HUNTINGTON'S DISEASE – CORE CONCEPTS AND CURRENT ADVANCES

Edited by **Nagehan Ersoy Tunali**

Huntington's Disease – Core Concepts and Current Advances
Edited by Nagehan Ersoy Tunali

Published by InTech

Janeza Trdine 9, 51000 Rijeka, Croatia

Publishing Process Manager Gorana Scerbe

Technical Editor Teodora Smiljanic

Cover Designer InTech Design Team

Additional hard copies can be obtained from orders@intechopen.com

Huntington's Disease – Core Concepts and Current Advances,
Edited by Nagehan Ersoy Tunali
p. cm.
ISBN 978-953-307-953-0

For EU product safety concerns:
IN TECH d.o.o., Prolaz Marije Krucifikse Kozulić 3, 51000 Rijeka, Croatia,
info@intechopen.com or visit our website at intechopen.com.

Meet the editor

Dr Nagehan Ersoy Tunalı received her Ph.D., M.Sc. and B.Sc. degrees in Molecular Biology and Genetics at Boğaziçi University, Istanbul (TR). Her PhD work involved "Molecular Analysis of Polyglutamine Diseases and Investigation of the Interaction Between Huntingtin and Nuclear Receptor Corepressor". She had the opportunity to gain experience in Huntington's Disease (HD) research at the University of Manchester (UK), CNR-Istituto di Medicina Sperimentale e Biotecnologie (IT) and the University of Wales College of Medicine (UK). Nagehan Ersoy Tunalı currently holds the position of Assistant Professor at Haliç University, Department of Molecular Biology and Genetics, Istanbul (TR) and Editor-in-Chief at Journal of Cell and Molecular Biology. Her research interests cover molecular diagnosis of polyglutamine repeat diseases, genetic modifiers of HD, localization and interactions of huntingtin, and the molecular mechanisms of excitotocity in HD. They are extended to genetic polymorphisms in kidney stone disease, and genetic susceptibilities and gene expression changes in urological cancers.

Contents

Preface

In the late 20th century the scientific community has witnessed a glorious outcome of an enviable long term collaboration among researchers working on Huntington's Disease. The invaluable efforts of the 58 international scientists and clinicians were eventuated in successful mapping of the disease gene to chromosome 4 in 1983. Being the first hereditary disease for which a DNA marker was used to localize the disease gene, HD has served as a model for mapping other genetic diseases. This achievement not only demonstrated the power of using linkage to DNA polymorphisms to approach genetic diseases, but also contributed to the concept of Human Genome Project.

Ten years later the gene was isolated and the genetic mutation causing HD was identified as the expansion in the number of CAG repeats in the first exon of the gene. Since that time, extensive research has been going on to decipher the changes in the molecular mechanisms caused by polyglutamines in the mutant protein product. Although there is only one gene and one mutation causing the disease, genotype-phenotype correlations and the molecular pathways involved were turned out to be extremely complex. One of the main complexities is that there is a huge amount of variation in the age of onset and the severity of symptoms among HD patients of the same CAG repeat size, which implicates the existence of genetic modifiers of the disease. The other is that, both gain of toxic function and loss of wild type function of the huntingtin protein are involved at the molecular level.

In the last almost 20 years many considerable achievements have been made and many questions found persuasive answers, however, we are still left with many missing pieces of the HD puzzle. There are currently no drugs available to cure the disease, which implies that we still have some way to go before completely understanding the neurodegenerative process in HD. In this regard, sharing of the experiences, the data, and the knowledge is of great importance to both the HD families and the scientific world.

This book, "Huntington's Disease - Core Concepts and Current Advances", was prepared to serve as a source of up-to-date information on a wide range of issues involved in Huntington's Disease. I believe that it will help the clinicians, health care providers, researchers, graduate students and life science readers to increase their understanding of the clinical correlates, genetic aspects, neuropathological findings, cellular and molecular events and potential therapeutic interventions involved in HD. The book not only serves reviewed fundamental information on the disease but also presents original research in several disciplines, which collectively provide comprehensive description of the key issues in the area.

Nagehan Ersoy Tunalı, PhD
Halic University, Faculty of Arts and Sciences,
Department of Molecular Biology and Genetics, Istanbul,
Turkey

Part 1

Cell Biology and Modeling of Huntington's Disease

1

Huntington's Disease: From the Physiological Function of Huntingtin to the Disease

Laurence Borgs[1,2], Juliette D. Godin[1,2],
Brigitte Malgrange[1,2] and Laurent Nguyen[1,2,3]
[1]*GIGA-Neurosciences,*
[2]*Interdisciplinary Cluster for Applied Genoproteomics (GIGA-R),*
University of Liège, C.H.U. Sart Tilman, Liège,
[3]*Wallon Excellence in Lifesciences and Biotechnology (WELBIO),*
Belgium

1. Introduction

Huntington's Disease (HD) is a progressive, fatal, autosomal dominant neurodegenerative disorder characterized by motor, cognitive, behavioural, and psychological dysfunction. HD symptoms usually appear at middle age. However, the disease can start earlier, and about 6% of HD patients develop juvenile forms (Foroud et al., 1999). Affecting approximately 1 in 10,000 people worldwide (Myers et al., 1993), the most obvious aspect of the pathology is a progressive neurodegeneration, particularly within the striatum (caudate and putamen). The massive loss of neurons in this region, normally responsible (among many things) for facilitation of volitional movement, is believed to lead to the characteristic motor dysfunctions of HD, such as uncontrolled limb and trunk movements, difficulty in maintaining gaze, and general lack of balance and coordination. The initial symptoms vary from person to person but the early stage of the disease is generally marked by involuntary movements of the face, fingers, feet or thorax associated with progressive emotional, psychiatric, and cognitive disturbances (Folstein et al., 1986). Psychiatric symptoms include depression, anxiety, apathy and irritability (Craufurd et al., 2001). In the later stages, HD is characterized by motor signs (mainly rigidity and akinesia), progressive dementia, or gradual impairment of the mental processes involved in comprehension, reasoning, judgment, and memory (Bachoud-Levi et al., 2001). Weight loss, alterations in sexual behaviour, and disturbances in the wake-sleep cycle are other characteristics of the disease and may be explained by hypothalamic dysfunction (Petersen et al., 2005). The patient usually dies within 10 to 20 years after the first symptoms appear, as there is currently no treatment to prevent or delay disease progression. As the disease progresses, there is general neuronal loss in several brain regions such as the cerebral cortex, the globus pallidus, the subthalamic nuclei, the substantia nigra, the cerebellum and the thalamus. Together with the neuronal loss, glial proliferation is observed (Vonsattel et al., 1985), although whether this proliferation is a cause or a consequence of the disease remains to be determined. The cause of HD is an expansion of CAG tract (encoding polyglutamine, polyQ) in exon 1 of the huntingtin gene (also called IT15 gene for Interesting Transcript) (HDCRG, 1993). The translated wild-type huntingtin protein is a 348-kDa protein containing a polymorphic stretch of 6 to 35 glutamine residues in its N-terminal domain. When the number of glutamine of huntingtin exceeds 36, it leads to the disease (HDCRG, 1993; Snell et al., 1993). The pathological mechanisms are not fully understood, but increasing evidences suggest that in addition to the gain of toxic properties, loss of wild-type huntingtin function also contributes to pathogenesis (Borrell-Pages et al., 2006).

2. Functions of wild-type huntingtin

Although the gene was discovered 18 years ago, the physiological role of the protein only has just begun to be understood. Huntingtin is ubiquitously expressed. Within neurons, huntingtin is found in the cytoplasm, within neurites and at synapses. It associates with various organelles and structures, such as clathrin-coated vesicles, endosomal and endoplasmic compartment, mitochondria, microtubules and plasma membrane (DiFiglia et al., 1995; Gutekunst et al., 1995; Kegel et al., 2005; Trottier et al., 1995a). Although mainly distributed in the cytoplasm, huntingtin is also detected in the nucleus (Hoogeveen et al., 1993; Kegel et al., 2002). Given its subcellular localization, huntingtin appears to contribute to various cellular functions in the cytoplasm and the nucleus. Consistent with this, huntingtin interacts with numerous proteins involved in gene expression, intracellular transport, intracellular signalling and metabolism (Borrell-Pages et al., 2006; Harjes & Wanker, 2003; S. H. Li & Li, 2004). An obvious feature of the huntingtin protein is the polyQ stretch at its NH2 terminus. To determine the contribution of the polyQ stretch to normal huntingtin function, a mice with a precise deletion of the short CAG triplet repeat encoding 7Q in the mouse HD gene - Hdh (DeltaQ/DeltaQ) - has been generated (Clabough & Zeitlin, 2006). Hdh (DeltaQ/DeltaQ) mice exhibit only a subtle phenotype, with slight defects in learning and memory tests suggesting that the polyQ tract is not required for essential function of huntingtin but instead may modulate the activity of huntingtin.

2.1 Huntingtin function during development and neurogenesis

Huntingtin is widely expressed in the early developing embryo where it plays an essential role in several processes including cell differentiation and neuronal survival. Inactivation of the mouse gene results in developmental retardation and embryonic lethality at E7.5 (Duyao et al., 1995; Nasir et al., 1995; Zeitlin et al., 1995). Null homozygous embryos (*Hdh-/-* mice) display abnormal gastrulation associated with increased apoptosis. It is known that the developmental defects observed in the *Hdh-/-* mice embryos derives from an inadequacy in the organization of extraembryonic tissue, possibly as a consequence of a disruption in the nutritive function of the visceral endoderm (Dragatsis et al., 1998). Additionally, huntingtin is essential for the early patterning of the embryo during the formation of the anterior region of the primitive streak (Woda et al., 2005). With the progression of embryonic development, experimental reductions of huntingtin levels below 50% cause defects in epiblast formation, the structure that will give rise to the neural tube, and profound cortical and striatal architectural anomalies (Auerbach et al., 2001; White et al., 1997). Defects in the formation of most of the anterior regions of the neural plate, specifically in the formation of telencephalic progenitor cells and the preplacodal tissue, have been recently described in the developing zebrafish with reduced huntingtin levels (Henshall et al., 2009). These data indicate that, in addition to its early extraembryonic function, huntingtin contributes to the formation of the nervous system at postgastrulation stages. Finally, specific inactivation of huntingtin in Wnt1 cell lineage leads to congenital hydrocephalus in mice further establishing a role for huntingtin in brain development (Dietrich et al., 2009).

A recent study specifically shows that huntingtin is involved in neurogenesis. Invalidation of huntingtin in murine cortical progenitors changes the nature of the division cleavages that lowers the pools of both apical and basal progenitors and promotes neuronal differentiation of daughter cells (Godin et al., 2010). This may explain previous observations showing that lowering the levels of huntingtin in mouse results, in addition to severe anatomical brain abnormalities, in ectopic masses of differentiated neurons near the striatum (White et al., 1997). Huntingtin localizes specifically at spindle poles during mitosis and associates with several component of the mitotic spindle (Caviston et al., 2007; Gauthier et al., 2004; Kaltenbach et al., 2007). Silencing of huntingtin in cells disrupts spindle orientation by modulating its integrity and disrupting the proper localization of several key components such as p150^{Glued} subunit of dynactin, dynein and the large nuclear mitotic apparatus (NuMA) protein (Godin et al., 2010).

2.2 Anti-apoptotic properties of huntingtin

Wild-type huntingtin is believed to have a pro-survival role. First the high level of apoptosis shown in knock-out mouse models suggests an anti-apoptotic function of wild-type huntingtin (Zeitlin et al., 1995). This has been corroborated in several *in vitro* and *in vivo* studies, demonstrating that expression of the full-length protein protected from a variety of apoptotic stimuli (Imarisio et al., 2008; Leavitt et al., 2001; Leavitt et al., 2006; Rigamonti et al., 2000; Rigamonti et al., 2001; Zuccato et al., 2001). Neuroprotection is enhanced with a progressive increase in the level of wild-type huntingtin, which indicates a gene-dosage effect (Leavitt et al., 2006). Several molecular mechanisms underlying the pro-survival activities of huntingtin have been elucidated. Wild-type huntingtin appeared to act downstream of mitochondrial cytochrome c release, preventing the activation of caspase-9 (Rigamonti et al., 2001) and caspase-3 (Rigamonti et al., 2000). Moreover, huntingtin physically interacts with active caspase-3 and inhibits its activity (Zhang et al., 2006). Huntingtin could also prevent the formation of the HIP1-HIPPI complex (huntingtin interacting protein 1 (HIP1)- HIP1 protein interactor (HIPPI)) and the subsequent activation of caspase-8 by sequestering HIP1 (Gervais et al., 2002). Finally, huntingtin exerts anti-apoptotic effects by binding to Pak2 (p21-activated kinase 2), which reduces the abilities of caspase-3 and caspase-8 to cleave Pak2 and convert it into a mediator of cell death (Luo & Rubinsztein, 2009).

2.3 Huntingtin and transcription

Huntingtin functions in transcription are well established. Huntingtin has been shown to interact with a large number of transcription factors such as the cAMP response-element binding protein (CREB)-binding protein (CBP) (McCampbell et al., 2000; Steffan et al., 2000), p53 (McCampbell et al., 2000; Steffan et al., 2000), the co-activator CA150 (Holbert et al., 2001) and the transcriptional co-repressor C-terminal binding protein (CtBP) (Kegel et al., 2002). In one hand, huntingtin acts as an activator of transcription. Huntingtin can bind to the transcriptional activator Sp1 (Specificity protein1) and the co-activator TAFII130 (TBP (TATA Box binding Protein) Associated Factor II 130) (Dunah et al., 2002). TAFII130 directly interacts with Sp1 and stimulates the transcriptional activation of genes. Huntingtin acts as a scaffold that links Sp1 to the basal transcription machinery, thus strengthening the bridge between the DNA-bound transcription factor Sp1 and the co-activator TAFII130 and, thereby, stimulating expression of target genes (Dunah et al., 2002). In addition, huntingtin binds to the transcriptional, repressor element-1 transcription/neuron restrictive silencer factors (REST/NRSFs), and therefore sequesters this complex in the cytoplasm (Zuccato et al., 2003). Huntingtin activates transcription by keeping REST/NRSF in the cytoplasm, away from its nuclear target, the neuron restrictive silencer element (NRSE), a consensus sequence found in many genes. Consistently, overexpression of huntingtin leads to an increase of the mRNAs transcribed from many RE1/NRSE-controlled neuronal genes (Zuccato et al., 2003; Zuccato et al., 2007). Huntingtin does not seem to interact with REST/NRSF directly, but rather belongs to a complex that contains HAP1 (Huntingtin associated protein 1), dynactin p150^{Glued} and RILP (REST/NRSF-interacting LIM domain protein), a protein that directly binds REST/NRSF and promotes its nuclear translocation (Shimojo, 2008). Huntingtin may therefore act in the nervous system as a general facilitator of neuronal gene transcription for a subclass of genes. In particular, huntingtin regulates the production of brain-derived neurotrophic factor protein (BDNF), a neurotrophin required for the survival of striatal neurons and for the activity of the cortico-striatal synapses (Charrin et al., 2005; Zuccato et al., 2001; Zuccato et al., 2003; Zuccato et al., 2007). This is supported by studies in zebrafish showing that loss of BDNF recapitulates most developmental abnormalities seen with huntingtin knockdown (Diekmann et al., 2009). Finally, it has been shown that the interaction of wild-type huntingtin with both HAP1 and mixed-lineage kinase 2 (MLK2) promotes the expression of NeuroD (Marcora et al., 2003), a basic helix–loop–helix transcription factor that is crucial for the development of the dentate gyrus of the hippocampus (M. Liu et al., 2000). In the other hand, huntingtin also promotes repression of gene transcription by binding to a repressor complex containing N-CoR and Sin3A. Such interaction is believed to favour the binding of N-CoR–Sin3a repressor complex to the basal transcription machinery and modulates transcriptional gene repression (Boutell et al., 1999). This hypothesis is supported by microarray analyses indicating an involvement of huntingtin in the regulation of the N-CoR–Sin3A-mediated transcription in HD transgenic mice (Luthi-Carter et al., 2000).

2.4 Huntingtin and intracellular transport

Huntingtin is predominantly found in the cytoplasm where it associates with vesicular structures and microtubules (DiFiglia et al., 1995; Gutekunst et al., 1995; Trottier et al., 1995b). Indeed, huntingtin associates with various proteins that play a role in intracellular trafficking (Harjes & Wanker, 2003; Kaltenbach et al., 2007). In particular, huntingtin interacts with dynein (Caviston et al., 2007) and the huntingtin-associated protein-1 (HAP1), a protein that associates with p150^{Glued} dynactin subunit, an essential component of the dynein/dynactin microtubule-based motor complex (Block-Galarza et al., 1997; Engelender et al., 1997; S. H. Li et al., 1998a; S. H. Li et al., 1998b; Schroer et al., 1996). Huntingtin and its interacting partner HAP1 are both anterogradely and retrogradely transported in axons at a speed characteristic for vesicles that move along microtubules (Block-Galarza et al., 1997). The first evidence of a role of huntingtin in intracellular transport came from a study in *Drosophila* showing that a reduction in huntingtin protein expression resulted in axonal transport defects in larval nerves and neurodegeneration in adult eyes (Gunawardena et al.,

2003). This was confirmed by further studies in mammals (Colin et al., 2008; Gauthier et al., 2004; Trushina et al., 2004). First it has been shown that wild-type huntingtin stimulates transport by binding with HAP1 and subsequently interacting with the molecular motors dynein/dynactin and kinesin (Engelender et al., 1997; Gauthier et al., 2004; S. H. Li et al., 1998b; McGuire et al., 1991). Huntingtin directly promotes the microtubule-based transport of BDNF and Ti-VAMP (tetanus neurotoxin-insensitive vesicle-associated membrane protein) vesicles in neurons through this interaction (Gauthier et al., 2004). Second, it has been shown that fast axonal trafficking of mitochondria was altered in mammalian neurons expressing less than 50% of wild-type huntingtin (Trushina et al., 2004). Accumulating or decreasing huntingtin in cells increases or reduces the speed of intracellular transport, respectively. Thus, this suggests that huntingtin is a processivity factor for the microtubule-dependent transport of vesicles (Colin et al., 2008; Gauthier et al., 2004). In particular, decreasing huntingtin levels in cells alters the interaction of the anterograde molecular motor kinesin with vesicles (Colin et al., 2008), whereas the direct interaction of huntingtin with dynein facilitates dynein-mediated vesicle motility (Caviston et al., 2007). Finally, phosphorylation of wild-type huntingtin at S421 is crucial to control the direction of vesicles in neurons (Colin et al., 2008). When phosphorylated, huntingtin recruits kinesin to the dynactin complex on vesicles and microtubules and therefore promotes anterograde transport. Conversely, when huntingtin is not phosphorylated, kinesin detaches and vesicles are more likely to undergo retrograde transport (Colin et al., 2008).

2.5 Huntingtin, endocytosis and synapses

Huntingtin interacts with many proteins that regulate exo- and endocytosis, such as the huntingtin-interacting protein 1 (HIP1) and 14 (HIP14), the HIP1-related protein (HIP1R), the protein kinase C, and the casein kinase substrate in neurons-1 (PACSIN1) (Engqvist-Goldstein et al., 2001; Kalchman et al., 1997; X. J. Li et al., 1995; Modregger et al., 2002; Singaraja et al., 2002; Wanker et al., 1997). Huntingtin is modified by the HIP14 protein, a palmitoyl-transferase involved in the sorting of many proteins from the Golgi region (Yanai et al., 2006). Huntingtin is important for the function of Rab11, a critical GTPase in regulating membrane traffic from recycling endosomes to the plasma membrane. The Rab11 nucleotide exchange activity is altered in cells depleted for huntingtin suggesting a role for huntingtin in Rab11 activation (X. Li et al., 2008). Huntingtin may also take part to the presynaptic complex through its interaction with HIP1, which has been associated with the presynaptic terminal (J. A. Parker et al., 2007). Furthermore, huntingtin can bind to PACSIN1/syndapin, syntaxin, and endophilin A, which collectively play a key role in synaptic transmission, as well as in synaptic vesicles and receptor recycling. Finally, wild-type huntingtin interacts with postsynaptic density 95 (PSD95; a protein located in the postsynaptic membrane) through its Src homology-3 (SH3) sequence, regulating the anchoring of *N*-methyl-d-aspartate (NMDA) and kainate (KA) receptors to the postsynaptic membrane (B. Sun et al., 2002). At the postsynaptic membrane, HAP1 binds Duo (the human orthologue of Kalirin) that is known to activate Rac1 signalling that plays an important role in the remodelling of the actin cytoskeleton (Colomer et al., 1997). Thus huntingtin might modulate Rac1 signalling and actin dynamics in dendrites via its interactions with HAP1 and PSD-95. This is further supported by the reported interaction of huntingtin with Cdc42-interacting protein 4 (CIP4) (Holbert et al., 2003) and FIP-2 (Hattula & Peranen, 2000), two proteins involved in actin dynamics and dendritic morphogenesis in the postsynaptic density.

3. Consequences of polyglutamine expansion of mutant huntingtin

The physiopathology of the Huntington Disease arises from aberrant interactions of mutant huntingtin, or its proteolytic fragments, with a wide set of cellular proteins and components. The extended stretch of polyglutamines (polyQ) causes huntingtin to acquire a non-native structural conformation, a common feature of mutant proteins associated with CAG-triplet repeat disorders (Muchowski, 2002). Misfolding of mutant huntingtin leads to both loss of huntingtin function and gain of novel properties, allowing it to engage in diverse aberrant interactions with multiple cellular components, thereby perturbing many cellular functions essential for neuronal homeostasis (Kaltenbach et al., 2007). This results in a combination of multiple physiopathological changes among which the most severe include protein aggregation, transcriptional deregulation and chromatin remodelling, impaired axonal transport, mitochondrial metabolism dysfunction, disruption of calcium homeostasis, excitotoxicity, and caspase activation.

3.1 Nuclear translocation of mutant huntingtin

The proteolytic cleavage of huntingtin into N-terminal fragments containing the polyQ stretch and their subsequent translocation to the nucleus is a key step of the disease. N-terminal fragments of mutant huntingtin are sufficient to reproduce HD pathology in animal models of the disease (Davies et al., 1997; Palfi et al., 2007; Schilling et al., 1999b). Proteolytic cleavage and nuclear translocation of mutant huntingtin are required to induce neurodegeneration (Saudou et al., 1998; Wellington et al., 2000b) and reducing polyQ-huntingtin cleavage decreases its toxicity and slows disease progression (Gafni et al., 2004; Wellington & Hayden, 2000). In addition, expression of truncated fragments of mutant huntingtin that contain the polyQ stretch results in an increased toxicity compare to expression of full length huntingtin with the same polyQ expansion suggesting that susceptibility to neuronal death is greater with decreasing protein length and increasing polyQ size (Hackam et al., 1998). Several proteases cleave huntingtin *in vitro* and *in vivo*, and the corresponding cleavage products have been found in the brain of patients and in murine models (Mende-Mueller et al., 2001). These proteases include caspase-1, -3, -6, -7 and -8 (Goldberg et al., 1996; Hermel et al., 2004; Wellington et al., 1998; Wellington et al., 2000b; Wellington et al., 2002), calpain (Bizat et al., 2003a; Gafni & Ellerby, 2002; Gafni et al., 2004; Goffredo et al., 2002; M. Kim et al., 2003; Y. J. Kim et al., 2001) and aspartic proteases (Lunkes et al., 2002). These different proteases can cleave huntingtin sequentially to produce N-terminal mutant fragments that are even more toxic and more susceptible to aggregation (Y. J. Kim et al., 2001; Ratovitski et al., 2009). Proteolytic cleavage depends on the length of the polyQ stretch within huntingtin, with pathological polyQ repeat-containing huntingtin being more efficiently cleaved than huntingtin containing polyQ repeats of non-pathological size (Gafni & Ellerby, 2002; B. Sun et al., 2002). Abnormal activation of these proteases could result from various insults received by HD neurons such as excessive levels of cytosolic Ca2+, reduced trophic support and activation of the apoptotic machinery. Once cleaved, N-terminal fragments of mutant huntingtin translocate into the nucleus. Small N-terminal huntingtin fragments interact with the nuclear pore protein translocated promoter region (Tpr), which is involved in nuclear export. PolyQ expansion alters this interaction compromising the export of the N-terminal fragments to the cytoplasm and increasing the nuclear accumulation of huntingtin (Cornett et al., 2005). Thus, intranuclear accumulation of N-terminal fragments of huntingtin may result of nuclear export rather than nuclear import dysfunctions. Finally, preventing huntingtin cleavage reduces neuronal toxicity and delays the onset of the disease (Gafni et al., 2004; Wellington et al., 2000a). Indeed, mutant huntingtin resistant to caspase-6 but not to caspase-3 cleavage does promote neuronal dysfunction and degeneration, indicating that the nature of the protease involved is critical for disease progression (Graham et al., 2006; Pouladi et al., 2009).

3.2 Aggregation and toxicity

The abnormal PolyQ tract of truncated mutant huntingtin changes the native structural protein conformation and consequently induces the formation of insoluble aggregates (Davies et al., 1997; Scherzinger et al., 1997). Aggregates are found in cytoplasm, nucleus and dendrites of affected neurons and appear with the onset of the disease when patients develop symptoms (DiFiglia et al., 1997). The exact mechanism for aggregation is still unclear but the SH3-containing Grb2-like protein (SH3GL3) protein interacts with the first exon of mutant huntingtin and promotes the formation of insoluble aggregates (Sittler et al., 1998). In the nucleus of neurons, N-terminal fragments of mutant huntingtin form intranuclear aggregates (NIIs) (DiFiglia et al., 1997; DiFiglia, 2002; Goldberg et al., 1996). Although it is well established that the nuclear localization of mutant huntingtin is required for neuronal death (Saudou et al., 1998), the toxicity of these nuclear aggregates is still being debated (Arrasate et al., 2004; Davies et al., 1997; Saudou et al., 1998). NIIs are not strictly correlated with neuronal death, as the highest percentage of NII-containing neurons is found in non-degenerating regions (Gutekunst et al., 1999; Kuemmerle et al., 1999). Also, NIIs are not correlated with cell death in neuronal models of HD *in vitro* or *in vivo* (M. Kim et al., 1999; Saudou et al., 1998; E. Slow, 2005; E. J. Slow et al., 2005), and the probability that a given neuron will die is lower when it contains inclusion bodies (Arrasate et al., 2004). The formation of NIIs may thus correspond to a protective mechanism that temporarily concentrates soluble and toxic huntingtin products to favour their degradation by the proteasome. Consistent with this is the suppression of aggregates accelerated polyQ-induced cell death caused by inhibition of the ubiquitination process (Arrasate et al., 2004; Saudou et al., 1998). Huntingtin aggregation could be facilitated by proteasomal chaperones such as Rpt4 and Rpt6, two subunits of the 19S proteasome (Rousseau et al., 2009). Studies using a conditional HD mouse model (in which silencing of mutant huntingtin expression leads to the disappearance of intranuclear aggregates (Yamamoto et al., 2000) showed that aggregates formation is a balance between the rate of huntingtin synthesis and its degradation by the proteasome (Martin-Aparicio et al., 2001). Therefore, over the course of the disease, the proteasome degradation system may become overloaded with an increasing number of misfolded and mutated proteins in the cell. As a consequence, the neurons may be progressively depleted of functional proteasomes, which will lead to a progressive accumulation of misfolded and abnormal proteins, further increasing the rate of protein aggregation (Jana et al., 2001; Waelter et al., 2001). Indeed, several components of the proteasome, such as its regulatory and catalytic subunits and ubiquitin conjugation enzymes, are also sequestered in these aggregates *in vitro* (Jana et al., 2001; Wyttenbach et al., 2000) and *in vivo* (Jana et al., 2001), resulting in the impairment of the ubiquitin–proteasome system (Bence et al., 2001).

3.3 Transcriptional deregulation

One consequence of mutant huntingtin is transcriptional deregulation. Nuclear huntingtin aggregates interfere with normal transcriptional control (Davies et al., 1997; DiFiglia et al., 1997). Comprehensive studies have shown a direct interference of mutant huntingtin with transcriptional complexes, altering levels of hundreds of RNA transcripts and leading to transcriptional deregulation (Hodges et al., 2006). It has been first proposed that mutant huntingtin establishes abnormal protein–protein interactions with several nuclear proteins and transcription factors, recruiting them into the aggregates and inhibiting their transcriptional activity. However, this hypothesis was disputed by findings in mice showing no significant differences in transcript levels of specific genes between NII-positive and NII-negative neurons (Sadri-Vakili et al., 2006). Whether the same is true in men is currently unknown. Subsequently, a large number of studies have deciphered molecular mechanisms underlying the transcriptional abnormalities in HD. These discoveries include demonstration of transcription factor sequestration, loss of protein-protein interaction and inhibition of enzymes involved in chromatin remodelling.

3.3.1 Sequestration of transcription factors

Numerous transcription factors have been reported to interact with polyQ huntingtin. Examples include TATA-binding protein (TBP) (Schaffar et al., 2004), CREB (cyclic-adenosine monophosphate (cAMP) response element (CRE) binding protein)-binding protein (CBP) (Schaffar et al., 2004; Steffan et al., 2000), specificity protein-1 (Sp1) (S. H. Li et al., 2002), and the TBP-associated factor (TAF)II130 (Dunah et al., 2002), all of which directly interact with mutant huntingtin through the expanded polyQ tail. Under pathological condition, TBP function is altered. Indeed, the interaction of TBP with huntingtin polyQ stretch leads to the sequestration of TBP into mutant huntingtin aggregates preventing TBP binding to DNA promoters (Friedman et al., 2008; Huang et al., 1998). CRE-mediated transcription is regulated by TAFII130, which is part of the basal transcriptional machinery and can abnormally interact with mutant huntingtin, rendering the transcriptional complex ineffective (Dunah et al., 2002). Mutant huntingtin could also alter CRE-mediated transcription through inhibition of CBP transcriptional activities. CBP plays a role in histone acetylation by acting as an acetyltransferase which opens the chromatin structure and exposes the DNA to transcription factors such as TAFII130, enhancing the CRE-mediated transcription. In the presence of mutant huntingtin, the interaction between huntingtin and CBP is enhanced leading to histone hypoacetylation and inhibition of CBP-mediated transcription (Cong et al., 2005; Steffan et al., 2000). One consequence of CBP inhibition is mitochondrial dysfunction (Quintanilla & Johnson, 2009). Mutant huntingtin-induced CBP inhibition leads to downregulation of PGC-α expression, a transcriptional co-activator that regulates the expression of genes involved in mitochondrial function such as the mitochondrial respiratory gene PPARγ thus impairing mitochondrial function that contributes to neuronal striatal cell death (Quintanilla & Johnson, 2009). Mutant huntingtin also represses the transcription of p53-regulated target genes through enhanced binding to p53 without any involvement of the polyQ stretch (Steffan et al., 2000). Sp1 is a regulatory protein that binds to guanine–cytosine boxes and mediates transcription through its glutamine-rich activation domains, which target components of the basal transcriptional complex, such as TAF130 (TFIID subunit) and TFIIF. Sequestration of Sp1 and TAFII130 into

NIIs leads to the inhibition of Sp1-mediated transcription (Dunah et al., 2002; S. H. Li et al., 2002). In addition, by interacting with TAFII130 or RAP30 (a TFIIF subunit), mutant huntingtin prevents the recruitment of TFIID into a functional transcriptional machinery (Dunah et al., 2002; Z. X. Yu et al., 2002). It has also been shown that the binding of Sp1 to specific promoters of susceptible genes is significantly decreased in transgenic HD mouse brains, striatal HD cells and human HD brains. This suggests that polyQ huntingtin dissociates Sp1 from target promoters, inhibiting the transcription of specific genes (Chen-Plotkin et al., 2006), such as the dopamine D2 receptor gene or nerve growth factor gene, two crucial gene in HD (Dunah et al., 2002).

3.3.2 Loss of transcription factor interaction

On the other hand, mutant huntingtin may also lose the ability to bind and interact with other transcription factors, as it is the case for the NRSE-binding transcription factors. The failure of mutant huntingtin to interact with REST / NRSF in the cytoplasm leads to its nuclear accumulation, where it binds to NRSE sequences and represses a large cohort of neuronal-specific genes containing the RE1/NRSE motif. This includes the *BDNF* gene, coding for a protein necessary for striatal neurons survival (Zuccato et al., 2003). Interestingly, *BDNF*-knockout models largely recapitulate the expression profiling of human HD (Strand et al., 2007), suggesting that striatal medium-sized spiny neurons suffer from similar insults in HD and BDNF-deprived environments. Analysis of human and mouse genome have identified more than 1800 RE1/NRSE sequences, suggesting that many other genes could be repressed by expression of mutant huntingtin (Bruce et al., 2004; Zuccato et al., 2003). By using a microarray-based survey of gene expression in a large cohort of HD patients and matched controls (Hodges et al., 2006), many genes whose expression is down-regulated in HD caudate are REST/NRSF target genes (Johnson & Buckley, 2009). These findings strongly support a model of strengthened REST/NRSF repression of target genes in HD brains. Besides REST/NRSF, mutation in huntingtin proteins impairs its interaction with the transcription repressor CtBP (Kegel et al., 2002) and N-CoR (Boutell et al., 1999) or the activator CA150 (Holbert et al., 2001), thereby impairing their activities.

3.3.3 Mutant huntingtin and chromatin structure

Regulation of gene expression results from the action of transcription factors and enzymes that modify chromatin structure. Histone acetyltransferases (HATs) favour gene transcription through the opening of chromatin architecture whereas histone deacetyltransferases (HDACs) repress gene transcription through chromatin condensation. Expanded polyQ huntingtin binds directly the acetyltransferase domain of CBP and p300/CBP associated factor (P/CAF), blocking their acetyltransferase activity (Cong et al., 2005; Steffan et al., 2001). This causes a condensed chromatin state and reduced gene transcription. These results indicate that reduced acetyltransferase activity might be an important component of polyglutamine pathogenesis. In accordance, HDAC inhibitors restore genes transcription and limit polyQ-induced toxicity in HD (Gardian et al., 2005; Steffan et al., 2001). Moreover, histone methylation promotes gene repression through chromatin condensation. Interestingly hypermethylation of histones has been found in HD patients and in several mouse models of HD (Gardian et al., 2005; Ryu et al., 2003). Finally, huntingtin can act directly on chromatin. Indeed, huntingtin binds to gene promoters *in vivo* in a polyQ-dependent manner suggesting that mutant huntingtin may modulate gene expression through abnormal interactions with genomic DNA, altering DNA conformation and transcription factor binding.

3.3.4 Post-transcriptional deregulation

Two independent studies have revealed that the microRNA (miRNA) machinery is perturbed in HD (Johnson et al., 2008; Packer et al., 2008). MiRNAs recognize complementary sequences located mostly in the untranslated 3'UTR sequence of target mRNAs and repress their transcription (Bartel, 2009). Recent data reveal that miRNAs are essential for neuronal survival and abnormal miRNAs expression is observed in the brain of HD patients (Johnson et al., 2008; S. T. Lee et al., 2011; Marti et al., 2010; Packer et al., 2008; Sinha et al., 2010). Among them, many miRNAs genes are targeted by REST. Accordingly, the expression of mir-7, mir-9, mir-22, mir-29, mir-124, mir-128, and mir-132, and mir-138 is downregulated in the brain of human patients and mouse models of HD (Johnson et al., 2008; S. T. Lee et al., 2011). The failure of mutant huntingtin to sequester REST in the cytoplasm (Zuccato et al., 2003) may thus lead to aberrant expression of miRNAs in HD. Downregulation of miRNAs correlates with increased expression level of many target mRNAs. Indeed, a recent study has revealed that the lack of TBP repression by mir-146a contributes to HD pathogenesis (Sinha et al., 2010). Moreover, it was reported that mir-132 downregulation in HD patients leads to higher levels of p250GAP expression, an inhibitor of the Rac/Rho family (Johnson et al., 2008). Mutant huntingtin also indirectly regulates the transcription of miRNA genes by destabilizing the interaction of Argonaute 2 with P-bodies, two key components of the miRNA-silencing pathway (Savas et al., 2008). These findings suggest that miRNA processing, as a whole, is impaired in HD.

3.4 Excitotoxicity

The loss of function of wild-type huntingtin engenders multiple cellular dysfunctions including an increase of pathological excitotoxicity, which is responsible for striatal neuronal injury. It has been described that huntingtin polyQ expansion correlates with hyperactivation of the ionotropic glutamate receptor *N*-methyl-d-aspartate (NMDA) resulting in a massive increase of intracellular $Ca2^{+}$ that activates in turn signalling pathways leading to cell death (Coyle & Puttfarcken, 1993; Fan & Raymond, 2007; Lipton & Rosenberg, 1994). Importantly, mutant huntingtin can also sensitize the inositol (1,4,5)-triphosphate receptor type 1 located in the membrane of the endoplasmic reticulum, promoting a further increase in intracellular $Ca2^{+}$ (Tang et al., 2003). Increased intracellular $Ca2^{+}$ concentration can have deleterious consequences including mitochondrial dysfunction, activation of the $Ca2^{+}$-dependent neuronal isoform of nitric oxide (NO) synthase, generation of NO and other reactive oxygen species, activation of $Ca2^{+}$-dependent proteases such as calpains, activation of phosphatases such as calcineurin and apoptosis (Fan & Raymond, 2007; Gil & Rego, 2008). Several molecular mechanisms underlying glutamate excitotoxicity have been elucidated. First polyQ expansion interferes with the ability of wild-type huntingtin to interact with PSD-95 (Section 2.5), resulting in the sensitization of NMDA (and KA) receptors and promoting glutamate-mediated excitotoxicity (Y. Sun et al., 2001). Second, mutant huntingtin can increase tyrosine phosphorylation of NMDA receptors, further promoting their sensitization (Song et al., 2003). Indeed, increased activity of Src family of tyrosine kinase induces phosphorylation of NMDA receptors and therefore stabilizes the receptors at the post-synaptic membrane by decreasing their binding to the clathrin adaptator protein 2 and limiting their endocytosis (B. Li et al., 2002; Roche et al., 2001; Vissel et al., 2001). Finally, synaptic function and neurotransmitter release are impaired when mutant huntingtin aggregates at the synapses (H. Li et al., 2003). Mutant huntingtin aggregates bind synaptic vesicles membranes and inhibits their uptake and release. The biochemical bases have not been yet elucidated. However, mutant huntingtin could impair the association of HAP1 with synaptic vesicles in axonal terminals (H. Li et al., 2003). Activation of pathways that lead to the production of excitotoxins in the brain is likely to have an impact in HD. Indeed, endogenous levels of the NMDA-receptor agonist quinoleic acid (QA, a product of tryptophan degradation generated along the kynurenine pathway) and of its bioprecursor, the free radical generator 3-hydroxykynurenine (3-HK) are increased in the striatum and cortex of early stage HD patients (Guidetti & Schwarcz, 2003; Guidetti et al., 2004) and in several mouse models of HD (Guidetti et al., 2006). This suggests that an increased generation of QA may contribute, at least in part, to excitotoxicity in HD. In accordance, inhibition of this pathway with a structural analogue of kynurenic acid, suppresses toxicity of a mutant huntingtin fragment (Giorgini et al., 2005). Another factor that can contribute to the vulnerability of striatal neurons to excitotoxicity is the capacity of the surrounding glial cells to remove extracellular glutamate from the synaptic cleft. In agreement, a decrease in the mRNA levels of the major astroglial glutamate transporter (GLT1) and the enzyme glutamine synthetase were detected in the striatum and cortex of R6/1 and R6/2 mouse models of HD (Lievens et al., 2001). In addition, mutant huntingtin has been shown to accumulate in the nucleus of glial cells in HD brains, decreasing the expression of GLT1 and reducing glutamate uptake (Shin et al., 2005). It remains unclear how GLT-1 expression is altered in presence of mutant huntingtin. The inhibition of GLT-1 could be huntingtin/Sp1 mediated. The GLT-1 promoter contain Sp1-binding site that are recognize by mutant huntingtin. In accordance, increasing striatal GLT1 expression by pharmacological treatment attenuates the neurological signs of HD in R6/2 mice, suggesting that a dysregulation of striatal glutamate uptake by glial cells may play a key role in HD (Miller et al., 2008). Beyond glutamate, other neuromodulators controlling the activity of the corticostriatal synapse can sensitize striatal neurons to excitotoxic stimuli. Adenosine (A) and A2 receptors (Tarditi et al., 2006; Varani et al., 2001), as well as cannabinoids (CB) receptors (Maccarrone et al., 2007; Marsicano et al., 2003), which are particularly abundant on the corticostriatal terminals, can enhance glutamate release upon activation. A crucial input to the striatum comes from the *substantia nigra pars compacta*, whose fibers represent the main striatal source of dopamine. Dopamine can directly regulate glutamate release from corticostriatal terminals by stimulating the D2 receptors (D2R) located on the cortical afferents (Augood et al., 1997; Cha et al., 1999; Huot et al., 2007).

3.5 Mitochondrial dysfunction and energy

Studies in HD patients and HD post-mortem tissue have given substantial evidences that bioenergetic defects may play a role in the pathogenesis of Huntington Disease: (1) A significant decrease in glucose uptake in the cortex and striatum of both pre-symptomatic and symptomatic HD patients (Antonini et al., 1996; Ciarmiello et al., 2006; Garnett et al., 1984; Grafton et al., 1990; Kuhl et al., 1982; Kuwert et al., 1990; Kuwert et al., 1993; Mazziotta et al., 1987); (2) A significant reduction in aconitase activity in the striatum and cerebral cortex (Tabrizi et al., 1999), that can be interpreted as an indirect indicator of ROS generation, mitochondrial dysfunction and excitotoxicity.; (3) A significant decrease in the activities of mitochondrial complexes II–III (Brennan et al., 1985; Browne et al., 1997; Butterworth et al., 1985; Gu et al., 1996; Mann et al., 1990) and IV in the striatum (Browne et al., 1997; Gu et al., 1996). Contradictory results have also been published regarding the activity of the mitochondrial complex I with an initial study showing a striking reduction in the activity of this complex and subsequent studies reporting no deficiencies in platelet mitochondrial function(Arenas et al., 1998; Gu et al., 1996; W. D. Parker, Jr. et al., 1990; Powers et al., 2007a; Powers et al., 2007b; Turner et al., 2007) ; (4) Increased production of lactate in the cerebral cortex and basal ganglia of HD patients (Jenkins et al., 1993; Koroshetz et al., 1997), suggestive of an elevated glycolytic rate; (5) A reduced phosphocreatine / inorganic phosphate ratio in skeletal muscle (Lodi et al., 2000) and a significant delay in the recovery of phosphocreatine levels after exercise (a direct measure of ATP synthesis) in HD patients (Saft et al., 2005); (6) Decreased mitochondrial ATP generation (Milakovic & Johnson, 2005; Seong et al., 2005); (7) Morphological and morphometric changes, as well as decreased membrane potential in mitochondria from lymphoblasts of HD patients (Panov et al., 2002; Squitieri et al., 2006); (8) Depletion of mitochondrial DNA in leukocytes from HD patients (C. S. Liu et al., 2008). In accordance with major defects in mitochondrial biogenesis, it has been shown that the administration of the mitochondrial cofactor coenzyme Q10 extended survival and delayed the development of motor deficits, weight loss, cerebral atrophy, and neuronal intranuclear inclusions in the transgenic mouse model of HD (Ferrante et al., 2002). However it is not clear whether mitochondrial dysfunctions are a cause or a consequence of HD.

Several molecular mechanisms have been suggested. Mutant huntingtin can bind directly to mitochondria (Choo et al., 2004; Orr et al., 2008; Panov et al., 2002), thereby enhancing mitochondria permeability that could lead to abnormal release of apoptotic factors (Panov et al., 2002; Sawa, 2001). Increased mitochondrial DNA mutations and deletions that can affect mitochondrial respiration have been detected in neurons of the cerebral cortex of HD patients (Acevedo-Torres et al., 2009; Horton et al., 1995). Mutant huntingtin induces an upregulation of the nuclear levels of p53 and an increase in its activity (Bae et al., 2005) both in HD transgenic mice and in HD patients. Interestingly, genetic deletion of p53 suppresses neurodegeneration in HD transgenic flies and neurobehavioral abnormalities of HD transgenic mice (Bae et al., 2005). Thus, it is likely that mutant huntingtin-induced increase in p53 activity induces further mitochondrial abnormalities that contribute to HD. Moreover, mitochondria fission could participate to polyQ-induced cell death in HD (Liot et al., 2009). Finally, mutant huntingtin also affects mitochondria motility within the cells (section 3.6), leading to mitochondria aggregates within neurites (Chang et al., 2006; Trushina et al., 2004). Mutant huntingtin may indirectly influence mitochondrial function via effects on the transcription of genes involved in the functioning and biogenesis of this organelle as seen in section 3.3.

3.6 Disruption of intracellular dynamics

Altered intracellular dynamics are likely to contribute to the development of the disease. This involves defects in axonal transport but also alterations of the secretory and endocytic pathways. Dysfunction of huntingtin directly impairs axonal transport. Expression of mutant huntingtin short fragments directly inhibits fast axonal transport in isolated giant squid axoplasm. Effects were greater with truncated polypeptides and occurred without detectable morphological aggregates (Szebenyi et al., 2003). Further study in primary culture of striatal neurons show that mutant huntingtin is unable to stimulate transport resulting in reduced BDNF support and in a higher susceptibility of striatal neurons to death (Gauthier et al., 2004). When huntingtin contains the pathological polyQ expansion, it interacts more strongly with HAP1 and p150^{Glued} (Gauthier et al., 2004), leading to detachment of the molecular motors from the microtubules and to a lower processivity of vesicles along the microtubules. Moreover, huntingtin in complex with HAP40 (Huntingtin associated protein 40) has been identified as a novel effector of the small guanosine triphosphatase Rab5, a key regulator of endocytosis (Pal et al., 2006). HAP40 mediates the recruitment of huntingtin to Rab5 onto early endosomes. HAP40 overexpression caused a drastic reduction of early endosomal motility through their displacement from microtubules and preferential association with actin filaments. Remarkably, in HD, endogenous HAP40 was up-regulated and endosome motility and endocytic activity were altered, suggesting that huntingtin/HAP40/Rab5 complex failed to regulate cytoskeleton-dependent endosome dynamics under pathological conditions. As well as nuclear aggregation, N-terminal huntingtin fragments form aggregates that accumulate in axonal processes and terminals (H. Li et al., 1999; H. Li et al., 2001; Sapp et al., 1997; Schilling et al., 1999a). Several studies have shown that N-terminal huntingtin polypeptide fragments containing the polyQ expansion cause axonal transport defects in cellular and *Drosophila* models of HD (Gunawardena et al., 2003; Szebenyi et al., 2003; Trushina et al., 2004). These aggregates physically block the circulating vesicles or organelles such as mitochondria but also titrates motor proteins, particularly p150^{Glued} and kinesin heavy chain (KHC), from other cargoes and pathways (Gunawardena et al., 2003; W. C. Lee et al., 2004).

3.7 Cell death

Cell death triggered by an apoptosis process is a common way for many neurodegenerative diseases, including HD. It has indeed been shown that huntingtin mutation engenders an activation of intrinsic apoptotic pathway implicating caspases in both HD patients and transgenic mouse models of HD (Hermel et al., 2004; Kiechle et al., 2002; Maglione et al., 2006; Ona et al., 1999; Sanchez et al., 1999; Wellington et al., 1998). Caspases activation leads to activation of factors that initiate the proteolytic destruction of cell. Several caspases including caspase-1, -3, -6, -7, -8 and -9 are transcriptionally up-regulated and activated in HD mouse models and human HD brain (Hermel et al., 2004; Kiechle et al., 2002; Maglione et al., 2006; Ona et al., 1999; Sanchez et al., 1999; Wellington et al., 1998). The activation of apoptotic signalling pathways causes the cytoplasmic release of cytochrome c, an intermediate protein associated with the membrane of mitochondria that can bind to caspases to activate the cell death process. Expression of cytochrome c is increased in HD striatal neurons (Kiechle et al., 2002; Wellington et al., 1998) or in excitotoxic lesion models of HD (Antonawich et al., 2002; Bizat et al., 2003b; Vis et al., 2001). In addition, another hallmark of apoptosis – the translocation of GlycerAldehyde 3-Phosphate DeHydrogenase (GAPDH) into the nucleus - has been observed in a transgenic mouse model of HD (Senatorov et al., 2003). Moreover, huntingtin possess a caspase-6 and caspase-3 cleavage site that enables its proteolytic cleavage. The resulting product is accumulated in the cells and facilitates the formation of insoluble and toxic aggregates, which can translocate into the nucleus and activate additional caspases (Wellington et al., 2000a). How mutant huntingtin induces apoptosis is still debated. In one hand, the polyQ expansion within mutant huntingtin reduces its ability to bind and thereby inhibit caspase-3 (Zhang et al., 2006). In the other hand, mutant huntingtin enhances caspase-8 activity that in turn activates caspase-3. Two models of caspase-8 activation have been proposed. First, mutant huntingtin could recruit caspase-8 into the aggregates, thus favouring its oligomerisation and its activation (Sanchez et al., 1999). In the other model, huntingtin binding to HIP1 is reduced by the polyQ expansion. The released HIP1 could then freely interacts with HIPPI and activates caspase-8 (Gervais et al., 2002; Zhang et al., 2006).

Some evidence suggests that autophagy may also mediate cell loss in HD. Autophagy is a bulk degradation process in which a portion of the cytosol and its content is enclosed by double-membrane structures named autophagosomes/autophagic vacuoles, which ultimately fuse with lysosomes for the degradation of the contents. Early studies showed increased numbers of autophagosome-like structures in the brain of HD patients (Davies et al., 1997; Kegel et al., 2000; Petersen et al., 2001; Qin et al., 2003; Roizin, 1979; Sapp et al., 1997; Tellez-Nagel et al., 1974). Furthermore, a positive correlation has been found between the number of autophagic vacuoles and the length of the polyglutamine expansion in HD lymphoblasts (Nagata et al., 2004). Mutant huntingtin induces endosomal and/or lysosomal activity (Kegel et al., 2000). In accordance an increased activity of the lysosomal proteases cathepsins D and H has been shown in the caudate nucleus of HD patients or in a cellular model of HD (del Toro et al., 2009; Mantle et al., 1995). Autophagy may represent an initial attempt of the HD cell to eliminate the mutant protein that over the course of the disease becomes overloaded, insufficient and dysfunctional, eventually resulting in cell degradation. Indeed, it was shown that the negative regulator of the autophagic pathway, mTOR (mammalian target of rapamycin), is sequestered into huntingtin-polyQ aggregates with subsequent inhibition of its kinase activity in HD cell models, transgenic mice, and patients' brain. This ultimately leads to the induction of autophagy and clearance of mutant huntingtin fragments, which protect cells from death (Ravikumar et al., 2004). Administration of chemical activators of autophagy or overexpression of genes implicated in autophagy enhances the clearance of mutant huntingtin, reduces aggregate formation, and improves the behavioural phenotype in HD mice, *Drosophila,* and *C.elegans* (Berger et al., 2006; Floto et al., 2007; Jia et al., 2007; Qin et al., 2003; Ravikumar et al., 2002; Ravikumar et al., 2004; Sarkar et al., 2007). In contrast, when the autophagy/lysosomal pathway is inhibited, soluble mutant huntingtin levels, aggregate formation, and toxicity increase (Ravikumar et al., 2002). Interestingly, posttranslational modifications of mutant huntingtin can modulate its clearance. First, clearance of mutant huntingtin can be achieved by acetylation at lysine residue 444 (Jeong et al., 2009). Increased acetylation at K444 facilitates trafficking of mutant huntingtin into autophagosomes, significantly improves clearance of the mutant protein by macroautophagy, and reverses the toxic effects of mutant huntingtin (Jeong et al., 2009). Second, phosphorylation of huntingtin by the inflammatory kinase IKK enhances its clearance by the proteasome and lysosome. In particular, phosphorylation of huntingtin increases clearance mediated by lysosomal-associated membrane protein 2A and Hsc70 (Thompson et al., 2009).

4. HD modeling

Histological analyses of post-mortem human HD brain samples gave limited information on molecular and cellular neurodegenerative mechanisms that lead to the disease. Thus, several animal models were developed to reproduce the neuropathology. These models have been very useful to discover novel pathological mechanisms that underlie the onset or the progression of the HD. However, they only partially reproduce features of the human disease and they are thus not appropriate to elaborate and evaluate novel therapies. This is the main reason why novel human based cellular models have recently been established.

4.1 Excitotoxic lesion models

KA is an excitatory amino acid and a non-*N*-methyl-*D*-aspartate (NMDA) glutamate receptor agonist. In the mammalian central nervous system, glutamic acid binds to its excitatory amino-acid receptors and promotes membrane depolarisation to favour transmission of synaptic information. Excessive or prolonged activation of glutamic acid receptors leads to damage and eventually excitotoxic death of the target neurons. Intra-striatal injection of KA in mice mimics many of neuropathological features of HD including specific striatal medium-sized neuronal loss (Coyle & Schwarcz, 1976; McGeer & McGeer, 1976). The modelling of HD using KA striatal injections revealed a toxic role for endogenous glutamate in the disease progression. However, KA intra-striatal injections do not perfectly reproduce features of HD because it also affects projection neurons and NADPH-positive neurons (Beal et al., 1985; Beal et al., 1986). The intra-striatal injection of the NMDA receptor agonist quinolinic acid (QA) reproduces even more faithfully the striatal lesions observed in HD by targeting a subset of medium spiny neurons - the GABAergic and substance P medium spiny neurons (Beal et al., 1986; Schwarcz & Kohler, 1983). The QA model has been successfully tested in primates with similar neuropathological lesions (Ferrante et al., 1993). The mitochondrial toxin, 3-nitropropionic acid (3-NP) is a mitochondrial inhibitor of succinate dehydrogenase that is able to mimic some mitochondrial dysfunction found in HD (Beal et al., 1993; Brouillet & Hantraye, 1995; Brouillet et al., 1999; Tunez & Santamaria, 2009). While the selective toxicity of 3-NP to striatal neurons remains unknown, its major advantage is that HD symptoms develop spontaneously after systemic administration (Reynolds et al., 1998). This model has been extended to non-human primates in which chronic systemic administration of 3-NP recapitulates behavioural, histological and neurochemical features of HD (Brouillet & Hantraye, 1995; Brouillet et al., 1995).

4.2 Genetics models of Huntington Disease

One major advance in HD research was the generation of various genetic mouse models of HD. These include knock-out, transgenic and knock-in models (Table 1).

4.2.1 Knock-out mice

Soon after the discovery of the *Hdh* gene, it was reported that homozygous deletion of the gene in mice was embryonically lethal (Duyao et al., 1995; Nasir et al., 1995; Zeitlin et al., 1995), which contrasts with the late onset of the human disease. Thus, these knock-out mice are not ideal models of HD, but they indicate that huntingtin has an essential role during embryonic development. Furthermore, huntingtin can rescue the knock-out phenotype, which indicates that the effect of the mutation is not primarily due to loss of function. Further analyses of conditional Hdh mice (Cre-loxp mouse) at different stages and in several tissues showed that conditional inactivation of Hdh in the adult mice forebrain results in progressive neurodegeneration (Dragatsis et al., 2000). This indicates that huntingtin is required postnatally for neuronal survival in cortex and striatum. A similar strategy was used to investigate the role of huntingtin in brain development, showing that the loss of huntingtin in Wnt1 cells results in congenital hydrocephalus associated with abnormalities in the choroid plexus and subcommissural organ (Dietrich et al., 2009).

Mice models	CAG expansion	Onset of symptoms	Survival	Nuclear inclusions	References
Transgenic mice (fragment of Human *IT15* gene)					
HD-N171-82Q	82Q	10 weeks	10-24 weeks	Nuclear inclusion in the cortex, striatum and amygdala	(Duan et al., 2003; Hersch & Ferrante, 2004; Schilling et al., 1999a)
R6/1	115Q	15-21 weeks	4-5 months	Nuclear and dendritic inclusions throughout the brain	(Davies et al., 1997; Mangiarini et al., 1996)
R6/2	144Q	4-5 weeks	2 months	Nuclear and dendritic inclusions throughout the brain	(Mangiarini et al., 1996)
Transgenic mice (full length of human *IT15* gene)					
YAC128	128Q	8-12 weeks	Normal life span	No inclusions	(Hodgson et al., 1999)
YAC 72	72Q	3 months	Normal life span	Inclusions in the striatum	(Hodgson et al., 1999)
Knock-in mice					
HdhQ80	80Q	No movement disorder	Normal life span	No inclusions	(Shelbourne et al., 1999)
HdhQ92	92Q	No movement disorder	Normal life span	Nuclear inclusions within the striatum	(Wheeler et al., 2000)
HdhQ111	111Q	No movement disorder	Normal life span	Nuclear inclusions within the striatum	(Wheeler et al., 2000)
Hdh$^{(CAG)150}$	150Q	60 weeks	Normal life span	Nuclear inclusions within the striatum	(Lin et al., 2001)

Table 1. This table summarizes the main characteristics of the most widely used mouse models of Huntington's Disease in fundamental and applied research. It is divided into three categories: transgenic mice bearing a fragment or full length of human IT15 gene and knock-in mice.

4.2.2 Transgenic models of Huntington Disease

In transgenic mouse models, the mutant gene, or part of it, is inserted randomly into the mouse genome, leading to the expression of a mutant protein in addition to the endogenous normal huntingtin.

Several transgenic mouse models of HD exist and are grouped in 2 categories: 1) mice expressing huntingtin N-terminal fragments, usually the first 1 or 2 exons of the human *huntingtin* gene that contain the polyQ expansion; 2) transgenic mice expressing the full-length human HD gene with an expanded polyQ tract.

The first transgenic mice models of HD include the insertion of a fragment of the human *IT15* gene coding for huntingtin. This widely used mutant mouse model, termed R6/2, contains a mutant N-terminus segment of the exon 1 of the human *IT15* gene encoding huntingtin with approximately 144 CAG expansions (Mangiarini et al., 1996). These transgenic mice exhibit progressive neurological features of human HD with choreiform-like movements and pathological cellular events such as inclusions formation at 4-5 weeks (J. Y. Li et al., 2005; Mangiarini et al., 1996). The neurological dysfunctions appear between 4-5 weeks and are followed by an early death around two-month old. However, anatomical analyses revealed that neuronal death was minimal compared to feature in human HD patients. R6/1 mutant mice that expressed a truncated *IT15* gene with around 115 CAG repeats (Davies et al., 1997) exhibit a more progressive course of disease probably due to the shorter CAG-repeats and lower expression rate of the mutant transgene, with death occurring within 4-5 months. Like in human feature of HD where the juvenile forms exhibiting high number of CAG repeats are the most dramatic, the severity of the neuropathological and neuroanatomical phenotype in mouse models of HD depends on the CAG repeat length. The N-171-82Q mouse model of HD contains a longer N-terminal fragment of huntingtin (exon 1 and exon2) with 82 CAG. In these mice, neuropathological features are more similar to human HD in that neurodegeneration is more prominent and seems more selective for the striatum (Duan et al., 2003; Hersch & Ferrante, 2004; Schilling et al., 1999a). All these transgenic mouse models represent a major benefit to study HD and each mouse model could provide information about specific biochemical abnormalities. Nevertheless, these models not faithfully reflect the neuropathological defaults observed in humans as the huntingtin fragment produce in these mutant mouse models may not be produced in the human brain.To overcome this problem, several transgenic mice with full-length human *IT15* gene were developed. Transgenic mice that express a full-length human HD cDNA with 48 or 89 CAG repeats manifested progressive behavioural motor dysfunctions with neuron loss in various brain areas including striatum and cerebral cortex but extremely rare nuclear inclusions (Reddy et al., 1998). Similar features were observed in YAC72 transgenic mice. YAC72 and YAC128 mice were developed with yeast artificial chromosome containing the full size *huntingtin* gene with 72 or 128 CAG repeats (Hodgson et al., 1999). The nuclear inclusions appear more gradually in YAC72 mice than in R6/2 models and cell loss appears limited to the striatum (Van Raamsdonk et al., 2005). YAC128 mice also show a progressive increase in total ventricular volume and a layer specific cortical atrophy, similarly to human HD patients (Carroll et al., 2011). Despite the fact that YAC mouse models closely recapitulate the region specific damage that occurs in HD, disease progression is slow (Hodgson et al., 1996).

All these models share some features with human HD. However some of them present divergences with human HD pathology. In HD patients, BDNF protein level is decreased in frontal cortex, striatum, cerebellum and substantia nigra (Zuccato et al., 2001). While cortical and striatal BDNF protein levels are reduced in the N171-82Q and R6/1 mice (Saydoff et al., 2006) like human feature of HD, they remain unchanged in R6/2 mice and conversely increase in the striatum and cerebellum of YAC72 transgenic mice (Seo et al., 2008). Moreover, a progressive age-dependant decrease of the ubiquitine proteasome system is observed in YAC72 transgenic mice, like in HD patient, but not in R6/2 mice (Seo et al., 2008). So animal models do not cover all aspects of HD but each model is valuable to study specific biochemical abnormalities.

4.2.3 Knock-in models

Knock-in mice are characterized by a progressive development of behavioural, pathological, cellular, and molecular abnormalities. These mouse models thus represent valuable tools to understand the early pathological events triggered by the mutation in humans.

Initially, knock-in models were disappointing because the first mice generated with an extended stretch of 50 or 80 CAG repeats into the endogenous mouse *Hdh* gene ((HdhQ50; HdhQ80)) showed no behavioural phenotypes or abnormalities (Shelbourne et al., 1999; White et al., 1997). Those mice don't exhibit huntingtin aggregates. However, in other knock-in mice (HdhQ92 and HdhQ111 and $Hdh^{(CAG)150}$, see below), microaggregates of huntingtin are detected in the brains of mice at 2–6 months and nuclear inclusions in older mice (10–18 months, depending on the model) in absence of cell death or abnormal behaviour (H. Li et al., 2000; Lin et al., 2001; Menalled et al., 2000; Wheeler et al., 2000). These findings suggest that neuronal dysfunction precedes cell death in HD and might be primarily responsible for early functional deficits. This correlates with the finding that subtle motor deficits precede by many years the appearance of behavioural symptoms and striatal atrophy in HD patients (smith, nature, 2000).

The phenotype describe for *Hdh* knock-ins with shorter repeats is less severe than for longer tracks suggesting that increase in repeats number produces mice with earlier age at onset that are close to human feature of HD. Wheeler and colleagues developed genetic knock-in mouse models of juvenile HD, HdhQ92 and HdhQ111, with expanded CAG repeats inserted into the murine *Hdh* gene. These mice present progressive neuropathological phenotype with specificity for striatal neurons and nuclear inclusions and insoluble aggregates (Wheeler et al., 2000). Finally the knock-in mouse model of HD $Hdh^{(CAG)150}$, with alleles of approximately 150 units, shows abnormalities, including late-onset behavioural, motor task deficit, activity disturbances and striatal injury similar to that found early in the course of human HD patients (Lin et al., 2001).

4.3 Human *in vitro* models of Huntington Disease

Besides *in vivo* animal models of HD, new *in vitro* culture models of human embryonic stem cells (hES) and human induced pluripotent stem cells (hIPS) have been developed and offer new hope to overcome the limitations of animal models.

Human ES cell lines (hES) are isolated from the inner cell mass of the embryo blastocyst (around 6 days post-fertilization) (Mateizel et al., 2010). hES cells maintain self-renewal ability and have the potential to differentiate into the three cell germ layers, endoderm, mesoderm and ectoderm. Under appropriate culture conditions, neural cell types of the central nervous system (CNS), including neurons, can be generated from hES. Such *in vitro* model represents an ideal tool for drug screening and also a promising source of neurons for cell replacement therapy in HD patients. It was actually reported that human neural precursors, derived from hES, transplanted into QA rat model of HD survived and underwent extensive migration and differentiation into DARPP32 medium spiny neurons (Aubry et al., 2008; Vazey et al., 2010). The pre-implantatory genetic diagnosis performed in embryos prior *in utero* implantation was a first step towards deriving pathological hES cell lines that carried mutations of HD (Mateizel et al., 2006; Mateizel et al., 2010; Niclis et al., 2009; Verlinsky et al., 2005). It has been described that HD-hES cells can efficiently differentiate into neurons (Niclis et al., 2009). Nevertheless, HD pathological hES cell lines represent a limiting source of information on the disease as they are very difficult to obtain for obvious questions of ethics and reproducibility.

A couple of years ago, a novel human cell model of the disease was generated. Human induced pluripotent stem cells (hIPS) were derived from skin fibroblasts of HD patients (Park et al., 2008b). Indeed, human adult somatic cells such as fibroblasts can be successfully converted into hIPS cells by expressing four genes linked to pluripotency (i.e. Oct4, klf4, c-myc and Sox2 or Oct4, Sox2, Lin28 and Nanog) (Park et al., 2008a; Park et al., 2008b; Takahashi et al., 2007; J. Yu et al., 2007). Like hES cells, hIPS cells are characterized by their ability to self-renew and pluripotency properties. In addition, hIPS can be efficiently differentiated into neural precursors, glia and neurons, including DARPP-32 medium spiny neurons (Boulting et al., 2011; Schwartz et al., 2008; Takahashi et al., 2007). Numerous biological variables including the number of CAG repetitions, the age of disease onset and the severity of the symptoms are likely to influence the response to drug treatment. Thus, the generation of patient-specific pluripotent stem cells will become a valuable resource to better characterize the physiopathological mechanisms of HD and further design the most appropriate drugs to treat each patient.

5. Conclusion

It is now well established that huntingtin is ubiquitously expressed from stem cells to mature neurons and thus plays sequential biological functions that contribute to both, the development and the homeostasis of the brain tissue. HD is a progressive neurodegenerative disorder that results from both, gain of toxic activities of polyQ huntingtin and loss of physiological functions of the corresponding wild-type protein. It is currently believed that the lack of huntingtin activity during brain development weakens neurons, which are then more susceptible to death induced by accumulation and aggregation of polyQ huntingtin. However, we still have no answer regarding the selective death of the subpopulation of striatal DARPP32+ neurons that occurs progressively as the disease worsens.

6. References

Acevedo-Torres, K., Berrios, L., Rosario, N., Dufault, V., Skatchkov, S., Eaton, M. J., Torres-Ramos, C. A. & Ayala-Torres, S. (2009). Mitochondrial DNA damage is a hallmark of chemically induced and the R6/2 transgenic model of Huntington's disease. *DNA Repair (Amst)*, Vol. 8, 1, pp.(126-136)

Antonawich, F. J., Fiore-Marasa, S. M. & Parker, C. P. (2002). Modulation of apoptotic regulatory proteins and early activation of cytochrome C following systemic 3-nitropropionic acid administration. *Brain Res Bull*, Vol. 57, 5, pp.(647-649)

Antonini, A., Leenders, K. L., Spiegel, R., Meier, D., Vontobel, P., Weigell-Weber, M., Sanchez-Pernaute, R., de Yebenez, J. G., Boesiger, P., Weindl, A. & Maguire, R. P. (1996). Striatal glucose metabolism and dopamine D2 receptor binding in asymptomatic gene carriers and patients with Huntington's disease. *Brain*, Vol. 119 (Pt 6), pp.(2085-2095)

Arenas, J., Campos, Y., Ribacoba, R., Martin, M. A., Rubio, J. C., Ablanedo, P. & Cabello, A. (1998). Complex I defect in muscle from patients with Huntington's disease. *Annals of Neurology*, Vol. 43, 3, pp.(397-400)

Arrasate, M., Mitra, S., Schweitzer, E. S., Segal, M. R. & Finkbeiner, S. (2004). Inclusion body formation reduces levels of mutant huntingtin and the risk of neuronal death. *Nature*, Vol. 431, 7010, pp.(805-810)

Aubry, L., Bugi, A., Lefort, N., Rousseau, F., Peschanski, M. & Perrier, A. L. (2008). Striatal progenitors derived from human ES cells mature into DARPP32 neurons in vitro and in quinolinic acid-lesioned rats. *Proc Natl Acad Sci U S A*, Vol. 105, 43, pp.(16707-16712)

Auerbach, W., Hurlbert, M. S., Hilditch-Maguire, P., Wadghiri, Y. Z., Wheeler, V. C., Cohen, S. I., Joyner, A. L., MacDonald, M. E. & Turnbull, D. H. (2001). The HD mutation causes progressive lethal neurological disease in mice expressing reduced levels of huntingtin. *Hum Mol Genet*, Vol. 10, 22, pp.(2515-2523.)

Augood, S. J., Faull, R. L. & Emson, P. C. (1997). Dopamine D1 and D2 receptor gene expression in the striatum in Huntington's disease. *Ann Neurol*, Vol. 42, 2, pp.(215-221)

Bachoud-Levi, A. C., Maison, P., Bartolomeo, P., Boisse, M. F., Dalla Barba, G., Ergis, A. M., Baudic, S., Degos, J. D., Cesaro, P. & Peschanski, M. (2001). Retest effects and cognitive decline in longitudinal follow-up of patients with early HD. *Neurology*, Vol. 56, 8, pp.(1052-1058)

Bae, B. I., Xu, H., Igarashi, S., Fujimuro, M., Agrawal, N., Taya, Y., Hayward, S. D., Moran, T. H., Montell, C., Ross, C. A., Snyder, S. H. & Sawa, A. (2005). p53 Mediates Cellular Dysfunction and Behavioral Abnormalities in Huntington's Disease. *Neuron*, Vol. 47, 1, pp.(29-41)

Bartel, D. P. (2009). MicroRNAs: target recognition and regulatory functions. *Cell*, Vol. 136, 2, pp.(215-233)

Beal, M. F., Marshall, P. E., Burd, G. D., Landis, D. M. & Martin, J. B. (1985). Excitotoxin lesions do not mimic the alteration of somatostatin in Huntington's disease. *Brain Res*, Vol. 361, 1-2, pp.(135-145)

Beal, M. F., Kowall, N. W., Ellison, D. W., Mazurek, M. F., Swartz, K. J. & Martin, J. B. (1986). Replication of the neurochemical characteristics of Huntington's disease by quinolinic acid. *Nature*, Vol. 321, 6066, pp.(168-171)

Beal, M. F., Brouillet, E., Jenkins, B. G., Ferrante, R. J., Kowall, N. W., Miller, J. M., Storey, E., Srivastava, R., Rosen, B. R. & Hyman, B. T. (1993). Neurochemical and histologic characterization of striatal excitotoxic lesions produced by the mitochondrial toxin 3-nitropropionic acid. *J Neurosci*, Vol. 13, 10, pp.(4181-4192)

Bence, N. F., Sampat, R. M. & Kopito, R. R. (2001). Impairment of the ubiquitin-proteasome system by protein aggregation. *Science*, Vol. 292, 5521, pp.(1552-1555.)

Berger, Z., Ravikumar, B., Menzies, F. M., Oroz, L. G., Underwood, B. R., Pangalos, M. N., Schmitt, I., Wullner, U., Evert, B. O., O'Kane, C. J. & Rubinsztein, D. C. (2006). Rapamycin alleviates toxicity of different aggregate-prone proteins. *Hum Mol Genet*, Vol. 15, 3, pp.(433-442)

Bizat, N., Hermel, J. M., Boyer, F., Jacquard, C., Creminon, C., Ouary, S., Escartin, C., Hantraye, P., Kajewski, S. & Brouillet, E. (2003a). Calpain is a major cell death effector in selective striatal degeneration induced in vivo by 3-nitropropionate: implications for Huntington's disease. *J Neurosci*, Vol. 23, 12, pp.(5020-5030)

Bizat, N., Hermel, J. M., Humbert, S., Jacquard, C., Creminon, C., Escartin, C., Saudou, F., Krajewski, S., Hantraye, P. & Brouillet, E. (2003b). In Vivo Calpain/Caspase Cross-talk during 3-Nitropropionic Acid-induced Striatal Degeneration: Implication of a calpain-mediated cleavage of active caspase-3. *J Biol Chem*, Vol. 278, 44, pp.(43245-43253)

Block-Galarza, J., Chase, K. O., Sapp, E., Vaughn, K. T., Vallee, R. B., DiFiglia, M. & Aronin, N. (1997). Fast transport and retrograde movement of huntingtin and HAP 1 in axons. *Neuroreport*, Vol. 8, 9-10, pp.(2247-2251)

Borrell-Pages, M., Zala, D., Humbert, S. & Saudou, F. (2006). Huntington's disease: from huntingtin function and dysfunction to therapeutic strategies. *Cell Mol Life Sci*, Vol. 63, 22, pp.(2642-2660)

Boulting, G. L., Kiskinis, E., Croft, G. F., Amoroso, M. W., Oakley, D. H., Wainger, B. J., Williams, D. J., Kahler, D. J., Yamaki, M., Davidow, L., Rodolfa, C. T., Dimos, J. T., Mikkilineni, S., MacDermott, A. B., Woolf, C. J., Henderson, C. E., Wichterle, H. & Eggan, K. (2011). A functionally characterized test set of human induced pluripotent stem cells. *Nat Biotechnol*, Vol. 29, 3, pp.(279-286)

Boutell, J. M., Thomas, P., Neal, J. W., Weston, V. J., Duce, J., Harper, P. S. & Jones, A. L. (1999). Aberrant interactions of transcriptional repressor proteins with the Huntington's disease gene product, huntingtin. *Hum Mol Genet*, Vol. 8, 9, pp.(1647-1655)

Brennan, W. A., Jr., Bird, E. D. & Aprille, J. R. (1985). Regional mitochondrial respiratory activity in Huntington's disease brain. *J Neurochem*, Vol. 44, 6, pp.(1948-1950)

Brouillet, E. & Hantraye, P. (1995). Effects of chronic MPTP and 3-nitropropionic acid in nonhuman primates. *Curr Opin Neurol*, Vol. 8, 6, pp.(469-473)

Brouillet, E., Hantraye, P., Ferrante, R. J., Dolan, R., Leroy-Willig, A., Kowall, N. W. & Beal, M. F. (1995). Chronic mitochondrial energy impairment produces selective striatal degeneration and abnormal choreiform movements in primates. *Proc Natl Acad Sci U S A*, Vol. 92, 15, pp.(7105-7109)

Brouillet, E., Conde, F., Beal, M. F. & Hantraye, P. (1999). Replicating Huntington's disease phenotype in experimental animals. *Prog Neurobiol*, Vol. 59, 5, pp.(427-468)

Browne, S. E., Bowling, A. C., MacGarvey, U., Baik, M. J., Berger, S. C., Muqit, M. M., Bird, E. D. & Beal, M. F. (1997). Oxidative damage and metabolic dysfunction in Huntington's disease: selective vulnerability of the basal ganglia. *Annals of Neurology*, Vol. 41, 5, pp.(646-653)

Bruce, A. W., Donaldson, I. J., Wood, I. C., Yerbury, S. A., Sadowski, M. I., Chapman, M., Göttgens, B. & Buckley, N. J. (2004). Genome-wide analysis of repressor element 1 silencing transcription factor/neuron-restrictive silencing factor (REST/NRSF) target genes. *PNAS*, Vol. 101, 28, pp.(10458-10463)

Butterworth, J., Yates, C. M. & Reynolds, G. P. (1985). Distribution of phosphate-activated glutaminase, succinic dehydrogenase, pyruvate dehydrogenase and gamma-glutamyl transpeptidase in post-mortem brain from Huntington's disease and agonal cases. *J Neurol Sci*, Vol. 67, 2, pp.(161-171)

Carroll, J. B., Lerch, J. P., Franciosi, S., Spreeuw, A., Bissada, N., Henkelman, R. M. & Hayden, M. R. (2011). Natural history of disease in the YAC128 mouse reveals a discrete signature of pathology in Huntington disease. *Neurobiol Dis*, Vol. 43, 1, pp.(257-265)

Caviston, J. P., Ross, J. L., Antony, S. M., Tokito, M. & Holzbaur, E. L. (2007). Huntingtin facilitates dynein/dynactin-mediated vesicle transport. *Proc Natl Acad Sci U S A*, Vol. 104, 24, pp.(10045-10050)

Cha, J. H., Frey, A. S., Alsdorf, S. A., Kerner, J. A., Kosinski, C. M., Mangiarini, L., Penney, J. B., Jr., Davies, S. W., Bates, G. P. & Young, A. B. (1999). Altered neurotransmitter receptor expression in transgenic mouse models of Huntington's disease. *Philos Trans R Soc Lond B Biol Sci*, Vol. 354, 1386, pp.(981-989)

Chang, D. T., Rintoul, G. L., Pandipati, S. & Reynolds, I. J. (2006). Mutant huntingtin aggregates impair mitochondrial movement and trafficking in cortical neurons. *Neurobiol Dis*, Vol. 22, 2, pp.(388-400)

Charrin, B. C., Saudou, F. & Humbert, S. (2005). Axonal transport failure in neurodegenerative disorders: the case of Huntington's disease. *Pathol Biol (Paris)*, Vol. 53, 4, pp.(189-192)

Chen-Plotkin, A. S., Sadri-Vakili, G., Yohrling, G. J., Braveman, M. W., Benn, C. L., Glajch, K. E., DiRocco, D. P., Farrell, L. A., Krainc, D., Gines, S., MacDonald, M. E. & Cha, J. H. (2006). Decreased association of the transcription factor Sp1 with genes downregulated in Huntington's disease. *Neurobiol Dis*, Vol. 22, 2, pp.(233-241)

Choo, Y. S., Johnson, G. V., MacDonald, M., Detloff, P. J. & Lesort, M. (2004). Mutant huntingtin directly increases susceptibility of mitochondria to the calcium-induced permeability transition and cytochrome c release. *Hum Mol Genet*, Vol. 13, 14, pp.(1407-1420)

Ciarmiello, A., Cannella, M., Lastoria, S., Simonelli, M., Frati, L., Rubinsztein, D. C. & Squitieri, F. (2006). Brain white-matter volume loss and glucose hypometabolism precede the clinical symptoms of Huntington's disease. *J Nucl Med*, Vol. 47, 2, pp.(215-222)

Clabough, E. B. & Zeitlin, S. O. (2006). Deletion of the triplet repeat encoding polyglutamine within the mouse Huntington's disease gene results in subtle behavioral/motor phenotypes in vivo and elevated levels of ATP with cellular senescence in vitro. *Hum Mol Genet*, Vol. 15, 4, pp.(607-623)

Colin, E., Zala, D., Liot, G., Rangone, H., Borrell-Pages, M., Li, X. J., Saudou, F. & Humbert, S. (2008). Huntingtin phosphorylation acts as a molecular switch for anterograde/retrograde transport in neurons. *Embo J*, Vol. 27, 15, pp.(2124-2134)

Colomer, V., Engelender, S., Sharp, A. H., Duan, K., Cooper, J. K., Lanahan, A., Lyford, G., Worley, P. & Ross, C. A. (1997). Huntingtin-associated protein 1 (HAP1) binds to a Trio-like polypeptide, with a rac1 guanine nucleotide exchange factor domain. *Human Molecular Genetics*, Vol. 6, 9, pp.(1519-1525)

Cong, S. Y., Pepers, B. A., Evert, B. O., Rubinsztein, D. C., Roos, R. A., van Ommen, G. J. & Dorsman, J. C. (2005). Mutant huntingtin represses CBP, but not p300, by binding and protein degradation. *Mol Cell Neurosci*, Vol. 4, pp.(560-571)

Cornett, J., Cao, F., Wang, C. E., Ross, C. A., Bates, G. P., Li, S. H. & Li, X. J. (2005). Polyglutamine expansion of huntingtin impairs its nuclear export. *Nat Genet*, Vol. 37, 2, pp.(198-204)

Coyle, J. T. & Schwarcz, R. (1976). Lesion of striatal neurones with kainic acid provides a model for Huntington's chorea. *Nature*, Vol. 263, 5574, pp.(244-246)

Coyle, J. T. & Puttfarcken, P. (1993). Oxidative stress, glutamate, and neurodegenerative disorders. *Science*, Vol. 262, 5134, pp.(689-695)

Craufurd, D., Thompson, J. C. & Snowden, J. S. (2001). Behavioral changes in Huntington Disease. *Neuropsychiatry Neuropsychol Behav Neurol*, Vol. 14, 4, pp.(219-226)

Davies, S. W., Turmaine, M., Cozens, B. A., DiFiglia, M., Sharp, A. H., Ross, C. A., Scherzinger, E., Wanker, E. E., Mangiarini, L. & Bates, G. P. (1997). Formation of neuronal intranuclear inclusions underlies the neurological dysfunction in mice transgenic for the HD mutation. *Cell*, Vol. 90, 3, pp.(537-548)

del Toro, D., Alberch, J., Lazaro-Dieguez, F., Martin-Ibanez, R., Xifro, X., Egea, G. & Canals, J. M. (2009). Mutant huntingtin impairs post-Golgi trafficking to lysosomes by delocalizing optineurin/Rab8 complex from the Golgi apparatus. *Mol Biol Cell*, Vol. 20, 5, pp.(1478-1492)

Diekmann, H., Anichtchik, O., Fleming, A., Futter, M., Goldsmith, P., Roach, A. & Rubinsztein, D. C. (2009). Decreased BDNF levels are a major contributor to the embryonic phenotype of huntingtin knockdown zebrafish. *J Neurosci*, Vol. 29, 5, pp.(1343-1349)

Dietrich, P., Shanmugasundaram, R., Shuyu, E. & Dragatsis, I. (2009). Congenital hydrocephalus associated with abnormal subcommissural organ in mice lacking huntingtin in Wnt1 cell lineages. *Hum Mol Genet*, Vol. 18, 1, pp.(142-150)

DiFiglia, M., Sapp, E., Chase, K., Schwarz, C., Meloni, A., Young, C., Martin, E., Vonsattel, J. P., Carraway, R., Reeves, S. A. & et al. (1995). Huntingtin is a cytoplasmic protein associated with vesicles in human and rat brain neurons. *Neuron*, Vol. 14, 5, pp.(1075-1081)

DiFiglia, M., Sapp, E., Chase, K. O., Davies, S. W., Bates, G. P., Vonsattel, J. P. & Aronin, N. (1997). Aggregation of huntingtin in neuronal intranuclear inclusions and dystrophic neurites in brain. *Science*, Vol. 277, 5334, pp.(1990-1993)

DiFiglia, M. (2002). Huntingtin Fragments that Aggregate Go Their Separate Ways. *Mol Cell*, Vol. 10, 2, pp.(224.)

Dragatsis, I., Efstratiadis, A. & Zeitlin, S. (1998). Mouse mutant embryos lacking huntingtin are rescued from lethality by wild-type extraembryonic tissues. *Development*, Vol. 125, 8, pp.(1529-1539)

Dragatsis, I., Levine, M. S. & Zeitlin, S. (2000). Inactivation of hdh in the brain and testis results in progressive neurodegeneration and sterility in mice. *Nat Genet*, Vol. 26, 3, pp.(300-306)

Duan, W., Guo, Z., Jiang, H., Ware, M., Li, X. J. & Mattson, M. P. (2003). Dietary restriction normalizes glucose metabolism and BDNF levels, slows disease progression, and increases survival in huntingtin mutant mice. *Proc Natl Acad Sci U S A*, Vol. 100, 5, pp.(2911-2916)

Dunah, A. W., Jeong, H., Griffin, A., Kim, Y. M., Standaert, D. G., Hersch, S. M., Mouradian, M. M., Young, A. B., Tanese, N. & Krainc, D. (2002). Sp1 and TAFII130 transcriptional activity disrupted in early Huntington's disease. *Science*, Vol. 296, 5576, pp.(2238-2243.)

Duyao, M. P., Auerbach, A. B., Ryan, A., Persichetti, F., Barnes, G. T., McNeil, S. M., Ge, P., Vonsattel, J. P., Gusella, J. F., Joyner, A. L. & et al. (1995). Inactivation of the mouse Huntington's disease gene homolog Hdh. *Science*, Vol. 269, 5222, pp.(407-410)

Engelender, S., Sharp, A. H., Colomer, V., Tokito, M. K., Lanahan, A., Worley, P., Holzbaur, E. L. & Ross, C. A. (1997). Huntingtin-associated protein 1 (HAP1) interacts with the p150Glued subunit of dynactin. *Hum Mol Genet*, Vol. 6, 13, pp.(2205-2212)

Engqvist-Goldstein, A. E., Warren, R. A., Kessels, M. M., Keen, J. H., Heuser, J. & Drubin, D. G. (2001). The actin-binding protein Hip1R associates with clathrin during early stages of endocytosis and promotes clathrin assembly in vitro. *J Cell Biol*, Vol. 154, 6, pp.(1209-1223.)

Fan, M. M. & Raymond, L. A. (2007). N-methyl-D-aspartate (NMDA) receptor function and excitotoxicity in Huntington's disease. *Prog Neurobiol*, Vol. 81, 5-6, pp.(272-293)

Ferrante, R. J., Kowall, N. W., Cipolloni, P. B., Storey, E. & Beal, M. F. (1993). Excitotoxin lesions in primates as a model for Huntington's disease: histopathologic and neurochemical characterization. *Exp Neurol*, Vol. 119, 1, pp.(46-71)

Ferrante, R. J., Andreassen, O. A., Dedeoglu, A., Ferrante, K. L., Jenkins, B. G., Hersch, S. M. & Beal, M. F. (2002). Therapeutic effects of coenzyme Q10 and remacemide in transgenic mouse models of Huntington's disease. *J Neurosci*, Vol. 22, 5, pp.(1592-1599.)

Floto, R. A., Sarkar, S., Perlstein, E. O., Kampmann, B., Schreiber, S. L. & Rubinsztein, D. C. (2007). Small molecule enhancers of rapamycin-induced TOR inhibition promote autophagy, reduce toxicity in Huntington's disease models and enhance killing of mycobacteria by macrophages. *Autophagy*, Vol. 3, 6, pp.(620-622)

Folstein, S. E., Leigh, R. J., Parhad, I. M. & Folstein, M. F. (1986). The diagnosis of Huntington's disease. *Neurology*, Vol. 36, 10, pp.(1279-1283)

Foroud, T., Gray, J., Ivashina, J. & Conneally, P. M. (1999). Differences in duration of Huntington's disease based on age at onset. *J Neurol Neurosurg Psychiatry*, Vol. 66, 1, pp.(52-56)

Friedman, M. J., Wang, C. E., Li, X. J. & Li, S. (2008). Polyglutamine expansion reduces the association of TATA-binding protein with DNA and induces DNA binding-independent neurotoxicity. *J Biol Chem*, Vol. 283, 13, pp.(8283-8290)

Gafni, J. & Ellerby, L. M. (2002). Calpain activation in Huntington's disease. *J Neurosci*, Vol. 22, 12, pp.(4842-4849.)

Gafni, J., Hermel, E., Young, J. E., Wellington, C. L., Hayden, M. R. & Ellerby, L. M. (2004). Inhibition of calpain cleavage of huntingtin reduces toxicity: accumulation of calpain/caspase fragments in the nucleus. *J Biol Chem*, Vol. 279, 19, pp.(20211-20220)

Gardian, G., Browne, S. E., Choi, D. K., Klivenyi, P., Gregorio, J., Kubilus, J. K., Ryu, H., Langley, B., Ratan, R. R., Ferrante, R. J. & Beal, M. F. (2005). Neuroprotective effects of phenylbutyrate in the N171-82Q transgenic mouse model of Huntington's disease. *J Biol Chem*, Vol. 280, 1, pp.(556-563)

Garnett, E. S., Firnau, G., Nahmias, C., Carbotte, R. & Bartolucci, G. (1984). Reduced striatal glucose consumption and prolonged reaction time are early features in Huntington's disease. *J Neurol Sci*, Vol. 65, 2, pp.(231-237)

Gauthier, L. R., Charrin, B. C., Borrell-Pages, M., Dompierre, J. P., Rangone, H., Cordelieres, F. P., De Mey, J., MacDonald, M. E., Lessmann, V., Humbert, S. & Saudou, F. (2004). Huntingtin controls neurotrophic support and survival of neurons by enhancing BDNF vesicular transport along microtubules. *Cell*, Vol. 118, 1, pp.(127-138)

Gervais, F. G., Singaraja, R., Xanthoudakis, S., Gutekunst, C. A., Leavitt, B. R., Metzler, M., Hackam, A. S., Tam, J., Vaillancourt, J. P., Houtzager, V., Rasper, D. M., Roy, S., Hayden, M. R. & Nicholson, D. W. (2002). Recruitment and activation of caspase-8 by the Huntingtin-interacting protein Hip-1 and a novel partner Hippi. *Nat Cell Biol*, Vol. 4, 2, pp.(95-105.)

Gil, J. M. & Rego, A. C. (2008). Mechanisms of neurodegeneration in Huntington's disease. *Eur J Neurosci*, Vol. 27, 11, pp.(2803-2820)

Giorgini, F., Guidetti, P., Nguyen, Q., Bennett, S. C. & Muchowski, P. J. (2005). A genomic screen in yeast implicates kynurenine 3-monooxygenase as a therapeutic target for Huntington disease. *Nat Genet*, Vol. 37, 5, pp.(526-531)

Godin, J. D., Colombo, K., Molina-Calavita, M., Keryer, G., Zala, D., Charrin, B. C., Dietrich, P., Volvert, M. L., Guillemot, F., Dragatsis, I., Bellaiche, Y., Saudou, F., Nguyen, L. & Humbert, S. (2010). Huntingtin is required for mitotic spindle orientation and mammalian neurogenesis. *Neuron*, Vol. 67, 3, pp.(392-406)

Goffredo, D., Rigamonti, D., Tartari, M., De Micheli, A., Verderio, C., Matteoli, M., Zuccato, C. & Cattaneo, E. (2002). Calcium-dependent Cleavage of Endogenous Wild-type Huntingtin in Primary Cortical Neurons. *J Biol Chem*, Vol. 277, 42, pp.(39594-39598.)

Goldberg, Y. P., Nicholson, D. W., Rasper, D. M., Kalchman, M. A., Koide, H. B., Graham, R. K., Bromm, M., Kazemi-Esfarjani, P., Thornberry, N. A., Vaillancourt, J. P. & Hayden, M. R. (1996). Cleavage of huntingtin by apopain, a proapoptotic cysteine protease, is modulated by the polyglutamine tract. *Nat Genet*, Vol. 13, 4, pp.(442-449)

Grafton, S. T., Mazziotta, J. C., Pahl, J. J., St George-Hyslop, P., Haines, J. L., Gusella, J., Hoffman, J. M., Baxter, L. R. & Phelps, M. E. (1990). A comparison of neurological, metabolic, structural, and genetic evaluations in persons at risk for Huntington's disease. *Ann Neurol*, Vol. 28, 5, pp.(614-621)

Graham, R. K., Deng, Y., Slow, E. J., Haigh, B., Bissada, N., Lu, G., Pearson, J., Shehadeh, J., Bertram, L., Murphy, Z., Warby, S. C., Doty, C. N., Roy, S., Wellington, C. L., Leavitt, B. R., Raymond, L. A., Nicholson, D. W. & Hayden, M. R. (2006). Cleavage at the caspase-6 site is required for neuronal dysfunction and degeneration due to mutant huntingtin. *Cell*, Vol. 125, 6, pp.(1179-1191)

Gu, M., Gash, M. T., Mann, V. M., Javoy-Agid, F., Cooper, J. M. & Schapira, A. H. (1996). Mitochondrial defect in Huntington's disease caudate nucleus. *Ann Neurol*, Vol. 39, 3, pp.(385-389)

Guidetti, P. & Schwarcz, R. (2003). 3-Hydroxykynurenine and quinolinate: pathogenic synergism in early grade Huntington's disease? *Adv Exp Med Biol*, Vol. 527, pp.(137-145)

Guidetti, P., Luthi-Carter, R. E., Augood, S. J. & Schwarcz, R. (2004). Neostriatal and cortical quinolinate levels are increased in early grade Huntington's disease. *Neurobiol Dis*, Vol. 17, 3, pp.(455-461)

Guidetti, P., Bates, G. P., Graham, R. K., Hayden, M. R., Leavitt, B. R., MacDonald, M. E., Slow, E. J., Wheeler, V. C., Woodman, B. & Schwarcz, R. (2006). Elevated brain 3-hydroxykynurenine and quinolinate levels in Huntington disease mice. *Neurobiol Dis*, Vol. 23, 1, pp.(190-197)

Gunawardena, S., Her, L. S., Brusch, R. G., Laymon, R. A., Niesman, I. R., Gordesky-Gold, B., Sintasath, L., Bonini, N. M. & Goldstein, L. S. (2003). Disruption of axonal transport by loss of huntingtin or expression of pathogenic polyQ proteins in Drosophila. *Neuron*, Vol. 40, 1, pp.(25-40)

Gutekunst, C. A., Levey, A. I., Heilman, C. J., Whaley, W. L., Yi, H., Nash, N. R., Rees, H. D., Madden, J. J. & Hersch, S. M. (1995). Identification and localization of huntingtin in brain and human lymphoblastoid cell lines with anti-fusion protein antibodies. *Proc Natl Acad Sci U S A*, Vol. 92, 19, pp.(8710-8714)

Gutekunst, C. A., Li, S. H., Yi, H., Mulroy, J. S., Kuemmerle, S., Jones, R., Rye, D., Ferrante, R. J., Hersch, S. M. & Li, X. J. (1999). Nuclear and neuropil aggregates in Huntington's disease: relationship to neuropathology. *J Neurosci*, Vol. 19, 7, pp.(2522-2534)

Hackam, A. S., Singaraja, R., Wellington, C. L., Metzler, M., McCutcheon, K., Zhang, T., Kalchman, M. & Hayden, M. R. (1998). The influence of huntingtin protein size on nuclear localization and cellular toxicity. *Journal of Cell Biology*, Vol. 141, pp.(1097-1105)

Harjes, P. & Wanker, E. E. (2003). The hunt for huntingtin function: interaction partners tell many different stories. *Trends Biochem Sci*, Vol. 28, 8, pp.(425-433)

Hattula, K. & Peranen, J. (2000). FIP-2, a coiled-coil protein, links huntingtin to Rab8 and modulates cellular morphogenesis. *Curr. Biol.*, Vol. 24, pp.(1603-1606)

HDCRG. (1993). A novel gene containing a trinucleotide repeat that is expanded and unstable on Huntington's disease chromosomes. *Cell*, Vol. 72, 6, pp.(971-983)

Henshall, T. L., Tucker, B., Lumsden, A. L., Nornes, S., Lardelli, M. T. & Richards, R. I. (2009). Selective neuronal requirement for huntingtin in the developing zebrafish. *Hum Mol Genet*, Vol. 18, 24, pp.(4830-4842)

Hermel, E., Gafni, J., Propp, S. S., Leavitt, B. R., Wellington, C. L., Young, J. E., Hackam, A. S., Logvinova, A. V., Peel, A. L., Chen, S. F., Hook, V., Singaraja, R., Krajewski, S., Goldsmith, P. C., Ellerby, H. M., Hayden, M. R., Bredesen, D. E. & Ellerby, L. M. (2004). Specific caspase interactions and amplification are involved in selective neuronal vulnerability in Huntington's disease. *Cell Death Differ*, Vol. 11, 4, pp.(424-438)

Hersch, S. M. & Ferrante, R. J. (2004). Translating therapies for Huntington's disease from genetic animal models to clinical trials. *NeuroRx*, Vol. 1, 3, pp.(298-306)

Hodges, A., Strand, A. D., Aragaki, A. K., Kuhn, A., Sengstag, T., Hughes, G., Elliston, L. A., Hartog, C., Goldstein, D. R., Thu, D., Hollingsworth, Z. R., Collin, F., Synek, B., Holmans, P. A., Young, A. B., Wexler, N. S., Delorenzi, M., Kooperberg, C., Augood, S. J., Faull, R. L., Olson, J. M., Jones, L. & Luthi-Carter, R. (2006). Regional and cellular gene expression changes in human Huntington's disease brain. *Hum Mol Genet*, Vol. 15, 6, pp.(965-977)

Hodgson, J. G., Smith, D. J., McCutcheon, K., Koide, H. B., Nishiyama, K., Dinulos, M. B., Stevens, M. E., Bissada, N., Nasir, J., Kanazawa, I., Disteche, C. M., Rubin, E. M. & Hayden, M. R. (1996). Human huntingtin derived from YAC transgenes compensates for loss of murine huntingtin by rescue of the embryonic lethal phenotype. *Human Molecular Genetics*, Vol. 5, 12, pp.(1875-1885)

Hodgson, J. G., Agopyan, N., Gutekunst, C. A., Leavitt, B. R., LePiane, F., Singaraja, R., Smith, D. J., Bissada, N., McCutcheon, K., Nasir, J., Jamot, L., Li, X. J., Stevens, M. E., Rosemond, E., Roder, J. C., Phillips, A. G., Rubin, E. M., Hersch, S. M. & Hayden, M. R. (1999). A YAC mouse model for Huntington's disease with full-length mutant huntingtin, cytoplasmic toxicity, and selective striatal neurodegeneration. *Neuron*, Vol. 23, 1, pp.(181-192)

Holbert, S., Denghien, I., Kiechle, T., Rosenblatt, A., Wellington, C., Hayden, M. R., Margolis, R. L., Ross, C. A., Dausset, J., Ferrante, R. J. & Neri, C. (2001). The Gln-Ala repeat transcriptional activator CA150 interacts with huntingtin: neuropathologic and genetic evidence for a role in Huntington's disease pathogenesis. *Proc Natl Acad Sci U S A*, Vol. 98, 4, pp.(1811-1816.)

Holbert, S., Dedeoglu, A., Humbert, S., Saudou, F., Ferrante, R. J. & Neri, C. (2003). Cdc42-interacting protein 4 binds to huntingtin: neuropathologic and biological evidence for a role in Huntington's disease. *Proc Natl Acad Sci U S A*, Vol. 100, 5, pp.(2712-2717)

Hoogeveen, A. T., Willemsen, R., Meyer, N., de Rooij, K. E., Roos, R. A., van Ommen, G. J. & Galjaard, H. (1993). Characterization and localization of the Huntington disease gene product. *Hum Mol Genet*, Vol. 2, 12, pp.(2069-2073)

Horton, T. M., Graham, B. H., Corral-Debrinski, M., Shoffner, J. M., Kaufman, A. E., Beal, M. F. & Wallace, D. C. (1995). Marked increase in mitochondrial DNA deletion levels in the cerebral cortex of Huntington's disease patients. *Neurology*, Vol. 45, 10, pp.(1879-1883)

Huang, C. C., Faber, P. W., Persichetti, F., Mittal, V., Vonsattel, J. P., MacDonald, M. E. & Gusella, J. F. (1998). Amyloid formation by mutant huntingtin: threshold, progressivity and recruitment of normal polyglutamine proteins. *Somat Cell Mol Genet*, Vol. 24, 4, pp.(217-233)

Huot, P., Levesque, M. & Parent, A. (2007). The fate of striatal dopaminergic neurons in Parkinson's disease and Huntington's chorea. *Brain*, Vol. 130, Pt 1, pp.(222-232)

Imarisio, S., Carmichael, J., Korolchuk, V., Chen, C. W., Saiki, S., Rose, C., Krishna, G., Davies, J. E., Ttofi, E., Underwood, B. R. & Rubinsztein, D. C. (2008). Huntington's disease: from pathology and genetics to potential therapies. *Biochem J*, Vol. 412, 2, pp.(191-209)

Jana, N. R., Zemskov, E. A., Wang, G. & Nukina, N. (2001). Altered proteasomal function due to the expression of polyglutamine- expanded truncated N-terminal huntingtin induces apoptosis by caspase activation through mitochondrial cytochrome c release. *Hum Mol Genet*, Vol. 10, 10, pp.(1049-1059.)

Jenkins, B. G., Koroshetz, W. J., Beal, M. F. & Rosen, B. R. (1993). Evidence for impairment of energy metabofism in vivo in Huntington's disease using localized lH NMR spectroscopy. *Neurology*, Vol. 43, pp.(2689-2695)

Jeong, H., Then, F., Melia, T. J., Jr., Mazzulli, J. R., Cui, L., Savas, J. N., Voisine, C., Paganetti, P., Tanese, N., Hart, A. C., Yamamoto, A. & Krainc, D. (2009). Acetylation targets mutant huntingtin to autophagosomes for degradation. *Cell*, Vol. 137, 1, pp.(60-72)

Jia, K., Hart, A. C. & Levine, B. (2007). Autophagy genes protect against disease caused by polyglutamine expansion proteins in Caenorhabditis elegans. *Autophagy*, Vol. 3, 1, pp.(21-25)

Johnson, R., Zuccato, C., Belyaev, N. D., Guest, D. J., Cattaneo, E. & Buckley, N. J. (2008). A microRNA-based gene dysregulation pathway in Huntington's disease. *Neurobiol Dis*, Vol. 29, 3, pp.(438-445)

Johnson, R. & Buckley, N. J. (2009). Gene dysregulation in Huntington's disease: REST, microRNAs and beyond. *Neuromolecular Med*, Vol. 11, 3, pp.(183-199)

Kalchman, M. A., Koide, H. B., McCutcheon, K., Graham, R. K., Nichol, K., Nishiyama, K., Kazemi-Esfarjani, P., Lynn, F. C., Wellington, C., Metzler, M., Goldberg, Y. P., Kanazawa, I., Gietz, R. D. & Hayden, M. R. (1997). HIP1, a human homologue of S. cerevisiae Sla2p, interacts with membrane-associated huntingtin in the brain. *Nat Genet*, Vol. 16, 1, pp.(44-53)

Kaltenbach, L. S., Romero, E., Becklin, R. R., Chettier, R., Bell, R., Phansalkar, A., Strand, A., Torcassi, C., Savage, J., Hurlburt, A., Cha, G. H., Ukani, L., Chepanoske, C. L., Zhen, Y., Sahasrabudhe, S., Olson, J., Kurschner, C., Ellerby, L. M., Peltier, J. M., Botas, J. & Hughes, R. E. (2007). Huntingtin interacting proteins are genetic modifiers of neurodegeneration. *PLoS Genet*, Vol. 3, 5, pp.(e82)

Kegel, K. B., Kim, M., Sapp, E., McIntyre, C., Castano, J. G., Aronin, N. & DiFiglia, M. (2000). Huntingtin expression stimulates endosomal-lysosomal activity, endosome tubulation, and autophagy. *J Neurosci*, Vol. 20, 19, pp.(7268-7278)

Kegel, K. B., Meloni, A. R., Yi, Y., Kim, Y. J., Doyle, E., Cuiffo, B. G., Sapp, E., Wang, Y., Qin, Z. H., Chen, J. D., Nevins, J. R., Aronin, N. & DiFiglia, M. (2002). Huntingtin is present in the nucleus, interacts with the transcriptional corepressor C-terminal binding protein, and represses transcription. *J Biol Chem*, Vol. 277, 9, pp.(7466-7476.)

Kegel, K. B., Sapp, E., Yoder, J., Cuiffo, B., Sobin, L., Kim, Y. J., Qin, Z. H., Hayden, M. R., Aronin, N., Scott, D. L., Isenberg, G., Goldmann, W. H. & DiFiglia, M. (2005). Huntingtin associates with acidic phospholipids at the plasma membrane. *J Biol Chem*, Vol. 280, 43, pp.(36464-36473)

Kiechle, T., Dedeoglu, A., Kubilus, J., Kowall, N. W., Beal, M. F., Friedlander, R. M., Hersch, S. M. & Ferrante, R. J. (2002). Cytochrome C and Caspase-9 Expression in Huntington 's Disease. *NeuroMolecular Medicine*, Vol. 1, 3, pp.(183-196)

Kim, M., Lee, H. S., LaForet, G., McIntyre, C., Martin, E. J., Chang, P., Kim, T. W., Williams, M., Reddy, P. H., Tagle, D., Boyce, F. M., Won, L., Heller, A., Aronin, N. & DiFiglia, M. (1999). Mutant huntingtin expression in clonal striatal cells: dissociation of inclusion formation and neuronal survival by caspase inhibition. *J Neurosci*, Vol. 19, 3, pp.(964-973)

Kim, M., Roh, J. K., Yoon, B. W., Kang, L., Kim, Y. J., Aronin, N. & DiFiglia, M. (2003). Huntingtin is degraded to small fragments by calpain after ischemic injury. *Exp Neurol*, Vol. 183, 1, pp.(109-115)

Kim, Y. J., Yi, Y., Sapp, E., Wang, Y., Cuiffo, B., Kegel, K. B., Qin, Z. H., Aronin, N. & DiFiglia, M. (2001). Caspase 3-cleaved N-terminal fragments of wild-type and mutant huntingtin are present in normal and Huntington's disease brains, associate with membranes, and undergo calpain-dependent proteolysis. *Proc Natl Acad Sci U S A*, Vol. 98, 22, pp.(12784-12789.)

Koroshetz, W. J., Jenkins, B. G., Rosen, B. R. & Beal, M. F. (1997). Energy metabolism defects in Huntington's disease and effect of coenzyme Q10. *Annals of Neurology*, Vol. 41, pp.(160-165)

Kuemmerle, S., Gutekunst, C. A., Klein, A. M., Li, X. J., Li, S. H., Beal, M. F., Hersch, S. M. & Ferrante, R. J. (1999). Huntington aggregates may not predict neuronal death in Huntington's disease. *Ann Neurol*, Vol. 46, 6, pp.(842-849)

Kuhl, D. E., Phelps, M. E., Markham, C. H., Metter, E. J., Riege, W. H. & Winter, J. (1982). Cerebral metabolism and atrophy in Huntington's disease determined by 18FDG and computed tomographic scan. *Ann Neurol*, Vol. 12, 5, pp.(425-434)

Kuwert, T., Lange, H. W., Langen, K. J., Herzog, H., Aulich, A. & Feinendegen, L. E. (1990). Cortical and subcortical glucose consumption measured by PET in patients with Huntington's disease. *Brain*, Vol. 113 (Pt 5), pp.(1405-1423)

Kuwert, T., Lange, H. W., Boecker, H., Titz, H., Herzog, H., Aulich, A., Wang, B. C., Nayak, U. & Feinendegen, L. E. (1993). Striatal glucose consumption in chorea-free subjects at risk of Huntington's disease. *J Neurol*, Vol. 241, 1, pp.(31-36)

Leavitt, B. R., Guttman, J. A., Hodgson, J. G., Kimel, G. H., Singaraja, R., Vogl, A. W. & Hayden, M. R. (2001). Wild-type huntingtin reduces the cellular toxicity of mutant huntingtin in vivo. *Am J Hum Genet*, Vol. 68, 2, pp.(313-324.)

Leavitt, B. R., Raamsdonk, J. M., Shehadeh, J., Fernandes, H., Murphy, Z., Graham, R. K., Wellington, C. L., Raymond, L. A. & Hayden, M. R. (2006). Wild-type huntingtin protects neurons from excitotoxicity. *J Neurochem*, Vol. 96, 4, pp.(1121-1129)

Lee, S. T., Chu, K., Im, W. S., Yoon, H. J., Im, J. Y., Park, J. E., Park, K. H., Jung, K. H., Lee, S. K., Kim, M. & Roh, J. K. (2011). Altered microRNA regulation in Huntington's disease models. *Exp Neurol*, Vol. 227, 1, pp.(172-179)

Lee, W. C., Yoshihara, M. & Littleton, J. T. (2004). Cytoplasmic aggregates trap polyglutamine-containing proteins and block axonal transport in a Drosophila model of Huntington's disease. *Proc Natl Acad Sci U S A*, Vol. 101, 9, pp.(3224-3229)

Li, B., Chen, N., Luo, T., Otsu, Y., Murphy, T. H. & Raymond, L. A. (2002). Differential regulation of synaptic and extra-synaptic NMDA receptors. *Nature Neuroscience*, Vol. 5, 9, pp.(833-834)

Li, H., Li, S. H., Cheng, A. L., Mangiarini, L., Bates, G. P. & Li, X. J. (1999). Ultrastructural localization and progressive formation of neuropil aggregates in Huntington's disease transgenic mice. *Hum Mol Genet*, Vol. 8, 7, pp.(1227-1236)

Li, H., Li, S. H., Johnston, H., Shelbourne, P. F. & Li, X. J. (2000). Amino-terminal fragments of mutant huntingtin show selective accumulation in striatal neurons and synaptic toxicity. *Nat Genet*, Vol. 25, 4, pp.(385-389)

Li, H., Li, S. H., Yu, Z. X., Shelbourne, P. & Li, X. J. (2001). Huntingtin aggregate-associated axonal degeneration is an early pathological event in Huntington's disease mice. *J Neurosci*, Vol. 21, 21, pp.(8473-8481.)

Li, H., Wyman, T., Yu, Z. X., Li, S. H. & Li, X. J. (2003). Abnormal association of mutant huntingtin with synaptic vesicles inhibits glutamate release. *Hum Mol Genet*, Vol. 12, 16, pp.(2021-2030)

Li, J. Y., Popovic, N. & Brundin, P. (2005). The use of the R6 transgenic mouse models of Huntington's disease in attempts to develop novel therapeutic strategies. *NeuroRx*, Vol. 2, 3, pp.(447-464)

Li, S. H., Gutekunst, C. A., Hersch, S. M. & Li, X. J. (1998a). Association of HAP1 isoforms with a unique cytoplasmic structure. *J Neurochem*, Vol. 71, 5, pp.(2178-2185)

Li, S. H., Hosseini, S. H., Gutekunst, C. A., Hersch, S. M., Ferrante, R. J. & Li, X. J. (1998b). A human HAP1 homologue. Cloning, expression, and interaction with huntingtin. *J Biol Chem*, Vol. 273, 30, pp.(19220-19227)

Li, S. H., Cheng, A. L., Zhou, H., Lam, S., Rao, M., Li, H. & Li, X. J. (2002). Interaction of huntington disease protein with transcriptional activator sp1. *Mol Cell Biol*, Vol. 22, 5, pp.(1277-1287.)

Li, S. H. & Li, X. J. (2004). Huntingtin-protein interactions and the pathogenesis of Huntington's disease. *Trends Genet*, Vol. 20, 3, pp.(146-154)

Li, X., Sapp, E., Valencia, A., Kegel, K. B., Qin, Z. H., Alexander, J., Masso, N., Reeves, P., Ritch, J. J., Zeitlin, S., Aronin, N. & Difiglia, M. (2008). A function of huntingtin in guanine nucleotide exchange on Rab11. *Neuroreport*, Vol. 19, 16, pp.(1643-1647)

Li, X. J., Li, S. H., Sharp, A. H., Nucifora, F. C., Jr., Schilling, G., Lanahan, A., Worley, P., Snyder, S. H. & Ross, C. A. (1995). A huntingtin-associated protein enriched in brain with implications for pathology. *Nature*, Vol. 378, 6555, pp.(398-402)

Lievens, J. C., Woodman, B., Mahal, A., Spasic-Boscovic, O., Samuel, D., Kerkerian-Le Goff, L. & Bates, G. P. (2001). Impaired glutamate uptake in the R6 Huntington's disease transgenic mice. *Neurobiol Dis*, Vol. 8, 5, pp.(807-821.)

Lin, C. H., Tallaksen-Greene, S., Chien, W. M., Cearley, J. A., Jackson, W. S., Crouse, A. B., Ren, S., Li, X. J., Albin, R. L. & Detloff, P. J. (2001). Neurological abnormalities in a knock-in mouse model of Huntington's disease. *Hum Mol Genet*, Vol. 10, 2, pp.(137-144.)

Liot, G., Bossy, B., Lubitz, S., Kushnareva, Y., Sejbuk, N. & Bossy-Wetzel, E. (2009). Complex II inhibition by 3-NP causes mitochondrial fragmentation and neuronal cell death via an NMDA- and ROS-dependent pathway. *Cell Death Differ*, Vol. 16, 6, pp.(899-909)

Lipton, S. A. & Rosenberg, P. A. (1994). Excitatory Amino Acids as a Final Common Pathway for Neurologic Disorders. *The New England Journal of Medicine*, Vol. 330, 9, pp.(613-622)

Liu, C. S., Cheng, W. L., Kuo, S. J., Li, J. Y., Soong, B. W. & Wei, Y. H. (2008). Depletion of mitochondrial DNA in leukocytes of patients with poly-Q diseases. *J Neurol Sci*, Vol. 264, 1-2, pp.(18-21)

Liu, M., Pleasure, S. J., Collins, A. E., Noebels, J. L., Naya, F. J., Tsai, M. J. & Lowenstein, D. H. (2000). Loss of BETA2/NeuroD leads to malformation of the dentate gyrus and epilepsy. *Proc Natl Acad Sci U S A*, Vol. 97, 2, pp.(865-870)

Lodi, R., Schapira, A. H., Manners, D., Styles, P., Wood, N. W., Taylor, D. J. & Warner, T. T. (2000). Abnormal in vivo skeletal muscle energy metabolism in Huntington's disease and dentatorubropallidoluysian atrophy. *Ann Neurol*, Vol. 48, 1, pp.(72-76)

Lunkes, A., Lindenberg, K. S., Ben-Haiem, L., Weber, C., Devys, D., Landwehrmeyer, G. B., Mandel, J. L. & Trottier, Y. (2002). Proteases acting on mutant huntingtin generate cleaved products that differentially build up cytoplasmic and nuclear inclusions. *Mol Cell*, Vol. 10, 2, pp.(259-269.)

Luo, S. & Rubinsztein, D. C. (2009). Huntingtin promotes cell survival by preventing Pak2 cleavage. *J Cell Sci*, Vol. 122, Pt 6, pp.(875-885)

Luthi-Carter, R., Strand, A., Peters, N. L., Solano, S. M., Hollingsworth, Z. R., Menon, A. S., Frey, A. S., Spektor, B. S., Penney, E. B., Schilling, G., Ross, C. A., Borchelt, D. R., Tapscott, S. J., Young, A. B., Cha, J. H. & Olson, J. M. (2000). Decreased expression of striatal signaling genes in a mouse model of Huntington's disease. *Hum Mol Genet*, Vol. 9, 9, pp.(1259-1271)

Maccarrone, M., Battista, N. & Centonze, D. (2007). The endocannabinoid pathway in Huntington's disease: a comparison with other neurodegenerative diseases. *Prog Neurobiol*, Vol. 81, 5-6, pp.(349-379)

Maglione, V., Cannella, M., Gradini, R., Cislaghi, G. & Squitieri, F. (2006). Huntingtin fragmentation and increased caspase 3, 8 and 9 activities in lymphoblasts with heterozygous and homozygous Huntington's disease mutation. *Mech Ageing Dev*, Vol. 127, 2, pp.(213-216)

Mangiarini, L., Sathasivam, K., Seller, M., Cozens, B., Harper, A., Hetherington, C., Lawton, M., Trottier, Y., Lehrach, H., Davies, S. W. & Bates, G. P. (1996). Exon 1 of the HD gene with an expanded CAG repeat is sufficient to cause a progressive neurological phenotype in transgenic mice. *Cell*, Vol. 87, 3, pp.(493-506)

Mann, V. M., Cooper, J. M., Javoy-Agid, F., Agid, Y., Jenner, P. & Schapira, A. H. (1990). Mitochondrial function and parental sex effect in Huntington's disease. *Lancet*, Vol. 336, 8717, pp.(749)

Mantle, D., Falkous, G., Ishiura, S., Perry, R. H. & Perry, E. K. (1995). Comparison of cathepsin protease activities in brain tissue from normal cases and cases with Alzheimer's disease, Lewy body dementia, Parkinson's disease and Huntington's disease. *Journal of the Neurological Sciences*, Vol. 131, 1, pp.(65-70)

Marcora, E., Gowan, K. & Lee, J. E. (2003). Stimulation of NeuroD activity by huntingtin and huntingtin-associated proteins HAP1 and MLK2. *Proc Natl Acad Sci U S A*, Vol. 100, 16, pp.(9578-9583)

Marsicano, G., Goodenough, S., Monory, K., Hermann, H., Eder, M., Cannich, A., Azad, S. C., Cascio, M. G., Gutierrez, S. O., van der Stelt, M., Lopez-Rodriguez, M. L., Casanova, E., Schutz, G., Zieglgansberger, W., Di Marzo, V., Behl, C. & Lutz, B. (2003). CB1 cannabinoid receptors and on-demand defense against excitotoxicity. *Science*, Vol. 302, 5642, pp.(84-88)

Marti, E., Pantano, L., Banez-Coronel, M., Llorens, F., Minones-Moyano, E., Porta, S., Sumoy, L., Ferrer, I. & Estivill, X. (2010). A myriad of miRNA variants in control and Huntington's disease brain regions detected by massively parallel sequencing. *Nucleic Acids Res*, Vol. 38, 20, pp.(7219-7235)

Martin-Aparicio, E., Yamamoto, A., Hernandez, F., Hen, R., Avila, J. & Lucas, J. J. (2001). Proteasomal-dependent aggregate reversal and absence of cell death in a conditional mouse model of Huntington's disease. *J Neurosci*, Vol. 21, 22, pp.(8772-8781.)

Mateizel, I., De Temmerman, N., Ullmann, U., Cauffman, G., Sermon, K., Van de Velde, H., De Rycke, M., Degreef, E., Devroey, P., Liebaers, I. & Van Steirteghem, A. (2006). Derivation of human embryonic stem cell lines from embryos obtained after IVF and after PGD for monogenic disorders. *Hum Reprod*, Vol. 21, 2, pp.(503-511)

Mateizel, I., Spits, C., De Rycke, M., Liebaers, I. & Sermon, K. (2010). Derivation, culture, and characterization of VUB hESC lines. *In Vitro Cell Dev Biol Anim*, Vol. 46, 3-4, pp.(300-308)

Mazziotta, J. C., Phelps, M. E., Pahl, J. J., Huang, S. C., Baxter, L. R., Riege, W. H., Hoffman, J. M., Kuhl, D. E., Lanto, A. B., Wapenski, J. A. & et al. (1987). Reduced cerebral glucose metabolism in asymptomatic subjects at risk for Huntington's disease. *N Engl J Med*, Vol. 316, 7, pp.(357-362)

McCampbell, A., Taylor, J. P., Taye, A. A., Robitschek, J., Li, M., Walcott, J., Merry, D., Chai, Y., Paulson, H., Sobue, G. & Fischbeck, K. H. (2000). CREB-binding protein sequestration by expanded polyglutamine. *Hum Mol Genet*, Vol. 9, 14, pp.(2197-2202.)

McGeer, E. G. & McGeer, P. L. (1976). Duplication of biochemical changes of Huntington's chorea by intrastriatal injections of glutamic and kainic acids. *Nature*, Vol. 263, 5577, pp.(517-519)

McGuire, M. A., Beede, D. K., Collier, R. J., Buonomo, F. C., DeLorenzo, M. A., Wilcox, C. J., Huntington, G. B. & Reynolds, C. K. (1991). Effects of acute thermal stress and amount of feed intake on concentrations of somatotropin, insulin-like growth factor (IGF)-I and IGF-II, and thyroid hormones in plasma of lactating Holstein cows. *J Anim Sci*, Vol. 69, 5, pp.(2050-2056.)

Menalled, L., Zanjani, H., MacKenzie, L., Koppel, A., Carpenter, E., Zeitlin, S. & Chesselet, M. F. (2000). Decrease in striatal enkephalin mRNA in mouse models of Huntington's disease. *Exp Neurol*, Vol. 162, 2, pp.(328-342)

Mende-Mueller, L. M., Toneff, T., Hwang, S. R., Chesselet, M. F. & Hook, V. Y. (2001). Tissue-specific proteolysis of Huntingtin (htt) in human brain: evidence of enhanced levels of N- and C-terminal htt fragments in Huntington's disease striatum. *J Neurosci*, Vol. 21, 6, pp.(1830-1837.)

Milakovic, T. & Johnson, G. V. (2005). Mitochondrial respiration and ATP production are significantly impaired in striatal cells expressing mutant huntingtin. *J Biol Chem*, Vol. 280, pp.(30773-30782)

Miller, B. R., Dorner, J. L., Shou, M., Sari, Y., Barton, S. J., Sengelaub, D. R., Kennedy, R. T. & Rebec, G. V. (2008). Up-regulation of GLT1 expression increases glutamate uptake and attenuates the Huntington's disease phenotype in the R6/2 mouse. *Neuroscience*, Vol. 153, 1, pp.(329-337)

Modregger, J., DiProspero, N. A., Charles, V., Tagle, D. A. & Plomann, M. (2002). PACSIN 1 interacts with huntingtin and is absent from synaptic varicosities in presymptomatic Huntington's disease brains. *Hum Mol Genet*, Vol. 11, 21, pp.(2547-2558)

Muchowski, P. J. (2002). Protein misfolding, amyloid formation, and neurodegeneration: a critical role for molecular chaperones? *Neuron*, Vol. 35, 1, pp.(9-12)

Myers, R. H., MacDonald, M. E., Koroshetz, W. J., Duyao, M. P., Ambrose, C. M., Taylor, S. A., Barnes, G., Srinidhi, J., Lin, C. S., Whaley, W. L. & et al. (1993). De novo expansion of a (CAG)n repeat in sporadic Huntington's disease. *Nature Genetics*, Vol. 5, 2, pp.(168-173)

Nagata, E., Sawa, A., Ross, C. A. & Snyder, S. H. (2004). Autophagosome-like vacuole formation in Huntington's disease lymphoblasts. *Neuroreport*, Vol. 15, 8, pp.(1325-1328)

Nasir, J., Floresco, S. B., O'Kusky, J. R., Diewert, V. M., Richman, J. M., Zeisler, J., Borowski, A., Marth, J. D., Phillips, A. G. & Hayden, M. R. (1995). Targeted disruption of the Huntington's disease gene results in embryonic lethality and behavioral and morphological changes in heterozygotes. *Cell*, Vol. 81, 5, pp.(811-823)

Niclis, J. C., Trounson, A. O., Dottori, M., Ellisdon, A. M., Bottomley, S. P., Verlinsky, Y. & Cram, D. S. (2009). Human embryonic stem cell models of Huntington disease. *Reprod Biomed Online*, Vol. 19, 1, pp.(106-113)

Ona, V. O., Li, M., Vonsattel, J. P., Andrews, L. J., Khan, S. Q., Chung, W. M., Frey, A. S., Menon, A. S., Li, X. J., Stieg, P. E., Yuan, J., Penney, J. B., Young, A. B., Cha, J. H. & Friedlander, R. M. (1999). Inhibition of caspase-1 slows disease progression in a mouse model of Huntington's disease. *Nature*, Vol. 399, 6733, pp.(263-267)

Orr, A. L., Li, S., Wang, C. E., Li, H., Wang, J., Rong, J., Xu, X., Mastroberardino, P. G., Greenamyre, J. T. & Li, X. J. (2008). N-terminal mutant huntingtin associates with mitochondria and impairs mitochondrial trafficking. *J Neurosci*, Vol. 28, 11, pp.(2783-2792)

Packer, A. N., Xing, Y., Harper, S. Q., Jones, L. & Davidson, B. L. (2008). The bifunctional microRNA miR-9/miR-9* regulates REST and CoREST and is downregulated in Huntington's disease. *J Neurosci*, Vol. 28, 53, pp.(14341-14346)

Pal, A., Severin, F., Lommer, B., Shevchenko, A. & Zerial, M. (2006). Huntingtin-HAP40 complex is a novel Rab5 effector that regulates early endosome motility and is up-regulated in Huntington's disease. *J Cell Biol*, Vol. 172, 4, pp.(605-618)

Palfi, S., Brouillet, E., Jarraya, B., Bloch, J., Jan, C., Shin, M., Conde, F., Li, X. J., Aebischer, P., Hantraye, P. & Deglon, N. (2007). Expression of mutated huntingtin fragment in the putamen is sufficient to produce abnormal movement in non-human primates. *Mol Ther*, Vol. 15, 8, pp.(1444-1451)

Panov, A. V., Gutekunst, C. A., Leavitt, B. R., Hayden, M. R., Burke, J. R., Strittmatter, W. J. & Greenamyre, J. T. (2002). Early mitochondrial calcium defects in Huntington's disease are a direct effect of polyglutamines. *Nat Neurosci*, Vol. 5, 8, pp.(731-736.)

Park, I. H., Arora, N., Huo, H., Maherali, N., Ahfeldt, T., Shimamura, A., Lensch, M. W., Cowan, C., Hochedlinger, K. & Daley, G. Q. (2008a). Disease-specific induced pluripotent stem cells. *Cell*, Vol. 134, 5, pp.(877-886)

Park, I. H., Zhao, R., West, J. A., Yabuuchi, A., Huo, H., Ince, T. A., Lerou, P. H., Lensch, M. W. & Daley, G. Q. (2008b). Reprogramming of human somatic cells to pluripotency with defined factors. *Nature*, Vol. 451, 7175, pp.(141-146)

Parker, J. A., Metzler, M., Georgiou, J., Mage, M., Roder, J. C., Rose, A. M., Hayden, M. R. & Neri, C. (2007). Huntingtin-interacting protein 1 influences worm and mouse presynaptic function and protects Caenorhabditis elegans neurons against mutant polyglutamine toxicity. *J Neurosci*, Vol. 27, 41, pp.(11056-11064)

Parker, W. D., Jr., Boyson, S. J., Luder, A. S. & Parks, J. K. (1990). Evidence for a defect in NADH: ubiquinone oxidoreductase (complex I) in Huntington's disease. *Neurology*, Vol. 40, 8, pp.(1231-1234)

Petersen, A., Larsen, K. E., Behr, G. G., Romero, N., Przedborski, S., Brundin, P. & Sulzer, D. (2001). Expanded CAG repeats in exon 1 of the Huntington's disease gene stimulate dopamine-mediated striatal neuron autophagy and degeneration. *Hum Mol Genet*, Vol. 10, 12, pp.(1243-1254.)

Petersen, A., Gil, J., Maat-Schieman, M. L., Bjorkqvist, M., Tanila, H., Araujo, I. M., Smith, R., Popovic, N., Wierup, N., Norlen, P., Li, J. Y., Roos, R. A., Sundler, F., Mulder, H. & Brundin, P. (2005). Orexin loss in Huntington's disease. *Hum Mol Genet*, Vol. 14, 1, pp.(39-47)

Pouladi, M. A., Graham, R. K., Karasinska, J. M., Xie, Y., Santos, R. D., Petersen, A. & Hayden, M. R. (2009). Prevention of depressive behaviour in the YAC128 mouse model of Huntington disease by mutation at residue 586 of huntingtin. *Brain*, Vol. 132, Pt 4, pp.(919-932)

Powers, W. J., Haas, R. H., Le, T., Videen, T. O., Hershey, T., McGee-Minnich, L. & Perlmutter, J. S. (2007a). Normal platelet mitochondrial complex I activity in Huntington's disease. *Neurobiol Dis*, Vol. 27, 1, pp.(99-101)

Powers, W. J., Videen, T. O., Markham, J., McGee-Minnich, L., Antenor-Dorsey, J. V., Hershey, T. & Perlmutter, J. S. (2007b). Selective defect of in vivo glycolysis in early Huntington's disease striatum. *Proc Natl Acad Sci U S A*, Vol. 104, 8, pp.(2945-2949)

Qin, Z. H., Wang, Y., Kegel, K. B., Kazantsev, A., Apostol, B. L., Thompson, L. M., Yoder, J., Aronin, N. & DiFiglia, M. (2003). Autophagy regulates the processing of amino terminal huntingtin fragments. *Hum Mol Genet*, Vol. 12, 24, pp.(3231-3244)

Quintanilla, R. A. & Johnson, G. V. (2009). Role of mitochondrial dysfunction in the pathogenesis of Huntington's disease. *Brain Res Bull*, Vol. 80, 4-5, pp.(242-247)

Ratovitski, T., Gucek, M., Jiang, H., Chighladze, E., Waldron, E., D'Ambola, J., Hou, Z., Liang, Y., Poirier, M. A., Hirschhorn, R. R., Graham, R., Hayden, M. R., Cole, R. N. & Ross, C. A. (2009). Mutant huntingtin N-terminal fragments of specific size mediate aggregation and toxicity in neuronal cells. *J Biol Chem*, Vol. 284, 16, pp.(10855-10867)

Ravikumar, B., Duden, R. & Rubinsztein, D. C. (2002). Aggregate-prone proteins with polyglutamine and polyalanine expansions are degraded by autophagy. *Hum Mol Genet*, Vol. 11, 9, pp.(1107-1117)

Ravikumar, B., Vacher, C., Berger, Z., Davies, J. E., Luo, S., Oroz, L. G., Scaravilli, F., Easton, D. F., Duden, R., O'Kane, C. J. & Rubinsztein, D. C. (2004). Inhibition of mTOR induces autophagy and reduces toxicity of polyglutamine expansions in fly and mouse models of Huntington disease. *Nat Genet*, Vol. 36, 6, pp.(585-595)

Reddy, P. H., Williams, M., Charles, V., Garrett, L., Pike-Buchanan, L., Whetsell, W. O., Jr., Miller, G. & Tagle, D. A. (1998). Behavioural abnormalities and selective neuronal loss in HD transgenic mice expressing mutated full-length HD cDNA. *Nat Genet*, Vol. 20, 2, pp.(198-202)

Reynolds, D. S., Carter, R. J. & Morton, A. J. (1998). Dopamine modulates the susceptibility of striatal neurons to 3-nitropropionic acid in the rat model of Huntington's disease. *J Neurosci*, Vol. 18, 23, pp.(10116-10127)

Rigamonti, D., Bauer, J. H., De-Fraja, C., Conti, L., Sipione, S., Sciorati, C., Clementi, E., Hackam, A., Hayden, M. R., Li, Y., Cooper, J. K., Ross, C. A., Govoni, S., Vincenz, C. & Cattaneo, E. (2000). Wild-type huntingtin protects from apoptosis upstream of caspase-3. *J Neurosci*, Vol. 20, 10, pp.(3705-3713)

Rigamonti, D., Sipione, S., Goffredo, D., Zuccato, C., Fossale, E. & Cattaneo, E. (2001). Huntingtin's neuroprotective activity occurs via inhibition of procaspase-9 processing. *J Biol Chem*, Vol. 276, 18, pp.(14545-14548.)

Roche, K. W., Standley, S., McCallum, J., Ly, C. D., Ehlers, M. D. & Wenthold, R. J. (2001). Molecular determinants of NMDA receptor internalization. *Nature Neuroscience*, Vol. 4, 8, pp.(794-802)

Roizin, L. (1979). The relevance of the structural co-factor (chemogenic lesion) in adverse and toxic reactions of neuropsychotropic agents. *Prog Neuropsychopharmacol*, Vol. 3, 1-3, pp.(245-257)

Rousseau, E., Kojima, R., Hoffner, G., Djian, P. & Bertolotti, A. (2009). Misfolding of proteins with a polyglutamine expansion is facilitated by proteasomal chaperones. *J Biol Chem*, Vol. 284, 3, pp.(1917-1929)

Ryu, H., Lee, J., Olofsson, B. A., Mwidau, A., Dedeoglu, A., Escudero, M., Flemington, E., Azizkhan-Clifford, J., Ferrante, R. J. & Ratan, R. R. (2003). Histone deacetylase inhibitors prevent oxidative neuronal death independent of expanded polyglutamine repeats via an Sp1-dependent pathway. *Proc Natl Acad Sci U S A*, Vol. 100, 7, pp.(4281-4286)

Sadri-Vakili, G., Menon, A. S., Farrell, L. A., Keller-McGandy, C. E., Cantuti-Castelvetri, I., Standaert, D. G., Augood, S. J., Yohrling, G. J. & Cha, J. H. (2006). Huntingtin inclusions do not down-regulate specific genes in the R6/2 Huntington's disease mouse. *Eur J Neurosci*, Vol. 23, 12, pp.(3171-3175)

Saft, C., Zange, J., Andrich, J., Muller, K., Lindenberg, K., Landwehrmeyer, B., Vorgerd, M., Kraus, P. H., Przuntek, H. & Schols, L. (2005). Mitochondrial impairment in patients and asymptomatic mutation carriers of Huntington's disease. *Mov Disord*, Vol. 20, 6, pp.(674-679)

Sanchez, I., Xu, C. J., Juo, P., Kakizaka, A., Blenis, J. & Yuan, J. (1999). Caspase-8 is required for cell death induced by expanded polyglutamine repeats. *Neuron*, Vol. 22, 3, pp.(623-633)

Sapp, E., Schwarz, C., Chase, K., Bhide, P. G., Young, A. B., Penney, J., Vonsattel, J. P., Aronin, N. & DiFiglia, M. (1997). Huntingtin localization in brains of normal and Huntington's disease patients. *Annals of Neurology*, Vol. 42, 4, pp.(604-612)

Sarkar, S., Perlstein, E. O., Imarisio, S., Pineau, S., Cordenier, A., Maglathlin, R. L., Webster, J. A., Lewis, T. A., O'Kane, C. J., Schreiber, S. L. & Rubinsztein, D. C. (2007). Small molecules enhance autophagy and reduce toxicity in Huntington's disease models. *Nat Chem Biol*, Vol. 3, 6, pp.(331-338)

Saudou, F., Finkbeiner, S., Devys, D. & Greenberg, M. E. (1998). Huntingtin acts in the nucleus to induce apoptosis but death does not correlate with the formation of intranuclear inclusions. *Cell*, Vol. 95, pp.(55-66)

Savas, J. N., Makusky, A., Ottosen, S., Baillat, D., Then, F., Krainc, D., Shiekhattar, R., Markey, S. P. & Tanese, N. (2008). Huntington's disease protein contributes to RNA-mediated gene silencing through association with Argonaute and P bodies. *Proc Natl Acad Sci U S A*, Vol. 105, 31, pp.(10820-10825)

Sawa, A. (2001). Mechanisms for neuronal cell death and dysfunction in Huntington's disease: pathological cross-talk between the nucleus and the mitochondria? *J Mol Med*, Vol. 79, 7, pp.(375-381.)

Saydoff, J. A., Garcia, R. A., Browne, S. E., Liu, L., Sheng, J., Brenneman, D., Hu, Z., Cardin, S., Gonzalez, A., von Borstel, R. W., Gregorio, J., Burr, H. & Beal, M. F. (2006). Oral uridine pro-drug PN401 is neuroprotective in the R6/2 and N171-82Q mouse models of Huntington's disease. *Neurobiol Dis*, Vol. 24, 3, pp.(455-465)

Schaffar, G., Breuer, P., Boteva, R., Behrends, C., Tzvetkov, N., Strippel, N., Sakahira, H., Siegers, K., Hayer-Hartl, M. & Hartl, F. U. (2004). Cellular toxicity of polyglutamine expansion proteins: mechanism of transcription factor deactivation. *Mol Cell*, Vol. 15, 1, pp.(95-105)

Scherzinger, E., Lurz, R., Turmaine, M., Mangiarini, L., Hollenbach, B., Hasenbank, R., Bates, G. P., Davies, S. W., Lehrach, H. & Wanker, E. E. (1997). Huntingtin-encoded polyglutamine expansions form amyloid-like protein aggregates in vitro and in vivo. *Cell*, Vol. 90, 3, pp.(549-558)

Schilling, G., Becher, M. W., Sharp, A. H., Jinnah, H. A., Duan, K., Kotzuk, J. A., Slunt, H. H., Ratovitski, T., Cooper, J. K., Jenkins, N. A., Copeland, N. G., Price, D. L., Ross, C. A. & Borchelt, D. R. (1999a). Intranuclear inclusions and neuritic aggregates in transgenic mice expressing a mutant N-terminal fragment of huntingtin. *Hum Mol Genet*, Vol. 8, 3, pp.(397-407)

Schilling, G., Wood, J. D., Duan, K., Slunt, H. H., Gonzales, V., Yamada, M., Cooper, J. K., Margolis, R. L., Jenkins, N. A., Copeland, N. G., Takahashi, H., Tsuji, S., Price, D. L., Borchelt, D. R. & Ross, C. A. (1999b). Nuclear accumulation of truncated atrophin-1 fragments in a transgenic mouse model of DRPLA. *Neuron*, Vol. 24, 1, pp.(275-286)

Schroer, T. A., Bingham, J. B. & Gill, S. R. (1996). Actin-related protein 1 and cytoplasmic dynein-based motility - what's the connection? *Trends Cell Biol*, Vol. 6, 6, pp.(212-215)

Schwarcz, R. & Kohler, C. (1983). Differential vulnerability of central neurons of the rat to quinolinic acid. *Neurosci Lett*, Vol. 38, 1, pp.(85-90)

Schwartz, P. H., Brick, D. J., Stover, A. E., Loring, J. F. & Muller, F. J. (2008). Differentiation of neural lineage cells from human pluripotent stem cells. *Methods*, Vol. 45, 2, pp.(142-158)

Senatorov, V. V., Charles, V., Reddy, P. H., Tagle, D. A. & Chuang, D. M. (2003). Overexpression and nuclear accumulation of glyceraldehyde-3-phosphate dehydrogenase in a transgenic mouse model of Huntington's disease. *Mol Cell Neurosci*, Vol. 22, 3, pp.(285-297)

Seo, H., Kim, W. & Isacson, O. (2008). Compensatory changes in the ubiquitin-proteasome system, brain-derived neurotrophic factor and mitochondrial complex II/III in YAC72 and R6/2 transgenic mice partially model Huntington's disease patients. *Hum Mol Genet*, Vol. 17, 20, pp.(3144-3153)

Seong, I. S., Ivanova, E., Lee, J. M., Choo, Y. S., Fossale, E., Anderson, M., Gusella, J. F., Laramie, J. M., Myers, R. H., Lesort, M. & MacDonald, M. E. (2005). HD CAG repeat implicates a dominant property of huntingtin in mitochondrial energy metabolism. *Hum Mol Genet*, Vol. 14, 19, pp.(2871-2880)

Shelbourne, P. F., Killeen, N., Hevner, R. F., Johnston, H. M., Tecott, L., Lewandoski, M., Ennis, M., Ramirez, L., Li, Z., Iannicola, C., Littman, D. R. & Myers, R. M. (1999). A Huntington's disease CAG expansion at the murine Hdh locus is unstable and associated with behavioural abnormalities in mice. *Hum Mol Genet*, Vol. 8, 5, pp.(763-774)

Shimojo, M. (2008). Huntingtin regulates RE1-silencing transcription factor/neuron-restrictive silencer factor (REST/NRSF) nuclear trafficking indirectly through a complex with REST/NRSF-interacting LIM domain protein (RILP) and dynactin p150 Glued. *J Biol Chem*, Vol. 283, 50, pp.(34880-34886)

Shin, J. Y., Fang, Z. H., Yu, Z. X., Wang, C. E., Li, S. H. & Li, X. J. (2005). Expression of mutant huntingtin in glial cells contributes to neuronal excitotoxicity. *J Cell Biol*, Vol. 171, 6, pp.(1001-1012)

Singaraja, R. R., Hadano, S., Metzler, M., Givan, S., Wellington, C. L., Warby, S., Yanai, A., Gutekunst, C. A., Leavitt, B. R., Yi, H., Fichter, K., Gan, L., McCutcheon, K., Chopra, V., Michel, J., Hersch, S. M., Ikeda, J. E. & Hayden, M. R. (2002). HIP14, a novel ankyrin domain-containing protein, links huntingtin to intracellular trafficking and endocytosis. *Hum Mol Genet*, Vol. 11, 23, pp.(2815-2828.)

Sinha, M., Ghose, J., Das, E. & Bhattarcharyya, N. P. (2010). Altered microRNAs in STHdh(Q111)/Hdh(Q111) cells: miR-146a targets TBP. *Biochem Biophys Res Commun*, Vol. 396, 3, pp.(742-747)

Sittler, A., Walter, S., Wedemeyer, N., Hasenbank, R., Scherzinger, E., Eickhoff, H., Bates, G. P., Lehrach, H. & Wanker, E. E. (1998). SH3GL3 associates with the Huntingtin exon 1 protein and promotes the formation of polygln-containing protein aggregates. *Mol Cell*, Vol. 2, 4, pp.(427-436)

Slow, E. (2005). Inclusions to the rescue? Neuroprotective role for huntingtin inclusions in HD. *Clin Genet*, Vol. 67, 3, pp.(228-229)

Slow, E. J., Graham, R. K., Osmand, A. P., Devon, R. S., Lu, G., Deng, Y., Pearson, J., Vaid, K., Bissada, N., Wetzel, R., Leavitt, B. R. & Hayden, M. R. (2005). Absence of behavioral abnormalities and neurodegeneration in vivo despite widespread neuronal huntingtin inclusions. *Proc Natl Acad Sci U S A*, Vol. 102, 32, pp.(11402-11407)

Snell, R. G., MacMillan, J. C., Cheadle, J. P., Fenton, I., Lazarou, L. P., Davies, P., MacDonald, M. E., Gusella, J. F., Harper, P. S. & Shaw, D. J. (1993). Relationship between trinucleotide repeat expansion and phenotypic variation in Huntington's disease. *Nat Genet*, Vol. 4, 4, pp.(393-397)

Song, C., Zhang, Y., Parsons, C. G. & Liu, Y. F. (2003). Expression of polyglutamine-expanded huntingtin induces tyrosine phosphorylation of N-methyl-D-aspartate receptors. *J Biol Chem*, Vol. 278, 35, pp.(33364-33369)

Squitieri, F., Cannella, M., Sgarbi, G., Maglione, V., Falleni, A., Lenzi, P., Baracca, A., Cislaghi, G., Saft, C., Ragona, G., Russo, M. A., Thompson, L. M., Solaini, G. & Fornai, F. (2006). Severe ultrastructural mitochondrial changes in lymphoblasts homozygous for Huntington disease mutation. *Mech Ageing Dev*, Vol. 127, 2, pp.(217-220)

Steffan, J. S., Kazantsev, A., Spasic-Boskovic, O., Greenwald, M., Zhu, Y. Z., Gohler, H., Wanker, E. E., Bates, G. P., Housman, D. E. & Thompson, L. M. (2000). The Huntington's disease protein interacts with p53 and CREB-binding protein and represses transcription. *Proc Natl Acad Sci U S A*, Vol. 97, 12, pp.(6763-6768)

Steffan, J. S., Bodai, L., Pallos, J., Poelman, M., McCampbell, A., Apostol, B. L., Kazantsev, A., Schmidt, E., Zhu, Y. Z., Greenwald, M., Kurokawa, R., Housman, D. E., Jackson, G. R., Marsh, J. L. & Thompson, L. M. (2001). Histone deacetylase inhibitors arrest polyglutamine-dependent neurodegeneration in Drosophila. *Nature*, Vol. 413, 6857, pp.(739-743.)

Strand, A. D., Baquet, Z. C., Aragaki, A. K., Holmans, P., Yang, L., Cleren, C., Beal, M. F., Jones, L., Kooperberg, C., Olson, J. M. & Jones, K. R. (2007). Expression profiling of Huntington's disease models suggests that brain-derived neurotrophic factor depletion plays a major role in striatal degeneration. *J Neurosci*, Vol. 27, 43, pp.(11758-11768)

Sun, B., Fan, W., Balciunas, A., Cooper, J. K., Bitan, G., Steavenson, S., Denis, P. E., Young, Y., Adler, B., Daugherty, L., Manoukian, R., Elliott, G., Shen, W., Talvenheimo, J., Teplow, D. B., Haniu, M., Haldankar, R., Wypych, J., Ross, C. A., Citron, M. & Richards, W. G. (2002). Polyglutamine repeat length-dependent proteolysis of huntingtin. *Neurobiol Dis*, Vol. 11, 1, pp.(111-122)

Sun, Y., Savanenin, A., Reddy, P. H. & Liu, Y. F. (2001). Polyglutamine-expanded huntingtin promotes sensitization of N-methyl-D- aspartate receptors via post-synaptic density 95. *J Biol Chem*, Vol. 276, 27, pp.(24713-24718.)

Szebenyi, G., Morfini, G. A., Babcock, A., Gould, M., Selkoe, K., Stenoien, D. L., Young, M., Faber, P. W., MacDonald, M. E., McPhaul, M. J. & Brady, S. T. (2003). Neuropathogenic forms of huntingtin and androgen receptor inhibit fast axonal transport. *Neuron*, Vol. 40, 1, pp.(41-52)

Tabrizi, S. J., Cleeter, M. W., Xuereb, J., Taanman, J. W., Cooper, J. M. & Schapira, A. H. (1999). Biochemical abnormalities and excitotoxicity in Huntington's disease brain. *Ann Neurol*, Vol. 45, 1, pp.(25-32)

Takahashi, K., Tanabe, K., Ohnuki, M., Narita, M., Ichisaka, T., Tomoda, K. & Yamanaka, S. (2007). Induction of pluripotent stem cells from adult human fibroblasts by defined factors. *Cell*, Vol. 131, 5, pp.(861-872)

Tang, T. S., Tu, H., Chan, E. Y., Maximov, A., Wang, Z., Wellington, C. L., Hayden, M. R. & Bezprozvanny, I. (2003). Huntingtin and huntingtin-associated protein 1 influence neuronal calcium signaling mediated by inositol-(1,4,5) triphosphate receptor type 1. *Neuron*, Vol. 39, 2, pp.(227-239)

Tarditi, A., Camurri, A., Varani, K., Borea, P. A., Woodman, B., Bates, G., Cattaneo, E. & Abbracchio, M. P. (2006). Early and transient alteration of adenosine A2A receptor signaling in a mouse model of Huntington disease. *Neurobiol Dis*, Vol. 23, 1, pp.(44-53)

Tellez-Nagel, I., Johnson, A. B. & Terry, R. D. (1974). Studies on brain biopsies of patients with Huntington's chorea. *J Neuropathol Exp Neurol*, Vol. 33, 2, pp.(308-332)

Thompson, L. M., Aiken, C. T., Kaltenbach, L. S., Agrawal, N., Illes, K., Khoshnan, A., Martinez-Vincente, M., Arrasate, M., O'Rourke, J. G., Khashwji, H., Lukacsovich, T., Zhu, Y. Z., Lau, A. L., Massey, A., Hayden, M. R., Zeitlin, S. O., Finkbeiner, S., Green, K. N., LaFerla, F. M., Bates, G., Huang, L., Patterson, P. H., Lo, D. C., Cuervo, A. M., Marsh, J. L. & Steffan, J. S. (2009). IKK phosphorylates Huntingtin and targets it for degradation by the proteasome and lysosome. *J Cell Biol*, Vol. 187, 7, pp.(1083-1099)

Trottier, Y., Devys, D., Imbert, G., Saudou, F., An, I., Lutz, Y., Weber, C., Agid, Y., Hirsch, E. C. & Mandel, J. L. (1995a). Cellular localization of the Huntington's disease protein and discrimination of the normal and mutated form. *Nat Genet*, Vol. 10, 1, pp.(104-110)

Trottier, Y., Lutz, Y., Stevanin, G., Imbert, G., Devys, D., Cancel, G., Saudou, F., Weber, C., David, G., Tora, L. & et al. (1995b). Polyglutamine expansion as a pathological epitope in Huntington's disease and four dominant cerebellar ataxias. *Nature*, Vol. 378, 6555, pp.(403-406)

Trushina, E., Dyer, R. B., Badger, J. D., 2nd, Ure, D., Eide, L., Tran, D. D., Vrieze, B. T., Legendre-Guillemin, V., McPherson, P. S., Mandavilli, B. S., Van Houten, B., Zeitlin, S., McNiven, M., Aebersold, R., Hayden, M., Parisi, J. E., Seeberg, E., Dragatsis, I., Doyle, K., Bender, A., Chacko, C. & McMurray, C. T. (2004). Mutant huntingtin impairs axonal trafficking in mammalian neurons in vivo and in vitro. *Mol Cell Biol*, Vol. 24, 18, pp.(8195-8209)

Tunez, I. & Santamaria, A. (2009). [Model of Huntington's disease induced with 3-nitropropionic acid]. *Rev Neurol*, Vol. 48, 8, pp.(430-434)

Turner, C., Cooper, J. M. & Schapira, A. H. (2007). Clinical correlates of mitochondrial function in Huntington's disease muscle. *Mov Disord*, Vol. 22, 12, pp.(1715-1721)

Van Raamsdonk, J. M., Pearson, J., Rogers, D. A., Bissada, N., Vogl, A. W., Hayden, M. R. & Leavitt, B. R. (2005). Loss of wild-type huntingtin influences motor dysfunction and survival in the YAC128 mouse model of Huntington disease. *Hum Mol Genet*, Vol. 14, 10, pp.(1379-1392)

Varani, K., Rigamonti, D., Sipione, S., Camurri, A., Borea, P. A., Cattabeni, F., Abbracchio, M. P. & Cattaneo, E. (2001). Aberrant amplification of A(2A) receptor signaling in striatal cells expressing mutant huntingtin. *Faseb J*, Vol. 15, 7, pp.(1245-1247.)

Vazey, E. M., Dottori, M., Jamshidi, P., Tomas, D., Pera, M. F., Horne, M. & Connor, B. (2010). Comparison of transplant efficiency between spontaneously derived and noggin-primed human embryonic stem cell neural precursors in the quinolinic acid rat model of Huntington's disease. *Cell Transplant*, Vol. 19, 8, pp.(1055-1062)

Verlinsky, Y., Strelchenko, N., Kukharenko, V., Rechitsky, S., Verlinsky, O., Galat, V. & Kuliev, A. (2005). Human embryonic stem cell lines with genetic disorders. *Reprod Biomed Online*, Vol. 10, 1, pp.(105-110)

Vis, J. C., Verbeek, M. M., de Waal, R. M., ten Donkelaar, H. J. & Kremer, B. (2001). The mitochondrial toxin 3-nitropropionic acid induces differential expression patterns of apoptosis-related markers in rat striatum. *Neuropathol Appl Neurobiol*, Vol. 27, 1, pp.(68-76)

Vissel, B., Krupp, J. J., Heinemann, S. F. & Westbrook, G. L. (2001). A use-dependent tyrosine dephosphorylation of NMDA receptors is independent of ion flux. *Nature Neuroscience*, Vol. 4, 6, pp.(587-596)

Vonsattel, J. P., Myers, R. H., Stevens, T. J., Ferrante, R. J., Bird, E. D. & Richardson, E. P., Jr. (1985). Neuropathological classification of Huntington's disease. *Journal of Neuropathology & Experimental Neurology*, Vol. 44, 6, pp.(559-577)

Waelter, S., Boeddrich, A., Lurz, R., Scherzinger, E., Lueder, G., Lehrach, H. & Wanker, E. E. (2001). Accumulation of mutant huntingtin fragments in aggresome-like inclusion bodies as a result of insufficient protein degradation. *Mol Biol Cell*, Vol. 12, 5, pp.(1393-1407.)

Wanker, E. E., Rovira, C., Scherzinger, E., Hasenbank, R., Walter, S., Tait, D., Colicelli, J. & Lehrach, H. (1997). HIP-I: a huntingtin interacting protein isolated by the yeast two-hybrid system. *Hum Mol Genet*, Vol. 6, 3, pp.(487-495)

Wellington, C. L., Ellerby, L. M., Hackam, A. S., Margolis, R. L., Trifiro, M. A., Singaraja, R., McCutcheon, K., Salvesen, G. S., Propp, S. S., Bromm, M., Rowland, K. J., Zhang, T., Rasper, D., Roy, S., Thornberry, N., Pinsky, L., Kakizuka, A., Ross, C. A., Nicholson, D. W., Bredesen, D. E. & Hayden, M. R. (1998). Caspase cleavage of gene products associated with triplet expansion disorders generates truncated fragments containing the polyglutamine tract. *J Biol Chem*, Vol. 273, 15, pp.(9158-9167)

Wellington, C. L. & Hayden, M. R. (2000). Caspases and neurodegeneration: on the cutting edge of new therapeutic approaches. *Clin Genet*, Vol. 57, 1, pp.(1-10)

Wellington, C. L., Leavitt, B. R. & Hayden, M. R. (2000a). Huntington disease: new insights on the role of huntingtin cleavage. *J Neural Transm Suppl*, Vol. 58, pp.(1-17)

Wellington, C. L., Singaraja, R., Ellerby, L., Savill, J., Roy, S., Leavitt, B., Cattaneo, E., Hackam, A., Sharp, A., Thornberry, N., Nicholson, D. W., Bredesen, D. E. & Hayden, M. R. (2000b). Inhibiting caspase cleavage of huntingtin reduces toxicity and aggregate formation in neuronal and nonneuronal cells. *J Biol Chem*, Vol. 275, 26, pp.(19831-19838)

Wellington, C. L., Ellerby, L. M., Gutekunst, C. A., Rogers, D., Warby, S., Graham, R. K., Loubser, O., van Raamsdonk, J., Singaraja, R., Yang, Y. Z., Gafni, J., Bredesen, D., Hersch, S. M., Leavitt, B. R., Roy, S., Nicholson, D. W. & Hayden, M. R. (2002). Caspase cleavage of mutant huntingtin precedes neurodegeneration in Huntington's disease. *J Neurosci*, Vol. 22, 18, pp.(7862-7872.)

Wheeler, V. C., White, J. K., Gutekunst, C. A., Vrbanac, V., Weaver, M., Li, X. J., Li, S. H., Yi, H., Vonsattel, J. P., Gusella, J. F., Hersch, S., Auerbach, W., Joyner, A. L. & MacDonald, M. E. (2000). Long glutamine tracts cause nuclear localization of a novel form of huntingtin in medium spiny striatal neurons in HdhQ92 and HdhQ111 knock- in mice. *Hum Mol Genet*, Vol. 9, 4, pp.(503-513)

White, J. K., Auerbach, W., Duyao, M. P., Vonsattel, J. P., Gusella, J. F., Joyner, A. L. & MacDonald, M. E. (1997). Huntingtin is required for neurogenesis and is not impaired by the Huntington's disease CAG expansion. *Nat Genet*, Vol. 17, 4, pp.(404-410)

Woda, J. M., Calzonetti, T., Hilditch-Maguire, P., Duyao, M. P., Conlon, R. A. & MacDonald, M. E. (2005). Inactivation of the Huntington's disease gene (Hdh) impairs anterior streak formation and early patterning of the mouse embryo. *BMC Dev Biol*, Vol. 5, pp.(17)

Wyttenbach, A., Carmichael, J., Swartz, J., Furlong, R. A., Narain, Y., Rankin, J. & Rubinsztein, D. C. (2000). Effects of heat shock, heat shock protein 40 (HDJ-2), and proteasome inhibition on protein aggregation in cellular models of Huntington's disease. *Proc Natl Acad Sci U S A*, Vol. 97, 6, pp.(2898-2903)

Yamamoto, A., Lucas, J. J. & Hen, R. (2000). Reversal of neuropathology and motor dysfunction in a conditional model of Huntington's disease. *Cell*, Vol. 101, 1, pp.(57-66)

Yanai, A., Huang, K., Kang, R., Singaraja, R. R., Arstikaitis, P., Gan, L., Orban, P. C., Mullard, A., Cowan, C. M., Raymond, L. A., Drisdel, R. C., Green, W. N., Ravikumar, B., Rubinsztein, D. C., El-Husseini, A. & Hayden, M. R. (2006). Palmitoylation of huntingtin by HIP14 is essential for its trafficking and function. *Nat Neurosci*, Vol. 9, 6, pp.(824-831)

Yu, J., Vodyanik, M. A., Smuga-Otto, K., Antosiewicz-Bourget, J., Frane, J. L., Tian, S., Nie, J., Jonsdottir, G. A., Ruotti, V., Stewart, R., Slukvin, II & Thomson, J. A. (2007). Induced pluripotent stem cell lines derived from human somatic cells. *Science*, Vol. 318, 5858, pp.(1917-1920)

Yu, Z. X., Li, S. H., Nguyen, H. P. & Li, X. J. (2002). Huntingtin inclusions do not deplete polyglutamine-containing transcription factors in HD mice. *Hum Mol Genet*, Vol. 11, 8, pp.(905-914.)

Zeitlin, S., Liu, J. P., Chapman, D. L., Papaioannou, V. E. & Efstratiadis, A. (1995). Increased apoptosis and early embryonic lethality in mice nullizygous for the Huntington's disease gene homologue. *Nat Genet*, Vol. 11, 2, pp.(155-163)

Zhang, Y., Leavitt, B. R., van Raamsdonk, J. M., Dragatsis, I., Goldowitz, D., MacDonald, M. E., Hayden, M. R. & Friedlander, R. M. (2006). Huntingtin inhibits caspase-3 activation. *Embo J*, Vol. 25, 24, pp.(5896-5906)

Zuccato, C., Ciammola, A., Rigamonti, D., Leavitt, B. R., Goffredo, D., Conti, L., MacDonald, M. E., Friedlander, R. M., Silani, V., Hayden, M. R., Timmusk, T., Sipione, S. & Cattaneo, E. (2001). Loss of huntingtin-mediated BDNF gene transcription in Huntington's disease. *Science*, Vol. 293, 5529, pp.(493-498.)

Zuccato, C., Tartari, M., Crotti, A., Goffredo, D., Valenza, M., Conti, L., Cataudella, T., Leavitt, B. R., Hayden, M. R., Timmusk, T., Rigamonti, D. & Cattaneo, E. (2003). Huntingtin interacts with REST/NRSF to modulate the transcription of NRSE-controlled neuronal genes. *Nat Genet*, Vol. 35, 1, pp.(76-83)

Zuccato, C., Belyaev, N., Conforti, P., Ooi, L., Tartari, M., Papadimou, E., MacDonald, M., Fossale, E., Zeitlin, S., Buckley, N. & Cattaneo, E. (2007). Widespread disruption of repressor element-1 silencing transcription factor/neuron-restrictive silencer factor occupancy at its target genes in Huntington's disease. *J Neurosci*, Vol. 27, 26, pp.(6972-6983)

2

Modeling Huntington's Disease: *in vivo, in vitro, in silico*

Nagehan Ersoy Tunalı
Haliç University, Department of Molecular Biology and Genetics, İstanbul Turkey

1. Introduction

Since the discovery of the Huntington's Disease (HD) gene (Huntington's Disease Collaborative Research Group, 1993) various research groups have aimed to discover the subcellular and tissue distribution of its mRNA and protein. The human HD gene is expressed ubiquitously in all human tissues as two major messenger RNA (mRNA) transcripts, 13.6 kb and 10.3 kb in length, which differ in the size of their 3′ UTRs due to differential polyadenylation (Trottier et al., 1995). HD mRNA is expressed in both neural and non-neural tissues with high levels of expression in brain and testis (Sharp et al. 1995; Strong et al. 1993). Northern blot and *in situ* hybridization analyses indicate that the two transcripts are expressed in many human tissues (heart, kidney, lungs, pancreas, muscles, liver, placenta) with higher expression of the longer transcript in the brain (Strong et al., 1993; Sharp et al., 1995). In the brain, highest levels were found in cerebral cortex and cerebellum, intermediate levels in the hippocampus and the lowest levels in the caudate nucleus and thalamus (Li et al., 1993). In addition to this, neuronal expression predominates over glial expression (Strong et al., 1993). No difference in the mRNA expression pattern between HD brains and controls was reported (Landwehrmeyer et al., 1995). The HD gene encodes a protein of 3144 amino acids with a molecular mass of 348 kDa, termed huntingtin (htt). The polyQ tract starts at residue 18 and is followed by a stretch of prolines. Similar to RNA studies, protein studies also indicate ubiquitous expression of htt in a variety of cells and tissues throughout the development and in the adult (Zeitlin et al., 1995) in both brain and peripheral tissue (Hoogeveen et al., 1993; Jou and Myers, 1995; Sharp et al., 1995). Normal huntingtin is widely distributed in the body, with the highest levels in the brain and testis. In HD patients, normal and mutant huntingtin have similar distribution and expression patterns (Sharp et al., 1995; Trottier et al., 1995).

The pathology of HD is restricted to the brain, medium spiny GABA-ergic striatal neurons are selectively lost (Graveland et al., 1985). The neuronal intranuclear inclusions (NII), which contain the N-terminal fragment htt, are accepted as neuropathological markers of HD (Davies et al., 1997; DiFiglia et al., 1997; Juenemann et al., 2011). Since htt has no known homologies to any other protein, it is not easy to assign its exact function(s) and therefore to identify the mechanisms involved in disease process. The molecular mechanism underlying HD pathogenesis has been explained by toxic gain of function of the mutant htt (Housman, 1995; Jacobsen et al., 2011). However, recent findings point out loss of function of the normal protein as a contributor to the disease process (Dragatsis et al., 2000; Zuccato et al., 2001).

The exact cellular and subcellular localization of htt should be regarded as a key to understand the tissue-specific death in HD and the underlying molecular mechanisms. In this regard, localization of both the endogenous and overexpressed htt has been investigated by several research groups in cell lines, animal models, and post-mortem patient tissues. However, there is still no certain agreement on the precise subcellular distribution of huntingtin. Although original observations indicated an exclusively cytoplasmic localization (DiFiglia et al., 1995; Gutekunst et al., 1995; Sharp et al., 1995), recently both normal and mutant htt have been reported in the nucleus (Bae et al., 2006; Havel et al., 2011; Kegel et al., 2002; Tanaka et al., 2006; Yan et al., 2011). Three putative NLS were identified, but later they were shown to be non-functional (Hackam et al., 1998; Xia et al., 2003); and it is still not very clear how htt is transported into the nucleus. A detailed analysis of the subcellular distribution of htt may provide suggestions for its possible roles.

The current debate about the precise localization of htt emerges mostly from the diversity of experimental setups and the methods used. In post-mortem studies, handling of the tissue and the method of fixation have important effects on the following staining pattern. In analyzing endogenous protein expression in cell lines by immunocytochemistry methods, cell type (neuronal/non-neuronal), fixation methods, and the specificity of the antibodies are the major determinants of the subsequent detection of localization. Much less controllable variables like culture conditions may also have important effects on cell cycle and growth. Any interruptions or changes in cell cycle programme may change the localization of the proteins (Martin-Aparico et al., 2002). Various antibodies directed to N- or C-terminal regions of htt have been used to detect htt in various cell lines, but the results are contraversial (DiFiglia et al., 1995; De Rooij et al., 1996, Wilkinson et al., 1999). In overexpression systems, localization of normal and mutant htt can be studied more extensively, since htt constructs of various sizes and polyQ lengths can be created. In this case, the associated tag may have substantial effects on the subcellular localization. Large tags may prevent nuclear localization of the proteins. In overexpression systems, size of the huntingtin construct and the associated repeat length proved to have major impacts on subcellular localization, and these two criteria should always be considered together, since the repeat length alone cannot determine the localization. Overexpressed htt can be visualized in fixed or live cells. To overcome the drawbacks of working with fixed cells, proteins can be fluorescently tagged and the transfected cells can be analyzed in their natural environments. In recent protein expression and localization studies, live cell analysis using fluorescent recombinant vectors have been the preferred method of choice. Use of laser scanning confocal microscopy (LSCM) adds more power, since it enables simultaneous multi-channel imaging of two or more fluorescent proteins, and cellular transfections can be analyzed in space and time.

Considering the above mentioned factors affecting htt localization, this study was constructed to establish the endogenous and overexpressed, wild type and mutant, full length (FL) and truncated htt localizations in cell lines and in HD mouse models. These various constructs, run under the same experimental conditions are expected to provide a full delineation of htt localization. In this regard, the expression pattern of endogenous and overexpressed htt was investigated in neuronal and non-neuronal cell lines (HEK 293, N2A, PC12, IMR32) and embryonic striatal neurons of R6/1 and HdhQ150 mice expressing truncated and FL htt, respectively. In addition to localization studies, 3D htt structure was analyzed using fold recognition model and *in silico* polyQ expansion mutations were created in the htt protein using VMD programme which will help to decipher the effects of the mutation on protein structure and function.

2. Endogenous htt expression

2.1 Maintenance of cell lines

Endogenous htt expression was investigated in HEK 293, N2A, PC12 and IMR32 cell lines using immunocytochemistry methods. All cell lines were purchased from ECACC and maintained in their respective growth medium at a density of 2-5x100.000 cells/ml, in a humid 37°C incubator supplied with 5% CO_2. HEK 293 and N2A cells were grown in culture flasks in MEM supplemented with penicillin/streptomycin (100 units ml^{-1}/100 µg ml^{-1}), glutamine (2 mM), 1X non-essential amino acids (NEAA) and 10%FBS. PC12 cells were grown in complete RPMI 1640 medium, consisting of RPMI 1640, penicillin/streptomycin (100 units ml^{-1}/100 µg ml^{-1}), glutamine (2 mM) and 10 %FBS. They were differentiated with NGF-β (100 ng/ml) when needed. IMR32 cells were grown adherent to culture flasks in complete RPMI 1640 containing penicillin/streptomycin (100 units ml^{-1}/ 100µg ml^{-1}), glutamine (2 mM) and 5% FBS. HEK 293, N2A and IMR32 cells were adherent and were passaged every four days. PC12 cells grow in suspension, but can be made adherent by coating the flasks with poly-D-lysine.

2.2 Immunostaining methods

One of the important steps in protein localization by immunostaining methods is the fixation of cell preparations. The cells should be appropriately fixed and permeabilized prior to staining. There are organic solvents and cross-linking fixatives available, both having advantages and disadvantages. Use of organic solvents may be regarded as much less toxic and time-saving since permeabilization is not required, rehydration in Phosphate Buffered Saline (PBS) prior to staining procedure is enough. However, cross-linking fixatives may fix the cells better on the slides, but cells will need permeabilization prior to staining. In this study both the organic solvents (methanol, acetone) and a cross-linking fixative (paraformaldehyde) were utilized in order to decipher their effects on staining patterns (Table 1). It was shown that fixation with organic solvents can mask epitope binding sites (Figure 1a), and paraformaldehyde fixations followed by Triton-X-100 permeabilization reveal better staining patterns (Fig. 1.b). In the framework of this study, 4% paraformaldehyde fixation at room temperature (RT) followed by 0.1% Triton-X-100 permeabilization was used to determine endogenous htt localizations.

Fixation	Permeabilization
4% paraformaldehyde, 15 mins at RT	0.1% Triton-X-100, 20 mins at RT
4% paraformaldehyde, 15 mins at RT	0.5% Triton-X-100, 20 mins at RT
Methanol at -20°C, 5 mins; air dry & rehydrate in 1XPBS	-
Acetone at -20°C, 10 mins, air dry & rehydrate in 1X PBS	-
Methanol/Acetone mix at -20°C, 10 mins, air dry & rehydrate in 1X PBS	-

Table 1. Cell fixation methods.

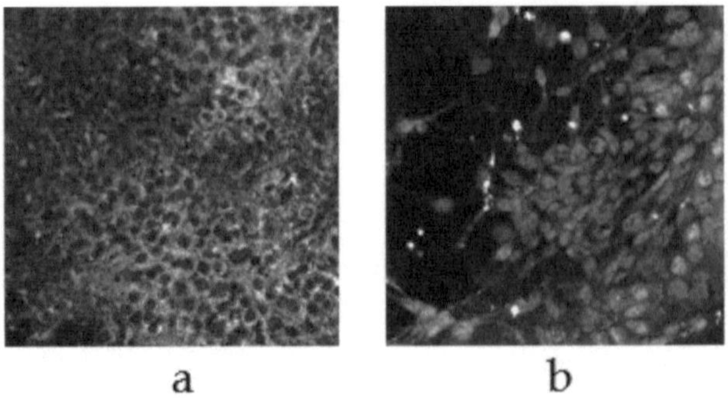

Fig. 1. R6/1 embryonic striatal cells, immunostained with the N675 antibody after (a) 100% MeOH and (b) 4% paraformaldehyde fixations.

Following fixation, cells were blocked in 10% serum of the host of the secondary antibody for 1 hr, and then incubated with the primary antibody diluted in 1% serum, for an hour at 37°C. Following washes in 1X PBS with three changes in 1 hr, secondary antibody in 1% serum was applied to the cells, and incubated for 1 hr at 37°C. Finally, cells were washed in 1X PBS and mounted with fluorsave reagent. Subcellular htt localization was determined using antibodies directed to different regions of the protein (Table 2)

Antibody	Description	Detection
N675	Rabbit polyclonal, gift from Dr. Lesley Jones (UK)	against amino acids 1-17
HDA	Mouse monoclonal, gift from Dr. Glenn Morris (UK)	against amino acids 997–1276
HDC	Mouse monoclonal, gift from Dr. Glenn Morris (UK)	against amino acids 2703–2911

Table 2. Primary antibodies used to localize endogenous htt.

2.3 Endogenous htt localization

N675 antibody should detect the first 17 amino acids of htt, just prior to the polyQ tract. Therefore, it should catch up the FL htt protein and any N-terminal cleavage products. HEK293 cells showed strong granular cytoplasmic staining with N675, and nuclear signal was restricted to a few small puncta (Fig.2.a). On the other hand, N2A cells exhibited strong diffuse nuclear staining (Fig.2.b,c). PC12 cells, when treated with NGF, showed one to two nuclear puncta, otherwise they showed a cytoplasmic staining pattern (Fig.2.d,e).

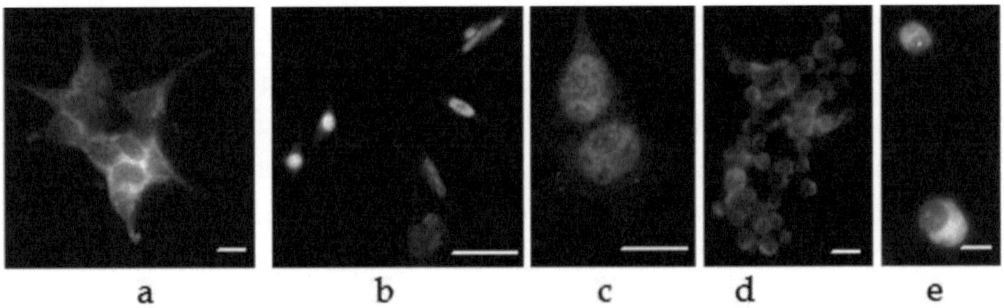

Fig. 2. HEK 293 (a, scale bar=20μm), N2A (b-c, scale bar=10μm), PC12 (d), NGF-treated PC12 (e, scale bar=20μm) cells immunostained with N675.

The localization and expression pattern of endogenous htt was further analyzed using another N-terminal antibody, HDA. This mouse monoclonal antibody was raised against the htt amino acids 997–1276, therefore it should detect the FL htt protein and any N-terminal cleavage products. HDA antibody caught very distinctive htt inclusions in HEK293 cells (Fig.3.a). However, N2A (Fig.3.b) and PC12 cells (Fig.3.c) showed homogenous nuclear localization, with a few puncta in higher expressing N2A cells. IMR32 cells generally demonstrated a nuclear expression with perinuclear inclusions in some cells (Fig.3.d, Fig4).

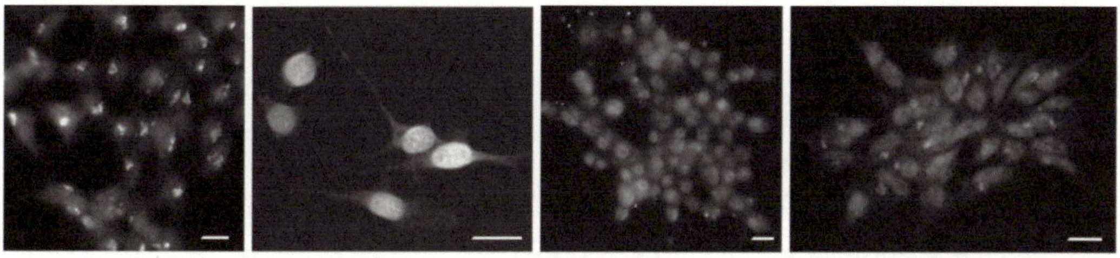

a b c d

Fig. 3. HEK 293 (a), N2A (b), NGF-treated PC12 (c), IMR32 (d) cells immunostained with HDA (scale bars=20μm for (a) and (c), 10μm for (b) and (d)).

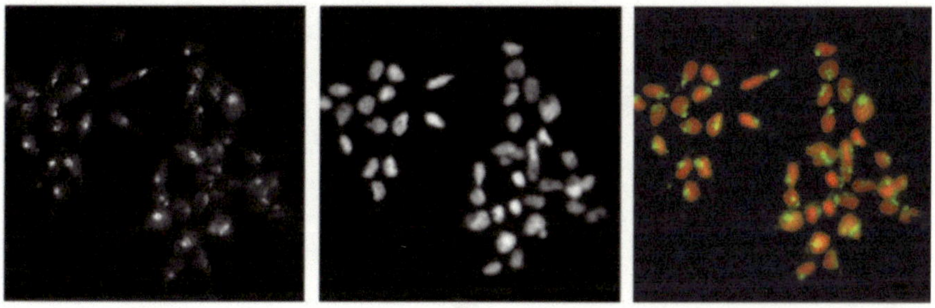

a b c

Fig. 4. Htt localization in IMR32 cells with HDA (a), nuclear staining with PI (b) and merged image (c).

The localization of endogenous htt also assessed by using a C-terminal antibody, HDC. This mouse monoclonal antibody was raised against amino acids 2703–2911, therefore expected to catch up FL htt and any C-terminal cleavage products. All cell types studied showed exclusively cytoplasmic localization and diffuse expression with the HDC antibody. HEK293 cells showed diffuse cytoplasmic localization, and occasionally, one inclusion per cell was noticed (Fig.5.a). In N2A cells, in addition to diffuse cytoplasmic staining, there was higher expression in the dendrites and nerve terminals (Fig.5.b.), cytoplasmic aggregates were noticed only in a few cells. In PC12 cells, htt expression was in the form of cytoplasmic punctates (Fig.5.c), and when treated with NGF, localization was extended to dendrites (Fig.5.d.). IMR32 cells showed diffuse cytoplasmic expression (Fig.5.e.).

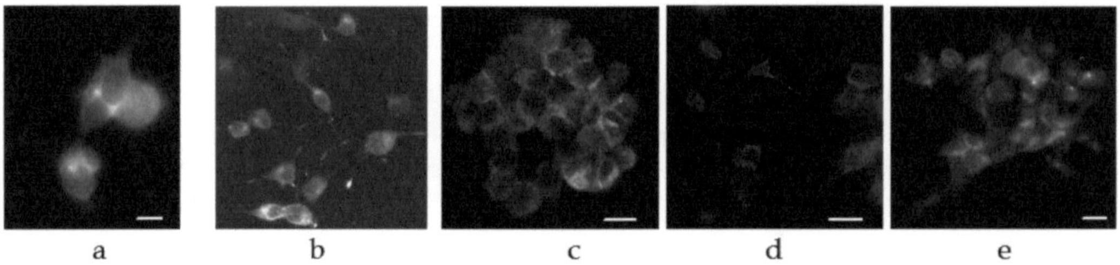

Fig. 5. HEK 293 (a), N2A (b), PC12 (c), NGF-treated PC12 (d), IMR32 cells (e) immunostained with HDC (scale bars=20μm).

3. Htt overexpression in HD cell models

3.1 Cloning full length huntingtin into pEYFP-C1

Green fluorescent protein (GFP), extracted from the jellyfish Aequorea victoria, is a widely used fluorescent reporter molecule. GFP and its variants can be expressed as fusion constructs with other proteins to monitor dynamic cellular processes. Since it does not require any additional substrates to emit light, it is ideal for *in vivo, in situ*, and real time protein expression and localization studies (Chalfie& Kain, 1998). In this study pEYFP-C1 and pEYFP-N1 vectors were used for cloning and expression studies of htt. Full length (FL) htt (10 kb) with normal and expanded CAG repeats were cloned into yellow fluorescent vector, pEYFP-C1. For this purpose, FL htt25Q and FL htt82Q sequences were first released from the pRcCMV vector (Cooper et al., 1998), which were generously provided by Dr. Christopher Ross (Johns Hopkins School of Medicine, USA), with BstZI and NotI restriction enzymes. FL htt was cut just before the ATG start codon and just after the polyA tail of htt, respectively. FL htt sequences were inserted into EcoRI-digested pEYFP-C1 vector (Fig.6.a). Ultracompetent Epicurian coli XL2 Blue cells were transformed with the ligation products and grown on kanamycin-containing agar plates at 37°C o/n. Bacterial colonies, which have taken up the ligation constructs, were selected first by filter hybridization and then with sequencing. After verification of the sequence frames, successful clones were maxi-preped.

3.2 Cloning truncated huntingtin into pEYFP-N1

The constructs containing only the exon1 of the htt gene with 23 and 65 CAG repeats in the pcDNA6c-myc/His vector were kind gifts from Dr.Mark Lesort (The University of Alabama at Birmingham. Truncated htt sequences were released from this vector with BamHI and XhoI restriction enzymes, and ligated into NheI- and XhoI-digested pEYFP-N1 vector (Fig.6.b). Ultracompetent Epicurian coli XL2 Blue cells were transformed with the ligations and were grown on kanamycin-containing agar plates at 37°C o/n. The transformed bacterial colonies were subjected to sequencing to confirm vector-insert junction sequences and the CAG repeat size.

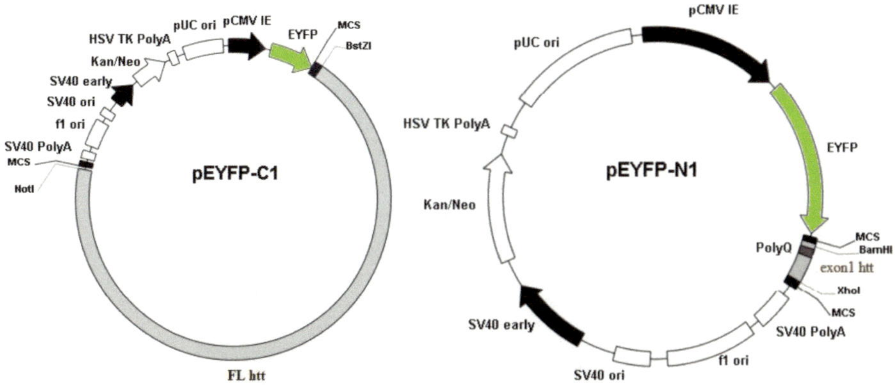

Fig. 6. Vector diagrams for FLhtt- pEYFPC1 (a) and Ex1htt- pEYFPN1 (b).

3.3 Transient transfection and microscopy

Transient transfections of cell lines with plasmid DNAs were carried out in 6-well plates, using Fugene6 (Roche) according to the instructions provided. Following transfection, localizations were visualized by both conventional fluorescence microscopy and LSCM. Living colors fluorescent proteins can be visualized with fluorescence microscopes, and can be independently distinguished using filter sets specific for each color. Images are acquired with a cooled charge-coupled device (CCD) camera. Live cells expressing fluorescent fusion proteins, or fixed cells prepared by immunocytochemistry were visualised under the 40X objective of the Zeiss Axiovert-S100 TV fluorescence microscope equipped with appropriate filters. The collected images were analyzed using Kinetic Imaging software. In confocal microscopy, images are produced by scanning the cells. When the light source is a laser beam, it is called LSCM. Scanning the object in x-, y-, and z-directions along the optical axis allows visualization of the object from all sides. In conventional fluorescence microscopy, co-localization of different proteins is performed using different filter sets sequentially. However, in LSCM, each detector is equipped with its filter sets to enable simultaneous multi-channel imaging of two or more fluorescent proteins. In this study, the BioRad 1024-MP laser scanning microscope system was used to analyse live and immunostained fixed cells. Zeiss Axiovert-S100 TV microscope was attached to the 1024 scan head. Live cell EYFP fluorescence was detected with a krypton-argon ion laser at 488 nm. Two dimensional (x, y) high resolution images (512 x 512) were collected with 40X, 1.3 NA oil immersion lens and filtered eight times with Kalman filter. For live cell analysis, three dimensional images were also captured for the purpose of spatial localization.

3.4 FL htt overexpression

HEK293, N2A and PC12 cells were transiently transfected with FL normal and mutant htt, in the form of YFP fusion constructs. Mutant htt originally contained 82 CAG repeats; however, the repeat size was contracted to 60 and expanded to 90 CAGs during bacterial transformation. Transfected cells were analyzed live, since derivatives of GFP expression vectors allow direct analysis of the cells, without the need for fixation or staining. As a first step, localization and expression of the pEYFPC1 vector itself, which encodes a 27 kDa fluorescent protein, was analyzed. Expression was mostly localized to the nucleus in all cell types examined and showed a diffuse expression pattern, excluding nucleoli. In addition to that, very weak expression in the cytoplasm and dendrites were noticed (Fig.7.).

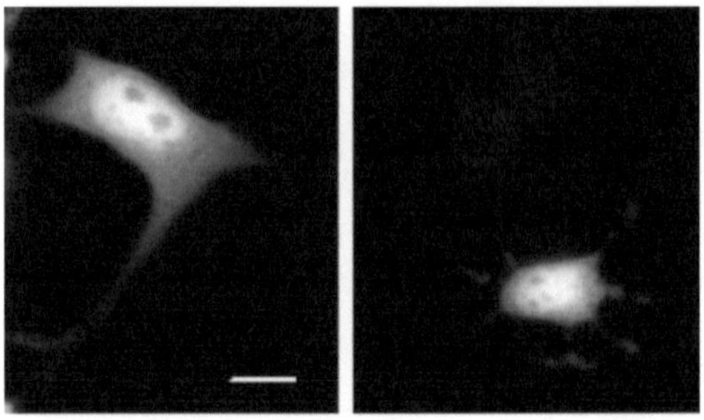

Fig. 7. pEYPC1 expression and localization in N2A cells (scale bar=10µm).

Full length wild type htt (FL htt25Q-YFP) expression was studied on a time scale in HEK293 cells. Transfected HEK 293 cells started to express htt–YFP fusion proteins four hours after transfection. As far as it was expressed, htt showed a diffuse cytoplasmic expression pattern. Transient transfections of wild type and mutant htt were analyzed for 72hr in HEK293, N2A, and PC12 cells, however, the overexpressed proteins did not change their cellular localization. Wild type htt-YFP expression was studied also with laser scanning confocal microscopy, which revealed a homogenous cytoplasmic expression in HEK293 cells (Fig.8.a). Cells expressing FL mutant htt-YFP also demonstrated diffuse cytoplasmic expression (Fig.8.b,c). However, more cells presented inclusions and apoptotic features, like big vacuoles, membrane blebbing and cellular dissociation (Fig.9.).

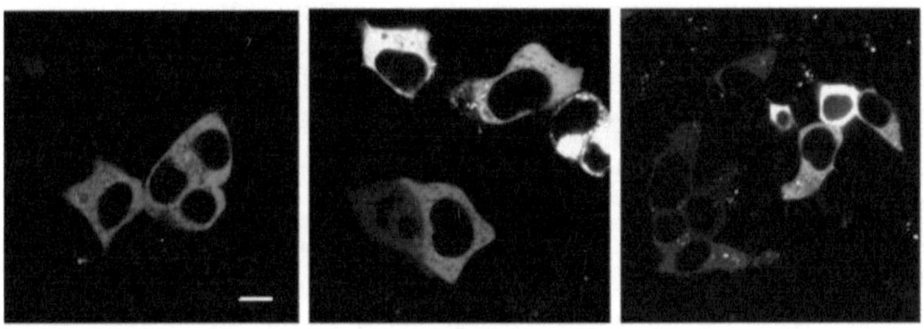

Fig. 8. LSCM images of FL htt25Q-YFP (a), FL htt60Q-YFP 8b), FL htt90Q-YFP (c) expression in HEK293 cells (scale bar=20µm).

Huntingtin expression was also studied in live N2A cells, transfected with htt25Q-YFP (Fig.10.a), htt60Q-YFP (Fig.10.b), and htt90Q-YFP (Fig.10.c) constructs. All plasmids exhibited a diffuse cytoplasmic expression pattern. In addition, htt was expressed in dendrites at high levels. Cells expressing the wild type htt were healthy; however those expressing mutant htt presented apoptotic features, like big vacuoles and membrane blebbing.

Live PC12 cells transfected with FL htt25Q-YFP showed diffuse cytoplasmic expression, and all were healthy (Fig.11.a). In cells treated with NGF, expression was extended to the dendrites; PC12 cells expressing mutant htt presented small cytoplasmic inclusions and failed to grow neurites when treated with NGF (Fig.11.b,c).

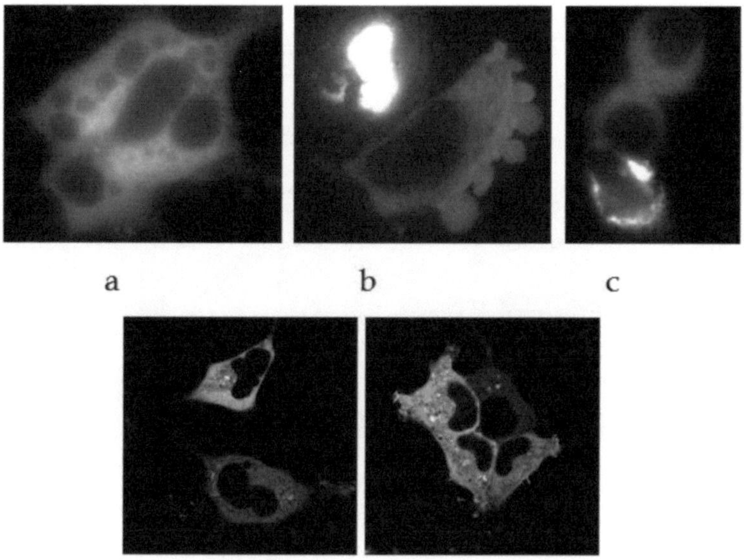

Fig. 9. FL htt60Q-YFP (a-c) and FL htt90Q-YFP (d,e) expressions in HEK cells as revealed by conventional fluorescence microscopy and LSCM, respectively.

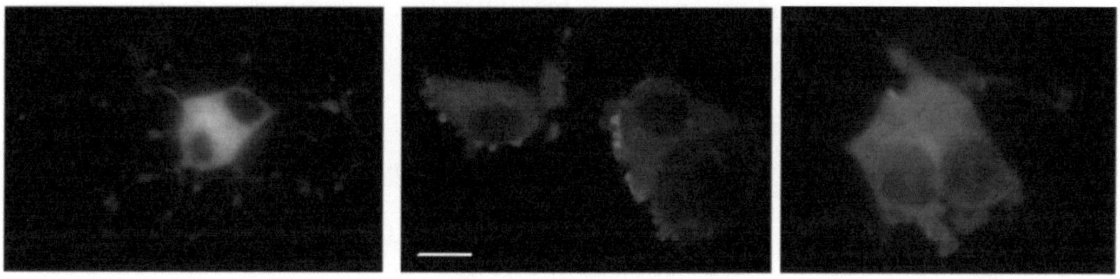

a b c

Fig. 10. a) FL htt25Q-YFP, b) FL htt60Q-YFP and c) FL htt90Q-YFP expressions in N2A cells (scale bar=10μm).

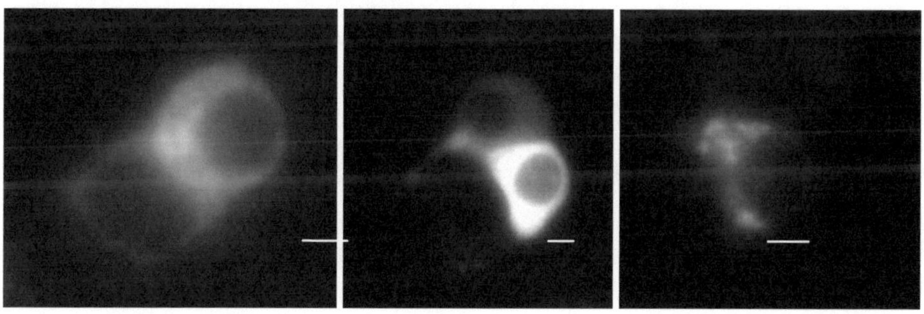

a b c

Fig. 11. a) FL htt25Q-YFP, b) FL htt60Q-YFP and c) FL htt90Q-YFP expressions in PC12 cells (scale bars=20μm).

In order to eliminate the possible effects of the 27 kDa fluorescent YFP tag on the localization of FL htt, HEK 293 (Fig.12.a), N2A (Fig.12.b), and PC12 cells (Fig.12.c) were transfected with untagged FL htt, and expression was detected with the N675 antibody. All transfected cells demonstrated diffuse cytoplasmic localization of both the wild type and mutant FL htt, as observed with FL htt-YFP constructs.

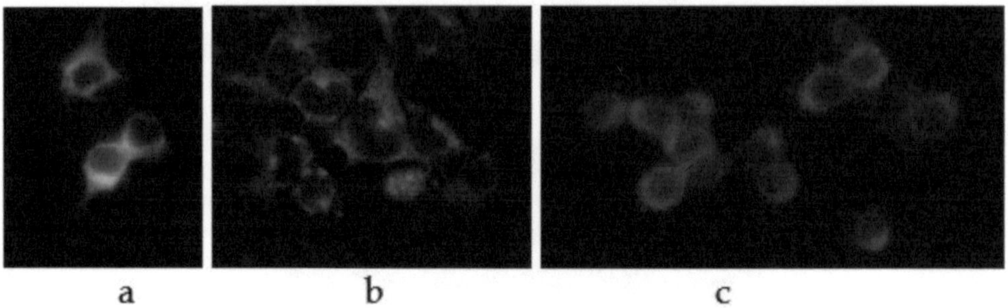

Fig. 12. a) HEK293, b) N2A, c) PC12 cells transfected with untagged FL htt.

The viability of the cells transiently transfected with the wild type and mutant htt constructs were studied by Trypan Blue Exclusion and MTT Assays. The experiments were performed in triplicates and repeated three separate times. The viability of the transfected cells are shown in Table 3. The significance of the toxicity caused by htt overexpression was calculated with the two-sample t-test. According to the results, FL htt90Q-YFP (p=0.039), but neither FL htt 25Q-YFP nor FL htt60Q-YFP caused significant cell death after 24 hr expression (at 95% CI, p=0.635 and p=0.255, respectively).

Construct	Viability (%)
Untransfected	88.0
pEYFP	85.9
FL htt25Q-YFP	74.8
FL htt60Q-YFP	65.8
FL htt90Q-YFP	62.4

Table 3. Viability of the cells transfected with htt constructs.

The metabolic activity of the cells was investigated with the colorimetric MTT assay. The significance of toxicity created by the wt and mutant htt constructs were calculated with the two sample t-test). The data has shown that mutant htt (p=0.008), but not wt htt (p=0.22), created significant toxicity in the cells in 24 hr.

3.5 Truncated htt overexpression in cell lines

In order to gain insight into the localization of truncated htt, HEK 293 and N2A cells were transfected with the truncated wild type and mutant htt, containing only the Exon 1 in pEYFP-N1 expression vector. Wild type truncated htt23Q showed more nuclear but also cytoplasmic diffuse expression pattern. In the nuclei, one to four htt inclusions were observed (Fig.13.a,b). Truncated htt65Q exhibited the same pattern; however the inclusions were of bigger size (Fig.13.c). The number and size of the inclusions were increased after 72hr transfection (Fig.13.b,d).

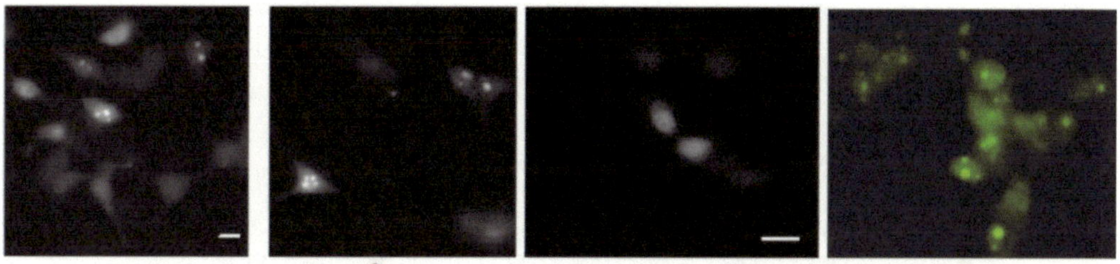

Fig. 13. Truncated htt23Q-YFP expression after 24hr (a) and 72hr (b) transfection in HEK293 cells; truncated htt65Q-YFP expression after 24hr (c) and 72hr (d) transfection in N2A cells (scale bars=15μm).

4. Htt expression in HD mouse models

The R6/1 and R6/2 mice were the first transgenic mouse models established to study HD. They both express only the exon 1 part of the human HD gene with 115 and 150 CAG repeats, respectively. The transgene is driven by the human huntingtin promoter. The transgene expression levels were identified as 31% and 75% of the endogenous huntingtin levels in the R6/1 and R6/2 mice, respectively (Mangiarini et al., 1996).

Endogenous and mutant htt expressions were analyzed in embryonic striatal neurons of two HD mouse models, R6/1 and HdhQ150. After breeding, pregnant female mice at E15 were sacrificed, embryos were removed and primary striatal cell cultures were prepared from E15 embryonic brains. The exact gestational stage of the embryos were determined by measuring the crown length. Embryos at E15 reach a crown length of 11-13 mm.

4.1 Huntingtin expression in R6/1 mouse model

Prior to analyze the htt expression pattern, primary striatal cells were stained for a widely expressed protein, β-tubulin III as a control for the protein expression level. Both wild type and R6/1 mutant embryonic striatal cells showed the same cytoplasmic localization and expression pattern (data not presented).

The N675 antibody showed a nuclear localization in both wild type (Fig.14.a) and mutant cells (Fig.14.b). The pattern of expression was occasionally punctate, but usually diffuse and homogenous, excluding nucleoli. Nuclear localization of htt was also demonstrated as co-stained with cytoplasmic β-tubulin III (Fig.15.).

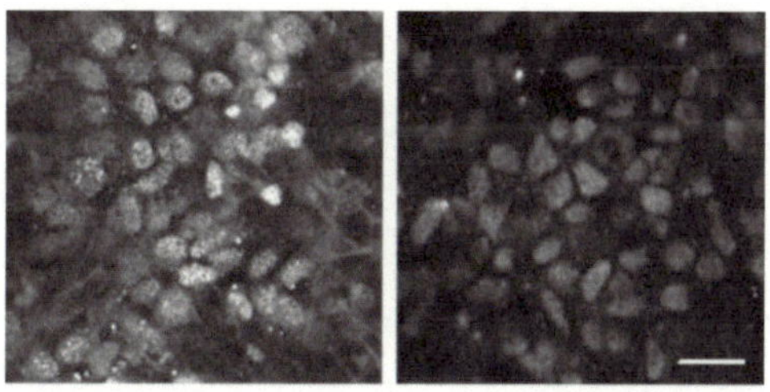

Fig. 14. LSCM images of N675-stained wt (a) and R6/1 (b) embryonic striatal cells (scale bar=20μm).

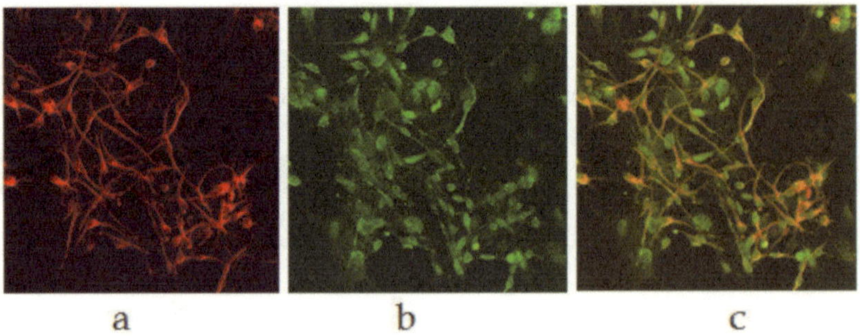

Fig. 15. Htt and β-tubulin III staining of embryonic striatal cells. a) β-tubulin III, b) htt, c) merged images.

Localization of htt was also assessed by the C-terminal antibody, HDC. This antibody revealed an exclusively cytoplasmic staining pattern of htt in both wild type (Fig.16.a) and R6/1 (Fig.16.b) primary striatal cells.

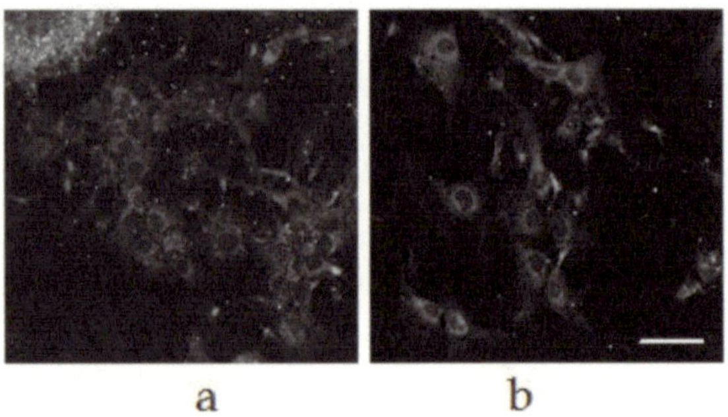

Fig. 16. LSCM images of HDC-stained wt (a) and R6/1 (b) primary striatal cells (scale bar=20μm).

4.2 Huntingtin expression in HdhQ150 mouse model

HdhQ150 knock-in mice carry FL mutant htt with 150 glutamine residues, therefore represent a perfect replica of the mutant human HD gene (Lin et al., 2001). Heterozygous male HdhQ150 mice were bred with heterozygous female HdhQ150 mice, and primary striatal cell cultures were prepared from E15 embryonic brains. Prior to analyze the htt expression pattern, striatal cells were stained with β-tubulin III as a control. Both wild type and mutant primary striatal cells showed the same cytoplasmic localization, and the expression levels were not different (data not presented here). Striatal neurons were identified by immunostaining the cells with an antibody against dopamine- and cyclic AMP-regulated phosphoprotein, DARPP-32. The cultured cells were all DARPP-32 positive in their nuclei after seven days (data not presented here).

Huntingtin localization and expression pattern were analyzed in wild type, heterozygous and homozygous embryos using antibodies directed to N- and C- terminal htt protein. With N675, primary striatal cells of both wild type (Fig.17.a) and mutant embryos (Fig.17.b,c) demonstrated either punctate or diffuse expression in the nuclei, excluding nucleoli. Nuclear localization (Fig.17.d) was verified by the nuclear counterstain, Draq5 (Fig.17.e).

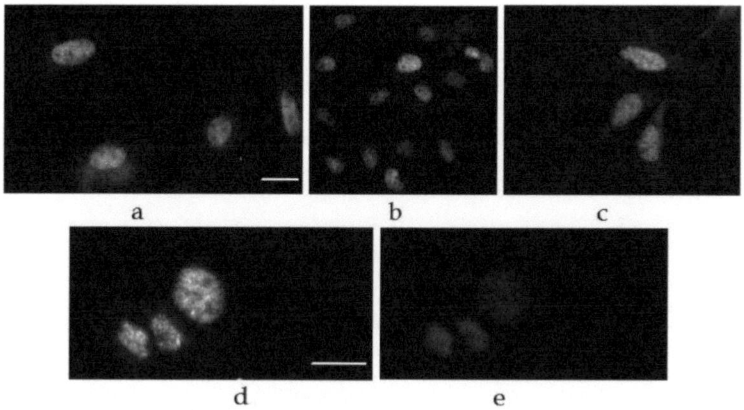

Fig. 17. N675 staining of HdhQ150 primary striatal cells (scale bars=10μm).

Localization of N-terminal htt was further assessed using another N-terminal antibody, HDA. This antibody detected htt in the nucleus, either diffuse or granular, in wild type (Fig.18.a), heterozygous (Fig.18.b) and homozygous mutant embryonic striatal cells (Fig.18.c).

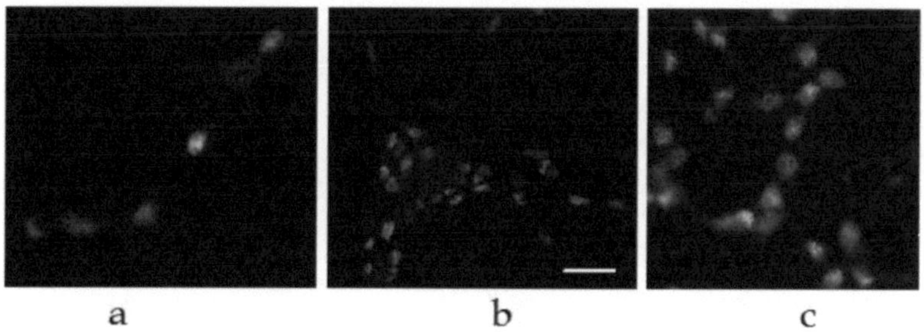

Fig. 18. HDA staining of HdhQ150 primary striatal cells (scale bar=20μm).

The antibody specific to the C-terminus of htt, HDC, revealed an exclusively cytoplasmic and diffuse staining pattern in both wild type (Fig.19.a) and heterozygous (Fig.19.b) and homozygous mutant (Fig.19.c) embryonic striatal cells.

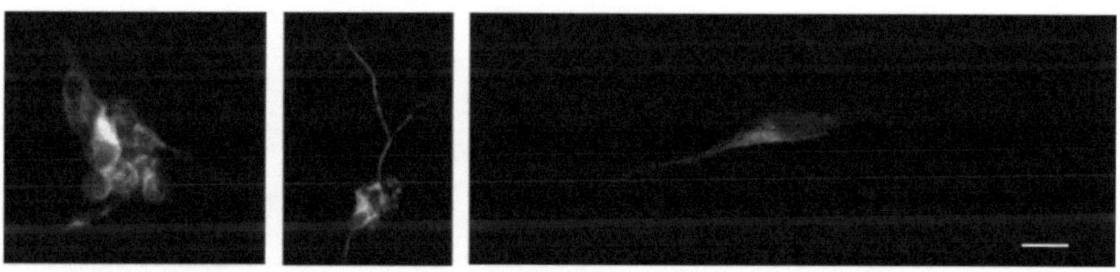

Fig. 19. HDC staining of HdhQ150 primary striatal cells (scale bar=20μm).

5. Western blotting and htt expression

Htt protein expressions were verified with SDS-PAGE and Western blotting. Protein samples extracted from transfected cell lines, quantified by Bradford Assay and subjected to SDS-PAGE electrophoresis. SDS-PAGE gels were blotted onto membranes for subsequent detection with the antibodies. Overexpressed htt was revealed with the antibodies N675 (Fig.20a), HDA (Fig.20.b) and HDC (Fig.20.c). Fluorescent fusion proteins were also detected with the EGFP antibody, specific for the derivatives of the GFP fluorescent protein (Fig.20.d).

Endogenous and overexpressed htt expressions were also investigated in cell lines and HdhQ150 mouse models which revealed different truncation products with different antibodies (Fig.21.)

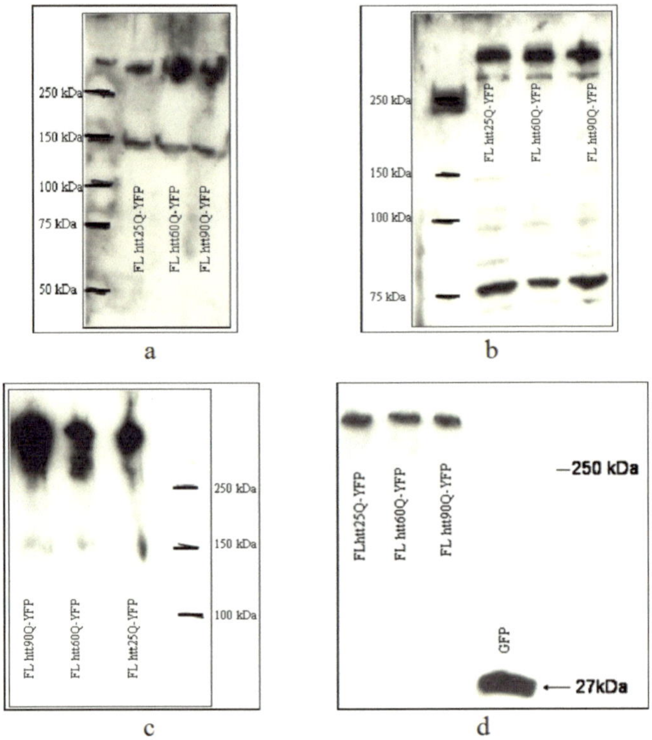

Fig. 20. Western blot of a 5% SDS-PAGE gel with FL htt-YFP expression in HEK293 cells after N675 (a), HDA (b), HDC (c) and anti-GFP antibodies.

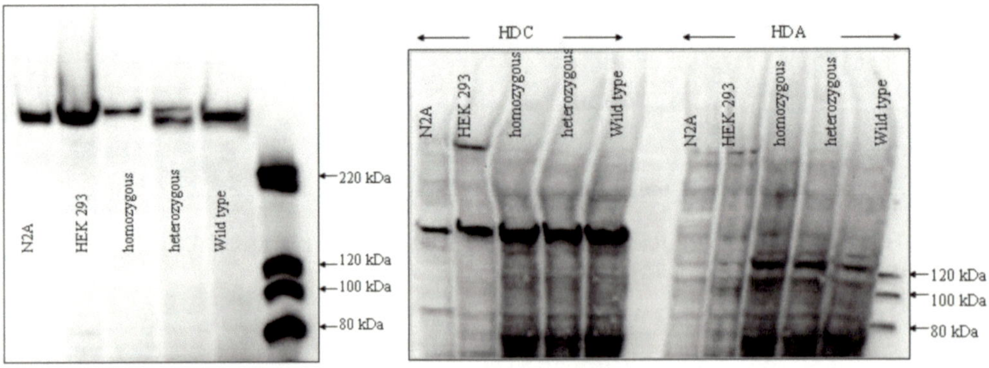

Fig. 21. Western blot of 3-8 % SDS-PAGE gels, comparing htt expression in cell lines and HdhQ150 embryos using N675 (a), HDA and HDC antibodies (b).

6. Modeling htt *in silico*

6.1 Building 3D htt model via fold recognition

The mechanisms leading to selective neurodegeneration can be explored via modeling the normal and mutant forms of the related proteins and analyzing their molecular structures. Most of the methods for protein structure prediction and modeling make use of the known structures of the homologous proteins. Htt shows no homology to any other counterparts, therefore it is not easy to identify its structure and assign its functions. However, since each protein folds into a unique 3D conformation, one should be able to predict its unknown structure using algorithms. Fold recognition method should be the method of choice when the protein of interest has no known homologues. Given a library of known structures, fold recognition determines which of them shares a folding pattern with the query protein, for which the sequence but not the structure is known (Lesk, A.M., 2008). A method for fold recognition is threading. The idea behind threading is to create many models for the query using many possible alignments between the known structures and the unknown protein. So, the threading method tries all possible folds and all possible alignments to establish the rough models. For successful fold recognition, the models should be scored and the best one should be selected. In addition to that, the scores should be calibrated to explore whether the rough model with the best score is likely to be correct.

There are programs available, like Jmol, Opendx, Rasmol, VMD and XCrySden, to determine the 3D structures of proteins. In this study VMD (Visual Molecular Dynamics) is used since the program is very user-friendly in constructing and analyzing the 3D molecular structures of the proteins. In addition to this, VMD shows α-helix and β- sheet structures, coils, turns and van der Waals bonds as well as protein sequence information, atomic arrangements and micromolecular details of the proteins (Gibas, C. & Jambeck, P., 2002). VMD programme, provided by the University of Illinois at Urbana-Champaign, creates 3D structures of the proteins that are saved as PDB files (Humphrey et al., 1996). Since the program produces more reliable and fair results with short amino acid sequences, htt protein was loaded to the program as sequences of 400 amino acids (aa). For each 400 aa sequence, five best models were retrieved (Table 4.) Using HHpred programme, the best model for the polyQ-bearing first 400 aa sequence was identified to be the 1WA5_B model (Fig.22).

Htt sequence	Best models
1st 400aa	1WA5_B , 1B3U_A, 1IBR_B, 1W9C_A , 1Q1S_C
2nd 400 aa	1PAI_A, a.86.1.1.1, 1AO7_E, 1LP9_E, 1UP6_E
3rd 400 aa	2GO2_A, 1X9D_A, 1EE5_A, 1Y2A_C ve 1Q1S_C
4th 400 aa	1EE5_A, 1WA5_B, 2F6H_X, 2GO2_A, 2F5U_A
5th 400 aa	a.118.1.14.3_A, 1ZEE_A, 1GAI, 1Y2A_C ,1Q1S_C
6th 400 aa	2GFP_A, 1IBR_B, 1U7G_A, a.118.5.1.1_A, 1Y2A_C
7th 400 aa	2F5U_A, 1RH5_A, 1IBR_B, 1YFM, 1C3C_A
Last 344 aa	1HZ4_A, a.118.4.1.1_A, 1XM9_A, 1N4M_A , 1Y2A_C

Table 4. Best models of htt.

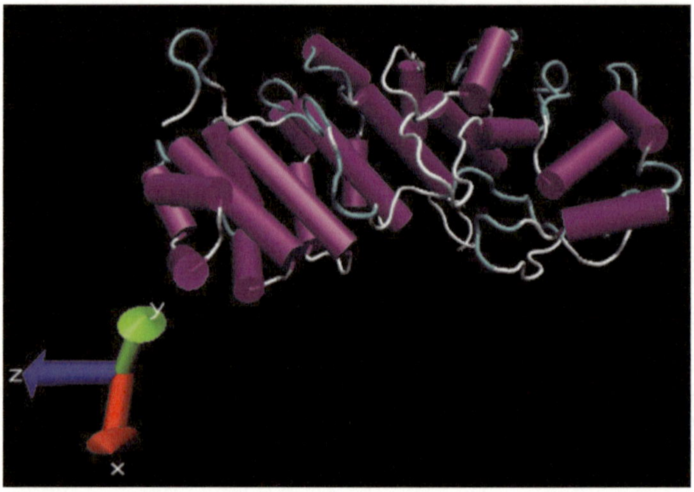

Fig. 22. The best model structure of the first 400 aa of htt.

The first 400 aa region of htt revealed parallel α-helices and no β-sheets in the initial parts of the structure, the turns and coils were normal in length. In the second 400 aa part,the number of α-helices decreased and β-sheets dominated, turn and coil structures were longer. The third 400 aa part showed long α-helices and very few β-sheets. In addition to that, turn and coil structures were considerably longer and α-helices tended to form tangles. the fourth 400 aa part resembles to the first part in terms of its α-helix and β-sheet content, however turns and coils were longer. In the fifth 400 aa region, there were much less but longer α-helices than that of the first part and they formed tangles. There were long coils and turns like that of the second part. Only one of the models have identified β-sheets in the fifth region. In the sixth 400 aa part of the protein, dense and parallel α-helices were found to dominate. In some models α-helices were considerably long. There were less number of α-helices which were long and organized as tangles. Some models indicated β-sheets, turn and coil structures were longer than that of other protein regions. In the last part of the protein, the density of the α-helices and the lengths of turns and coils were normal.

6.2 Modeling htt mutation *in silico*

In this part of the study, *in silico* polyQ expansion mutations were created by adding extra glutamine repeats to the first 400 aa part of the protein. Wild type normal htt protein is accepted to contain 23Qs. In humans, 27-35 CAGs show meiotic instability and 36-39 repeats are considered to show incomplete penetrance. Repeats above 40 definitely cause HD .In order to represent these stages, polyQ region of htt was made expanded with 10,13, 14, 15, 16, 20, 25 and 30 additional glutamines, which result in mutant proteins of 33, 36, 37, 38, 39, 43, 48 and 53Q. The mutant models were compared to the best model of the first 400 aa structure, 1WA5_B. According to the results, there is no significant structural change in the mutant proteins of 33, 36, 37 and 38Q. However, htt with 39 or more glutamines have shown conformational changes. Especially turn and coil regions were found to be longer and increased in number, and α-helices were found to be shorter. One previous study has reported increased α-helices and β-sheets, which was thought to be correlated with increased tendency to aggregate (Marchut, A.J.& Hall, C.K., 2007). As a second step, models of only the polyQ regions of 36Q and 53Q proteins were constructed and compared to the polyQ region of the 1WA5_B model (Fig.23). It was noticed that the turn and coil structures were increased in number and α-helices were shorter. In this situation, protein may gain non-covalent interactions within itself or with other proteins and aggregate in the form of twisted β-sheets.

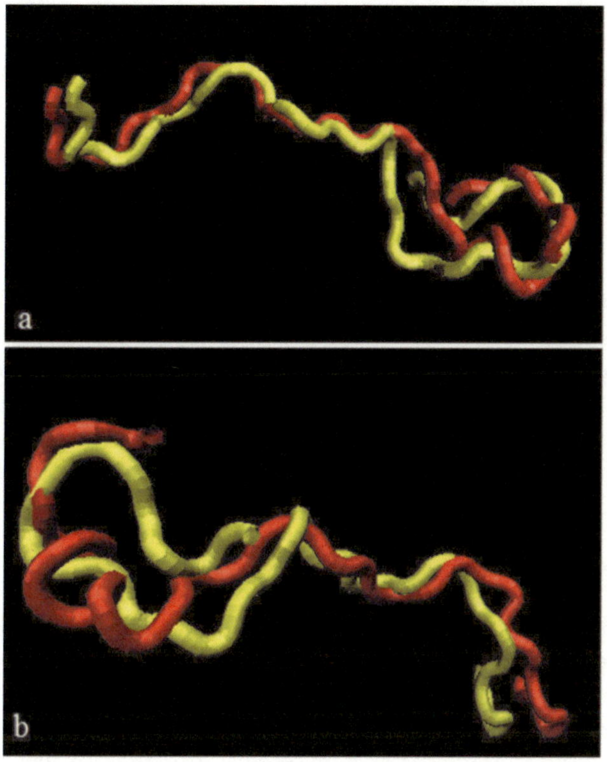

Fig. 23. Structural comparison of the polyQ regions of 1WA5_B to that of mutant 36Q (a) and 53Q (b). Wild type polyQ region is indicated in yellow, mutant polyQs are in red color.

Conformational change due to CAG repeat expansion results in toxic gain of function in the mutant protein and leads to cellular toxicity. Toxic gain of function, in turn, results in loss of function of the normal htt (Cattaneo et al., 2001; Chen et al., 2003). This interplay starts inevitable neurodegeneration processes. Increase in the number of turns and coils, together with shortened α-helices may result in folding of the protein within itself or favors its interactions with other polyQ proteins and its aggregation as β-pleated sheets. Research findings have identified aggregates in the form of β-sheets *in vitro* and in human HD brain (Scherzinger et al., 1997)., which can be explained by the previously proposed polar zipper model (Perutz et al., 1994). Distortion in the α-helix structure and formation of β-sheets were also proved in other neurodegenerative disorders. It is apparent that β-sheets render the protein more susceptible to aggregate formation (Goedert, M., 1999; Murray et al., 2001). Lengthened turn and coil and shortened α-helix structures significantly change the conformation of the protein. In this new conformation, conserved protein regions may become susceptible to cleavage by proteases which results in proteolysis, production of toxic htt fragments and regional pathology. On the other hand, conformational change may also cause deficiency in normal htt proteolysis, in that case toxicity comes into play due to inreased half life of the protein. It has been shown that aggregate-forming mutant htt is compatible with variable β-sheet/ β-turn model by using N-terminal htt in mammalian cell cultures and cortical neurons (Poirier et al., 2005). This proves that the resulting mutant structure is toxic to the cells. On the other hand, these intercellular aggregates can protect the cells against toxicity by isolating the mutant proteins in the cell (Rajagopalan, S. & Andersen, J.K., 2001). Overall, expanded polyQ leads to a conformational change by favoring lengthened turns and coils and shortened α-helices, which may lead to improper folding and aggregation of the mutant protein. Then, apoptosis follows when the cellular concentrations of the proteins are distorted.

7. Conclusion

Since the identification of the HD gene and its protein product, localization of both the endogenous and overexpressed htt has been investigated by several research groups in cell lines, animal models, and post-mortem patient tissues. However, after almost eighteen years, there is still no agreement on the precise distribution and function of htt. In the framework of this study, localization and expression patterns of endogenous and overexpressed htt were investigated in a variety of neuronal and non-neuronal cell lines (HEK 293, N2A, PC12, IMR32) and in embryonic striatal projection neurons expressing FL and truncated htt.

Endogenous htt was localized in cells using antibodies directed to N- and C-terminal regions of htt. The results demonstrated that htt is proteolysed as a cell type specific manner. Using the N-terminal antibody, localization of htt was found to be nuclear in neuronal cells, but cytoplasmic in the others. This suggests that htt might be processed differently in neuronal and non-neuronal cells. In the neurons, cleavage products might be small enough to enter into the nucleus by passive diffusion, or they might interact with proteins involved in nuclear functions and actively transported. In other words, htt cleavage and subsequent transportation to the nucleus might be required for the nuclear activities in neuronal cells. In addition, inclusions detected with the HDA antibody but not with N675, imply a different pool of N-terminal truncation products. In this study endogenous htt was shown to be exclusively cytoplasmic using the C-terminal antibody, with a diffuse and homogenous expression pattern. In neuronal cells localization was extended to the dendrites and concentrated in the nerve endings, suggesting a role in neurotransmission.

Overexpressed FL wild type and mutant htt with YFP tags were localized in the cytoplasm, regardless of the cell type. The possible effects of the 27 kDa YFP on the localization of proteins were analyzed by using untagged htt constructs, however the localizations were shown to be the same. All cells examined expressed htt four hours after transfection, and have never entered into the nucleus in three days, after which they started to die. In addition, htt was shown to be localized to dendrites in neuronal cells, which again suggests a role in synaptic transmission. On the other hand, expression of full length mutant htt containing 60Q and 90Q was shown to be toxic to the cells, as verified with the cell viability assays. Cells transfected with the mutant constructs showed apoptotic features like membrane blebbing, large vacuoles and cellular dissociation. In addition, PC12 cells have failed to grow neurites when stimulated with NGF, which proves that mutant htt has a considerable effect on cellular growth. Proteolytic cleavage of overexpressed htt was shown on Western blots. Detection of overexpressed htt proteins in HEK 293 cells with N- and C-terminal antibodies has revealed a 150 kDa htt fragment as well as the FL protein. Apparently, this large fragment cannot enter the nucleus by passive diffusion. On the other hand, differential processing of htt in different cell types is apparent on the gels. Overexpressed truncated wild type and mutant htt proteins show more nuclear, but also cytoplasmic localization. A few small nuclear inclusions were noticed with the wild type protein, but the mutant htt forms more and bigger inclusions, which increase in number and size in time.

Htt localization was also assessed in primary striatal neurons of HD mouse models expressing the FL and exon1 fragment of the protein. In both models, N-terminal antibodies recognized nuclear htt either diffuse or in the form of punctates, and the C-terminal antibody recognized htt exclusively in the cytoplasm. The localization and expression patterns were the same for striatal cells expressing wild type, heterozygous and homozygous mutant huntingtin. This implies that, in wild type and mutant embryonic striatal projection neurons, htt is cleaved and processed in similar ways.

Apart from evaluating htt localization in cell and animal models, 3D structures of normal and mutant htt were identified in order to correlate any conformational changes to disease pathology. The first 400 aa region of htt revealed parallel α-helices and no β-sheets in the initial parts of the structure, the turns and coils were normal in length. In silico polyQ expansion mutations were resulted in increased number of turn and coil structures and shorter α-helices in the polyQ region of htt. In this new conformation the protein may gain non-covalent interactions, fold improperly, resist degradation and aggregate in the form of β-sheets, which in turn depletes the soluble protein counterparts whose intracellular concentrations are crucial. *In silico* conformational changes due to expanded polyQ give clues about the pathogenic mechanisms of still unexplored neurodegeneration processes. Modeling mutant disease proteins *in silico* helps to predict possible changes in its conformation. Use of this information together with *in vivo* and *in vitro* protein localization data will help to explore the functions of the disease protein and the mechanisms involved in disease pathogenesis. In HD, where the molecular details of the neurodegeneration processes seem to be highly complex, concurrent evaluation of *in vivo, in vitro* and *in silico* data should better enlighten the way to discover selective neurodegeneration and ultimately to disease treatment.

8. Acknowledgment

The work presented here was supported by University of Wales College of Medicine (Cardiff, United Kingdom), Boğaziçi University (İstanbul, Turkey) and Haliç University (İstanbul, Turkey).

9. References

Bae, B.I., Hara, M.R., Cascio M.B., Wellington, C.L., Hayden, M.R., Ross, C.A., Ha, H.C., Li, X.J., Snyder, S.H., Sawa, A. (2006). Mutant huntingtin: nuclear translocation and cytotoxicity mediated by GAPDH. *Proc Natl Acad Sci USA,* Vol.103, No.9, (Feb 2006), pp.3405-9, ISSN: 0027-8424

Cattaneo, E., Rigamonti,D., Goffredo, D., Zuccato, C., Squitieri, F. & Sipione, S. (2001). Loss of normal huntingtin function: new developments in Huntington's disease research. *Trends in Neurosciences,* Vol.24, No.3, (March 2001), pp.182-188, ISSN: 0166-2236.

Chalfie, M. & Kain, S. (eds). (1998). *Green Fluorescent Protein Properties, Applications, and Protocols,* Wiley-Liss, Inc., ISBN: 047117839X, 9780471178392, New York

Chen, Y.W., (2003). Local Protein Unfolding and Pathogenesis of Polyglutamine-Expansion Diseases. *Proteins: Structure, Function, and Genetics,* Vol.51, No.1, (April 2003), pp.68-73, ISSN (printed): 0887-3585. ISSN (electronic): 1097-0134

Cooper, J. K., Schilling, G., Peters, M. F., Herring, W. J., Sharp, A. H., Kaminsky, Z., Masone, J., Khan, F. A., Delanoy, M., Borchelt, D. R., Dawson, V. L., Dawson, T. M.& Ross, C. A. (1998). Truncated N-Terminal Fragments of Huntingtin with Expanded Glutamine Repeats form Nuclear and Cytoplasmic Aggregates in Cell Culture. *Hum. Mol. Genet.,* Vol. 7, No. 5, (May 1998), pp.783-790, Online ISSN 1460-2083 - Print ISSN 0964-6906

Davies, S., Turmaine, M., Cozens, B.A., DiFiglia, M., Sharp, A.H., Ross, C.A., Scherzinger, E., Wanker, E.E., Mangiarini, L. & Bates, G.P. (1997). Formation of Neuronal Intranuclear Inclusions Underlies the Neurological Dysfunction in Mice Transgenic for the HD Mutation. *Cell,* Vol.90, No.3, (Aug1997), pp.537-548, ISSN 0092-8674

De Rooij, K. E., Dorsman, J. C., Smoor , M. A., Den Dunnen, J. T. & Van Ommen, G. J. B. (1996). Subcellular Localization of the Huntington's Disease Gene Product in Cell Lines by Immunofluorescence and Biochemical Subcellular Fractionation. *Hum. Mol. Genet.,* Vol. 5, No. 8, (Aug 1996), pp. 1093-1099, ISSN 0964-6906

DiFiglia, M., Sapp, E., Chase, K., Schwarz, C., Meloni, A., Young, C., Martin, E., Vonsattel, J. P., Carraway, R., Reeves, S. A., Boyce, F. M. & Aronin, N. (1995). Huntingtin is a Cytoplasmic Protein Associated with Vesicles in Human and Rat Brain Neurons. *Neuron,* Vol. 14, No.5, (May 1995), pp.1075-1081, ISSN 0896-6273

DiFiglia, M., Sapp, E., Chase, K.O., Davies, S.W., Bates, G.P., Vonsattel, J.P. & Aronin, N. (1997). Aggregation of Huntingtin in Neuronal Intranuclear Inclusions and Dystrophic Neurites in Brain. *Science,* Vol. 277, No.5334, (Sep 1997), pp. 1990-1993, ISSN: 0036-8075

Dragatsis, I., Levine, M. S. & Zeitlin, S. (2000). Inactivation of Hdh in the Brain and Testis Results in Progressive Neurodegeneration and Sterility in Mice. *Nat. Genet.,* Vol. 26, No. 3, (Nov 2000), pp. 300-306, ISSN: 1061-4036

Gibas,C.& Jambeck,P. (2002). *Einführung in die Praktische Bioinformatik. 1st edition,* O'Reilly Media, ISBN-10: 3897212897, ISBN-13: 978-3897212893, Köln

Goedert, M. (1999). Filamentous nevre cell inclusions in neurodegenerative diseases: tauopathies and α- synucleinopathies. *Philosophical Transactions of Royal Society London B*, Vol.354, No.1386, (June 1999), pp.1101-1118, ISSN: 0080-4622

Graveland, G. A., Williams, R. S. & DiFiglia, M. (1985). Evidence for degenerative and regenerative changes in neostriatal spiny neurons in Huntington's disease. *Science*, Vol. 227, No. 4688, (Feb 1985), pp. 770-773, ISSN: 0036-8075

Gutekunst, C. A., Levey, A. I., Heilman, C. J., Whaley, W. L., Yi, H., Nash, N. R., Rees, H. D., Madden, J. J. & Hersch, S. M. (1995). Identification and Localisation of Huntingtin in Brain and Human Lymphoblastoid Cell Lines with Anti-Fusion Protein Antibodies. *Proc. Natl. Acad. Sci. USA*, Vol. 92, No. 19, (Sep 1995), pp.8710-8714, ISSN: 0027-8424

Hackam, A.S., Singaraja, R., Wellington, C. L., Metzler, M., McCutcheon, K., Zhang, T., Kalchman, M. & Hayden, M. R. (1998). The influence of Huntington protein size on nuclear localisation and cellular toxicity. *J. Cell Biol.*, Vol. 141, No. 5,(Jun 1998), pp. 1097-1105, ISSN: 0021-9525

Havel, L.S., Wang, C.E., Wade, B., Huang, B., Li, S. & Li, X.J. (2011). Preferential accumulation of N-terminal mutant huntingtin in the nuclei of striatal neurons is regulated by phosphorylation. *Hum Mol Genet.*, Vol.20, No.7, (1Apr 2011), pp.1424-1437, ISSN: 0964-6906

Hoogeveen, A.T., Willemsen, R., Meyer, N., de Rooij, K.E., Roos, R.A., van Ommen, G.J. & Galjaard, H. (1993). Characterization and localization of the Huntington disease gene product. *Hum Mol Genet.*, Vol.2, No.12, (Dec 1993), pp.2069-73, ISSN: 0964-6906

Housman D. (1995). Gain of Glutamines, Gain of Function? *Nat. Genet.*, Vol. 10, No. 1, pp. 3-4, ISSN: 1061-4036

Humphrey, W., Dalke, A., Schulten, K. (1996). Visual Molecular Dynamics. *J.Molec. Graphics*, Vol.14, No.1, pp.33-38, ISSN 0263-7855

Huntington's Disease Collaborative Research Group. (1993). A Novel Gene Containing a Trinucleotide Repeat That is Expanded and Unstable on Huntington's Disease Chromosomes. *Cell*. Vol. 26, No. 72 (6), (Mar 1993), pp. 971-983, ISSN: 0092-8674

Jacobsen, J.C., Gregory, G.C., Woda, J.M., Thompson, M.N., Coser, K.R., Murthy, V., Kohane, I.S., Gusella, J.F., Seong, I.S., MacDonald, M.E., Shioda, T. & Lee, J.M. (2011). HD CAG-correlated gene expression changes support a simple dominant gain of function. *Hum Mol Genet.*, Vol.20, No.14, (15 Jul 2011), pp.2846-60. ISSN: 0964-6906

Jou, Y.S & Myers, R.M. (1995). Evidence from antibody studies that the CAG repeat in the Huntington disease gene is expressed in the protein. *Hum Mol Genet*, Vol. 4, No. 3, (Mar 1995), pp.465-469, ISSN: 0964-6906

Juenemann, K., Weisse, C., Reichmann, D., Kaether, C., Calkhoven, C.F. & Schilling, G. (2011). Modulation of mutant huntingtin N-terminal cleavage and its effect on aggregation and cell death. *Neurotox Res.*, Vol.20, No.2, (Aug 2011), pp.120-133, ISSN: 0892-0362

Kegel, K. B., Kim, M., Sapp, E., McIntyre, C., Castano, J. G., Aronin, N. & DiFiglia, M. (2002). Huntingtin is Present in the Nucleus, Interacts with the Transcriptional Corepressor C-Terminal Binding Protein, and Represses Transcription. *J. Biol. Chem.*, Vol. 277, No. 9, (Mar 2002), pp. 7466-7476. ISSN: 0021-9258

Landwehrmeyer, G. B., Mcneil, S. M., Dure, L. S., Ge P., Aizawa, H., Huang, Q., Ambrose, C. M., Duyao, M. P., Bird, E. D., Bonilla, E., De Young, M., Avila-Gonzales, A. J., Wexler, N. S., DiFiglia, M., Gusella, J. F., MacDonald, M. E., Penney, J. B., Young, A. B. & Vonsattel, J. P. (1995). Huntington's Disease Gene: Regional and Cellular Expression in Brain of Normal and Affected Individuals. *Ann. Neurol.*, Vol. 37, No. 2, pp. 218-230, ISSN: 0364-5134

Lesk, A.M. (2008). *Introduction to Bioinformatics*, Oxford University Press Inc. (*3rd edition*), ISBN 978-0-19-920804-3, New York .

Li, S. H., Schilling, G., Young, W. S. 3rd, Li, X.J., Margolis, R.L., Stine, O.C., Wagster, M.V., Abbott, M.H., Franz, M.L., Ranen, N.G., Folstein, S.E. , Hedreen, J.C.& Ross, C.A. (1993). Huntington's disease gene (IT15) is widely expressed in human and rat tissues. *Neuron,* Vol.11, No.5, (Nov 1993), pp.985-93, ISSN: 0896-6273

Lin, C. H., Tallaksen-Greene, S., Chien, W.M., Cearley, J.A., Jackson, W.S., Crouse, A.B., Ren, S., Li, X.J., Albin, R.L., Detloff, P.J.(2001) Neurological abnormalities in a knock-in mouse model of Huntington's disease. *Hum Mol Genet.* Vol. 10, No.2, (March 2001), pp.137-44. , ISSN: 0964-6906

Mangiarini, L., Sathasivam, K., Seller, M., Cozens, B., Harper, A., Hetherington, C.,Lawton, M., Trottier, Y., Lehrach, H., Davies, S.W.& Bates, G.P. (1996). Exon 1 of the HD gene with an expanded CAG repeat is sufficient to cause a progressive neurological phenotype in transgenic mice. *Cell* Vol.87 pp.493–506, ISSN 0092-8674

Marchut, A.J. & Hall, C.K. (2007). Effects of Chain Length on the Aggregation of Model Polyglutamine Peptides: Molecular Dynamics Simulations. *Proteins: Structure, Function, and Bioinformatics,* Vol.66, No.1, (Jan 2007), pp.96-109, ISSN: 1097-0134

Martin-Aparicio, E., Avila, J. & Lucas, J. L. (2002). Nuclear Localization of N-terminal Mutant Huntingtin is Cell Cycle Dependent. *European Journal of Neuroscience*, Vol. 16, No. 2, (Jul 2002), pp. 355-359, Online ISSN: 1460-9568

Murray, IVJ., Lee, VM.-Y. & Trojanowski, JQ. (2001). Synucleinopathies: a pathological and molecular review. *Clin Neurosci Res.,* Vol.1, No.6, (December 2001), pp.445-455, ISSN: 1566-2772.

Perutz, M. F., Johnson, T., Suzuki, M. &. Finch, J. T. (1994). Glutamine Repeats as Polar Zippers: Their Possible Role in Inherited Neurodegenerative Diseases. *Proc. Natl. Acad. Sci. USA*, Vol. 91, pp. 5355-5358, ISSN: 0027-8424

Poirier, M.A., Jiang, H. & Ross, C.H. (2005). A structure-based analysis of huntingtin mutant polyglutamine aggregation and toxicity: evidence for a compact beta-sheet structure, *Hum. Mol. Genet.,* Vol.14, No.6, (15 March 2005), pp.765-77, ISSN: 0964-6906

Rajagopalan, S. & Andersen, J.K. (2001). Alpha synuclein aggregation: Is it the toxic gain of funtion responsible for neurodegeneration in Parkinson's disease? *Mechanisms of Ageing and Development*, Vol.122, No.14, (30 Sep 2001), pp.1499-1510, ISSN: 0047-6374

Scherzinger,E., Lurz,R., Turmaine,M., Mangiarini, L., Hollenbach, B., Hasenbank, R., Bates, G.P., Davies, S.W., Lehrach, H. & Wanker, E.E. (1997). Huntigtin-Encoded Polyglutamine Expansions Form Amyloid-like Protein Aggregates In Vitro and In Vivo. *Cell,* Vol.90, No.3, (August 1997), pp.549-558, ISSN: 0092-8674

Sharp, A. H., Loev, S. J., Schilling, G., Li, S. H., Li X. J., Bao, J., Wagster, M. V., Kotzuk, J. A., Steiner, J. P., Lo, A., Hedreen J. S. & Ross, C. A. (1995). Widespread Expression of Huntington's Disease Gene (IT15) Protein Product. *Neuron.* Vol. 14, No.5, (May 1995), pp. 1065-1074, ISSN: 0896-6273

Strong, T. V., Tagle, D. A., Valdes, J. M., Elmer, L. W., Boehm, K., Swaroop, M., Kaatz, K. W., Collins, F. S. & Albib, R. L. (1993). Widespread Expression of the Human and Rat Huntington's Disease Gene in Brain and Nonneural Tissues. *Nature Genetics*. Vol. 5, No. 3, (Nov 1993), pp. 259-265, ISSN: 1061-4036

Tanaka, Y., Igarashi, S., Nakamura, M., Gafni, J., Torcassi, C., Schilling, G., Crippen, D., Wood, J.D., Sawa, A., Jenkins, N.A., Copeland, N.G., Borchelt, D.R., Ross, C.A. & Ellerby, L.M. (2006). Progressive Phenotype and Nuclear Accumulation of an Amino-Terminal Cleavage Fragment in a Transgenic Mouse Model with Inducible Expression of Full-length Mutant Huntingtin. *Neurobiol. Dis.,* Vol. 21, No.2, (Feb 2006), pp. 381-391, ISSN: 0969-9961.

Trottier, Y., Devys, D., Imbert, G., Saudou, F., An, I., Lutz, Y., Weber, C., Agid, Y., Hirsch, E. C. & Mandel, J. L. (1995). Cellular Localization of the Huntington's Disease Protein and Discrimination of the Normal and Mutated Form. *Nature Genetics*. Vol. 10, (May 1995), pp. 104-110, ISSN: 1061-4036

Wilkinson, F. L., Man, N. T., Manilal, S. B., Thomas, P., Neal, J. W., Harper, P. S., Jones, A. L. & Morris, G. E. (1999). Localization of Rabbit Huntingtin Using a New Panel of Monoclonal Antibodies. *Molecular Brain Research,* Vol. 69, No. 1, (May 1999), pp. 10-20, ISSN: 0169-328X

Wyttenbach, A., Swartz, J., Kita, H., Thykjaer, T., Carmichael, J., Bradley, J., Brown, R., Maxwell, M., Schapira, A., Orntoft, T. F., Kato, K. & Rubinsztein, D. C. (2001). Polyglutamine Expansions Cause Decreased Cre-Mediated Transcription and Early Gene Expression Changes Prior to Cell Death in an Inducible Cell Model of Huntington's Disease. *Hum. Mol. Genet.,* Vol. 10, No. 17, (Jun 2001), pp. 1829-1845, ISSN: 0964-6906

Xia, J., Lee, D.H., Taylor, J., Vandelft, M. & Truant, R. (2003). Huntingtin contains a highly conserved nuclear export signal. *Hum. Mol. Genet.,* Vol.12, No.12, (Jun 2003), pp.1393-1403, ISSN: 0964-6906

Yan, Y., Peng, D., Tian, J., Chi, J., Tan, J., Yin, X., Pu, J., Xia, K. & Zhang, B. (2011). Essential sequence of the N-terminal cytoplasmic localization-related domain of Huntingtin and its effect on Huntingtin aggregates. *Sci China Life Sci.,*Vol.54, No.4, (Apr 2011), pp.342-350, ISSN: 1674-7305

Zeitlin, S., J. P. Liu, D. L. Chapman, V. E. Papaioannu & Efstratiadis, A. (1995) Increased Apoptosis and Early Embryonic Lethality in Mice Nullizygous for the Huntington's Disease Gene Homologue. *Nature Genetics,* Vol. 11, pp. 155-163, ISSN: 1061-4036

Zuccato, C., Ciammola, A., Rigamonti D., Leavitt, B. R., Goffredo, D., Conti, L., MacDonald, M. E., Friedlander, R. M., Silani, V., Hayden, M. R., Timmusk, T., Sipione, S. & Cattaneo, E. (2001). Loss of Huntingtin-Mediated BDNF Gene Transcription in Huntington's Disease. *Science*, Vol. 293, No. 5529, (Jul 2001), pp. 493-498, ISSN: 0036-8075

3

Molecular Mechanism of Huntington's Disease – A Computational Perspective

Giulia Rossetti[1] and Alessandra Magistrato[2]
[1]*German Research School for Simulation Science, FZ-Juelich and RWTH,*
[2]*CNR-IOM-National Simulation Center c/o,*
International School for Advanced Studies (SISSA/ISAS), Trieste,
[1]*Germany*
[2]*Italy*

1. Introduction

1.1 The Huntington's Disease

Huntington's Disease (HD) is a devastating autosomal dominant[a] neurodegenerative human disease for which there is currently no cure. HD is characterized by progressive motor, cognitive, and psychiatric symptoms (Huntington 1872). The gene responsible for HD (HTT) encodes the ubiquitously expressed Huntingtin protein (Htt) (MacDonald et al. 1993) (Fig. 1). Human Htt is essential for brain development (Reiner et al. 2003), although its exact biological function is unknown. This protein is located mostly in the cytoplasm, but a small amount of Htt is also present in the nucleus (Kegel et al. 2002). Moreover, the protein can dynamically travel back and forth between the two cellular compartments (Kegel et al. 2002). Htt may be associated also with the plasma membrane, the endocytic and autophagic vesicles, the endosomal compartments, the endoplasmic reticulum, the Golgi apparatus, mitochondria and microtubules (Kegel et al. 2002; Caviston et al. 2007; Kegel et al. 2005; Rockabrand et al. 2007; Strehlow et al. 2007; Atwal et al. 2007). Htt is a large, multidomain protein (3144 aa and molecular weight 348 kDa) for which structural information at atomic resolution is not available (Zuccato et al. 2010). Htt has been proposed to be an elongated super-helical solenoid with a diameter of ~ 200 Å (Li et al. 2006) (Fig. 1). The best-characterized part of the protein is the Exon 1 (Ex1), which consists of the following regions: N17 (the 17-amino acid-long N-term), the variable polyQ stretch (less than 36 Qs in healthy individuals (Mangiarini et al. 1996)), and a polyProline (polyP)-rich region (Fig. 1). Ex1, with an extended polyQ tract (m-Ex1), is sufficient to cause HD-like pathology in animal models (Mangiarini et al. 1996). Moreover, expression of m-Ex1 is sufficient to cause the typical formation of Htt aggregates found in brains of HD patients (Mangiarini et al. 1996; Bates et al. 1998; Davies et al. 1997). Hence, investigations of structural, dynamical and kinetic properties of m-Ex1 may help to understand key aspects of the disease. The amino acids of N17 are highly conserved (100% similarity) in all vertebrate species (Tartari et al. 2008), and

---

[a]Autosomal dominant conditions are achieved in cases in which a mutated gene from one parent is sufficient to cause a disease, in spite of the presence of a normal gene inherited from the other parent

N17 was originally believed to be unstructured (Perutz 1999). However, mutational analysis *in vivo* and Circular Dicroism (CD) (Atwal et al. 2007; Thakur et al. 2009) spectroscopy and NMR (Thakur et al. 2009) on peptides *in vitro* pointed out that this polypeptide may be an amphipathic α-helix with membrane-associating properties with respect to the endoplasmic reticulum (Atwal et al. 2007). The polyQ stretch begins at the 18th amino acid in human Htt (MacDonald et al. 1993).[b]

In 1994, Max Perutz (Perutz 1994) suggested, for the first time, that the physiological function of polyQ was to bind transcription factors containing also a polyQ region. Consistently, it was later shown that the polyQ tract is a key regulator of Htt binding to its partners (Harjes & Wanker, 2003). This hypothesis is supported by the presence of HEAT repeats (Fig. 1) along the Htt sequence, which favor protein-protein interactions (Andrade & Bork 1995). Moreover, the polyQ region may have flexible and multifunctional structures, which can assume specific conformations and different activities, depending on its binding partners, on its sub-cellular location, and on time of maturation in a given cell type and tissue (Kim et al. 2009; Zuccato et al. 2010).

The polyQ region is followed by a polyP tract (Tartari et al. 2008). This latter may affect the stability of the polyQ segment by keeping it soluble (Bhattacharyya et al. 2006; Steffan et al. 2004). Hence, it may protect polyQ against its conformational collapse. In addition, the polyP may also work also as a protein-interaction domain. Consistent with these hypotheses, structural data provided hints that the polyQ's aggregation and toxicity are influenced by the COOH-terminal polyP region (Kim et al. 2009; Bhattacharyya et al. 2006).

1.2 Mutated huntingtin in HD and the role of polyQ

The causative mutation of Htt is an abnormal expansion of CAG trinucleotide repeats within the coding sequence of the gene. The expansion leads to an elongated stretch of Q residues beyond the first 17 amino acids (MacDonald et al. 1993). In healthy individuals the number of Qs repeats is 35 or fewer, with 17–20 repeats found most commonly (Myers 2004; Housman 1995; Leavitt et al. 1999; Leavitt et al. 2001).[c]

Most adult-onset HD cases feature a mutated form of the protein (m-Htt) with 40–50 Qs. Expansions of 50 and more repeats generally cause the juvenile form of the disease (Gusella & MacDonald 2000). There is a strong negative correlation between the age of HD's onset and the number of Qs (Gusella & MacDonald 2000). Usually, the longer is the polyQ tract, the earlier is the age of the onset (Ross 1995; Gusella & MacDonald 2000). However, for a

---

[b]In 2008, the first Htt orthologs multi-alignment provides evidence that the polyQ is an ancient acquisition of Htt (Tartari et al. 2008). Its appearance dates back to sea urchin in which a NHQQ sequence is present, which consists of a group of four hydrophilic amino acids that can be considered bio-chemically comparable to the four glutamines (QQQQ) found in fish, amphibians, and birds (Tartari et al. 2008). The polyQ has then expanded gradually in mammals to become the longest and most polymorphic polyQ in humans (Tartari et al. 2008). One possible hypothesis is that wild-type Htt function, during development, may arise from the binding of different sets of interactors: many proteins in the cells contain a polyQ tract, in particular transcription factors and transcriptional regulators (Cha 2007).

[c]Repeats between 27 and 35 are rare and are not associated with disease. However, they are meiotically unstable and can expand into the disease range of 36 and above, when transmitted through the paternal line. Incomplete penetrance has been observed in individuals with 36– 41 repeats, but the estimates of penetrance for this group are imprecise.

given Q length there is a large variation in the age of onset, and the number of Qs by itself has poor predictive power on the age of HD's onset (Imarisio et al. 2008). m-Htt abnormally interacts with other proteins (Sapp et al. 1999) and causes brain damage (Borrell-Pages et al. 2006) producing oxidative stress, excitotoxic processes and metabolism deregulation (Grunewald & Beal 1999; Sapp et al. 1999).[d]

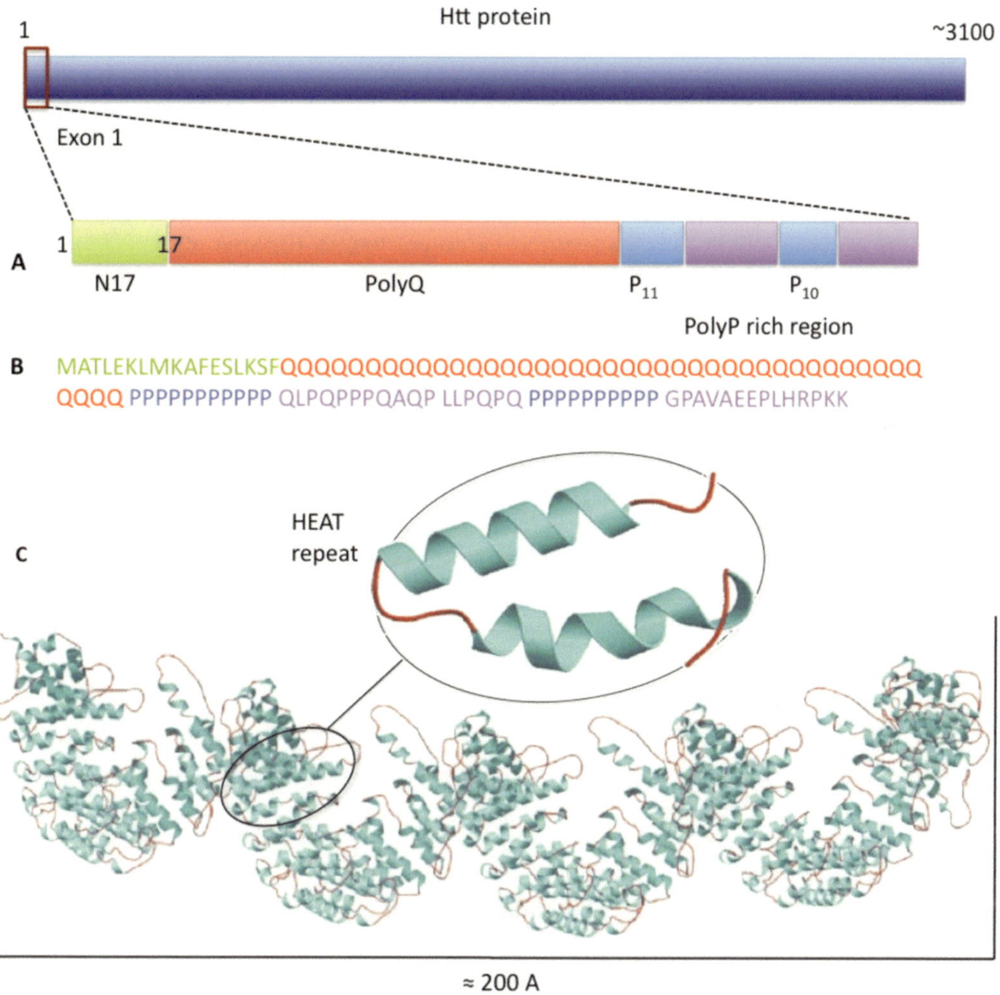

Fig. 1. Htt and Ex1: A) Scheme of Ex1 regions with B) the corresponding primary sequence; C) Proposed model of Htt. The HEAT domain is magnified. HEAT repeats are ~40 amino-acid domains, which fold in two anti-parallel α-helices forming a hairpin (Palidwor et al. 2009). Htt features 16 of these repeats (Andrade & Bork 1995; Palidwor et al. 2009; Li et al. 2006) organized in 4 clusters (Tartari et al. 2008).

[d]The expression of long Q tracts alone, in the context of an N-terminal fragment or full-length Htt protein was shown to disrupt a wide variety of biological functions in cells and model organisms (Johnson and Davidson 2010; Mangiarini et al. 1996; Zoghbi and Orr 2000).

Neuronal intra-nuclear and intra-cytoplasmic inclusions rich in polyQs are the pathological hallmarks of HD (Davies et al. 1997; Trottier et al. 1995). Inclusions are believed to be toxic since they can provoke a physical block of axonal transport between the cell body and the synaptic terminal as well as the recruitment of other polyQ-containing proteins, mainly transcription factors. These latter, interacting with m-Htt, may lose their physiological function, leading to cell death (Gunawardena et al. 2003; Lee et al. 2004; Li et al. 2001; Parker et al. 2001). However, inclusions may also result from an attempt of the cells to proteolytically degrade or inactivate m-Htt (Kuemmerle et al. 1999; Saudou et al. 1998). This alternative proposal is supported by the fact that cells forming Htt inclusions have an improved survival with respect to those not forming them (Arrasate et al. 2004). Accordingly, there is little correlation between inclusions burden and the areas of the brain most affected in HD (Gutekunst et al. 1999; Kuemmerle et al. 1999). The formation of polyQ rich inclusions proceeds through steps that generate different aggregated species, whose populations and stabilities may increase with polyQ length (Bulone et al. 2006). Among these different species are the nuclei, the oligomers, the protofibrils and, finally, the large fibers, which form the microscopic aggregates found in neurons (Ross & Poirier 2004). Unfortunately, the exact degree of toxicity of each species is not known.

Few information is available for protofibrils and fibers (Zuccato et al. 2010), while several efforts were been done to characterize oligomeric species. Indeed, oligomers may be highly reactive toward cellular environment because they have a large surface area with respect to volume ratios, as compared with larger inclusions. This reactivity may be correlated with toxicity (Ross & Poirier 2004; Ross & Poirier 2005; Nagai et al. 2007; Truant et al. 2008). Recent studies highlighted that the oligomers could be formed in several ways such as via N-terminal or direct polyQ interactions (Legleiter et al. 2010; Olshina et al. 2010; Ramdzan et al. 2010). However, the oligomers may also not be the pathway thought which the formation of polyQ larger inclusions takes place (Ross & Tabrizi 2011).

1.3 Role of the flanking regions

N17 and polyP modulate toxicity of m-Htt Ex1 (Truant et al. 2008; Atwal et al. 2007; Bhattacharyya et al. 2005).[e] Indeed, deletion of the proline-rich (P-rich) region in m-Ex1 fragments greatly increases their toxicity in yeast. These m-Ex1 fragments are otherwise innocuous (Dehay & Bertolotti 2006). Therefore, the P-rich region appears to be protective against the effects of expanded polyQ (Bhattacharyya et al. 2005).

N17, present in all mouse models of HD, was shown to modulate the toxicity of m-Htt in a structure-dependent manner (Truant et al. 2008; Atwal et al. 2007). A single point mutation in the middle of N17 was shown to disrupt the possibility to obtain a helical structure, completely abrogating any visible aggregates of m-Htt (Truant et al. 2008; Atwal et al. 2007). Indeed, the initial phases of the aggregation process seem to be accelerated by hydrophobic

[e] Also sequences exogenous to Ex1 modulate aggregation. In the yeast toxicity model, the positioning of flag-tags on the expression constructs modulate toxicity and the nature of aggregated protein (Duennwald et al. 2006). Another group observed modulation of polyQ aggregation by the use of structured chimeras with the cellular retinoic-acid binding protein in *E. coli* (Ignatova et al. 2007). Finally, it has shown that also some purification tag, such as the glutathione S-transferase fusion does affect the aggregation dynamics of polyQ (Perutz 1994).

interactions within an amphipathic α-helical structure of N17 (Thakur et al. 2009). Accordingly, the deletion of this region strongly reduces polyQ aggregation *in vitro* (Thakur et al. 2009). These results suggest that the regions outside the polyQ tract may interact with each other, influencing aggregation (Truant et al. 2008; Zuccato et al. 2010).

The ***first*** proposed polyQ aggregation pathway was mediated only by aggregation of the polyQ stretches (Bates 2003; Ross & Poirier 2004; Wanker 2000). It displayed the kinetics of nucleated-growth polymerization with a prolonged lag-phase required to form an aggregation nucleus, followed by a fast extension phase during which additional polyQ monomers rapidly joined the growing aggregate.

The ***second,*** recently proposed, aggregation pathway comes from Wetzel's group (Kar et al. 2011; Thakur et al. 2009). This depends mainly on N17 and involves several intermediates. In particular, the aggregation process may be characterized by the formation of oligomers having N17 in their core and polyQ sequences exposed on the surface. As the polyQ length increases, the structure decompacts and oligomers or protofibrils rearrange into amyloid-like structures capable of rapidly propagating via monomer addition (Kar et al. 2011; Thakur et al. 2009). The importance of the flanking regions suggests other therapeutic targets for polyQ-mediated neurodegeneration related to N17 or polyP, rather than polyQ itself.

2. Computational studies of Huntington Disease

In the following paragraph we provide an overview of the computational studies present in the literature carried out on the different aspects of HD such as the structure of the oligomers, the factors driving the formation and determining the thermodynamic stability of amyloidogenic aggregates and the role of the flanking regions on aggregation mechanism. All these studies are grouped on the basis of the topic and of the computational methods employed to address them.

2.1 Structural models of polyQ oligomers – Atomistic simulations

Many aspects of the HD's onset mechanism could be elucidated obtaining structural information at atomistic level on polyQ aggregates. However, detailed structural information are difficult to obtain experimentally as short wild-type polyQ tracts are insoluble at the high concentrations required for crystallographic or NMR studies (Truant et al. 2008).[f] In contrast, structural information at atomic level of resolution can be provided by computational approaches (Moroni et al. 2009; Miller et al. 2010). Simulations, in fact, can offer insights into structural and dynamical properties of polyQ peptides of different lengths, shapes and oligomeric states. Computational studies of neurodegenerative diseases can be carried out via classical molecular dynamics (MD) (Miller et al. 2010; Ma & Nussinov 2006). In this method, the atoms move according to the Newton's law on a predefined

[f]A simple search within the PDB (http://www.rcsb.org/) reveals that polyQ tracts present in a variety of normal cellular proteins are annotated as 'unstructured' or have to be removed to facilitate crystallization (Truant et al. 2008). Only one structure exists of the N-terminal part of Htt with 17Qs, obtained by a fusion with maltose-binding protein. It features polyQ stretch that can adopt an α-helical, random-coil, or an extended-loop conformation (Kim et al. 2009).

potential energy surface. Namely, interatomic interactions are described via empirical force fields. However, MD simulations in explicit water allow accessing a time scale limited to hundreds of ns. Thus, to simulate relevant biological processes, occurring on longer time scales, they have to be combined with enhanced sampling computational techniques (Christ et al 2010; Laio & Gervasio 2008) or it is, otherwise, necessary to use simplified interaction potentials (Ma & Nussinov 2006; Tozzini 2010; Miller et al. 2010).

Several structural models of the aggregated polyQ units were proposed with geometries compatible with available experimental information (electron microscopy and X-ray data). These models were based on Perutz's suggestion that Q side chains, being similar to the amino acid backbone units, could establish an H-bond network (Perutz 1999; Perutz 1994). In fact, Perutz, initially interpreted the X-ray diffraction data of polyQ aggregates as a polar zipper model, and later reinterpreted them as a circular β-helix model in which polyQ tracts can form turns composed by 20 res, with the Q side chains alternatively inside and outside the water filled nanotube (Fig. 2). According to this model, polyQ aggregates would be stabilized by H-bond interactions between main and side chain atoms (Fig. 2). In fact, the β-sheets were proposed to be antiparallel so that an amine group of one side chain could H-bond with the carbonyl group of the side chain belonging to the following turn (Esposito et al. 2008). Consistent with these X-ray diffraction data is also triangular β-helix model (Stork et al. 2005; Raetz & Roderick 1995), which is formed by turns of 16 residues, and the Atkins's model, in which the H-bond network of the Q side chains allows for high-density packing (Sikorski & Atkins 2005). Although several models were proposed for the polyQ aggregates, this issue is highly debated and it is still not clear which is the most common structure present in the aggregates.

Many studies of the proposed models for the polyQ aggregates were preformed via classical MD simulations (Perutz 1994; Perutz et al. 2002; Sikorski & Atkins 2005; Sunde & Blake 1997; Sunde et al. 1997; Sharma et al. 2005), providing valuable insights on their stabilities (Sikorski & Atkins 2005; Esposito et al. 2008; Stork et al. 2005; Armen et al. 2005; Finke et al. 2004; Finke & Onuchic 2004; Marchut & Hall 2006, 2006, 2007; Merlino et al. 2006; Zanuy et al. 2006; Ogawa et al. 2008). Among these, several studies investigated the dependence of the structural stability of the circular β-helix, as well as of other possible structures, on the Q length (Stork et al. 2005; Merlino et al. 2006; Ogawa et al. 2008; Rossetti et al. 2008; Hajime et al. 2008; Khare et al. 2005), leading sometimes to conflicting views. For example classical MD studies showed that β-helices with three turns were unstable with circular geometries, being, instead, stable in a triangular β-helix shape. Moreover, these studies pointed out that two-coiled triangular polyQ β-helices, which were individually unstable, became, instead, stable upon dimerization. This suggested that the formation of the initial aggregation seed of huntingtin amyloids requires dimers of at least 36 Qs (Stork et al. 2005). A subsequent study verified the stability of the circular β-helix model by performing MD simulations on polyQ fragments of different lengths. The results pointed out that circular β-helix models maintained a regular structure during the MD run, only when containing more than 40 Qs (Merlino et al. 2006). Moreover, a different MD study showed that the stability of the circular β-helix structure increased with an increasing number of Q, reaching the maximal stability above 30 Qs (Ogawa et al. 2008). In contrast to these computational results, annular units smaller than the circular β-helix model were detected experimentally and confirmed by other computational studies

(Marchut & Hall 2006a, 2006b, 2007; Papaleo & Invernizzi 2011). A more recent MD study proposed a systematic investigation of structural characteristic of polyQ strands in the early stage of nucleation, considering left handed circular, right handed rectangular, left and right handed triangular β-helices of different lengths (Zhou et al. 2011). These simulations showed that left handed triangular and right handed rectangular conformations were stable when they had at least three turns, preserving a high degree of the β-sheet content during the simulation. The stability of the systems increased with an increasing number of rungs, but it was insensitive to the number of Qs in each polyQ fragment (Zhou et al. 2011). Classical MD simulations were also performed for the cross-β-spine steric zipper model (Esposito et al. 2008), a motif found for the GNNQQNY peptide, an heptapetide present in the N-terminal prion-determining domain of the yeast protein Sup35. The simulations revealed that this kind of polyQ assemblies were very stable. In fact, the H-bonds between either parallel or antiparallel β-sheets, greatly affected the high stability of these structures, with a large contribution coming from the Q side chains H-bonds.

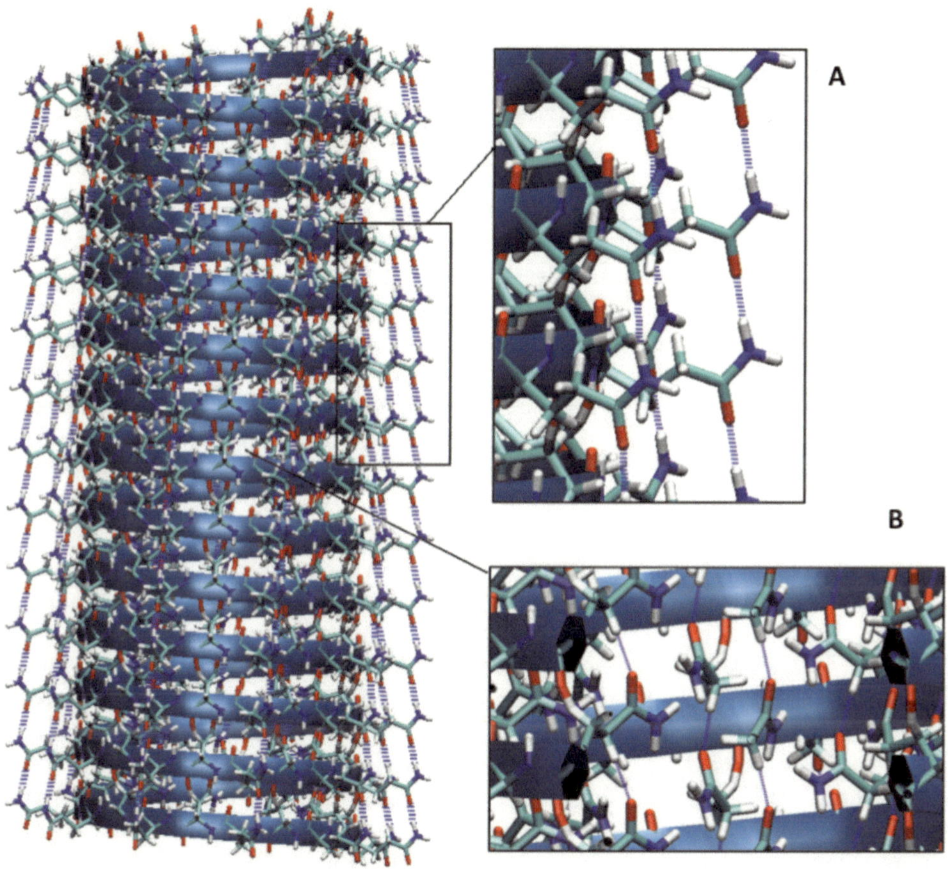

Fig. 2. Molecular view of circular β-helix structure with particular of: A) external and B) internal H-bond network. A similar H-bond network is observed also for the triangular β-helix structure.

In summary, all these computational studies suggested that several polymorphic forms of polyQ oligomers can exist and that all of them, when present as monomers, become more stable with an increasing number of Qs or upon aggregation with other polyQ tracts. The polymorphism of polyQ structures may possibly result in different pathways leading to the formation of toxic oligomers and fibrils.

2.2 Cooperativity of polyQ H-bonds – *ab inito* and Hybrid QM/MM simulations

Computer simulations can also help in elucidating the role of the peculiar electronic properties of the Q side chains in the formation of polyQ the aggregates. β-sheets are ubiquitous in protein structures and in aggregates of amyloidogenic proteins (Tartari et al. 2008). Thus, understanding the electronic factors contributing to the thermodynamic stability of β-sheets is of fundamental importance in neurodegeneration (Rossetti et al. 2010). In the past, several research groups tried to address experimentally and computationally whether the formation of β-sheets is cooperative. Cooperativity in H-bonding exists from a structural and from an energetic point of view when the strength of H-bonds and the thermodynamic stability of the H-bonding structures, respectively, increase non-linearly with an increasing number of H-bonds. From an electronic point of view H-bond cooperativity depends on the polarization of the electronic clouds of adjacent molecules or strands. If, polarization effects are present, a rearrangement of electronic structure takes place. This aspect is clearly not accounted in force field calculations, which are grounded on predefined parameters of the potential energy. Thus, despite the successes of classical MD simulations in the study of neurodegenerative diseases, there are cases in which the use of effective potentials may be not accurate enough. In these cases a more sophisticated computational approach is provided by static or dynamics *ab initio* calculations, typically based on Density Functional Theory (DFT). In static DFT calculations the electronic structure problem is solved parametrically for nuclear configurations generated by minimization schemes. Instead, in dynamic DFT calculations the atoms move according to the Newton's law on a potential energy surface, which is evaluated from electronic structure calculations (Spiegel & Magistrato 2006; Carloni et al. 2002). Clearly, *ab initio* schemes, requiring to solve the electronic structure problem for different nuclear configurations, are much more computationally demanding than classical MD, limiting the size of the systems studied to hundreds of atoms and the time scale accessible to *ab initio* MD to tens of ps (Carloni et al. 2002). Since biological systems treated in their environment comprise several hundreds of thousands of atoms, they are clearly too large to be treated with a full *ab initio* description. An alternative approach to treat these systems relies on hybrid quantum-classical (QM/MM) MD simulations. The QM/MM MD approach combines classical with *ab initio* MD. In this approach the system under investigation is divided into two different regions, allowing to concentrate the computational efforts of the electronic structure calculations (QM part) to the part in which the force field can fail. The rest of the system is, instead, treated with empirical force fields in a computationally more efficient manner (Spiegel & Magistrato 2006). This allows to extend the size of the systems studied to hundreds of thousands of atoms, although the computational cost of the QM part still limits the time scale accessible to few tens of ps.

The role of cooperative effect (CE) of H-bonds between different strands of amyloidogenic aggregates was investigated by performing static DFT calculations on the known molecular structure of the heptameric peptide GNNQQNY. This study showed that the strength of H-bonds between layers of fibrils increased nonlinearly up to four layers. Moreover, it showed that the H-bonding interactions within the β-sheets of the amyloid structure were cooperative, with contributions to the binding energy from several layers away within the fibril (Tsemekhman et al. 2007). Other studies carried out on polyAlanine (A), polySerine, polyValine homo-polymers showed that H-bonding and dipole–dipole interactions were strengthened through CEs, contributing to the stability of the secondary structures (Horvath et al. 2004, 2005; Varga & Kovács 2005; Improta & Barone 2004; Improta et al. 2001; Improta et al. 2001; Wieczorek & Dannenberg 2003). In these studies the influence of the side chains on the thermodynamic stability of the investigated structures was also verified. These results highlighted the presence of cooperativity in the C=O..H-N H-bond, which was an important source of long-range interactions (Horvath et al. 2005). Instead, DFT calculations performed on polyGlycine (G) of different lengths and conformations showed the influence of long-range effects on the stability of different conformers (Improta & Barone 2004; Improta et al. 2001; Improta et al. 2001). Moreover, long-range interactions were shown to contribute considerably to the stability of the β-sheet structures, with appreciable effects on the molecular geometry. It was also shown that the H-bond length was very sensitive to long-range interactions (Horvath et al. 2004, 2005). Finally, a DFT study carried out also on polyG peptides showed that repeating H-bonds either in parallel and antiparallel β-sheets were not cooperative in terms of enthalpy contribution in the direction parallel to strand elongation. CEs existed, instead, in the perpendicular direction and they depended on the number of residues in each strand (Zhao & Wu 2002). Thus, all these studies suggested that H-bond CEs exist in homopolypeptides in different conformations, including the β-sheet structure. Since the Q side chain resembles an animoacid backbone, it is likely that CEs will greatly contribute to H-bond cooperativity and to the thermodynamic stabilization of the polyQ β-sheets. This aspect was addressed recently by performing *ab initio* and QM/MM MD calculations (Rossetti et al. 2010).

2.3 Aggregation properties and the role of exon-1- coarse grain models and enhanced sampling techniques

The details of molecular mechanism leading from the association of monomeric polyQs to the formation of mature fibrils remain highly debated. As mentioned above, Wetzel at al. suggested initially a nucleated grow polymerization model based only on the polyQ peptides to connect the disordered monomers to the highly ordered β-sheet structures present in the fibrils (Papaleo & Invernizzi 2011).

Several computational studies were carried out to shed light on the aggregation properties of polyQ chains, on the formation of the initial aggregation nucleus and on the role of the regions flanking the polyQ tract in m-Ex1 (Papaleo & Invernizzi 2011). A common methodology employed in these studies was the use of coarse grain (CG) models, in which groups of atoms are described as a single bead (Tozzini 2010). In the simplest model only three types of beads, namely hydrophobic, hydrophilic and neutral, are considered. Other CG models instead have special terms to account for H-bond formation, which is crucial for aggregation (Ma & Nussinov 2006). CG methods usually allow to extend the time scale accessible to MD simulations, loosing, at the same time, the accuracy of an atomistic description. Sometimes atomistic simulations with the use of implicit solvent models give an alternative to the CG models. However, an explicit treatment of the solvent is of crucial importance for a correct characterization of the folding and of the aggregation properties of polypeptides (Papaleo & Invernizzi 2011). Conversely, atomistic simulations can be used in combination with methods, which allow to extend the time scale of the simulations, enhancing the sampling of the underlying free energy surface (Laio & Gervasio 2008; Christ et al. 2010; Biarnes et al. 2011; Sugita & Okamoto 2000; Bussi et al. 2006). These latter warrant the accuracy of the atomistic description, but they are computationally very demanding and, in most cases, not yet at the stage of being able to simulate the folding and/or the aggregation of peptides of the biologically relevant lengths (Rohrig et al. 2006; Miller et al. 2010). Among the aggregation studies carried out so far for HD, a force filed based monte carlo simulation study in implicit solvent (Vitalis et al. 2009) investigated the free energy cost associated with the formation of ordered β-sheet structures in dependence an increasing number of Qs in the single monomer (Vitalis et al. 2009). This work reported the free energy costs to form structures with a high β-sheet content consistent with literature data. However, an increase of this free energy cost with an increasing chain length was observed, in contrast to previous interpretation of kinetic data. Moreover, the authors suggested that β-sheet formation may be an attribute of peptide-rich phases characterized by high molecular weight aggregates rather than monomers or oligomers (Vitalis et al. 2009).

Discrete Molecular Dynamics (DMD), an efficient MD method based on a simplified interparticle potential (Miller et al. 2010), was employed to show that the cooperativity in the folding of a chimeric monomer (composed by Chymotrypsin inhibitor 2 with an inserted polyQ repeat) decreased for peptides with polyQ lengths above the pathogenic threshold. Moreover, it was demonstrated that the dominant mode for dimer formation was inter Qs H-bonding (Barton et al. 2007). The aggregation of model polyQ peptides was also investigated via MD simulations with a simplified model of polyQ (Marchut & Hall 2006a, 2006b, 2007). This model accounted for the most important types of intra- and inter-molecular interactions, namely H-bond and hydrophobic interactions, allowing the folding process to take place within the time scale of the simulation (Marchut & Hall 2007). In this study the folding of isolated polyQ tracts of non-pathogenic and pathogenic lengths, and the folding and the aggregation of systems of polyQ peptides of various lengths were investigated. The isolated polyQ peptides formed some backbone–backbone H-bonds, although the hydrogen bond content (HBC) was markedly lower than that of an ordered β-sheet structure. In the multi-chains simulations, instead, ordered aggregates with significant β-sheet and random coil characters were observed at intermediate and high temperatures, respectively. Interestingly, the temperature at which the peptides underwent the transition from amorphous to ordered aggregates and from ordered aggregates to random coils increased with increasing polyQ length. More recently, the aggregation of polyQ peptides of different lengths was addressed via replica exchange (RE) MD and a simplified force filed. REMD simulations combine several MD simulations at elevated temperature to generate a variety of conformational ensembles with a Monte Carlo like conformational selection (Bussi et al. 2006). Thus, this method allows to explore the conformational space of peptide aggregation and folding, instead of getting trapped in local minima (Sugita & Okamoto 2000). In this work REMD was applied to study the aggregation kinetics of the polyQ monomers and dimers with chain lengths from 30 to 50 residues. The results showed that for the monomers a structural change from an α-helical structure to random coil occurs with no formation of a β-strand. For dimers, instead, starting from random coils there was the initial formation of antiparallel β-sheets, of circular and of triangular β-helices, which may lead to the formation of toxic oligomers and fibrils (Laghaei & Mousseau 2010).

As stated previously, the sequences flanking the polyQ tract have been recently demonstrated to have a key role on aggregation mechanism (Thakur et al. 2009). However, structural information on these segments are lacking. Thus, more recent computational studies employed the classical MD method in combination with enhanced sampling techniques to investigate the complex free energy landscape for the folding and for the aggregation of the N terminal region of Htt (Ex1 or N17). These studies are of crucial importance to understand how the misfolding and aggregation of polyQ tract(s) is affected by the flanking sequences.

Classical MD studies in combination with simulated tempering and folding at home infrastructure were employed to study the thermodynamics of N17 (Kelley et al. 2009). In these simulations N17 was found to be highly helical, although adopting two different and seemingly stable states. The most populated state was a two-helix bundle, although a significant percentage of structures still assumed the conformation of a single straight helix. Since N17 was demonstrated to be involved in the rate-limiting step for the formation of the initial aggregation nucleus, two possible mechanisms for the nucleating event were proposed in this study. These are based on a transition between the two-helix and single-helix state of N17 and on the interactions between the N17 and the polyQ tract (Kelley et al. 2009). Moreover, a recent Monte Carlo simulation study, along with circular dichroism experiments, described the effect of N17 on polyQ conformations and intermolecular interactions. This study showed that N17 and polyQ domains were increasingly disordered as the polyQ length increased in N17-polyQs peptides. In contrast with experimental suggestions (Thakur et al. 2009), N17 suppressed the intrinsic propensity of the polyQ tracts to aggregate by forming incipient micellar structures adopted by N17 segments. Instead, increasing the polyQ length the degree of intermolecular association increased, becoming mainly governed by the associations between polyQ tracts (Williamson et al. 2010). Finally, a systematic DMD study, in combination with the RE method, was carried out on monomeric Ex1 with the full flanking regions on a variant of Ex1 missing the polyP region, which is hypothesized to prevent aggregation, and on an isolated polyQ peptide. For each of these three constructs, polyQ tracts of pathogenic and non-pathogenic lengths were considered. Interestingly, the study showed a correlation between the length polyQ tract and the probability to form a misfolded state rich in β-sheets. Furthermore, it showed that N17 more likely adopted a β-sheet rather than an α-helix conformation as the length of the polyQ tract increased. Finally, this study demonstrated that the polyP region formed polyP type II helices, decreasing the probability of the polyQ to form a state rich in β-sheets (Lakhani et al. 2010). More recently, enhanced sampling techniques were employed to predict the conformational properties of N17 fragments in water solution and to shed light on its crucial role in Htt aggregation (Kar et al. 2011).

3. Selected applications

In the following paragraph we present three selected examples taken from our work in which we explain in detail how computer simulations can be employed to address the different, still unclear, aspects of the HD's onset mechanism.

3.1 The HD threshold and the structural stability of toxic conformers

In this work we have addressed one of the questions lengthily debated concerning the polyQ length-dependent toxicity threshold. One hypothesis suggested that the length dependent toxicity of HD was based on a specific structural transition, occurring only when the polyQ tract is above 36 amino acids (structural transition hypothesis). Consistent with this hypothesis, an anti-polyQ monoclonal antibody was observed, which was able to specifically recognize the expanded toxic polyQ tracts. This suggested the existence of a generic conformational epitope formed only above a certain polyQ length (Trottier et al. 1995; Kaltenbach et al. 2007; Sugaya et al. 2007). The presence of such abnormally folded protein, which can aggregate and form fibrillar structures, could highlight similarities between HD and other neurodegenerative diseases such as Alzheimer's, Parkinson's D, and prion disorders (Ross & Poirier 2004). Hence, several efforts were done for "hunting the elusive toxic polyQ conformer" (Trottier et al. 1995). This hypothesis was, however, challenged by various experimental evidences. First, polyQ fragments shorter than the disease threshold were also shown to aggregate, adopting similar structures to those of peptides longer than threshold (Klein et al. 2007; Masino et al. 2002; Bennett et al. 2002; Tanaka et al. 2001) and to exhibit toxicity in an eukaryote organism (*Caenorhabditis elegans)* (Morley et al. 2002). Second, it was shown that polyQ length influences the stability of the initial aggregation seed and that, in turn, it may affect the kinetics of its formation (Chen et al. 2002). The kinetics of the elongation phase is, instead, independent of the polyQ length (Chen et al. 2002). Furthermore, it was suggested that the toxicity of mut-Htt may be simply due to the fact that polyQ tracts are inherently toxic sequences, whose deleterious effects gradually increase with their length (Klein et al. 2007). We investigated the influence of the polyQ length on the structural stability of monomers and oligomers by performing atomistic MD simulations on different β-helical models featuring a number of Qs below and well beyond the disease threshold (Rossetti et al. 2008). We considered the circular (Raetz & Roderick 1995) and triangular β-helices (Stork et al. 2005; Perutz et al. 2002) as shapes of the oligomers, since these are the only models consistent with the 'structural threshold hypothesis'. Thus, we studied two large monomeric models based on the circular β-helix (labeled as P from Perutz, who introduced this model in 2002 (Perutz et al. 2002)), and on the triangular β-helix model (labeled as T). This latter was constructed starting from the regularly shaped coils of UDP-N-acetyl glucosamine acyltransferase (Raetz & Roderick 1995) (Protein Data Bank entry: 1LXA) (Berman 2000), and replacing each residue with a glutamine. The circular and triangular β-helix models contained 266 and 179 residues and each turn was composed by 20 and 18 Qs, respectively. Both the T and the P models were composed by a single polyQ chain and had a number of Qs well above that observed at physiological conditions. In addition, we considered different oligomeric models built starting from the single chain P and T systems: (i) 4 oligomers in circular β-helix conformation composed by 4, 3, 2, 1 monomers (each composed by 40 Qs). These were named P_{AD}, P_{AC}, P_{AB}, P_{A}, respectively. (ii) 4 oligomers in triangular β-helix conformation with respectively 4, 3, 2, 1 monomers (each composed by 36 Qs). These models were symbolized by T_{AD}, T_{AC}, T_{AB}, T_{A} , respectively. (iii) One oligomer in circular β-helix conformation composed by 8 monomers each containing 25 Qs residues. The model was named P_{AH25}. (iv) Finally, we considered 4 small monomeric models in circular β-helix conformation composed by 25, 30, 35, 40 Qs and symbolized by P_{25}, P_{30}, P_{35}, P_{40}, respectively.

These models were chosen to perform a systematic study that allowed us to validate or discard the 'structural threshold hypothesis'. Moreover, by varying systematically the size of the polyQ units in these models, and considering both the monomeric and oligomeric states, our calculations shed light on the dependence of the stability of β-helical structures upon the number of monomers. Finally, considering both the P and T helical structures, our findings became independent of the structural model chosen. To simplify the discussion we defined qualitatively the Structural Stability (SS) as a quantity which increased (i) with the compactness of the structure, as measured by the plots of the RMSD of backbone atoms, as well as the gyration radius (*R*g) versus time and (ii) with the HBC, defined as the total number of H-bonds formed within the structural models, divided by the total number of H-bond donor functionalities. Our MD simulations at finite temperature and in aqueous solution pointed out that the two different β-helix shapes influenced only the β-sheet content. In particular, the T helix displayed a larger number of residues in random coil conformation than the P one. However, the HBC as well as the SS of the two shapes were comparable (Fig. 3). Moreover, we demonstrated that SS did not depend on the number of Qs in the monomers. In fact, oligomers composed by 4 monomers of 40 Qs and by 8 monomers of 25 Qs had similar SS. Consistent with our results an NMR study revealed no structural difference between aggregates formed by short and long polyQ peptides (Klein et al. 2007). We also showed for the first time that the SS of polyQ oligomers was not affected by the shape. We suggested, instead, that only the number of monomers - thus, the concentration in an (*in vivo* or *in vitro*) experiment - contributes to the overall stability of the oligomers. This may be due to the additive contribution of the single monomer in the H-bond network formed between backbone atoms (Fig. 3).

Conversely, the H-bonds formed between Q side-chains influenced mainly the stability of the single isolated monomers (Fig. 4). In fact, the isolated monomer with Q length above the disease threshold, P_{40}, was characterized by a larger number of β-sheet content and HBC, with respect to shorter monomers (Fig. 4). This latter depended mainly on side chain H-bonds, and thus, on the number of Qs. Therefore, if the Q length was lower than that of the disease threshold, the β-stranded monomers were unstable and, hence, they might aggregate with lower probability (Fig. 4), consistent with experimental findings.

In conclusion, our data discarded the structural threshold hypothesis. However, interpreting our findings on the basis of the whole landscape of available experimental data (Klein et al. 2007), we suggested that the observed length-dependent toxicity threshold may be explained by a faster aggregation kinetics, occurring for longer polyQ tracts.

3.2 Hydrogen bonding cooperativity in polyQ β-sheet investigated by *ab initio* and QM/MM MD simulations

Perutz was the first to show that the polyQ tracts may form β-sheet based structures, which are able to establish tighter interactions with increasing polyQ length. Therefore, the correlation between the strength of the polyQ chain and the strength of the interactions may be a key aspect at the basis of the correlation between polyQ length and severity of disease (Perutz 1994). Peruz suggested that this may be due to the Q side chain, which, having the same chemical characteristics of an amino acid backbone, can form a network of H-bonds involving both the main and the side chain atoms (Perutz 1994; Klein et al. 2007; Perutz &

Windle 2001)[g]. CE in H-bonding is very important for both the structure and the energetic of polypeptide systems. As the presence of CE was demonstrated for other homopolymers, it is likely that this effect may be particularly relevant in the peculiar H-bond network of polyQ, playing a key role in Htt/Ex1 misfolding and aggregation (Perutz 1994; Perutz & Windle 2001). Most of the studies carried out on the structural stability of polyQ oligomers

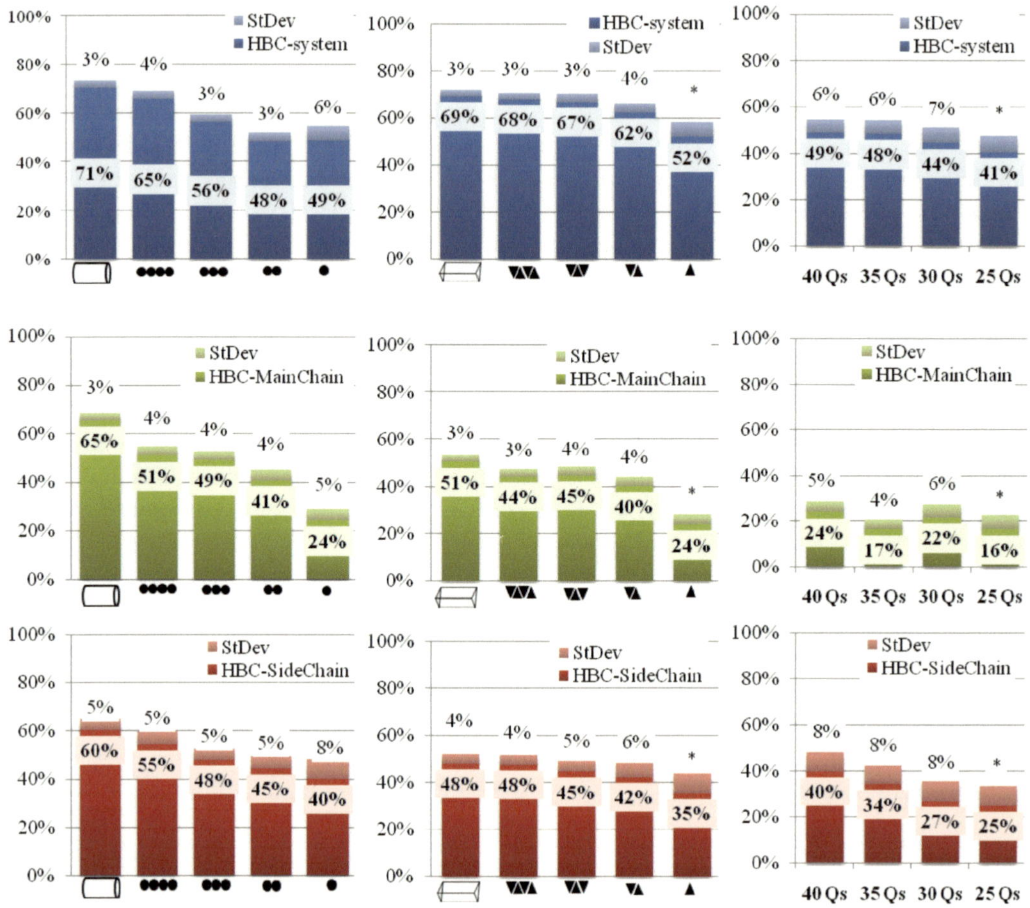

Fig. 3. HBC in P and T series given as percentage of H-bonds in the P series (left column), in the T series (middle column) and in the monomeric series (right column). Blue, green and red histograms represent the total, the main chain and the side chain HBC, respectively. The solid cylinder refers to the P model and the black circles refer to monomers of 40 Qs in circular β-helix. The triangular cylinder refers to the T models and the black triangles refer to monomers of 36Qs in triangular β-helix shape. Monomers of different lengths are indicated with the number of Qs present in the chain.

---

[g] Consistently with this hypothesis, aggregates of protein are not seen in proteins expressing polyasparagine, an amino acid that differs from glutamine even by only one methyl group (Oma et al. 2004).

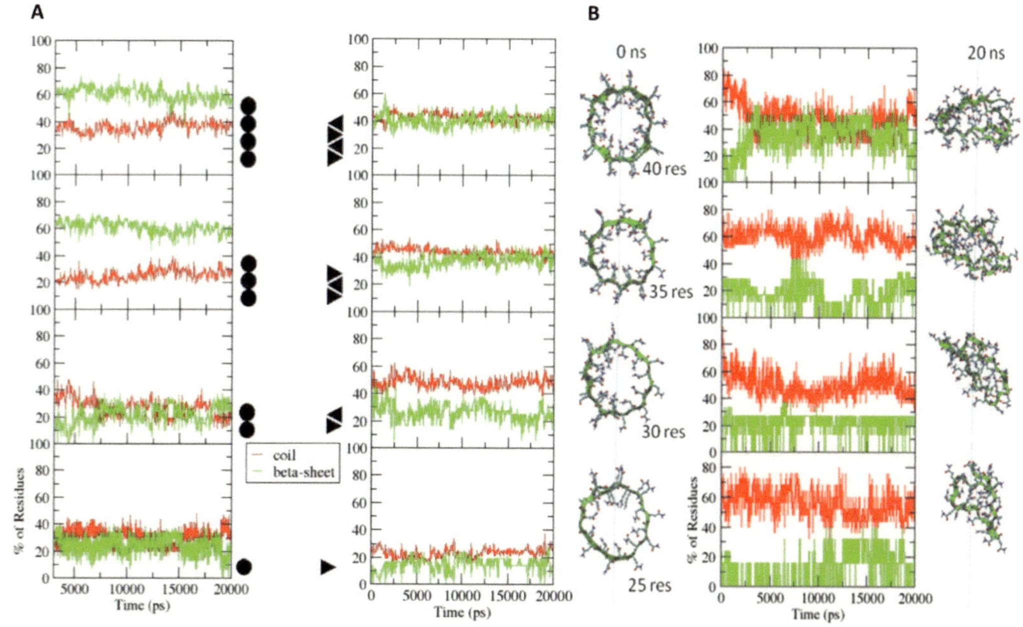

Fig. 4. β-sheet (green) and random coil conformation (red) of oligomers in circular (left panel) and triangular (middle panel) β-helix conformations, symbolized as black circles and triangles, respectively (A) and of P_{40}, P_{35}, P_{30}, P_{25} monomers (B) (right panel). Monomers of different lengths are indicated with the number of Qs present in the chain. On the left and right side of the graph in B the initial and final (after 20 nanoseconds (ns) MD) geometries of each monomer are shown. Water is not shown for clarity.

were achieved by classical MD calculations, which, not dealing with electronic polarizability, were not suitable to characterize the presence of CE in this kind of aggregates. CE was investigated on other polypeptides (Tsemekhman et al. 2007; Varga & Kovács 2005; Horvath et al. 2004, 2005; Improta & Barone 2004; Improta et al. 2001; Improta et al. 2001; Zhao & Wu 2002; Wieczorek & Dannenberg 2003; Viswanathan et al. 2004; Scheiner & Kar 2005) with the application of first principle methods. However, here we provide a summary of the first study in which the presence and the importance of CE in the H-bonds of Qs side chains was verified with DFT approaches (Rossetti et al. 2008). We performed first principles DFT-PBE (Benedek et al. 2005; Morozov et al. 2004; Perdew et al 1996) calculations on polyQ peptides of increasing complexity, assembled in parallel[h] β-sheets (Tsemekhman et al. 2007; Koch et al. 2005; Beke et al. 2006; Perczel et al. 2005). In order to carry out this study we used different models[i] (labeled as ***N***x*n* hereafter), which differed from each other for the number of strands (***N***=1, 2, 3, 4) and/or for the number of Qs in each strand (*n*=1, 2, 3, 4).[j] The resulting 16 models ranged from 29 to 320 atoms. Furthermore, to verify the contribution of the polyQ side chains to CE we also considered a series of models where we varied the initial Q side chains conformations putting them in a position in which they could not H-

[h]CE turns out to be stronger in parallel β-sheets (like the systems considered here) than in anti-parallel ones (Koch et al. 2005).
[i]The models were built using HyperChem 8.0 program (Hypercube)
[j]Each polypeptide is terminated by the addition of -NCH3 and -OCCH3 groups.

bond with the adjacent strand and, a series of models built with polyA. Finally, to check the role of solvent and temperature effects on polypeptide conformation (Scheiner & Kar 2005) we performed 2 *ps* of hybrid DFT/MM MD calculations on a large a β-helix nanotube (8 turns of 20 Qs) in aqueous solution (Perutz et al. 2002; Berendsen et al. 1995; CPMD; 2002; van der Spoel et al. 2005). In this case, we considered three models in which the QM part included the *4x4*, *3x4* and *4x3* moieties.

Although circular β-helix is only one of the possible polyQ structures (Sikorski & Atkins 2005; Zanuy et al. 2006), we investigated it since we demonstrated by classical MD studies that the structural stability of the polyQ oligomers was independent from the β-sheet shape (Rossetti et al. 2008). In this study, in fact, we aimed at providing a qualitative description of CE. Quantitative predictions would, instead, require an investigation on a variety of proposed structures. CE on β-sheet strands may be present in patterns *perpendicular* to the peptide elongation ($\perp$ CE) or *parallel* to it (||CE) (Fig. 5A). When $\perp$ CE is present a decrease in H-bond length should be observed with an increasing number of strands. Moreover, in ∞CE the H-bonds at the center of the pile should be shorter than at the rim. Consistent with the presence of $\perp$ CE, in all models considered the H-bond distances of both the backbone and the side chains decreased with an increasing number of strands. In addition, H-bond lengths turned out to be shorter at the center of H-bonded chains than at the rim when at least three H-bonds were piled up in the perpendicular direction (***N***= 4). This feature was observed both for the side chains and the backbone. Interestingly, it was observed that the backbone dipoles along the same column (H-bond in the perpendicular direction) of β-strands had the same orientations (in contrast to those of the adjacent column) and could, therefore, sum up increasing the polarization of the systems (Zhao & Wu 2002). However, in the peculiar case of polyQ β-strands, the Q side chains counterbalanced this polarization, affecting the H-bonds of the backbone. As a result, in the columns where the H-bond dipole orientations were enhanced by similar side chain H-bond dipole orientations, a $\perp$ CE was present. This resulted in a H-bond shorter at the center of the column. On the other hand, when neighboring side chain columns had H-bond dipoles oriented in opposite directions (with respect to the column considered), the inner H-bond was not the shortest of the column (Fig. 5 D and E). This explains why for the backbone the $\perp$ CE, namely the fact that H-bonds are shorter at the center of the pile and not at the rim, was visible only by tacking averages (Fig 5D).

A different type of CE is that parallel to peptide elongation (||CE). When ||CE is present, a shortening of the central H-bond lengths between two adjacent strands takes place. This is usually not present in β-sheets due to the alternative orientation of backbone H-bond dipoles along the strands. However, the dipoles associated with the Q side chains added up in a coherent way for the central H-bonds between two strands (Fig. 5E). This occurred at position 2 in $\boldsymbol{N} \times 2$ series, at positions 2 and 3 in $\boldsymbol{N}\times 3$ series and at positions 2, 3, and 4 in $\boldsymbol{N} \times 4$ series. Thus, these H-bonds were shorter than those of the rim. As expected, in the calculations in which the side chains were impaired of H-bonding or Qs were replaced by As, the ||CE was not observed. QM/MM MD calculations qualitatively reproduced the H-bonds trends of the corresponding *in vacuo* models. However, in these simulations H-bond lengths were larger and the side chains formed mostly H-bonds with the solvent. These differences were probably due to the presence of the solvent and to temperature effects, which were completely neglected in the *in vacuo* calculations. These calculations suggested that environmental effects influence only the magnitude of CE in H-bonding, while the qualitative trend was the same of that found in the *in vacuo* calculations.

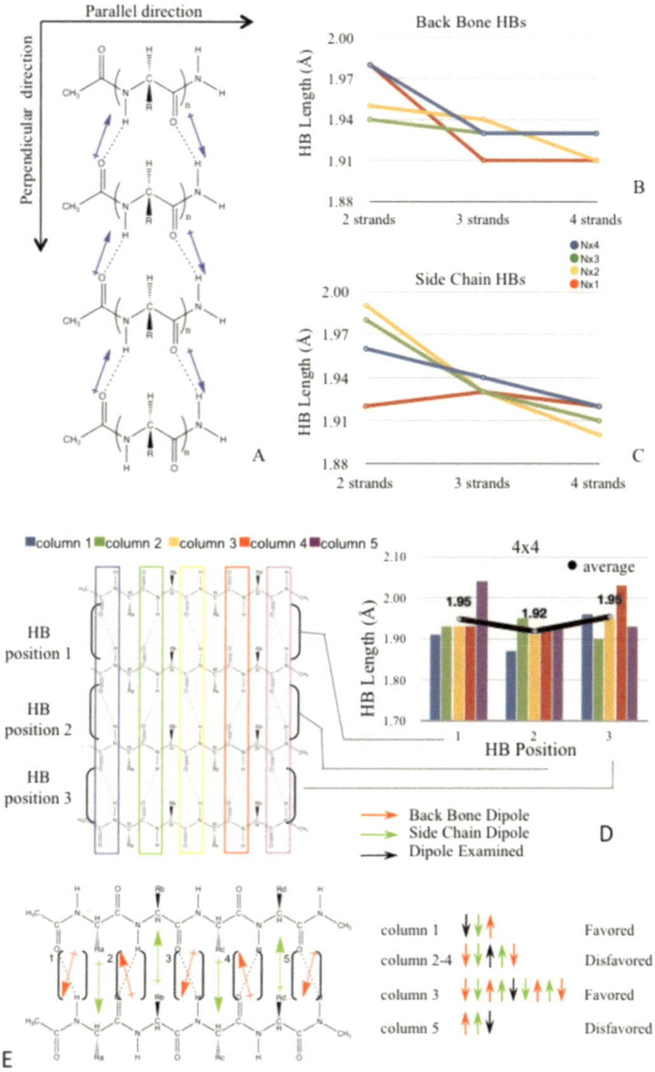

Fig. 5. Cooperative Effect - A) Definition of CE in H-bonds parallel (||) and perpendicular (⊥) directions to peptides elongation. B)-C) Structural aspects of CE: B) Backbone CE (⊥CE): Mean values of H-bond lengths of the backbone atoms versus the number of strands for each series of n Qs. C) Side chains CE (⊥CE): Mean values of H-bond lengths for the side chain atoms versus the number of strands for each series of n Q. D)-E) ⊥CE in system 4x4. D) In the histograms: H-bond length of backbone for different positions inside each strand as a function of the position across the different strands. Color of the histogram corresponds to the H-bonds circled on the left picture of B. The black line connects the mean values over the rows. E) Orientation of dipoles associated with the H-bonds for the 4x4 system.

Consistent with these results, in the simulations in which the Q side chains were impaired to H-bond and in the models in which the Qs were replaced by As, the H-bond at the center of the polyQ chain was longer than at the rim. The H-bond lengths, instead, continued to decrease with the number of piled strands even in these systems.

Finally, we also calculated the stabilization energy associated with the formation of H-bonds between the different strands of the systems *in vacuo*. To this end we defined the stabilization energy *per strand* (ΔE_N) as the energy associated with the addition of the $\mathbf{N}^{th}$ Q strand to the Q_{N-1} strands ($E_{N\times n}$), minus the formation energy of the N isolated strand ($\Delta E_N = E_{N\times n} - N \cdot E_{1\times n}$). In this definition, $E_{N\times n}$ is the energy of a system containing N strands and belonging to the n series; while $E_{1\times n}$ is the energy associated with an isolated strand containing n Qs. In practice this is the energy of a strand containing n Qs isolated from long-range effects. We also introduced the stabilization energy per H-bond (ΔE_{HB}) as Δ_{EN} divided by the number of H-bonds (n_{HB}) in each system ($\Delta E_{HB} = \Delta E_N / n_{HB}$).

Our study showed that ΔE_{HB} decreased nonlinearly with the number of strands (Fig. 6). ΔE_{HB} ranged from -5.0 kcal/mol in the smallest system to -6.5 kcal/mol in the larger system, suggesting that a CE existed and that for the present systems this was at most of 1.5 kcal/mol per H-bond. Clearly this stabilization energy was smaller for models containing A residues and with Qs side chains rotated to impair H-bonding.

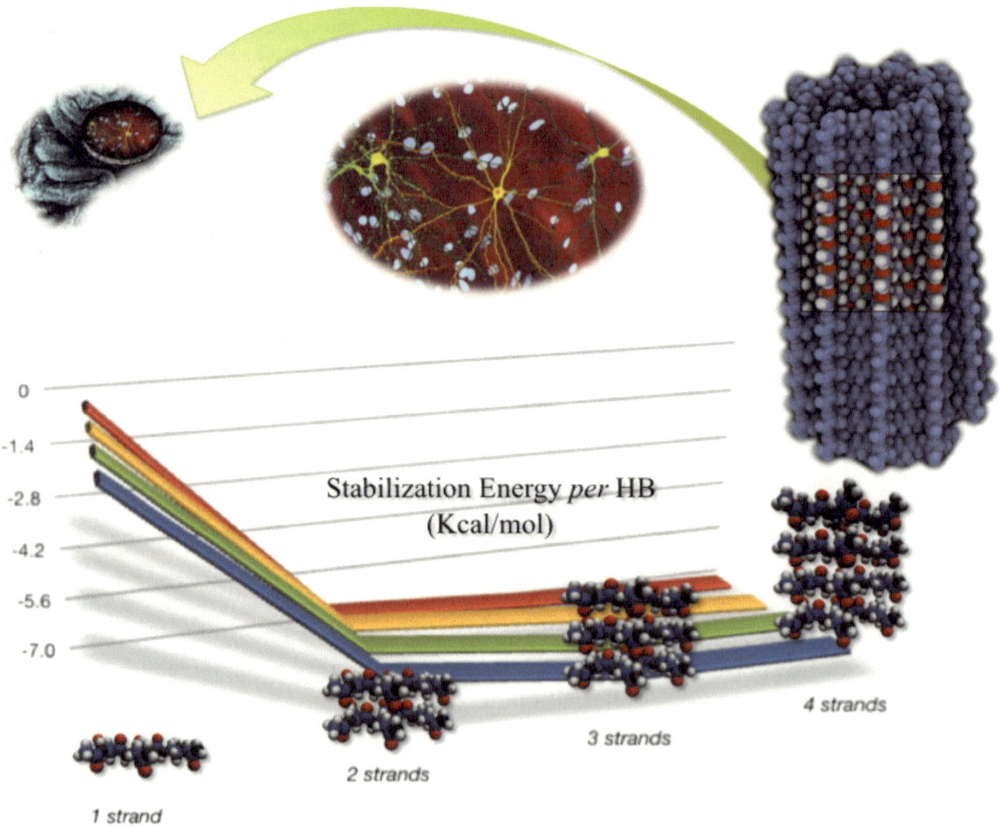

Fig. 6. Cooperative effect calculated as stabilization energy per H-bond.

3.3 Conformation of N17 in aqueous solution investigated by bias-exchange metadynamics

As stated in the introduction, recently, *in vivo* (Truant et al. 2008; Duennwald et al. 2006; Aiken et al. 2009), in cell (Ignatova et al. 2007; Cornett et al. 2005; Lakhani et al. 2010), *in vitro* (Rockabrand et al. 2007; Kim et al. 2009; Williamson et al. 2010) and *in silico* (Lakhani et al. 2010) studies showed that N17 modulates Htt fibrillation. This might arise by a variety of mechanisms, including changes in subcellular localization, nucleation of aggregation and/or interaction with cellular partners (Truant et al. 2008).

Understanding the influence of N17 on the aggregation mechanisms of polyQ in HD highly depends on structural information. Different spectroscopic techniques such as NMR (Thakur et al. 2009), CD (Thakur et al. 2009; Williamson et al. 2010) and FRET (Thakur et al. 2009) showed that N17 in aqueous solution adopts predominantly a random-coil structure with transient helical conformations (Thakur et al. 2009). Thus, N17 in solution can exist in equilibrium between different conformations. As these experimental techniques can provide only information on averages between the populations of different conformers, the secondary and the tertiary structure contents of the different N17 conformers remain not known. In this selected example (Rossetti G 2011) the Bias Exchange Metadynamics (BEM) was adopted to describe the thermodynamics and the kinetics of N17 in aqueous solution and at room temperature (Piana & Laio 2007). The BEM method relies on a combination of metadynamics and replica exchange. (Piana & Laio 2007; Laio & Gervasio 2008; Laio & Parrinello 2002). Metadynamics is a powerful algorithm used for accelerating rare events. In this scheme the system is described by a set of collective variables (CVs) and its normal evolution in the space of the CVs is biased by a history-dependent potential that forces the system to escape from local minima. This potential is, later, used to reconstruct the underlying free energy surface. Metadynamics, however, is effective only to explore few reaction coordinates as its performance decreases enormously with an increasing number of CVs. Typically, RE method is performed between replica of the system at different temperature as this latter is adopted to enhance the phase-space exploration (Piana et al. 2008; Sugita & Okamoto 2000). An example of RE metadynamics exists, in which a metadynamics run is performed in replicas of the system at different temperature (Bussi et al. 2006). However, in BEM exchanges are performed between replicas of the system at the same temperature, but using different CVs. This allows to extend the metadynamics approach to a virtually unlimited number of variables, becoming very effective for protein folding (Marinelli et al. 2009).

In this study BEM was employed to predict the free energy landscape of N17[k] in aqueous solution (Rossetti et al. 2008).[j] Our results showed that N17 populated four main kinetic basins, which interconverted on the second time-scale.[l] In each basin these were several possible clusters and an attractor, which was the lowest free energy cluster of the basin (Rossetti et al. 2011). The most populated basin (about 75%) was a random coil, with an extended flat exposed hydrophobic surface (B2, in Fig. 7). The latter may be crucial for the

[k]N17's extended coil conformation was built with the Modeller 9v8 program (Sánchez and Sali 1997). The D and K residues were considered to be in their ionized state.

[j]The first three (CV_1, CV_2, CV_3) count the number of hydrophobic contacts, of C contacts, and of backbone hydrogen bonds. CV_4 and CV_5 monitor the helical content in the whole and central part of the peptide. CV_6 is the dihedral correlation between successive dihedrals.

[l]The free energy of each cluster is estimated by a weighted-histogram approach (Kumar et al. 1992).

role of N17 in Htt oligomerization because such surface may create a hydrophobic seed around which the flanking polyQ tract can collapse (Truant et al. 2008; Ross & Tabrizi 2011; Ross et al. 2003) and promote hydrophobic-force driven associations between Htt N-terminal fragments (Thakur et al. 2009; Colby et al. 2004; Angeli et al. 2010; Tam et al. 2009). The other significantly populated basins, B1 and B3 (Fig. 7) assumed an amphipathic helical conformation, from residues 1 to 11 and from 1 to 7, respectively. Such conformation may

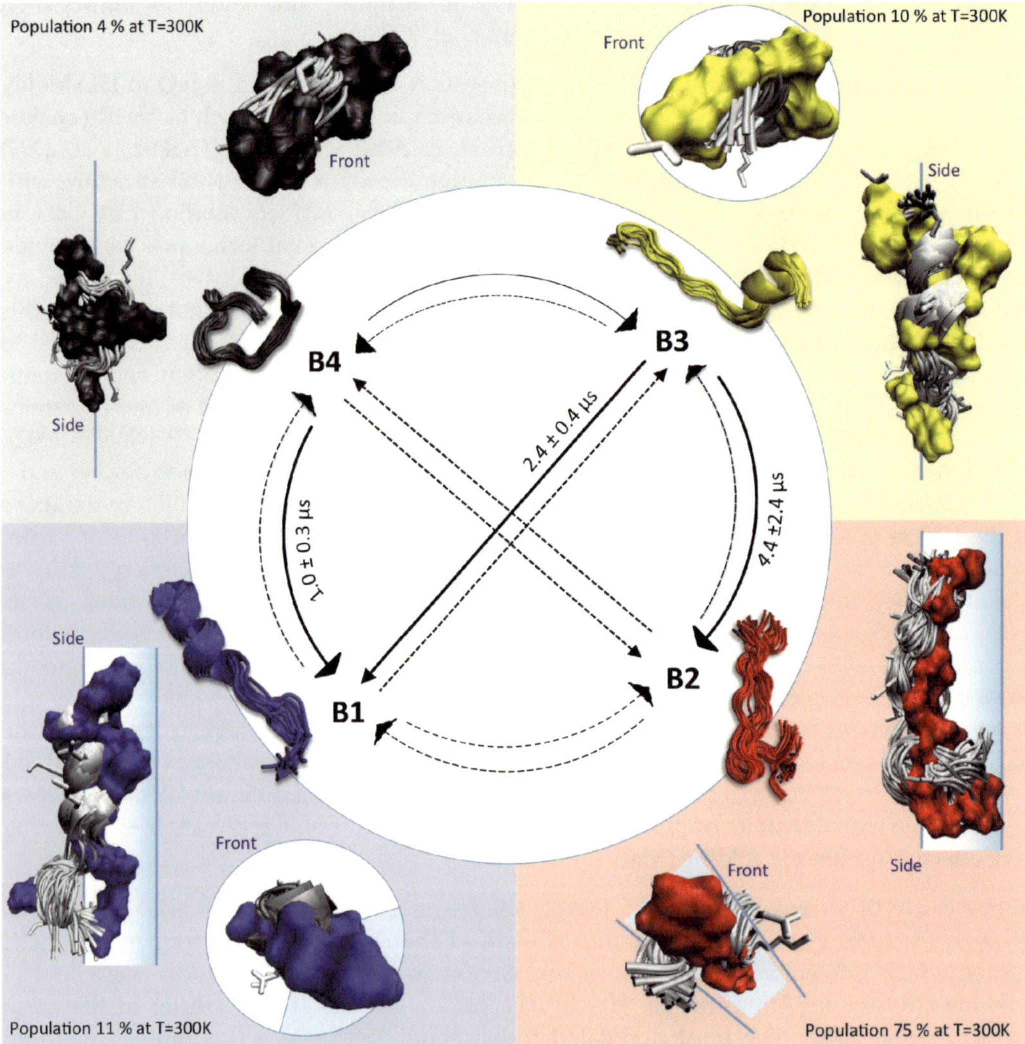

Fig. 7. The four basins (B1-B4) of N17 in aqueous solution. B1-B4 are characterized by their population and by their attractor. This latter is defined as the cluster with lowest free energy in the basin. Only the attractors' structures and their correspondent views of the hydrophobic side chain distribution are shown for clarity. The attractors' structures of B1 are colored in red, those of B2 in blue, those of B3 in yellow and those of B4 in grey. The calculated interconversion rates along with their corresponding statistical errors are reported. Dotted arrows are used for rates > 2 ms.

Modification	Q tract	In vivo	In cell	In vitro	Helix Propensity	Hydro phobicity	**SS pred**: secondary structure prediction **Burial_25:** burial, less than 25% solvent accesibility **Burial 5:** burial, less than 5% exposure **Reliability** of prediction accuracy, ranges from 0 to 9, bigger is better.
MATLEKLMKAFESLKSF	-	-	-	-	4.36	0.05	Sequence : MATLEKLMKAFESLKSF SS pred : - HHHHHHHHHHHHHHHH Burial_25 : ---B--BB-BB--B--B Burial_5 : ------B---------- Reliability: 73799999999999998
MATAAAAAAAFESLKSF (Tam et al. 2009)	103	!	-	!	3.06	-0.09	Sequence : MATAAAAAAAFESLKSF SS pred : -- HHHHHHHHHHHHHHH Burial_25 : ---B--BB-BB--B--B Burial_5 : ----------------- Reliability: 98489999999999998
MATLEKPMPAFESLKSF (Tam et al. 2009)	103	-	-	!	0.22	-0.02	Sequence : MATLEKPMPAFESLKSF SS pred : ----------HHHHHHH Burial_25 : ---B---B-BB--B--B Burial_5 : ----------------- Reliability: 99846777746899998
Polar to A MATLAALMAAFESLKSF (Tam et al. 2009)	103	↓	-	↓	6.01	0.10	Sequence : MATLAALMAAFESLKSF SS pred : - HHHHHHHHHHHHHHHH Burial_25 : ---BBBBBBB--B--B Burial_5 : ------B---------- Reliability: 73799999999999998
Non polar to A MATAEKAAKAFESLKSF (Tam et al. 2009)	103	!	-	↓	2.26	-0.14	Sequence : MATAEKAAKAFESLKSF SS pred : --- HHHHHHHHHHHHHH Burial_25 : ---B--BB-BB--B--B Burial_5 : ----------------- Reliability: 99768999999999998
MATLEKLMKAFEDLKDF *(Gu et al. 2009)*	97	↓	-	↓	4.31	-0.02	Sequence : MATLEKLMKAFEDLKDF SS pred : - HHHHHHHHHHHHHHHH Burial_25 : ---B--BB-BB--B--B Burial_5 : ------B---------- Reliability: 73799999999999987
MAALEKLMKAFESLKSF *(Aiken et al. 2009)*	46	↓	↓	-	6.26	0.04	Sequence : MAALEKLMKAFESLKSF SS pred : -- HHHHHHHHHHHHHHH Burial_25 : ---B--BB-BB--B--B Burial_5 : ------B---------- Reliability: 70899999999999987

↓ = decrease; ! = block; H=helix ; B = buried

Table 1. Bioinformatic calculations on N17. Calculated (Improta et al. 2001) helix propensity, hydrophobicity and number of buried residues of N17 and non-amyloidogenic mutants. In all calculations, pH, temperature and ionic strength were assumed to be the same as in the calculations, namely 7, 300K, 0.1M.

facilitate the binding on N17's target surface. This was consistent with the proposal that N17 assumes a helical fold by binding to a variety of cellular partners (Thakur et al. 2009; Colby et al. 2004; Angeli et al. 2010; Tam et al. 2009). This aspect, in turn, may have an impact on the formation of fibrils (Tam et al. 2009; Gu et al. 2009). The last basin B4 was characterized, instead, by a very small population and assumed a globular compact coiled structure. A variety of mutants of N17 are non-amyloidogenic (Tam et al. 2009; Gu et al. 2009). As for most of them, the mutation changes the nature of the residue from apolar to polar or vice versa, the result of the mutation is, probably, a reduction of the large content of amphiphatic conformations of N17 (Tam et al. 2009; Gu et al. 2009).[m] Consistently, the calculated folding propensity of these mutants differed significantly from that of N17 (Tab. 1). This hold true even if only a single point mutation was introduced (the N17(T3A) peptide) (Tam et al. 2009; Gu et al. 2009). In conclusion, changes in the relative population of the different basins induced by a change of amphiphaty may substantially affect the propensity of N17 mutants to form fibrils as observed experimentally (Rossetti et al. 2011).

4. Conclusions and perspectives

Our review clearly remarks the importance of computer simulations in complementing and interpreting experimental findings in neurodegenerative diseases. However, although computer simulations techniques are becoming more and more powerful to investigate these biological problems and the computer power continues to increase enormously, several aspects still limit the effective application of computational methods to a detailed understanding of the polyQ aggregation mechanism (Papaleo & Invernizzi 2011). The limited time scale accessible to full atomistic simulations requires the use of enhanced sampling algorithms to explore the conformational space of the folding and of the aggregation of Htt fragments, or to get insights into the physico-chemical determinants at the basis of this mechanism. However, these computational techniques are very demanding from the computational point of view and not yet capable of simulating the aggregation of long biologically relevant peptides. CG models may be suitable to study larger systems and to explore longer time scale than force filed based MD. However, they are lack of an atomistic description, which may be crucial to correctly describe the complex aggregation processes at the basis of neurodegenerative diseases. In this respect the development of accurate multiscale approaches based on a combination of CG and force field methods may be useful to overcome the limitations of both methodologies (Moroni et al. 2009; Tozzini 2010; Neri et al. 2005). From the experimental point of view, instead, limitations to a complete understanding of HD's onset mechanism are given by the lack of the crystallographic structure of the entire Htt protein, as well as by the lack of a detailed mapping of its interacting partners proteins. This latter aspect is becoming to be addressed both experimentally and computationally (Angeli et al. 2010; Rossetti et al. 2011). Moreover, theoretical and experimental studies demonstrated that several different aggregation pathways exist, resulting in different oligomeric and fibrillar structures of comparable stabilities. Probably, the dominant morphology of the aggregates is determined by the species having the lowest barrier to form the initial nucleation seed, rather than the largest

[m]The helix propensity, hydrophobicity and number of buried residues of N17 as well as those of the mutants in Tab 1 were estimated using the AGADIR ((Lacroix, et al. 1998) at http://agadir.crg.es/), PEPINFO (Sweet and Eisenberg 1983) at http://emboss.sourceforge.net/ and JPRED3 (Cole et al. 2008) at http://www.compbio.dundee.ac.uk/, respectively.

thermodynamic stability. However, the relative importance of kinetic and thermodynamics factors in amyloids grow is still a highly debated issue (Papaleo & Invernizzi 2011). Due to the increased potentialities of both experimental and computational approaches a synergistic effort should be immensely useful to unravel the toxicity mechanism of protein aggregation and, in particular, to further clarify several unclear aspects of the HD mechanism (Ma & Nussinov 2006; Miller et al. 2010). This may be also of help to identify and design specific molecules to hamper polyQ aggregation (Robertson & Bottomley 2010).

5. Acknowledgments

The authors thank Prof. P. Carloni, Dr. A. Pastore, Prof. F. Persichetti, Prof. A. Laio and P. Cossio as they have contributed to the selected applications presented here.

6. References

Aiken, C. T., J. S. Steffan, C. M. Guerrero, H. Khashwji, T. Lukacsovich, D. Simmons, J. M. Purcell, K. Menhaji, Y. Z. Zhu, K. Green, F. Laferla, L. Huang, L. M. Thompson, and J. L. Marsh. 2009. Phosphorylation of threonine 3: implications for Huntingtin aggregation and neurotoxicity. *J Biol Chem* 284 (43):29427-36. 1083-351X (Electronic) 0021-9258 (Linking).

Andrade, M. A., and P. Bork. 1995. Heat Repeats in the Huntingtons-Disease Protein. *Nature Genetics* 11 (2):115-116. 1061-4036.

Angeli, S., J. Shao, and M. I. Diamond. 2010. F-actin binding regions on the androgen receptor and huntingtin increase aggregation and alter aggregate characteristics. *PLoS ONE* 5 (2):e9053.

Armen, R. S., B. M. Bernard, R. Day, D. O. V. Alonso, and V. Daggett. 2005. Characterization of a possible amyloidogenic precursor in glutamine-repeat neurodegenerative diseases. *Proc Nat Acad Sci USA* 102 (38):13433-13438. 0027-8424.

Arrasate, M., S. Mitra, E. S. Schweitzer, M. R. Segal, and S. Finkbeiner. 2004. Inclusion body formation reduces levels of mutant huntingtin and the risk of neuronal death. *Nature* 431 (7010):805-810. 0028-0836.

Atwal, R. S., J. Xia, D.Pinchev, J. Taylor, R. M. Epand, and R. Truant. 2007. Huntingtin has a membrane association signal that can modulate huntingtin aggregation, nuclear entry and toxicity. *Hum Mol Gen* 16 (21):2600-15. Online ISSN 1460-2083 - Print ISSN 0964-6906.

Barton, S., R. Jacak, S. D. Khare, F. Ding, and N. V. Dokholyan. 2007. The length dependence of the polyQ-mediated protein aggregation. *J Biol Chem* 282 (35):25487-92. 0021-9258 (Print) 0021-9258 (Linking).

Bates, G. 2003. Huntingtin aggregation and toxicity in Huntington's disease. *Lancet* 361 (9369):1642-4. 0140-6736 (Print) 0140-6736 (Linking).

Bates, G. P., L. Mangiarini, and S. W. Davies. 1998. Transgenic Mice in the Study of Polyglutamine Repeat Expansion Diseases. *Brain Pathology* 8 (4):699-714. 1750-3639.

Beke, T., I. G. Csizmadia, and A. Perczel. 2006. Theoretical study on tertiary structural elements of beta-peptides: nanotubes formed from parallel-sheet-derived assemblies of beta-peptides. *J Am Chem Soc* 128 (15):5158-67. 0002-7863 (Print) 0002-7863 (Linking).

Benedek, N. A., I. K. Snook, K. Latham, and I. Yarovsky. 2005. Application of numerical basis sets to hydrogen bonded systems: a density functional theory study. *J Chem Phys* 122 (14):144102. 0021-9606 (Print).

Bennett, M. J., K. E. Huey-Tubman, A. B. Herr, A. P. West, S. A. Ross, and P. J. Bjorkman. 2002. A linear lattice model for polyglutamine in CAG-expansion diseases. *Proc Nat Acad Sci USA* 99 (18):11634-11639. 0027-8424.

Berendsen, H. J. C., D. van der Spoel, and R. van Drunen. 1995. GROMACS: a message-passing parallel molecular dynamics implementation. *Comput Phys Commun.*

Berman, H. 2000. The protein data bank: A retrospective and prospective. *Biophys J* 78 (1):267A-267A. 0006-3495.

Bhattacharyya, A. M., A. K. Thakur, and R. Wetzel. 2005. polyglutamine aggregation nucleation: thermodynamics of a highly unfavorable protein folding reaction. *Proc Natl Acad Sci U S A* 102 (43):15400-5. 0027-8424 (Print) 0027-8424 (Linking).

Bhattacharyya, A., A. K. Thakur, V. M. Chellgren, G. Thiagarajan, A. D. Williams, B. W. Chellgren, T. P. Creamer, and R. Wetzel. 2006. Oligoproline Effects on Polyglutamine Conformation and Aggregation. *J Mol Biol* 355 (3):524-535. 0022-2836.

Biarnes, X., S. Bongarzone, A. Vargiu, P. Carloni, and P. Ruggerone. 2011. Molecular motions in drug design: the coming age of the metadynamics method. *J Comput Aid Mol Des* 25 (5):395-402. 0920-654.

Borrell-Pages, M., D. Zala, S. Humbert, and F. Saudou. 2006. Huntington's disease: from huntingtin function and dysfunction to therapeutic strategies. *Cell Mol Life Sci* 63 (22):2642-2660. 1420-682.

Bulone, D., L.Masino, D. J. Thomas, P. L. San Biagio, and A. Pastore. 2006. The Interplay between PolyQ and Protein Context Delays Aggregation by Forming a Reservoir of Protofibrils. *PLoS ONE* 1 (1):e111.

Bussi, G., F. L. Gervasio, A. Laio, and M. Parrinello. 2006. Free-Energy Landscape for Œ≤ Hairpin Folding from Combined Parallel Tempering and Metadynamics. *J Am Chem Soc* 128 (41):13435-13441. 0002-7863.

Carloni, P., U. Rothlisberger, and M. Parrinello. 2002. The Role and Perspective of Ab Initio Molecular Dynamics in the Study of Biological Systems. *Accounts Chem Res* 35 (6):455-464. 0001-4842.

Caviston, J. P, J. L. Ross, S. M. Antony, M. Tokito, and E. L. F. Holzbaur. 2007. Huntingtin facilitates dynein/dynactin-mediated vesicle transport. *P Natl Acad Sci Usa* 104 (24):10045-50.

Cha, J.H. J. 2007. Transcriptional signatures in Huntington's disease. *Prog Neurobiol* 83 (4):228-48. 0301-0082.

Chen, S. M., V. Berthelier, J. B. Hamilton, B. O'Nuallain, and R. Wetzel. 2002. Amyloid-like features of polyglutamine aggregates and their assembly kinetics. *Biochem* 41 (23):7391-7399. 0006-2960.

Christ, C. D., A. E. Mark, and W. F. van Gunsteren. 2010. Basic ingredients of free energy calculations: A review. *J Comput Chem* 31 (8):1569-1582. 1096-987X.

Colby, D. W., Y.J. Chu, J.P. Cassady, M. Duennwald, H. Zazulak, J.M. Webster, A. Messer, S. Lindquist, V.M. Ingram, and K.D. Wittrup. 2004. Potent inhibition of huntingtin and cytotoxicity by a disulfide bond-free single-domain intracellular antibody. *Proc Nat Acad Sci USA* 101 (51):17616-17621.

Cole, C., J. D. Barber, and G. J. Barton. 2008. The Jpred 3 secondary structure prediction server. *Nucleic Acids Res* 36:W197-W201.

Cornett, J., F. Cao, C.E. Wang, C. A. Ross, G. P. Bates, S. H. Li, and X. J. Li. 2005. Polyglutamine expansion of huntingtin impairs its nuclear export. *Nat Genet* 37 (2):198-204.

CPMD 3.11.1. Copyright IBM Corp 1990-2008.

Davies, S. W., M. Turmaine, B. A. Cozens, M. DiFiglia, A.H. Sharp, C. A. Ross, E. Scherzinger, Erich E. Wanker, Laura Mangiarini, and Gillian P. Bates. 1997. Formation of Neuronal Intranuclear Inclusions Underlies the Neurological Dysfunction in Mice Transgenic for the HD Mutation. *Cell* 90 (3):537-548. 0092-8674.

Dehay, B., and A. Bertolotti. 2006. Critical role of the proline-rich region in Huntingtin for aggregation and cytotoxicity in yeast. *J Biol Chem* 281 (47):35608-35615.

Duennwald, M. L., S. Jagadish, P. J. Muchowski, and S. Lindquist. 2006. Flanking sequences profoundly alter polyglutamine toxicity in yeast. *Proc Nat Acad Sci USA* 103 (29):11045-11050.

Esposito, L., A. Paladino, C. Pedone, and L. Vitagliano. 2008. Insights into structure, stability, and toxicity of monomeric and aggregated polyglutamine models from molecular dynamics simulations. *Biophys J* 94 (10):4031-4040. 0006-3495.

Finke, J. M., and J. N. Onuchic. 2004. Simulations exploring the structural ensemble in the folding of proteins and amyloid peptides. Biophys J 86 (1):340A-340A. 0006-3495.

Finke, J. M., M. S. Cheung, and J. N. Onuchic. 2004. A Structural Model of Polyglutamine Determined from a Host-Guest Method Combining Experiments and Landscape Theory. *Biophys J* 87 (3):1900-1918. 0006-3495.

Grunewald, T., and M. F. Beal. 1999. Bioenergetics in Huntington's Disease. *Ann NY Acad Sci* 893 (1):203-213. 1749-6632.

Gu, X., E. R. Greiner, R. Mishra, R. Kodali, A. Osmand, S. Finkbeiner, J. S. Steffan, L. M. Thompson, R. Wetzel, and X. W. Yang. 2009. Serines 13 and 16 Are Critical Determinants of Full-Length Human Mutant Huntingtin Induced Disease Pathogenesis in HD Mice. *Neuron* 64 (6):828-840. 0896-6273.

Gunawardena, S., L.S. Her, R. G. Brusch, R. A. Laymon, I. R. Niesman, B. Gordesky-Gold, L. Sintasath, N. M. Bonini, and L. S. B. Goldstein. 2003. Disruption of Axonal Transport by Loss of Huntingtin or Expression of Pathogenic PolyQ Proteins in Drosophila. *Neuron* 40 (1):25-40. 0896-6273.

Gusella, J. F., and M. E. MacDonald. 2000. Molecular genetics: unmasking polyglutamine triggers in neurodegenerative disease. *Nat Rev Neurosci* 1 (2):109-15. 1474-1776.

Gutekunst, C. A., S. H. Li, H. Yi, J. S. Mulroy, S. Kuemmerle, R. Jones, D. Rye, R. J. Ferrante, S. M. Hersch, and X. J. Li. 1999. Nuclear and neuropil aggregates in Huntington's disease: Relationship to neuropathology. *J Neurosci* 19 (7):2522-2534. 0270-6474.

Hajime, O., N. Miki, W. Hirofumi, E. B. Starikov, M. Rothstein Stuart, and T. Shigenori. 2008. Molecular dynamics simulation study on the structural stabilities of polyglutamine peptides. *Comput Biol Chem* 32 (2):102-110. 1476-9271.

Harjes P., E.E. Wanker 2003. The hunt for huntingtin function: interaction partners tell many different stories. *Trends Biochem Sci* 28:425-433.

Horvath, V., Z. Varga, and A. Kovacs. 2004. Long-range effects in oligopeptides. A theoretical study of the beta-sheet structure of Gly(n) (n=2-10). *J Phys Chem A* 108 (33):6869-6873. 1089-5639.

Horvath V, Varga Z, Kovacs. 2005. Substituent effects on long-range interactions in the β-sheet structure of oligopeptides. *J. Mol. Struct. (Theochem.)* 755 (1-3):247-251. 0166-1280.

Housman, D. 1995. Gain of glutamines, gain of function? *Nat Genet* 10 (1):3-4. 1061-4036.

Huntington, G. 1872. On chorea. *The Medical and Surgical Reporter* 26 (15):317-321.

HyperChem 8.0, 1115 NW 4th St. Gainesville, FL 32608 (USA).

Ignatova, Z., A. K. Thakur, R. Wetzel, and L. M. Gierasch. 2007. In-cell aggregation of a polyglutamine-containing chimera is a multistep process initiated by the flanking sequence. *J Biol Chem* 282 (50):36736-43.

Imarisio, S., J. Carmichael, V. Korolchuk, C.W. Chen, S. Saiki, C. Rose, G. Krishna, J. E. Davies, E. Ttofi, B. R. Underwood, and D. C. Rubinsztein. 2008. Huntington's disease: from pathology and genetics to potential therapies. *Biochem J* 412 (2):191-209.

Improta, R., V. Barone, K. N. Kudin, and G. E. Scuseria. 2001. Structure and conformational behavior of biopolymers by density functional calculations employing periodic boundary conditions. I. The case of polyglycine, polyalanine, and poly-alpha-aminoisobutyric acid in vacuo. *J Am Chem Soc* 123 (14):3311-3322. 0002-7863.

Improta, R., and V. Barone. 2004. Assessing the reliability of density functional methods in the conformational study of polypeptides: The treatment of intraresidue nonbonding interactions. *J Comput Chem* 25 (11):1333-1341. 1096-987X.

Improta, R., V. Barone, K. N. Kudin, and G. E. Scuseria. 2001. The conformational behavior of polyglycine as predicted by a density functional model with periodic boundary conditions. *J Chem Phys* 114 (6):2541-2549. 0021-9606.

Johnson, C. D, and B. L. Davidson. 2010. Huntington's disease: progress toward effective disease-modifying treatments and a cure. *Hum Mol Genet* 19 (R1):R98-R102. Online 1460-2083 - Print 0964-6906.

Kaltenbach, L. S., E. Romero, R. R. Becklin, R. Chettier, R. Bell, A. Phansalkar, A. Strand, C. Torcassi, J. Savage, A. Hurlburt, G.H. Cha, L. Ukani, C.L. Chepanoske, Y. Zhen, S. Sahasrabudhe, J. Olson, C. Kurschner, L. M. Ellerby, J. M. Peltier, J. Botas, and R. E. Hughes. 2007. Huntingtin interacting proteins are genetic modifiers of neurodegeneration. *PLoS Genet* 3 (5):e82.

Kar, K., M. Jayaraman, B. Sahoo, R. Kodali, and R. Wetzel. 2011. Critical nucleus size for disease-related polyglutamine aggregation is repeat-length dependent. *Nat Struct Mol Biol*. 1545-9993.

Kegel, K. B., A. R. Meloni, Y. Yi, Y. J. Kim, E. Doyle, B. G. Cuiffo, E. Sapp, Y. Wang, Z.H. Qin, J. D. Chen, J. R. Nevins, N. Aronin, and M. DiFiglia. 2002. Huntingtin is present in the nucleus, interacts with the transcriptional corepressor C-terminal binding protein, and represses transcription. *J Biol Chem* 277 (9):7466-76. 0021-9258.

Kegel, K. B., E. Sapp, J. Yoder, B. Cuiffo, L. Sobin, Y. J. Kim, Z.H. Qin, M. R. Hayden, N. Aronin, D. L. Scott, G. Isenberg, W. H. Goldmann, and M. DiFiglia. 2005. Huntingtin associates with acidic phospholipids at the plasma membrane. *J Biol Chem* 280 (43):36464-73. 0021-9258.

Kelley, N. W., X. Huang, S. Tam, C. Spiess, J. Frydman, and V. S. Pande. 2009. The Predicted Structure of the Headpiece of the Huntingtin Protein and Its Implications on Huntingtin Aggregation. *J Mol Biol* 388 (5):919-927. 0022-2836.

Khare, S. D., F. Ding, K. N. Gwanmesia, and N. V. Dokholyan. 2005. Molecular origin of polyglutamine aggregation in neurodegenerative diseases. *PLoS Comp Biol* 1 (3):230-5.

Kim, M. W., Y. Chelliah, S. W. Kim, Z. Otwinowski, and I. Bezprozvanny. 2009. Secondary Structure of Huntingtin Amino-Terminal Region. *Structure (London, England : 1993)* 17 (9):1205-1212. 0969-2126.

Klein, F., A. Pastore, L. Masino, G. Zederlutz, H. Nierengarten, M. Ouladabdelghani, D. Altschuh, J. Mandel, and Y. Trottier. 2007. Pathogenic and Non-pathogenic Polyglutamine Tracts Have Similar Structural Properties: Towards a Length-dependent Toxicity Gradient. *J Mol Biol* 371 (1):235-244. 0022-2836.

Koch, O., M. Bocola, and G. Klebe. 2005. Cooperative effects in hydrogen-bonding of protein secondary structure elements: A systematic analysis of crystal data using Secbase. *Proteins: Struct, Fun, and Bio* 61 (2):310-317. 1097-0134.

Kuemmerle, S, C A Gutekunst, A M Klein, X J Li, S H Li, M F Beal, S M Hersch, and R J Ferrante. 1999. Huntington aggregates may not predict neuronal death in Huntington's disease. *Ann Neurol* 46 (6):842-9.

Kumar, S, D Bouzida, RH Swendsen, Peter A Kollman, and J.M. Rosenberg. 1992. THE weighted histogram analysis method for free-energy calculations on biomolecules.1. The Method. *J Comput Chem* 13 (8):1011-1021. 1096-987X.

Lacroix, E, A R Viguera, and L Serrano. 1998. Elucidating the folding problem of alpha-helices: local motifs, long-range electrostatics, ionic-strength dependence and prediction of NMR parameters. *J Mol Biol* 284 (1):173-91. 0022-2836.

Laghaei, R., and N. Mousseau. 2010. Spontaneous formation of polyglutamine nanotubes with molecular dynamics simulations. *J Chem Phys* 132 (16):165102. 0021-9606.

Laio, A., and F. L. Gervasio. 2008. Metadynamics: a method to simulate rare events and reconstruct the free energy in biophysics, chemistry and material science. *Rep Prog Phys* 71 (12):126601.

Laio, A., and M. Parrinello. 2002. Escaping free-energy minima. *P Natl Acad Sci Usa* 99 (20):12562-6.

Lakhani, Vinal V., Feng Ding, and Nikolay V. Dokholyan. 2010. Polyglutamine Induced Misfolding of Huntingtin Exon1 is Modulated by the Flanking Sequences. *PLoS Comput Biol* 6 (4):e1000772.

Leavitt, B. R., J. A. Guttman, J. G. Hodgson, G. H. Kimel, R. Singaraja, A. Wayne Vogl, and Michael R. Hayden. 2001. Wild-Type Huntingtin Reduces the Cellular Toxicity of Mutant Huntingtin In Vivo. *Am J Hum Genet* 68 (2):313-324. 0002-9297.

Leavitt, B. R., C. L. Wellington, and M. R. Hayden. 1999. Recent Insights into the Molecular Pathogenesis of Huntington Disease. *Semin Neurol* 19 (04):385,395. 0271-8235.

Lee, W.C. M., M. Yoshihara, and J. T. Littleton. 2004. Cytoplasmic aggregates trap polyglutamine-containing proteins and block axonal transport in a Drosophila model of Huntington's disease. *P Natl Acad Sci Usa* 101 (9):3224-9.

Legleiter, J., E. Mitchell, G. P. Lotz, E. Sapp, C. Ng, M. DiFiglia, L. M. Thompson, and P. J. Muchowski. 2010. Mutant huntingtin fragments form oligomers in a polyglutamine length-dependent manner in vitro and in vivo. *J Biol Chem* 285 (19):14777-90. 0021-9258.

Li, H, S H Li, Z X Yu, P Shelbourne, and X J Li. 2001. Huntingtin aggregate-associated axonal degeneration is an early pathological event in Huntington's disease mice. *J Neurosci* 21 (21):8473-81. 0270-6474.

Li, W., L. C. Serpell, W. J. Carter, D. C. Rubinsztein, and J. A. Huntington. 2006. Expression and characterization of full-length human huntingtin, an elongated HEAT repeat protein. *J Biol Chem* 281 (23):15916-22. 0021-9258.

Ma, B. and R. Nussinov. 2006. Simulations as analytical tools to understand protein aggregation and predict amyloid conformation. *Curr Opin Chem Biol* 10 (5):445-452. 1367-5931.

MacDonald, M. E., C. M. Ambrose, M. P. Duyao, R. H. Myers, C. Lin, L. Srinidhi, G. Barnes, S. A. Taylor, M. James, N. Groot, H. MacFarlane, B. Jenkins, M. A. Anderson, N. S. Wexler, J. F. Gusella, G. P. Bates, S. Baxendale, H. Hummerich, S. Kirby, M. North, S. Youngman, R. Mott, G. Zehetner, Z. Sedlacek, A. Poustka, A.M. Frischauf, H. Lehrach, A. J. Buckler, D. Church, L. Doucette-Stamm, M. C. O'Donovan, L. Riba-Ramirez, M. Shah, V. P. Stanton, S. A. Strobel, K. M. Draths, J. L. Wales, P. Dervan, D. E. Housman, M. Altherr, R. Shiang, L. Thompson, T. Fielder, J. J. Wasmuth, D. Tagle, J. Valdes, L. Elmer, M. Allard, L. Castilla, M. Swaroop, K. Blanchard, F. S. Collins, R. Snell, T. Holloway, K. Gillespie, N. Datson, D. Shaw, and P. S. Harper. 1993. A novel gene containing a trinucleotide repeat that is expanded and unstable on Huntington's disease chromosomes. *Cell* 72 (6):971-983. 0092-8674.

Mangiarini, L., K. Sathasivam, M. Seller, B. Cozens, A. Harper, C. Hetherington, M. Lawton, Y. Trottier, H. Lehrach, S. W. Davies, and G. P. Bates. 1996. Exon 1 of the HD gene with an expanded CAG repeat is sufficient to cause a progressive neurological phenotype in transgenic mice. *Cell* 87 (3):493-506. 0092-8674.

Marchut, A. J., and C. K. Hall. 2006a. Side-chain interactions determine amyloid formation by model polyglutamine peptides in molecular dynamics simulations. *Biophys J* 90 (12):4574-4584. 0006-3495.

Marchut A.J., C. K. Hall 2006b. Spontaneous formation of annular structures observed in molecular dynamics simulations of polyglutamine peptides. *Comput Biol Chem* 30 (3):215-218. 1476-9271.

Marchut A.J., C. K. Hall 2007. Effects of chain length on the aggregation of model polyglutamine peptides: Molecular dynamics simulations. *Proteins Struct Funct Bioinf* 66 (1):96-109. 0887-3585.

Marinelli, F., F. Pietrucci, A. Laio, and S. Piana. 2009. A kinetic model of trp-cage folding from multiple biased molecular dynamics simulations. *PLoS Comp Biol* 5 (8):e1000452.

Masino, L., G. Kelly, K. Leonard, Y. Trottier, and A. Pastore. 2002. Solution structure of polyglutamine tracts in GST-polyglutamine fusion proteins. *Febs Lett* 513 (2-3):267-272. 0014-5793.

Merlino, A., L. Esposito, and L. Vitagliano. 2006. Polyglutamine repeats and beta-helix structure: Molecular dynamics study. *Proteins Struct Funct Bioinf* 63 (4):918-927. 0887-3585.

Miller, Y., B. Ma, and R. Nussinov. 2010. Polymorphism in Alzheimer A beta Amyloid Organization Reflects Conformational Selection in a Rugged Energy Landscape. *Chem Rev* 110 (8):4820-4838. 0009-2665.

Morley, J. F., H. R. Brignull, J. J. Weyers, and R. I. Morimoto. 2002. The threshold for polyglutamine-expansion protein aggregation and cellular toxicity is dynamic and influenced by aging in Caenorhabditis elegans. *Proc Nat Acad Sci USA* 99 (16):10417-10422. 0027-8424.

Moroni, E., G. Scarabelli, and G. Colombo. 2009. Structure and sequence determinants of aggregation investigated with molecular dynamics. *Front Biosci* 14:523-539. 1093-4715.

Morozov, A. V., T. Kortemme, K. Tsemekhman, and D. Baker. 2004. Close agreement between the orientation dependence of hydrogen bonds observed in protein structures and quantum mechanical calculations. *Proc Natl Acad Sci U S A* 101 (18):6946-51. 0027-8424 (Print).

Myers, R. 2004. Huntington's Disease Genetics. *NeuroRX* 1 (2):255-262.

Nagai, Y., T. Inui, H. A. Popiel, N. Fujikake, K. Hasegawa, Y. Urade, Y. Goto, H. Naiki, and T. Toda. 2007. A toxic monomeric conformer of the polyglutamine protein. *Nat Struct Mol Biol* 14 (4):332-340. 1545-9993.

Neri, M., C. Anselmi, M. Cascella, A. Maritan, and P. Carloni. 2005. Coarse-grained model of proteins incorporating atomistic detail of the active site. *Phys Rev Lett* 95 (21):218102.

Ogawa, H., M. Nakano, H. Watanabe, E. B. Starikov, S. M. Rothstein, and S. Tanaka. 2008. Molecular dynamics simulation study on the structural stabilities of polyglutamine peptides. *Comput Biol Chem* 32 (2):102-110. 1476-9271.

Olshina, M A, L M Angley, Y M Ramdzan, J Tang, M F Bailey, A F Hill, and D M Hatters. 2010. Tracking Mutant Huntingtin Aggregation Kinetics in Cells Reveals Three Major Populations That Include an Invariant Oligomer Pool. *J Biol Chem* 285 (28):21807-21816. 0021-9258.

Oma, Y., Y. Kino, N. Sasagawa, and S. Ishiura. 2004. Intracellular localization of homopolymeric amino acid-containing proteins expressed in mammalian cells. *J Biol Chem* 279 (20):21217-22. 0021-9258.

Palidwor, G. A, S. Shcherbinin, M. R. Huska, T. Rasko, U. Stelzl, A. Arumughan, R. Foulle, P. Porras, L. Sanchez-Pulido, E. E. Wanker, and M. A. Andrade-Navarro. 2009. Detection of alpha-rod protein repeats using a neural network and application to huntingtin. *PLoS Comp Biol* 5 (3):e1000304.

Papaleo, E., and G. Invernizzi. 2011. Conformational Diseases: Structural Studies of Aggregation of Polyglutamine Proteins. *Curr Comput-Aid Drug.* no. 7 (1):23-43.

Parker, J A, J B Connolly, C Wellington, M Hayden, J Dausset, and C Neri. 2001. Expanded polyglutamines in Caenorhabditis elegans cause axonal abnormalities and severe dysfunction of PLM mechanosensory neurons without cell death. *P Natl Acad Sci Usa* 98 (23):13318-23.

Perczel, A., Z. Gaspari, and I. G. Csizmadia. 2005. Structure and stability of beta-pleated sheets. *J Comput Chem* 26 (11):1155-1168. 1096-987X.

Perdew, J. P., K. Burke, and M. Ernzerhof. 1996. Generalized Gradient Approximation Made Simple. *Phys Rev Lett* 77 (18):3865-3868. 0031-9007 (Print).

Perutz, M. 1994. Polar Zippers - Their Role in Human-Disease. *Protein Sci* 3 (10):1629-1637. 0961-8368.

Perutz, M. F. 1999. Glutamine repeats and neurodegenerative diseases: molecular aspects. *Trends Biochem Sci* 24 (2):58-63. 0968-0004.

Perutz, M. F., J. T. Finch, J. Berriman, and A. Lesk. 2002. Amyloid fibers are water-filled nanotubes. *Proc Nat Acad Sci USA* 99 (8):5591-5595.

Perutz, M. F., and A. H. Windle. 2001. Cause of neural death in neurodegenerative diseases attributable to expansion of glutamine repeats. *Nature* 412 (6843):143-144. 0028-0836.

Piana, S, and A. Laio. 2007. A Bias-Exchange Approach to Protein Folding. *J Phys Chem B* 111 (17):4553-4559. 1520-6106.

Piana, S., A. Laio, F. Marinelli, M. Van Troys, D. Bourry, C. Ampe, and J. C. Martins. 2008. Predicting the Effect of a Point Mutation on a Protein Fold: The Villin and Advillin Headpieces and Their Pro62Ala Mutants. *J Mol Biol* 375 (2):460-470. 0022-2836.

Raetz, C. R. H., and S. L. Roderick. 1995. A Left-Handed Parallel beta Helix in the Structure of UDP-N-Acetylglucosamine Acyltransferase. *Science* 270 (5238):997-1000. 0036-8075.

Ramdzan, Y. M., R. M. Nisbet, J. Miller, S. Finkbeiner, A. F. Hill, and D. M. Hatters. 2010. Conformation Sensors that Distinguish Monomeric Proteins from Oligomers in Live Cells. *Chem. & Biol* 17 (4):371-379. 1074-5521.

Reiner, Anton, Ioannis Dragatsis, Scott Zeitlin, and Daniel Goldowitz. 2003. Wild-type huntingtin plays a role in brain development and neuronal survival. *Mol Neurobiol* 28 (3):259-275. 0893-7648.

Robertson, A. L., and S. P. Bottomley. 2010. Towards the Treatment of Polyglutamine Diseases: The Modulatory Role of Protein Context. *Curr Med Chem* 17 (27):3058-3068. 0929-8673.

Rockabrand, E., N. Slepko, A. Pantalone, V. N Nukala, A. G. Kazantsev, J. L. Marsh, P. G. Sullivan, J. S. Steffan, S. L. Sensi, and L. M. Thompson. 2007. The first 17 amino acids of Huntingtin modulate its sub-cellular localization, aggregation and effects on calcium homeostasis. *Hum Mol Gen* 16 (1):61-77. 0964-6906.

Rohrig, U. F., A. Laio, N. Tantalo, M. Parrinello, and R. Petronzio. 2006. Stability and structure of oligomers of the Alzheimer peptide A beta(16-22): From the dimer to the 32-mer. *Biophys J* 91 (9):3217-3229. 0006-3495.

Ross, C. A., and M. A. Poirier. 2004. Protein aggregation and neurodegenerative disease. *Nature Med*:S10-S17. 1078-8956.

Ross, C. 1995. When more is less: Pathogenesis of glutamine repeat neurodegenerative diseases. *Neuron* 15 (3):493-496. 0896-6273.

Ross, C. A, M. A. Poirier, E. E. Wanker, and Mario Amzel. 2003. Polyglutamine fibrillogenesis: The pathway unfolds. *Proc Nat Acad Sci USA* 100 (1):1-3.

Ross, C. A., and S. J. Tabrizi. 2011. Huntington's disease: from molecular pathogenesis to clinical treatment. *Lancet Neurol* 10 (1):83-98. 1474-4422.

Ross, C., and M. Poirier. 2005. What is the role of protein aggregation in neurodegeneration? *Nature Rev Mol Cell Biol* 6 (11):891-898. 1471-0072.

Rossetti G, Angeli S, Magistrato A, Diamod M, Carloni P. 2011. Actin binding by Htt blocks intracellular aggregation. *sumitted to Plos One.*

Rossetti, G:, Pilar C:, A. Laio, and P. Carloni. 2011. Conformations of the Huntingtin N-term in aqueous solution from atomistic simulations. *Febs Letters* 585 (19):3086-3089. 0014-5793.

Rossetti, G., A. Magistrato, A. Pastore, and P. Carloni. 2010. Hydrogen Bonding Cooperativity in polyQ beta-Sheets from First Principle Calculations. *J Chem Theory Comput* 6 (6):1777-1782. 1549-9618.

Rossetti, G., A. Magistrato, A. Pastore, F. Persichetti, and P. Carloni. 2008. Structural Properties of Polyglutamine Aggregates Investigated via Molecular Dynamics Simulations. *J Phys Chem B* 112 (51):16843-16850. 1520-6106.

Sánchez, R, and A Sali. 1997. Evaluation of comparative protein structure modeling by MODELLER-3. *Proteins* Suppl 1:50-8.

Sapp, E., J. Penney, A. Young, N. Aronin, J. P. Vonsattel, and M. DiFiglia. 1999. Axonal transport of N-terminal huntingtin suggests early pathology of corticostriatal projections in Huntington disease. *J Neuropathol Exp Neurol* 58 (2):165-173. 0022-3069.

Saudou, F., S. Finkbeiner, D. Devys, and M. E. Greenberg. 1998. Huntingtin Acts in the Nucleus to Induce Apoptosis but Death Does Not Correlate with the Formation of Intranuclear Inclusions. *Cell* 95 (1):55-66. 0092-8674.

Scheiner, S., and T. Kar. 2005. Effect of Solvent upon CH...O Hydrogen Bonds with Implications for Protein Folding. *J Phys Chem B* 109 (8):3681-3689. 1520-6106.

Sharma, D., L. M. Shinchuk, H. Inouye, R. Wetzel, and D. A. Kirschner. 2005. Polyglutamine homopolymers having 8-45 residues form slablike beta-crystallite assemblies. *Proteins Struct Funct Bioinf* 61 (2):398-411. 0887-3585.

Sikorski, P., and E. Atkins. 2005. New model for crystalline polyglutamine assemblies and their connection with amyloid fibrils. *Biomacromolecules* 6 (1):425-432. 1525-7797.

Spiegel, K., and A. Magistrato. 2006. Modeling anticancer drug-DNA interactions via mixed QM/MM molecular dynamics simulations. *Org Biomol Chem* 4 (13):2507-2517. 1477-0520.

Steffan, J. S, N. Agrawal, J. Pallos, E. Rockabrand, L. C Trotman, N. Slepko, K. Illes, T. Lukacsovich, Y.Z. Zhu, E. Cattaneo, P. P. Pandolfi, L. M. Thompson, and J. L. Marsh. 2004. SUMO modification of Huntingtin and Huntington's disease pathology. *Science* 304 (5667):100-4. 0036-8075.

Stork, M., A. Giese, H. A. Kretzschmar, and P. Tavan. 2005. Molecular dynamics simulations indicate a possible role of parallel beta- helices in seeded aggregation of poly-Gln. *Biophys J* 88 (4):2442-2451. 0006-3495.

Strehlow, A. N. T., J. Z Li, and R. M. Myers. 2007. Wild-type huntingtin participates in protein trafficking between the Golgi and the extracellular space. *Hum Mol Genet* 16 (4):391-409. 1460-2083.

Sugaya, K., S. Matsubara, Y. Kagamihara, A. Kawata, and H. Hayashi. 2007. Polyglutamine Expansion Mutation Yields a Pathological Epitope Linked to Nucleation of Protein Aggregate: Determinant of Huntington's Disease Onset. *PLoS ONE* 2 (7):e635.

Sugita, Y., and Y. Okamoto. 2000. Replica-exchange multicanonical algorithm and multicanonical replica-exchange method for simulating systems with rough energy landscape. *Chem Phys Lett* 329 (3-4):261-270. 0009-2614.

Sunde, M., and C. Blake. 1997. The structure of amyloid fibrils by electron microscopy and X-ray diffraction. *Adv Protein Chem* 50:123-159. 0065-3233.

Sunde, M., L. C. Serpell, M. Bartlam, P. E. Fraser, M. B. Pepys, and C. C. F. Blake. 1997. Common core structure of amyloid fibrils by synchrotron X-ray diffraction. *J Mol Biol* 273 (3):729-739. 0022-2836.

Sweet, R. M., and D. Eisenberg. 1983. Correlation of sequence hydrophobicities measures similarity in three-dimensional protein structure. *J Mol Biol* 171 (4):479-488. 0022-2836.

Tam, S., C. Spiess, W. Auyeung, L. Joachimiak, B. Chen, M. A. Poirier, and J. Frydman. 2009. The chaperonin TRiC blocks a huntingtin sequence element that promotes the conformational switch to aggregation. *Nat Struct Mol Biol* 16 (12):1279-1285. 1545-9993.

Tanaka, M., I. Morishima, T. Akagi, T. Hashikawa, and N. Nukina. 2001. Intra- and intermolecular beta-pleated sheet formation in glutamine-repeat inserted myoglobin as a model for polyglutamine diseases. *J Biol Chem* 276 (48):45470-45475. 0021-9258.

Tartari, Marzia, Carmela Gissi, Valentina Lo Sardo, Chiara Zuccato, Ernesto Picardi, Graziano Pesole, and Elena Cattaneo. 2008. Phylogenetic comparison of huntingtin homologues reveals the appearance of a primitive polyQ in sea urchin. *Mol Biol Evol* 25 (2):330-8. 1537-1719.

Thakur, A. K., M. Jayaraman, R. Mishra, M. Thakur, V. M. Chellgren, I.J. L. Byeon, D. H. Anjum, R. Kodali, T. P. Creamer, J. F. Conway, A. M. Gronenborn, and R. Wetzel. 2009. Polyglutamine disruption of the huntingtin exon 1 N terminus triggers a complex aggregation mechanism. *Nat Struct Mol Biol* 16 (4):380-389. 1545-9993.

Tozzini, V. 2010. Multiscale Modeling of Proteins. *Accounts Chem Res*. no. 43 (2):220-230. doi: 10.1021/ar9001476

Trottier, Y., Y. Lutz, G. Stevanin, G. Imbert, D. Devys, G. Cancel, F. Saudou, C. Weber, G. David, L. Tora, Y. Agid, A. Brice, and J. L. Mandel. 1995. Polyglutamine Expansion as a Pathological Epitope in Huntingtons-Disease and 4 Dominant Cerebellar Ataxias. *Nature* 378 (6555):403-406. 0028-0836.

Truant, R., R. S. Atwal, C. Desmond, L. Munsie, and T. Tran. 2008. Huntington's disease: revisiting the aggregation hypothesis in polyglutamine neurodegenerative diseases. *FEBS J* 275 (17):4252-4262. 1742-4658.

Tsemekhman, K., L. Goldschmidt, D. Eisenberg, and D. Baker. 2007. Cooperative hydrogen bonding in amyloid formation. *Protein Sci* 16 (4):761-764. 1469-896X.

van der Spoel, D., E. Lindahl, B. Hess, G. Groenhof, A. E. Mark, and H. J. Berendsen. 2005. GROMACS: Fast, flexible, and free. *J Comput Chem* 26 (16):1701-1718. 1096-987X.

Varga, Z., and A. Kovács. 2005. Hydrogen bonding in peptide secondary structures. *Int J Quantum Chem* 105 (4):302-312. 1097-461X.

Viswanathan, R., A. Asensio, and J. J. Dannenberg. 2004. Cooperative Hydrogen-Bonding in Models of Antiparallel b-Sheets. *J Phys Chem A* 108 (42):9205-9212. 1089-5639.

Vitalis, A., N. Lyle, and R. V. Pappu. 2009. Thermodynamics of ≤-Sheet Formation in Polyglutamine. *Biophys J* 97 (1):303-311. 0006-3495.

Wanker, E E. 2000. Protein aggregation and pathogenesis of Huntington's disease: mechanisms and correlations. *Biol Chem* 381 (9-10):937-942. 1431-6730.

Wieczorek, R., and J. J. Dannenberg. 2003. H-bonding cooperativity and energetics of alpha-helix formation of five 17-amino acid peptides. *J Am Chem Soc* 125 (27):8124-9. 0002-7863 (Print).

Williamson, T. E., An. Vitalis, S. L. Crick, and R. V. Pappu. 2010. Modulation of Polyglutamine Conformations and Dimer Formation by the N-Terminus of Huntingtin. *J Mol Biol* 396 (5):1295-1309. 0022-2836.

Zanuy, D., K. Gunasekaran, A. M. Lesk, and R. Nussinov. 2006. Computational study of the fibril organization of polyglutamine repeats reveals a common motif identified in beta-helices. *J Mol Biol* 358 (1):330-345. 0022-2836.

Zhao, Y. L., and Y. D. Wu. 2002. A theoretical study of beta-sheet models: is the formation of hydrogen-bond networks cooperative? *J Am Chem Soc* 124 (8):1570-1. 0002-7863.

Zhou, Z., J. Zhao, H. Liu, J. W. Wu, K. Liu, C. Chuang, W. Tsai, and Y. Ho. 2011. The Possible Structural Models for Polyglutamine Aggregation: A Molecular Dynamics Simulations Study. *J Biomol Struct Dyn* 28 (5):743-758. 0739-1102.

Zoghbi, H. Y., and H. T. Orr. 2000. Glutamine repeats and neurodegeneration. *Annu Rev Neurosci* 23:217-247.

Zuccato, C, M Valenza, and E. Cattaneo. 2010. Molecular Mechanisms and Potential Therapeutical Targets in Huntington's Disease. *Phys Rev* 90 (3):905-981.

Part 2

Neuropathological Mechanisms and Biomarkers in Huntington's Disease

4

Biomarkers for Huntington's Disease

Jan Kobal[1], Luca Lovrečič[2] and Borut Peterlin[2]
[1]*University Medical Center Ljubljana and University Psychiatric Hospital Ljubljana, Department of Neurology,*
[2]*University Medical Center Ljubljana, Department of Obstetrics and Gynecology, Slovenia*

1. Introduction

The core clinical features of Huntington's Disease (HD) were outlined by George Huntington in 1872 (Huntington 1872). Like nowadays , in George Huntington's time no cure for HD was yet available. However, genetic testing for HD that is now available can reliably predict the individuals at risk that will develop the disease. In such premanifest individuals slowing down the disease process may potentially delay the onset of disease symptoms. Therefore, there is an increasing need of finding the markers for the disease progression in premanifest HD individuals.

A biomarker is defined as an attribute of the disease that is objectively measured and evaluated as an indicator of normal biological processes, pathogenic processes or pharmacological response to a therapeutic intervention (Biomarkers definitions working group 2001). In adult HD mouse it has been demonstrated that stopping the expression of mutant Huntingtin may reverse the clinical and pathological phenotype (Yamamoto et al 2000). However, treatment trials expected to modify disease progression remain confined to population of manifest HD patients until reliable markers of disease process progression can be found for the premanifest HD gene carriers. Clinical measures may be used as primary endpoints and we will first focus on them. In our opinion, a comprehensive neurological and physical examination of premanifest HD gene carriers represents a reliable way towards identification of potential clinical biomarkers.

2. Clinical biomarkers for HD

A broad consensus exists among clinicians that a clinical diagnosis of HD can be made with certainty only in the presence of specific motor disorders. Thus, fixing the onset of the motor disorder in this way is a more or less reproducible method to conduct age at onset surveys or genotype-phenotype correlation studies (Kremer 2002). The most complete technique of assessing the early signs and symptoms of HD is to follow up a cohort of at risk individuals for an extended period of time. The most instructive follow-up study continues to be the one of the Venezuelan HD kindred (Penney et al 1990). It was performed prior to identification of the gene; nevertheless its conclusions are still valid. It demonstrated that patients pass through a transitional state from the normal presymptomatic phase to the time at which the diagnosis can clearly be made on neurological examination. The study revealed that there was no single presenting sign or symptom in HD. In the earliest phases there was an insidious and slow deterioration of intellectual functions as well as mild personality change. The clear appearance of extrapyramidal signs such as chorea, hypokinesia, rigidity or dystonia indicates a phase on the disease progression, not the beginning of the disease. Prior to these signs however, most individuals will display minor motor abnormalities (Penney et al 1990). These minor abnormalities include general restlessness, abnormal eye movements, or impaired optokinetic nystagmus, hyperreflexia, impaired finger tapping or rapid alternating hand movements, and excessive and inappropriate movements of the fingers, hands, or toes during emotional stress as well as mild dysarthria. Minor abnormalities usually precede the obvious signs of extrapyramidal dysfunction by at least 3 years. Persons with a completely normal neurological examination have only a 3 per cent chance of being diagnosed as clinically manifest HD patients within the next 3 years (Penney et al 1990). A retrospective assessment of the affected individuals has revealed that minor involuntary movements are among the earliest symptoms experienced and that soon by those mental and emotional symptoms, including sadness, depression, irritability, and episodes of verbal and physical abuse may develop (Kirkwood et al 2001). Various research groups have revealed that so-called asymptomatic gene carriers statistically display subtle cognitive defects; such subtle cognitive deficits may precede motor abnormalities by years (Campodonico et al 1996, Lawrence et al 1998). However, it is important to realize that individuals with expanded repeats due to HD mutation may perform just as well or better than matched controls. Only when an individual is close to the estimated age of onset, as predicted by cytosine-adenine-guanine (CAG) repeat length (Brinkman et al 1997) that minor deficits in selected cognitive domain may become apparent (Campodonico et al 1996).

2.1 Unified Huntington's Disease rating scale

Clinical biomarkers are standardised clinical tests and rating scales that measure progression of HD phenotype. In order to provide a comprehensive assessment of motor performance, cognitive functioning, behavioral and psychiatric problems and functional status of an individual the United Huntington's Disease Rating Scale (UHDRS) was developed by the Huntington Study Group (Huntington Study group 1996). It enables a comprehensive, rapid, and efficient survey that is highly sensitive to disease progression over relatively short periods of time, such as 1 year. Using the UHDRS clinical score subtle motor abnormalities were found in premanifest HD subjects and were increasing with the proximity of the predicted time of clinical diagnosis (Biglan et al 2009). Although UHDRS is a standard assessment of disease progression it does not encompass every possible manifestation of HD. Special techniques have been developed to detect subtle premanifest clinical abnormalities that may lead to the development of new potential clinical biomarkers. HD progression may additionally be tracked by clinical techniques of oculomotor assessment (Klöppel et al 2008), tapping test (Andrich et al 2007), and gait analysis (Rao et al 2005).

2.2 Cognitive impairment

Subtle cognitive changes are present already in presymptomatic gene carriers (Kirkwood et al 2001, Craufurd & Snowden 2002); they become evident close to onset and early in the course of the disease and grow to be more severe as the disease evolves (Campodonico et al 1996; Brandt & Butters 1986). Cognitive changes therefore have the potential to identify premanifest HD gene carriers close to the onset of the disease. Asides to Clinical psychological tests encompassed in the UHDRS (Verbal fluency, Symbol digit and Stroop test) other neuropsychological test batteries may be used for the purpose. However, the natural history of HD-related cognitive impairment is still not completely understood. Executive tests, combined with neuroimaging techniques have provided new evidence of cognitive abnormalities in HD; abnormal connectivity between basal ganglia and cortical areas has been suggested (Montoya et al 2006).

3. Positron emission tomography

Prior to HD gene identification the transitional state in HD development was proven to be accompanied by changes in metabolic rates of glucose as seen on positron emission tomography (PET) (Grafton et al 1992). After identification of the HD gene Huntington's disease Collaborative Research Group 1993) longitudinal follow up studies of identified presymptomatic gene carriers were started. Using serial 11C-SCH 23390 and 11C-raclopride PET striatal dopamine D1 and D2 receptor binding was followed in a group of HD gene carriers of which 4 were in transitional state (Andrews et al 1999). The affected subjects showed mean annual reductions of 5.0 and 3.0 per cent loss of striatal dopamine D1 an D2 binding, respectively, while presymptomatic HD gene carriers showed mean annual reductions of 2.0 and 4.0 per cent, respectively. In mutation negative group no loss of dopamine binding was detected. The rate of loss of striatal dopamine D2 receptors correlated with CAG repeat length in presymptomatic HD gene carriers. Longitudinal studies have shown a mean annual decrease in dopamine D2 receptor binding of 5-6 per cent in HD patients and of around 4 per cent in premanifest HD gene carriers (Pavese et al 2003) Microglial activation was observed in the striatum of both HD patients and presymptomatic HD gene carriers by reduced binding of 11C-raclopride (Pavese et al 2006). The correlation with probability of time of onset was also shown in presymptomatic HD. (Tai et al 2007). Two-stage PET scanning method was applied to a cohort of presymptomatic and symptomatic HD individuals; this technique enables better visualization of anatomic structures and might potentially serve as a useful biomarker in the future. (Tomasi et al 2011).

PET scanning therefore shows promise for early visualization and quantification of pathological abnormalities in HD and therefore may be helpful in finding new potential biomarkers. There are however a number of weaknesses which limit usefulness of this technique. The cost is high and availability limited, scanning is time-consuming, radioactive ligands are difficult to manipulate. PET scanning also is susceptible to neuroleptic abuse which is common in HD patients.

4. Magnetic resonance imaging

Volumetric magnetic resonance imaging (MRI) enables estimation of brain region volumes. T1 volumetric MRI is the standard MRI technique most often used also in HD; however, other standard MRI techniques may provide useful information as well.

Longitudinal studies have shown significantly faster brain atrophy in early HD patients (Aylard et al 1997) and in presymptomatic gene carriers as far as 11 years from the predicted onset (Aylard et al 2004). Longitudinal assessment of striatal volumes thus seems to hold capacity of providing potential biomarkers. The use of T1 weighted combined to diffusion-weighted scans seem to provide good information about the nature, and topographic specificity of brain changes in pre-HD individuals (Stoffers et al 2010). Basal ganglia are parts of the brain that are most affected by atrophy in HD patients, however, atrophy of other parts of the brain also takes place early in the course of the disease. Measurements of larger brain volumes may thus be more precise and less susceptible to local changes. Quantitatively, most of pathology in HD is extrastriatal and relative contributions to disease manifestation by striatal atrophy are not known. Without effective treatment techniques it is not possible to validate whether change in MRI striatal volumes can serve as an effective surrogate endpoint (Aylard 2007) Also, basal ganglia are closely interconnected to many parts of human brain and their atrophy may be contributed to different clinical pathology.

4.1 Brain volumes measurements

Using a semi-automated MRI volumetric technique Rosas et al proved that numerous extrastriatal brain areas are atrophied (Rosas et al 2002). Using an automated MRI technique they further managed to demonstrate regional cortical thinning in early HD patients (Rosas et al 2003). In premanifest HD gene carriers selective thinning of cortical parts was found that correlated positively with changes in cognition measured by the cognitive part of UHDRS (Rosas et al 2005). Further analyses revealed a significant association between regional cortical thinning and total functional capacity which is the leading primary outcome measure in neuroprotection trials (Rosas et al 2008). Progression of HD was evaluated by a longitudinal follow up volumetric MRI analysis and efficient measurement of the volume changes was performed within 15 years from the estimated onset of the clinical disease (Aylard et al 2011).

The boundary shift integral (BSI) is a semi- automated method by which changes in the brain volume can be calculated from registered 2-year interval scan pairs. Using BSI, Wild et al have demonstrated that whole-brain atrophy was significantly faster in early HD patients than in control subjects, and accelerated atrophy during the course of the disease was noted (Wild et al 2010).

Voxel-Based Morphometry (VBM) is an automated technique for analysis of series of MRI scans. VBM identifies variably affected brain regions in different stages of HD (Kassubek et al 2004) Longitudinal studies using this technique are promising (Tabrizi et al 2011).

The aforementioned MRI techniques are potentially useful in identifying the regions that may serve as biomarkers of disease progression in prevention trials.

4.2 Functional MRI

Functional MRI (fMRI) identifies subtle changes in regional blood flow during increased neuronal activity to identify brain regions active during performance of a specific task. Early abnormalities due to neuronal dysfunction can be detected. Neuronal dysfunction in early phase of the disease is potentially reversible which increases value of this technique. Thus, fMRI as a functional technique may reveal early functional pathology and may not require longitudinal measurements like morphometric methods. Several fMRI studies have demonstrated regional functional abnormalities in early HD (Georgiu-Karistanis et al 2007). A study conducted in presymptomatic HD gene carriers alterations in cortical functional activity have been shown to correlate with the time of onset (Paulsen et al 2004) A study comparing data obtained from volumetric MRI and fMRI found that regions with altered activity were not those experiencing the most atrophy (Gavazzi et al 2007) While fMRI technique may represent a useful biomarker there are also weak points for its general use. Technical equipment is more demanding than conventional MRI and expertise of the technique is required.

4.3 Molecular MRI techniques

Diffusion tensor imaging (DTI) is an MRI technique developed from standard diffusion weight imaging (DWI) technique which applies the ability of water molecules to diffuse along axons and produces maps of white matter tracts. It can detect abnormalities in myelin which would appear normal on conventional MRI. In HD, the regions of decreased fractional isotropy (FI, measure of axonal organization) compared to controls were detected. Rosas et al found the regions that correlated to cognitive performance in presymptomatic HD gene carriers; more widespread lesions were detected in manifest HD (Rosas et al 2008). DTI shows promise to become a biomarker capable of detecting changes in HD earlier than other imaging techniques (Magnotta et al 2009) although not many studies have been performed and its potential remains to be tested.

MR spectroscopy is capable of noninvasive quantification of the biochemical composition of brain tissue. Lower neuronal markers (N-acetylaspartate) levels were shown in presymptomatic and early HD, whereas glial cell markers (myo-inositol) were increased (Surrock et al 2010). Elevated lactate and reduced creatine levels were shown in the striatum of presymptomatic HD gene carriers and early HD patients (Reynolds et al 2005). The technique is capable of detecting biochemical changes in the central nervous system and promises to be helpful in a potential biomarker discovery; its utility, however, is limited by long scan times, small number of molecules it can accurately detect, and comparatively low sensitivity.

5. Molecular biomarkers

Various molecular biomarkers can also be obtained from peripheral blood, urine and cerebrospinal fluid (CSF). Ideally, biomarkers obtained from body fluids would be expected to reflect pathologic changes in CNS. Such a substance is normally not present in the blood, but in HD gene carriers/patients it leaks across blood-brain barrier and becomes detectable. However, mutant huntingtin is expressed ubiquitously over all body tissues, therefore molecular changes detected in body fluids may reflect peripheral processes promoted by mutant huntingtin. In this way, biomarkers obtained from CSF could reflect CNS pathology more precisely (Huang et al 2011).

Candidate biomarkers obtained from body fluids can be divided in metabolic, endocrine, markers of oxidative stress, and markers obtained from signalling pathways.

5.1 Metabolic biomarkers

Due to ubiquitous expression of huntingtin in addition to neurological features peripheral deficits may be detected. HD-associated differences in metabolite levels in peripheral blood were detected by Underwood et al that identified a pro-catabolic pattern of metabolic changes, present even in presymptomatic HD gene carriers (Underwood et al 2006). Another research group found decreasing levels of branched chain amino acids in presymptomatic HD gene carriers and clinically manifest HD patients in different stages of the disease compared to controls. The levels were found to correlate with CAG repeat length and UHDRS motor score (Mochel et al 2007). Uric acid, a known antioxidant agent that was found to be connected with the progression of Parkinson's disease has been investigated as a putative biomarker/modifiable agent that could slow down HD progression (Auinger et al 2010).

5.2 Endocrine biomarkers

Several features of early HD like weight loss, depression, disturbed sleep cycle could be due to hypothalamic dysfunction. Undeniably, loss of hypothalamic cells has been found in HD patients (Petersen et al 2006). Endocrine disturbances that may track disease progression have been identified. Urinary cortisol levels increase progressively with the advancing disease in HD patients (Bjorquist et al 2006). Still other potential endocrine biomarkers are under investigation in clinically expresed HD (Hult et al 2010).

Endocrine changes are of interest as potential biomarkers to track disease, yet endocrine features are susceptible to influence of drugs such as neuroleptics and antidepressants, and psychiatric pathology such as depression which may occur in early HD.

5.3 Oxidative stress biomarkers

Mitochondrial dysfunction has recently been shown in HD patients and presymptomatic HD gene carriers. (Saft et al 2005). Other markers of oxidative stress and metabolism are under investigation (Chen 2011). Mutant huntingtin and its cleavage products as the immediate cause of neuronal dysfunction and death in HD are being investigated as potential biomarkers (Moskovitch-Lopatin 2010).

5.4 Signalling pathways biomarkers

A significant decrease in brain-derived neurotrophic factor (BDNF), an agent that promotes survival of neurons was found in the serum of symptomatic HD patients (Ciammmola et al 2007). Augmentation of neurotrophic gene products such as BDNF could present a potential therapeutic target in HD (Ross & Shoulson 2009). BDNF is an interesting potential biomarker of disease progression, however, it does not cross the blood-brain barrier. The balance of central and peripheral contributions to altered serum BDNF in HD requires further study.

Abnormalities of the endocannabinoid system were observed in premanifest HD gene carriers as well as in manifest HD patients (Fernanadez-Ruiz et al 2009). Adenosine A2 receptors were found to increased density and affinity in different stages of manifest HD as well as in presymptomatic HD gene carriers (Varani et al 2007).

6. Autonomic nervous system function as a putative biomarker in HD

Our research started with the study of autonomic nervous system function (ANS) in presymptomatic HD gene carriers and symptomatic HD patients. Based on a standardized clinical ANS questionnaire (Turkka 1987) a group of 33 patients was enrolled, among them 8 presymptomatic HD gene carriers. Symptomatic patients were classified according to the

Shoulsoh and Fahn's HD disability scale (Shoulson & Fahn 1979) to mildly affected group and moderately /severely affected which were evaluated together. Mostly, an increase in the ANS function , especially of the sympathetic part, was observed in presymptomatic HD gene carriers and mildly affected HD patients, and a decrease in the ANS function in moderately/severely affected patients was observed (Kobal et al 2004). Our further research was modified by observing a presymptomatic HD gene carrier in whom choreatic movements appeared after suffering from chronic subdural hematoma that compromised the cerebral cortex, but not the basal ganglia (Kobal et al 2007) (Fig. 1).

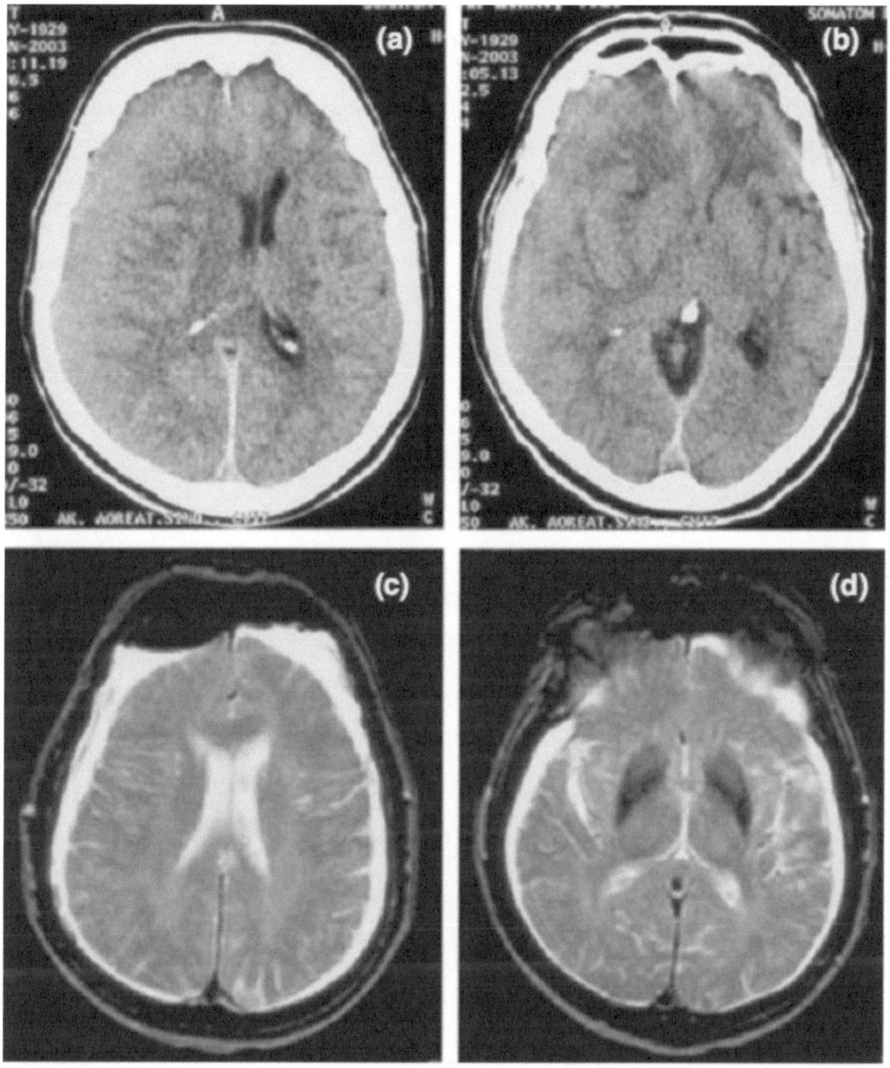

Fig. 1. Computed tomography scan of the head in a patient with chorea discovering isodense bilateral chronic hematomas expanding over the entire right hemisphere and over the left parietooccipitotemporal cortex (a, b). Proton density weighted Magnetic resonance imaging 6 days after surgical evacuation and after reappearance of the initially regressed chorea showed no structural abnormality in the basal ganglia (c, d).

We hypothesized that early autonomic dysfunction could be due to imbalance in the central ANS centres and conducted further research in this direction. In the next study we enlarged the number of presymptomatic HD gene carriers and early manifest HD patients, which were clinically evaluated by UHDRS clinical scale. ANS tests to challenge higher-order ANS centres like mental stress and the cold pressor test were introduced. Attenuated response to simple mental arithmetic test was shown in a group of 14 presymptomatic HD gene carriers and 11 early symptomatic HD patients (Fig. 2). The response to late phase of cold pressor test in the same patients was exaggerated (Fig. 3).

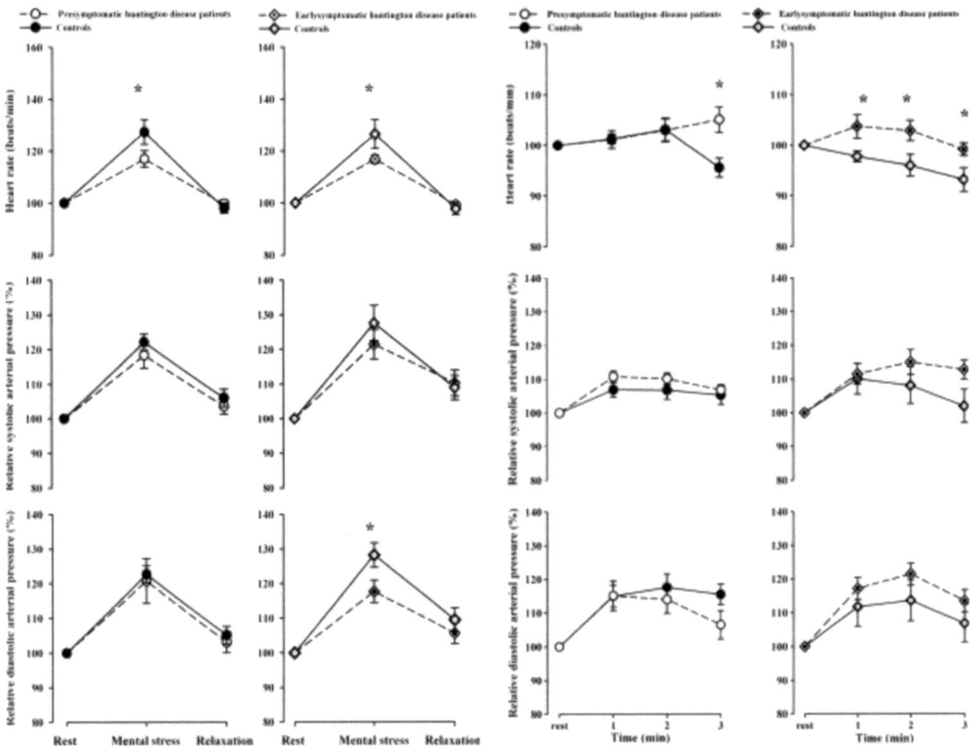

Fig. 2. (left): Arterial pressure and heart rate values during a simple mental arithmetic test. Values are expressed as percentage of resting arterial pressure and heart beat rates. * statistically significant differences between the groups (p<0.05). Fig. 2. (right): Heart rate and arterial pressure values during cold pressure test. Values are expressed as percentage of resting heart rate and arterial pressure values. * statistically significant differences between the groups (p<0.05).

The results were in favour of highest-order cortical ANS centres hypofunction which, according to the concept of central autonomic network organization, could lead to hyperfunction of hypothalamus and lower order central autonomic centres (Kobal et al 2010, Melik et al-in print). Our findings were in line with the findings of a recent study on thalamic metabolism in preclinical HD. Thalamic metabolism was elevated at baseline, but fell to subnormal levels in the pre-HD subjects who developed symptoms (Feigin et al 2007). A recent survey found significantly more gastrointestinal, urinary cardiovascular and sexual problems in group of HD patients. In premanifest HD group swallowing problems and light-headness on standing up were prominent (Aziz et al 2010)

The ANS function could potentially represent a useful biomarker in HD however, further cross-sectional as well as longitudinal studies are needed. Drawbacks of these methods are that they are unspecific; they also may show variable intersubject response and are sensitive to use of drugs with anticholinergic effect such as neuroleptics which are commonly used in HD patients.

7. "Omic" biomarkers for HD

7.1 Use of "Omic" biomarkers in clinical practice

Complete sequencing of the human genome has launched a new era of systems biology referred to as »omics«. The term refers to the comprehensive analysis of biological systems and a variety of omics subdisciplines are acknowledged. Through genomics new approaches to monitor diseases are becoming available. New technologies are capable of defining large sets of biomarkers systematically in biological samples (Bell 2004), and provide an analytical approach to investigation of all the products of the genome at messenger RNA or protein level at once. These methodologies are capable of generating data on multiple biomarkers that vary quantitatively very early in the disease, in response to disease onset, progression or therapeutic intervention and may provide sets of prognostic factors (Schadt et al 2003). The development of biomarkers for prognostic use in diseases with asymptomatic phases is particularly challenging and can be time-consuming, as they must be validated and monitored in long-term clinical outcomes (Frank & Hargreaves 2003).

Microarray analysis has significantly augmented the throughput of genomic studies and haemogenomic approach has been proposed; several examples of potential microarray-based biomarkers in blood have already been described. Peripheral blood is an easily accessible tissue, and specific gene expression signatures have been shown to exist in a wide variety of diseases where no obvious clinical phenotype in blood is present, such as tuberous sclerosis, neurofibromatosis, Down syndrome, multiple sclerosis, etc (Tang et al 2004, Bomprezzi et al 2003, Achiron et al 2004).

Another two important goals of genomics and genetics in clinical practice, besides diagnosis and staging, are to improve therapeutic efficacy and reduce drug toxicity (Evans & Relling 2004). The field of pharmacogenomics encircles the role of genes in an individual's response to drugs and comprises a broad area of basic drug discovery research, the genetic basis of drug responses, pharmacodynamics, pharmacokinetics and metabolism. The implications for the development of new drugs and clinical patient management are huge. For example, in multiple sclerosis it has already been shown that differences in the gene expression profiles of treatment-responsive and treatment-non-responsive patients are present and detectable (Sturzebecher et al 2003). In cancer therapy, several studies have been published where expression signatures indicated the response to certain treatment (Glinsky et al 2005, Rosenwald et al 2002). In neurodegenerative diseases like HD, where no effective therapies are available and symptoms progression is relatively slow, biomarkers for therapy response monitoring are important.

7.2 Transcriptomic research in HD

The research on HD has mainly been focused on the nervous system, only few studies have reported on muscle or other tissues. Our previous results show (Borovecki et al 2005) that expression changes of many genes were present and detectable in blood of HD patients when compared to healthy controls. Not only were these changes present in HD patients with clinical symptoms but disturbances in gene expression were detected in presymptomatic mutation carriers as well (Figure 3). The analysis of gene expression changes was performed on 2 different microarray platforms (Affymetrix, Amersham) in 12 symptomatic patients and 10 healthy controls, as well as in 5 presymptomatic mutation carriers and 4 healthy controls. Ten times more probes were differentially expressed in symptomatic HD group than in presymptomatic HD group when compared to controls. Interestingly, Amersham detected up to 4 times more probes as differentially expressed in the symptomatic group than Affymetrix microarrays, whereas this was not the case in the presymptomatic group (Table 1).

Based on this study, in which we have shown that gene expression changes in HD are detectable in blood of HD patients, research in the field of disease progression and novel therapy response in human HD may focus on this easily accessible tissue as well. Further work in the proposed direction is needed to provide clues to these implications.

	SYMPTOMATIC GROUP		**PRESYMPTOMATIC GROUP**	
***p*-value**	**Affymetrix**	**Amersham**	**Affymetrix**	**Amersham**
0.05	5267	12159	1369	939
0.01	3133	9685	392	490
0.005	2546	8678	223	314
0.001	1646	6579	58	75
0.0005	1366	5815	31	42
0.0001	884	4191	3	8
0.00005	740	3599	2	6

Table 1. Numbers of differentially expressed probes between HD group and healthy controls. Numbers of changed probes are shown for each microarray platform separately with respect to different *p*-values.

Only one study on gene expression changes in HD in human brain samples has been reported so far (Hodges et al 2006). The expression in three distinct brain regions from symptomatic HD patients was analyzed. We compared those with our expression results in blood of symptomatic HD patients on Affymetrix platform to compare expression changes in brain and blood (Table 2, Fig. 3). When using the same statistical measures ($p<0.001$) the greatest expression changes were observed in the caudate nucleus, followed by blood > BA4 cortex > cerebellum. 30% of probe sets changed in blood were also significantly changed in caudate samples suggesting that HD specific changes might be detectable in blood. There was not a single probe set differentially expressed in all four tested tissues and 47 probe sets were significantly changed in blood, caudate and BA4 motor cortex, two of the more affected areas of brain in HD. These findings imply that similar cellular processes are disturbed in the caudate nucleus and blood cells, although there is no clinical phenotype in blood of HD patients. The latter may be due to the fact that unlike neurons, the life span of lymphocytes is short and the turn over rapid.

	Blood	Caudate	BA4 Cortex	Cerebellum
Number of changed probe sets	1646	5225	963	340

Table 2. Numbers of probe sets differentially expressed in blood and three brain regions in symptomatic HD patients ($p<0.001$).

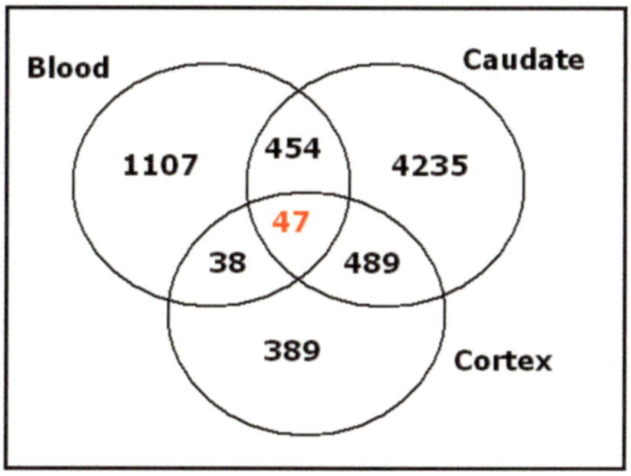

Fig. 3. Numbers of over-lapping differentially expressed probe sets in blood, caudate and BA4 cortex ($p<0.001$).

7.3 Age-at-onset prediction

Since the gene expression changes were present already in presymptomatic stages, expression profiles might be used to refine the prediction of disease onset. There is a long time period before HD manifests itself through clinical symptoms. Due to available mutational testing, one can learn earlier about his/her gene mutation status. After the diagnosis, prediction of age at disease onset is most important information for mutation carriers and their relatives. Two ways to speculate about the age of onset have been accepted so far, both quite insensitive - number of CAG repeats and polymorphisms of modifier genes. Age of onset may presently only be given in a rather wide range - for example a mutation carrier with 42 CAG repeats will most probably develop symptoms between 35-57 years of age. Clearly, this is of no use to someone to plan the future. Brain-imaging studies have been trying to add some sensitivity to the prediction of disease onset. Study of basal ganglia volume showed that atrophy of basal ganglia occured gradually, beginning years before symptoms onset (Aylard 2007). Mutation carriers who were close to the onset of HD as predicted by CAG repeat numbers had smaller volumes of basal ganglia than subjects far from onset for all structures except the caudate nucleus. Mutation carriers who were far from the onset had smaller basal ganglia volumes than healthy controls for all structures except the globus pallidus. A functional MR imaging (fMRI) study showed differences in the groups of mutation carriers far or close to the predicted age of onset when compared to healthy controls (Paulsen et al 2004). The group close to the onset had significantly less activation in subcortical regions than control subjects and the group far from the predicted onset had an intermediate degree of activation. Despite the mentioned findings there have not been any definite measurements or protocols for determining the age of onset proposed so far.

Our results suggest that gene expression in blood of HD mutation carriers is disturbed long before the onset of symptoms (some of tested mutation carriers were as young as 20 years with CAG repeat lengths of 41, suggesting the start of the disease between 40-50 years of age). These results imply that gene expression changes in blood might potentially be used not only to monitor the disease progression but to help predict the age of disease onset as well. Also, expression changes correlated with the disease progression in the symptomatic stages of HD, and they might be valuable in determining the progression of specific symptoms in the advanced stages.

7.4 Potential biomarkers of disease progression

Whole genome transcriptome analysis might define also biomarkers for disease progression. Using whole genome gene expression data from our study (Borovecki et al 2005) we have performed additional analysis to select a potential biomarker set - set of genes to be useful as a biomarker and test their expression with another independent method, quantitative RTPCR (QRT-PCR). To narrow down the list of differentially expressed genes, additional criteria for selecting genes of interest were implemented. We have selected top 12 candidate genes that had the best reproducibility of expression changes when tested with QRT-PCR in presymptomatic/symptomatic HD patients and healthy controls.

To make a study more stringent we validated the 12 gene set on an independent set of HD samples and controls (Fig. 4). Expression of individual genes increased with disease progression from the presymptomatic to advanced symptomatic stage, but the differences did not reach statistical significance. While only 6 genes, ANXA, MARCH7, CAPZA1, HIF1A, TAF7 and YPEL5, were significantly upregulated in the presymptomatic (P) group ($p<0.05$), 10 genes were significantly upregulated in the symptomatic (S) group ($p<0.05$) (PCNP and SF3B1 were not significant) and 11 genes were significantly upregulated in the late symptomatic (LS) group ($p<0.05$) (SF3B1 was not significant).

This study provided confirmatory evidence of significant gene expression changes in blood of HD patients published by Borovecki et al. Expression of the 12 genes appeared higher in the advanced symptomatic group of patients compared to the presymptomatic group, but these stage-dependent differences in expression did not reach statistical significance. In order to investigate predictive performance of the gene set, we examined logistic regression machine learning algorithm on our dataset. Proposed classifier reached overall positive predictive value of 78% with 82% sensitivity and 53% specificity for HD with respect to healthy control. In addition, the potential of gene set to discriminate between presymptomatic and symptomatic patients was evaluated using the logistic regression algorithm. The results showed overall positive predictive value of 85% with relatively high sensitivity (83%), but with low specificity (50%). A possible explanation for low specificity may be the unequal distribution of cases in our dataset (14 presymptomatic and 47 symptomatic cases) and small set of training cases (Lovrecic et al 2009).

While high specificity and sensitivity are generally desirable for diagnostic biomarkers, these parameters are not essential in diseases such as HD where the diagnosis is already known and the intended use of biomarkers is to primarily monitor disease progression. As a potential marker of disease progression, the 12-gene set showed promising overall positive predictive value and sensitivity (85% and 83%, respectively), but with relatively low specificity (50%). Nevertheless, our results suggest that the 12-gene set may be of better clinical value compared to individual genes as a marker of disease progression in HD. Moreover, we hypothesize that including more altered genes in the gene set may further enhance its clinical applicability.

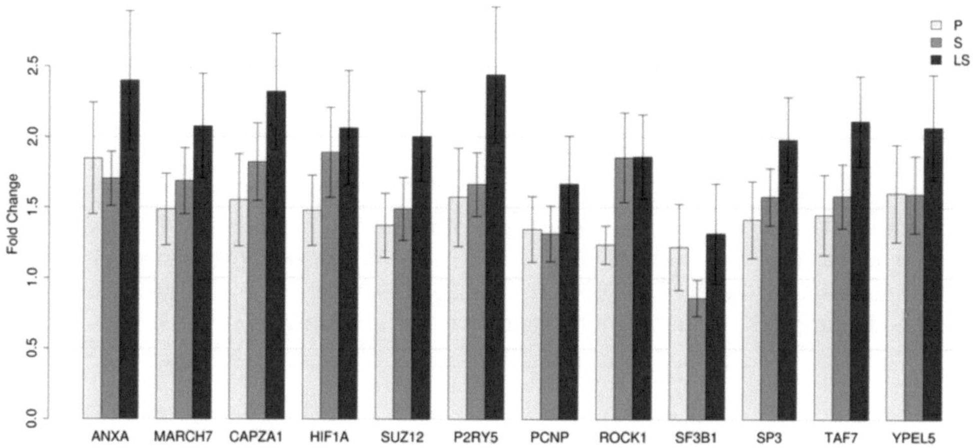

Fig. 4. Expression fold changes of 12 genes in different stages of HD.

The upregulation of expression of the 12 previously selected genes (8) was validated in the new cohort of HD patients. Bars represent fold increase in mRNAs in HD patients relative to healthy controls. Interval lines represent the (average fold change) x (2^{SEM}-1). P-presymptomatic HD mutation carriers, S-symptomatic patients, LS-late symptomatic patients: ANXA-annexin A1; *MARCH7*-membrane-associated ring finger 7; *CAPZA1*-capping protein muscle Z-line, alpha 1; ***HIF1A***-hypoxia-inducible factor 1, alpha subunit; SUZ12-suppressor of zeste 12 homolog; P2RY5-purinergic receptor P2Y, G-protein coupled, 5; PCNP-PEST proteolytic signal containing nuclear protein; ROCK1-Rho-associated, coiled-coil containing protein kinase 1; SF3B1-splicing factor 3b, subunit 1; SP3-Sp3 transcription factor; TAF7-TAF7 RNA polymerase II, TATA box binding protein (TBP)-associated factor; YPEL5-Yippee-like 5.

In addition, another study of global gene expression in lymphoblastic cell lines from HD patients failed to identify any significant changes in gene expression (Runne et al 2007) that were observed previously (Borovecki et al 2005). It is therefore evident that multiple independent validation studies will be required to evaluate potential clinical applicability of a putative biomarker. While development of novel hemogenomic approaches to non-invasively monitor disease progression showed promise, it remains unclear whether the observed changes in blood gene expression will be sufficiently robust to serve as biomarkers of disease. A combination of genomic, metabolomic and proteomic approaches may be required, in combination with neuroimaging, to successfully identify biomarkers of disease progression in HD and probably other neurodegenerative diseases.

7.5 Elucidation of pathophysiological processes in HD

Analysis of whole genome transcriptome might also give us insights into the disturbed pathways and processes involved in disease onset and progression. Although multiple pathological mechanisms by which mutant htt causes neuronal dysfunction have been proposed and studied in detail (Harjes & Wanker 2003), the exact molecular mechanisms how mutant htt induces cell death are not understood.

Using whole genome gene expression data from our study (Borovecki et al 2005) we have performed additional bioinformatic analysis of microarray data with freely available Onto-Tools (Draghici et al 2003) and Gene set enrichment analysis (GSEA) (Subramanian et al 2005) software. Three separate comparisons were done: 1) all HD samples compared to healthy control samples (HDvsC); 2) symptomatic HD samples compared to healthy control samples (SvsC); 3) presymptomatic HD samples compared to healthy control samples (PvsC). When looking for enriched gene set with GSEA, additional comparison was done - symptomatic HD samples compared to presymptomatic HD samples (SvsP). Onto- Express was used to more thoroughly characterize the sets of functionally related differentially expressed genes (Draghici et al 2003, Khatri et al 2002). The tool classified genes according to two Gene-Ontology (GO) categories: biological process and molecular function (Fig. 5).

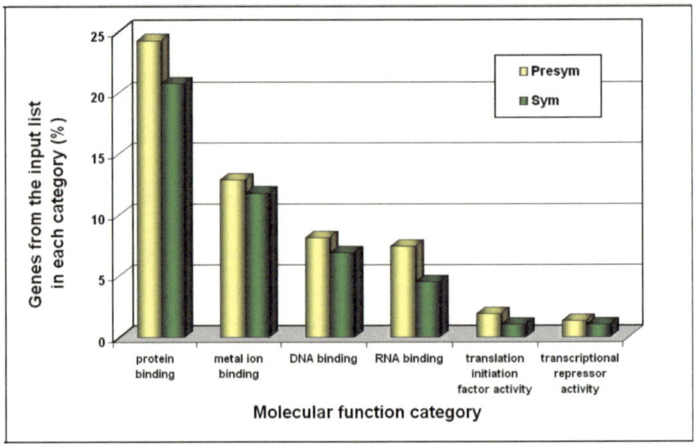

Fig. 5. Common molecular function categories changed in presymptomatic (Presym) HD mutation carriers and symptomatic (Sym) HD patients.

Molecular function category	Category rank*	Genes in category (%)
Structural constituent of ribosome	5	2.9
Unfolded protein binding	6	2.6
GTPase activity	7	2.3
Protein transporter activity	9	1.9
Protein heterodimerization activity	10	1.6
RNA polymerase II transcription factor activity	12	1.3
Hydrogen-transporting ATPase activity	13	1.3
Translation elongation factor activity	14	1
Helicase activity	15	1
Ubiquitin conjugating enzyme activity	16	1

* Category rank from PvsC comparison.

Table 3. Molecular function categories specifically changed only in presymptomatic HD mutation carriers.

Using Onto-tools we may propose that two novel mechanisms are disturbed in early stages of HD. Molecular function categories "metal ion binding" and "helicase activity" have been shown to be disrupted already in presymptomatic HD mutation carriers (Fig. 5, Table 3). Another interesting finding was that many more molecular function and biological process categories were disturbed at the gene expression level in presymptomatic than in symptomatic group (Table 3).

Using another method, Gene set enrichment analysis (GSEA) two additional mechanisms in terms of gene sets were significantly upregulated in the presymptomatic group - lipid metabolism with adipocyte function (Nadler et al 2000) and gene set linked to expression changes in major depressive disorder (Aston et al 2005). Also, our results have confirmed most of the previously described potential pathogenetic mechanisms to be disturbed at gene expression level using two completely different approaches.

Many hypotheses on HD pathogenesis have been investigated, but none has been able to decipher the basis of what goes wrong first. Since HD primarily affects the brain, majority of the research on the pathogenesis has been done on neuronal cells or tissue. We used a different approach in two aspects - we included presymptomatic mutation carriers that gave us an insight into the early changes in HD, and our analyses were done on blood cells which appear not to be affected in HD. Possibly, if their life span were longer, as is the case with neuronal cells, blood cells would also become affected. One study reported that lymphoblasts isolated from HD patients showed increased stress-induced apoptotic cell death, suggesting their abnormal function, but no apparent clinical phenotype was found present (Sawa et al 1999). These are more reasons to believe that the changes present in blood cells are early changes characteristic of HD. Moreover, the analysis of gene expression changes in presymptomatic mutation carriers separately might lead to explanation of some primarily disturbed mechanisms specific for HD.

Metal ion binding category was disturbed already in presymptomatic disease stage and only scarce data are currently available on metal ions in neurodegenerative diseases affecting basal ganglia (Dexter et al 1991, Moos & Morgan 2004). The results of previous studies have suggested that metal ions might contribute to neurodegenerative process. More studies are needed to elucidate the importance of metal ions in HD, but our results suggest that related mechanisms are disturbed already in presymptomatic disease stages. Interestingly, the analysis of molecular function and biological process GO categories have shown that many more categories are changed specifically in presymptomatic HD stages implying that many processes are active and changed in comparison to healthy controls before the onset of clinical symptoms. Since they are not present in symptomatic stages of HD they might exhibit the measures that cells are undertaking to counteract to mutation driven disturbances and efforts to execute normal processes appropriately.

In addition, expression results from our study showed that gene set consisting of genes controlling lipid metabolism and signal transduction (Nadler et al 2000) was specifically changed in the group of presmyptomatic HD mutation carriers. Disturbed lipid metabolism and an adipocyte function have been previously reported in R6/2 mice, where a defect in fat breakdown by adipocytes was suggested (Fain et al 2001). No results on human samples have been available so far. Our results suggest that this might be one of the mechanisms disturbed early in the human HD pathogenesis. The second gene set significantly enriched in presymptomatic group was previously defined in a study of expression changes in brain in major depressive disorder (Aston et al 2005) where a disruption in the expression of genes involved in neurodevelopment, signal transduction, synaptic function and cell communication was shown. Psychiatric symptoms usually precede motor impairment in HD for a few years and depression is one of them. Perhaps this might be the explanation for discovered enrichment of this gene set specifically in the presymptomatic HD group.

8. Conclusion

We conclude that identification of easily obtainable, reliable and robust biomarkers of Huntington's Disease progression will be important for development and evaluation of future therapies (Weir DW et al 2011). Specific pathogenic mechanisms can be readily proven by clinical, neuroimaging, and/or biochemical biomarkers. Peripheral blood due to its easy accessibility might be a representative tissue for genomic HD specific changes investigation and a potential tissue of choice for monitoring the course of the disease. However, without the effective treatment techniques it is not yet possible to validate which biomarker can serve as an effective surrogate endpoint for the disease process modification.

9. References

Achiron A, Gurevich M, Magalashvili D, et al. Understanding autoimmune mechanisms in multiple sclerosis using gene expression microarrays: treatment effect and cytokinerelated pathways. Clin Dev Immunol 2004; 11(3-4): 299-305.

Andrews TC, Weeks RA, Turjanski N et al. Huntington's disease progression. PET and clinical observations. Brain 1999; 122: 2353-2363.

Andrich J, Saft C, Ostholt N, et al. Assessment of simple movements and progression of Huntington's disease. J Neurol Neurosurg Psychiatry 2007; 78(4): 398-403.

Aston C, Jiang L,Sokolov BP. Transcriptional profiling reveals evidence for signaling and oligodendroglial abnormalities in the temporal cortex from patients with major depressive disorder. Mol Psychiatry 2005; 10(3): 309-22.

Auinger P, Kierbutz K, McDermott MP. The relationship between uric acid levels and Huntington's disease progression. Mov Disord 2010; 25(2): 224-228.

Aylard EH. Change in MRI striatal volumes as a biomarker in preclinical Huntington's disease. Brain Res Bull 2007; 72(2-3): 152-8.

Aylard EH, Li Q, Stine OC et al. Longitudinal change in basal ganglia volume in patients with Huntington's disease. Neurology 1997; 48(2): 394-399.

Aylard EH, Nopopulos PC, Ross CA. Longitudinal change in regional brain volumes in prodromal Huntington disease. J Neurol Neurosurg Psychiatry 2011; 82(4): 405-410.

Aylard EH, Rosenblatt A, Field K et al. Onset and rate of striatal atrophy in preclinical Huntington disease. Neurology 2004; 63(1): 66-72.

Aziz NA, Anguelova GV, Marinus J et al. Autonomic symptoms in patients and pre-manifest mutation carriers of Huntington's disease. Eur J Neurol 2010; 17: 1068-1074.

Bell J. Predicting disease using genomics. Nature 2004; 429(6990): 453-6.

Biglan KM, Ross CA, Langbehn DR et al. Motor abnormalities in premanifest persons with Huntington's disease: The PREDICT-HD study. Mov Disord 2009; 24(12): 1763-1772.

Biomarkers definitions working group. Biomarkers and surrogate endpoints: preferred definitions and conceptual framework. Clin Pharmacol Ther 2001; 69(3): 89-95.

Bjorquist M, Petersen A, Bacos K et al. Progressive alterations in the hypothalamic-pituitary-adrenal axis in the R6/2 transgenic mouse model of Huntington's disease. Hum Mol Genet 2006; 15(10): 1713-1721.

Bomprezzi R, Ringner M, Kim S, et al. Gene expression profile in multiple sclerosis patients and healthy controls: identifying pathways relevant to disease. Hum Mol Genet 2003; 12(17): 2191-9.

Borovecki F, Lovrecic L, Zhou J et al. Genome-wide expression profiling of human blood reveals biomarkers for Huntington's disease. PNAS 2005 2; 102(31) : 11023-8. Epub 2005.

Brandt J, Butters N. The neuropsychology of Huntington's disease. Trends in Neurosciences 1986; 9: 118-120.

Brinkman RR, Mezei MM, Thielmann J, Almquist E, Hayden MR. The likelihood of being affected with Huntington disease by a particular age for a specific CAG size. Am J Hum Genet 1997; 60: 1202-1210.

Campodonico JR, Codori AM, Brandt J. Neuropsychological stability over two years in asymptomatic carriers of the Huntington's disease mutation. J Neurol Neurosurg Psychiatry 1996; 61: 621-624.

Chen Chiung-Mei. Mitochondrial dysfunction, metabolic deficits, and increased oxidative stress in Huntington's disease. Chang Gung Med J 2011; 34: 135-152.

Ciammmola A, Sassone J, Canella M et al. Low brain-derived neurotrophic factor (BDNF) levels in serum of Huntington's disease patients. Am J Med Genet B Neuropsychiatr Genet 2007; 144(4)574-577.

Craufurd D, Snowden J. Neuropsychological and neuropsychiatric aspects of Huntington's disease. In: Bates G, Harper P. Huntington's disease. Third edition, Oxford University Press 2002, New York: 62-110.

Dexter DT, Carayon A, Javoy-Agid F, et al. Alterations in the levels of iron, ferritin and other trace metals in Parkinson's disease and other neurodegenerative diseases affecting the basal ganglia. Brain 1991; 114 (Pt 4): 1953-75.

Draghici S, Khatri P, Bhavsar P, et al. Onto-Tools, the toolkit of the modern biologist: Onto-Express, Onto-Compare, Onto-Design and Onto-Translate. Nucleic Acids Res 2003; 31(13): 3775-81.

Draghici S, Khatri P, Martins RP, Ostermeier GC,Krawetz SA. Global functional profiling of gene expression. Genomics 2003; 81(2): 98-104.

Evans WE,Relling MV. Moving towards individualized medicine with pharmacogenomics. Nature 2004; 429(6990): 464-8.

Fain JN, Del Mar NA, Meade CA, et al. Abnormalities in the functioning of adipocytes from R6/2 mice that are transgenic for the Huntington's disease mutation. Hum Mol Genet 2001; 10(2): 145-52.

Feigin A, Tang C, Ma Y et al. Thalamic metabolism and symptom onset in preclinical Huntington's disease. Brain 2007; 130: 2858-2867.

Fernanadez-Ruiz J. The endocannabinoid system as a target for the treatment of motor dysfunction. Br J Pharmacol 2009; 156:1029-1040.

Frank R, Hargreaves R. Clinical biomarkers in drug discovery and development. Nat Rev Drug Discov 2003; 2(7): 566-80.

Gavazzi C, Nave RD, PetralliR et al. Combining functional and structural brain magnetic resonance imaging in Huntington disease. J Comput Assist Tomogr 2007; 31(4): 574-580.

Georgiu-Karistanis N, Sritharan A, Farrow M et al. Increased cortical recruitment in Huntington's disease using a Simon task. Neuropsychologia 2007; 45(8): 1791-1800.

Glinsky GV, Berezovska O,Glinskii AB. Microarray analysis identifies a death-fromcancer signature predicting therapy failure in patients with multiple types of cancer. J Clin Invest 2005; 115(6): 1503-21.

Grafton ST, Mazziota JC, Pahl JJ et al. Serial changes of glucose cerebral metabolism and caudate size in persons at risk for Huntington's disease. Arch Neurol 1992; 49: 1161-1167.

Harjes P,Wanker EE. The hunt for huntingtin function: interaction partners tell many different stories. Trends Biochem Sci 2003; 28(8): 425-33.

Hodges A, Strand AD, Aragaki AK, et al. Regional and cellular gene expression changes in human Huntington's disease brain. Hum Mol Genet 2006; 15(6): 965-77.

Huang YC, Wu JR, Tseng MY et al. Increased prothrombin, Apolipoprotein A-IV, and Haptoglobulin in the cerebrospinal fluid of patients with Huntington's disease. PLoS ONE 2011; 6(1): e15809.

Hult S, Shultz K, Soylu R, Petersen A. Hypothalamic and neuroendocrine changes in Huntington's disease. Curr Drug Targets 2010; 11(10): 1237-49.

Huntington G. On chorea. Med Surg Rep 1872; 26: 317-321.

Huntington Study group. Unified Huntington's disease rating scale: reliability and consistency. Movement Disorder 1996; 11: 136-142.

Huntington's disease Collaborative Research Group. A novel gene containing a trinucleotide repeat that is expanded and unstable on Huntington's disease chromosomes. Cell 1993: 72: 971-983.

Kassubek J, Gaus W, Landwehrmeyer GB. Evidence for more widespread cerebral pathology in early HD; an MRI based morphometric analysis. Neurology 2004; 60(10): 1615-1620.

Khatri P, Draghici S, Ostermeier GC, Krawetz SA. Profiling gene expression using ontoexpress. Genomics 2002; 79(2): 266-70.

Kirkwood SC, Su JL, Conneally P, Foroud T. Progression of symptoms in early and middle stages of Huntington disease. Arch Neurol 2001; 58: 273-278.

Klöppel S, Draganski B, Golding CV et al. White matter connections reflect changes in voluntary-guided saccades in pre-symptomatic Huntington's disease. Brain 2008;131(1): 196-204.

Kobal J, Bosnjak R, Milosevic Z, Mesec A, Bajrovic FF. Choreatic movements first appear in Huntington's disease associated with brain cortex lesion due to subdural hematoma. Eur J Neurol 2007; 14:e3-4.

Kobal J, Meglic B, Mesec A, Peterlin B. Early sympathetic hyperactivity in Huntington's disease. Eur. J Neurol 2004; 11: 842-848.

Kobal J, Melik Z, Cankar K et al. Autonomic dysfunction in presymptomatic and early symptomatic Huntington's disease. Acta Neurol Scand 2010; 121(3): 392-399.

Kremer B. Clinical neurology of Huntington's disease. Diversity in unity, unity in diversity. In: Bates G, Harper P. Huntington's disease. Third edition, Oxford University Press 2002, New York: 28-61.

Lawrence AD, Hodges JR, Rosser AR et al. Evidence for specific cognitive deficits in preclinical Huntington's disease. Brain 1998; 121; 1329-1341.

Lovrecic L, Kastrin A, Kobal J, Pirtosek Z, Krainc D, Peterlin B.Gene expression changes in blood as a putative biomarker for Huntington's disease. Mov Disord. 2009 Nov 15; 24 (15): 2277-81.

Magnotta VA, Kim J, Koschik T et al. Diffusion tensor imaging in preclinical Huntington disease. Brain Imaging Behav 2009; 3(1): 77-84.

Melik Z, Kobal J, Cankar K, Strucl M. Microcirculation response to local cooling in patients with Huntington's disease. J Neurol-in print.

Mochel F, Charles P, Seguin F et al. Early energy deficit in Huntington disease: identification of plasma biomarkers traceable during disease progression. PLoS One 2007; 25(7): e674.

Montoya A, Price BH, Menear M, Lepage M. Brain imaging and cognitive dysfunctions in Huntington's disease. J Psychiaty Neurosci 2006;31(1): 21-29.

Moos T,Morgan EH. The metabolism of neuronal iron and its pathogenic role in neurological disease: review. Ann N Y Acad Sci 2004; 1012: 14-26.

Moskovitch-Lopatin M, Weiss A, Rosas HD et al. Optimization of an HTRF assay for the detection of soluble mutant huntingtin in human buffy coats: A potential biomarker in blood for Huntington disease. PLoS Curr. 2010; 2: RRN1205.

Nadler ST, Stoehr JP, Schueler KL, et al. The expression of adipogenic genes is decreased in obesity and diabetes mellitus. Proc Natl Acad Sci U S A 2000; 97(21): 11371-6.

Paulsen JS, Zimbelman JL, Hinton DR et al. fMRI biomarker of early neuronal dysfunction in presymptomatic Huntington disease. AJNR 2004; 25: 1715-1721.

Pavese N, Andrews TC, Brooks DJ et al. Progressive striatal and cortical dopamine receptor dysfunction in Huntington's disease. Brain 2003; 126(5): 1127-1135.

Pavese N, Andrews TC, Grhard A et al. Microglial activation correlates with severity in Huntington disease: a clinical and PET study. Neurology 2006; 66(11): 1638-1643.

Penney JB Jr, Young AB, Shoulson I, Starosta-Rubenstein S et al. Huntington's disease in Venezuela: 7 years of follow-up on symptomatic and asymptomatic individuals. Movement Dis 1990; 5: 93-99.

Petersen A, Björquist M. Hypothalamic-endocrine aspects in Huntington's disease. Eur J Neurosci 2006; 24(4): 961-967.

Rao AK, Quinn L, Marder KS. Reliability of spatiotemporal gait outcome measures in Huntington's disease. Mov Disord 2005; 20(8): 1033-1037.

Reynolds NC Jr, Prost RW, Mark LP. Heterogeneity in 1H-MRS profiles of presymptomatic and early manifest Huntington's disease. Brain Res 2005;1031(1): 82-89.

Rosas HD, Hevelone ND, Zaleta AK et al. Regional cortical thinning in preclinical Huntington disease and its relationship to cognition. Neurology 2005; 65(5): 745-747.

Rosas HD, Koroshetz WJ, Chen YI et al. Evidence of more widespread cerebral pathology in early HD: an MRI-based morphometric analysis. Neurology 2003; 60(10): 1615-1620.

Rosas HD, Liu AK, Hersch S et al. Regional and progressive thinning of the cortical ribbon in Huntington's disease. Neurology 2002; 58(5): 695-701.

Rosas HD, Salat DH, Stephanie E Lee et al. Cerebral cortex and the clinical expression of Huntington's disease: complexity and heterogeneity. Brain 2008; 131: 1057-1068.

Rosas HD, Tuch D, Hevelone N et al. Diffusion tensor imaging in presymptomatic and early Huntington's disease: selective white matter pathology and its relationship to clinical measures. Mov Disord 2006; 21(7): 1043-1047.

Rosenwald A, Wright G, Chan WC, et al. The use of molecular profiling to predict survival after chemotherapy for diffuse large-B-cell lymphoma. N Engl J Med 2002; 346(25): 1937-47.

Ross CA, Shoulson I. Huntington,s disease: pathogenesis, biomarkers, and approaches to experimental therapeutics. Parkinsonism Relat Disord 2009; 15 Suppl 3: S135-138.

Runne H, Kuhn A, Wild EJ, et al. Analysis of potential transcriptomic biomarkers for Huntington's disease in peripheral blood. Proc Natl Acad Sci U S A 2007; 104(36): 14424-14429.

Saft C, Zange J, Andrich J et al. Mitochondrial impairment in patients and asymptomatic mutation carriers of Huntington's disease. Mov Disord 2005; 130:2585-2567.

Sawa A, Wiegand GW, Cooper J, et al. Increased apoptosis of Huntington disease lymphoblasts associated with repeat length-dependent mitochondrial depolarization. Nat Med 1999; 5(10): 1194-8.

Schadt EE, Monks SA, Drake TA, et al. Genetics of gene expression surveyed in maize, mouse and man. Nature 2003; 422(6929): 297-302.

Shoulson I, Fahn S. Huntington's disease: clinical care and evaluation. Neurology 1979; 29: 1-3.

Stoffers D, Sheldon S, Kuperman JM et al. Contrasting gray and white matter changes in preclinical Huntington disease: An MRI study. Neurology 2010; 74: 1208-1216.

Sturzebecher S, Wandinger KP, Rosenwald A, et al. Expression profiling identifies responder and non-responder phenotypes to interferon-beta in multiple sclerosis. Brain 2003; 126(Pt 6): 1419-29.

Subramanian A, Tamayo P, Mootha VK, et al. Gene set enrichment analysis: a knowledge-based approach for interpreting genome-wide expression profiles. Proc Natl Acad Sci U S A 2005; 102(43): 15545-50.

Surrock A, Laule C, Decolongon J et al. Magnetic resonance spectroscopy biomarkers in premanifest and early Huntington disease. Neurology 2010; 75(19): 1702-1710.

Tabrizi SJ, Scahill RI, Durr A et al. Biological and clinical changes in premanifest and early stage Huntington's disease in the TRACK-HD study: the 12 month longitudinal analysis. Lancet Neurol 2011; 10: 31-42.

Tai YF, Pavese N, Gerhard A et al. Microglial activation in presymptomatic Huntington's disease gene carriers. Brain 2007; 150(pt7): 1759-1766.

Tang Y, Schapiro MB, Franz DN, et al. Blood expression profiles for tuberous sclerosis complex 2, neurofibromatosis type 1, and Down's syndrome. Ann Neurol 2004; 56(6): 808-14.

Tomasi G, Bartoldo A, Cobelli C et al. Global-two-stage filtering of clinical PET parametric maps: application to [(11)C]-®-PK11195. Neuroimage 2011;55(3):942-953.

Turkka JT. Correlation of the severity of autonomic dysfunction to cardiovascular reflexes and to plasma noradrenaline levels in Parkinson's disease. Eur Neurol 1987; 26: 203-210.

Underwood BR, Broadhurst D, Warwick BD et al. Huntington disease patients and transgenic mice have similar pro-catabolic serum metabolite profiles. Brain 2006;129:877-886.

Varani K, Bachoud-Levi AC, Mariotti C et al. Biological abnormalities of peripheral A(2A) receptors in a large representation of polyglutamine disorders and Huntington's disease stages. Neurobiol Dis 2007; 27(1): 36-43.

Weir DW, Sturrock A, Leavitt BR. Development of biomarkers for Huntington's disease. Lancet Neurol 2011; 10: 573-90.

Wild EJ, Henley SM, Hobbs NZ et al. Rate and acceleration of whole-brain atrophy in premanifest and early Huntington's disease. Mov Disord 2010, 25(7); 888-895.

Yamamoto A, Lucas JJ, Hen R. Reversal of neuropathology and motor dysfunction in a conditional model of Huntington's disease. Cell 2000; 101(1): 57-66.

5

Quinolinate Accumulation in the Brains of the Quinolinate Phosphoribosyltransferase (QPRT) Knockout Mice

Shin-Ichi Fukuoka[1], Rei Kawashima[1,2,3], Rei Asuma[1], Katsumi Shibata[4] and Tsutomu Fukuwatari[4]
[1]*Department of Chemistry and Biological Science, College of Science and Engineering, Aoyama Gakuin University, Chuo-ku, Sagamahara-shi, Kanagawa,*
[2]*Department of Gastroenterology, Research Institute, National Center for Global Health and Medicine, Shinjuku-ku, Tokyo,*
[3]*Department of Biochemistry, Graduate School of Medical Sciences, Kitasato University, Minami-ku, Sagamahara-shi, Kanagawa,*
[4]*Department of Life Style Studies, School of Human Cultures, The University of Shiga, Hassaka-cho, Hikone-shi, Shiga,*
Japan

1. Introduction

The kynurenine pathway (KP) is the main route of L-tryptophan catabolism, thus resulting in the production of the essential pyridine nucleotide, nicotinamide adenine dinucleotide (NAD^+) (Figure 1) (Stone, 1993). Quinolinic acid (QA) is one of the KP metabolites, which are synthesized from the essential amino acid tryptophan (Trp). QA is a potent endogenous excitotoxin of neuronal cells that acts as N-methyl-D-aspartate (NMDA) receptor agonist. Quinolinate phosphoribosyltransferase (QPRT) is the only enzyme that degrades QA in mammalian cells, so the concentration of QA is modulated directly by the QPRT activity. QA is an endogenous excitotoxin acting on N-methyl-D-aspartate receptors (NMDARs) which leads to pathological and neurochemical features similar to those observed in HD. Neurons expressing high levels of NMDARs are lost early from the striatum of individuals affected with Huntington's Disease (HD), and injection of NMDA receptor agonists, such as QA, into the striatum of rodents or non-human primates mimics the pattern of neuronal damage observed in HD. When QA is loaded into rat brains by autodialysis, the striatal region is specifically severely damaged (Schwarcz & Köhler, 1983). An autoradioreceptor assay showed that the number of NMDA glutamate receptors in patients of HD was reduced by 93% (Young et al., 1988), thus supporting the hypothesis that an endogenous agonist of the receptor is primarily responsible for the neural degradation associated with the disease. Unlike kainate or ibotenate, QA is thought to be the only physiological agonist for the NMDARs involved in the disorder (Stone et al., 1981). Thus, a dysfunction of QA metabolism in the human brain has been postulated to be involved in the pathogenesis of such neurodegenerative disorders as epilepsy, Alzheimer's Disease (AD) and HD (the "quinolinate hypothesis") (Schwarcz et al., 1986).

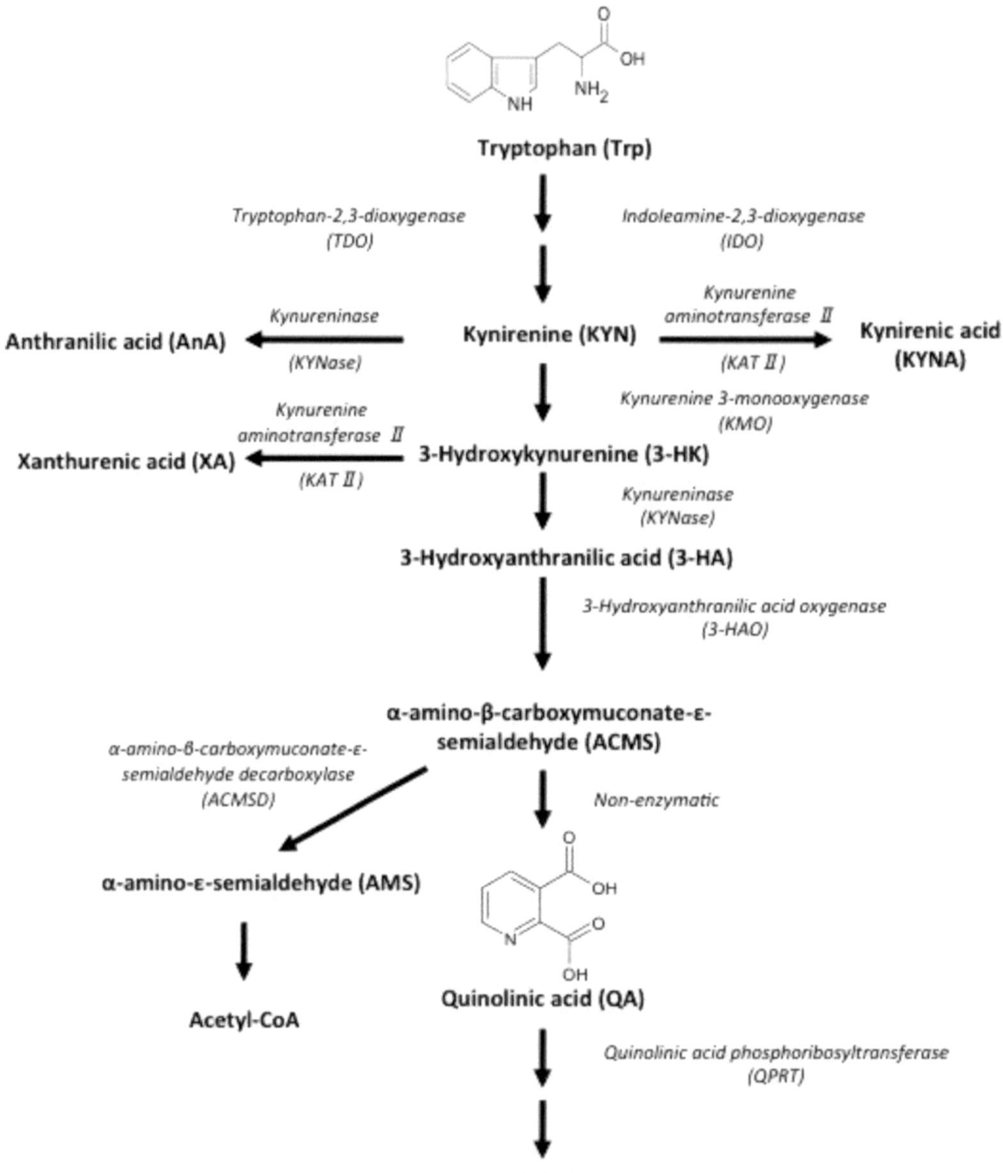

Fig. 1.

However, this hypothesis has not yet been corroborated by the measurement of endogenous QA in neurodegenerative disorders. In this study, we generated QPRT gene deficient mice (QPRT knockout mice) to investigate this hypothesis *in vivo*. We succeeded in detecting the endogenous QA accumulation-induced neurodegenerations in the striatum of middle-aged

QPRT knockout (KO) mice by an immunohistochemical analysis. In KO mice, the expression levels of KP enzymes and NMDA receptor (NMDAR) subunits were altered compared to those of wild type (WT) mice. The expression of the NR2B subunit was also significantly increased in middle-aged KO mice. The results of biochemical analyses indicated that QA tended to exert NMDAR-mediated excitotoxiciy in the brains of these mice. We observed behavioral disorders in QPRT KO mice using two behavioral tests. Many previous studies have demonstrated that disturbances in gait are symptomatic of Parkinson's Disease (PD) and HD. Gait abnormalities in PD include a shortened stride length. HD also shows gait abnormalities include changes in stride length (Koller & Trimble, 1985). We therefore measured the stride length of KO mice based on these studies. We found that the aged QPRT KO mice displayed shortening of their strides compared to the WT group. In contrast, the middle-aged QPRT KO mice did not exhibited any significant gait abnormalities. Fernagut et al. reported that the stride length is a reliable index of motor disorders due to basal ganglia dysfunction in mice (Carter et al., 1999). Our findings suggest that the striatal neuronal lesions in the QPRT KO mice progressed with age, such that the younger mice had not yet developed sufficient basal ganglia dysfunction to result in a change in gait. The shortening of strides may be an event that occurs during the later stage of neurodegeneration.

2. Generation of QPRT gene deficient models

2.1 Construction of the QPRT gene targeting vector

Based on the genomic information obtained previously and using genomic clones of the 129 Svj mouse QPRT gene, a 2.9 kb 50 homologous recombination region including a portion of exon 2, intron 3, exon 3, and a portion of intron 4 was amplified by PCR. A PGK-βgeo selection marker cassette was ligated and subcloned into the targeting vector with the MC1-diphtheria toxin A gene to select against nonhomologous recombination (Figure 2a).

2.2 Generation of the QPRT disrupted mice

The constructed targeting vector was introduced into a 129 Svj mouse ES cell line (Genome Systems) by electroporation. ES cells were selected in media containing G418, and the surviving cells were purified by dilution to obtain single clones. Homologous recombination in the ES cells was confirmed by a Southern blotting analysis using a probe localized 50 methods. The positive clone was injected into C57BL/6N mouse blastocysts to obtain chimeric mice that transmitted the mutation through the germline. Mice were bred and maintained using standard mouse husbandry procedures. The detailed physiological and biochemical analyses of the QPRT gene deficient mice will be published elsewhere.

2.3 Validation studies

Genomic DNA extracted from mice tails was used for genotyping PCR. Figure 2b is an electropherogram of the mouse DNA amplification products. Figure 2b is an electropherogram of the mouse DNA amplification products. WT, heterozygous (HZ) and KO mice showed distinct band patterns.

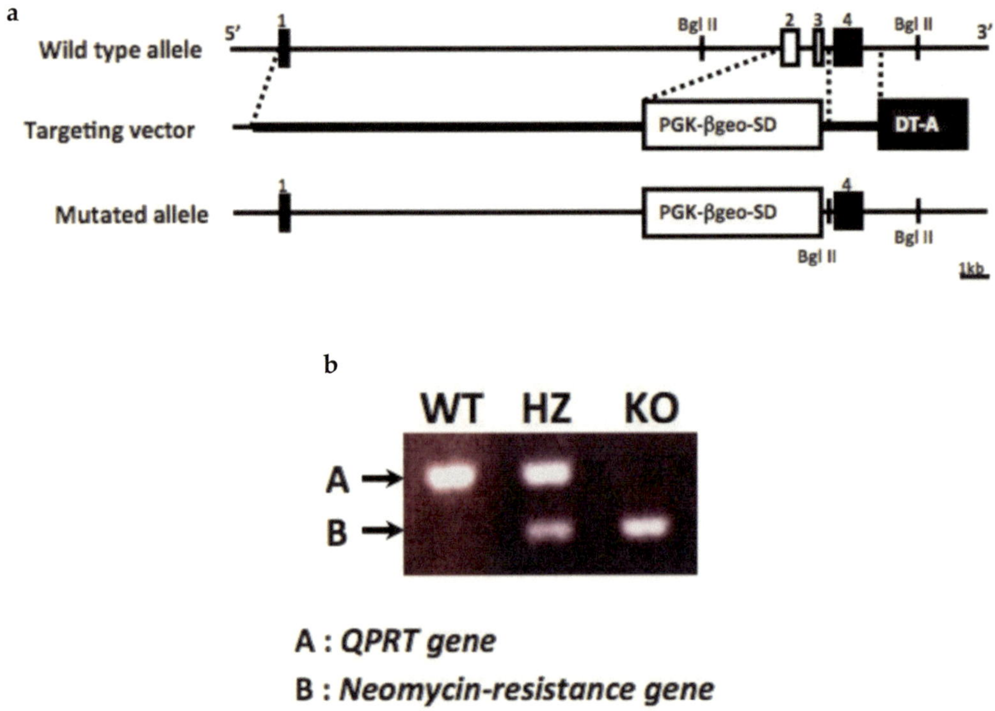

Fig. 2. Generation of QPRT gene deficient mice. WT, wild type mice; KO, QPRT KO mice. (a) The targeting strategy used for QPRT gene disruption. Exons are represented as *numbered boxes*. DT-A, diphtheria toxin-A. (b) An agarose gel showing genotyped PCR amplicons. Genomic DNA extracted from mouse tails were used for PCR with primer pairs for the QPRT and PGK-βgeo genes. The product sizes are 398 and 221 base pairs (bp), respectively.

3. Morphological analysis

3.1 The presence of QA in the brains

The QA in the brain of middle-aged (18-week-old) WT and QPRT KO mice was stained using an anti-QA antibody (Figure 3). The histochemical analyses were performed with frozen-section tissues prepared from brains fixed with 4% paraformaldehyde in phosphate-buffered saline (PBS). The frozen sections were excised and embedded in O. C. T. compound for cryosectioning, then dried and treated with 5% BSA for 30 min at room temperature. The tissues were incubated with 5 μg/ml rabbit anti-QA antibody (Ab) (Sigma, MO) for 12 hours at 4°C, followed by 0.5 μg/ml secondary biotinylated donkey anti-rabbit IgG Ab (Molecular Probes, OR) for 2 hours at room temperature. The immunostaining was visualized with a VECTASTAIN ABC kit (Vector Laboratories, CA) using DAB as the chromagen. The images were captured with a fluorescence microscope (Axioplan2; Carl Zeiss Inc., Jena, Germany) equipped with a CCD camera. The staining intensities were determined by using the Image-J software program to measure the stained areas in each striatum of sections after the experiment. Stained cells existed in both groups, especially in their striatum. There was no consistent pattern of labeling with regard to specific cell types. Quantification of QA staining intensities suggested that there were high QA levels in the middle-aged KO mouse striatum. In the striatum, KO mice showed approximately two times the amount of QA staining intensity compared to WT mice (Data not shown).

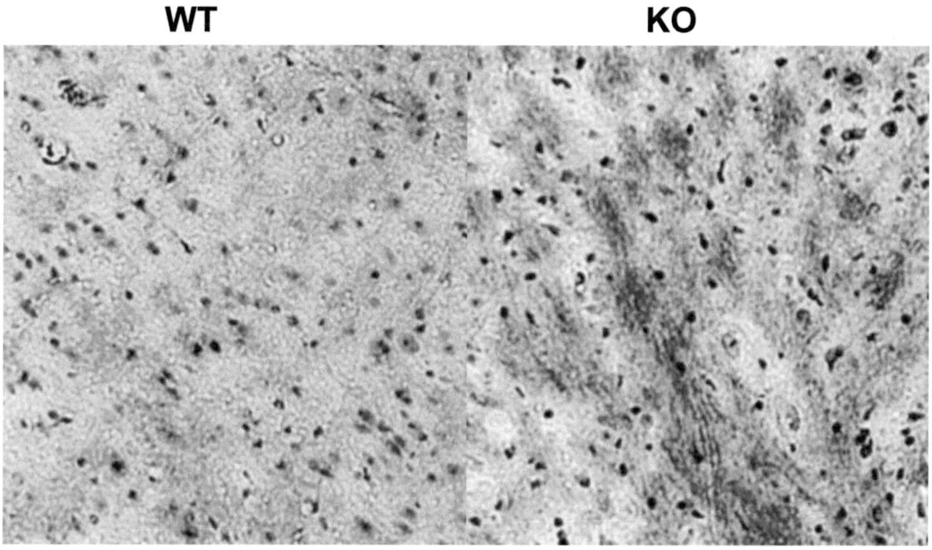

Fig. 3. Detection of QA by immunostaining of the striatum of WT and QPRT KO mice. WT, wild type mice; KO, QPRT KO mice. The frozen sections of middle-aged WT and KO brains were labeled by anti-QA polyclonal antibodies. The QA-positive cells were stained in the striatum of WT (a, left) and KO (a, right).

3.2 Detection of neurodegeneration

According to the results of the morphological analysis using the QA antibody, we prepared sections from middle-aged WT mice and QPRT KO mice to detect neuronal degeneration by Fluoro Jade C staining (Figure 4).

To detect neurodegeneration, the tissues were treated with fluoro-jade C (Histo-Chem Inc.; Jefferson, AR) according to the previously described method. The slides bearing frozen cut tissue sections were first immersed in a basic alcohol solution consisting of 1% sodium hydroxide in 80% ethanol for 5 min. They were then rinsed for 2 min in 70% ethanol, for 2 min in distilled water, and then incubated in 0.06% potassium permanganate solution for 10 min. Slides were subsequently transferred for 10 min to a 0.0001% solution of Fluoro-Jade C dissolved in 0.1% acetic acid vehicle. The slides were then rinsed through three changes of distilled water for 1 min per change. Excess water was drained onto a paper towel, and the slides were air dried on a slide warmer at 50°C for at least 5 min. The air dried slides were then cleared in xylene for at least 1 min and then coverslipped with entellan new (Merck Inc., Japan) non-fluorescent mounting media.

The neurons in the striatum of the KO mouse brains were labeled in their cell bodies. In contrast, there were few stained neurons in the WT mouse brain sections. These results indicate that neurodegeneration occurred remarkably in the middle-aged QPRT KO mouse striatum, but not in those of WT mice.

Fig. 4. Neuronal degeneration in the WT and QPRT KO mouse striatum stained with Fluoro-Jade C. WT, wild type mice; KO, QPRT knockout mice. 25μm frozen brain sections of middle-aged WT (right) and KO (right) mice were used for immunofluoresence studies with Fluoro-Jade C staining. The degenerated neurons were labeled in their cell bodies.

4. Detection of gene expressions

4.1 The gene expression of kynurenine pathway enzymes in the striatum of WT and QPRT KO mice

We detected the mRNA expression levels of metabolic enzymes in the KP (Figure 1) to clarify the role of metabolism in the QPRT KO mice. To determine the mechanism of KP metabolism in the middle-aged (14 ~ 22-week-old) and aged (68-week-old) mouse striatum, we analyzed the gene expression levels of KP enzymes by real-time PCR using primers for IDO, TDO, KATII, KYNase, KMO, 3-HAO and ACMSD. The total RNA was extracted from the mouse striatum using TRIZOL (Invitrogen, CA). The purity of RNA was confirmed by spectrophotometer readings at 260/280 nm. Total RNA was reverse-transcribed with the PrimeScript RT reagent kit (TaKaRa) and amplified by PCR. For KP enzyme genes (Figure 1), the following primers were used; IDO, 5'-TTCTTCTTAGAGTCAGCTCCCC (sense) and TCACAGAGACCAGACCATTCAC-3' (antisense); TDO, 5'-AAGAGGAACAGATGGCAGAG (sense) and TCGTCGTTCACCTTTACTCA-3' (antisense); KAT II, 5'-CGGTTTGAAGA CGACTTGA (sense) and TTGGGTGGGTAGTTGACAGT-3' (antisense); KYNase, 5'-AGCCCATGAGAAAGAAATAG (sense) and TGCCGCTTTGGAGTAG-3' (antisense); KMO, 5'-CGCGATCATGCCCTCTA (sense) and GGACCCAAGGACAAAGAGTC-3' (antisense); 3-HAO, 5'-TTGAGTGGTTGAGAGCTGTCAC (sense) and GGCTATGGCTG TTAGAAGATCG-3' (antisense); ACMSD, 5'-GGTACATGCCTCTTACATCAGC (sense) and GCTATCCTAGAGCTTGCTATGC-3' (antisense); QPRT, 5'-GCTCCTGTTACCCCCTACAACC (sense) and GGATGCAAAATTGAGGCCCGGG-3' (antisense). GAPDH was used in each reaction as an internal standard. For quantitative analysis, the SYBR Premix Ex Taq™ (TaKaRa) was used according to the manufacturer's instructions in a LightCycler® 480 Real-Time PCR System (Roche, Basel, Schweiz). The conditions for the reaction were as follows: 5 s at 95°C and 20 s at 60°C for 40 cycles.

In the middle-aged mice, the mRNA levels of ACMSD in the striatum of QPRT KO mice were lower than those in the WT group ($P = 0.036$). In contrast, no significant differences in the mRNA expression levels were seen in other KP enzymes (Figure 5).

On the other hand, the aged groups showed the opposite results for ACMSD expression. The mRNA levels of ACMSD in the QPRT KO mice were increased significantly compared to those of WT mice ($P = 0.0088$). However, in both the middle-aged and aged groups, there were no significant changes in any of the other KP enzymes between the WT and KO mice.

4.2 The gene expression levels of NMDAR subunits in the striatum of WT and QPRT KO mice

We next investigated the effects of QPRT deletion on the glutamatergic pathway. The mRNA expression of NMDAR subunits in the striatum of middle-aged and aged WT and QPRT KO mice were analyzed by real-time PCR (Figure 6). In the middle-aged QPRT KO group, the NR2B subunit mRNA expression level increased approximately two-fold compared to the WT group ($P = 0.002$). In addition, the NR2A subunit genes in the KO mouse striatum showed a tendency toward an increased expression compared to the WT group. However, there were no significant differences for any other subunits.

In the aged KO mouse group, the mRNA expression levels of the NR1, NR2A and NR2B subunits were significantly higher than those of the WT group ($P = 0.016$, 0.049 and 0.044, respectively). KO mice also showed two-fold increased expression of the NR2D subunit ($P = 0.015$). However, the NR2C subunits did not show any significant differences in expression between WT and KO mice.

These results about NMDAR expression levels suggested that there were similar propensities with regard to about the expression of NR2A and NR2B subunits in the middle-aged and aged groups, but that more subunits were affected by QPRT deficiency in the aged group.

5. Discussion

In this study, we generated QPRT gene deficient mice (QPRT knockout mice) and confirmed that the mRNA and protein expression of QPRT were not detected in the tissues of the QPRT KO mice. Therefore, it is expected that endogenous QA cannot be degraded by QPRT in these mice, allowing for the possible accumulation of QA. According to the "quinolinate hypothesis" and other previous studies utilizing animal models of neurodegenerative disorders, the accumulation of QA was associated with remarkable abnormal phenotypes such as defects in growth and development, such as were observed in a mouse model of Huntington's disease (HD) (Dellen, 2008). However, the QPRT KO mice exhibited normal phenotypes. In our long-term follow-up study, the changes in the body weight and intake of food were almost the same in the QPRT KO mice as in the wild type (WT) mice (data not shown). These results made us wonder why the QPRT KO mice did not show any effects in their growth and development. We assumed that there were two possibilities that needed to be investigated. First, there was the possibility that the QA concentrations in the brains of QPRT KO mice were decreased to nontoxic levels due to changes in the expression levels of kynurenine pathway metabolic enzymes. The second possibility was that there might have been a change in the mechanism of NMDARs-mediated QA excitotoxicity in the presence of excessive QA. Based on these possibilities, we investigated the mechanisms of QA accumulation, degradation and excitotoxicity in QPRT KO mice.

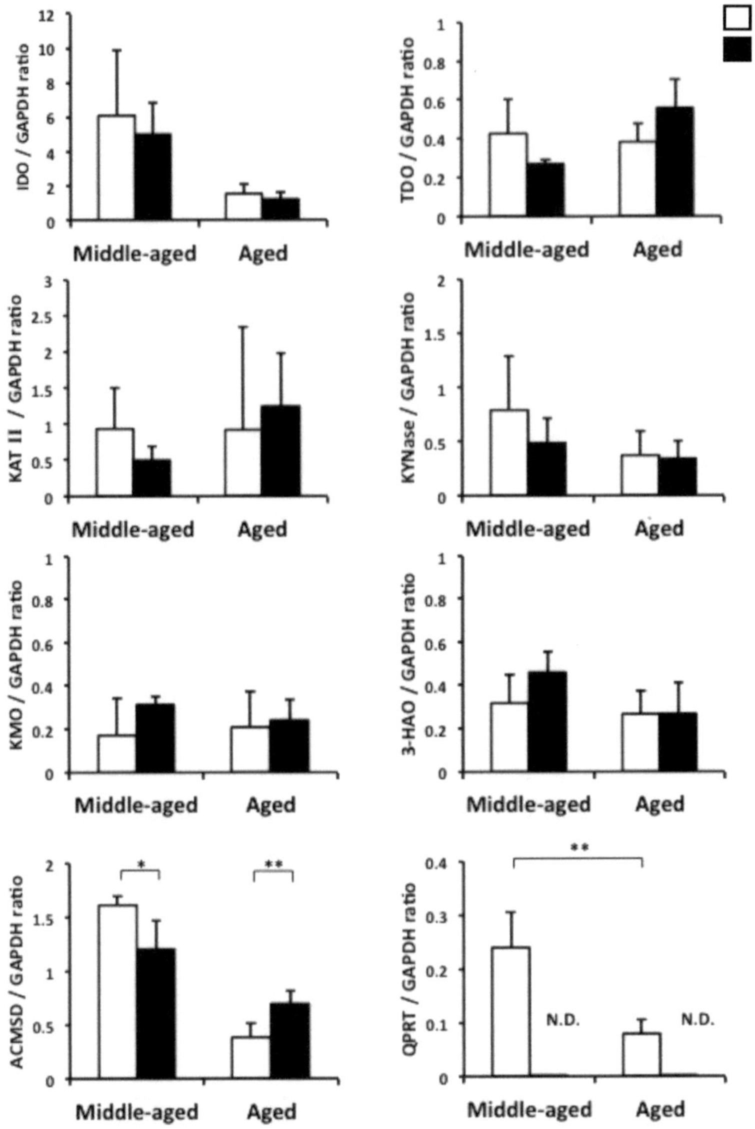

Fig. 5. The mRNA expression levels of KP enzymes in the striatum tissue samples of WT and QPRT KO mice. WT, wild type mice; KO, QPRT knockout mice. Total RNA extracted from middle-aged (14~22-week-old) and aged (68-week-old) mouse striatum samples was subjected to real time quantitative RT-PCR. The figures show the gene expression levels of KP enzymes IDO (middle-aged,WT: n = 4, KO: n = 8, aged, WT: n = 3, KO: n = 6), TDO (middle-aged, WT: n = 4, KO: n = 6; aged, WT: n = 3, KO: n = 5), KATII (middle-aged,WT: n = 3, KO: n = 6; aged, WT: n = 3, KO: n = 6), KYNase (middle-aged,WT: n = 4, KO: n = 6; aged, WT: n = 3, KO: n = 6), KMO (middle-aged,WT: n = 3, KO: n = 4; aged, WT: n = 3, KO: n = 6), 3-HAO (middle-aged,WT: n = 3, KO: n = 7; aged, WT: n = 3, KO: n = 6), ACMSD (middle-aged,WT: n = 3, KO: n = 6; aged, WT: n = 3, KO: n = 6) and QPRT (middle-aged,WT: n = 3, KO: n = 7; aged, WT: n = 3, KO: n = 6). The values are shown as the ratios of KP enzymes / GAPDH (internal standard). The data are presented as the means ± S.D. *, $p < 0.05$, **, $p < 0.01$. (Student's *t*-test).

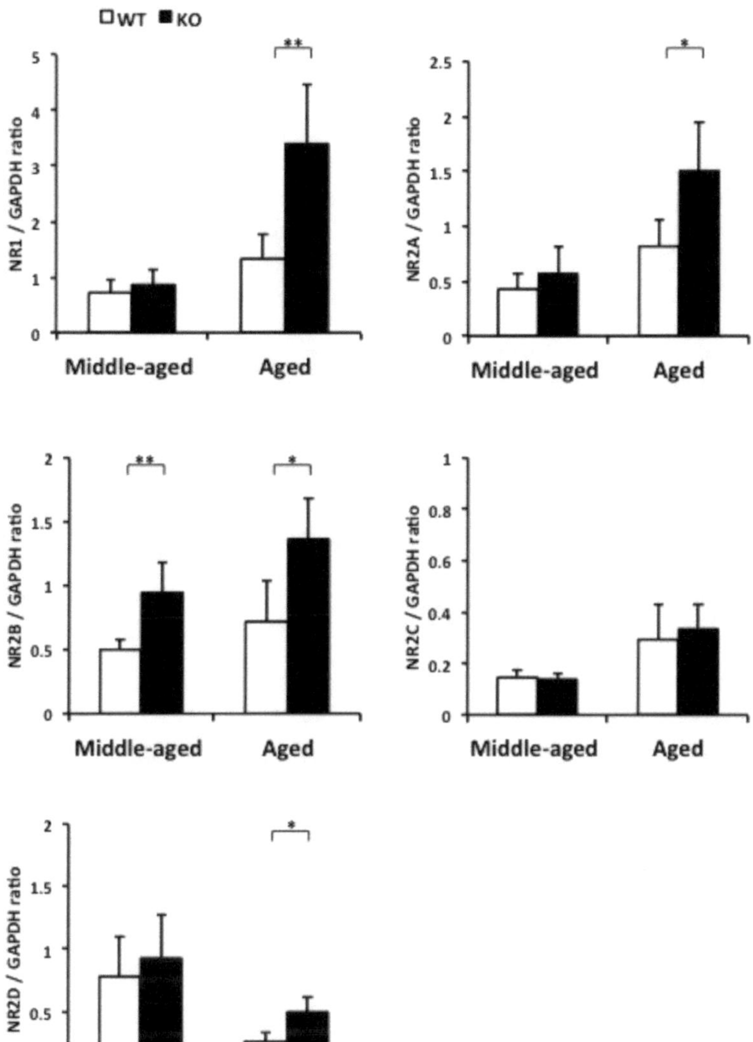

Fig. 6. The mRNA expression levels of NMDAR subunits in striatum tissue samples from WT and QPRT KO mice. WT, wild type mice; KO, QPRT knockout mice. Total RNA was extracted from middle-aged (14~22-week-old) and aged (68-week-old) mouse striatum samples, and the mRNA expression levels of NMDAR subunits was determined by real time quantitative RT-PCR. The figures show the gene expression levels of NMDAR subunits NR1 (middle-aged,WT: n = 6, KO: n = 7; aged, WT: n = 3, KO: n = 6), NR2A (middle-aged,WT: n = 5, KO: n = 7; aged, WT: n = 3, KO: n = 6), NR2B (middle-aged,WT: n = 5, KO: n = 8; aged, WT: n = 3, KO: n = 6), NR2C (middle-aged,WT: n = 5, KO: n = 8; aged, WT: n = 3, KO: n = 6), and NR2D (middle-aged,WT: n = 5, KO: n = 7; aged, WT: n = 3, KO: n = 6). The values are shown as the ratio of NMDAR subunits/GAPDH (internal standard). The data are presented as the means ± S.D. *,$p < 0.05$. (Student's *t*-test).

We first tried to visualize the intrastriatal QA in mouse brain sections to clarify whether QA accumulated in the brains of the QPRT KO mice. In human and rat brains, QA is present at concentrations in the high nanomolar range (Wolfensberger, 1983). An HD study revealed that QA levels were increased (by 300-400%) in the neostriatum of human with early stage HD compared to controls (Guidetti, 2004)]. By an immunohistochemical analysis using an anti-QA polyclonal antibody, we observed that the number of QA-positive cells in the striatum of middle-aged QPRT KO mice was higher than in WT mice. This result suggested that QA does indeed accumulate in the QPRT KO mice. An *in vitro* study showed that prolonged exposure of rat organotypic cortico-striatal cultures to as little as 100 nM QA results in characteristic excitotoxic damage (Whetsell & Schwarcz, 1989).

Using immunohistochemical detection of QA, we tried to validate the existence of neurodegenerarion in the striatum of middle-aged QPRT KO mice. Several previous studies demonstrated that selective striatal neuronal damage occurs in the striatum of HD patients. Therefore, we and others have postulated that a pathological elevation of QA levels may produce excitotoxic neurodegeneration in HD. We used fluoro-jade C staining for detection of neuronal degeneration. As a result, there were an increased number of degenerated neurons in the striatum of QPRT KO mice compared to WT mice. These results were consistent with QA accumulation in the striatum of the QPRT KO mice. We hypothesized that the neuronal degeneration in the striatum of KO mice might have been induced by the high levels of QA due to the QPRT deficiency.

We revealed significant differences in the gene expression levels of KP enzymes between WT and QPRT KO mice. The expression levels of ACMSD, which degrades the QA precursor α-amino-β-carboxymuconate-ε-semialdehyde (ACMS) in the striatum of middle-aged QPRT KO mice were lower than those of WT mice. In middle-aged QPRT KO mice, this change in KP metabolism might promote QA production, because QA is non-enzymatically derived from ACMS (Figure 1). This is consistent with the results of our immunohistochemical staining for QA. However, the results in the aged groups showed the opposite, with an increase in expression in the QPRT KO mice compared to the WT mice. This may reflect the acquisition of a defense mechanism against accumulation of QA in the aged QPRT KO mice, and we expected that this mechanism was established during the aging process. A reciprocal relationship between ACMSD mRNA and enzymatic activity was described by previous studies; the fluctuation of hepatic ACMSD mRNA expression was followed by that of ACMSD activity (Tanabe et al., 2002), and in the mouse brain, the changes in ACMSD expression at the message levels are shown to be highly correlated to those at the enzyme activity levels. This suggests that the quantification of the message levels with the real-time PCR technique is useful to address the regulation of ACMSD expression and QA levels (Fukuoka et al., 2002). Although we have not measured the enzymatic activities in the different groups of mice, based on the previous studies, we speculated that the mRNA expression levels of ACMSD observed in our present study were reflective of the enzymatic activity.

Because QA is known as a selective N-methyl-D-aspartate receptor (NMDAR) agonist, we determined the mRNA expression levels of NMDAR subunits to evaluate the mechanism underlying the neurotoxicity of endogenous QA.

The expression of functional recombinant NMDARs in mammalian cells requires the co-expression of at least one NR1 and one NR2 subtype. The stoichiometry of NMDARs has not yet been established conclusively, but the current consensus is that NMDARs are tetramers that most often incorporate two NR1 and two NR2 subunits of the same or different subtypes. In the case of middle-aged QPRT KO mice, only the NR2B subunit expression levels were significantly increased compared to the levels in WT mice. In contrast, the expression levels of both the NR2A and NR2B subunits were increased in the aged QPRT KO mice compare to the aged WT mice. Previous studies have elucidated that the pharmacological and functional properties of NMDARs depend heavily on the NR2 subunit composition. Moreover, other groups have confirmed that the critical factor affecting the NMDAR activity is the subunit composition: NR2B-containing NMDARs promote neuronal death, while NR2A-containing NMDARs promote neuronal survival. Heng et al. created double-mutant mice by crossing a murine genetic model of HD to a transgenic mouse overexpressing the NMDAR–NR2B subunits (Tang et al., 1999; Heng et al., 2007) and their recent study showed that the double-mutant mice exhibited a significant decrease in striatal neuron number and striatal volume. This result demonstrated that the overexpression of the NR2B subunit leads to the degeneration of striatal neurons. Based on these previous studies and our present findings, we believe that the high sensitivity of striatal neurons in the middle-aged mice to damage was due to their high expression of the NR2B subunit. Similarly, we believe that the aged QPRT KO mice expressed high levels of the NR2A subunit as an adaptive neuroprotective mechanisms.

6. Conclusion

In summary, our study detected the presence of endogenous QA accumulation and QA-induced neurodegeneration in the striatum of middle-aged QPRT KO mice. Our results raised the possibility that QPRT KO mice are able to be used as a model of endogenous QA accumulation mimicking various human neurodegenerative conditions. Although it was difficult to demonstrate the "quinolinate hypothesis" in previous *in vivo* studies, the new QPRT KO mouse model will therefore be a useful model for further investigating this hypothesis.

7. Acknowledgements

The authors thank Professor Tomoko Tashiro (Molecular and Neurobiology Laboratory, Aoyama Gakuin University, Kanagawa, Japan) for a lot of beneficial advice and her assistance with the florescence microscope and the cryostat; Assistant Professor Takayuki Negishi (Molecular and Neurobiology Laboratory, Aoyama Gakuin University, Kanagawa, Japan) for many helpful instructions, especially about neurobiology; Dr. Yasuhiro Arii (Mukogawa Woman's University, Hyogo, Japan) for his important advice about this study; and CALPIS Co., Inc. (Tokyo, Japan) for assistance with the Light Cycler® 480 Real-Time PCR System.

8. References

Carter, R.J. et al. (1999) Characterization of progressive motor deficits in mice transgenic for the human Huntington's disease mutation., J Neurosci 19 pp3248-3257.

Dellen, A.V. (2008) Wheel running from a juvenile age delays onset of specific motor deficits but does not alter protein aggregate density in a mouse model of Huntington's disease., BMC Neurosci 9 p34.

Fukuoka, S. et al. (2002) Identification and expression of a cDNA encoding human alpha-amino-beta-carboxymuconate-epsilon-semialdehyde decarboxylase (ACMSD). A key enzyme for the tryptophan-niacine pathway and "quinolinate hypothesis". J Biol Chem 277pp35162-35167.

Guidetti, P. (2004) Neostriatal and cortical quinolinate levels are increased in early grade Huntington's disease., Neurobiol Dis 17 pp455-461.

Heng, M.Y. et al. (2007) Longitudinal evaluation of the Hdh(CAG)150 knock-in murine model of Huntington's disease., J Neurosci 27 pp8989-8998.

Koller, W.C. & Trimble, J. (1985) The gait abnormality of Huntington's disease., Neurology 35 pp1450-1454.

Schwarcz, R. & Köhler, C. (1983) Differential vulnerability of central neurons of the rat to quinolinic acid., Neurosci Lett 38 pp85-90.

Schwarcz, R. et al. (1986) Quinolinic acid: a pathogen in seizure disorders?, Adv Exp Med Biol 203 pp697-707.

Stone, T.W. (1993) Neuropharmacology of quinolinic and kynurenic acids., Pharmacol Rev 45 pp309-379.

Stone, T.W. et al. (1981) Activity of the enantiomers of 2-amino-5-phosphono-valeric acid as stereospecific antagonists of excitatory aminoacids., Neuroscience 6 pp2249-2252.

Tanabe, A. et al. (2002) Expression of rat hepatic 2-amino-3-carboxymuconate-6-semialdehyde decarboxylase is affected by a high protein diet and by streptozotocin-induced diabetes., J Nutr 132 pp1153-1159.

Tang, Y.P. et al. (1999) Genetic enhancement of learning and memory in mice., Nature 401 pp63-69.

Whetsell, W.O. & Schwarcz, R. (1989) Prolonged exposure to submicromolar concentrations of quinolinic acid causes excitotoxic damage in organotypic cultures of rat corticostriatal system., Neurosci Lett 97 pp271-275.

Wolfensberger, M. (1983) Identification of quinolinic acid in rat and human brain tissue., Neurosci Lett 41 pp247-252.

Young, A.B. et al. (1988) NMDA receptor losses in putamen from patients with Huntington's disease., Science 241 pp981-983.

6

Alterations in Expression and Function of Phosphodiesterases in Huntington's Disease

Robert Laprairie*, Greg Hosier*, Matthew Hogel
and Eileen M. Denovan-Wright
Department of Pharmacology, Dalhousie University, Canada

1. Introduction

Cyclic AMP (cAMP, cyclic 3′, 5′-adenosine monophosphate) was first identified as a signalling molecule in 1958 (Rall & Sutherland, 1958), however it was not until 1962 that the enzyme responsible for hydrolysis of cAMP was identified and named phosphodiesterase (PDE; Butcher & Sutherland, 1962). Shortly afterwards, cyclic GMP (cGMP, cyclic 3′, 5′-guanosine monophosphate) was identified as another important second messenger that was hydrolyzed by PDE (Ashman et al., 1963). PDEs inactivate cAMP or cGMP by hydrolizing the 3′ cyclic phosphate bond of the cyclic nucleotide in question (Bender & Beavo 2006). Through molecular cloning and sequencing, it is now known that mammalian PDEs are encoded by 21 distinct genes (Bender & Beavo, 2006). These 21 genes encode protein isoforms of which variants can exist through the use of multiple transcription start sites and alternative mRNA splicing (Bender & Beavo, 2006). The 21 identified isoforms have been grouped into 11 families based on similarities in amino acid sequence, structure and function.

1.1 Phosphodiesterase are key regulators of cyclic nucleotide signalling cascades

cAMP is formed from ATP by adenylyl cyclase (Fig. 1; Rall & Sutherland, 1958). Adenylyl cyclase is a membrane-bound enzyme that can be activated by the Gα subunit, as well as the βγ subunit of the G-protein family, by calcium, and by protein kinase C (Tang & Ziboh, 1991, Iyengar, 1993). Once formed, cAMP activates protein kinase A. Protein kinase A is a tetrameric protein composed of two catalytic subunits and two regulatory subunits (Johnson & Jameson, 2000; Johnson et al., 2001). Two cAMP molecules bind to each regulatory subunit, which results in the release of the active catalytic subunits. Protein kinase A is known to phosphorylate proteins involved in cell signalling, apoptosis, ion channel regulation, osmotic homeostasis, and protein trafficking (reviewed by Shabb, 2001). Protein kinase A can also enter the nucleus where it is known to phosphorylate cAMP-response element binding (CREB) protein (Delghandi et al., 2005). Phosphorylated CREB stimulates transcription of genes related to cell signalling and proliferation such as brain-derived

*Co-first Authors

neurotrophic factor (Delghandi et al., 2005). In addition to signalling through protein kinase A, cAMP can directly alter ion channel conductance (reviewed by Wang et al., 2007).

PDEs also regulate cGMP signalling cascades. The cGMP pathway is activated by nitric oxide, which is produced by nitric oxide synthase (Francis et al. 2010). Nitrous oxide activates guanylyl cyclase, which can be membrane-bound or cytosolic. Guanyl cyclase converts GTP to cGMP, which can go on to activate protein kinase G (Francis et al., 2010). The cGMP pathway, regulates smooth muscle relaxation (Walter, 1984), synaptic plasticity (Kleppisch & Feil, 2009), and regulation of platelet aggregation (Walter, 1984).

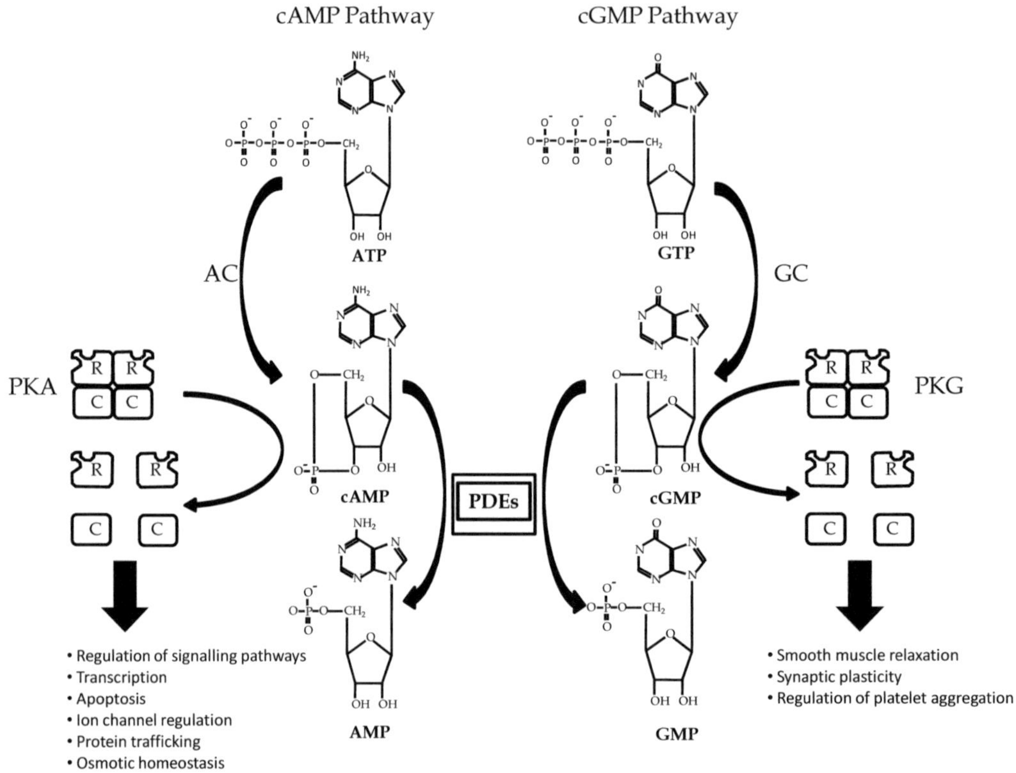

Fig. 1. PDEs regulate the cyclic nucleotide signalling pathways. The cAMP pathways is activated by adenylyl cyclase (AC) which converts ATP to cAMP. cAMP binds to the regulatory subunits (R) of protein kinase A (PKA), causing the release of the catalytic subunits (C). The cGMP pathway functions in a similar manner to the cAMP pathway. Guanylyl cyclase (GC) catalyzes the conversion of GTP to cGMP which activates protein kinase G (PKG). PDEs eliminate active cAMP and cGMP by hydrolyzing the molecules to their inactive AMP and GMP forms.

1.2 Phosphodiesterase isoforms are grouped into families based on similarities in catalytic and regulatory domains

All mammalian PDE isoforms share a conserved catalytic domain consisting of approximately 270 amino acids located in the C-terminal half of the protein (Degerman et al., 1997; Fig. 2). The catalytic domain is more similar within an individual PDE family

(>80% amino acid identity) than between different PDE families (~25-40% identity). Isoforms within PDE families 1, 2, 3, 4, 10 and 11 have dual specificity for both cAMP and cGMP, while PDEs within families 7 and 8 specifically hydrolyze cAMP and PDEs within families 5, 6 and 9 specially hydrolyze cGMP. The molecular basis for cAMP, cGMP, or cAMP/cGMP selectivity is believed to rely on a "glutamine switch" within the PDE catalytic domain which refers to an invariant glutamine that takes an orientation that favours binding of either cAMP or cGMP based on the presence of surrounding amino acid residues (Zhang et al., 2004).

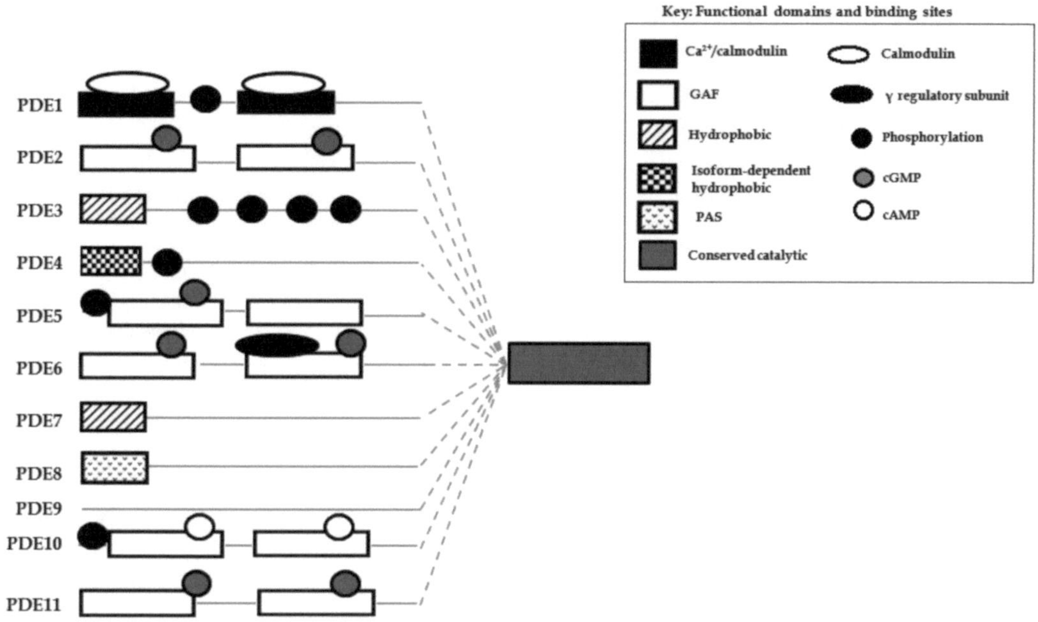

Fig. 2. Structural differences between the various PDE families. The different structural subunits that make up the individual PDE families help to dictate catalytic and regulatory specificity, as well as subcellular localization of the various PDEs.

The N-terminal portions of PDEs are widely divergent and contain functional domains that confer many of the regulatory and localization properties specific to the different PDE families (Degerman et al., 1997; Fig. 2). Isoforms of PDE families 2, 5, 6, and 7 contain two GAF domains (named after the proteins in which these domains are found: cGMP-specific phosphodiesterases, adenylyl cyclases and transcriptional activator of formate metabolism). Binding of cGMP, or cAMP in the case of PDE10, to the GAF domain stimulates enzymatic activity (Bender & Beavo, 2006). Isoforms of the PDE1 family share common dual Ca^{2+}/Calmodulin binding sites, which, when bound by Ca^{2+}/Calmodulin, stimulates enzymatic activity(Bender & Beavo, 2006). PDE3 isoforms contain hydrophobic domains near the N-terminus, which are believed to localize these enzymes to the plasma membrane (Degerman et al., 1997). PDE3 isoforms are also unique in that they are inhibited by cGMP, though the functional domain responsible for this has not been identified (Degerman et al., 1997). PDE6 isoforms contain an inhibitory subunit (γ), which must be removed to stimulate catalytic activity (Bender & Beavo, 2006). PDE7 isoforms contain hydrophobic localization domains (Bender & Beavo, 2006). PDE8 isoforms contain PAS domains (named after the three proteins in which it occurs: period circadian protein, aryl hydrocarbon receptor nuclear translocator protein, single-minded protein), and REC, or receiver, domains, which are believed to function as environmental sensors (Bender & Beavo, 2006). Members of PDE families 1, 3, 4, 5, and 10 also contain phosphorylation sites which are known to play a role in activating or inhibiting enzymatic activity depending on the phosphorylation site in question (Bender & Beavo, 2006).

1.3 Subcellular localization of PDE isoforms plays an important role in compartmentation of cyclic nucleotide signalling

An important idea to come about in the last few years regarding cyclic nucleotide signalling is that of compartmentation of cAMP and cGMP (Bender & Beavo, 2006). Unique localization and protein-protein interaction domains allow PDEs isoforms to localize to specific areas of the cell which allows for compartmentation of cyclic nucleotides (Bender & Beavo, 2006). Because adenyl cyclase and some proportion of guanyl cyclase is membrane-bound, localization of PDEs to the membrane plays an important role in controlling cyclic nucleotide signalling. As previously discussed, PDE3A and PDE3B contain hydrophobic domains that can localize these proteins to the membrane. Hydrophobic domains are also found in PDE2A2, PDE2A3, and PDE4A1 (Bender & Beavo, 2006). Arrestin binding domains in PDE4 isoforms are also known to allow PDE4 isoforms to localize to arrestin / β- adrenergic receptor complexes where they can breakdown cyclic nucleotides and inhibit β-adrenergic receptor signalling (Baillie et al., 2003).

Subcellular localization of individual isoforms can also change through regulatory mechanisms. This is exemplified with PDE10A2 in medium spiny projection neurons (Fig. 3). When cAMP levels are low PDE10A2 is palmitoylated and becomes associated with vesicles or the plasma membrane (Charych et al., 2010). Once at the plasma membrane, PDE10A2 is trafficked to dendritic processes throughout the neuron where it may serve to regulate intracellular signalling cascades associated with dopaminergic and glutamatergic synapses (Charych et al., 2010). When levels of cAMP increase however PKA becomes activated which leads to phosphorylation of PDE10A2. Phosphorylation of PDE10A2 inhibits palmitoylation, which results in the cytosolic accumulation of PDE10A2 in the cell body where it can normalize cAMP levels through its catalytic activity (Charych et al., 2010). Consequently, subcellular localization of PDEs plays an important role in compartmentation of cyclic nucleotide signalling.

1.4 Conclusions

PDEs regulate cyclic nucleotide signalling through breakdown of cAMP and cGMP. Multiple PDE isoforms are expressed in mammals which differ in catalytic, regulatory and subcellular localization properties. Unique regulatory and localization properties allow for fine tuning of cyclic nucleotide levels through compartmentation of specific PDE isoforms. Properties of isoforms derived from each of the 21 PDE genes encoded in mammals are summarized in Table 1.

PDE isoforms	Preferred substrate	Regulatory properties	Subcellular localization
PDE1 ***A,B,C***	cAMP/cGMP	Ca2+/Calmodulin-activated	Cytosolic
PDE2 ***A***	cAMP/cGMP	GAF+	*A1*: Cytosolic *A2, A3*: Membrane bound
PDE3 ***A,B***	cAMP/cGMP	cGMP-inhibited	*A:* Membrane-bound or cytoplasmic[1] *B:* Membrane-associated
PDE4 ***A,B,C,D***	cAMP/cGMP	UCR may play as yet unknown role	*A,B*: Membrane-associated C: Cytosolic D: Membrane-bound or cytoplasmic[1]
PDE5 ***A***	cGMP	GAF+	Cytosolic
PDE6 ***A,B,C***	cGMP	Inhibited by γ subunit; GAF+	*A,B*: Membrane associated, but becomes cytosolic after association with δ subunit C: Cytosolic
PDE7 ***A,B***	cAMP	Unknown	A1:Cytosolic A2: Membrane-bound
PDE8 ***A,B***	cAMP	PAS and REC environmental sensors	Cytosolic
PDE9 ***A***	cGMP	No known regulatory domains	*A1*:Nuclear *A5*: Cytosolic
PDE10 ***A***	cAMP/cGMP	GAF+	*A1,A3*: Cytosolic *A2*: Cytosolic when cAMP levels are high, membrane associated when cAMP levels low
PDE11 ***A***	cAMP/cGMP	GAF+	Cytosolic

[1] -Depends on splice variant and cell type.

Table 1. Properties of PDE isoforms. Adapted from Bender & Beavo (2006), Lugnier (2006), and Kleppisch & Feil (2009). Abbreviations: GAF, cGMP-activated PDEs, adenylyl cyclase, and transcriptional activator of formate metabolism; UCR, upstream conserved region; PAS, period circadian protein, aryl hydrocarbon receptor nuclear translocator protein, single-minded protein; REC, receiver.

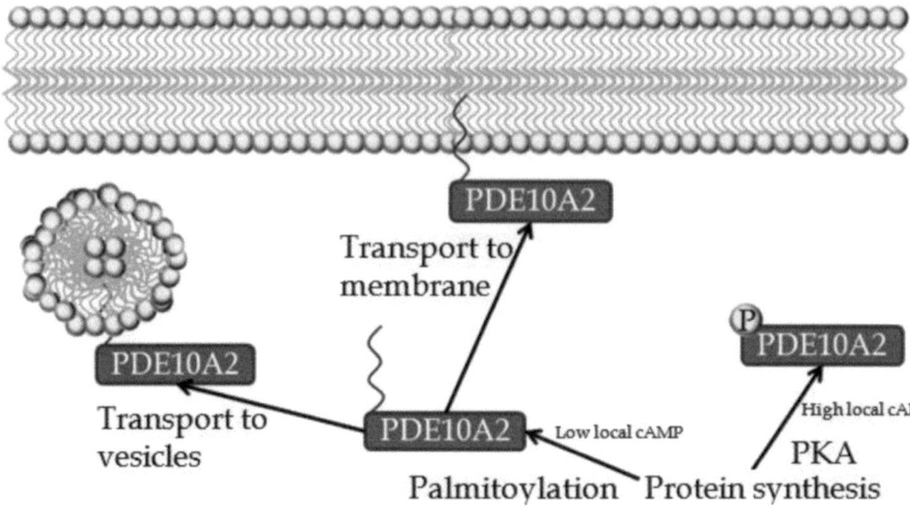

Fig. 3. Proposed model for the regulation of PDE10A2 localization in neurons in response to fluctuations in cAMP. PDE10A2 protein is synthesized in the cytoplasm. High levels of cAMP activate PKA to cause phosphorylation, and thus activation, of PDE10A2. During periods of low cAMP, PDE10A2 is palmitoylated and becomes associated with vesicles or the plasma membrane. Once at the plasma membrane, PDE10A2 is trafficked to dendritic processes throughout the neuron.

2. Phosphodiesterase isoforms have unique tissue distributions which can change during normal physiological processes

Of the 21 encoded PDE isoforms, only a small sub-set is expressed in any cell type. This cell-specific expression gives rise to unique distributions of PDE isoforms across tissues. Evidence suggests that expression of PDE isoforms changes during normal development and aging. Because different isoforms display distinct catalytic, regulatory, and subcellular localization properties, tissue-specific expression of PDE isoforms provides a mechanism to finely tune cyclic nucleotide levels within an organism.

2.1 Phosphodiesterase isoforms have unique tissue distributions

PDE isoforms have unique tissue distributions in the central nervous system (CNS) and non-nervous tissue. Tissue-specific expression of PDE isoforms was first noted in studies examining mRNA and protein expression of individual isoforms using northern blot (Fidock et al., 2002; Loughney et al., 1996), *in situ* hybridization (Prickaerts et al., 2002), western blot (Sadhu et al., 1999), and immunohistochemistry (Vandeput et al., 2007). Since then, heterogeneous tissue distribution of PDE isoform transcripts has been conclusively shown using quantitative reverse transcription (qRT) polymerase chain reaction (PCR) to quantify PDE isoform expression profiles in 12 distinct CNS and 12 distinct non-nervous tissues (Lakics et al., 2010). The PDE isoforms that are highly expressed in tissues are summarized in Table 2, while the relative distribution of highly expressed PDE isoforms across tissues is summarized in Table 3. It appears that individual tissues typically express between one and four PDE isoforms at high levels (Table 2), and individual PDE isoforms may be expressed at high levels in multiple tissues of both the CNS and non-nervous tissues (Table 3).

CNS		*Non-nervous tissue*	
Tissue	**Highly Expressed PDE Isoforms**	**Tissue**	**Highly Expressed PDE isoforms**
fCT	2A	THY	8B
pCT	2A	ADR	2A
tCT	2A	LIV	2A, 3B, 8A
HIP	2A	PAN	3A,5A, 8A
CAU	1B, 10A	STO	5A
SN	1C, 4B,	INT	2A, 4B, 5A, 9A
NAC	1B, 2A,	HEA	1C, 3A
CER	4A, 4B, 9A,10A	MUS	4B, 4D
THA	1C, 4B,	KID	1A, 4D, 9A
HPT	1C, 4B	BLA	5A
DRG	1C, 2A,5A, 9A	LUN	5A
SPI	4B	SPL	2A

Table 2. Highly expressed PDE isoforms within a given tissue based on quantitative reverse transcription PCR data reported by Lakics et al. (2010). PDE isoforms were considered highly expressed if mRNA levels were 60% or higher relative to the most highly expressed isoform within a given tissue. fCT, frontal cortex; pCT, parietal cortex; tCT, temporal cortex; HIP, hippocampus; CAU, caudate; SN, substantia nigra; NAC, nucleus accumbens; CER, cerebellum; THA, thalamus; HPT, hypothalamus; DRG, dorsal root ganglion; SPI, spinal cord; THY, thyroid; ADR, adrenal gland; LIV, liver; PAN, pancreas; STO, stomach; INT, intestine; HEA, heart; MUS, skeletal muscle; KID, kidney; BLA, bladder; LUN, lung; SPL, spleen.

Predominant isoform	**Sites of high expression**	
	CNS	***Non-nervous tissue***
1B	CAU	
2A	fCT, pCT, tCT, HIP, CAU, NAC	SPL
3A		HEA
4B	fCT, pCT, HIP, CAU, SN, NAC, THA, HPT, SPI	SPL
5A		BLA, LUN
7B	CAU	
8B		THY
9A	CAU, CER, DRG	KID, BLA, SPL
10A	CAU	
11A	DRG	THY, LIV, PAN, MUS

Table 3. Sites of high expression for predominant PDE isoforms based on quantitative reverse transcription PCR data from Lakics et al. (2010). Expression was considered high if mRNA levels were 60% or greater relative to other sites measured. Members of the PDE6 family are not shown because PDE6 isoforms are only expressed at appreciable levels in retina, which was not tested in this study. fCT, frontal cortex; pCT, parietal cortex; tCT, temporal cortex; HIP, hippocampus; CAU, caudate; SN, substantia nigra; NAC, nucleus accumbens; CER, cerebellum; THA, thalamus; HPT, hypothalamus; DRG, dorsal root ganglion; SPI, spinal cord; THY, thyroid; ADR, adrenal gland; LIV, liver; PAN, pancreas; STO, stomach; INT, intestine; HEA, heart; MUS, skeletal muscle; KID, kidney; BLA, bladder; LUN, lung; SPL, spleen.

2.2 Expression of phosphodiesterase isoforms can change during development and aging

Expression of PDE isoforms is dynamic and can change during different physiological processes such as development and aging. Prickaerts and colleagues (2002) showed that PDE5 mRNA is expressed in cerebellar Purkinje cells of rat brains only on and after postnatal (P) day 10, whereas PDE9A mRNA is present at 15 days gestation and several postnatal stages (P0, P5, P10, P21) until adulthood, thus providing an example of altered expression of PDE isoforms during development. Changes to PDE isoform expression during aging have also been documented, as PDE5 is significantly decreased in old compared to young adult rat brains, while expression of PDE 9 is higher in old compared to young rat brains (Prickaerts et al., 2002). Decreases in PDE4 expression have also been reported in the in the aging brain, as PDE4A mRNA levels are reduced in striatum of old compared to young mice (Hebb et al., 2004). Other PDEs including PDE1B and PDE10A do not appear to change with age (Hebb et al., 2004).

3. Expression of phosphodiesterase isoforms PDE1B, PDE4A, and PDE10A is decreased in Huntington's Disease

Huntington's Disease (HD) is caused by the inheritance of a mutant *huntingtin* gene containing an expanded CAG repeat region, which codes for an expanded polyglutamine (polyQ) region in the mutant huntingtin (mHtt) protein (reviewed by Zuccato et al., 2010). The CAG repeat length of mHtt is inversely correlated with the age of HD symptom onset. mHtt is cleaved by caspase enzymes (Graham et al., 2010). The resulting, truncated, amino-terminus of mHtt (N-mHtt) translocates to the nucleus. It is thought that the nuclear, soluble, N-mHtt interferes with transcription and thus effects gene expression, cell function, and survival (Hermel et al., 2004). The neurodegeneration observed during HD progression is tissue- and cell-specific, such that the medium spiny neurons of the striatum (caudate/putamen) are most severely affected (reviewed by Zuccato et al., 2010). Transcriptional dysregulation is a major component of HD pathogenesis. N-mHtt is thought to interfere with the assembly of the transcriptional machinery, either through the sequestration of certain transcription factors, or through the inappropriate binding and interactions with co-factors and transcription factors at the site of transcription initiation. Because transcription is dysregulated by N-mHtt during HD pathogenesis, several research groups have examined and identified the subset of genes whose expression is altered in the presence of N-mHtt to determine how altered gene expression might contribute to this disease. The identification of dysregulated genes in HD has been completed primarily with mouse models of HD and tissue from human patients suffering HD.

Several transgenic mouse models of HD exist, which can be broadly categorized as models over-expressing mHtt or knock-in models expressing mHtt within the mouse *huntingtin* locus at physiologically accurate levels (Heng et al., 2008). Of the over-expression mouse models of HD, the mouse N171-82Q, R6, and rat *huntingtin* cDNA models express N-mHtt containing between 82 (N171-82Q) and ~144 (R6/2) CAG repeats and maintain a full complement of wild-type, mouse Htt. HD symptom progression and neurodegeneration in these models is more rapid than in knock-in models of HD (Heng et al., 2008). Two distinct transgenic HD lines are derived from the R6 model: R6/1 and R6/2. R6/1 mice express N-mHtt with ~113 CAG repeats and begin to exhibit HD motor symptoms at approximately 13 weeks of age. R6/2 mice expression N-mHtt with ~144 CAG repeats and begin to exhibit HD motor symptoms at approximately 8 weeks of age (Heng et al., 2008). Two other over-expression models, the mouse YAC128 and the HD cDNA models, express the full-length mHtt containing 128 CAG repeats. Disease progression is more rapid in these models than in the mouse knock-in models, but less rapid than in rodent models over-expressing N-mHtt (Heng et al., 2008). Also, the degree of neurodegeneration, as observed following euthanasia, is less severe in the full-length over-expression models than those models over-expressing N-mHtt (Heng et al., 2008). Knock-in mouse models of HD, including the Hdh/Q72 - 80 and Q111 - 150, express exon 1 of the human mutant *huntingtin* transgene in the mouse *huntingtin* locus. HD motor symptom onset, cognitive decline, and decreased socializing behaviours are delayed in these mouse models, relative to other rodent models of HD (Heng et al., 2008). Striatal cell loss, gross brain atrophy, and the size and number of neuronal intranuclear inclusions are also less prominent in mouse knock-in models of HD. Of the transgenic mouse models of HD, the R6 lines have been extensively studied because they display many behavioural and physiological changes associated with HD progression over a short period of time (Heng et al., 2008). R6/1 mice begin to exhibit motor symptoms related to the pathophysiology of HD between 15 and 18 weeks of age. R6/2 mice begin to exhibit HD-like symptoms between 8 and 9 weeks of age (Mangiarini et al., 1996). These symptoms include increased spontaneous locomotor activity, increased esacape latency in the Morris water maze, spatial learning deficits, and progressive rotarod deficit (Cha et al., 1998).

3.1 Phosphodiesterase 1B mRNA levels decrease in the striatum of transgenic Huntington`s Disease mice prior to symptom onset

PDE1B mRNA levels are lower in fully symptomatic, 12 week-old, R6/2 HD transgenic mice compared to age-matched wild-type mice, as demonstrated by microarray analysis (Luthi-Carter et al., 2004). Subsequent microarray analyses of PDE1B mRNA expression in symptomatic R6/1 mice, N171-82Q HD transgenic mice, a rat model of N-mHtt over-expression, and cDNA derived from the mRNA of symptomatic HD patients, all provide evidence for an N-mHtt-dependent decrease in PDE1B expression (Luthi-Carter et al., 2002; Desplats et al., 2006; Crocker et al., 2006; Nguyen et al., 2008). To determine when changes were first detected, and how CAG repeat length effected the rate or relative decline, PDE1B mRNA expression was measured in the striatum of the R6/1 and R6/2 HD transgenic mice, and wild-type mice, using *in situ* hybridization (Hebb et al., 2004; Fig. 4). An analysis of background-corrected optical density for PDE1B mRNA hybridization in coronal sections of mouse striatum revealed significant differences between genotypes and across ages. mRNA expression of PDE1B was reduced in R6/2 mice, relative to wild-type and R6/1 mice by 4 weeks of age. A significant decline in PDE1B was detectable by 10 week in R6/1 mice compared with wild-type. After the initial decline in transcript level observed in R6/1 and R6/2 mice, no further decline occurred. Therefore, PDE1B mRNA expression decreases in the presence of N-mHtt in the R6/1 and R6/2 transgenic mouse models of HD prior to motor symptom onset (Hebb et al., 2004). Decreased PDE1B expression may be a direct effect of expression of N-mHtt, or represent a compensatory mechanism during disease progression.

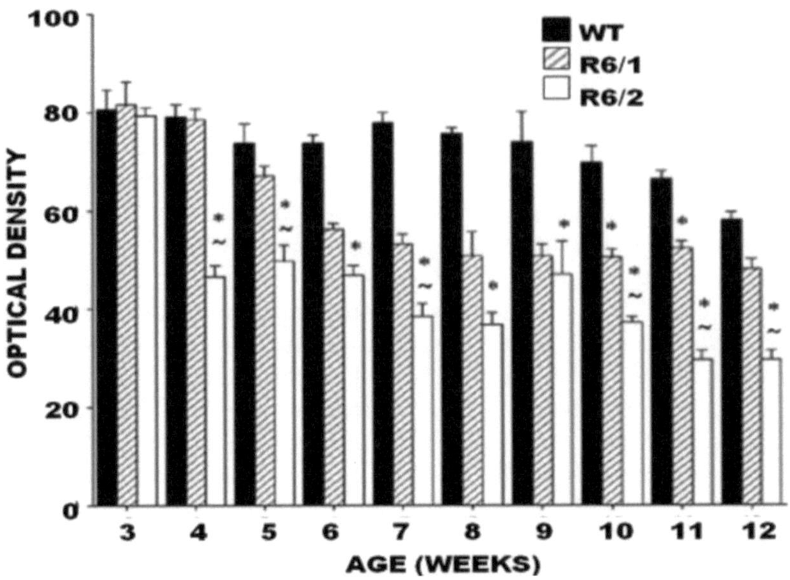

Fig. 4. PDE1B expression decreases in the striatum of R6/1 and R6/2 HD transgenic mice prior to symptom onset. This figure depicts the optical density of PDE1B mRNA *in situ* hybridization in the lateral striatum of wild-type (WT), R6/1, and R6/2 mice. In both R6/1 and R6/2 mice, there is an N-mHtt- and age-dependent decrease in PDE1B mRNA. Data represents means ± S.E.M. for $n = 4$ of each genotype and of mice as indicated. * $P < 0.01$, significant difference from age-matched WT. $P < 0.01$, ~ significant difference from age-matched R6/1.

3.2 Phosphodiesterase 10A mRNA and protein levels decrease in the striatum of transgenic Huntington`s Disease mice prior to symptom onset

PDE10A mRNA is expressed in the striatum, nucleus acumbens, and olfactory tubercle of R6 and wild-type mice. PDE10A mRNA distribution through the rostral-caudal axis of the mouse striatum is uniform (Fig. 5A). PDE10A mRNA expression is decreased in the striatum of R6/2 mice, relative to wild-type mice, by 4 weeks of age, as determined by *in situ* hybridization (Fig. 5A; Hebb et al., 2004). PDE10A mRNA levels continue to decline until reaching a new steady-state level, which is approximately 25% of that found in age-matched wild-type, by 9 weeks of age. Expression of PDE10A begins to decline between 6 and 7 weeks of age in R6/1 mice and continues to decline over the next 5 weeks, until reaching a new steady-state level of approximately 50% that of wild-type (Fig. 5B and C). Overall, three conclusions can be formed from *in situ* analysis of PDE10A expression in R6 mice. First, PDE10A mRNA levels do not normally change significantly within the striatum from 3 to 30 weeks of age. This conclusion suggests there is no effect of age on the expression of PDE10A. Second, PDE10A mRNA levels decline, and reach a final steady-state level, in an N-mHtt-dependent manner prior to symptom onset in both R6 lines. Third, the rate of PDE10A mRNA expression's decline is dependent upon the CAG repeat length of the mutant *huntingtin* transgene, as demonstrated by the more rapid rate of PDE10A mRNA in R6/2 mice, which express the *huntingtin* gene with a greater repeat length than R6/1 mice.

The N-mHtt-dependent decrease in PDE10A expression in the striatum of R6 mice was measured using western blot. PDE10A protein levels decrease in R6/2 mice, relative to wild-type, at 9 weeks of age and continue to decline until 15 weeks of age, when they achieve a new steady-state level (Fig. 5D; Hebb et al., 2004). In R6/1 mice a decrease in protein abundance is observed at 9 weeks of age, and the decrease continues until 18 weeks of age. The pattern of protein and mRNA decrease is similar in both R6 lines in that a significant decrease in levels is detected prior to or during motor symptom onset and the decline continues until a new steady state is achieved. In the case of PDE10A protein, the decrease is delayed, which is likely caused by a relatively long protein half-life.

As is observed in R6 transgenic mouse models of HD, PDE10A protein expression is decreased in human patients suffering from HD. PDE10A protein expression was analyzed in post-mortem human tissue from the caudate, nucleus acumbens, and putamen of grade 3 HD patients using western blot. When equal amounts of protein from healthy and HD patients were resolved by SDS-PAGE and probed with an anti-PDE10A antibody, little PDE10A could be detected in protein samples derived from HD patients (Fig. 6). These results demonstrate that PDE10A protein levels are decreased in HD, relative to age-matched healthy individuals (Hebb et al., 2004).

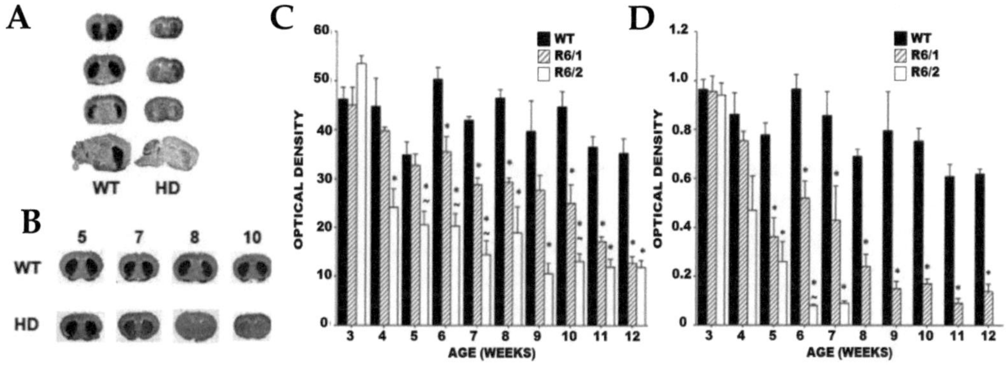

Fig. 5. PDE10A mRNA and protein levels decrease in the striatum of R6/1 and R6/2 HD transgenic mice prior to symptom onset. Panel A depicts PDE10A mRNA hybridization through the rostral-caudal axis of 10 week-old R6/2 HD and wild-type (WT) mice, with the bottom section shown in the sagittal plane. In panel B, the striatum-specific decline in PDE10A mRNA in R6/2 HD mice, relative to wild-type, is apparent by 7 weeks of age. Panel C depicts the optical density of PDE10A mRNA *in situ* hybridization in the lateral striatum of wild-type, R6/1, and R6/2 mice. In both R6/1 and R6/2 mice there is an N-mHtt- and age-dependent decrease in PDE10A mRNA. Panel D depicts the optical density of PDE10A protein from a western blot membrane for protein derived from striata of wild-type, R6/1, and R6/2 mice. Data represents means ± S.E.M. for $n = 4$ of each genotype and of mice as indicated. * $P < 0.01$, significant difference from age-matched wild-type. $P < 0.01$, ~ significant difference from age-matched R6/1.

The decreased steady state levels of PDE10A2 in the R6 mouse striatum are caused by an altered rate of transcriptional initiation, rather than an alteration in mRNA stability (Hu et al., 2004; Gomez et al., 2006). A comparison of the human and mouse PDE10A2 promoters reveals a high degree of conservation with respect to the presence and relative positions of several *cis*-regulatory elements. Altered expression of PDE10A2 may be a direct consequence of N-mHtt acting upon the transcriptional machinery present at the promoter of this gene (Hu et al., 2004).

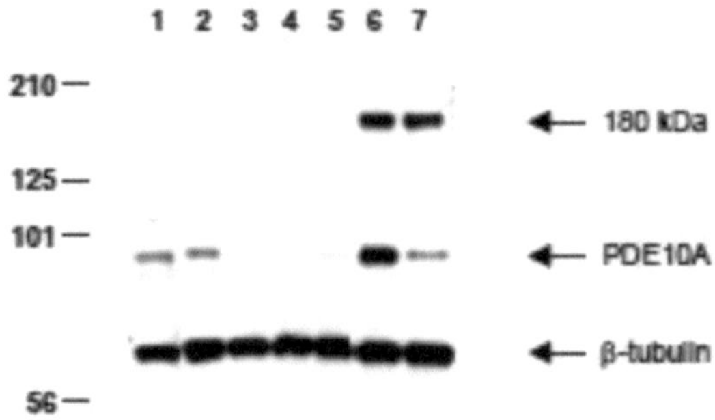

Fig. 6. PDE10A protein levels are lower in post-mortem samples from the caudate, nucleus accumbens, and putamen of patients with grade 3 HD, compared to age-matched controls. Lanes 1 and 2 represent 1 µg of protein derived from the caudate and nucleus accumbens of 66 and 53 year-old non-HD males, respectively. Lanes 3 - 5 represent 1 µg of protein derived from the caudate, nucleus accumbens, and putamen of grade 3 HD patients of 52, 67, and 48 year-old HD females (3 and 4) and a male (5). Lane 6 represents 1 µg of protein from the striatum of wild-type mice, which was included as a positive control. Lane 7 represents 1 µg of protein from the striatum of 12 wk-old R6/1 mice.

3.3 Phosphodiesterase 4A mRNA levels decrease in the striatum with age independently of mutant huntingtin

PDE4A expression is higher in the cortex than the striatum of wild-type and R6 mice. The optical density of cortical PDE4A mRNA has been measured by *in situ* hybridization. PDE4A mRNA abundance in the cortex declines as the animals age, and is significantly greater in wild-type mice relative to R6/2 at 3 weeks of age (Fig. 7). The decline in PDE4A mRNA expression is correlated with age, but not with the expression of N-mHtt, although it is possible that PDE4A mRNA begins to decline in R6/2 transgenic mice before it begins to decline in wild-type and R6/1 mice (Hebb et al., 2004).

3.4 Conclusions

PDE1B and 10A mRNA and protein levels are decreased in R6 mice relative to age-matched wild-type mice prior to motor symptom onset. PDE4A mRNA levels decline with increasing age. Decreased PDE1B and 10A expression in the R6 mouse models of HD is dependent upon expression of N-mHtt. Greater polyQ repeat length within the N-mHtt, such as in the fragment expressed in R6/2 mice relative to R6/1 mice, leads to earlier decreases in PDE1B and 10A expression. The lifespan of the R6 HD transgenic mouse models is summarized in figure 8. In the case of PDE10A2, decreased mRNA expression is the result of transcriptional interference by N-mHtt at the PDE10A2 promoter. Collectively, these data suggest expression of N-mHtt causes a decrease, at the level of transcription, in abundance of PDE1B, 10A, and possibly 4A, in the striatum (caudate/putamen). The functional consequence of decreased PDE expression in HD remains unclear.

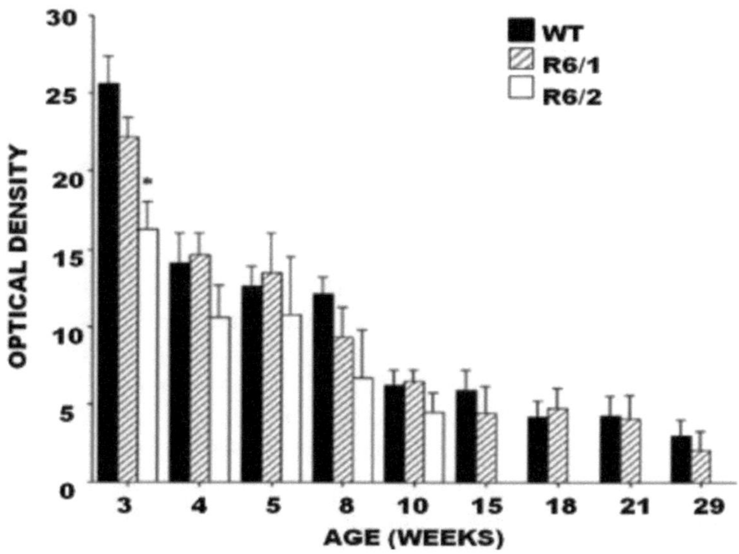

Fig. 7. PDE4A mRNA levels decrease with age. This figure depicts the optical density of PDE4A mRNA *in situ* hybridization in the cortex of WT, R6/1, and R6/2 HD transgenic mice. In all genotypes, there is a decrease in striatal PDE4A mRNA expression with increasing age. Data represents means ± S.E.M. for $n = 4$ of each genotype and of mice as indicated. * $P < 0.001$, significant difference from age-matched WT.

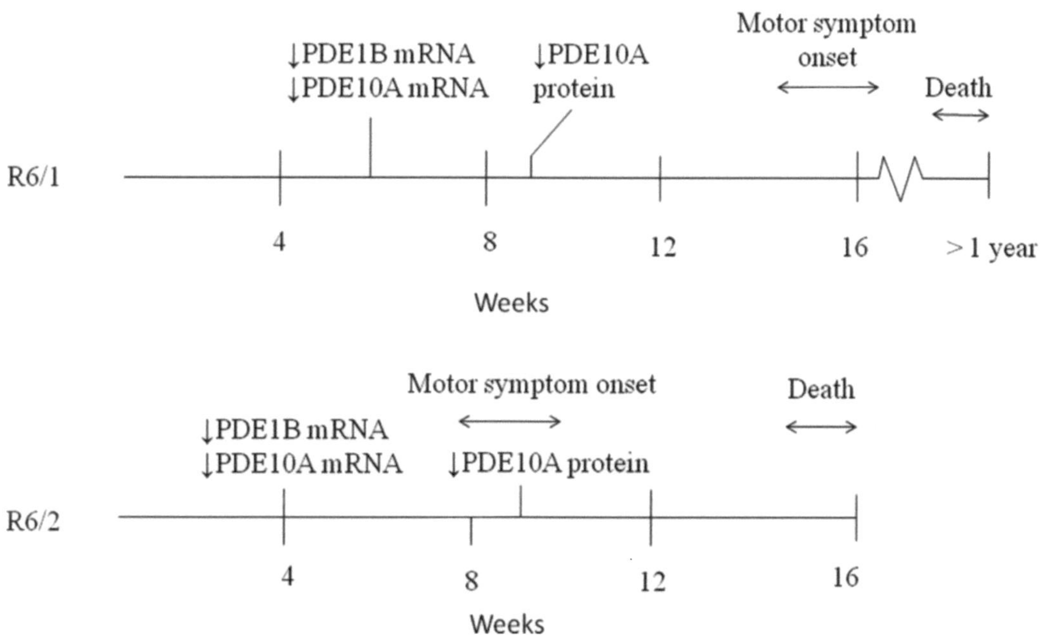

Fig. 8. Time-line of motor symptom onset, life span, and changes in PDE expression in R6/1 and R6/2 HD transgenic mice.

4. Impaired function of phosphodiesterase isoforms is associated with various pathological conditions of the central nervous system

HD progression is associated with distinct pathological changes in motor control and behaviour. Transgenic rodent models of HD that express mHtt, in part or whole, recapitulate many of the symptoms associated with HD and often experience cell-specific decreases in PDE1B and 10A mRNA expression (Table 4). However, the precise role of decreased PDE expression in these models is difficult to determine. Genetic knock-out of specific PDE isoforms, such as 1B, 4, 10A, and 11A, in mice causes phenotypic changes that often resemble the symptom profile of transgenic rodent models of HD (Kleppisch & Feil, 2009). Moreover, several mutations in specific PDE4, 6, 8, 10, and 11 isoforms are associated with disorders of the central nervous system, such as schizophrenia and major depressive disorder (Esposito et al., 2009). By comparing and contrasting the phenotype of HD to PDE knock-out models and other central nervous system disorders where PDE expression or activity are dysregulated, certain hypotheses can be made regarding the consequence of decreased PDE expression in HD.

4.1 Genetic knock-out of specific phosphodiesterases causes distinct behavioural phenotypes

Knock-out studies in which the expression of a specific PDE is eliminated by gene ablation or mutation reveal how changes in catalytic activity or expression of specific PDEs may contribute to disease pathophysiology. Mice lacking PDEs 1B, 1C, 4B, 4D, 6B, 9A, 10A, and 11A have been generated. These mice exhibit behaviours that resemble some behaviour associated with schizophrenia, major depressive disorder, hyperkinesias, and HD.

4.1.1 Phosphodiesterase 1B knock-out causes hyper-locomotion and spatial learning deficits in mice

PDE1B mRNA and protein are highly expressed in the striatum relative to other brain regions (Table 3). Expression of PDE1B decreases prior to symptom onset in R6 mouse models of HD. PDE1B knock-out mice were generated using homologous recombination to remove exons 2 – 13 of the mouse PDE1B gene (Reed et al., 2002). Knock-out of PDE1B is associated with increased locomotor activity, increased dopamine receptor-mediated phosphporylation of dopamine and cAMP-regulated neuronal phosphoprotein (DARPP-32), and performance deficits in spatial learning tasks. DARPP-32 is expressed in medium spiny projection neurons where it is phosphorylated and activated by protein kinase A following dopamine receptor-mediated cAMP production. Upon activation, DARPP-32 inhibits protein phosphatase 1, and thus facilitates phosphorylation and activation of pro-survival proteins. Double knock-out mice lacking PDE1B and DARPP-32 do not differ in phenotype from PDE1B null mice, suggesting that PDE1B activity upstream of DARPP-32 represents the major modulatory pathway for cyclic nucleotide messenger systems in the striatum (Ehrman et al., 2006). PDE1B knock-out mice show increased dopamine turnover and decreased serotonin levels in the striatum, and model depression-like behaviours such as decreased pleasure-seeking activity (Siuciak et al., 2007). These data indicate enhancement of cyclic nucleotide second messenger systems by PDE1B knock-out causes significant changes in locomotion and dopamine-mediated signal transduction within the striatum. Consequently, PDE1B null mice recapitulate the increased spontaneous locomotor activity and depression observed in HD mouse models and patients. Importantly during HD progression, DARPP-32 mRNA and protein levels decrease in an N-mHtt dependent manner (Gomez et al., 2006). This indicates that treatments that alter PDE1B activity may be limited by defects in downstream DARPP-32 levels or activity.

4.1.2 Complete phosphodiesterase 4B, or conditional phosphodiesterase 4D, knock-out produces a schizophrenia-like phenotype in mice

The dual-specificity PDE4 isoforms, including PDE4A, B, and D, are expressed in the cerebral cortex and amygdala (Siuciak et al., 2007). In HD, PDE4A mRNA declines with age and may contribute to changes in mood and behaviour observed during disease progression (Hebb et al., 2004). PDE4B knock-out mice were generated by homologous recombination of exons 3 - 6, which ablated the catalytic subunit of the mouse PDE4B gene (Jin et al., 1999). Specific knock-out of PDE4B in mice reduces prepulse inhibition, which is considered a mouse behavioural model of schizophrenia. The prepulse inhibition test utilizes a series of paired stimuli to determine whether an animal is capable of filtering external stimuli. Mice with normal executive function have a reduced response to the second of two, paired, stimuli, relative to the first. Mice exhibiting schizophrenia-like symptoms have a heightened response to the second stimuli due to a deficit in the ability to filter external stimuli. Mice lacking PDE4B are defective in their response to prepulse inhibition, and have decreased baseline locomotor activity (Siuciak et al., 2007). In addition, PDE4B null mice display anxiogenic-like behaviour, as measured by decreased head-dips in the hole board test, reduced transitions into the light side of a light-dark chamber, and decreased exploration of an open field. PDE4B null mice do not display changes in memory or nociception (Zhang et al., 2002). PDE4B knock-out mice display impaired reversal learning in the Morris water maze, but no differences in spatial memory or fear conditioning, relative to wild-type littermates (Rutten et al., 2009). Taken together, these data illustrate that PDE4B expression in the cortex and amygdala contributes to control of locomotion and anxiety-like behaviours and that other PDEs do not compensate for loss of PDE4B function.

PDE4D expression is more abundant in the cerebral cortex and hippocampus than other brain regions. Loss of PDE4D is associated with behaviours that mimic the effects of antidepressants, although the precise role of PDE4D in MDD pathophysiology is unclear. Mice lacking PDE4D display increased mobility in the forced swim and tail-suspension tests, indicative of antidepressant-like behaviours in mice (Zhang et al., 2002). These data demonstrate PDE4 regulate susceptibility to psychoses and changes in mood. Depressive symptoms, such as anhedonia and decreased socializing behaviour are observed in mouse models of HD and may result from the decline in expression of certain PDE4 isoforms.

4.1.3 Phosphodiesterase 10A knock-out reduces spontaneous locomotor activity and increases social interactions in mice

PDE10A is highly expressed in the medium spiny projection neurons of the striatum. Protein and mRNA expression of PDE10A is decreased in the striatum (caudate/putamen) in human patients with, and mouse models of, HD. Knock-out of PDE10A in mice causes increased escape latency in the Morris water maze, impaired conditioned avoidance behaviour, reduced spontaneous locomotor activity, increased social interaction, and increased levels of striatal cAMP, relative to wild-type mice. PDE10A knock-out does not induce anxiety-, or depression-like behaviours, or produce altered nociception (Siuciak et al., 2006). Further, hyper-locomotion associated with amphetamine treatment is absent in PDE10A knock-out mice

(Siuciak et al., 2006). Siuciak and colleagues (2006) concluded that inhibition of PDE10A may represent a novel therapeutic approach to the treatment of schizophrenia. PDE10A knock-out mice recapitulate the reduced spontaneous locomotor activity characteristic of late-stage HD rigidity, and increased escape latency in the Morris water maze, but differ in that mouse models of HD display decreased, not increased, social interaction behaviours (Table 4).

Species	Model	Behavioural phenotype	Decreases in PDE expression	Reference
Mouse	N171-82Q	• Increased spontaneous locomotor activity • Progressive accelerated rotarod deficit beginning at 12 weeks	• PDE2A (6 weeks)	Schilling et al. (1999) Yu et al. (2003) Runne et al. (2008)
Mouse	R6/1 and R6/2	• Increased spontaneous locomotor activity • Progressive accelerated rotarod deficit beginning at 5 and 12 weeks • Increased escape latency in the MWM • Spatial learning deficit	• PDE1B (4 weeks R6/2, 10 weeks R6/1) • PDE4A (with aging) • PDE10A (4 weeks R6/2, 6 weeks R6/1)	Cha et al. (1998) Meade et al. (2002) Ribchester et al. (2004) Hebb et al. (2004)
Mouse	*HD* cDNA	• Hyperactivity (12 weeks) • Decreased baseline motor activity (24 weeks)	• PDE1B, 10A (14 weeks)	Reddy et al. (1998) Thomas et al. (2008)
Mouse	YAC 128	• Increased spontaneous locomotor activity (12 weeks) • Decreased baseline motor activity (48 weeks) • Increased escape latency in the MWM	• PDE1B, 10A (12 weeks)	Benn et al. (2007) Mazarei et al. (2010)
Rat	*huntingtin* cDNA	• Cognitive decline (40 weeks) • Increased spontaneous locomotor activity	• PDE1B, PDE10A (12 weeks)	Nguyen et al. (2008) Cao et al. (2006)
Mouse	Hdh/Q72 - 80	• Increased aggression, decreased socializing behaviours • Anhedonia	• None reported	Kennedy et al. (2005)
Mouse	Hdh/Q111 - 150	• Decreased baseline motor activity (24 weeks) • Hyperactivity (4 weeks)	• None reported	Wheeler et al. (2002)

Table 4. Behavioural phenotypes and decreases in PDE expression observed in rodent models of HD. Phenotypes for specific transgenic rodent models of HD are described with the approximate time at which behaviours become present where possible. Changes in PDE expression were determined via microarray and subsequently confirmed by quantitative polymerase chain reaction.

4.1.4 Phosphodiesterase 11A knock-out produces a schizophrenia-like phenotype in mice

PDE11A mRNA is expressed in the hippocampus CA1, subiculum, amygdalohippocampal area, and dorsal root ganglia, as demonstrated by *in situ* hybridization (Kelly et al., 2010). PDE11A knock-out mice were generated by creating a missense mutation in the catalytic subunit of the protein, which caused it to be non-functional. PDE11A knock-out mice exhibit hyperactivity in an open field test, deficits in social odour recognition and social avoidance behaviours, enlarged lateral ventricles, and increased CA1 activity. Overall, this knock-out mouse model displays symptoms that are thought to be like some symptoms seen in psychotic patients.

In contrast, humans homozygous for loss-of-function mutations in PDE11A were more likely to suffer major depressive disorder than those with normal levels of PDE11A expression (Wong et al., 2006). These studies highlight the essential differences between mouse models and human disorders. In both cases though, deficits in social behaviours were present, which suggests PDE11A function is required for normal socialization processes. Microarray analysis of gene expression in tissue derived from rodent models of HD demonstrate that PDE11A expression is not changed in HD (Cha et al., 1998). The phenotype of PDE11A knock-out mice does, however, resemble the hyperactivity and social avoidance behaviours observed in rodent models of HD.

In conclusion, altered expression of PDE1B, 4, 10A, or 11A appear to change behaviour in similar manners in rodent models of HD and PDE knock-out mice. The phenotypes associated with genetic knock-out of specific PDEs are summarized in table 5.

Species	Model	Mutant gene/ gene locus	Associated phenotype	References
Mouse	Knock-out	PDE1B	• Increased locomotor activity • Increased dopamine receptor-mediated phosphporylation of DARPP-32 • Spatial learning deficit • Reduced pleasure-seeking activity	Reed et al. (2002)
Mouse	Knock-out	PDE4B	• Decreased baseline motor activity • Exaggerated locomotor response to amphetamine	Siuciak et al. (2007)
Mouse	Knock-out	PDE4D	• Increased mobility in the forced swim and tail-suspension tests	Zhang et al. (2002)
Mouse	Knock-out	PDE10A	• Increased escape latency in Morris water maze • Impaired conditioned avoidance learning • Reduced spontaneous locomotor activity	Siuciak et al. (2006)
Human	SNPs	PDE11A	• Major Depressive Disorder	Wong et al. (2006)
Mouse	Knock-out	PDE11A	• Hyperactivity • Deficits in social avoidance behaviours • Enlarged lateral ventricles	Kelly et al. (2010)

Table 5. CNS phenotypes related to ablation and mutations of PDE genes in mice.

4.2 Mutations in phosphodiesterases and their interacting proteins are associated with schizophrenia

Schizophrenia is a neurological disorder described by a range of behavioural, attention, sensory, and executive function-based deficits (Ebix Inc. Animated Dissection of Anatomy for Medicine, [A.D.A.M.], 2010). Individuals with schizophrenia may experience psychoses, delusions, and hallucinations, collectively known as positive symptoms, as well as feelings of depression and social isolation, described as negative symptoms. This disorder affects approximately 24 million people worldwide (A.D.A.M., 2010). Schizophrenia is complex in that both the underlying cause and pathogenesis are highly variable when individuals suffering schizophrenia are compared. Several environmental factors, such as prenatal stress, infection, and substance abuse contribute, or predispose individuals, to developing schizophrenia (A.D.A.M., 2010). Genetic factors also play a role in the disorder's etiology. Specific mutations in PDEs, and the proteins they interact with, are an example of these genetic factors. The symptom profiles of schizophrenia and HD overlap in several respects. First, the behavioural changes associated with both disorders are highly variable. Second, schizophrenia and HD are both associated with symptoms of depression and social withdrawal. Third, individuals with schizophrenia may exhibit hyperactivity and individuals with HD present with choreic movements, which may be neurologically related to hyperactivity (Siuciak et al., 2007). In this section the role of PDEs and their interacting partners in the etiology of schizophrenia will be summarized. We will demonstrate the important role these enzymes play in the central nervous system and how a dysregulation of their activity can contribute to schizophrenic disorders.

Several authors have reported up-regulation of PDE5 protein in post-mortem tissue samples from patients with schizophrenia, particularly those with prominent negative symptoms (Akhondzadeh et al., 2011). PDE1C and PDE8B mRNA are up-regulated in post-mortem samples derived from the lateral cerebellum of patients with schizophrenia (Fatemi et al., 2009). The precise cause of this up-regulation is unknown, but the data demonstrate schizophrenia pathogenesis is associated with dysregulation of several PDE families and isoforms.

PDE10A has garnered significant attention as a potential therapeutic target for schizophrenia. As previously described, the PDE10A variant, PDE10A2, displays differential sub-cellular localization depending on local cAMP level. PDE10A2 localizes to the membrane and is transported along dendritic processes by palmitoylation at cysteine 11 (Charych et al., 2010). Protein kinase A is activated by high cAMP and phosphorylates PDE10A2 at threonine 16, which interferes with trafficking of PDE10A2 to the membrane. The authors postulate that differential dopamine signalling, as observed in schizophrenia, in the direct and indirect striatal output pathways, would change cAMP levels and thus localization and activation of PDE10A2. Their model of dopamine-dependent PDE10A2 localization and activity is summarized in figure 2.

Mutations in the disrupted-in-schizophrenia-1 protein are considered strong genetic risk factors for the development of schizophrenia. Specifically, mutation of glutamine 31 to leucine (Q31L) or leucine 100 to proline (L100P) in the N-terminal region of this protein are associated with depression-like and schizophrenia-like phenotypes in mutant mice, respectively (Lipina et al., 2011). Disrupted-in-schizophrenia-1 protein exists in a protein complex with glycogen synthase kinase-3 and PDE4B in the rat dorso-lateral prefrontal cortex and hippocampus. This complex localizes to the synapse in primary mouse hippocampal cultured neurons (Lipina et al., 2011). In protein extracts derived from the hippocampus or dorso-lateral prefrontal cortex of L100P mice, disrupted-in-schizophrenia-1 protein -PDE4B binding was reduced by 75% and disrupted-in-schizophrenia-1 protein - glycogen synthase kinase-3 binding was reduced by 50%, relative to protein extracts derived from mice with wild-type disrupted-in-schizophrenia-1 protein. Similarly, disrupted-in-schizophrenia-1 protein -PDE4B binding was reduced by 50%, and disrupted-in-schizophrenia-1 protein –glycogen synthase kinase-3 binding by 75%, in Q31L mouse models of depression. The group hypothesized that disrupted-in-schizophrenia-1 protein acts as a scaffold to integrate and down-regulate the signalling pathways of PDE4B and glycogen synthase kinase-3. Sub-threshold, doses of the glycogen synthase kinase-3 inhibitor TDZD-8 and the PDE4 inhibitor rolipram effectively treat depression- and schizophrenia-like symptoms in both mutant mouse strains, as demonstrated by measuring pre-pulse inhibition deficit and mobility in the forced swim test. The authors conclude that disrupted-in-schizophrenia-1 protein mutations produce inappropriate interactions with PDE4B. The result of these inappropriate reactions was an inability to converge PDE4B and glycogen synthase kinase-3 signalling pathways contributing to schizophrenia-like phenotypes in mice. These data demonstrate the proper signalling of PDE4 isoforms is required for normal executive function, mood, and behaviour.

Mutations in the PDE4B gene itself have also been examined for associations with schizophrenia. The existence of PDE4B gene variants was examined in a population of 169 Caucasian patients taking antipsychotic medication. Two PDE4B variants associated with tardive dyskinesia and two additional variants associated with female-specific tardive dyskinesia were discovered (Souza et al., 2011). However, correction for multiple testing eliminated these variants as being truly genetically associated with the tardive dyskinesia observed in schizophrenia. In contrast, a similar study examined variations in the PDE4B gene in 837 individuals with schizophrenia and 1473 controls (Kahler et al., 2010). They found four variants in the PDE4B3 isoform nominally associated with schizophrenia in females, and four additional single nucleotide polymorphisms associated with positive symptom scores according to Positive And Negative Symptoms Scale (PANSS) testing of patients. Similar results were found in the PDE4B gene in a Japanese population, lending further support to the theory that certain PDE4B variants have a positive association with schizophrenia (Numata et al., 2009). Up-regulation of PDE4A and 4B mRNA has been observed in the frontal cortex of patients with schizophrenia (Fatemi et al., 2009). Overall, mutation of PDE4 isoforms, or changes in the level of expression of PDE4, is associated with changes in mood and behaviour related to schizophrenia. Therefore, decreased PDE4 expression during HD progression may contribute to changes in mood and behaviour as well.

4.3 Changes in phosphodiesterase 4 mRNA expression, but not allelic variability of phosphodiesterases, is implicated in major depressive disorder

Major depressive disorder is a neurological disorder characterized by emotional, attentional, sensory, and executive function-based deficits (A.D.A.M., 2011). Individuals with major depressive disorder may experience feelings of sadness, loss, anger, or frustration that persist for extended periods of time such that these feelings interfere with their normal ability to function and be productive. Major depressive disorder affects approximately 8 –

12% of all people at some point during their lives (A.D.A.M., 2011). Approximately 40% of patients suffering from HD exhibit symptoms of depression (A.D.A.M., 2011). The Hdh mouse models of HD exhibit anhedonia and decreased socializing behaviours, which are considered to be analogous to human depression (Kennedy et al., 2005). Genetic factors, or heritability, contribute 40 - 50% to the probability a person will suffer major depressive disorder (Numata et al., 2009). Changes in the mRNA expression or the activity of certain PDEs can contribute to major depressive disorder etiology.

PDE4B mRNA expression, single nucleotide polymorphisms, and haplotype variants were examined in a large Japanese population (655) suffering major depressive disorder (Numata et al., 2009). No significant correlation between allelic variation and major depressive disorder was found. PDE4B is most likely implicated in the pathophysiology of major depressive disorder because of the differential mRNA expression observed in animal models and human patients suffering from major depressive disorder. During HD progression, mRNA expression of PDE4A declines in a cell-specific manner in the cortex (Hebb et al., 2004). Depressive symptoms often observed in individuals suffering HD, and mouse models of HD, may therefore be explained by a decline in PDE4 expression.

Other authors have analyzed associations between allelic variation in PDE1A, 8A, 9A, and 11A and major depressive disorder (Wong et al., 2006; reviewed by Esposito et al., 2009). Nominally significant allelic associations between these PDEs and major depressive disorder have been found. However, independent analyses of these data, or attempts to replicate these findings in other populations, have failed to demonstrate significance. Of these, only one demonstrated a significant association between an inactivating mutation in PDE11A and individuals with adrenocortical hyperplasia and major depressive disorder (reviewed by Esposito et al., 2009).

4.4 Conclusions

Transgenic mouse models of HD display locomotor and cognitive deficits, including early-symptomatic increased spontaneous locomotor activity, late-symptomatic hypoactivity, and increased escape latency in the Morris water maze. These mouse models also display the depression-like phenotypes of anhedonia and decreased socializing behaviours (Heng et al., 2008). Similarly, genetic ablation of PDE1B, 4, 10A, or 11A is associated with specific locomotor and cognitive declines. Increased locomotor activity, spatial learning deficits, and reduced pleasure-seeking activity are observed in PDE1B knock-out mice (Reed et al., 2002). Increased escape latency and reduced spontaneous locomotor activity are observed in PDE10A knock-out mice (Siuciak et al., 2006). These data suggest that the consequence of decreased PDE expression, as observed in HD, may be a change in motor control and mood.

Human patients suffering from HD experience spontaneous choreic movements early in disease progression, rigidity late in disease progression, and symptoms of depression. Schizophrenia and major depressive disorder are two disorders of the central nervous system where PDE expression and/or catalytic activity are dysregulated. Therefore changes in PDE1B, 4, and 10A may play a contributing factor in the pathogenesis of HD and other central nervous system disorders. Expression and function of PDE4B and PDE11A in the amygdala, cortex, and hippocampus is critical to maintain normal cognitive function and social interaction. PDE4D also appears to be involved in social interaction and depression-like behaviour. Expression of PDE8B is up-regulated in the cortex and hippocampus of Alzheimer's Disease patients compared to age-matched controls (Pérez-Torres et al., 2003). Although microarray analyses suggest no significant changes in the expression of PDE4B, 4D, 8A, or 11A mRNA, it is interesting to note that other central nervous system disorders are associated with cell-specific changes in PDE expression, which may contribute to their pathophysiology.

5. Pharmacological inhibition of phosphodiesterases in the central nervous system

Because multiple PDE isoforms are expressed in the central nervous system (Tables 2 and 3) and individual isoforms are tightly coupled to specific physiological functions, pharmacological inhibitors may be used to treat pathological conditions of the central nervous system without a high likelihood of causing non-specific side effects. PDEs represent a logical target for competitive inhibition because concentrations of their substrate (cAMP and cGMP) are low (>1µM to 10 µM; Koyanagi et al., 1998). This means that competition with endogenous substrate could be achieved using low concentrations of PDE inhibitors. However, PDE isoforms share similar structure in the catalytic domain, which makes design and development of truly selective competitive inhibitors difficult. Within the active site of all PDE isoforms studied to date, 11 invariant residues have been identified which maintain a consistent arrangement between isoforms and are believed to be important for catalytic activity (Manallack et al., 2005). Nevertheless, multiple PDE competitive inhibitors have been developed which show some degree of isoform-specificity as demonstrated by a lower half-maximal inhibitory concentration (IC_{50}) for one isoform relative to other isoforms. For treatment of central nervous system disorders, competitive PDE inhibitors must also effectively cross the blood-brain-barrier. This section will review the pharmacological profile of PDE competitive inhibitors that show some degree of selectivity for individual PDE isoforms and have well known effects in the central nervous system.

5.1 Papaverine, TP-10 and MP-10 are selective competitive inhibitors of PDE10A

Papaverine is an opium alkaloid that was first isolated in 1848 from poppies, or *Papaver somniferum*, from which the name "papaverine" is derived (Hollman, 2005). Medical use of papaverine was first suggested in 1914 for treatment of hypertension and angina (Hollman, 2005). Papaverine was shown to competitively inhibit PDE10A following quantification of IC_{50} values in mice (Siuciak et al., 2006). The IC_{50} of papaverine for PDE10A is 36 nM, which is between 9 and 52-fold lower than the IC_{50} for the next most easily inhibited isoform, PDE4 (Siuciak et al., 2006).

TP-10 and MP-10 are two PDE10A competitive inhibitors that were developed by the pharmaceutical company Pfizer in 2008. The IC_{50} of TP-10 and MP-10 for PDE10A is approximately 0.3 nM and 0.18 nM respectively. This is between 3,333 and 10,000-fold lower than the IC_{50} for the 18 other PDE isoforms tested (Schmidt et al., 2008), which makes these compounds more selective for PDE10A than papaverine. TP-10 and MP-10 are more potent than papaverine as 3.2 mg/kg of TP-10 administered sub-cutaneously produced a 3- and 3.5-fold increase in extracellular cAMP and cGMP respectively in rat striatum (Schmidt et al., 2008). The dose of papaverine required to achieve a similar effect was 56 mg/kg (Siuciak et al., 2006).

5.2 Rolipram is a selective competitive inhibitor of PDE4 isoforms

Rolipram is a PDE4 competitive inhibitor originally developed as an antidepressant (Kehr et al., 1985). The IC_{50} of rolipram is approximately 500 nM for PDE4 in mice (Bader et al., 2006), which is approximately 24-fold lower than the IC_{50} for PDE10A (Bader et al., 2006). Within the PDE4 family, rolipram seems to inhibit PDE 4A most effectively as rolipram inhibited immunopurified PDE4A activity in U937 human histiocytic lymphoma cells with an IC_{50} of approximately 3 nM, compared to IC_{50} values of approximately 130 nM and 240 nM for PDE4B and PDE4D respectively (Bader et al., 2006). Inhibition of PDE4B and PDE4D is known to contribute to the antidepressant effects of rolipram, as the ability of rolipram to alleviate depression-like behaviours, is partially lost in PDE4B and PDE4D knock-out mice (Zhang et al., 2002; Siuciak et al., 2007).

5.3 Sildenafil is used for inhibition of PDE5A in the periphery, but also has effects in the central nervous system through inhibition of PDE5A and possibly PDE6

Sildenafil (trade name Viagra) was first developed by the pharmaceutical company Pfizer for treatment of angina, hypertension, and erectile dysfunction (Boolell et al., 1996). The IC_{50} of sildenafil for PDE5 is 3 nM as measured in human corpus cavernosum (Ballard et al., 1998). This is between 80- and 8500-fold lower than the IC_{50} for PDEs 1-4 (Ballard et al., 1998). Anti-angina, -hypertension, and –erectile dysfunction effects of sildenafil are mediated by inhibition of PDE5 which is enriched in smooth muscle of the lungs and corpus callosum (Boolell et al., 1996). PDE5 is also expressed in the brain and growing evidence suggests that orally delivered sildenafil has effects in the central nervous system, as inhibition of PDE5 in the brain is associated with improved object recognition memory in rats (Prickaerts et al., 2002) and altered event-related brain potentials in humans (Schultheiss et al., 2001). The IC_{50} of sildenafil for PDE6 is 9–fold greater than the IC_{50} for PDE5 (Ballard et al., 1998). Despite a higher IC_{50} for PDE6 than PDE5, inhibition of PDE6 in the retina is believed to contribute to visual disturbances reported in a minority of patients taking sildenafil (Marmor & Kessler, 1999). IC_{50} values of sildenafil for PDEs 7-11 have not yet been reported. Taken together, evidence suggests that sildenafil inhibits PDE5 and PDE6 in the central nervous system in addition to inhibiting PDE5 in the periphery.

6. Pharmacological inhibition of phosphodiesterase activity is useful in the treatment of several neurological disorders

Phosphodiesterase inhibitors are used for the treatment of embolism, thrombocytosis, inflammation, decreased cerebral blood flow, heart failure, asthma, chronic obstructive pulmonary disease, and erectile dysfunction. PDE inhibitors exhibit antidepressant, and nootropic (*i.e.* memory enhancing), properties, which has led to the development of central nervous system-specific PDE inhibitors for the treatment of several neurological disorders.

6.1 Phosphodiesterase inhibitors improve cognitive and sensorimotor deficits in schizophrenia

Inhibition of PDE4B, 5, and 10A, enzymes has been investigated as a potential therapeutic means of reducing psychoses. In particular, inhibitors of these enzymes improve attentional and sensorimotor deficits, as well as socializing deficits, in animal models of schizophrenia.

Several research groups have investigated the clinical efficacy of PDE inhibitors for the treatment of both positive and negative symptoms associated with schizophrenia. Two common rodent models of schizophrenia have been used to examine the effect of PDEs. Dopamine receptor-agonist treated mice and rats exhibit stereotypy and hyperactivity, which are behaviours thought to model the positive symptoms of schizophrenia in rodents. The other pharmacological rodent model of schizophrenia is the phencyclidine-treated mouse or rat. Phencyclidine acts as an N-methyl-D-aspartic acid (NMDA) receptor antagonist. Phencyclidine-treated rodents exhibit hyperactivity, prepulse inhibition, and anhedonia and are considered to be rodent models of the positive and negative symptoms of schizophrenia. The most well-known PDE inhibitor studied for use in schizophrenia is rolipram. Rolipram improves cognition, memory, and prepulse inhibition deficits in dopamine receptor agonist-treated mice. PDE4B activity and regulation are disrupted in the disrupted-in-schizophrenia-1 protein-L100P transgenic mouse model of schizophrenia (Lipina et al., 2011). Treatment of disrupted in schizophrenia-1 protein-L100P mice with the PDE4-specific inhibitor rolipram (0.1 mg/kg) corrects the deficit in prepulse inhibition and hyperactivity without producing overt side effects. Rolipram also reduces psychoses and improves attentional deficits in patients with chronic schizophrenia. However, rolipram has been discontinued as a treatment for schizophrenia because its use is associated with nausea, emesis, weight loss, and acute insomnia. These adverse effects are observed following treatment with all known PDE4 inhibitors. Acute insomnia, as it pertains to PDE4 inhibition, describes an inability to sleep consistently for less than 1 month during drug use (reviewed by Zhang et al., 2002). In addition to the adverse effects observed generally for PDE4 inhibitors, rolipram causes gastrointestinal pain and cardiac arrhythmia (Zhang et al., 2002).

PDE10A was first identified as a "druggable" target for the treatment of schizophrenia in 1999 (Itoh et al., 2011). The PDE10A-selective inhibitor papaverine has gained attention as a possible treatment for schizophrenia because of its neuroprotective actions. Papaverine induces NGF-dependent neurite outgrowth in PC12 neuroblastoma cells (Itoh et al., 2011). However, another PDE10A-selective inhibitor, MP-10, has no effect on neurite outgrowth in this model. Therefore, the effect of papaverine on neurite outgrowth may not be mediated by inhibition of PDE10A. Other PDE10A-selective inhibitors, such as the imidazol[1,5-a]pyridol[3,2-e]pyrazines, are effective at reducing stereotypy and hyperactivity in rats treated with phencyclidine or dopamine receptor agonists (Itoh et al, 2011). The highly selective PDE10A inhibitors MP-10 and TP-10 decrease hyperactivity, attenuate conditioned avoidance responses, recover prepulse inhibition deficits, and improve social odour recognition and novel object recognition in methamphetamine- or phencyclidine-treated rats (Kahler et al., 2010). Pfizer began a Phase I, placebo-controlled, randomized, double-blind, parallel assignment, safety/efficacy clinical trial for MP-10 in 2007. This trial demonstrated that MP-10 has a clearance of 4 mL min^{-1} kg^{-1}, a half-life of 14 h, high oral bioavailability and low pharmacokinetic variability. Pfizer began a Phase II clinical trial for MP-10 with an anticipated end date of May 2008 and a primary end point of significant improvement for patients suffering schizophrenia on the Positive And Negative Symptoms Scale (PANSS). Unfortunately, this trial has been discontinued and the Pfizer website does not provide a clear statement regarding the reason for the trial being discontinued. Two Phase II clinical trials, utilizing the PDE4 inhibitor dipyridamole are ongoing at the University of Maryland and Hospital Espirita de Porto Alegre (United States National Institutes of Health [NIH],

2011). Pfizer has also disclosed a patent for PQ-10, a papaverine-like PDE10A inhibitor. However, this compound appears to inhibit PDE10A and the cardiac-specific PDE3A isoforms, and can cause hypotension (Kahler et al., 2010).

The PDE5-selective inhibitor has been shown to improve socializing behaviours in animal models of schizophrenia. Sildenafil has been utilized as an adjunct therapy to risperidone for the treatment of patients suffering chronic schizophrenia (Akhondzedah et al., 2011). Forty patients were treated in a double-blind fashion with risperidone (6 mg/day), and sildenafil (75 mg/day) or placebo, for 8 weeks. Patients receiving sildenafil experienced a significant improvement compared with those given risperidone alone when symptoms were measured by positive and negative symptoms scale (PANSS). Importantly, no negative side effects were reported. The Massachusetts General Hospital recently completed a Phase IV clinical trial for the use of sildenafil on improving cognitive functioning, verbal memory, fluency, attention, spatial memory, motor speed, executive function, and reducing the incidence of psychoses and withdrawal symptoms in patients with schizophrenia. Their study utilized single daily doses of sildenafil (50 or 100 mg) for 12 days. Results have not yet been published (NIH, 2011).

6.2 Phosphodiesterase inhibition improves deficits in social interaction and mood associated with depression

Evidence from animal models and clinical trials demonstrating PDE inhibition could alleviate the negative symptoms of schizophrenia led investigators to explore the utility of these compounds in major depressive and bipolar disorders. Inhibition of certain PDEs can effectively improve depression-related symptoms in animal models. Moreover, several PDE inhibitors are currently being tested in clinical trials to confirm their efficacy in treating human depression. The animal model and clinical trial data concerning the antidepressant effects of PDE inhibitors add support to the utility of PDE inhibitors in treating the depressive symptoms observed among individuals with HD.

Isoforms of PDE4, particularly PDE4A and D, are expressed in the hippocampus and frontal cortex, which are areas classically considered to be the mediators of antidepressant drug effects. PDE4 inhibitors, such as rolipram, were first investigated for their antidepressant properties in animal models 27 years ago (Zhang et al., 2002). Treatment of rodents with the drug reserpine is used as a model of depression. Reserpine causes hypothermia in rodents, which is reversed by antidepressant drugs, such as tricyclic antidepressants. Rolipram is capable of reversing reserpine-induced hypothermia, and reducing immobility in the forced swim test. More recent animal models of depression include the learned helplessness model and the serotonin-depletion model. These animals are more likely to engage in pleasure-seeking activity (*i.e.* seek a mate) when treated with rolipram. Rolipram's antidepressant activities have been confirmed by several clinical trials, yet the drug has not been marketed because it causes emesis and gastrointestinal complications. Despite these complications, rolipram effectively demonstrates the utility of PDE4 inhibition for treatment of depression. As an antidepressant, rolipram is 30 times more potent than the tricyclic antidepressants imipramine or desipramine, and the effects of rolipram are potentiated by serotonin-selective reuptake inhibitors, suggesting serotonin-selective reuptake inhibitors and PDE inhibitors might be useful for the combinatorial treatment of depression (reviewed by Zhang et al., 2002). A new Phase II clinical trial is ongoing by the NIH to determine the efficacy of low-dose rolipram for the treatment of major depressive disorder. The expected date of completion for this trial is December 2011. In this trial patients will receive rolipram, or placebo, for 3 years. During this time, symptoms of major depressive disorder will be monitored, brain PDE4 levels will be measured by PET scan, and the possible correlation of PDE4 level and major depressive disorder symptoms will be explored.

Inhibitors of PDE5 isoforms may represent equally promising means of treating depression. A double-blind, placebo-controlled clinical trial examined the usefulness of sildenafil for the treatment of mild-to-moderate, previously untreated, depression in men suffering erectile dysfunction (Kennedy et al., 2005). A total of 202 men were recruited for the trial, which lasted 6 weeks, and volunteers were treated with 50 mg sildenafil once daily, or placebo. Patients treated with sildenafil had significantly improved scores for the Beck Depression Inventory-II (BDI-II) and the erectile dysfunction domains, which are questionnaires to determine the depressive state and sexual satisfaction of a patient, respectively. Similar results were reported in an earlier, larger clinical trial, where the Self-Esteem And Relationship (SEAR) questionnaire was employed (Moncada et al., 2009). Inhibitors of PDE4 and 5 isoforms demonstrate robust antidepressant effects in animal models and clinical trials in men (Moncada et al., 2009).

PDE10A inhibition has also been investigated for antidepressant properties. Hypothermia is not reversed in reserpine-treated mice subsequently treated with papaverine or TP-10. Moreover, PDE10A knock-out mice do not differ from wild-type litter mates when they are tested in the forced swim test, which is considered a well-established test for depression in rodents (Siuciak et al., 2006).

6.3 Phosphodiesterase inhibition may be useful for cognitive enhancement in Alzheimer's Disease

Inhibitors of PDE isoforms have been investigated for their antidepressant and cognitive enhancement properties. Because of the ability of these compounds to enhance cognition and improve memory, they were tested in mouse models of Alzheimer's Disease. Treatment of transgenic Alzheimer's Disease mice (Tg2576), which over-express amyloid β precursor protein, with the PDE5 inhibitor sildenafil improves memory function, as measured in the Morris water maze, and significantly increases brain-derived neurotrophic factor levels in the hippocampus (Cuadrado-Tejedor et al., 2011). Brain-derived neurotrophic factor expression is decreased in human patients suffering from Alzheimer's Disease and major depressive disorder (Cuadrado-Tejedor et al., 2011). Brain-derived neurotrophic factor-mediated signal transduction is associated with increased protein kinase A activity, increased levels of phosphorylated CREB, and increased CREB-dependent gene expression (Zuccato et al., 2010). Brain-derived neurotrophic factor mRNA and protein levels are also decreased in HD. Decreased brain-derived neurotrophic factor is thought to contribute to decreased cell survival and gross atrophy during HD progression (Zuccato et al., 2010). Thus, the finding that sildenafil use might increase brain-derived neurotrophic factor levels suggests this drug might effectively delay neurodegeneration in disorders of the central nervous system such as HD.

Rolipram has also been investigated for Alzheimer's Disease treatment because of its nootropic properties. Initial evidence from pre-clinical models suggested inhibition of PDE4 isoforms would be logical in the context of Alzheimer's Disease because 1) they improved deficits in long-term memory in mouse models of Alzheimer's Disease, 2) they improved neurogenesis and pre-synaptic plasticity in mouse models of neurodegeneration, and 3) they evoked potent anti-inflammatory responses in mice challenged with lipopolysaccharide and other inflammatory agents (reviewed by Esposito et al., 2009). Moreover, expression of PDE4B is up-regulated in hippocampal neurons and microglia of mouse models of Alzheimer's Disease and autopsied Alzheimer's Disease human tissue (Cuadrado-Tejedor et al., 2011). Long-term potentiation, protein kinase A activity, and CREB phosphorylation are improved in hippocampal slice cultures derived from Alzheimer's Disease transgenic mice treated with rolipram (Cuadrado-Tejedor et al., 2011). Similarly, retention in the passive avoidance test is decreased, and an improvement in the Morris water maze test for memory is observed, in mice given amyloid β protein injections to the CA1 hippocampal region and treated with rolipram for 32 days (Cheng et al., 2010). Furthermore, coronal sections of hippocampus derived from mice treated with rolipram express phosphorylated CREB at significantly higher levels than mice that were not given rolipram. Despite these promising results, rolipram *per se* is not a useful drug for the treatment of Alzheimer's Disease because it causes emesis, insomnia, and cardiac arrhythmia. Current research in the area of PDE inhibitors for use in Alzheimer's Disease is focused on developing PDE4-selective inhibitors that do not cause emesis. Merck conducted a Phase II clinical trial for the use of their PDE4 inhibitor, MK-0952, on 55-year or older patients with mild-to-moderate Alzheimer's Disease in 2008. The trial has been completed and no significant improvements in patient cognitive function or memory were reported (NIH, 2011). One Phase II clinical trial, recently completed by Exonhit therapeutics, is attempting to establish the usefulness of the drug etazolate as a therapy for the treatment of Alzheimer's Disease. Etazolate is a PDE4 inhibitor and adenosine receptor antagonist. Data have not yet been published from this trial (NIH, 2011).

PDE1 inhibitors have shown nootropic properties in the treatment of Alzheimer's Disease, where PDE1 expression does not change. PDE1 inhibitors were first utilized clinically as vasodilators and anti-inflammatory agents. In 2003, the PDE1 inhibitor vinpocetine was investigated for its actions as a nootropic for therapeutic use in Alzheimer's Disease and dementia. Three clinical trials were conducted, all of which demonstrated significant cognitive improvement in patients treated with vinpocetine relative to those treated with placebo (reviewed by Szatmari & Whitehouse, 2003). However, the number of patients treated for a period greater than 6 months was too small for any conclusions to be drawn. The major side effect reported was agranulocytosis, a serious condition that reduces immune function. This potential adverse effect was considered too significant a risk to the treatment group. Consequently, no subsequent clinical investigations have been conducted.

6.4 Phosphodiesterase inhibition improves motor function in neurodegenerative disorders and traumatic central nervous system injury

PDE inhibition produces beneficial motor effects in animal models of neurodegenerative disorders such as Parkinson's Disease, and multiple sclerosis, as well as spinal cord injury, and ischemic-stroke. PDE7 inhibition, using the selective inhibitor S14, was shown to improve motor function in the lipopolysaccharide model of Parkinson's Disease, possibly due to protection of dopaminergic neurons in the substantia nigra (Picconi et al., 2011).

Long-term levodopa use by Parkinson's Disease patients causes involuntary jerky movements of the arms and/or head, which is described as dyskinesia. Dyskinesia also occurs in mice given 6-hydroxydopamine lesions to the substantia nigra and chronically treated with levopdopa. Dyskinesia is reduced in these mice if they are administered striatal injections of zaprinast (PDE 5,6,9,11 inhibitor) or UK-343664 (PDE5 inhibitor) twice per day for 21 days (Picconi et al., 2011).

PDE5 inhibition by subcutaneous sildenafil treatment (10 mg/kg, once per day) for 8 days reduced the incidence and severity of abnormal movements in experimental autoimmune encephalomyelitis mice, a model of multiple sclerosis (Picconi et al., 2011). Sildenafil treatment in this study also improved neuropathology, as shown by reductions in 1) demyelination and axonal loss in spinal cord, 2) inflammatory cell infiltration, and 3) microglia activation (Picconi et al., 2011). PDE4 and PDE5 inhibition seem to play a beneficial role in motor recovery after traumatic injury of the central nervous system. Various selective PDE5 inhibitors have been shown to promote neurogenesis and improve motor recovery after cerebral ischemia (Picconi et al., 2011). Rolipram treatment also improves motor recovery following spinal cord injury in rats (Picconi et al., 2011).

6.5 Inhibitors of phosphodiesterases 2, 3, and 5, used in the treatment of peripheral disorders, have deleterious effects on the central nervous system

Several PDE inhibitors are used to treat disorders in the periphery. Although the main site of action for these inhibitors is peripheral tissue, several of these drugs effect central nervous system function either directly if they are blood-brain barrier penetrant, or indirectly via their effect on blood flow. The PDE2/3-selective inhibitor anagrelide is used to treat severe cases of essential thrombocytosis and is not considered blood-brain barrier penetrant (Sadhu et al., 1999). However, anagrelide is known to cause migraine headaches and dizziness. These effects are thought to stem from the vasodilation and decreased blood pressure often associated with chronic use of anagrelide.

Perhaps the most widely known and studied peripheral PDE inhibitors with actions in the central nervous system are PDE5-selective inhibitors, such as sildenafil. As mentioned above, sildenafil - and similar drugs - are commonly used to treat clinical erectile dysfunction. They also demonstrate nootropic, antidepressant, and motor control-enhancing properties in clinical trials for, and animal models of, depression, Alzheimer's Disease, and multiple scleorsis. However, sildenafil is also known to cause headaches in approximately 10% of all users, sudden hearing loss, and anterior optic neuropathy (NIH, 2011). Sildenafil-associated headaches are thought to be the result of altered blood flow. Sudden hearing loss and anterior optic neuropathy, however, are caused by PDE5 or 6 inhibition in the cochlear nerve and eye, respectively (NIH, 2011; see section 3.3). The Pfizer corporation recently completed a Phase III observational trial to determine the prevalence of anterior optic neuropathy in men using sildenafil for clinical erectile dysfunction. Their study found that the risk of developing anterior optic neuropathy for users of sildenafil was significant and posed a major risk to chronic users of sildenafil (odds ratio 5.73, prevalence of 0.35 per 1000 individuals). Pfizer is currently recruiting for a Phase IV clinical trial of sildenafil to explore the prevalence of anterior optic neuropathy in greater detail. The Eli Lily company is conducting a parallel trial with another PDE5-selective inhibitor. Pfizer has also conducted clinical trials to determine the efficacy of sildenafil as a treatment of Miniere's Disease. In their Phase II clinical trials, Pfizer found sildenafil did not significantly improve symptoms associated with Meniere's Disease, relative to placebo (NIH, 2011).

PDE inhibition has profound effects on physiology in the central nervous system and periphery. Preclinical data from animal models of schizophrenia, depression, Alzheimer's Disease, Parkinson's Disease, and multiple sclerosis suggest PDE 4, 5, and 10A inhibitors effectively improve deficits in socializing behaviour, cognitive function, and locomotor activity. Clinical data from schizophrenia, depression, and Alzheimer's Disease trials confirm the beneficial effects of PDE4 and 5 inhibitors on locomotor control, cognitive function, and mood. Unfortunately, PDE1, 4, 5, and 10A inhibitors are known to have certain adverse side effects, such as emesis in the case of rolipram, or anterior optic neuropathy in the case of sildenafil.

Given the efficacy of PDE inhibitors at improving locomotor and cognitive deficits, entertaining their potential utility in HD appears logical. Schizophrenia, depression, and Alzheimer's Disease share certain symptoms with HD, such as altered locomotor control, mood, or neurodegeneration. Therefore, the utility of PDE inhibitors has been investigated as a potential treatment of HD.

7. It is uncertain whether pharmacological inhibition of phosphodiesterases is beneficial for treatment of Huntington's Disease

Because inhibition of specific PDE isoforms has been shown to effectively reduce motor, cognitive, and emotional changes associated with several neurological disorders such as, Parkinson's Disease, schizophrenia, Alzheimer's Disease, and major depressive disorder, inhibition of PDEs may improve motor, cognitive and emotional changes that occur in HD. Despite data that PDE10A and PDE1B mRNA and protein levels are decreased early in HD and PDE4A mRNA levels decrease with aging, cAMP levels have been shown to be reduced in striatum of pre-symptomatic STHdh Q111/111 transgenic HD mice (Gines, 2003), and reduced cAMP levels have been observed in cerebral spinal fluid of HD patients (Cramer et al., 1984), and post-mortem caudate of HD patients (Gines, 2003). It has also been reported that phosphorylation of CREB, a transcription factor that is phosphorylated by cAMP-dependent protein kinase A, is decreased in models of HD. Consequently, it has been theorized that the decreased PDE10A, PDE1B, and PDE4A levels present in HD progression may represent compensatory changes that occur in response to even earlier decreases in cAMP levels (Kleiman et al., 2011). By this paradigm, inhibition of PDEs early in disease progression may represent a valid therapeutic approach to elevate cAMP levels and overcome changes to gene expression, which may contribute to HD pathogenesis. It is unknown at this point whether loss of PDEs is compensatory or pathogenic, however inhibition of PDE4 and PDE10A has been tested in animal models of HD.

7.1 PDE4 inhibition using rolipram shows beneficial effects in animal models of Huntington's Disease, however adverse effects associated with PDE4 inhibitors may limit their usefulness in Huntington's Disease

The first experiments to test whether pharmacological inhibition of PDEs was beneficial in HD used the PDE4 inhibitor rolipram. Rolipram treatment (1.5 mg/kg, intra-peritoneal injection once per day) for 2 and 8 weeks in rats that had received striatal lesions by direct quinolinic acid injection to the striatum, a model used to recapitulate striatal degeneration that occurs in HD, resulted in decreased striatal cell loss as well as increased levels of phosphorylated CREB in the striatum (DeMarch et al., 2007). In a follow up study, Demarche and colleagues (2008) tested effects of rolipram treatment in the R6/2 transgenic mouse model of HD. Rolipram treatment (1.5 mg/kg, intra-peritoneal injection once per day) beginning at 4 weeks of age and continuing until euthanasia increased survival on a Kaplan-Meyer curve by approximately 1.5 weeks, reduced gross brain atrophy, increased the number of surviving striatal neurons, reduced microglia activation, reduced the size and number of neuronal intranuclear inclusions, increased phosphorylated CREB levels in striatal and cortical neurons, and increased brain-derived neurotrophic factor levels in striatal cells (DeMarch et al., 2008). A later study from the same group extended previous findings by examining changes to the parvalbuminergic interneurons (Giampà et al., 2009). Parvalbuminergic interneurons display reduced CREB phosphorylation, and reductions in levels of a transcriptional co-activator CREB binding protein, which is believed to contribute to HD pathogenesis (Nucifora et al., 2001). In support of a beneficial effect of PDE4 inhibition in this study, rolipram treatment was found to increase the number of parvalbumin interneurons and normalize levels of CREB binding protein, a transcriptional co-activator that is thought to be inhibited by N-mHtt. Rolipram treatment also increased motor activity in an open field test, and increased the time spent on the rotarod. Taken together, these studies support a beneficial effect of rolipram treatment in rat chemical-lesion and transgenic mouse models of HD.

Although rolipram has shown beneficial effects for treatment of HD in animal models, clinical use of rolipram and other PDE4 inhibitors is not being pursued due to high incidence of adverse effects such as emesis, gastrointestinal problems, and insomnia (Giampà et al., 2010). Instead HD researchers have turned to other PDEs as a therapeutic target. The main candidate currently being investigated is PDE10A.

7.2 Inhibition of PDE10A produces conflicting results in mouse models of Huntington's Disease

PDE10A has been pursued as a pharmacological target for treatment of HD because PDE10A is selectively expressed in the caudate, which degenerates in HD (Hebb et al., 2004; Giampà et al., 2010; Lakics et al., 2010). To date, two studies have directly tested effects of PDE10A inhibitors in animal models of HD. In the first study, R6/1 and wild-type mice were treated with papaverine (20 mg/kg subcutaneous once daily, 30 minutes before behavioural testing) beginning at 8 weeks and continuing for 14 days. Wild-type mice displayed significantly increased anxiety-like behaviours using the light-dark test, but anxiety-like behaviour was absent in R6/1 mice. Reduced CREB protein levels in striatum of R6/1 and wild-type mice were also reported as shown by western blot and densitometric analysis. In the same study, effects of chronic papaverine treatment (20 mg/kg subcutaneous once daily for 42 days, given 30 mins before testing) in wild-type mice were also examined. Papaverine treatment led to distinct motor deficits, mild cognitive deficits, and anxiety-like behaviour as measured by rotarod, Morris Water Maze, and light-dark test respectively. In contrast, Giampa and colleagues (2010) showed that PDE10A inhibition using TP-10 (1.5 mg/kg intra-peritoneal injection, once daily) beginning at 4 weeks of age and continuing until euthanasia improved symptoms related to HD in R6/2 mice without producing deficits in wild-type mice. TP-10 treatment was shown to improve motor deficits as shown by delayed development of hind paw clasping, increased time spent on rotarod, and increased distance travelled in an open field test. TP-10 treatment also decreased neurodegeneration, as shown by increased striatal and cortical neuron number, decreased number of neuronal intranuclear inclusions, and reduced microglia activation. Additionally, TP-10 treatment was shown to increase levels of phosphorylated CREB and brain-derived neurotrophic factor in the striatum and cortex of R6/2 mice. Differences between the results reported by Giampa et al. (2010) and Hebb et al. (2004) could be due to differences in pharmacological properties of papaverine and TP-10 or methodological differences such as behavioural tests used and age at which treatment was started. Consequently, effects of PDE10A inhibition in genetic mouse models of HD are not clear.

Other studies provide indirect evidence that PDE10A inhibition may be beneficial for treatment of HD. Threlfell and colleagues (2009) report that inhibition of PDE10A using striatal infusion of papaverine or TP-10, or systemic administration of TP-10, increases the probability that medium spiny projection neurons will depolarize in response to cortical input as shown by single-unit extracellular recordings performed in the dorsal striatum of anesthetised rats. Loss of medium spiny neurons is believed to contribute to HD pathophysiology, so enhancement of medium spiny neurons responsiveness to cortical input represents a potentially beneficial effect of PDE10A inhibition in HD. Kleiman and colleagues (2011) observed changes in gene expression associated with chronic PDE10A inhibition that were predicted to provide neuroprotective effects in models of HD. Microarray analysis of RNA obtained from wild-type mice treated with TP-10 (25 mg/kg administered by oral galvage once per day for 18 days) showed down-regulation of mRNAs encoding histone deacetylase 4, follistatin, and claspin mRNAs in the striatum. Down-regulation of these mRNAs has been predicted to provide neuroprotection in HD (Hughes et al., 1999; Freudenreich and Lahiri, 2004; Thomas et al., 2008). In this study, no differences in gene expression were observed in PDE10A knock-out mice treated with TP-10, thus indicating that the effect of TP-10 on gene expression is selective for PDE10A. Kleiman and colleagues (2011) also showed that CREB-mediated transcription was significantly increased in striatum of wild-type mice treated with TP-10 (3.2 mg/kg subcutaneously for 1 week) using *in vivo* imaging of the bioluminescence produced from a CRE driven luciferase lenti-viral vector. Taken together, these studies provide indirect evidence that inhibition of PDE10A may be beneficial in HD.

8. Conclusions

PDE10A and 1B mRNA and protein levels are decreased early in HD (Hebb et al., 2004). PDE4A mRNA levels decrease with aging (Hebb et al., 2004). Impaired function of PDE isoforms is associated with various pathological conditions of the central nervous system, so decreases in PDE10A and 1B may contribute to HD pathology. However, cAMP levels are reduced prior to symptom onset in rodent models of HD and it has been theorized that decreases in PDE10A and 1B may represent compensatory changes that occur in response to even earlier decreases in cAMP levels (Keliman et. al., 2011). If this is true, then inhibition of PDEs early in disease progression represents a valid therapeutic approach to elevate cAMP levels and overcome changes in gene expression, which may contribute to HD pathogenesis. In support of this view, PDE4 inhibition via rolipram has shown beneficial results in mouse models of HD (DeMarch et al., 2007; Giampa et al., 2009). Additionally, PDE10A inhibition by TP-10 treatment at 4 weeks of age reduces behavioural and cellular changes associated with HD progression in R6/2 mice (Giampa et al., 2010). Indirect evidence also supports a beneficial effect of PDE10A inhibition for treatment of HD, as TP-10 treatment increases the probability that medium spiny projection neurons fire in response to cortical input (Threfell et al., 2009) and both genetic ablation of PDE10A and TP-10 treatment result in gene expression changes that are predicted to be neuroprotective in HD (Kleiman et al., 2011). However, PDE10A2 mRNA expression is decreased at the level of transcription by N-mHtt in R6 mouse models of HD (Hu et al., 2004; Gomez et al., 2006). N-mHtt interacts with, and interferes with, the normal function of transcription factors and co-factors required for the appropriate expression of PDE10A2 (Hu et. al., 2004). These data specifically argue against the hypothesis that decreased PDE expression represents a compensatory mechanism on the part of the cell during HD progression. PDE10A inhibition by papaverine produces marked cognitive and motor deficits in wild-type mice, while having no beneficial effect in R6/1 transgenic HD mice (Hebb et. al., 2004). Moreover, PDE10A knock-out mice display cognitive deficits, including increased escape latency in the Morris water maze, and reduced spontaneous locomotor activity (Siuciak et al., 2006). Ablation of PDE1B is associated with increased spontaneous locomotor activity and reduced pleasure-seeking behaviour. The phenotype of PDE10A and 1B knock-out mice resembles those observed in several transgenic rodent models of HD. Taken together, these data indicate that decreased expression of PDEs 1B, 4A, and 10A may play a pathogenic role in HD.

HD progression is also associated with decreased expression of DARPP-32 and brain-derived neurotrophic factor. Evidence from PDE1B knock-out mice suggests that PDE1B is the major up-stream regulator of DARPP-32 activity in medium spiny projection neurons (Reed et al., 2002). Consequently, PDE1B inhibition in the context of HD, may have limited efficacy as DARPP-32 levels and activity are decreased. Decreased brain-derived neurotrophic factor is associated with decreased cell survival in Alzheimer's Disease and HD, and neuronal atrophy in major depressive disorder (Zuccato et al., 2010). Sildenafil may induce expression of brain-derived neurotrophic factor. However, PDE5 is not expressed at high levels in the caudate/putamen, so the effect of sildenafil may have little benefit in the treatment of HD neuronal cell loss.

To date, the majority of data regarding changes in PDE expression, or the efficacy of PDE inhibition in HD, have been collected in the R6 mouse model. The R6 transgenic mouse model of HD is limited in several respects. R6 mice over-express an N-terminal fragment of mHtt (Mangiarini et al., 1996). HD progression is accelerated by N-mHtt over-expression and certain aspects of HD pathophysiology, such as behavioural changes, may not be observed (Cha et al., 1998). Other, more physiologically accurate, transgenic mouse models of HD recapitulate the longitudinal progression of this disorder. For example, the Hdh/Q model, which is a knock-in mouse model expressing exon 1 of the human *huntingtin* gene containing 72 - 150 CAG repeats within the mouse *huntingtin* locus (Wheeler et al., 2002), display locomotor symptoms resembling those observed in human patients suffering HD, as well as decreased socializing behaviours, and anhedonia (Kennedy et al., 2005). Future studies that examine longitudinal changes in cAMP levels and PDE expression in the striatum of HD mice may elucidate whether decreased PDE1B, 4A, and 10A expression is compensatory or pathogenic. If PDE reductions are shown to be compensatory in HD knock-in models, clinical trials could be conducted to determine whether PDE inhibitors could delay HD symptom onset without producing adverse side effects. PDE1 and 4 inhibitors cause agranulocytosis and emesis, respectively, and consequently may not be practical for use in treating HD. PDE10A inhibitors are not known to cause adverse side effects and may represent the most logical target in such clinical trials for the safe and effective treatment of HD.

9. Acknowledgements

Support was provided by: Canadian Institute of Health Research, Nova Scotia Health Research Foundation, and Huntington Society of Canada. Figures reproduced from Hebb et al., 2004 were used with permission from *Neuroscience*.

10. References

Akhondzadeh S., Ghayyoumi R., Rezaei F., Salehi B., Modabbernia A.H., Maroufi A., Esfandiari G.R., Naderi M., Ghebleh F., Tabrizi M., & Rezazadeh S.A. (2011). Sildenafil adjunctive therapy to risperidone in the treatment of the negative symptoms of schizophrenia: a double-blind randomized placebo-controlled trial. *Psychopharmacology*, Vol. 213, No. 4, (Feb 2011), pp. 809 - 815.

Ashman D.F., Lipton R., Melicow M.M., & Price T.D. (1963). Isolation of adenosine 3', 5'-monophosphate andguanosine 3', 5'-monophosphate from rat urine. *Biochemical and Biophysical Research Communications*, Vol. 11,No. 1, (May 1963), pp. 330 - 334.

Bader S., Korholt A., Snippe H., & Van Haastert P.J.M. (2006). DdPDE4, a novel cAMP-specific phosphodiesterase at the surface of dictyostelium cells. *The Journal of Biological Chemistry*, Vol. 281, No. 29, (Jul 2006), pp. 20018 - 20026.

Baillie G.S., Sood A., McPhee I., Gall I., Perry S.J., Lefkowitz R.J., & Houslay M.D. (2003). beta-Arrestin-mediated PDE4 cAMP phosphodiesterase recruitment regulates beta-adrenoceptor switching from Gs to Gi. *Proceedings of the National Academy of Sciences in the United States of America*, Vol. 100, No. 3, (Feb 2003), pp.940 - 945.

Ballard S.A., Gingell C.J., Tang K., Turner L.A., Price M.E., & Naylor A.M. (1998). Effects of sildenafil on therelaxation of human corpus cavernosum tissue in vitro and on the activities of cyclic nucleotide phosphodiesterase isozymes. *The Journal of Urology*, Vol 159., No. 6, (Jun 1998), pp. 2164 - 2171.

Bender A.T., & Beavo J.A. (2006). Cyclic Nucleotide Phosphodiesterases: Molecular Regulation to Clinical Use. *Pharmacological Reviews*, Vol. 58, No. 3, (Sep 2006), pp. 488 - 520.

Benn C.L., Slow E.J., Farrell L.A., Graham R., Deng Y., Hayden M.R., & Cha J.H. (2007). Glutamate receptor abnormalities in the YAC128 transgenic mouse model of Huntington's disease. *Neuroscience*, Vol. 147, No. 2, (Jun 2007), pp. 354 - 372.

Boolell M., Allen M.J., Ballard S.A., Gepi-Attee S., Muirhead G.J., Naylor A.M., Osterloh I.H., & Gingell C. (1996). Sildenafil: an orally active type 5 cyclic GMP-specific phosphodiesterase inhibitor for the treatment of penile erectile dysfunction. *International Journal of Impotence Research*, Vol. 8, No. 2, (Jun 1996), pp. 47 - 52.

Butcher R.W., & Sutherland E.W. (1962). Adenosine 3',5'-phosphate in biological materials. I. Purification andproperties of cyclic 3',5'-nucleotide phosphodiesterase and use of this enzyme to characterize adenosine 3',5'-phosphate in human urine. *The Journal of Biological Chemistry*, Vol. 237, No. 1, (Apr 1962), pp. 1244 - 1250.

Cao C., Temel Y., Blokland A., Ozen H., Steinbusch H.W., Vlamings R., Nguyen H.P., von Horston S., Schmitz C.,Visser-Vandewalle V. (2006). Progressive deterioration of reaction timer performance and choreiformsymptoms in a new Huntington's disease transgenic rat model. *Behavioural Brain Research*, Vol. 170, No. 2, (Jun 2006), 257 - 261.

Cha J.H., Kosinski C.M., Kerner J.A., Alsdorf S.A., Mangiarini L., Davies S.W., Penney J.B., Bates G.P., & Young A.B.(1998). Altered brain neurotransmitter receptors in transgenic mice expressing a portion of an abnormal human huntingtin disease gene. *Proceedings of the National Academy of Sciences in the United States of America*, Vol. 95, No. 11, (May 1998), pp. 6480 - 6485.

Charych E.I., Jiang L.-X., Lo F, Sullivan K., & Brandon N.J. (2010). Interplay of palmitoylation and phosphorylation in the trafficking and localization of phosphodiesterase 10A: implications for the treatment of schizophrenia. *The Journal of neuroscience*, Vol. 30, No. 27, (Jul 2010), pp. 9027 - 9037.

Cheng Y.F., Wang C., Lin H.B., Li Y.F., Huang Y., Xu J.P., & Zhang H.T. (2010). Inhibition of phosphodiesterase-4 reverses memory deficits produced by Aβ25-35 or Aβ1-40 peptide in rats. *Psychopharmacology*, Vol. 212, No.2, (Oct 2010) pp. 181 - 191.

Cramer H., Warter J.M., & Renaud B. (1984). Analysis of neurotransmitter metabolites and adenosine 3',5'-monophosphate in the CSF of patients with extrapyramidal motor disorders. *Advances in Neurology*, Vol. 40, No. 1, (Jan 1984), pp. 431 - 435.

Crocker S.F., Costain W.J., & Robertson H.A. (2006). DNA microarray analysis of striatal gene expression in symptomatic transgenic Huntington's mice (R6/2) reveals neuroinflammation and insulin associations.*Brain Research*, Vol. 1088, No. 1, (May 2006), pp. 176 - 186.

Cuadrado-Tejedor M, Hervias I, Ricobaraza A, Puerta E, Pérez-Roldán JM, García-Barroso C, Franco R, Aguirre N, García-Osta A. (2011). Sildenafil restores cognitive function without affecting Aß burden in an Alzheimer's disease mouse model. *British Journal of Pharmacology*, E-pub ahead of print. (May 2011).

DeMarch Z., Giampa C, Patassini S., Bernardi G., & Fusco F.R. (2008). Beneficial effects of rolipram in the R6/2 mouse model of Huntington's disease. *Neurobiology of Disease*, Vol. 30, No. 3, (Jun 2008), pp. 375 - 387.

DeMarch Z., Giampa C., Patassini S., Martorana A., Bernardi G., & Fusco F.R. (2007). Beneficial effects of rolipram in a quinolinic acid model of striatal excitotoxicity. *Neurobiology of Disease*, Vol. 25, No. 2, (Feb 2007), pp. 266 - 273.

Degerman E., Belfrage P., & Manganiello V.C. (1997). Structure, localization, and regulation of cGMP-inhibited phosphodiesterase (PDE3). *The Journal of Biological Chemistry*, Vol. 272. No. 11, (Mar 1997), pp. 6823 - 6826.

Delghandi M.P., Johannessen M., & Moens U. (2005). The cAMP signalling pathway activates CREB through PKA, p38 and MSK1 in NIH 3T3 cells. *Cellular Signalling*, Vol. 17, No. 11, (Nov 2005), pp. 1343 - 1351.

Desplats P.A., Kass K.E., Gilmartin T., Stanwood G.D., Woodward E.L., Head S.R., Sutcliffe J.G., & Thomas E.A.(2006). Selective deficits in the expression of striatal-enriched mRNAs in Huntington's disease. *Journal of Neurochemistry*, Vol. 96, No. 3, (Feb 2006), pp. 743 - 757.

Ebix Inc. (2011) Schizophrenia, Major depressive disorder, and Alzheimer's disease. In: *Animated Dissection of Anantomy for Medicine (A.D.A.M.)*, Aug 12, 2011, Available from: <http://www.adam.com/healthsolutions.aspx >

Ehrman L.A., Williams M.T., Schaefer T.L., Gudelsky G.A., Reed T.M., Fienberg A.A., Greenberg P., & Vorhees C.V. (2006). Phosphodiesterase 1B differentially modulates the effects of methamphetamine on locomotor activityand spatial learning through DARPP32-dependent pathways: evidence from PDE1B-DARPP32 double-knockout mice. *Genes Brain and Behaviour*, Vol. 5, No. 7, (Oct 2006), pp. 540 - 551.

Esposito K., Reierson G.W., Luo H.R., Wu G.S., Licinio J., Wong M.L. (2009). Phosphodiesterase genes and antidepressant treatment response: a review. *Annals of Medicine*, Vol. 41, No. 3, (Jan 2009), pp. 177 - 185.

Fatemi S.H., Reutiman T.J., Folsom T.D., & Lee S. (2009). Phosphodiesterase-4A expression is reduced in cerebella ofpatients with bipolar disorder. *Psychiatry and Genetics*, Vol. 18, No. 6, (Dec 2008), pp. 282 - 288.

Fidock M., Miller M., & Lanfear J., (2002). Isolation and differential tissue distribution of two human cDNAs endocing PDE1 splice variants. *Cell Signalling*, Vol. 14, No. 1, (Jan 2002), pp. 53 - 60.

Francis S.H., Busch J.L., Corbin J.D., & Sibley D. (2010). cGMP-dependent protein kinases and cGMP phosphodiesterases in nitric oxide and cGMP action. *Pharmacology Reviews*, Vol. 62, No. 3, (Sep 2010), pp. 525 - 563.

Freudenreich C.H., & Lahiri M. (2004). Structure-forming CAG/CTG repeat sequences are sensitive to breakage in the absence of Mrc1 checkpoint function and S-phase checkpoint signaling: implications for trinucleotide repeat expansion diseases. *Cell Cycle*, Vol. 3, No. 11, (Nov 2004), pp. 1370 - 1374.

Giampa C., Laurenti D., Anzilotti S., Bernardi G., Menniti F.S., & Fusco F.R. (2010). Inhibition of the striatal specific phosphodiesterase PDE10A ameliorates striatal and cortical pathology in R6/2 mouse model of Huntington's disease. *Public Library of Science: One*, Vol. 5, No. 15, (Oct 2010), pp. e13417.

Giampa C., Middei S., Patassini S., Borreca A., Marullo F., Laurenti D., Bernardi G., Ammassari-Tuele M., & Fusco F.R. (2009). Phosphodiesterase type IV inhibition prevents sequestration of CREB binding protein, protects striatal parvalbumin interneurons and rescues motor deficits in the R6/2 mouse model of Huntington's disease. *The European Journal of Neuroscience*, Vol. 29, No. 5, (Mar 2009), pp. 902 - 910.

Gines S., Seong I.S., Fossale E., Ivanova E., Trettel F., Gusella J.F., Wheeler V.C., Persichetti F., & MacDonald M.E. (2003). Specific progressive cAMP reduction implicates energy deficit in presymptomatic Huntington's disease knock-in mice. *Human Molecular Genetics*, Vol. 12, No. 5, (Mar 2003), pp. 497 - 508.

Gomez G.T., Hu H., McCaw E.A., & Denovan-Wright E.M. (2006). Brain-specific factors in combination with mutanthuntingtin induce gene-specific transcriptional dysregulation. *Molecular and Cellular Neuroscience*, Vol. 31, No. 4, (Apr 2006), pp. 661 - 675.

Graham R.K., Deng Y., Carroll J., Vaid K., Cowan C., Pouladi M.A., Metzler M., Bissada N., Wang L., Faull R.L., GrayM., Yang X.W., Raymond L.A., & Hayden M.R. (2010). Cleavage at the 586 amino acid caspase-6 site inmutant huntingtin influences caspase-6 activation in vivo. *Journal of Neuroscience*, Vol. 30, No. 45, (Nov 2010), 15019 - 15029.

Hebb A.L.O., Robertson H.A., & Denovan-Wright E.M. (2004). Striatal phosphodiesterase mRNA and protein levels are reduced in Huntington's disease transgenic mice prior to the onset of motor symptoms. *Neuroscience*, Vol. 123, No. 4, (Jan 2004) pp. 967 - 981.

Heng M.Y., Detloff P.J., & Albin R.L. (2008). Rodent genetic models of Huntington disease. *Neurobiological Disorders*, Vol. 32, No. 1, (Oct 2008), pp. 1 - 9.

Hermel E., Gafni J., Propp S.S., Leavitt B.R., Wellington C.L., Young J.E., Hackman A.S., Logvinova A.V., Peel A.L., Chen S.F., Hook V., Singaraja R., Krajewski S., Goldsmith P.C., Ellerby H.M., Hayden M.R., Bredesen D.E., &Ellerby L.M. (2004). Specific caspase interactions and amplification are involved in selective neuronal vulnerability in Huntington's disease. *Cell Death & Differentiation*, Vol. 11, No. 4, (Apr 2004), pp. 424 - 428.

Hollman A. (2005). Plants and the Heart. *Dialogues in Cardiovascular Medicine*, Vol. 10, No. 4, (Jan 2005), pp. 259 - 263.

Hughes, P.E., Alexi, T., Williams C.E., Clark R.G., & Gluckman P.D. (1999). Administration of recombinant human Activin-A has powerful neurotrophic effects on select striatal phenotypes in the quinolinic acid lesion model of Huntington's disease. *Neuroscience*, Vol. 92, No. 1, (Jan 1999), pp. 197 - 209.

Hu H., McCaw E.A., Hebb A.L., Gomez G.T., & Denovan-Wright E.M. (2004). Mutant huntingtin affects the rate oftranscription of striatum-specific isoforms of phosphodiesterase 10A. *European Journal of Neuroscience*, Vol.20, No. 12, (Dec 2004), pp. 3351 - 3361.

Itoh K., Ishima T., Kehler J., & Hashimoto K. (2011). Potentiation of NGF-induced neurite outgrowth in PC12 cells by papaverine: role played by PLC-gama, IP3 receptors. *Brain Research*, Vol. 1377, No. 4, (Mar 2011), pp. 32 - 40.

Iyengar R. (1993). Molecular and functional diversity of mammalian Gs-stimulated adenylyl cyclases. *Journal of Federation of American Societies for Experimental Biology*, Vol. 7, No. 9, (Jun 1993), pp. 768 - 775.

Johnson D.A., Akamine P., Radzio-Andzelm E., Madhusadan M, & Taylor S.S. (2001). Dyanamincs of cAMP-dependent protein kinase. *Chemical Reviews*, Vol. 101, No. 8, (Aug 2001), pp. 2243 - 2270.

Johnson W., & Jameson J.L. (2000). Role of Ets2 in cyclic AMP regulation of the human chronic gonadotropin beta promoter. *Molecular and Cellular Endocrinology*, Vol. 165, No. 1 - 2, (Jul 2000), pp. 17 - 24.

Jin S.L., Richard F.J., Kuo W.P., D'Ercole A.J., & Conti M. (1999). Impaired growth and fertility of cAMP-specific phosphodiesterase PDE4B-deficient mice. *Proceedings of the National Academy of Science in the United States of America*, Vol. 96, No.21, (Oct 1999), pp. 1998 - 2003.

Kähler A.K., Otnaess M.K., Wirgenes K.V., Hansen T., Jönsson E.G., Agartz I., Hall H., Werge T., Morken G., Mors O., Mellerup E., Dam H., Koefod P., Melle I., Steen V.M., Andreassen O.A., & Djurovic S. (2010). Association study of PDE4B gene variants in Scandinavian schizophrenia and bipolar disorder multicenter case-control samples. *American Journal of Medical Genetics B: Neuropsychiatric Genetics*, Vol. 1538, No. 1, (Jan 2010), pp. 86 - 96.

Kehr W., Debus G., & Neumeister R. (1985). Effects of rolipram, a novel antidepressant, on monoamine metabolismin rat brain. *Journal of Neural transmission*, Vol. 63, No. 1, (Jan 1985), pp. 1 - 12.

Kennedy L., Shelbourne P.F., & Dewar D. (2005). Alterations in dopamine and benzodiazepine receptor binding precede overt neuronal pathology in mice modelling early Huntington's disease pathogenesis. *Brain Research*, Vol. 1039, No. 1 - 2, (Mar 2005), pp. 14 - 21.

Kelly M.P., Logue S.F., Brennan J., Day J.P., Lakkaraju S., Jiang L., Zhong X., Tam M., Sukoff Rizzo S.J., Platt B.J., Dwyer J.M., Neal S., Pulito V.L., Agostino M.J., Grauer S.M., Navarra R.L., Kelley C., Comery T.A., Murrills R.J., Houslay M.D., & Brandon N.J. (2010). Phosphodiesterase 11A in brain is enriched in ventral hippocampus and deletion causes psychiatric disease-related phenotypes. *Proceedings of the National Academy of Sciences in the United States of America*, Vol. 107, No. 18, (May 2010), 8457 - 8462.

Kleiman R.J., Kimmel L.H., Bove, S.E., Lanz T.A., Harms, J.F., Romegialli A., Miller K.S., Willis A., des Etages S., Kuhn M, & Schmidt C.J. (2011). Chronic Suppression of Phosphodiesterase 10A Alters Striatal Expression of Genes Responsible for Neurotransmitter Synthesis , Neurotransmission , and Signaling Pathways Implicated in Huntington ' s Disease. *The Journal of Pharmacology and Experimental Therapeutics*, Vol. 336. No. 1, (Jan 2011), pp. 64 - 76.

Kleppisch T., & Feil R. (2009). cGMP signalling in the mammalian brain: role in synaptic plasticity and behaviour. *Handbook of Experimental Pharmacology*, Vol. 1, No. 191, (Jan 2009), pp. 549 - 579.

Koyanagi M., Suga H., Hoshiyama D, Ono K., Iwabe N., Kuma K., & Miyata T. (1998). Ancient gene duplication anddomain shuffling in the animal cyclic nucleotide phosphodiesterase family. *Federation of European BiochemicalSocieties: letters*, Vol. 436, No. 3, (Oct1998), pp. 323 - 328.

Lakics V., Karran E.H., & Boess F.G. (2010). Quantitative Comparison of Phosphodiesterase mRNA distribution in human brain and peripheral tissues. *Neuropharmacology*, Vol. 59, No. 6, (Nov 2010), pp. 367 - 374.

Lipina T.V., Wang M., Liu F., & Roder J.C. (2011). Synergistic interactions between PDE4B and GSK-3: DISC1 mutantmice.*Neuorpharmacology*, E-pub ahead of print, (Mar 2011).

Loughney K., Martins T.J., Harris E.A., Sadhu K., Hicks J.B., Sonnenburg W.K., Beavo J.A., & Ferguson K. (1996).Isolation and characterization of cDNAs corresponding to two human calcium, calmodulin-regulated, 3',5'-cyclic nucleotide phosphodiesterases. *The Journal of Biological Chemistry*, Vol. 271, No. 2, (Jan 1996), pp. 796 - 806.

Lugnier C. (2006). Cyclic nucleotide phosphodiesterase (PDE) superfamily: a new target for the development of specific therapeutic agents. *Pharmacology & Therapeutics*, Vol. 109, No. 3, (Mar 2006), pp. 366 - 398.

Luthi-Carter R., Hanson S.A., Strand A.D., Bergstrom D.A., Chun W., Peters N.L., Woods A.M., Chan E.Y.,Kooperberg C., Krainc D., Young A.B., Tapscott S.J., & Olson J.M. (2002). Dysregulation of gene expressionin the R6/2 model of polyglutamine disease: parallel changes in muscle and brain. *Human MolecularGenetics*, Vol. 15, No. 11, (Aug 2002), pp. 1911 - 1926.

Luthi-Carter R., Apostol B.L., Dunah A.W., DeJohn M.M., Farrell L.A., Bates G.P., Young A.B., Standaert D.G., Thompson L.M., & Cha J.H. (2004). Complex alteration of NMDA receptors in transgenic Huntington's disease mouse brain: analysis of mRNA and protein expression, plasma membrane association, interactingproteins, and phosphorylation. *Neurobiological disorders*, Vol. 14, No. 3, (Dec 2004), pp. 624 - 636.

Manallack D.T., Hughes R.A., & Thompson P.E. (2005). The next generation of phosphodiesterase inhibitors: structural clues to ligand and substrate selectivity of phosphodiesterases. *Journal of Medicinal Chemistry*, (May2005), pp. 3449 - 3462.

Mangiarini L., Sathasivam K., Seller M., Cozens B., Harper A., Hetherington C., Lawton M., Trottier Y., Lehrach H.,Davies S.W., & Bates G.P. (1996). Exon 1 of the HD gene with an expanded CAG repeat is sufficient to causea progressive neurological phenotype in transgenic mice. *Cell*, Vol. 87, No. 3, (Nov 1996), pp. 493 - 506.

Marmor M.F., & Kessler R. (1999). Sildenafil (Viagra) and ophthalmology. *Survey of Ophthalmology*, Vol. 44, No. 2, (Sep-Oct 1999), pp. 153 - 162.

Mazarei G., Neal S.J., Becanovic K., Luthi-Carter R., Simpson E.M., & Leavitt B.R. (2010). Expression analysis of novel striatal-enriched genes in Huntington's disease. *Human Molecular Genetics*, Vol. 19, No. 4, (Feb 2010), pp. 609 - 622.

Meade C.A., Deng Y.P., Fusco F.R., Del Mar N., Hersch S., Goldowitz D., & Reiner A. (2002). Cellular localization anddevelopment of neuronal intranuclear inclusions in striatal and cortical neurons in R6/2 transgenic mice. *The Journal of Comparative Neurology*, Vol. 449, No. 3, (Jul 2002), pp. 241 - 269.

Moncada I., Martínez-Jabaloyas J.M., Rodriguez-Vela L., Gutiérrez P.R., Giuliano F., Koskimaki J., Farmer I.S., Renedo V.P., & Schnetzler G. (2009). Emotional changes in men treated with sildenafil citrate for erectile dysfunction: a double-blind, placebo-controlled clinical trial. *Journal of Sexual Medicine*, Vol. 6, No. 12, (Dec 2009), pp. 3469 - 3477.

Nguyen H.P., Metzger S., Holzmann C., Koczan D., Thiesen H.J., von Hörsten S., Riess O., & Bonin M. (2008). Age-dependent gene expression profile and protein expression in a transgenic rat model of Huntington's disease.*Proteomics Clinical Applications*, Vol. 2, No. 12, (Dec 2008), pp. 1638 - 1650.

Nucifora F.C., Sasaki M., Peters M.F., Huang H., Cooper J.K., Yamada M., Takahashi H., Tsuji S., Troncoso J., Dawson V.L., Dawson T.M., Ross, C.A. (2001). Interference by huntingtin and atrophin-1 with CBP-mediated transcription leading to cellular toxicity. *Science*, Vol. 291, No. 5512, (Mar 2001), pp. 2423 -2428.

Numata S., Iga J., Nakataki M., Tayoshi S., Taniguchi K., Sumitani S., Tomotake M., Tanahashi T., Itakura M., Kamegaya Y., Tatsumi M., Sano A., Asada T., Kunugi H., Ueno S., & Ohmori T. (2009). Gene expression and association analyses of the phosphodiesterase 4B (PDE4B) gene in major depressive disorder in the Japanese population. *American Journal of Medical Genetics B: Neuropsychiatric Genetics*, Vol. 1508, No. 4, (Jun 2009), pp. 527 - 534.

Pérez-Torres S., Cortés R., Tolnay M., Probst A., Palacios J.M., & Mengod G. (2003). Alterations on phosphodiesterase type 7 and 8 isozyme mRNA expression in Alzheimer's disease brains examined by in situ hybridization. *Experimental Neurology*, Vol. 182, No. 2, (Aug 2003), pp. 322 - 334.

Picconi B., Bagetta V., Ghiglieri V., Paillè V., Di Filippo M., Pendolino V., Tozzi A., Giampà C., Fusco F.R., Sgobio C., & Calabresi P. (2011). Inhibition of phosphodiesterases rescues striatal long-term depression and reduces levodopa-induced dyskinesia. *Brain*, Vol. 134, No. 2, (Dec 2010), pp. 357 - 387.

Prickaerts J., van Staveren W.C.G., Sik A., Markerink-van Ittersym M., Niewohnen U., van der Staay F.J., Blokland A., & de Vente J. (2002). Effects of two selective phosphodiesterase type 5 inhibitors, sildenafil and vardenafil, on object recognition memory and hippocampal cyclic GMP levels in the rat. *Neuroscience*, Vol. 113, No. 2, (Feb 2002), pp. 351 - 361.

Rall T.W., & Sutherland E.W. (1958). Formation of cyclic adenine ribonucleotide by tissue particles. *The Journal of Biological Chemistry*, Vol. 232, No. 1, (Oct 1957), pp. 1065 - 1076.

Reddy P.H., Williams M., Charles V., Garrett L., Pike-Buchanan L., Whetsell W.O. Jr., Miller G., & Tagle D.A. (1998). Behavioural abnormalities and selective neuronal loss in HD transgenic mice expressing mutated full-length HD cDNA. *Nature Genetics*, Vol. 20, No. 2, (Oct 1998), pp. 198 - 202.

Reed T.M., Repaske D.R., Snyder G.L., Greengard P., & Vorhees C.V. (2002). Phosphodiesterase 1B knock-out miceexhibit exaggerated locomotor hyperactivity and DARPP-32 phosphorylation in response to dopamine agonists and display impaired spatial learning. *Journal of Neuroscience*, Vol. 22, No. 12, (Jun 2002), pp. 5188 -5197.

Ribchester R.R., Thomson D., Wood N.I., Hinks T., Gillingwater T.H., Wishart T.M.,Court F.A., & Morton A.J. (2004). Progressive abnormalities in skeletal muscle and neuromuscular junctions of transgenic mice expressing theHuntington's disease mutation. *European Journal of Neuroscience*, Vol. 20, No. 11, (Dec 2004), pp. 3092 - 3144.

Runne H., Regulier E., Kuhn A., Zala D., Gokce O., Perrin V., Sick B., Aebischer P, Deglon N., & Luthi-Carter R.(2008). Dysregulation of gene expression in primary neuron models of Huntington's disease shows that polyglutamine-related effects on the striatal transcriptome may not be dependent on brain circuitry. *Journal of Neuroscience*, Vol. 28, No. 39, (Sep 2008), pp. 9723 - 9731.

Rutten K., Van Donkelaar E.L., Ferrington L., Blokland A., Bollen E., Steinbusch H.W., Kelly P.A., Prickaerts J.H.(2009). Phosphodiesterase inhibitors enhance object memory independent of cerebral blood flow andglucose utilization in rats. *Neuropsychopharmacology*, Vol. 34, No. 8, (Jul 2009), pp 1914 - 1925.

Sadhu K., Hensley K., Florio V.A., & Wolda S.L. (1999). Differential expression of the cyclic GMP-stimulated phosphodiesterase PDE2A in human venous and capillary endothelial cells. *Journal of Histochemistry and Cytochemistry*, Vol. 47, No. 7, (Jul 1999), pp. 895 - 906.

Schilling G., Becker M.W., Sharp A.H., Jinnah A.H., Duan K., Kotzuk J.A., Slunt H.H., Ratovitski T., Cooper J.K., Jenkins N.A., Copeland N.G., Price D.L., & Borchelt D.R. (1999). Intranuclear inclusions and neuritic aggregates in transgenic mice expressing a mutant N-terminal fragment of huntingtin. *Human Molecular Genetics*, Vol. 8, No. 3, (Mar 1999), pp. 397 - 407.

Schmidt C.J., Chapin D.S., Cianfrogna J., Corman M.L., Hajos M., Harms J.F., Hoffman W.E., Lebel L.A., McCarthy S.A., Nelson F.R., Prouix-LaFrance C., Majchrzak M.J., Ramirez A.D., Schmidt K, Seymour P.A., Siuciak J.A., Tingley F.D. 3rd, Williams R.D., Verhoest P.R., & Menniti F.S. (2008). Preclinical Characterization of Selective Phosphodiesterase 10A Inhibitors: A New Therapeutic Approach to the Treatment of Schizophrenia. *The Journal of Pharmacology & Experimental Therapeutics*, Vol. 352, No. 2, (May 2008), pp. 681 - 690.

Schultheiss D., Muller S.V., Nager W., Stief C.G., Schlote N., Jonas U., Asvestis C., Johannes S., & Munte T.F. (2001). Central effects of sildenafil (Viagra) on auditory selective attention and verbal recognition memory in humans: a study with event-related brain potentials. *World Journal of Urology*, Vol. 19, No. 1, (Feb 2001), pp. 46 - 50.

Shabb J.B. (2001). Physiological substrates of cAMP-dependent protein kinase. *Chemical Reviews*, Vol. 101, No. 8, (Aug2001), pp. 2381 - 2411.

Siuciak J.A., Chapin D.S., Harms J.F., Lebel L.A., McCarthy S.A., Chambers L., Shrikhande A., Wong S., Menniti F.S., & Schmidt C.J. (2006). Inhibition of the striatum-enriched phosphodiesterase PDE10A: a novel approach to the treatment of psychosis. *Neuropharmacology*, Vol. 51, No. 2 (Aug 2006), pp. 386 - 396.

Siuciak J.A., Chapin D.S., McCarthy S.A., & Martin A.N. (2007). Antipsychotic profile of rolipram: efficacy in rats and reduced sensitivity in mice deficient in the phosphodiesterase-4B (PDE4B) enzyme. *Psychopharmacology*, Vol. 192, No. 3, (Jun 2007), pp. 415 - 424.

Szatmari S.Z., & Whitehouse P.J. (2003). Vinpocetine for cognitive impairment and dementia. *Cochrane Database of Systematic Reviews*, Vol. 1, No. 1, (Jan 2003), pp. CD003119.

Souza R.P., Meltzer H.Y., Lieberman J.A., Voineskos A.N., Remington G., & Kennedy G.L. (2011). Prolactin as a biomarker for treatment response and tardive dyskinesia in schizophrenia subjects: old thoughts revisited from a genetic perspective. *Human Psychopharmacology*, E-pub ahead of print. (Feb 2011).

Tang W, & Ziboh V.A. (1991). Phorbol ester inhibits 13-cis-retinoic acid-induced hydrolysis of phosphatidylinositol4,5-bisphosphate in cultured murine keratinocytes: a possible negative feedback via protein kinase C-activation. *Cellular Biochemical Function*, Vol. 9, No. 3, (Jul 1991), pp. 183 - 191.

Thomas E.A., Coppola G., Desplats P.A., Tang B., Soragni E., Burnett R., Gao F., Fitzgerald K.M., Borok J.F., Herman D., Geschwind D.H., & Gottesfeld J.M. (2008). The HDAC inhibitor 4b ameliorates the disease phenotype and transcriptional abnormalities in Huntington's disease transgenic mice. *Proceedings of the National Academy of Sciences in the United States of America*, Vol. 105, No. 40, (Oct 2008), pp. 15564 - 15569.

Threlfell S, Sammut S, Menniti FS, Schmidt CJ, West AR. (2009). Inhibition of Phosphodiesterase 10A Increases the Responsiveness of Striatal Projection Neurons to Cortical Stimulation. *Journal of Pharmacology & Experimental Therapeutics*, Vol. 328, No. 3, (Mar 2009), pp. 785 - 795.

United States National Institutes of Health. (2011). Clinical trials for Phosphodiesterase inhibitors. In: *Clinicaltrials.gov*, Aug 12, 2011, Available from: <http://clinicaltrials.gov/>

Vandeput F., Wolda S.L., Krall J., Hambleton R., Uher L., McCaw K.N., Radwanski P.B., Florio V., & Movsesian M.A. (2007). Cyclic nucleotide phosphodiesterase PDE1C1 in human cardiac myocyces. *The Journal of Biological Chemistry*, Vol. 282, No. 45, (Nov 2007), pp. 32749 - 32757.

Walter U. (1984). cGMP-regulated enzymes and their possible physiological functions. *Advances in Cyclic Nucleotide and Protein Phosphorylation Research*, Vol. 17, No. 1, (Jan 1984), pp. 249 - 258.

Wang Z., Jiang Y., Lu L., Huang R., Hou Q., & Shi F. (2007). Molecular mechanisms of cyclic nucleotide-gated ion channel gating. *Journal of Genetics and Genomics*, Vol. 34, No. 6, (Jun 2007), pp. 477 - 485.

Wheeler V.C., Gutekunst C.A., Vrbanac V., Lebel L.A., Schilling G., Hersch S., Friedlander R.M., Gusella J.F., Vonsattel J.P., Borchelt D.R., & MacDonald M.E. (2002). Early phenotypes that presage late-onset neurodegenerative disease allow testing modifiers in Hdh CAG knock-in mice. *Human Molecular Genetics*, Vol. 11, No. 6, (Mar 2002), 633 - 640.

Wong M.L., Wheelan F., Deloukas P., Whittaker P., Delgado M., Cantor R.M., McCann S.M., & Licinio J. (2006). Phosphodiesterase genes are associated with susceptibility to major depression and antidepressant treatment response. *Proceedings of the National Academy of Sciences in the United States of America*, Vol. 103, No. 41, (Oct 2006), pp. 15124 - 15129.

Yu Z.X., Li S.H., Evans J., Pillarisetti A., Li H., & Li X.J. (2003). Mutant huntingtin causes context-dependent neurodegeneration in mice with Huntington's disease. *Journal of Neuroscience*, Vol. 23, No. 6, (Mar 2003), pp. 2193 - 2202.

Zhang, H.-T., Huang, Y., Jin S.-L., Frith S.A., Suvarna N., Conti M., & O'Donnell J.M. (2002). Antidepressant-like profile and reduced sensitivity to rolipram in mice deficient in the PDE4D phosphodiesterase enzyme. *Neuropsychopharmacology*,Vol. 27, No. 4, (Oct 2002), pp. 587 - 595.

Zhang KYJ, Card GL, Suzuki Y, Artis DR, Fong D, Gillette S, Hsieh D, Neiman J, West BL, Zhang C, Milburn MV, Kim S-H, Schlessinger J, Bollag G. (2004) A Glutamine Switch Mechanism for Nucleotide Selectivity by Phosphodiesterases. *Molecular cell*, Vol. 15, No. 2, (July 2004), pp. 79 - 86.

Zuccato C., Valenza M., & Cattaneo E. (2010). Molecular mechanisms and potential therapeutic targets in Huntington's disease. *Physiology Reviews*, Vol. 90, No. 1, (Jul 2010), pp. 905 - 981.

Part 3

Cognitive Dysfunction in Huntington's Disease

7

Cognition in Huntington's Disease

Tarja-Brita Robins Wahlin[1,2] and Gerard J. Byrne[2]
[1]*Department of Neurobiology, Care Sciences and Society, Karolinska Institutet, Stockholm,*
[2]*School of Medicine, The University of Queensland, Brisbane,*
[1]*Sweden*
[2]*Australia*

1. Introduction

Huntington's Disease (HD) is an autosomal dominant, neurodegenerative disease. It is characterized by severe involuntary motor dysfunction, so-called choreic movements, neurological and psychiatric symptoms and cognitive impairments that lead to dementia (Bates et al., 2002). Genetic markers for the gene that causes HD were identified in 1983, located on the short arm of chromosome four (Gusella et al., 1983). Ten years later in 1993 the gene was cloned (Huntington's Disease Collaborative Research Group, 1993). HD was thus from the mid-1980s one of the first diseases where it was possible to predict whether an asymptomatic individual had inherited the genetic markers and would therefore become ill in the future. The clinical diagnosis of HD is based on the presence of motor symptoms and a positive mutation analysis, or on neurological and psychiatric symptoms in patients with a family history of HD.

In almost half of HD cases clinical onset is indicated with psychiatric symptoms such as depression, anxiety and aggressive outbursts (Close Kirkwood et al., 2002a; Julien et al., 2007; van Duijn et al., 2007). Sometimes onset of the disease presents with schizophrenic or manic-like symptoms (Julien et al., 2007; Shiwach, 1994). However, the initial indication of onset is often in the form of subtle cognitive impairment before manifest neurological or psychiatric symptoms occur. All patients become demented over the course of the disease (Brandt et al., 1984). The first symptoms occur most frequently in the 45 to 50 age bracket, although age of onset ranges from 2 to 80 years (Roos et al., 1991). The average life expectancy after clinical onset is 15-17 years (Roos et al., 1993). In the juvenile form of Huntington's Disease onset occurs before age 20 (5-10% of cases) and approximately 25% of HD debuts after age 50, some at age 70 or older (Kremer, 2002). There is currently no specific treatment to cure or delay the disease.

2. General aspects of the disease

2.1 Nomenclature

Huntington's Disease has been described in varying ways throughout history. Christian Lund described HD or Anundsjö disease in Norwegian in 1860 (Orbeck, 1960) and a young American doctor, George Huntington, published a description of HD in 1872 which is still largely valid (Huntington, 1872). The disease has since carried his name and is also called Huntington's Chorea (from the Greek, χορός, dance and khoreia, chorea). The Westphal variant of Huntington's Disease manifests in muscular rigidity and hypokinesia in young adults, usually between 20-30 years. The correct term today is Huntington's Disease.

2.2 Prevalence

The prevalence of HD in many countries is not established and estimates differ considerably from country to country (Harper, 2002). The prevalence in Western Europe is estimated at approximately 3-7 per 100 000 depending on city and country. For example, the prevalence is well mapped in England and varies in the range 2.5-9.95 per 100 000 (Harper, 2002). North American prevalence is estimated at 4.1 to 8.4 per 100,000 inhabitants. Many countries have no information or only sporadic information on the prevalence of HD. The prevalence is lower among indigenous populations in Africa (e.g. 0.01 per 100 000 in South Africa) and Asia (0.7 per 100 000 in Japan and 0.4 per 100 000 in Hong Kong). Areas with notably high prevalence of HD are found in Tasmania, Australia (17:100 000) (Conneally, 1984) and in the Lake Maracaibo district of Venezuela (Young et al., 1986).

2.3 Cause and heredity

HD (OMIM 143100) is an autosomal dominant neurodegenerative disease caused by a mutation in the short arm of chromosome 4 (4p16.3) (Gusella et al., 1983) (Figure 1).

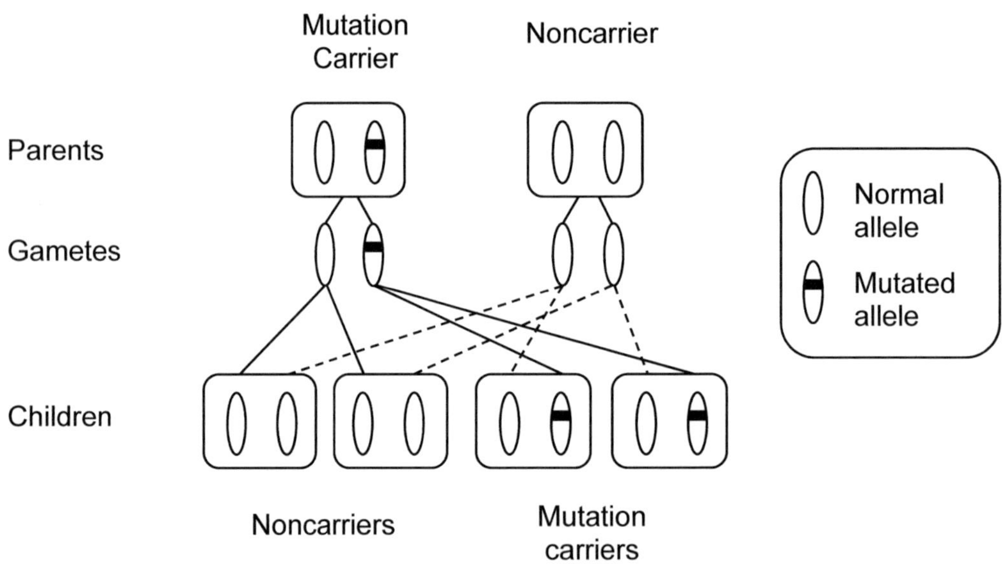

Fig. 1. Autosomal dominant inheritance

2.4 CAG sequences

A mutation in the huntingtin (HTT) gene causes an increase in the number of trinucleotide CAG (Cytosine, Adenosine, Guanine) repetitions, always 36 or more for individuals with HD (Huntington's Disease Collaborative Research Group, 1993). A person with normal function has between 9 and 35 repetitions of the CAG sequence. Repetitions from 36 to 39 are characterized by reduced penetrance, therefore an individual in this range will not automatically develop the characteristic symptoms of HD during his lifetime, but children of such an individual are still at risk (Rubinsztein et al., 1996). A sequence of 40 CAG repetitions or more has full penetrance. There is also a negative correlation between the number of CAG repetitions and age at onset, but this does not explain all variation in the age at onset, which means that other factors possibly interact and determine when disease symptoms will appear (see table 1).

CAG sequences	
≤ 28	Normal function
29-35	The individual will not develop HD, but the next generation inherits the risk of developing the disease.
36-39	Reduced penetrance; some individuals will develop HD and development is generally late in life. The next generation inherits the risk of developing the disease. A number of non-symptomatic cases in older individuals with 36-39 CAG sequences have been reported.
≥ 40	Full penetrance; all individuals will develop the disease. Higher CAG sequences provides earlier disease onset (negative correlation). Juvenile HD manifests itself most often in people with ≥ 60 CAG repetitions.

Table 1. Number of CAG sequences and risk of onset.

The number of CAG repetitions tends to be extended at the formation of the sex cells and takes place primarily when the gene is inherited from the father (Kehoe et al., 1999). Inheritance via the male line leads to cases of earlier onset and also more deleterious disease outcome, so called anticipation (Ridley et al., 1988). Estimated age at onset can be calculated using a regression equation, although not always accurately (Langbehn et al., 2004; Langbehn et al., 2009; Rubinsztein et al., 1997). Spontaneous mutations are very rare and explain only 0.1 % of the cases of the disease. However, it is reported that about 8% of HD patients do not have an affected family member (Almqvist et al., 2001; Siesling et al., 2000).

2.5 Neuropathological changes

A widespread, selective neuropathology is found in HD, with cell loss and atrophy. The changes are strikingly selective in their effect on specific brain cell types and particular brain structures. Medium γ-aminobutyric acid (GABA) spiny neurons are the neuronal cells primarily affected, mainly in the caudate nucleus and putamen. The cortex is less affected and the cerebellum is relatively spared. The HD gene product, a very large 350 kDa protein, termed *huntingtin*, is believed to have a toxic effect which leads to cellular dysfunction and eventual death of neurons (Huntington's Disease Collaborative Research Group, 1993). The exact mechanism of the toxic effect is still poorly understood. Early neuropathological changes are seen selectively in the striatum, where 90% of neuronal cells are medium spiny projection neurons (MSP neurons). Loss of projection neurons in the caudate nucleus is the dominant neuropathological change. Death of neuronal cells continues gradually in layers 3, 5 and 6 of the cortex, the substantia nigra and the CA1 region of the hippocampus. Loss of enkephalin-withholding MSP neurons in the striatum, which indirectly controls voluntary and related movements, constitutes the neurobiological basis for HD chorea. The preferential involvement of the indirect pathway of basal ganglia-thalamocortical circuitry is believed to be the cause of chorea (Paulsen et al., 2005a). Fronto-striatal circuitry linking the striatum with frontal lobes is also affected. In addition, changes in the substantia nigra, hippocampus, hypothalamus and selectively in the cortex and white matter are found.

2.6 Chemical changes

Profound atrophy in large parts of the brain is seen in the final stage of the disease. Neuronal loss leads to reduction of neurotransmitters such as γ-aminobutyric acid (GABA), glutamate, glutamic acid decarboxylase (GAD), peptides (e.g. enkephalin) and acetylcholine (choline acetyltransferase, ChAT) in the striatum. On the other hand, there are increases in serotonin levels, while serotonin receptor density decreases. A reduction in postsynaptic D_1 and D_2 dopamine receptors and in the dopamine transporter DAT in the striatum also has the potential to explain the cognitive impairments of HD patients (Antonini et al., 1996; Backman et al., 1997). The complex and multifarious symptoms of HD have been attributed to these neuropathological and neurochemical changes (Walker, 2007).

3. Clinical picture and progress

3.1 Clinical picture

The clinical picture includes severe motor dysfunction, cognitive decline leading to dementia and neurological and psychiatric symptoms. Symptoms of HD vary from patient to patient and although all symptoms may be present, some symptoms are more dominant during different phases. Cognitive impairments occur early in the disease, exacerbated when manifest disease progresses and causes reductions in everyday functions. Affected cognitive domains include psychomotor speed, language, memory and executive functions; and later in the disease visuospatial abilities are also affected (Robins Wahlin et al., 2010; Robins Wahlin et al., 2007).

3.2 Motor symptoms

Severe locomotor dysfunction with hyperkinesia characterizes HD. These involuntary movements are seen first in the fingers and toes, then in the trunk. Approximately 10% of all patients with HD may, however, have the juvenile onset or Westphal variant of HD with symptoms of hypokinesia and rigidity similar to Parkinson's disease (Bittenbender & Quadfasel, 1962; Bruyn, 1962). Difficulties with balance occur, with exaggerated, fidgeting motor action and a tendency to violent involuntary movements. HD patients often walk with a dance-like gait with legs widely separated to compensate for the lack of balance and control. The symptoms may cause the patient to appear to be intoxicated by alcohol. Almost all patients manifest irregularly timed, randomly distributed and abrupt choreatic movements (Barbeau et al., 1981). They may keep their hands in their pockets to limit uncontrollable arm actions. Facial musculature is also affected with characteristic chorea of the face showing in the form of pouting of the lips, lifting of the eyebrows, frowning and nodding head movements. Eye movements become disturbed at an early stage, with jerky action and the patient has difficulty focusing the eyes on moving objects. Fine motor skills decline, characterized by clumsiness and problems with grasping and holding objects. A patient may be diagnosed when subtle neurologic symptoms are identifiable as disturbed tongue and eye movements. Dysarthria is found early, while dysphasia is common in the final stage. For about half of patients the extrapyramidal motor symptoms manifest at clinical onset (Mattsson, 1974). Patients with later age of onset, 50-70 years of age, debut with involuntary movements, walking difficulties and dysphasia. These patients usually have a slower and more benign development of the pathological processes compared with patients with a younger age of onset (see Table 2).

Prodromal phase – Early Signs	**Manifest and clinical phase -- Signs & symptoms**	**Dementia Phase -- Late in the disease**
Agitation Egocentricity, persistence	Myotonic dystrophy Myoclonus Problems initiating movements	Rigidity Decreasing involuntary movements
Irritability, aggressiveness, anger	Increasing involuntary movements	Grave or diminishing chorea
Apathy	Choreatic manifestations; writhing, jerky movements	Increase in falls
Anxiety	Balance and gait difficulty	Inability to walk
Uninhibited behaviour	Problems with fine motor skills (such as shoe-laces)	Developmental Dyspraxia
Impaired impulse control	Problem with swallowing; danger of inhalation	Dysphagia
Euphoria	Slowed voluntary movements	Bradykinesia
Abnormal eye movements	Inability to control the speed and force of movements, clumsiness	Difficulty in swallowing & eating
Sadness	Dyskinesia	Neglected nutrition
Depression	General weakness	Wheelchair bound
Suicidal Ideation	Weight loss	Weight loss
Slowness of speech	Speech impairments; slurred speech & phonological impairment, difficulty with pronunciation	Dysphasia, serious speech impairments, mutism
Motion	Problems with daily living activities (ADL)	Inability to manage ADL
Psychological denial	Muscle stiffness	Incontinence
Symptom searching (mutation carriers)	Delusions, hallucinations	Evident regression

Table 2. Clinical signs and symptoms.

3.3 Behavioural changes and psychiatric disorders

About half of patients debut with affective disorders or psychiatric symptoms (Mattsson, 1974). These may occur before other clear symptoms manifest and can be very difficult to manage. They sometimes dominate the clinical picture. Around 72-98% of HD individuals develop significant neuropsychiatric problems, including both affective psychoses and non-affective psychoses (Mendez, 1994; Paulsen et al., 2001; van Duijn et al., 2007). Major depression (Larsson et al., 2006) and manic episodes also occur (van Duijn et al., 2008). In the manic phase, presentation is the same as for bipolar disorder. As Huntington noted, socially deviant behavior occurs when the individual fails to recognize or register their divergent behavior (Huntington, 1872). Hallucinations of hearing, smell, sight, taste and touch may be present in HD. The most common neuropsychiatric symptoms are reported to be dysphoria (69%), agitation (67%), irritability (65%), apathy (56%), anxiety (52%), disinhibition (35%) and euphoria (31%) (Paulsen et al., 2001). The unusually diverse manifestations of the disease have made diagnosis difficult to determine, especially before DNA testing (Tost et al., 2004)(see Table 2).

3.4 Depression

General sadness, depression and anxiety are frequently displayed early in the disease course (Larsson et al., 2006). Apathy and irritability (33% to 76%) are also amongst the first symptoms (van Duijn et al., 2007). When depression occurs it is often characterized by hopelessness, guilt and shame (Baudic et al., 2006; Kessler, 1987). The suicide rate in people with HD is twice that of the normal population (Robins Wahlin et al., 2000) and suicide risk is highest in the context of disease onset (Paulsen et al., 2005b). It is not known whether depression is an integral part of the disease or a response to the knowledge of the severity of the disease in the patient's future, or possibly a combination of the two. The affective disorder may be an explicit manifestation of brain damage. Depression can also be an expression of grief and anxiety, as HD patients are aware that their children may await the same fate that they are facing (Bird, 1999; Paulsen et al., 2001).

3.5 Cognitive impairments and dementia

The cognitive symptoms of HD vary from patient to patient and although several symptoms can be present, some dominate more than others through the different phases (see Table 3). Cognitive signs manifest early in the disease, exacerbated when manifest disease progresses causing reduced ability to perform everyday functions. Affected cognitive domains include psychomotor speed, language, memory, executive functions and later also visuospatial abilities (Lawrence et al., 2000; Robins Wahlin et al., 2010; Robins Wahlin et al., 2007; Snowden et al., 2002; Stout et al., 2011). The cognitive deterioration can be divided into three main phases, depending on the disease progress: prodromal phase, clinical phase and dementia phase. The cognitive phases are associated with reductions in the total functional capacity (Total Functional Capacity scale, TFC) in the areas of occupational activity, finances, domestic chores, activities of daily living (ADL) and increasing care needs (Beglinger et al., 2010; Paulsen, 2010; Shoulson & Fahn, 1979).

4. The prodromal phase

4.1 The prodromal phase and early signs

Neurological symptoms may not be detected in this phase and therefore it is called the prodromal, preclinical or presymptomatic phase. Patients often report memory difficulties, concentration and attention problems or psychosomatic symptoms before the disease can be diagnosed definitively (Verny et al., 2007). Changes in behavior in relation to either family or friends are very subtle. Increasing difficulty managing emotions sometimes leads to aggressive outbursts in surroundings where the outbursts may not seem warranted. These are sometimes referred to as "catastrophic reactions" (Almqvist et al., 1999). Patients may seek out physicians due to stress or psychosomatic symptoms such as gastrointestinal problems or insomnia, which need symptomatic treatment. Difficulties arise in managing tasks at home and work and such difficulties are only understood with hindsight after diagnosis.

Prodromal phase – Early signs	**Manifest and clinical phase - Sign & symptoms**	**Dementia Phase -- Late in the disease**
Psycho-motor slowness ▼ ▼	Clear psycho-motoric slowness ▼ ▼ ▼	Impaired cognitive functions ▼ ▼ ▼
Executive functions: Concentration ▼ Initiation ▼ ▼ Attention ▼ Flexible thinking ▼ Logical thinking ▼ Simultaneous capacity ▼ Judgement ▼	Executive functions: Concentration ▼ ▼ Initiation ▼ ▼ ▼ Attention ▼ ▼ Flexible thinking ▼ ▼ Logical thinking ▼ ▼ Simultaneous capacity ▼ ▼ Judgement ▼ ▼	Executive functions: Concentration ▼ ▼ ▼ Initiation ▼ ▼ ▼ Attention ▼ ▼ ▼ Flexible thinking ▼ ▼ ▼ Logical thinking ▼ ▼ ▼ Simultaneous capacity ▼ ▼ ▼ Judgement ▼ ▼ ▼
Slightly impaired verbal flow ▼	Markedly impaired verbal flow ▼ ▼	Impaired verbal flow or mutism ▼ ▼ ▼
Declining working memory ▼ ▼	Clearly impaired working memory ▼ ▼ ▼	Greatly impaired working memory ▼ ▼ ▼
Reductions in episodic memory: Encoding ▼ Retrieval; search strategies ▼ Learning ▼ ▼ Recognition (minor problem)	Significant reductions in episodic memory: Encoding ▼ ▼ Retrieval; search strategies ▼ ▼ Learning ▼ ▼ ▼ Recognition ▼	Severe reductions in episodic memory: Encoding ▼ ▼ ▼ Retrieval; search strategies ▼ ▼ Learning ▼ ▼ ▼ Recognition ▼ ▼
Prospective memory difficulties (remember appointments) ▼	Prospective memory difficulties (forgets to pay bills) ▼ ▼	Inability to access prospective memory ▼ ▼ ▼
Mild visuospatial difficulties ▼	Notable visuospatial difficulties ▼	Greatly reduced visuospatial ability ▼ ▼

▼ Mild signs and symptoms; ▼ ▼ Moderate disturbance; ▼ ▼ ▼ Grave disorder

Table 3. Neuropsychological characteristics in Huntington's Disease.

4.2 The prodromal cognitive disorder

The first cognitive changes occur approximately 12-15 years before clinical (motor) onset of the disease (Paulsen et al., 2008; Robins Wahlin et al., 2007; Stout et al., 2011). These prodromal changes are characterized by reduced executive functions which present as alterations in flexibility, reasoning and verbal fluency (Larsson et al., 2008). Lack of logical thinking and difficulties in the skills of decision-making, initiative, attention and planning are very early signs (Lemiere et al., 2004; van Walsem et al., 2009). The ability to conduct complex reasoning, to perform tasks in sequence and to demonstrate simultaneous capacity (dual tasking) all deteriorate gradually (Stout et al., 2011). Linguistic features work relatively well, but syntactic complexity and verbal fluency decrease (Larsson et al., 2008). Working memory and attention show early impairments (Verny et al., 2007). Episodic memory impairment begins with active learning difficulties and inability to apply effective search strategies for information (Montoya et al., 2006a; Solomon et al., 2007; Verny et al., 2007). Memory tasks where only recognition is required exhibit better performance, suggesting a greater problem with retrieval than with encoding of information. Learning new skills becomes more difficult and takes longer (see Figure 2) (van Walsem et al., 2009; Verny et al., 2007). Reduced learning ability and concentration cause problems in occupational, financial and domestic functioning. Mutation carriers in the prodromal phase may need to be referred for comprehensive neuropsychological assessment to determine their capacity for employment, driving of motor vehicles and decision-making (Beglinger et al., 2010).

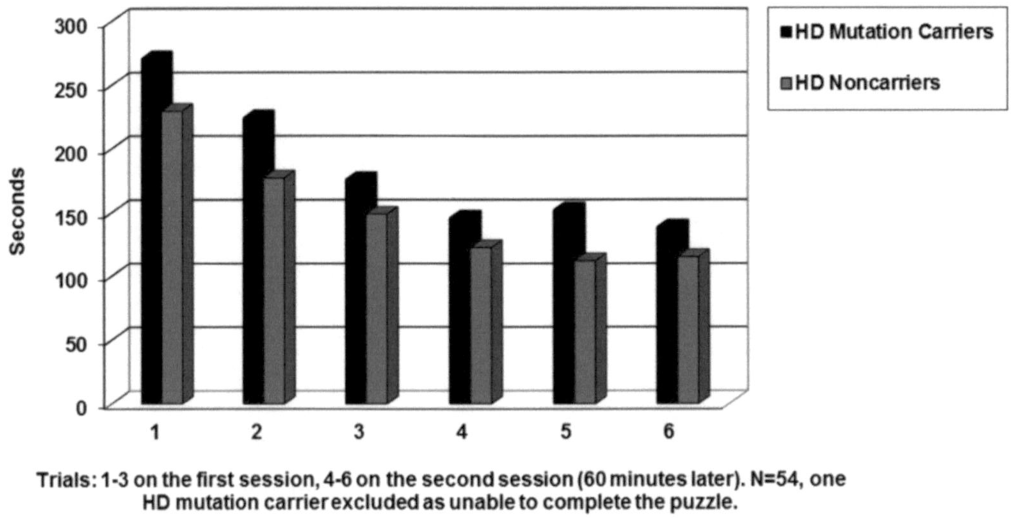

Fig. 2. Performance time on the Tower of Hanoi Puzzle Across Huntington's Disease Groups (Robins Wahlin et al., 2007).

5. Clinical phase

5.1 Clinical neurological and psychiatric disorder

In the clinical phase the neurological symptoms emerge, including the involuntary movements that are characteristic of HD. Many patients lose weight, despite maintaining or sometimes even increasing their food intake. The motor symptoms vary in intensity depending on the degree of mental tension and activation. At the beginning of the clinical phase the involuntary movements are so subtle that patients do not notice them. Later on, usually after about 10 years, the chorea causes grave disability. Over this period there is also a marked decline in executive functioning and a diminishing ability to get organized in everyday situations (Paulsen, 2010; Paulsen et al., 2001). Thinking skills and social and emotional functioning deteriorate. Depression, social isolation and denial of symptoms (anosognosia) are characteristic also in this phase. The critical period for suicide is precisely connected with the stage of illness (Paulsen et al., 2005a) when understanding is maintained and the disease's debilitating symptoms are seen as a threat for the future. Cognitive slowness in combination with reduced attention and failure to notice or correct errors eventually cause the patient to lose their driving licence and their employment (Beglinger et al., 2010). The behavioral problems lead to increased impulsivity, aggressiveness and sometimes even hypersexuality. Marital relationships are commonly strained.

5.2 Clinical cognitive disorder

As the disease progresses, neuropsychological skills requiring executive functioning deteriorate further, including skills such as concentration, patience and stamina. Reduced ability for abstract thinking becomes more apparent and the patient exhibits a greater degree of concrete thinking. Judgment, discrimination and ability to plan are increasingly reduced. Sometimes the patient becomes apathetic. Their semantic memory is initially only slightly reduced (Robins Wahlin et al., 2010) and they continue to recognize close relatives and familiar surroundings. On the other hand, episodic and prospective memory decline, for example remembering future tasks such as appointments (Lundervold et al., 1994b). Working memory continues to deteriorate. Memory deficits consist of a generalized impaired ability to learn new information and retrieve old knowledge, in other words learning curves show a low, flat line (Butters et al., 1994; Verny et al., 2007). Phonemic and semantic fluency becomes decidedly slower (e.g. FAS and categories). Vocabulary decreases, the patient becomes taciturn, distinctions in meaning are lost and the patient has difficulty keeping up with discussions. Because parietal functions are connected to the striatum, visuospatial difficulties are manifested at this stage (Lundervold & Reinvang, 1991). Deteriorating visuospatial skills as well as psychomotor slowness create major difficulties and patients at this stage are strongly advised not to drive a car. Reductions in visuospatial functions can be tapped by missed details and distorted relationships when the patient copies shapes (e.g. Rey-Osterrieth Complex Figure, see Figure 3 and 4) (Osterrieth, 1944; Rey, 1941).

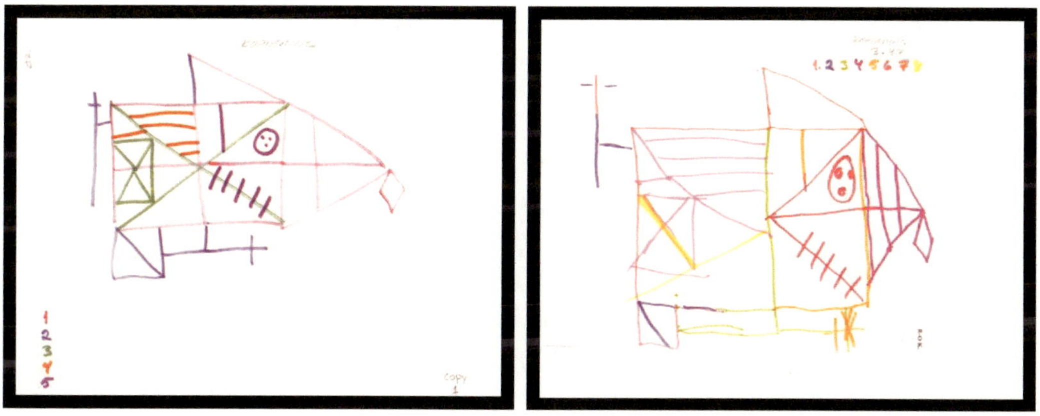

Fig. 3. and 4. Examples of organized (left panel) and disorganized (right panel) Rey-Osterrieth Complex Figures (ROCF) as copied by mutation carriers of HD. The left ROCF was drawn by a female mutation carrier with nine years to estimated disease onset and the ROCF right figure was drawn by a female carrier with one year to estimated disease onset, although the latter patient had not yet been diagnosed by a neurologist as having manifest HD.

6. Dementia phase

6.1 The neuropsychiatric disorder in dementia phase

Dementia symptoms in HD differ in part from what is seen in other dementias such as Alzheimer's disease in that the three great "As" (Aphasia, Apraxia, Agnosia) do not dominate in the early symptomatic stage, however, these symptoms manifest in the dementia phase. Increasing dementia expresses in passivity, loss of non-verbal communication and increasing apathy. Lack of awareness of loss of ability frequently occurs. Patients with HD are extremely slow and restricted in both movement and speech during the depressive and apathetic phases (Paulsen et al., 2001). Lack of initiative may cause them to cease taking care of their own health and hygiene. The sleep cycle is often disturbed, probably because of hypothalamic dysfunction (Soneson et al., 2010). Explosive eruptions occur, with emotional lability and even catastrophic reactions requiring extra resources from daily caregivers (Almqvist et al., 1999; Decruyenaere et al., 2005). In the final stage of the disease, the patient is often bedridden, has lost nearly all functions in everyday activities and is in need of 24 hour care (Zakzanis, 1998).

6.2 Cognitive disturbance in dementia phase

All intellectual functions are severely reduced in the final stage (Lundervold et al., 1994a; Zakzanis, 1998). Inability to mobilize knowledge combined with a lack of motivation presents as generalized dementia (Redondo-Verge, 2001). Evidence of general intellectual sluggishness and semantic slowness is substantial. It may take up to one minute to express a word or idea. As the disease progresses, speech and linguistic ability becomes increasingly difficult and eventually the patient may become mute. Severe dementia is dominant in the final stage (Zakzanis, 1998).

6.3 Psychological and practical implications

HD is often associated with shame and guilt. The disease is viewed as a problematic psychiatric disease because early cognitive and psychiatric symptoms produce abnormal behavior (Paulsen et al., 2001; Robins Wahlin et al., 2000). Symptoms of chorea (involuntary movements) have relatively little impact on the person's functional ability, but unfortunately attract attention and often give the wrong impression that the patient is intoxicated. Neurological signs such as unsteadiness, grimacing and twitching of the face and symptoms of mental illness may be seen as shameful by the victim and his family. Stories about patients being hidden, rejected or repudiated exist in many families. Many times the disease is a great family secret and associated with a lot of denial (Deckel & Morrison, 1996). There is a fear of contracting the disease and its insidious onset leads to intense trawling for early signs which can then be interpreted as symptoms that the manifest phase has begun (Robins Wahlin, 2007).

7. Assessment and diagnosis

7.1 The neuropsychological assessment

Verbal episodic memory, working memory, executive function, verbal fluency and psychomotor speed should be investigated early in the prodromal phase of HD as impairments occur before manifest symptoms are visible (Kirkwood et al., 2000; Verny et al.,

2007). A simple test battery is recommended that includes tests of a variety of cognitive abilities including verbal fluency and memory tasks. The Unified Huntington's Disease Rating Scale (UHDRS) is specially developed for HD and contains a brief cognitive test battery (Huntington Study Group, 1996). The UHDRS is designed to be administered each time the patient comes to follow-up appointments. UHDRS takes approximately 30-45 minutes to implement and includes the following neuropsychological tests which are relatively simple to administer.

1. Letter Fluency (Phonemic) Controlled Oral Word Association Test (COWAT, FAS)
2. Dementia Rating Scale
3. Hopkins Verbal Learning Test (HVLT)
4. Symbol Digit Modalities Test (Digit Symbol in Wechsler Adult Intelligence Scale, WAIS).
5. Stroop Test.
   (a) Colour Naming
   (b) Word Reading
   (c) Interference
6. Trail Making Test A & B
7. Category Fluency (Semantic) Animals

If there is insufficient time for the entire test battery, assessment can be shortened to 10-15 minutes by examining verbal fluency (FAS, animals, see Figures 5 and 6), speed and executive functions (Symbol Digit, Stroop Test, TMT A & B) (Lezak et al., 2004; Strauss et al., 2006). The full test battery is suitable for annual examination which track the mutation carrier's functional levels for employment and holding a driver's licence while in the prodromal phase. Learning effects of the UHDRS are minimal, except for the HVLT, which is available in six parallel versions (Beglinger et al., 2010; Brandt & Benedict, 2001). The HVLT consists of an episodic memory task with 12 words, with four words from each of three semantic categories. The parallel versions avoid learning effects in annual patient evaluations (Solomon et al., 2007; Woods et al., 2005).

7.2 In-depth clinical neurological/neuropsychological examination

In recent years, quantified neurological examinations (Folstein et al., 1983) have been supplemented and/or replaced by the Unified Huntington's Disease Rating Scale (UHDRS) for examination of motor functions in HD (Huntington Study Group, 1996). The Total Functional Capacity (TFC) scale is used internationally as a neurological functional scale that includes areas such as activities of daily living (ADL), occupational activities, financial management, living chores and care needs (Paulsen et al., 2010; Shoulson & Fahn, 1979). The total functional scale (0-13) score correlates highly with cognitive and motor impairments (Kremer, 2002). Great care is needed in the selection of neuropsychological test battery instruments, in order to ensure high sensitivity in detecting prodromal evidence of HD. Although there are now a number of published studies of cognitive impairment in prodromal HD, it has not yet been established whether there is a characteristic pattern of impaired cognitive function prior to the onset of the motor symptoms of HD. Two studies report a pattern of prodromal cognitive function and intelligence as measured by the Wechsler Adult Intelligence Scale (Wechsler, 1997), with Robins Wahlin reporting significant differences in functioning across the various WAIS-R subtests and Verbal, Performance and Full Scale IQ scores (see Figure 7 and 8) (de Boo et al., 1997; Robins Wahlin et al., 2010).

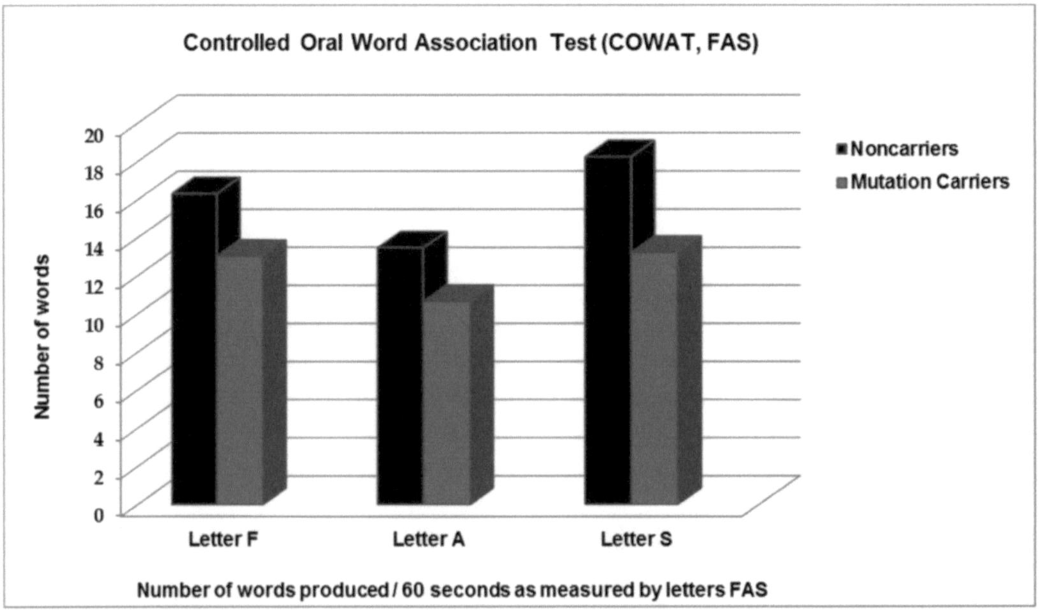

Fig. 5. Phonemic fluency as measured by the letters FAS in a Swedish sample of prodromal mutation carriers (n=29) and noncarriers (n=34) of HD (Larsson et al., 2008).

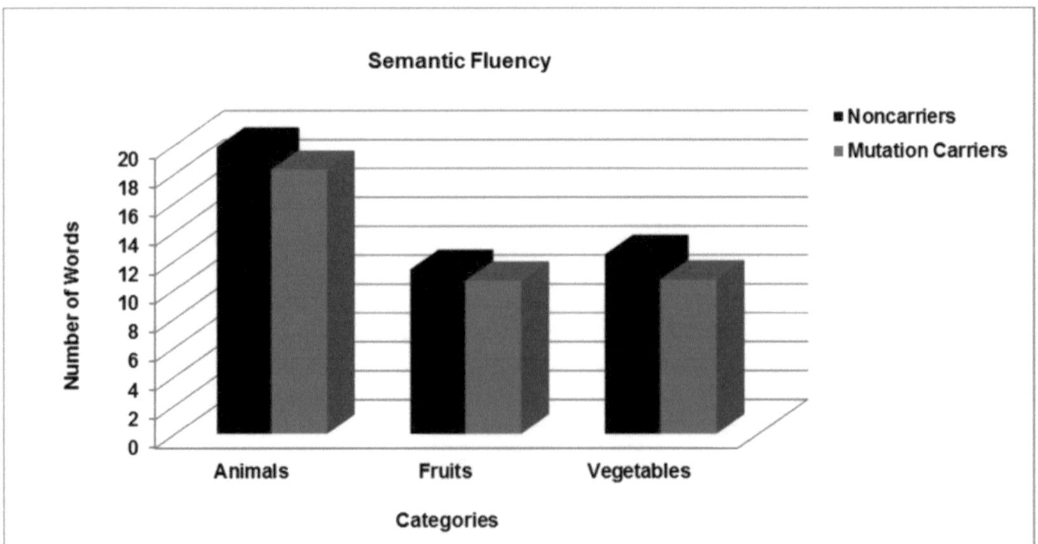

Fig. 6. Semantic fluency as measured by the categories Animals, Fruits and Vegetables in a Swedish sample of prodromal mutation carriers (n=29) and noncarriers (n=34) of HD (Larsson et al., 2008).

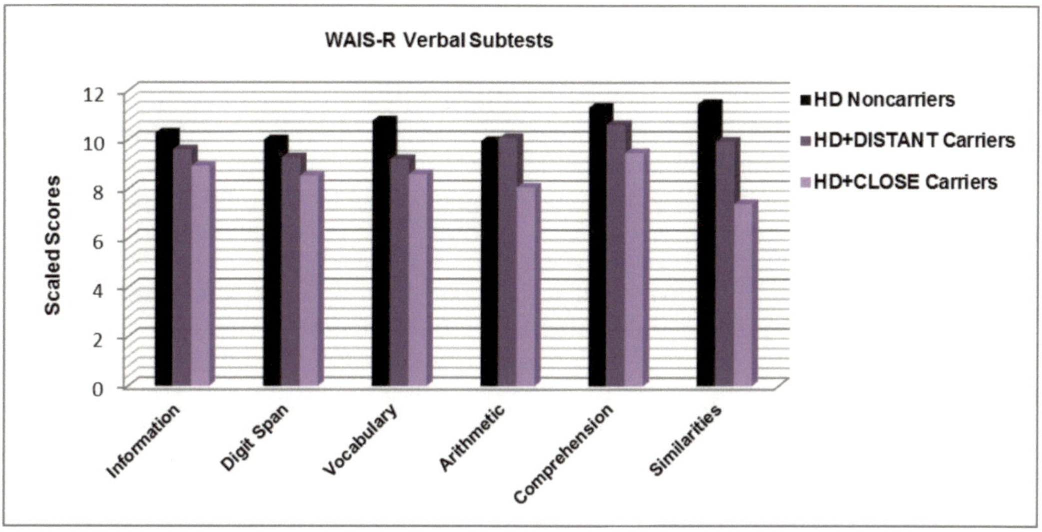

Fig. 7. Weighted scores in the WAIS-R Verbal subtests in noncarriers (HD noncarriers, n=35), mutation carriers with 12 or more years to disease onset (HD DISTANT onset, n=15) and mutation carriers with less than 12 years to disease onset (HD CLOSE onset, n=15) for Huntington's Disease (Robins Wahlin et al., 2010).

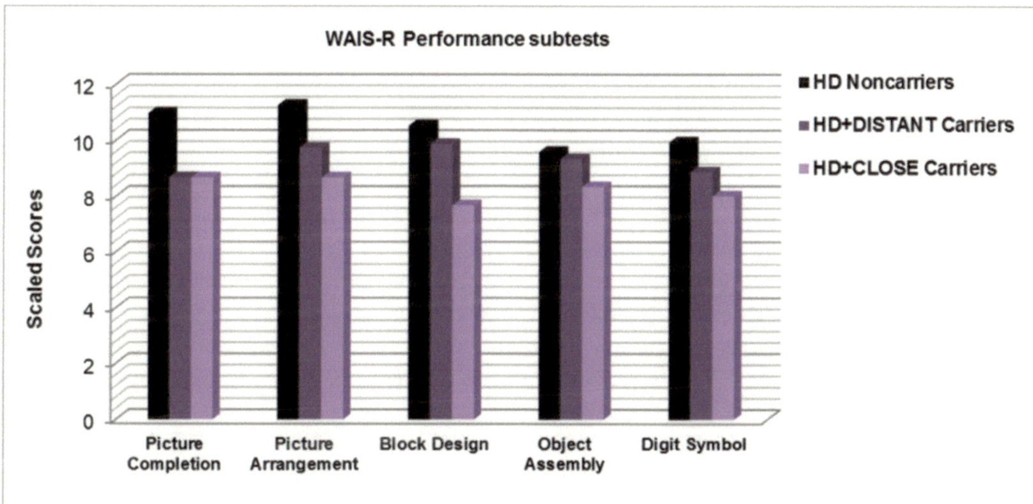

Fig. 8. Weighted scores in the WAIS-R Performance subtests in noncarriers (HD noncarriers, n=35), mutation carriers with 12 or more years to disease onset (HD DISTANT onset, n=15) and mutation carriers with less than 12 years to disease onset (HD CLOSE onset, n=15) for Huntington's Disease (Robins Wahlin et al., 2010).

7.3 Comprehensive psychometrics

Neuropsychological testing is recommended every two years to identify early signs of cognitive disabilities in mutation carriers, with the test battery covering a wide range of cognitive functions. *Verbal fluency* must always be included and it is worth noting that *phonemic fluency* shows earlier reductions than *semantic fluency* (Larsson et al., 2008). This pattern indicates that the frontal functions are involved at an early stage in disease progress. The most common categories of semantic fluency are animals, professions, vegetables and means of transport, for which there are normative data (Strauss et al., 2006). Numerous tests are available to test *Episodic Memory* with both verbal (words, sentences) and non-verbal materials (faces, objects, spatial positions and geometric shapes) (Lezak et al., 2004; Strauss et al., 2006). Furthermore, both free recall and recognition tests can be administered directly after the learning opportunity and also after longer time intervals (Lezak et al., 2004). Tests of Vocabulary and Information (WAIS, Figure 7) are useful to study *Semantic memory* (Wechsler, 1997). *Short-term memory* can be examined in both verbal (Digit Span, see Figure 7) and non-verbal (Corsi Block) tasks, while *procedural memory* can be studied with the Tower of London or Tower of Hanoi test (Figure 2). Appropriate tests for *executive functions* and *psychomotor speed* are the Stroop and the TMT A & B. The Wisconsin Card Sorting Test is less appropriate because it is time-consuming and has been shown to be not sensitive to HD (Grant & Berg, 1948; Milner, 1963). Block Design (WAIS, Figure 8), the Rey-Osterrieth Complex Figure test (ROCF, Figure 3 and 4) (Osterrieth, 1944; Rey, 1941)] and Mental Rotations (Vanderberg, 1971) are sensitive for *visuospatial features* (Robins Wahlin et al., 2007). *Mental tempo* is best noted during the neuropsychological investigations. Digit Symbol (WAIS, Figure 8) and Dots (Ekberg & Hane, 1984) or equivalent, are particularly straightforward to administer and provide reliable measures of psychomotor slowing in HD.

8. Genetic testing

8.1 Guidelines for genetic testing

The disclosure of the huntingtin gene mutation and trinucleotide CAG repeats in HD has provided new opportunities to determine diagnosis prior to the onset of motor symptoms. At risk persons may choose to find out whether they have inherited the mutation for the disease. International guidelines for genetic testing are (a) 18 years or older, (b) 50% risk of HD, (c) 25% risk of HD if parent is deceased or if the parent with a 50% risk is participating in the genetic test process (Broholm et al., 1994; Nance et al., 2003; World Federation of Neurology: Research Committee. Research Group on Huntington's chorea, 1989). The most common reasons for genetic testing are to obtain certainty for genetic status (77%), general planning for the future and family (38%), or in the interests of the children (Robins Wahlin et al., 2000). Information about genetic testing for HD is given in the clinical genetics department (or equivalent) in larger hospitals or specialized centers for HD. Prenatal diagnosis can be provided when the parents are considering termination of the pregnancy if the result shows that the fetus carries the mutation (Decruyenaere et al., 2007; Simpson & Harper, 2001). In some countries prenatal diagnosis, including IVF, is covered by public health insurance. However, approximately 35% of couples decide not to have children after undertaking genetic testing (Decruyenaere et al., 2007). In practice, however, only a small minority of at risk individuals elects to undergo genetic testing for HD (Harper et al., 2000; Robins Wahlin et al., 2000).

8.2 Imaging studies in HD and prodromal HD

Neuroimaging studies such as MRI, CT, SPECT and PET provide extra support for prodromal and manifest diagnosis of HD and are also valuable tools for studying disease progression (Antonini et al., 1996; Aylward, 2007; Montoya et al., 2006b; Paulsen, 2010; Paulsen et al., 2004). Volume of basal ganglia structures is reduced long before neuropsychological deficiencies can be demonstrated (Aylward et al., 1994). Recent MRI studies indicate that atrophy begins in the caudate nucleus approximately 11 years before clinical disease debut (Aylward et al., 2004). The putamen atrophies 9 years before manifest HD, followed by the cortico-striato-thalamocortical network and the frontal lobes which are affected later in the disease process (Paulsen, 2010).

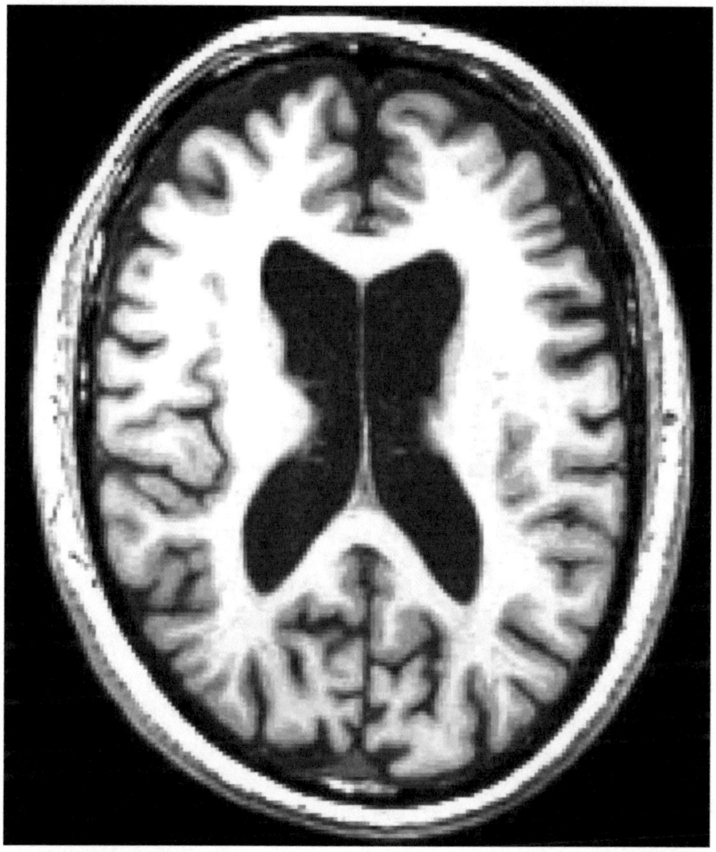

Fig. 9. MRI scan showing atrophy of the brain in Huntington's Disease (courtesy of Joakim Tedroff and Mouna Esmaeilzadeh).

Volumetric MRI data also demonstrate associations with cognitive impairments. The volume of the caudate nucleus and putamen has a strong relationship with decline in executive functions. Cortical matter degenerates later (Paulsen, 2010). PET studies have shown that other factors that contribute to cognitive impairments in HD are reduced dopamine transporter (DAT) and dopamine receptor density (D_1 and D_2) in the caudate and the putamen due to cell death (Antonini et al., 1998; Bäckman et al., 1997; Feigin et al., 2007).

A functional MRI study has shown reduced blood-oxygen-level dependent (BOLD) activity in the left dorsolateral prefrontal cortex in preclinical HD mutation carriers (Wolf et al., 2011).

8.3 Diagnostic significance

Genetic testing in recent years has provided possibilities for early diagnosis of HD and it is assumed that as the testing becomes more widely available more at risk persons will seek this information (Harper et al., 2000). There is, however, minimal research into the impact of early diagnosis on persons who are at risk of developing HD (Robins Wahlin, 2007). Some of those undergoing genetic testing will already have mild cognitive impairment that might interfere with their ability to process the results of testing. More importantly, support services are still inadequate for people who learn early in adulthood that they will develop the disease 10-40 years later. Mutation carriers and family are very concerned about cognitive impairments and dementia, since symptoms in HD are multifaceted and it is difficult to specify exactly how cognitive decline will begin. The provision of factual information to mutation carriers and family might greatly reduce fear and anxiety. This information can be provided as feedback and support in the context of neuropsychological testing or during early visits to neurologists. Effective information about the cognitive deficits in the disease and a better understanding of the long prodromal phase and time course of HD would be beneficial for affected persons and family. Counseling and support may offer opportunities for the development of psychological compensation strategies (Robins Wahlin et al., 2010; Walsh, 1999). Evaluation of therapeutic interventions in HD requires a detailed neurological/neuropsychological examination based on cognitive and clinical methods. Blood workup is indicated to exclude other diseases affecting cognitive and everyday functions.

9. Therapy and approach

9.1 Treatment and care

There are currently no effective, disease-modifying treatments for HD that might prevent its onset or slow down its progress (Mestre et al., 2009a, 2009b). In addition, the cognitive deficits found in HD are not susceptible to treatment. However, early detection of HD offers the potential in the future for prodromal diagnosis and early use of medication thereby possibly modifying its course. Treatment with donepezil (a cholinesterase inhibitor used primarily for symptomatic treatment in Alzheimer's disease) has not been shown to improve motor performance or cognition in HD. Recent studies of treatment with an omega-3 fatty acid (ethyl-eicosapentaenoic acid) and vitamin E did not show any beneficial effects (Feigin et al., 1996; Mestre et al., 2009a). However, the anti-dopaminergic, monoamine depleting agent, tetrabenazine, has been shown to be effective for the treatment of involuntary movements, but not for cognitive impairment or depression (Mestre et al., 2009b). Symptomatic treatment with small doses of conventional neuroleptics (eg haloperidol) and atypical neuroleptics (eg olanzapine), can be used to treat psychotic symptoms and outbursts of aggression (Grove et al., 2000; Warby et al., 2007). Valproic acid has been used to treat myoclonic hyperkinesia and anti-parkinsonian medication can improve hypokinesis and rigidity. However, medication containing L-dopa can also increase the severity of choreatic movements. Benzodiazepines can reduce irritability and aggressiveness, as well as help with anxiety and sleep disorders. Neuropsychiatric problems are treated in the usual way in the early stages of HD and such treatment can be of great benefit to the patient and the family in crisis. If the patient has delusions or hallucinations, contact with a psychiatrist is desirable, since treatment with neuroleptics is often complex. Side effects of neuroleptics may be unfavorable for the patient's cognition and this needs to be balanced in treatment. Depression should always be treated with medication and if possible with psychological therapy (such as cognitive behavior therapy) or regular clinical contact. Antiepileptic treatment is sometimes indicated, especially in juvenile HD.

9.2 Clinical relevance and the approach to cognitive handicap

Cognitive impairments in the prodromal phase of HD often reduce the working lives of mutation carriers and may even lead to major crises. The inertia, reduced psychomotor speed and cognitive decline, especially affecting attention and visuospatial cognition (Lawrence et al., 2000), may lead to issues with driving ability, thus causing social and mobility problems (Beglinger et al., 2010). A formal on-road driving assessment may be needed for those individuals with preclinical or clinical HD who dispute the clinician's assessment of their likely driving ability. Through a thorough explanation of any cognitive disabilities, patients can understand and learn to adapt to the signs of approaching HD and to some extent compensate for the lost abilities. The development of the disease and cognitive disabilities requires annual follow-up as a minimum and compassionate measures in explaining, communicating and organizing supportive interventions, thus providing higher quality of life for struggling HD patients and families (Robins Wahlin et al., 1997).

9.3 Specific approach for the mutation testing

Choosing to have oneself tested should always be an individual choice and a thorough psychosocial evaluation should precede any such testing (Copley et al., 1995). A genetic test should never be imposed on an individual from any direction, especially as genetic discrimination against HD persons has been noted, although fortunately it is rare (Harper et al., 2004; Robins Wahlin, 2007). From 1993 onwards, direct mutation analysis has been used in most developed countries. This has allowed precise determination of an individual's mutation status (Huntington's Disease Collaborative Research Group, 1993). Prenatal diagnosis is available as well as prenatal exclusion testing if the prospective parents do not wish to know their genetic status (Warby et al., 2007). In some countries, as for instance in Belgium, prenatal exclusion testing using linkage analysis, pre-implantation genetic diagnosis exclusion testing and non-disclosure is available for at risk persons who want to exclude the mutation in their offspring but do not want to know their own carrier status (Decruyenaere et al., 2007). International consensus argues that children and young people under 18 years of age should not be tested for the HD mutation (Warby et al., 2007; 1989). Testing of minors is not ethically defensible because it removes the individual's option to know or not know and may cause stigma within the family and society (Robins Wahlin, 2007). It may also have serious educational and career consequences.

9.4 Psychological support during genetic testing

Since there is currently only limited symptomatic relief available, it is important for genetic counseling to occur both before and after disclosure of the results of genetic testing (Hedera, 2001). Many patients feel ambivalent about knowing their risk status (Robins Wahlin, 2007).

The request is often for an immediate and quick investigation but frequently the at risk person does not attend the appointment, only to request a new one later. Changing the at risk status to mutation carrier or noncarrier status is a psychologically complex process, which should not be rushed. The candidate's defence mechanisms and history of stress tolerance are key variables in this difficult process. After the discovery of their genetic status all candidates experience an immediate period of adaptation (Robins Wahlin, 2007). Many mutation carriers undergo a process of denial and a long period of significantly increased stress and depression, leading to long-term sick leave (Paulsen et al., 2010). Sometimes even those who find out that they do not carry the mutation have great difficulty adapting to their new genetic status, which may show itself in survivor guilt and even require therapy or counseling (Robins Wahlin et al., 1997). Psychosomatic symptoms of stress, anxiety and depression generally require symptomatic treatment and a therapy contact can be envisaged (Robins Wahlin et al., 2000).

9.5 Support actions

Personality changes and odd behavior associated with HD may also lead to relationship difficulties within the family (Close Kirkwood et al., 2002b). Catastrophic reactions to the cognitive impairment that is integral to HD and other psychosocial crises require professional support for the whole family (Robins Wahlin, 2007). Psychological support is critical when HD is complicated by depression and suicidal thoughts (Paulsen et al., 2005b; Robins Wahlin et al., 2000). Since HD has major consequences for the entire family, the therapeutic team needs to look beyond the individual patient. Psychological and social care planning can be adapted according to the different phases of illness seen in HD (Walker, 2007). Different stages require varying degrees of assistance from psychiatrists, neurologists, social workers, guardians, dietitians, speech therapists, dental hygienists, physiotherapists, occupational therapists or clinical psychologists (Paulsen et al., 2005a). In the later stages of the illness, the HD patient often has difficulty eating, swallowing and managing their personal hygiene. Incontinence and balance problems make the patient dependent on total care (Warby et al., 2007; World Federation of Neurology: Research Committee. Research Group on Huntington's chorea, 1989). Management of basic ADL functions requires a personal carer in the affected family or nursing home, as the ADL management becomes increasingly important in the late HD stages. As family members witness severe cognitive decline in the affected person, their own predisposition to the mutation constitutes an additional strain. Siblings, children and grandchildren often becomes isolated from the rest of society and need additional support (Dewhurst et al., 1970).

9.6 Diagnostic significance and cognitive testing

Genetic testing for HD has in recent years provided opportunities for early diagnosis and evidence to date suggests that demand for genetic testing will increase in the future. However, knowledge is still inadequate about the psychological and physical impact of early diagnostics on people who are at risk of HD and those who learn that they will develop the disease. Family members and mutation carriers are often concerned about cognitive impairments and dementia. As the signs and symptoms in HD are multifaceted and clinicians cannot specify how and when the early indications will show, provision of factual information to carrier and family will greatly reduce fear and anxiety. This information to the patient can be provided by means of feedback and support in the context of neuropsychological testing by a psychologist or during medical visits. Diagnostic information about the symptomatic and cognitive profile provides the best understanding of the disease to the carrier and family. To evaluate therapeutic interventions requires a detailed neurological/neuropsychological examination based on cognitive, clinical and sometimes even experimental methods. Significant progress in understanding cognition in prodromal HD has reinforced the need for HD to be viewed as both a cognitive and motor disease. Cognition needs to be viewed as a component of quality of life and this recognition by clinicians will aid in the treatment of the long prodromal stages of the disease.

9.7 Future directions

HD is an uncommon but devastating condition for which there is currently no effective disease-modifying treatment. Symptomatic treatments are only modestly effective for some manifestations of the illness. Subtle cognitive impairment commences well before motor manifestations in many patients, complicating social and occupational functioning. Dementia is inevitable if the person with HD lives long enough. Because of the autosomal dominant pattern of HD inheritance, the ability to definitively identify mutation carriers and the availability of mouse models with various phenotypes, the future of HD research should be positive. Clinical trials of putative disease-modifying treatments in preclinical HD mutation carriers are needed. Such trials will need to assess HD carriers serially with both neuropsychological and neuroimaging tests, as well as other more specific biomarkers (Weir et al., 2011). Public and philanthropic funds will be needed for the development of therapies for HD as its low-prevalence status is unlikely to drive sufficient entrepreneurial interest. However, further research into the slowly progressive cognitive impairment found in people with HD might serve as a useful model for other neurodegenerative conditions with less certain etiological factors.

10. References

Almqvist, E. W., Bloch, M., Brinkman, R., et al. (1999). A worldwide assessment of the frequency of suicide, suicide attempts, or psychiatric hospitalization after predictive testing for Huntington disease. *American Journal of Human Genetics, 64*(5), 1293-1304, ISSN: 0002-9297

Almqvist, E. W., Elterman, D. S., MacLeod, P. M., et al. (2001). High incidence rate and absent family histories in one quarter of patients newly diagnosed with Huntington disease in British Columbia. *Clinical Genetics, 60*(3), 198-205, ISSN: 0009-9163

Antonini, A., Leenders, K. L., & Eidelberg, D. (1998). [11C]raclopride-PET studies of the Huntington's disease rate of progression: relevance of the trinucleotide repeat length. *Annals of Neurology, 43*(2), 253-255, ISSN: 0364-5134

Antonini, A., Leenders, K. L., Spiegel, R., et al. (1996). Striatal glucose metabolism and dopamine D2 receptor binding in asymptomatic gene carriers and patients with Huntington's disease. *Brain, 119*(6), 2085-2095, ISSN: 0006-8950

Aylward, E. H. (2007). Change in MRI striatal volumes as a biomarker in preclinical Huntington's disease. *Brain Research Bulletin, 72*(2-3), 152-158, ISSN: 0361-9230

Aylward, E. H., Brandt, J., Codori, A. M., et al. (1994). Reduced basal ganglia volume associated with the gene for Huntington's disease in asymptomatic at-risk persons. *Neurology, 44*(5), 823-828, ISSN: 0028-3878

Aylward, E. H., Sparks, B. F., Field, K. M., et al. (2004). Onset and rate of striatal atrophy in preclinical Huntington disease. *Neurology, 63*(1), 66-72, ISSN: 1526-632X

Bäckman, L., Robins Wahlin, T.-B., Lundin, A., et al. (1997). Cognitive deficits in Huntington's disease are predicted by dopaminergic PET markers and brain volumes. *Brain, 120,* 2207-2217, ISSN 0006-8950

Backman, L., Robins-Wahlin, T. B., Lundin, A., et al. (1997). Cognitive deficits in Huntington's disease are predicted by dopaminergic PET markers and brain volumes. *Brain, 120 (Pt 12),* 2207-2217, ISSN: 0006-8950

Barbeau, A., Duvoisin, R. C., Gerstenbrand, F., et al. (1981). Classification of extrapyramidal disorders. Proposal for an international classification and glossary of terms. *Journal of the Neurological Sciences, 51*(2), 311-327, ISSN: 0022-510X

Bates, G., Harper, P. S., & Jones, L. (eds.) (2002) *Huntington's Disease, Third Edition Oxford Monographs on Medical Genetics 45,* Oxford University Press, ISBN 0-19-851060-8, Oxford

Baudic, S., Maison, P., Dolbeau, G., et al. (2006). Cognitive impairment related to apathy in early Huntington's disease. *Dementia and Geriatric Cognitive Disorders, 21*(5-6), 316-321, ISSN: 1420-8008

Beglinger, L. J., O'Rourke, J. J., Wang, C., et al. (2010). Earliest functional declines in Huntington disease. *Psychiatry Research, 178*(2), 414-418, ISSN: 0165-1781

Bird, T. D. (1999). Outrageous fortune: the risk of suicide in genetic testing for Huntington disease. *American Journal of Human Genetics, 64*(5), 1289-1292, ISSN: 0002-9297

Bittenbender, J. B., & Quadfasel, F. A. (1962). Rigid and akinetic forms of Huntington's chorea. *Archives of Neurology, 7,* 275-288, ISSN: 0003-9942

Brandt, J., & Benedict, R. H. B. (2001). *Hopkins verbal learning test – revised.* Lutz, FL: Psychological Assessment Resources.

Brandt, J., Strauss, M. E., Larus, J., et al. (1984). Clinical correlates of dementia and disability in Huntington's disease. *Journal of Clinical Neuropsychology, 6*(4), 401-412, ISSN: 0165-0475

Broholm, J., Cassiman, J. J., Crauford, D., et al. (1994). Guidelines for the molecular genetics predictive test in Huntington's Disese. *Neurology, 44,* 1533-1536, ISSN: 0028-3878

Bruyn, G. W. (1962). Thiopropazate dihydrochloride (Dartal) in the treatment of Huntington's chorea. *Psychiatria, Neurologia, Neurochirurgia, 65,* 430-438, ISSN: 0033-2666

Butters, N., Salmon, D., & Heindel, W. C. (1994). Specificity of the memory deficits associated with basal ganglia dysfunction. *Revue Neurologique, 150*(8-9), 580-587, ISSN: 0035-3787

Close Kirkwood, S., Siemers, E., Viken, R. J., et al. (2002a). Evaluation of psychological symptoms among presymptomatic HD gene carriers as measured by selected MMPI scales. *Journal of Psychiatric Research, 36*(6), 377-382, ISSN: 0022-3956

Close Kirkwood, S., Siemers, E., Viken, R. J., et al. (2002b). Longitudinal personality changes among presymptomatic Huntington disease gene carriers. *Neuropsychiatry, Neuropsychology, & Behavioral Neurology, 15*(3), 192-197, ISSN: 0894-878X

Conneally, P. M. (1984). Huntington disease: genetics and epidemiology. *American Journal of Human Genetics, 36*(3), 506-526, ISSN: 0002-9297

Copley, T. T., Wiggins, S., Dufrasne, S., et al. (1995). Are we all of one mind? Clinicians' and patients' opinions regarding the development of a service protocol for predictive testing for Huntington disease. Canadian Collaborative Study for Predictive Testing for Huntington Disease. *American Journal of Medical Genetics, 58*(1), 59-69, ISSN: 0148-7299

de Boo, G. M., Tibben, A., Lanser, J. B., et al. (1997). Intelligence indices in people with a high/low risk for developing Huntington's disease. *Journal of Medical Genetics, 34*(7), 564-568, ISSN: 0022-2593

Deckel, A. W., & Morrison, D. (1996). Evidence of a neurologically based "denial of illness" in patients with Huntington's disease. *Archives of Clinical Neuropsychology, 11*(4), 295-302, ISSN: 0887-6177

Decruyenaere, M., Evers-Kiebooms, G., Boogaerts, A., et al. (2005). Partners of mutation-carriers for Huntington's disease: forgotten persons? *European Journal of Human Genetics, 13*(9), 1077-1085, ISSN: 1018-4813

Decruyenaere, M., Evers-Kiebooms, G., Boogaerts, A., et al. (2007). The complexity of reproductive decision-making in asymptomatic carriers of the Huntington mutation. *European Journal of Human Genetics, 15*(4), 453-462, ISSN: 1018-4813

Dewhurst, K., Oliver, J. E., & McKnight, A. L. (1970). Socio-psychiatric consequences of Huntington's disease. *British Journal of Psychiatry, 116*(532), 255-258, ISSN: 0007-1250

Ekberg, K., & Hane, M. (1984). Test battery for investigating functional disorders--the TUFF battery. *Scandinavian Journal of Work, Environment and Health, 10 Suppl 1*, 14-17, ISSN: 0355-3140

Feigin, A., Kieburtz, K., Como, P., et al. (1996). Assessment of coenzyme Q10 tolerability in Huntington's disease. *Movement Disorders, 11*(3), 321-323, ISSN: 0885-3185

Feigin, A., Tang, C., Ma, Y., et al. (2007). Thalamic metabolism and symptom onset in preclinical Huntington's disease. *Brain: A Journal of Neurology, 130*(Part 11), 2858-2867, ISSN: 0006-8950

Folstein, S. E., Jensen, B., Leigh, R. J., et al. (1983). The measurement of abnormal movement: methods developed for Huntington's disease. [Research Support, U.S. Gov't, P.H.S.]. *Neurobehavioral Toxicology and Teratology, 5*(6), 605-609., ISSN: 0275-1380

Grant, D. A., & Berg, E. A. (1948). A behavioral analysis of degree of reinforcement and ease of shifting to new responses in a Weigl-type card-sorting problem. *Journal of Experimental Psychology, 38*(4), 404-411, ISSN: 0022-1015

Grove, V. E., Jr., Quintanilla, J., & DeVaney, G. T. (2000). Improvement of Huntington's disease with olanzapine and valproate. *New England Journal of Medicine, 343*(13), 973-974, ISSN: 0028-4793

Gusella, J. F., Wexler, N. S., Conneally, P. M., et al. (1983). A polymorphic DNA marker genetically linked to Huntington's disease. *Nature, 306*(5940), 234-238, ISSN: 0028-0836

Harper, P. S. (2002). The epidemiology of Huntington's disease. In *Huntington's disease,* G. Bates, P. S. Harper & A. L. Jones, pp. 159-197, Oxford University Press, ISBN 0-19-851060-8, Oxford

Harper, P. S., Gevers, S., de Wert, G., et al. (2004). Genetic testing and Huntington's disease: issues of employment. *Lancet Neurology, 3*(4), 249-252, ISSN: 1474-4422

Harper, P. S., Lim, C., & Craufurd, D. (2000). Ten years of presymptomatic testing for Huntington's disease: the experience of the UK Huntington's Disease Prediction Consortium. *Journal of Medical Genetics, 37*(8), 567-571, ISSN: 1468-6244

Hedera, P. (2001). Ethical principles and pitfalls of genetic testing for dementia. *Journal of Geriatric Psychiatry and Neurology, 14*(4), 213-221, ISSN: 0891-9887

Huntington Study Group. (1996). The Unified Huntington's Disease Rating Scale: reliability and consistency. *Movement disorders, 11*(2), 136-142 ISSN: 0885-3185

Huntington, G. (1872). On chorea. George Huntington, M.D. *Journal of Neuropsychiatry Clinical Neuroscience, 15*(1), 109-112, ISSN: 0895-0172

Huntington's Disease Collaborative Research Group. (1993). A novel gene containing a trinucleotide repeat that is expanded and unstable on Huntington's disease chromosomes *Cell, 72*, 971-983, ISSN: 0092-8674

Julien, C. L., Thompson, J. C., Wild, S., et al. (2007). Psychiatric disorders in preclinical Huntington's disease. *Journal of Neurology, Neurosurgery, and Psychiatry, 78*(9), 939-943, ISSN: 1468-330X

Kehoe, P., Krawczak, M., Harper, P. S., et al. (1999). Age of onset in Huntington disease: sex specific influence of apolipoprotein E genotype and normal CAG repeat length. *Journal of Medical Genetics, 36*(2), 108-111, ISSN: 0022-2593

Kessler, S. (1987). Psychiatric implications of presymptomatic testing for Huntington's disease. *American Journal of Orthopsychiatry, 57*(2), 212-219, ISSN: 0002-9432

Kirkwood, S. C., Siemers, E., Bond, C., et al. (2000). Confirmation of subtle motor changes among presymptomatic carriers of the Huntington disease gene. *Archives of Neurology, 57*(7), 1040-1044, ISSN: 0003-9942

Kremer, B. (2002). Clinical neurology of Huntington's disease; Diversity in unity, unity in diversity. In *Huntington's disease,* G. Bates, P. S. Harper & A. L. Jones, pp. 28-61, Oxford University Press, ISBN 0-19-851060-8, Oxford.

Langbehn, D. R., Brinkman, R. R., Falush, D., et al. (2004). A new model for prediction of the age of onset and penetrance for Huntington's disease based on CAG length. *Clinical Genetics, 65*(4), 267-277, ISSN: 0009-9163

Langbehn, D. R., Hayden, M. R., & Paulsen, J. S. (2009). CAG-repeat length and the age of onset in Huntington disease (HD): A review and validation study of statistical approaches. *American Journal of Medical Genetics Part B: Neuropsychiatric Genetics*(153B), 397-408, ISSN: 1552-485X

Larsson, M. U., Almkvist, O., Luszcz, M. A., et al. (2008). Phonemic fluency deficits in asymptomatic gene carriers for Huntington's disease. *Neuropsychology, 22*(5), 596-605, ISSN: 0894-4105

Larsson, M. U., Luszcz, M. A., Bui, T. H., et al. (2006). Depression and suicidal ideation after predictive testing for Huntington's disease: a two-year follow-up study. *Journal of Genetic Counseling, 15*(5), 361-374, ISSN: 1059-7700

Lawrence, A. D., Watkins, L. H., Sahakian, B. J., et al. (2000). Visual object and visuospatial cognition in Huntington's disease: implications for information processing in corticostriatal circuits. *Brain, 123*(Pt 7), 1349-1364, ISSN: 0006-8950

Lemiere, J., Decruyenaere, M., Evers-Kiebooms, G., et al. (2004). Cognitive changes in patients with Huntington's disease (HD) and asymptomatic carriers of the HD mutation – a longitudinal follow-up study. *Journal of Neurology, 251*(8), 935-942, ISSN: 0340-5354

Lezak, M. D., Howieson, D. B., & Loring, D. W. (2004). *Neuropsychological assessment.* (Fourth ed.). Oxford University Press, ISBN 987-0-19-5111-21-7, New York.

Lundervold, A. J., & Reinvang, I. (1991). Neuropsychological findings and depressive symptoms in patients with Huntington's disease. *Scandinavian Journal of Psychology, 32*(3), 275-283, ISSN: 0036-5564

Lundervold, A. J., Karlsen, N. R., & Reinvang, I. (1994a). Assessment of 'subcortical dementia' in patients with Huntington's disease, Parkinson's disease, multiple sclerosis and AIDS by a neuropsychological screening battery. *Scandinavian Journal of Psychology, 35*(1), 48-55, ISSN: 0036-5564

Lundervold, A. J., Reinvang, I., & Lundervold, A. (1994b). Characteristic patterns of verbal memory function in patients with Huntington's disease. *Scandinavian Journal of Psychology, 35*(1), 38-47, ISSN: 0036-5564

Mattsson, B. (1974). Huntington's chorea in Sweden. *Acta Psychiatrica Scandinavica. Supplementum, 255*, 221-235, ISSN: 0065-1591

Mendez, M. F. (1994). Huntington's disease: update and review of neuropsychiatric aspects. *International Journal of Psychiatry in Medicine, 24*(3), 189-208, ISSN: 0091-2174

Mestre, T., Ferreira, J., Coelho, M. M., et al. (2009a). Therapeutic interventions for disease progression in Huntington's disease. *Cochrane Database of Systematic Reviews*(3), CD006455, ISSN: 1469-493X

Mestre, T., Ferreira, J., Coelho, M. M., et al. (2009b). Therapeutic interventions for symptomatic treatment in Huntington's disease. *Cochrane Database of Systematic Reviews*(3), CD006456, ISSN: 1469-493X

Milner, B. (1963). Effects of different brain lesions on card sorting. *Archives of Neurology, 9*, 90-100, ISSN: 0003-9942.

Montoya, A., Pelletier, M., Menear, M., et al. (2006a). Episodic memory impairment in Huntington's disease: a meta-analysis. *Neuropsychologia, 44*(10), 1984-1994, ISSN: 0028-3932

Montoya, A., Price, B. H., Menear, M., et al. (2006b). Brain imaging and cognitive dysfunctions in Huntington's disease. *Journal of Psychiatry & Neuroscience, 31*(1), 21-29, ISSN: 1180-4882

Nance, M., Myers, R. H., Wexler, A., et al. (Producer). (2003). *Genetic Testing for Huntington's disease; It s relevance and implications.* Revised HDSA Guidelines. Retrieved from http://www.hdsa.org/images/content/1/1/11884.pdf.

Orbeck, A. L. (1960). [Lund-Huntington chorea]. *Tidsskrift for Den Norske Laegeforening, 80*, 95-96, ISSN: 0029-2001

Osterrieth, P. A. (1944). Le test de copie d'une figure complexe; contribution à l'étude de la perception et de la mémoire. *Archives de psychologie*, ISSN: 0003-9640

Paulsen, J. S. (2010). Early detection of Huntington's disease. *Future Neurology, 5*(1), 85-104, ISSN: 1479-6708

Paulsen, J. S., Hoth, K. F., Nehl, C., et al. (2005a). Critical periods of suicide risk in Huntington's disease. *American Journal of Psychiatry, 162*(4), 725-731, ISSN: 0002-953X

Paulsen, J. S., Langbehn, D. R., Stout, J. C., et al. (2008). Detection of Huntington's disease decades before diagnosis: the Predict-HD study. *Journal of Neurology, Neurosurgery, and Psychiatry, 79*(8), 874-880, ISSN: 1468-330X

Paulsen, J. S., Nehl, C., Hoth, K. F., et al. (2005b). Depression and stages of Huntington's disease. *Journal of Neuropsychiatry & Clinical Neurosciences, 17*(4), 496-502, ISSN: 0895-0172

Paulsen, J. S., Ready, R. E., Hamilton, J. M., et al. (2001). Neuropsychiatric aspects of Huntington's disease. *Journal of Neurology, Neurosurgery, and Psychiatry, 71*(3), 310-314, ISSN: 0022-3050

Paulsen, J. S., Wang, C., Duff, K., et al. (2010). Challenges assessing clinical endpoints in early Huntington disease. *Movement Disorders*, ISSN: 1531-8257

Paulsen, J. S., Zimbelman, J. L., Hinton, S. C., et al. (2004). fMRI biomarker of early neuronal dysfunction in presymptomatic Huntington's Disease. *AJNR American Journal of Neuroradiology, 25*(10), 1715-1721, ISSN: 0195-6108

Redondo-Verge, L. (2001). [Cognitive deterioration in Huntington disease]. *Revista de Neurología, 32*(1), 82-85, ISSN: 0210-0010

Rey, A. (1941). L'examen psychologique dans les cas d'encéphalopathie traumatique.(Les problems.). *Archives de psychologie*, ISSN: 0003-9640

Ridley, R. M., Frith, C. D., Crow, T. J., et al. (1988). Anticipation in Huntington's disease is inherited through the male line but may originate in the female. *Journal of Medical Genetics, 25*(9), 589-595., ISSN: 0022-2593

Robins Wahlin, T.-B. (2007). To know or not to know: a review of behaviour and suicidal ideation in preclinical Huntington's disease. *Patient Education and Counseling, 65*(3), 279-287, ISSN: 0738-3991

Robins Wahlin, T.-B., Backman, L., Lundin, A., et al. (2000). High suicidal ideation in persons testing for Huntington's disease. *Acta Neurologica Scandinavica, 102*(3), 150-161, ISSN: 0001-6314

Robins Wahlin, T.-B., Larsson, M. U., Luszcz, M. A., et al. (2010). WAIS-R features of preclinical Huntington's disease: implications for early detection. *Dementia and Geriatric Cognitive Disorders, 29*(4), 342-350, ISSN: 1421-9824

Robins Wahlin, T.-B., Lundin, A., & Dear, K. (2007). Early cognitive deficits in Swedish gene carriers of Huntington's disease. *Neuropsychology, 21*(1), 31-44, ISSN: 0894-4105

Robins Wahlin, T.-B., Lundin, A., Backman, L., et al. (1997). Reactions to predictive testing in Huntington disease: case reports of coping with a new genetic status. *American Journal of Medical Genetics, 73*(3), 356-365, ISSN: 0148-7299

Roos, R. A., Hermans, J., Vegter-van der Vlis, M., et al. (1993). Duration of illness in Huntington's disease is not related to age at onset. *Journal of Neurology, Neurosurgery, and Psychiatry., 56*(1), 98-100., ISSN: 0022-3050

Roos, R. A., Vegter-van der Vlis, M., Hermans, J., et al. (1991). Age at onset in Huntington's disease: effect of line of inheritance and patient's sex. *Journal of Medical Genetics, 28*(8), 515-519, ISSN: 0022-2593

Rubinsztein, D. C., Leggo, J., Chiano, M., et al. (1997). Genotypes at the GluR6 kainate receptor locus are associated with variation in the age of onset of Huntington disease. *Proceedings of the National Academy of Sciences of the United States of America, 94*(8), 3872-3876, ISSN: 0027-8424

Rubinsztein, D. C., Leggo, J., Coles, R., et al. (1996). Phenotypic characterization of individuals with 30-40 CAG repeats in the Huntington disease (HD) gene reveals HD cases with 36 repeats and apparently normal elderly individuals with 36-39 repeats. *American Journal of Human Genetics, 59*(1), 16-22, ISSN: 0002-9297

Shiwach, R. (1994). Psychopathology in Huntington's disease patients. *Acta Psychiatrica Scandinavica, 90*(4), 241-246, ISSN: 0001-690X

Shoulson, I., & Fahn, S. (1979). Huntington disease: clinical care and evaluation. *Neurology, 29*(1), 1-3, ISSN: 0028-3878

Siesling, S., Vegter-van de Vlis, M., Losekoot, M., et al. (2000). Family history and DNA analysis in patients with suspected Huntington's disease. *Journal of Neurology, Neurosurgery and Psychiatry, 69*(1), 54-59, ISSN: 0022-3050

Simpson, S. A., & Harper, P. S. (2001). Prenatal testing for Huntington's disease: experience within the UK 1994-1998. *Journal of Medical Genetics, 38*(5), 333-335, ISSN: 1468-6244

Snowden, J. S., Craufurd, D., Thompson, J., et al. (2002). Psychomotor, executive, and memory function in preclinical Huntington's disease. *Journal of Clinical and Experimental Neuropsychology 24*(2), 133-145, ISSN: 1380-3395

Solomon, A. C., Stout, J. C., Johnson, S. A., et al. (2007). Verbal episodic memory declines prior to diagnosis in Huntington's disease. *Neuropsychologia, 45*(8), 1767-1776, ISSN: 0028-3932

Soneson, C., Fontes, M., Zhou, Y., et al. (2010). Early changes in the hypothalamic region in prodromal Huntington disease revealed by MRI analysis. *Neurobiology of Disease*, ISSN: 1095-953X

Stout, J. C., Paulsen, J. S., Queller, S., et al. (2011). Neurocognitive signs in prodromal huntington disease. *Neuropsychology, 25*(1), 1-14, ISSN: 1931-1559

Strauss, E., Sherman, E. M. S., & Spreen, O. (2006). *A compendium of neuropsychological tests. Administration, Norms, and Commentary* (Third ed.). Oxford University Press, ISBN 0195159578 | ISBN-13: 9780195159578, New York.

Tost, H., Wendt, C. S., Schmitt, A., et al. (2004). Huntington's disease: phenomenological diversity of a neuropsychiatric condition that challenges traditional concepts in neurology and psychiatry. *American Journal of Psychiatry, 161*(1), 28-34, ISSN: 0002-953X

van Duijn, E., Kingma, E. M., & van der Mast, R. C. (2007). Psychopathology in verified Huntington's disease gene carriers. *Journal of Neuropsychiatry and Clinical Neurosciences, 19*(4), 441-448, ISSN: 0895-0172

van Duijn, E., Kingma, E. M., Timman, R., et al. (2008). Cross-sectional study on prevalences of psychiatric disorders in mutation carriers of Huntington's disease compared with mutation-negative first-degree relatives. *Journal of Clinical Psychiatry, 69*(11), 1804-1810, ISSN: 1555-2101

van Walsem, M. R., Sundet, K., Retterstøl, L., et al. (2009). A double blind evaluation of cognitive decline in a Norwegian cohort of asymptomatic carriers of Huntington's disease. *Journal of Clinical and Experimental Neuropsychology*, ISSN: 1744-411X

Vanderberg, S. G. (1971). *A test of three-dimensional spatial visualization based on the Shepard-Metzler "mental rotation" study*. University of Colorado. Boulder.

Verny, C., Allain, P., Prudean, A., et al. (2007). Cognitive changes in asymptomatic carriers of the Huntington disease mutation gene. *European Journal of Neurology, 14*(11), 1344-1350, ISSN: 1468-1331

Walker, F. O. (2007). Huntington's disease. *Lancet, 369*(9557), 218-228, ISSN: 0140-6736

Walsh, A. (1999). Presymptomatic testing for Huntington's disease: the role of genetic counseling. *Medicine & Health Rhode Island, 82*(5), 168-170, ISSN: 1086-5462

Warby, S., Graham, R., & Hayden, M. (2010). *Huntington disease.* GeneReviews, Retrieved from <http://www.ncbi.nlm.nih.gov/bookshelf/br.fcgi?book=gene&part=huntington>.

Wechsler, D. (1997). Wechsler Adult Intelligence Scale-III (WAIS-III). *New York: Psychological Corporation.*

Weir, D. W., Sturrock, A., & Leavitt, B. R. (2011). Development of biomarkers for Huntington's disease. *Lancet Neurology, 10*(6), 573-590, ISSN: 1474-4465

Wolf, R. C., Sambataro, F., Vasic, N., et al. (2011). Longitudinal functional magnetic resonance imaging of cognition in preclinical Huntington's disease. *Experimental Neurology,* ISSN: 1090-2430

Woods, S. P., Scott, J. C., Conover, E., et al. (2005). Test-retest reliability of component process variables within the Hopkins Verbal Learning Test-Revised. *Assessment, 12*(1), 96-100, ISSN: 1073-1911

World Federation of Neurology: Research Committee. Research Group on Huntington's chorea. Ethical issues policy statement on Huntington's disease molecular genetics predictive test, 94, Pub. L. No. 1-3 327-332 (1989 Dec).

Young, A. B., Shoulson, I., Penney, J. B., et al. (1986). Huntington's disease in Venezuela: neurologic features and functional decline. *Neurology, 36*(2), 244-249, ISSN: 0028-3878

Zakzanis, K. K. (1998). The subcortical dementia of Huntington's disease. *Journal of Clinical and Experimental Neuropsychology, 20*(4), 565-578, ISSN: 1380-3395

8

Early Dysfunction of Neural Transmission and Cognitive Processing in Huntington's Disease

Michael I. Sandstrom, Sally Steffes-Lovdahl, Naveen Jayaprakash, Antigone Wolfram-Aduan and Gary L. Dunbar
Central Michigan University
USA

1. Introduction

Huntington's disease (HD) is one of many deteriorative brain diseases, a class of disease in which neurons progressively die. In its final stages, HD robs patients of the dignity of their humanity; denying control of basic movements necessary for communication, facial expression and personal accomplishment. A means to test for the mutation has been available since 1993, when the Huntington's Disease Collaborative Research Group exposed the huntingtin gene and characterized the nature of the mutation process. Despite this, children of patients often avoid determining their genotype because such a diagnosis is currently merely bleak without hope of remedy, and because of legitimate fears of employment discrimination or difficulties maintaining health insurance given the legal definition of "pre-existing condition." In the absence of promising treatments or prospects for cures the devastating loss of muscular control during the final stages of disease progression is ominous. It is therefore not uncommon for HD patients to become aware of their own disease rather late into its progression when motor symptoms begin to emerge. As these movement symptoms arise they may be effectively masked by compensatory behavioral strategies. In time, however, these compensatory tactics fail to keep up with the advancing choreic movements which eventually dominate and negate purposeful motor control.

The regions of the brain that are most susceptible to neuron death in HD, in a manner that correlates with motoric symptom severity, are the cerebral cortex, and the caudate and putamen nuclei of the basal ganglia (Young et al., 1986; Halliday et al., 1998). At first glance, it may seem that halting or preventing progressive neuron death within these affected areas would provide an adequate therapeutic strategy for HD. While efforts to do this are indeed under way (see Mattson & Furukawa, 1996 or Mattson, 2000 for review; Leyva et al., 2010; Niatsetskaya et al., 2010), this approach has, in and of itself, proven insufficient. At best, efforts to block apoptosis-generating mechanisms in HD patients have delayed symptom onset at early stages, yet have failed to ward off motor symptom onset (*Vitamin E-related Antioxidant D-α-tocopherol* - Peyser et al., 1995; *Creatine* - Verbessem et al., 2003; *Coenzyme* Q_{10} - Huntington Study Group, 2001). Although higher dose studies are currently ongoing with these compounds, it remains unclear whether enticing benefits observed *in vitro* (e.g. Wang et al., 2005; Hoffstrom et al., 2010), or with animal models (Ferrante et al., 2000; Dedeoglu et al. 2002; van Raamsdonk, 2005a) will manifest in human clinical trials (see Delanty & Dichter, 2000, or Wang et al., 2010 for broader reviews of treatment efforts). The physiological perspective represents a plausible theoretical viewpoint that may explain the rather disappointing clinical results of cell preservation efforts. Preventing neuronal death may perpetuate neurons, but are these preserved neurons in HD patients capable of carrying out their prescribed roles sufficiently, given their diseased state at the time of treatment? Efforts to merely prevent neuronal death by increasing ATP synthesis, antioxidants, or other anti-apoptosis remedies are unlikely to provide sufficient benefit to patients if neurons are already malfunctioning. Furthermore, if malfunctioning neurons that are maintained by treatments nevertheless fail to engage their appropriate roles, then their survival may be disruptive. It would seem that the key to rectifying HD will require not only maintaining neuron survival, but also their proper physiology. Beyond total cell counts, protected neurons must be able to respond appropriately to afferent signals, sensitivity modulations, and engage in or disengage from longer-term plastic changes in a normal manner. This chapter will focus on the functional disruptions of neural transmission and related cognitive processing at very early disease stages. Therefore, *presymptomatic HD* (pre-HD) will be defined as HD-related malfunctions arising prior to the emergence of diagnosable motor abnormalities described by Paulson (2008).

2. Primary or compensatory mechanisms?

Neurons, either as individual cells or as part of an integrated nervous system, continually attempt to compensate for disruptive influences and maintain a dynamic equilibrium. When signals become weak, receptor sensitivity is boosted to compensate. When energy utilization is high, extra synapses are created to maintain signals at reduced cost. These compensatory responses are known as *plasticity* and they are at work not only in response to damaging or disruptive influences but also to support learning and memory formations or the process of forgetting when information becomes less applicable (Lee et al., 2004; Fusi et al., 2007). Several famous neuroscientists offered early descriptions of the mechanisms underlying plasticity. Among these, the one who received the most recognition in this arena was Donald Hebb, who was a student of Karl Lashley and subsequently collaborated with Wilder Penfield (Brown & Milner, 2003). Hebb's main contribution was his theoretical description of a modifiable synapse that supports extended increases in synaptic strength when specific conditions are met; a process known as *long term potentiation* or LTP (Hinton, 2003; Milner, 2003). Recently, it has become popular to refer to a myriad of neuronal modification processes as "Hebbian" when they lead to either synaptic LTP or its opposite, *long term depression* or LTD (Massey & Bashir, 2007; McBain & Kauer, 2009).

It is evident that the activity of neurons during pre-HD stages is distorted. In addition, various affected brain systems attempt to compensate for the mutation-related malfunctions, particularly during these earlier stages. These simultaneous processes present a substantial challenge to neuroscientists attempting to unravel the neurophysiological mysteries of HD. To make things yet more challenging, compensations in one system may compromise another system. If researchers could localize and reverse primary malfunctions, it may be possible to control the spread of these potentially maladaptive compensatory adjustments.

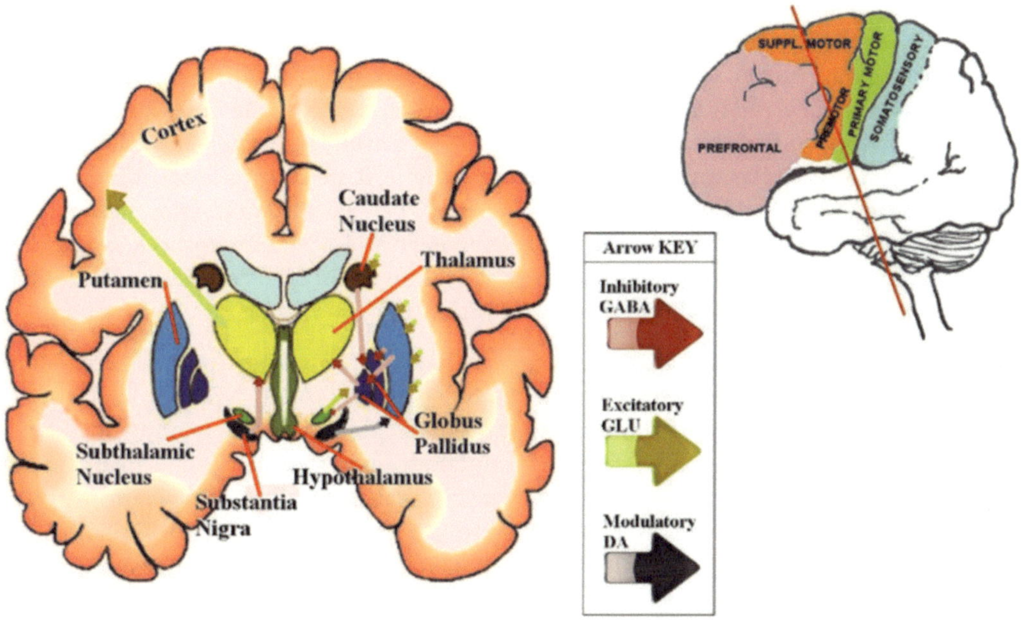

Fig. 1. Basal Ganglia Circuitry. The input regions (caudate, putamen, subthalamic nucleus) are generally conceptualized as receiving converging excitatory input from the cortex. Within these regions, modulatory DA input arising from the substantia nigra tailors the responses of the majority efferent MSNs. The subthalamic nucleus contributes excitatory input to the globus pallidus while the centromedian and intralaminar nuclei of the thalamus send excitatory input to the caudate and putamen. The caudate and putamen contribute sequential inhibitory signals through the globus pallidus and the substantia nigra reticulata, converging inhibitory signals on the thalamus. These converging inhibitory signals modulate thalamic relay neurons which return excitatory signals into the frontal cortex based on amassed inhibition or disinhibition. The thalamocortical targets are mostly the supplementary motor and premotor areas for movement, but other cortico-basal ganglia-cortical loops interact with other regions of prefrontal cortex involved in behavior planning. Slightly modified version of basal ganglia image reprinted with permission Courtesy of the Dana Foundation, Copyright 2007, all rights reserved.

To fully understand pre-HD, it is necessary to provide background about the cerebral circuitry where malfunctions begin to appear. HD is primarily a disorder of the basal ganglia, so we'll begin by describing the primary associated nuclei and connections of this system. The key associated neurotransmitters are glutamate (GLU), gamma aminobutyric acid (GABA), acetylcholine (ACh), adenosine (ADN), nitric oxide (NO), dopamine (DA), serotonin (5-HT), endocannabinoids, and various cotransmitter neuropeptides. As diagrammed in Figure 1, the basal ganglia are generally conceptualized first by orienting to the primary input regions, the *caudate* and *putamen* nuclei (these are indistinct in experimental animals and referred to as a combined "striatum"). The whole of the cortical mantle, along with centromedian and intralaminar thalamic nuclei, send excitatory GLU projections to these structures where they converge on both the GABAergic *medium spiny* *neurons (*MSNs) and local interneurons containing GABA, ACh, or NO, along with various neuropeptides. Following local integration, the MSNs project to the globus pallidus or the substantia nigra reticulata, which both harbor GABAergic neurons that feed forward to the thalamus where they modulate thalamic relay neurons that feed back to the cortex. DA originates from the substantia nigra compacta, releasing the highest levels of this neurotransmitter into the caudate and putamen where it modulates local MSN activity, along with 5-HT originating from the dorsal raphe nucleus. Thalamic relay neurons that close the "motor loop" feed back to the supplementary and premotor cortices, while those that close the "cognitive loop" feed back to the prefrontal cortex (see Middleton & Strick, 2000, for explanation of loops).

Given the high convergence of axon terminals, and associated astrocytes (see Pascual et al., 2005), releasing so many different neurotransmitters (GLU, GABA, NO, DA, 5-HT, ADN) onto MSNs, the complexity of their modulation seems to present a wide window for error in ordinary conditions. Additionally, if these modulatory inputs begin to send inappropriate signals (as will be discussed below) it is surprising that this system continues to process movement signals for as long as it does before motor symptoms begin. Striatal MSNs represent the majority (approximately 90%) of neurons in the striatum responsible for relaying processed information to subsequent basal ganglia stations. Since striatal MSNs are most notably vulnerable in HD, it will be important to place these neurons into proper context.

Figure 2 depicts a model striatal MSN with a subset of notable afferent influences. These neurons are influenced by nitric oxide arising from GABA interneurons that synapse on dendritic spine necks (Kubota & Kawaguchi, 2000), by ACh arising from cholinergic interneurons that are generally understood to be the "tonically active" striatal neurons (Wilson et al., 1990), and by adenosine (ADN) arising from both local neurons and astrocytes in a nonsynaptic but activity-dependent manner (Delaney & Geiger, 1998; Pascual et al., 2005; Pajski & Venton, 2010). Also, recent findings have elevated endocannabinoids to a prominent position in striatal synaptic processing, as these compounds tend to be released by MSNs in response to GLU and DA stimulation and provide feedback to CB1 receptors located on GLU-releasing axon terminals (Matyas et al., 2006; Uchigashima et al., 2007; Lovinger, 2010).

The history of basal ganglia exploration was profoundly influenced for many years by the pioneering work of Charles Gerfen, who was the first to expose distinctions between two prominent circuitry pathways emanating from the rat striatum: the *striatonigral* and the *striatopallidal* pathways (Gerfen & Young, 1988; Gerfen, 1992a). Thus, striatal MSNs were understood to send efferent axons from the striatum *either* to the substantia nigra pars reticulata (includes internal pallidum in humans) *or* the external globus pallidus, but not both. Within this seminal work, Gerfen and others delineated several important distinctions between these two GABAergic efferent pathways, such as differential neuropeptide expression, DA receptor expression (Gerfen, 1992b), and more recently, differential muscarinic receptor expression (Acquas & DiChiara, 2002). These data have been foundational to many speculations regarding the function of the basal ganglia and its role in selecting behavioral actions.

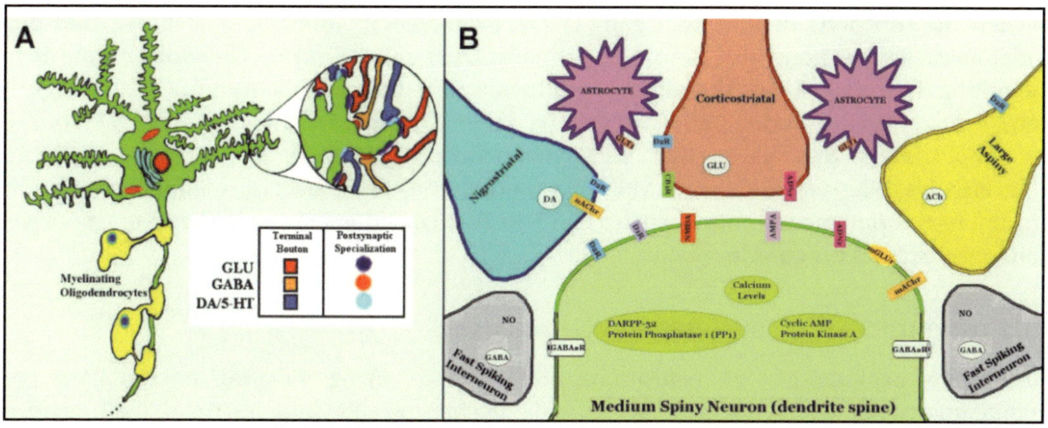

Fig. 2. Synaptic Interactions on a Medium Spiny Neuron. A. The dendritic spines on MSN surfaces are postsynaptic specializations that expand the input surface area and often act as synaptic compartments. Excitatory GLU (red) inputs, arising from corticostriatal or thalamostriatal afferents, tend to converge on the distal regions of these spines. Modulatory inputs of nigrostriatal DA or raphe-striatal 5-HT (blue) are either found juxtaposed near GLU inputs or on the main dendritic branch, where they can best modulate the excitation. GABA interneurons tend to synapse closer to the soma and the main dendritic branch, providing a shunting capacity that can truncate excitation. In response to input disruptions, remaining terminals often shift their positions to adjust their efficacy in driving MSN activity. B. Several modulatory inputs surround a dendritic spine simultaneously. The timing of transmitter arrival, as well as the specific combinations that arrive, are critical to both the initial response and the longer-term consequences of transmission. The many dendritic spines increase available surface area for all the synaptic structures. Astrocytes surrounding synapses circumscribe GLU and GABA terminals, containing synaptic overflow, while DA and acetylcholine (ACh) can diffuse over wider distances. Important intracellular response elements include the dopamine and cyclic AMP-regulated phosphoprotein weighing 32 kDa (DARPP-32) which typically inhibits protein phosphatase 1 (PP1), calcium levels, and the cyclic AMP produced by adenylate cyclase which regulates protein kinase A. The depicted receptors include: dopaminergic D1 (D1R) and D2 (D2R); glutamatergic n-methyl-D-aspartate (NMDA), α-amino-3-hydroxy-5-methyl-4-isoxazolepropionic (AMPA), and metabotropic (mGLUr); cholinergic nicotinic (nAChr) and muscarinic (mAChr); GABA (GABAaR, the GABAbR are also present but not shown); cannabinoid (CB1R, are also expressed on MSN terminals); and adenosine (ADNr, also expressed on MSN terminals).

However, despite recent experiments using the new cyclic recombinase expression (CRE) technology (Matamales et al., 2009; Valjent et al., 2009; Bateup et al., 2010) that have demonstrated a clear separation of these striatal efferent pathways in mice, substantial populations of MSNs in rats (30%) and primates (80%) project to both the nigral/internal pallidum and the external globus pallidus simultaneously (Kawaguchi, et al., 1990; Wu et al., 2000; Levesque & Parent, 2005; Fino & Venance, 2010). Therefore, while differences in receptor expression may yet yield distinct striatal neuron subpopulation responses in rats, primates, and humans, the importance of strictly *distinct* efferent pathways emanating from the striatum is less clear.

When the functions of the basal ganglia are evaluated in humans, it is likely that our anatomical connections are more like primates than rats or mice. Therefore, while it is tempting to speculate that the different pathways, often referred to as *direct* (striatonigral) and *indirect* (striatopallidal) pathways may be distinctly impacted within rodent HD models, it would seem less likely that such distinctions remain in the human condition. Nevertheless, the sensitivities of MSNs to modulation and plasticity within the striatum (caudate and putamen for humans) are likely to bear considerable resemblance to responses and mechanisms exposed in rodents.

3. Presymptomatic neurotransmitter release

Collective explorations of neurotransmitter release using *in vivo* microdialysis or voltammetry-related techniques can provide useful insights into neuronal malfunctions predating motor symptom expression in HD. Investigations into whether neurons are stimulated sufficiently to release, whether neurotransmitter availability/storage in vesicles provides sufficient quantities upon release, and whether neurotransmitters are removed appropriately to terminate the postsynaptic response, have all indicated that different neurotransmitter systems can exhibit unique malfunctions. Because of the common associations between the excitatory neurotransmitter GLU, excitotoxicity, and apoptosis in the HD brain, the glutamatergic striatal afferents were the first system to be investigated for early stage HD-associated problems using available rodent models (Greenamyre, 1986). A full account of all rodent HD models is beyond the scope of this chapter, but a brief description is pertinent to make a key point. The transgenic rodent (rat or mouse) has become a popular model of HD, and perhaps the most popular to date would be the R6/2 mouse that harbors exon 1 of the human mutated huntingtin gene with approximately 150 CAG repeats, exhibiting symptoms by 8 weeks (Carter et al., 1999). This is to be distinguished from the *knock-in* 150 mice (KI-150) that harbor full-length mutant but murine-based huntingtin genes inserted into the genome in a manner that replaces the endogenous gene but maintains endogenous expression control. These KI-150 mice take up to 80 weeks to exhibit symptoms despite harboring a similar number of CAG repeats (Heng et al., 2007). The key point here would be that when expression control presumably minimizes protein creation, symptom severity is dampened providing greater windows for exploring presymptomatic stages.

As previously indicated (Figure 2), striatal MSNs are heavily innervated by glutamatergic afferents arising from the cortex (corticostriatal) and the thalamus (thalamostriatal). The convergence of this input represents the primary excitatory drive for all striatal neurons, with the majority of glutamatergic synapses targeting the tips of dendritic spines. Striatal neurons recorded in slice preparations (Berretta, 2008; Stern et al., 1998; Wilson & Kawaguchi, 1996) and stationary awake animals (Sandstrom & Rebec, 2003; Wilson, 1993, 2004) are mostly silent. In turn, bursts of activity arise in intact animals during bouts of movement, and are believed to be generated by correlated surges among glutamatergic afferents (Wilson, 2004). Evidence indicates that striatal MSNs are stimulated by increased amounts of GLU beginning at presymptomatic stages, both because synaptic transport mechanisms begin to fail (Estrada-Sanchez et al., 2009; Brustovetsky et al., 2004; Behrens et al., 2002; Lievens et al., 2001), and as a result of altered control of glutamate release (authors' observations; Estrada-Sanchez et al., 2009; Nicniocaill et al., 2001). In fact, findings of increased resistance to excitotoxicity in the presymptomatic R6/2 mouse suggest early compensations occur to diminish GLU sensitivity somewhat (Qian et al., 2011; Estrada-Sanchez et al., 2010; Hansson et al., 1999). Furthermore, treatments that decrease glutamate transporter expression enhance glutamate-related neurotoxicity (Estrada-Sanchez et al., 2010). Conversely, boosting the expression of astrocyte-based GLT-1 glutamate transporter using an antibiotic drug called ceftriaxone, which increases glutamate reuptake in R6/2 mice, not only diminished motor-symptom expression, such as paw clasping and ballistic twitching, but also seemed to improve cognitive processing, as indicated in plus-maze activity (Sari et al., 2010; Miller et al., 2008a). These deficits in GLU control seem to extend into the frontal cortex as well (Hassel et al., 2008; Behrens et al., 2002) which would be expected to distort information processing more profoundly in both motor and cognitive domains by affecting both primary and downstream targets.

Microdialysis is a technique commonly used to measure extracellular release of neurotransmitters and is a versatile technique in that it can measure several different neurotransmitter substances at once. The basic technique requires pumping a solution that closely approximates cerebrospinal fluid across a small semi-permeable membrane-enclosed probe tip that is placed inside the brain region of interest. Via diffusion, neurotransmitter substances released will enter the semi-permeable membrane tip, pass into an output line, and accumulate in collection vials. The contents of these vials are then analyzed for neurotransmitter content using high performance liquid chromatography (HPLC) and the levels of neurotransmitter measured represent snapshots of release activity that took place during the experiment. Collections can be taken at intervals throughout the experiment to indicate the changes that take place over time.

Our laboratory has investigated release activity in, perhaps, the most valid model of pre-HD, the KI-150 mouse. As described previously, this knock-in model provides a wide window of presymptomatic development. They also lack even subtle motor symptoms until at least 14 weeks of age. With this model, we found disrupted GLU release control at the earliest age ever observed, prior to the onset of cognitive deficits. We performed a within-subject set of experiments that began following weaning with multi-stage training of an operant task during the next 3 weeks, from 6-9 weeks of age. Training culminated in a task that required animals to alternate bar-pressing between two levers in an operant box (Med Associates, St. Albans, VT) in a left-right-left-right-left sequence. The reward was a sucrose pellet for each successful sequence accomplished. Once trained, 9-10 week old animals were then challenged to adopt the reverse sequence (right-left-right-left-right); a challenge intended to test behavioral flexibility after habitual behavior had been established.

In vivo microdialysis was performed during this challenge, and during a task-free period that followed, in order to measure neurotransmitter changes elicited within the mouse striatum. After two recovery days, these same animals were then subject to a second microdialysis experiment, during which high-level (80mM, normally 2.9mM) potassium-containing artificial cerebrospinal fluid was pumped across the membrane. This technique is a popular method, used to generate local excitation of neurons and terminals within the region of interest (Nicniocaill et al., 2001; Tossman et al., 1986). Potassium typically flows through neuronal membranes easily, and when it is provided extracellularly this depolarizes the local neurons and afferent axons. Although no statistically significant genotype distinction was found in operant-stimulated GLU levels, potassium-stimulated

GLU was significantly increased among the homozygote mice, by comparison to the remaining genotypes (see Figure 3).

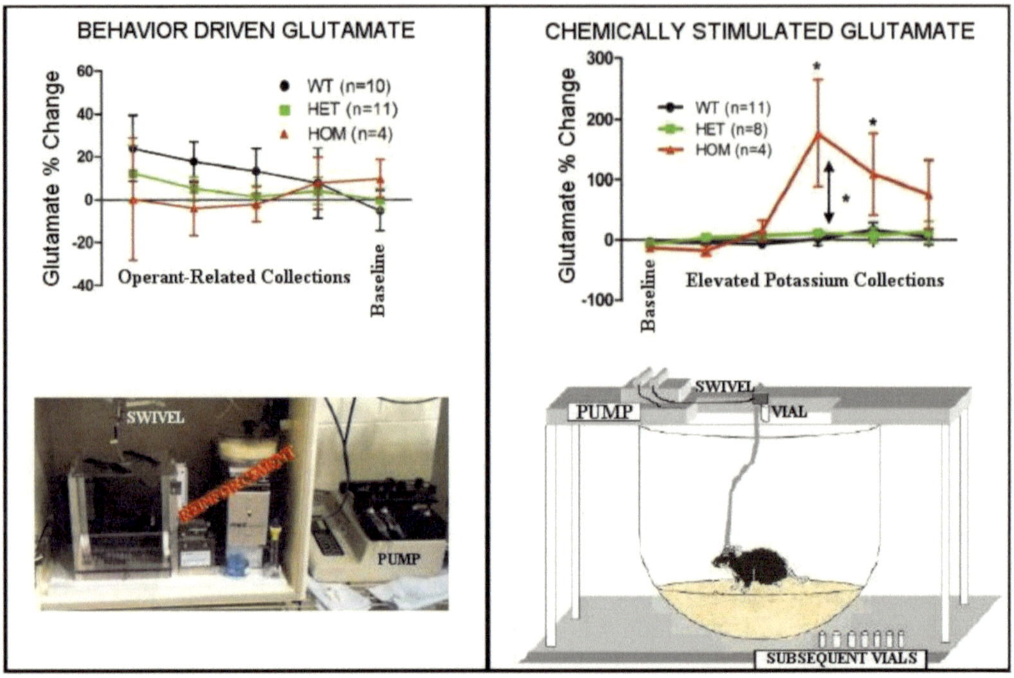

Fig. 3. Glutamate Microdialysis with Knock-In Mice. Two sequential microdialysis sessions were performed with freely-moving KI-150 mice. In the first session, measurements of GLU were taken during operant behavior, while these measures were taken in the second session during elevated potassium stimulation. GLU levels measured in the first session were not significant, while a significant difference was found between homozygote mice and both their heterozygote and wild-type littermates in response to striatal potassium stimulation. The homozygote animals averaged 160% increases in glutamate during potassium stimulation (from 2.9-mM to 80-mM) while no increases in GLU responses to behavioral stimulation were observed. These groups showed no genotype distinctions in the number of reinforcers earned when required to reverse their bar-pressing pattern. No genotype-related movement deficits were exhibited by these animals as measured by open field activity or grip strength measures, indicating they were presymptomatic. At this early age (10-11 weeks) these knock-in animals showed no signs of motoric pathology in the hallmark longitudinal study that used far more extensive batteries of tests to thoroughly assess deficit onset (Heng et al., 2007).

The data depicted in Figure 3 also relates to a common theme with neurochemical measurements in the context of deteriorative disease: deficits in release control often require stimulation to reveal an existing abnormality. This relates to the compensatory mechanisms previously described. In our experiments for example, it would seem that engaging in operant bar-pressing behavior for food reward was not sufficient to expose an underlying problem with glutamate control, while the 80 millimolar elevated potassium concentrations in the extracellular fluid apparently was. Within the striatum, it seems the large majority of synaptic GLU removal, post-release, is accomplished by transporters expressed by astrocytes (Lee & Pow, 2010). Energy is required for astrocytes to accomplish GLU transport (Azarias et al., 2011), and a corresponding decline in the astrocytic expression of GLU transporters seems to reach a point where the striatum can no longer keep up with the behavior-related surges of GLU necessary to generate striatal bursting activity (Wilson, 2004). It seems this occurs despite observations that astrocytes initially proliferate in HD, perhaps as a compensatory strategy (Faideau et al., 2010). It is also relevant to pre-HD that developmental changes seem to take place between youth and later adulthood regarding astrocytic participation in GLU clearance. Apparently astrocytes adopt a greater role in this clearance at later ages, when HD symptoms are more profound (Thomas et al., 2011a).

Another neurotransmitter that appears to become disrupted in early HD is dopamine (DA). The deficit in DA release is not as profound in HD as it is in Parkinson's disease (PD), which is known to result primarily from the deterioration of DA-producing neurons. Whenever a neurotransmitter system is dampened, it produces functional loss, either within the same system (movement-related), or within an alternate system that also depends on that neurotransmitter (cognitive or strategic). Evidence from animal models indicates that the capacity to release DA is substantially reduced in the context of both huntingtin mutations (Ortiz et al., 2010, 2011; Tang et al., 2007; Johnson et al., 2006, 2007; Yohrling et al., 2003), and the mitochondria-compromising neurotoxin, 3-nitroproprionic acid (3-NP), also used to model HD (Kraft et al., 2009). These 3-NP findings suggest that compromised DA may arise in part from a cellular energy deficit. Yohrling and colleagues (2003) also looked at loss of tyrosine hydroxylase, the rate limiting enzyme in DA production, in the substantia nigra of postmortem HD brains and found over 30% loss, which would functionally compromise DA availability. Therefore, ironically, despite the implications of DA overactivity that may arise from the clinical effectiveness of dampening this neurotransmitter with currently FDA-approved drugs such as tetrabenazine (Guay, 2010; de Tommaso et al., 2011), animal research indicates DA function is, in fact *reduced,* even prior to motor symptom expression (Ortiz et al., 2010, 2011; Johnson et al., 2006; Bibb et al., 2000).

When DA is compromised its modulatory effect is diminished, resulting in abnormal synaptic plasticity in the striatum and also in the frontal cortex, which is known to receive less dopaminergic innervation than the striatum (Cummings et al., 2006). This altered modulatory influence may contribute to the cognitive, strategic, or behavioral-flexibility-type symptoms that arise early in HD (Walker et al., 2008; Montoya et al., 2006; Paulsen & Conybeare, 2005; van Raamsdonk et al., 2005b; Nieoullon, 2002). From this presumption follow expected speculations as to whether stimulating the DA system may provide certain functional benefits. This approach has been attempted with R6/2 mice using methamphetamine in combination with levodopa treatment, which would be expected to promote both dopamine availability and release. However, while short-term improvement of some motor symptoms were found, animals treated with this regimen eventually exhibited increased problems on the rotarod, indicating loss of movement coordination as well as a shortened life-spans (Hickey et al., 2002).

It is important to recognize a distinct but equally important concept that has emerged from DA exploration: DA transmission aggravates oxidative processes that promote neuronal death in the context of the huntingtin mutation (Deyts et al., 2009; Charvin et al., 2005). Thus, interfering with DA transmission may be a rational choice for treatment, since it would serve to slow deteriorative processes, despite the repercussions for plasticity. In fact, direct research with two neuroleptics: haloperidol (Charvin et al., 2008) and tetrabenazine (Wang et al., 2010), demonstrated a neuroprotective effect on huntington mutation-bearing striatal neurons in isolation. Conversely, stimulation with DA receptor agonists tends to promote neuronal death (Tang et al., 2007). Furthermore, Paoletti and colleagues (2008) compared striatal cells from mouse models, harboring either 7 or 111 CAG repeats in their huntingtin genes, for vulnerability to DA and NMDA glutamate receptor stimulation. They found that the cells with 111 CAG repeats in huntingtin were killed more readily by stimulation with D1 DA receptor agonists, and cell vulnerability was enhanced when this stimulation was combined with NMDA receptor stimulation.

Thus, the presymptomatic rise in extracellular glutamate would be expected to enhance the destructive potential of DA. The Paoletti (2008) study also found that this combination of stimuli leads to intracellular molecular events that activate apoptotic or programmed cell death genes in a manner similar to what occurs in the brains of HD patients. If indeed this is the case in HD patients, treatment strategies may pose a Faustian bargain: "*Would you be willing to compromise your cognitive function in order to delay the loss of movement control*?" Such a dichotomous choice is not necessarily inevitable, but a DA-blocking treatment should not be offered without exploring the consequences. Ideally, a treatment will emerge that balances both the sensitivities to, and the need for, DA transmission.

Even in the absence of DA-compromising medication, the earliest cognitive decline exhibited by HD patients is in set shifting, where subjects must abandon attention to one strategy for another, also referred to as behavioral flexibility (Lawrence et al., 1996; Ho et al., 2003). In fact, direct correlations have been found between pre-HD patients' success with tasks requiring this sort of strategic flexibility and DA activity or control, as measured by positron emission tomography (PET) receptor or transporter binding assessment (Bäckman et al., 1997; Lawrence et al., 1998b). These correlations of pre-HD DA activity and cognitive flexibility are evaluated with tasks such as the Tower of Hanoi and Wisconsin Card Sort.

PET scan-related binding studies using radiolabeled raclopride (a D2 receptor antagonist) also reveal DA receptor malfunctions in pre-HD in both the striatum and the cortex that correlate with cognitive deficits (Pavese et al., 2003, 2010; van Oostrom et al., 2009). Unfortunately, it is practically impossible to distinguish binding to presynaptic versus postsynaptic DA D2 receptors in these regions using PET technology. However, reductions in presynaptic D2 receptors may well suggest either decreased dopaminergic terminals or decreased autoreceptor expression, both of which would distort DA release. The aforementioned study that found a correlation between performance on the Tower of Hanoi task and DA function assessed binding to DA transporters that are largely expressed on DA terminals (Bäckman et al., 1997), suggesting at least some of the reductions in D2 binding may result from terminal loss. Thus, it appears clear that suppressing DA in HD is simultaneously neuroprotective, and yet, more disruptive to cognitive processing; representing a dilemma that will need to be resolved in future treatment efforts.

4. Presymptomatic neuronal activity and plasticity dysfunction

In the context of pre-HD, when abnormal synaptic GLU levels linger for prolonged periods and a diminished DA modulation exists, MSNs must adapt to maintain normal function. To picture the various converging synaptic influences surrounding MSNs, refer back to Figure 2. In the context of the dramatic modulatory influences present at the MSN synapse, a new appreciation of spike-timing dependent plasticity has been developed that describes different ramifications when signals arrive before or after action potentials in the striatum. The known response patterns were recently reviewed by Fino and Venance (2010), who describe an impressive precision in the sensitivities to input timing exhibited by striatal neurons.

Apparently, striatal neurons exhibit differences in long-term reactions to input (as dramatic as LTP versus LTD) that depend upon whether modulatory inputs arrive before or after action potentials. Impressively, these responses can be realized with subthreshold membrane currents. Of course, GABA release at the terminal regions of MSNs (internal/external globus pallidus, substantia nigra reticulata) necessitates action potential generation. *In vitro* findings indicate that membrane currents occurring before thresholds are reached can influence the direction of subsequent responses. Given that experiments performed with freely-moving animals typically assess action potentials without appreciating sub-threshold activity, a great deal of information may be processed by striatal neurons that is likely to be missed in those experiments that typically rely on extracellular recording (Kiyatkin & Rebec, 1996). This subthreshold activity that can contribute timed depolarizations is nevertheless important for establishing extended response tendencies.

As mentioned, striatal neurons tend to exhibit low spontaneous action potential generation in healthy awake animals in the absence of spontaneous behavior (Sandstrom & Rebec, 2003). Therefore, the large majority of modulatory influences are likely to occur at subthreshold membrane potentials. This expands the potential for disruptive malfunctions resulting from the huntingtin mutation as even minor disruptions to transmitter release will distort pre-threshold synaptic currents.

Given the initial increases in GLU that may surge to higher endogenous levels, it makes sense that there seems to be a presymptomatic sensitivity to NMDA and quinolinic acid (a potent NMDA receptor agonist) among MSNs from YAC128 mice, a popular model with extended pre-HD periods (Graham et al., 2009). Later, at symptomatic stages, these same mice show resistance to quinolinic acid-induced excitotoxicity. This later-stage resistance to GLU excitotoxicity has also been observed in other HD mouse models (Starling et al., 2005; Zeron et al., 2002; Levine et al., 1999). Although this may represent a gradual decrease in sensitivity to GLU stimulation, recordings of striatal neuronal activity demonstrate hyperactivity in symptomatic transgenic R6/2 mice (Rebec et al., 2006) indicating corticostriatal malfunction. In fact, this study treated R6/2 mice with high amounts of systemic ascorbate which seemed to ameliorate the observed deficiency of endogenous striatal ascorbate, and subsequently reduced striatal hyperactivity. This effect suggests that the ascorbate contributes substantially to GLU uptake transport, and can help diminish detrimental excitatory drive when present.

Changes in responses to NMDA stimulation and AMPA stimulation among cortical neurons also takes place in the R6/2 model, and these are more easily seen when neurons are observed in isolation (André et al., 2006). This may be a general type of response that occurs when control of GLU levels or activity is compromised, whereby neurons decrease sensitivity to accommodate the increased basal GLU levels. Interestingly, striatal neuron hyperactivity is not a common feature of all HD mouse models, as the R6/2 mouse shows this but the knock-in 140 (KI-140) mouse does not (Miller et al., 2008b). Also, striatal hyperactivity is not observed in the transgenic rat model (51-CAG; Miller et al., 2010). However, the capacity to generate coordinated afferent bursts into the striatum, as measured by coordination of firing patterns between pairs of striatal neurons, was disrupted in all these models (Miller et al., 2011). It is likely that this disrupted cortical input originates from aberrant activity within the cortex (Dorner et al., 2009), which was also evidenced in terms of a lack of synchrony between pairs of neurons in the prefrontal cortex of both R6/2 and KI-140 mice (Walker et al., 2008).

It is important to note that these demonstrations of changes in cortical and striatal activity were shown in symptomatic animals. The contributions of cortical disruptions to the time course of behavioral deficit expression make sense when considering the presymptomatic loss of GLU regulation. Cummings and colleagues (2009) found that cortical activity became disrupted with larger and more frequent excitatory postsynaptic potentials in several animal models including the R6/2, YAC128, and KI-140 lines, largely in the presymptomatic stages. In addition, more frequent inhibitory postsynaptic potentials within the striatum could easily disrupt the coordination of cortical input to the striatum. Laforet and colleagues (2001) evaluated the pathological cortical and striatal alterations that precede HD symptoms in both humans and animal models and concluded the contributions of cortical malfunction must be critical.

Human data indicate that cortical metabolic dysfunction occurs among HD patients before brain-scan indications of pathology manifest in the striatum (Rosas et al., 2005; Paulsen et al., 2004; Sax et al., 1996). Combined with evidence that knock-in mouse models with lower CAG repeat numbers lack both cortical neuronal changes and later behavioral changes (Wheeler et al., 2000), while longer repeat containing knock-in models exhibit moderate cortical involvement and moderate behavioral changes (Lin et al., 2001), these data strongly implicate early cortical malfunctions preceding striatal malfunctions and perhaps contributing to their development. Research with restricted expression models, where mutated huntingtin is expressed only in striatal MSNs, show striatal NMDA sensitivities, but these animals do not seem to develop behavioral deficits in the normal progression. This suggests cortical expression of huntingtin mutations are also necessary for pathology (Gu et al., 2007; although see Thomas et al., 2011b).

A related complication that has commanded recent attention is the increased stimulation of extrasynaptic NMDA receptors (expressed outside the postsynaptic zone), arising in pre-HD. Stimulation of these extrasynaptic NMDA receptors seems to elevate apoptotic cascades, while synaptic NMDA stimulation serves to prevent this and maintain neuronal health (Milnerwood et al., 2010; Okamoto et al., 2009; Li et al., 2004). Apparently, the majority of extrasynaptic NMDA receptors in the striatum contain the NR2B subunit that seems to confer a disruptive influence on both MSN survival (Okamoto et al., 2009) and neuronal responses and plasticity, including a tendency to decrease CREB signaling (Milnerwood et al., 2010; Leveille et al., 2008; Hardingham et al., 2002). With the lack of synergy between cortical inputs, increasing chaotic nature of cortical impulses, and the diminished control of the GLU released via transporter malfunctions, it is not surprising that GLU spill-over into extrasynaptic domains increases as HD pathology advances.

MSNs normally exhibit hyperpolarized membrane potentials and decreased input resistances, both of which reduce their tendency to produce action potentials in response to sporadic and temporally uncoordinated input. This leads to their relative silence in healthy animals in the absence of movement. Findings from electrophysiological explorations of both presymptomatic and symptomatic R6/2 mouse striatal slices show both increased input resistance when stimulated directly and decreased paired-pulse facilitation when stimulated indirectly and repetitively via the cortex (Klapstein et al., 2001). It seems that decreased inwardly rectifying potassium channel expression may account for a corresponding depolarization of the resting membrane potential that occurs in MSNs of these mice, but neither of these are observed during the presymptomatic stage (Ariano et al., 2005). These combined data therefore also suggest that cortical neuronal malfunctions may precede striatal changes.

The most direct demonstration of presymptomatic cortical contributions during the development of HD in an animal model (R6/2 and R6/1 mice; Cepeda, 2003) indicated a progressive decrease in spontaneous currents in striatal neurons, along with increased generation of large synaptic current events that occur prior to symptom expression. This decreased spontaneous current generation is counter-intuitive, given the data described above that indicated reuptake transport control over GLU release is lost in pre-HD mouse models (as early as 10-11 weeks, among KI-150 mice; Figure 3). Nonetheless, a consequence of increased input resistance among MSNs in HD would be increased excitability, and a decreased rheobase (current intensity necessary to reach action potential) expressed in MSNs of pre-HD mice (Klapstein et al., 2001).

Interestingly, a recent series of elegant experiments were performed with pre-HD 51-CAG rats in which *in vivo* electrophysiological measures (taken while animals were under pentobarbital anesthesia) correlated with operant-task deficits in time appreciation (Höhn et al., 2011). The major electrophysiological finding in that study was an increased theta-burst generated LTD among homozygote rats, as demonstrated by before-and-after input-output curves, *despite the lack of any changes in paired-pulse facilitation*. It remains unclear why increased plasticity within the striatum would be responsible for the observed correlated behavioral deficit among homozygote rats in time appreciation. This presymptomatic deficit was exposed by challenging rats to recognize differences in the duration of signals and to respond differentially to short and long signals. The ability to discern differences in the length of these signals could be made more challenging by shortening the long signal, and homozygote rats exhibited more difficulty when this was done than wild type rats.

This sort of compromised appreciation of elapsed time also seems to present among pre-HD patients (Rowe et al., 2010; Paulsen et al., 2008; Beste et al., 2007). In fact, in an extensive review, Matell and Meck (2004) support the hypothesis that a primary role of the basal ganglia depends on the timing of coincident cortical inputs, and subsequent integration that occurs on MSNs, to generate conscious appreciation of time. Their modeling of expected oscillations in striatal neurons, should the cortical input become increasingly varied, seems to predict what was found to occur in actual oscillations recorded in an HD mouse model by Cepeda and colleagues (2003). The lack of synergy in cortical activity, chaotic nature of cortical impulses, diminished GLU control, and disruptions in plasticity, based on extrasynaptic NMDA stimulation, could easily underlie this sort of behavioral disruption.

Generalizations of DA effects on neurons in both the striatum and cortex are more useful in appreciating the changes likely to become relevant during pre-HD. DA has been shown to elicit an increased signal-to-noise ratio when applied in the context of recording from single striatal neurons at the time of stimulation by GLU iontophoresis in freely-moving rats

(Kiyatkin & Rebec, 1996). It is interesting that convergent stimulation of striatal neurons, arising during motor activity, was enhanced by DA iontophoresis, while spontaneous activity unrelated to movement was usually suppressed (Pierce & Rebec, 1995). From this perspective, DA seemed to promote behavior-related striatal activity and diminish the spontaneous activity generally not seen in intact rats that sit quietly, while otherwise awake (Sandstrom & Rebec, 2003). Increases in MSN input resistance, along with increased sensitivities of NMDA receptors and malfunctioning cortical input may lead to increased spontaneous striatal activity in freely-moving symptomatic R6/2 mice (Rebec et al., 2006), as well as reduced coordination of striatal activity (Miller et al., 2011).

One rather mysterious finding presented clearly and convincingly in a review by Fino and Venance (2010), is that LTP can be induced by pairings of corticostriatal input activity to MSN impulses in the order of *post-pre*, while LTD tends to result from pairings of the same activities in the order of *pre-post* (see also Fino et al., 2005). Even more intriguing would be the implications of sub-threshold EPSPs on later firing tendencies seeming to follow the same rules (Fino & Venance, 2010). It is challenging to justify these findings with what might be expected if striatal learning is accomplished by promoting the sensitivity of neurons commonly activated by corticostriatal input in the typical manner. It would seem that in circumstances involving spontaneous activity that is present prior to coordinated efforts of corticostriatal terminals converging on MSNs, the order of action potentials would emphasize *post-pre* relations, while the reverse (*pre-post*) would be the natural order of activity across these terminals, subsequently causing striatal action potentials. If the natural order leads to depression of MSN sensitivity (*pre-post*), while neurons abnormally active before coordinated inputs from corticostriatal terminals become more sensitive (*post-pre*), this would decrease the sensitivity of synapses that function properly and increase the sensitivity of those that do not. DA, which is not likely to be present in quantities as high as those found *in vivo* in the slices of these experiments, seems to do the opposite, promoting activity arising from coordinated corticostriatal input (behavior related), and diminishing unrelated activity (Pierce et al., 1995). If the capacity to release DA declines during pre-HD, it is therefore easy to imagine that both disrupted plasticity and increased noise in striatal transmission could corrupt contributions of the basal ganglia in behavior processing.

Investigations of DA receptors indicate a selective initial, presymptomatic vulnerability of either D2 receptor expression or D2 expressing striatal neurons in human HD patients, followed by progressive loss of D1 receptors at the intermediate and late stages of the disease (Glass et al., 2000). This pattern does not seem to occur in animal models, which seem to diminish both D1 and D2 dopamine receptors more readily (Cha et al., 1998). As has been suggested, DA also participates in inducing pathological responses among striatal neurons which are manifested in electrophysiological explorations of neurons in several mouse models of HD (André et al., 2011). Experiments with HD mouse models have revealed that during pre-HD, specific neurons that can be distinguished as expressing *either* D1 or D2 receptors exhibit abnormal NMDA-related plasticity more exclusively among the D1 expressing neurons, and that this seems to be normalized by reducing DA presence. As these same mice seem to exhibit deficiencies in DA signaling (Bibb et al., 2000), and the recent experiments demonstrating plasticity disruptions were performed in slices where maintenance of endogenous-like levels of extracellular dopamine is dubious, it may be premature to promote DA suppression as a therapeutic strategy, even though it is currently the only FDA-approved treatment for HD.

5. Presymptomatic cognitive dysfunction

The area where problems begin to appear in pre-HD is not in engaging in a chosen behavior, but rather in making the original decision about the best behavior to select. This is especially evident when an individual is faced with an array of seemingly relevant information. Given that there are often multiple strategies to effective problem-solving, choosing the best strategy depends on processing varied input, referencing past experience, and creative thinking. Patients, themselves, often do not recognize their own cognitive limitations; since most daily tasks do not involve being challenged to constantly switch strategies in order to keep up with changing scenarios. In fact, an interesting recent study, involving both pre-HD patients and their regular companions, explored the general perception of apathy, disinhibition, and executive dysfunction, using a modified version of the Frontal System Behavior Scale (FrSBe, Grace & Malloy, 2001). This study demonstrated that even the more severe diagnosis-predicting perceptions of problems were not as readily recognized by patients as they were by their companions (Duff et al., 2010). The lack of awareness that surrounds these early cognitive symptoms can complicate their exploration, evaluation, and intervention development.

As HD develops, there is a progressive decline in internal time assessment, attention, executive function, and short-term memory (Rowe et al., 2010; Beste et al., 2007; Bourne et al., 2006; Paulsen & Conybeare, 2005; Ho et al., 2003). These early cognitive declines arise in pre-HD expressed by patients (Rowe et al., 2010; Ho et al., 2003; Lemiere et al., 2002; Snowden et al., 2001; Kirkwood et al., 2000) and animal models (Höhn et al., 2011; Trueman et al., 2007, 2008; van Raamsdonk et al., 2005b). The diminished appreciation of elapsed time recently shown in HD animal models (Höhn et al., 2011) seems to recapitulate similar problems exhibited by HD patients, who present difficulties when they are required to maintain consistent self-pacing (Rowe et al., 2010).

The most commonly referenced aspect of cognitive decline exhibited in early HD is executive function, manifested as cognitive inflexibility. This cognitive inflexibility, defined as an inability to coordinate the most effective strategic response, or adaptation to apparent changes in circumstances, is related to expressions of apathy that are not easily perceived by preclinical patients but are recognized by their regular companions (Duff et al., 2010). It is easy to imagine that when circumstances become complex and patients become overloaded, frustration and disappointment lead to irritation or apathy, which can be difficult for companions (Quarrell, 2008; Bourne et al., 2006). Assessments of cognitive inflexibility in pre-HD typically requires the use of specialized tasks, such as the Wisconsin Card Sorting and Tower of London tasks (Brandt et al., 2008), whereby patients are challenged to routinely shift strategies, depending on circumstances. Complicated and multifaceted tasks are otherwise rare, and, as such, pre-HD patients are typically able to cope in their day-to-day functions.

Associating difficulties with cognitive flexibility with the known malfunctions of the basal ganglia in pre-HD can be complex. It is perhaps helpful to employ a simplified version of the information-processing circuit, which would proceed as follows (see Figure 1): (1) the whole cortical mantle including both sensations and emotions converge into the striatum (caudate and putamen in humans) which then proceeds to internally process signals within the basal ganglia way-stations, converting the original signals into a modulatory feedback that is directed back towards the frontal cortex where "executive functions" (strategic behavior decisions) are performed; (2) as the strategies are chosen, the frontal cortex attempts to hold the relevant information in short-term memory and adjusts the plan according to all available appreciated circumstances until finally feeding it forward to primary motor cortex; (3) from there the final decision is generated and sent to lower motor systems to be engaged, but these commands can still be vetoed even after initial movements begin by engaging antagonist muscles (e.g. when a batter starts a swing in baseball only to stop before committing when it becomes clear the ball will fly wide).

Pre-HD patients, when examined with functional magnetic resonance imaging, show metabolic malfunctions in the striatum (Kuwert et al., 1993), the frontal cortex (Wolf et al., 2007), and even the thalamus, all of which are interconnected (Feigin et al., 2007). The degree to which these alterations are evident depends on the cognitive load at the time of measurement (Wolf et al., 2008). To demonstrate this, Wolf and colleagues (2008) challenged pre-HD patients with a working memory task which would be expected to activate the prefrontal cortices (Kane & Engle, 2002). As the working memory load was increased, pre-HD patients presented decreased correlations in activity between the frontal cortices and their striatal activities, similar to the above-described findings of decreased coordination of cortical activity in experimental animals (Cummings et al., 2009; Walker et al., 2008). Similarly, in animal models of pre-HD, cortical neuron control seems to be diminished, largely by the lack of sufficient local inhibition, resulting in uncoordinated activity patterns (Cummings et al., 2009). Cortical neuropathology, and even some minor tissue deterioration observed in terms of thinning, clearly begins to arise during pre-HD (Kipps et al., 2005), as well as at the very beginning of symptom expression (Beglinger et al., 2005), correlating with apparent cognitive difficulties.

While the Wolf (2008) study found a disconnect between activity in the cortex and striatum in tasks requiring working memory, complex planning tasks, that are more related to executive function and cognitive flexibility, also challenge pre-HD patients. Two extensive studies demonstrated that testing pre-HD patients just prior to motoric symptom expression, using tests such as the Wisconsin Card Sorting task, resulted in greater difficulties than both mutation-free and pre-HD subjects who were further from motor symptom expression (Brandt et al., 2008; Snowden et al., 2002). In the first of these studies (Snowden et al., 2002), the data indicated that working memory malfunctions may arise earlier in the disease progression than problems with executive function. Further explorations of executive function suggest that DA plays a pivotal role within both the prefrontal cortex and the striatum. Tests of behavioral flexibility using rats in operant chambers demonstrated that pharmacological manipulations of DA receptor activity, within either the prefrontal cortex (Winter et al., 2009) or the ventral striatum (Haluk & Floresco, 2009) during ongoing behavior, disrupts the animals' capacities to switch patterns of behavior to obtain more reinforcers.

Interesting experiments performed to track the activity of DA-producing neurons in behaving monkeys, suggest that DA neuron activity depends more on the reward-predicting value of cues than on rewards themselves (Waelti et al., 2001). The general tendency when recording from DA neurons, in either the ventral tegmental area (VTA) or substantia nigra pars compacta (SNpc), is that phasic firing increases initially occur at the time of reward delivery, but in time become associated with stimuli that predict reward rather than the rewards themselves. As this change occurs (when they no longer indicate only the reward) this likely promotes changes in striatal or frontal cortex DA levels that increase via acquired associations with the predictive aspect of cues, which themselves would suggest appropriate behavior strategies.

Given the previously-described enhancements to the coordinated signals and diminishments to uncoordinated signals revealed in studies using DA iontophoresis in intact animals (Kiyatkin & Rebec, 1996; Pierce & Rebec, 1995), these phasic increases in firing would allow DA signals to enhance the neuronal responses in targeted areas by eliminating background noise in a normally-functioning system. Thus, intact DA modulations should be expected to enhance recognition of faulty or maladaptive behavior patterns or at least promote cortically coordinated patterns. Perhaps for this reason, unmedicated Parkinson's patients who are tested early in their disease progression (as DA diminishes well before movement deficits emerge in PD) exhibit impairment in "set-shifting" tasks that require cognitive flexibility (Owen et al., 1992). The previously described declines in presymptomatic DA release capacity exposed in HD animal models (Ortiz et al., 2010, 2011) would predict a DA-related deficit in behavioral flexibility in HD patients. Such deficits were found by Lawrence and his colleagues (1998b), who showed that pre-HD mutation carriers were impaired on cognitive tests in a manner that correlated with DA receptor binding levels measured by PET scans. A positive correlation was found across HD patients and control subjects between success on the Tower of Hanoi task and DA transporter binding (Bächman, 1997). The bottom line of these findings were that HD patients had lower DAT binding, which predicts lower release capacities, as these transporters are expressed on DA neuron terminals in the caudate and putamen (reduced release capacity = reduced success). Set-shifting deficits exhibited by pre-HD patients (Lawrence et al., 1998a) are also likely to depend upon early DA malfunctions.

Another pre-HD difficulty exhibited involves the appreciation of emotion. The area of social emotion appreciation that is most affected by HD was originally believed to be disgust recognition (Hennenlotter et al., 2004; Sprengelmeyer et al., 2006). However, in a more recent study (Johnson et al., 2007), the deficits in emotional processing were broadened to involve the recognition of all negative emotions (i.e., anger, disgust, fear, and sadness). These emotion perception issues are intimately connected to cognitive flexibility, as the same population of DA neurons shown to fire in accordance to reward-predictability (Waelti et al., 2001), also project throughout the limbic system, including the amygdala and prefrontal cortex (Salgado-Pineda et al., 2005) which provide critical support for perception of emotion. The critically important role emotion plays in cognitive processing is well documented (see Damasio, 1996, 1999) and its role in the disruptive cognitive processing observed in pre-HD patients provides a fertile area of research that promises to deliver further insights into the etiology and potential treatments for early stage HD.

6. Conclusions

Despite the discovery of the gene primarily responsible for HD, it is fair to say that our understanding of its etiology is largely preliminary. This is because the gene and corresponding protein seem to be incredibly complex and involved in multiple aspects of neuronal physiology. Delineating HD-related neurophysiological deficits will necessitate appreciation of events that occur before neuronal death and the onset of motoric symptoms. Determining the primary, pre-compensatory malfunctions will likely suggest treatment strategies that can target and alleviate these without becoming entangled in compensation cascades. Furthermore, coordination and normalization of neuronal activity in key brain regions such as the frontal cortex, caudate, and putamen would seem to require restoration of healthy GLU management. Experiments with ceftriaxone show promise in that regard along with other efforts to boost GLU reuptake.

The DA system in HD represents a greater puzzle since there are clearly pros and cons to the currently FDA approved strategies that mostly diminish DA in the earlier stages of HD, which may alleviate emerging motor symptoms but may also aggravate cognitive dysfunction. As such, it is important to consider the cognitive domain in the context of neuronal activity and transmission deficits, since these circuits seem to show changes before motoric disruptions emerge. If sensitivity to DA or NMDA transmission could be diminished, these systems could be normalized far more effectively. Attempts to decipher the dynamic transmission interactions and elucidate the role of mutant huntingtin should continue in parallel to testing potential treatments in animal models of HD.

In this chapter, we have identified several key sources of physiological disruptions and integrated them into a theoretical framework to help explain the early expressions of cognitive malfunction in this disease. Isolating the physiological disruptions underlying pre-HD is critical for devising more effective treatments. Until it becomes possible to repair damaged or mutated genes, the most effective therapies will be those that help relevant neuron populations resume their normal roles and compensate for the extensive dysfunction driven by abnormal huntingtin protein physiology.

It is likely that both the preliminary malfunctions, such as cognitive decline and the later-stage loss of movement control depend upon similar physiological alterations within the same neuronal populations. However, potential treatments given during earlier stages of HD should be more efficacious as they will benefit from greater neuron numbers that would be available prior to widespread neuron death. As such, investigations into the early pre-HD may provide the greatest hope of effectively slowing the progress of this devastating disease.

7. Acknowledgements

Support for this work was provided by the Field Neurosciences Institute, the Central Michigan University Neuroscience Programs, the John G. Kulhavi Professorship in Neuroscience, and the Central Michigan University Office of Research and Sponsored Programs.

8. References

Acquas, E., & DiChiara, G. (2002). Dopamine-acetylcholine interaction. Chapter 15, In: *Handbook of Experimental Pharmacology: Subseries Dopamine in the CNS*, Part 154/2, G. Dichiara (Ed.), pp. 85-115, Springer-Verlag, ISBN 978-3-540-42720-9, Berlin.

André, V.M., Cepeda, C., Venegas, A., Gomez, Y., & Levine, M.S. (2006). Altered cortical glutamate receptor function in the R6/2 model of Huntington's disease. *Journal of Neurophysiology*, Vol. 95, No. 4, (April 2006), pp. 2108-2119, ISSN 0022-3077

André, V.M., Fisher, Y.E., & Levine, M.S. (2011). Altered balance of activity in the striatal direct and indirect pathways in mouse models of Huntington's disease. *Frontiers in Systems Neuroscience*, Vol. 5, No. 46, pp. 1-11, ISSN 1662-5137

Ariano, M.A., Cepeda, C., Calvert, C.R., Flores-Hernandez, J., Hernandez-Echeagaray, E., Klapstein, G.J., Chandler, S.H., Aronin, N., DiFiglia, M., & Levine, M.S. (2005). Striatal potassium channel dysfunction in Huntington's disease transgenic mice. *Journal of Neurophysiology*, Vol. 93, No. 5, (May 2005), pp. 2565–2574, ISSN 0022-3077

Azarias, G., Perreten, H., Lengacher, S., Poburko, D., Demaurex, N., Magistretti, P.J., & Chatton, J.Y. (2011). Glutamate transport decreases mitochondrial pH and modulates oxidative metabolism in astrocytes. *Journal of Neuroscience*, Vol. 31, No. 10, (March 2011), pp. 3550-3559, ISSN 0270-6474

Bäckman, L., Robins-Wahlin, T.-B., Lundin, A., Ginovart, N., & Farde, L. (1997). Cognitive deficits in Huntington's disease are predicted by dopaminergic PET markers and brain volumes. *Brain*, Vol. 120, No. 12, (Dec 1997), pp. 2207-2217, ISSN 0006-8950

Bateup, H.S., Santini, E., Shen, W., Birnbaum, S., Valjent, E., Surmeir, D.J., Fisone, G., Nestler, E.J., & Greengard, P. (2010). Distinct subclasses of medium spiny neurons differentially regulate striatal motor behaviors. *Proceedings of the National Academy of Sciences, USA*, Vol. 107, No. 33, (August 2010), pp. 14845-14850, ISSN 0027-8424

Beglinger, L.J., Nopoulos, P.C., Jorge, R.E., Langbehn, D.R., Mikos, A.E, Moser, D.J., Duff, K., Robinson, R.G., & Paulsen, J.S. (2005). White matter volume and cognitive dysfunction in early Huntington's disease. *Cognitive and Behavioral Neurololgy*, Vol. 18, No. 2, (June 2005), pp. 102-107, ISSN 1543-3633

Behrens, P.F., Franz, P., Woodman, B., Lindenberg, K.S., & Landwehrmeyer, G.B. (2002). Impaired glutamate transport and glutamate-glutamine cycling: downstream effects of the Huntington mutation. *Brain*, Vol. 125, No. 8, (August 2002), pp. 1908-1922, ISSN 0006-8950

Berretta, N., Nistico, R., Bernardi, G., & Mercuri, N.B. (2008). Synaptic plasticity in the basal ganglia: A similar code for physiological and pathological conditions. *Progress in Neurobiology*, Vol. 84, No. 4, (April 2008), pp. 343-362, ISSN 0301-0082

Beste, C., Saft, C., Andrich, J., Müller, T., Gold, R., & Falkenstein, M. (2007). Time processing in Huntington's disease: a group control study. *PLoS One*, Vol. 2, No. 12, (December 2007), pp. e1263, ISSN 1932-6203

Bibb, J.A., Yan, Z., Svenningsson, P., Snyder, G.L., Pieribone, V.A., Horiuchi, A., Nairn, A.C., Messer, A., & Greengard, P. (2000). Severe deficiencies in dopamine signaling in presymptomatic Huntington's disease mice. *Proceedings of the National Academy of Sciences, USA*, Vol. 97, No. 12, (June 2000), pp. 6809-6814, ISSN 0027-8424

Bourne, C., Clayton, C., Murch, A., & Grant, J. (2006). Cognitive impairment and behavioral difficulties in patients with Huntington's disease. *Nursing Standard*, Vol. 20, No. 35, (May 2006), pp. 41-44, ISSN 0029-6570

Brandt, J., Inscore, A.B., Ward, J., Shpritz, B., Rosenblatt, A., Margolis, R.L., & Ross, C.A. (2008). Neuropsychological deficits in Huntington's disease gene carriers and correlates of early "conversion." *Journal of Neuropsychiatry and Clinical Neuroscience*, Vol. 20, No. 4, (January 2008), pp. 466-472, ISSN 1545-7222

Brown, R.E., & Milner, P.M. (2003). The legacy of Donald O. Hebb: more than the Hebb synapse. *Nature Reviews Neuroscience*, Vol. 4, No. 12, (Dec 2003), pp. 1013-1019, ISSN 1471-003X

Brustovetsky, T., Purl, K., Young, A., Shimizu, K., & Dubinsky, J.M. (2004). Dearth of glutamate transporters contributes to striatal excitotoxicity. *Experimental Neurology*, Vol. 189, No. 2, (October 2004), pp. 222-230, ISSN 0014-4886

Carter, R.J., Lione, L.A., Humby, T., Mangiarini, L., Mahal, A., Bates, G.P., Dunnett, S.B., & Morton, A.J. (1999), Characterization of progressive motor deficits in mice transgenic for the human Huntington's disease mutation, *Journal of Neuroscience*, Vol. 19, No. 8, (April 1999), pp. 3248-3257, ISSN 0270-6474

Cepeda, C., Hurst, R.S., Calvert, C.R., Hernández-Echeagaray, E., Nguyen, O.K., Jochoy, E., Christian, L.J., Ariano, M.A., & Levine, M.S. (2003). Transient and progressive electrophysiological alterations in the corticostriatal pathway in a mouse model of Huntington's disease. *Journal of Neuroscience*, Vol. 23, No. 3, (February 2003), pp. 961-969, ISSN 1529-2401.

Cha, J.H.J., Frey, A.S., Alsdorf, S.A., Kerner, J.A., Kosinski, C.M., Mangiarini, L., Penney, J.B.Jr., Davies, S.W., Bates, G.P., & Young, A.B. (1999). Altered neurotransmitter receptor expression in transgenic mouse models of Huntington's disease. *Philosophical Transactions of the Royal Society B: Biological Sciences*, Vol. 354, No. 1386, (June 1999), pp. 981-989, ISSN 1471-2970

Charvin, D., Roze, E., Perrin, V., Deyts, C., Betuing, S., Pages, C., Regulier, E., Luthi-Carter, R., Brouillet, E., Deglon, N., & Caboche, J. (2008). Haloperidol protects striatal neurons from dysfunction induced by mutated huntingtin in vivo. *Neurobiology of Disease*, Vol. 29, No. 1, (January 2008), pp. 22-29, ISSN 0969-9961

Charvin, D., Vanhoutte, P., Pages, C., Borrelli, E., & Caboche, J. (2005). Unraveling a role for dopamine in Huntington's disease: the dual role of reactive oxygen species and D2 receptor stimulation. *Proceedings of the National Academy of Sciences, USA*, Vol. 102, No. 34, (August 2005), pp. 12218-12223, ISSN 0027-8424.

Cummings, D.M., Andre, V.M., Uzgil, B.O., Gee, S.M., Fisher, Y.E., Cepeda, C., & Levine, M.S. (2009). Alterations in cortical excitation and inhibition in genetic mouse models of Huntington's disease. *Journal of Neuroscience*, Vol. 29, No. 33, (August 2009), pp. 10371-10386, ISSN 1529-2401

Cummings, D.M., Milnerwood, A.J., Dallerac, G.M., Waights, V., Brown, J.Y., Vatsavayai, S.C., Hirst, M.C, & Murphy, K.P. (2006). Aberrant cortical synaptic plasticity and dopaminergic dysfunction in a mouse model of Huntington's disease. *Human Molecular Genetics*, Vol. 15, No. 19, (October 2006), pp. 2856-2868, ISSN 0964-6906

Damasio, AR. (1996). The somatic marker hypothesis and the possible functions of the prefrontal cortex. *Philosophical Transactions of the Royal Society of London. Series B, Biological Sciences*, Vol. 351, No. 1346, (October 1996), pp. 1413-1420, ISSN 0962-8436

Damasio, A.R. (1999). *The Feeling of What Happens: Body and Emotion in the Making of Consciousness*. Harcourt Brace and Company, ISBN 0-15-1000369, New York.

de Tommaso, M., Serpino, C., & Sciruicchio, V. (2011). Management of Huntington's disease: role of tetrabenazine. *Therapeutics and Clinical Risk Management*, Vol. 7, (March 2011), pp. 123-129, ISSN 1178-203X

Dedeoglu, A., Kubilus, J.K., Jeitner, T.M., Matson, S.A., Bogdanov, M., Kowall, N.W., Matson, W.R., Cooper, A.J., Ratan, R.R., Beal, M.F., Hersch, S.M., & Ferrante, R.J. (2002). Therapeutic effects of cystamine in a murine model of Huntington's disease. *Journal of Neuroscience,* Vol. 22, No. 20, (October 2002), pp. 8942-8950, ISSN 1529-2401.

Delaney, S.M., & Geiger, J.D. (1998). Levels of endogenous adenosine in rat striatum. II. Regulation of basal and N-methyl-D-aspartate-induced levels by inhibitors of adenosine transport and metabolism. *Journal of Pharmacology and Experimental Therapeutics,* Vol. 285, No. 2, (May 1998), pp. 568-572, ISSN 0022-3565

Delanty, N., & Dichter, M.A., (2000). Antioxidant therapy in neurologic disease. *Archives of Neurology,* Vol. 57, No. 9, (September 2000), pp. 1265-1270, ISSN 0003-9942.

Deyts, C., Galan-Rodriguez, B., Martin, E., Bouveyron, N., Roze, E., Charvin, D., Caboche, J., & Betuing, S. (2009). Dopamine D2 receptor stimulation potentiates PolyQ-Huntingtin-induced mouse striatal neuron dysfunctions via Rho/ ROCK-II activation. *PloS One,* Vol. 4, No. 12, (December 2009), e8287, ISSN 1932-6203

Dorner, J.L., Miller, B.R., Klein, E.L., Murphy-Nakhnikian, A., Andrews, R.L, Barton, S.J., & Rebec, G.V. (2009). Corticostriatal dysfunction underlies diminished striatal ascorbate release in the R6/2 mouse model of Huntington's disease. *Brain Research,* Vol. 1290, (September 2009), pp. 111-120, ISSN 1872-6240

Duff, K., Paulsen, J.S., Beglinger, L.J., Langbehn, D.R., Wang, C., Stout, J.C., Ross, C.A., Aylward, E., Carlozzi, N.E., Queller, S., & general Predict-HD Investigators of Huntington Study Group. (2010). "Frontal" behaviors before the diagnosis of Huntington's disease and their relationship to markers of disease progression: Evidence of early lack of awareness. *Journal of Neuropsychiatry and Clinical Neuroscience,* Vol. 22, No. 2, (Spring 2010), pp. 196-207, ISSN 1545-7222

Estrada-Sanchez, A.M., Montiel, T., & Massieu, L. (2010). Glycolysis inhibition decreases the levels of glutamate transporters and enhances glutamate neurotoxicity in the R6/2 Huntington's disease mice. *Neurochemical Research,* Vol. 35, No. 8, (August 2010), pp. 1156-1163, ISSN 1573-6903.

Estrada-Sanchez, A.M., Montiel, T., Segovia, J., & Massieu, L. (2009). Glutamate toxicity in the striatum of the R6/2 Huntington's disease transgenic mice is age-dependent and correlates with decreased levels of glutamate transporters. *Neurobiology of Disease,* Vol. 34, No. 1, (April 2009), pp. 78-86, ISSN 1095-953X.

Faideau, M., Kim, J., Cormier, K., Gilmore, R., Welch, M., Auregan, G., Dufour, N., Guillermier, M., Brouillet, E., Hantraye, P., Deglon, N., Ferrante, R.J., & Bonvento, G. (2010). In vivo expression of polyglutamine-expanded huntingtin by mouse striatal astrocytes impairs glutamate transport: a correlation with Huntington's disease subjects. *Human Molecular Genetics,* Vol. 19, No. 15, (August 2010), pp. 3053-3067, ISSN 1460-2083

Feigin, A., Tang, C., Ma, Y., Mattis, P., Zgaljardic, D., Guttman, M., Paulsen, J.S., Dhawan, V., & Eidelberg, D. (2007). Thalamic metabolism and symptom onset in preclinical Huntington's disease. *Brain,* Vol. 130, Pt 11, (November 2007), pp. 2858-2867, ISSN 1460-2156

Ferrante, R.J., Andreassen, O.A., Jenkins, B.G., Dedeoglu, A., Kuemmerle, S., Kubilus, J.K., Kaddurah-Daouk, R., Hersch, S.M., & Beal, M.F. (2000). Neuroprotective effects of creatine in a transgenic mouse model of Huntington's disease. *Journal of Neuroscience*, Vol. 20, No. 12, (June 2000), pp. 4389-97, ISSN 0270-6474.

Fino, E., Glowinski, J., & Venance, L. (2005). Bidirectional activity-dependent plasticity at cortico-striatal synapses. *Journal of Neuroscience*, Vol. 25, No. 49, (December 2005) pp. 11279-11287, ISSN 1529-2401

Fino, E., & Venance, L. (2010). Spike-timing dependent plasticity in the striatum. *Frontiers in Synaptic Neuroscience*, Vol. 2, No. 6, (June 2010) pp.1-10, ISSN 1663-3563

Fusi, S., Asaad, W.F., Miller, E.K., & Wang, X.J. (2007). A neural circuit model of flexible sensorimotor mapping: learning and forgetting on multiple timescales. *Neuron*, Vol. 54, No. 2, (April 2007), pp. 319-33, ISSN 0896-6273

Gerfen, C.R. (1992a). The neostriatal mosaic: Multiple levels of compartmental organization in the basal ganglia. *Annual Review of Neuroscience*, Vol. 15, (March 1992), pp. 285-320, ISSN 0147-006X

Gerfen, C.R. (1992b). D1 and D2 dopamine receptor regulation of striatonigral and striatopallidal neurons. *Seminars in Neuroscience*, Vol. 4, No. 2, (April 1992), pp. 109-118, ISSN 1044-5765

Gerfen, C.R., & Young, W.S. (1988). Distribution of striatonigral and striatopallidal peptidergic neurons in both patch and matrix compartments: an in situ hybridization histochemistry and fluorescent retrograde tracing study. *Brain Research*, Vol. 460, No. 1, (September 1988), pp. 161-167, ISSN 0006-8993

Glass, M., Dragunow, M., & Faull, R.L. (2000). The pattern of neurodegeneration in Huntington's disease: a comparative study of cannabinoid, dopamine, adenosine, and GABA(A) receptor alterations in the human basal ganglia in Huntington's disease. *Neuroscience*, Vol. 97, No. 3, (May 2000), ISSN 0306-4522

Grace, J., & Malloy, G.J. (2001). *Frontal Systems Behavior Scale (FrSBe): Professional Manual*. Psychological Assessment Resources, ISBN 987-654-321, Lutz, Florida.

Graham, R.K., Pouladi, M.A., Joshi, P., Lu, G., Deng, Y., Wu, N.-K., Figueroa, B.E., Metzler, M., André, V.M., Slow, E.J., Raymond, L., Friedlander, R., Levine, M.S., Leavitt, B.R., & Hayden, M.R. (2009). Differential susceptibility to excitotoxic stress in YAC128 mouse models of Huntington disease between initiation and progression of disease. *Journal of Neuroscience*, Vol. 29, No. 7, (February 2009), pp. 2193-2204, ISSN 1529-2401.

Greenamyre, J.T. (1986). The role of glutamate in neurotransmission and in neurologic disease. *Archives of Neurology*, Vol. 43, No. 10, (October 1986), pp. 1058-1063, ISSN 0003-9942

Gu, X., Andre, V.M., Cepeda, C., Li, S.H., Li, X.J., Levine, M.S., & Yang, X.W. (2007). Pathological cell-cell interactions are necessary for striatal pathogenesis in a conditional mouse model of Huntington's disease. *Molecular Neurodegeneration*, Vol. 2, (April 2007), pp. 8, ISSN 1750-1326

Guay, D.R. (2010). Tetrabenazine, a monoamine-depleting drug used in the treatment of hyperkinetic movement disorders. *American Journal of Geriatric Pharmacotherapy*, Vol. 8, No. 4, (August 2010), pp. 331-373, ISSN 1876-7761.

Halliday, G.M., McRitchie, D.A., Macdonald,V., Double, K.L., Trent, R.J. & McCusker, E. (1998). Regional specificity of brain atrophy in Huntington's disease. *Experimental Neurology*, Vol. 154, No. 2, (December 1998), pp. 663-672, ISSN 0014-4886

Haluk, D.M., & Floresco, S.B. (2009). Ventral striatal dopamine modulation of different forms of behavioral flexibility. *Neuropsychopharmacology*, Vol. 34, No. 8, (July 2009), pp. 2041-2052, ISSN 1740-634X.

Hansson, O., Petersen, A., Leist, M., Nicotera, P., Castilho, R.F., & Brundin, P. (1999). Transgenic mice expressing a Huntington's disease mutation are resistant to quinolinic acid-induced striatal excitotoxicity. *Proceedings of the National Academy of Sciences, USA*, Vol. 96, No. 15, (July 1999), pp. 8727-8732, ISSN 0027-8424

Hardingham, G.E., Fukunaga, Y., & Bading, H. (2002). Extrasynaptic NMDARs oppose synaptic NMDARs by triggering CREB shut-off and cell death pathways. *Nature Neuroscience*, Vol. 5, No. 5, (May 2002), pp. 405-414, ISSN 1097-6256

Hassel, B., Tessler, S., Faull, R.L., & Emson, P.C. (2008). Glutamate uptake is reduced in prefrontal cortex in Huntington's disease. *Neurochemical Research*, Vol. 33, No. 2, (February 2008), pp. 232-237, ISSN 0364-3190

Heng, M.Y., Tallaksen-Greene, S.J., Detloff, P.J., & Albin, R.L. (2007), Longitudinal evaluation of the Hdh(CAG)150 knock-in murine model of Huntington's disease, *Journal of Neuroscience*, Vol. 27, No. 34, (August 2007), pp. 8989-8998, ISSN 1529-2401

Hennenlotter, A., Schroeder, U., Erhard, P., Haslinger, B., Stahl, R., Weindl, A., von Einsiedel, H.G., Lange, K.W., & Ceballos-Baumann, A.O. (2004). Neural correlates associated with impaired disgust processing in pre-symptomatic Huntington's disease. *Brain*, Vol. 127, Pt 6, (June 2004), pp 1446–1453, ISSN 0006-8950.

Hickey, M.A., Reynolds, G.P., & Morton, A.J. (2002). The role of dopamine in motor symptoms in the R6/2 transgenic mouse model of Huntington's disease. *Journal of Neurochemistry*, Vol 81, No. 1, (April 2002), pp. 46-59, ISSN 0022-3042.

Hinton, G. (2003). The ups and downs of Hebb synapses. *Canadian Psychology*, Vol. 44, No. 1, (February 2003), pp. 10-13, ISSN 0708-5591

Ho, A.K., Sahakian, B.J., Brown, R.G., Barker, R.A., Hodges, J.R., Ane, M.N., Snowden, J., Thompson, J., Esmonde, T., Gentry, R., Moore, J.W., & Bodner, T. (2003). Profile of cognitive progression in early Huntington's disease. *Neurology*, Vol. 61, No. 12, (December 2003), pp. 1702-1706, ISSN 1526-632X.

Hoffstrom, B.G., Kaplan, A., Letso, R., Schmid, R.S., Turmel, G.J., Lo, D.C., & Stockwell, B.R. (2010). Inhibitors of protein disulfide isomerase supress apoptosis induced by misfolded proteins, *Nature Chemical Biology*. Vol. 6, No. 12, (December 2010), pp. 900-906, ISSN 1552-4469

Höhn, S., Dallérac, G., Faure, A., Urbach, Y.K., Nguyen, H.P., Riess, O., von Hörsten, S., Le Blanc, P., Desvignes, N., El Massioui, N., Brown, B.L., & Doyére, V. (2011). Behavioral and in vivo electrophysiological evidence for presymptomatic alteration of prefrontostriatal processing in the transgenic rat model for Huntington disease. *Journal of Neuroscience*, Vol. 31, No. 24, (June 2011), pp. 8986-8997, ISSN 1529-2401

Huntington Study Group. (2001). A randomized placebo-controlled trial of coenzyme Q10 and Remacemide in Huntington's disease. *Neurology*, Vol. 57, No. 3, (August 14), pp. 397-404, ISSN 0028-3878

Huntington's Disease Collaborative Research Group. (1993). A novel gene containing a trinucleotide repeat that is expanded and unstable on Huntington's disease chromosomes. *Cell*, Vol. 72, No. 6, (March 1993), pp. 971-983, ISSN 0092-8674

Johnson, M..A, Rajan, V., Miller, C.E., & Wightman, R.M. (2006). Dopamine release is severely compromised in the R6/2 mouse model of Huntington's disease. *Journal of Neurochemistry*, Vol. 97, No. 3, (May 2006), pp. 737-746, ISSN 0022-3042

Johnson, S.A., Stout, J.C., Solomon, A.C., Langbehn, D.R., Aylward, E.H., Cruce, C.B., Ross, C.A., Nance, M., Kayson, E., Julian-Baros, E., Hayden, M.R., Kieburtz, K., Guttman, M., Oakes, D., Shoulson, I., Beglinger, L., Duff, K., Penziner, E., Paulsen, J.S., & Predict-HD Investigators of Huntington Study Group. (2007). Beyond disgust: impaired recognition of negative emotions prior to diagnosis in Huntington's disease. *Brain*, Vol. 130, Pt 7, (July 2005), pp. 1732-1744, ISSN 1460-2156

Kane, M.J., & Engle, R.W. (2002). The role of prefrontal cortex in working memory capacity, executive attention, and general fluid intelligence: An individual-differences perspective. *Psychonomic Bulletin & Review*, Vol. 9, No. 4, (December 2002), pp. 637-671, ISSN 1069-9384

Kawaguchi, Y., Wilson, C.J., & Emson, P.C. (1990). Projection subtypes of rat neostriatal matrix cells revealed by intracellular injection of biocytin. *Journal of Neuroscience*, Vol. 10, No. 10, (October 1990), pp. 3421-3438, ISSN 0270-6474

Kipps, C.M., Duggins, A.J., Mahant, N., Gomes, L., Ashburner, J., & McCusker, E.A. (2005). Progression of structural neuropathology in preclinical Huntington's disease: a tensor based morphometry study. *Journal of Neurology, Neurosurgery, and Psychiatry*, Vol. 76, No. 5, (May 2005), pp. 650-655, ISSN 0022-3050

Kirkwood, S.C., Siemers, E., Hodes, M.E., Conneally, P.M., Christian, J.C., & Foroud, T. (2000). Subtle changes among presymptomatic carriers of the Huntington's disease gene. *Journal of Neurology, Neurosurgery, and Psychiatry*, Vol. 69, No. 6, (December 2000), pp (773-779), ISSN 0022-3050

Kiyatkin, E.A., & Rebec, G.V. (1996). Modulatory action of dopamine on acetylcholine-responsive striatal and accumbal neurons in awake, unrestrained rats. *Brain Research*, Vol. 713, No. 1-2, (March 1996), pp. 70-78, ISSN 0006-8993

Klapstein, G.J., Fisher, R.S., Zanjani, H., Cepeda, C., Jokel, E.S., Chesselet, M.F., & Levine, M.S. (2001). Electrophysiological and morphological changes in striatal spiny neurons in R6/2 Huntington's disease transgenic mice. *Journal of Neurophysiology*, Vol. 86, No. 6, (December 2001), pp. 2667–2677, ISSN 0022-3077.

Kraft, J.C., Osterhaus, G.L., Ortiz, A.N., Garris, P.A., & Johnson, M.A. (2009). In vivo dopamine release and uptake impairments in rats treated with 3-nitropropionic acid. *Neuroscience*, Vol. 161, No. 3, (July 2009), pp. 940-949, ISSN 1873-7544.

Kubota, Y., & Kawaguchi, Y. (2000). Dependence of GABAergic synaptic areas on the interneuron type and target size. *Journal of Neuroscience*, Vol. 20, No. 1, (January 2000), pp. 375-386, ISSN 1529-2401

Kuwert, T., Lange, H.W., Boecker, H., Titz, H., Herzog, H., Aulich, A., Wang, B.C., Nayak, U., & Feinendegen, L.E. (1993). Striatal glucose consumption in chorea-free subjects at risk of Huntington's disease. *Journal of Neurology*, Vol. 241, No. 1, (November 1993), pp 31-36, ISSN 0340-5354.

Laforet, G.A., Sapp, E., Chase, K., McIntyre, C., Boyce, F.M., Campbell, M., Cadigan, B.A., Warzecki, L., Tagle, D.A., Reddy, P.H., Cepeda, C., Calvert, C.R., Jokel, E.S., Klapstein, G.J., Ariano, M.A., Levine, M.S., DiFiglia, M., & Aronin, N. (2001). Changes in cortical and striatal neurons predict behavioral and electrophysiological abnormalities in a transgenic murine model of Huntington's disease. *Journal of Neuroscience*, Vol. 21, No. 23, (December 2001), pp. 9112-9123, ISSN 1529-2401.

Lawrence, A.D., Hodges, J.R., Rosser, A.E., Kershaw, A., ffrench-Constant, C., Rubinsztein, D.C., Robbins, T.W., & Sahakian, B.J. (1998a). Evidence for specific cognitive deficits in preclinical Huntington's disease. *Brain*, Vol. 121, Pt 7, (July 1998), pp.1329-1341, ISSN 0006-8950.

Lawrence, A.D., Sahakian, B.J., Hodges, J.R., Rosser, A.E., Lange, K.W., & Robbins, T.W. (1996). Executive and mnemonic functions in early Huntington's disease. *Brain*, Vol. 119, Pt 5, (October 1996), pp.1633-1645, ISSN 0006-8950

Lawrence, A.D., Weeks, R.A., Brooks, D.J., Andrews, T.C., Watkins, L.H.A., Harding, A.E., Robbins, T.W., & Sahakian, B.J. (1998b). The relationship between striatal dopamine receptor binding and cognitive performance in Huntington's disease. *Brain*, Vol. 121, Pt 7, (July 1998), pp. 1343-1355, ISSN 0006-8950

Lee, A., & Pow, D.V. (2010). Astrocytes: Glutamate transport and alternate splicing of transporters. *International Journal of Biochemistry & Cell Biology*, Vol. 42, No. 12, (December 2010), pp.1901-1906., ISSN 1878-5875

Lee, J.L.C., Everitt, B.J., & Thomas, K.L. (2004). Independent cellular processes for hippocampal memory consolidation and reconsolidation. *Science*, Vol. 304, No. 5672, (May 2004), pp. 839-843, ISSN 1095-9203

Lemiere, J., Decruyenaere, M., Evers-Kiebooms, G., Vandenbussche, E., & Dom, R. (2002). Longitudinal study evaluating neuropsychological changes in so-called asymptomatic carriers of the Huntington's disease mutation after 1 year. *Acta Neurologica Scandinavica*, Vol. 106, No. 3, (September 2002), pp. 131-141, ISSN 0001-6314

Leveille, F., El Gaamouch, F., Gouix, E., Lecocq, M., Loner, D., Nicole, O., & Buisson, A. (2008). Neuronal viability is controlled by a functional relation between synaptic and extrasynaptic NMDA receptors. *FASEB Journal*, Vol. 22, No. 12, (December 2088), pp. 4258-4271, ISSN 1530-6860

Levesque, M., & Parent, A. (2005). The striatofugal fiber system in primates: a reevaluation of its organization based on single-axon tracing studies. *Proceedings of the National Academies of Science, USA*, Vol. 102, No. 33, (August 2005), pp. 11888-11893, ISSN 0027-8424

Levine, M.S., Klapstein, G.J., Koppel, A., Gruen, E., Cepeda, C., Vargas, M.E., Jokel, E.S., Carpenter, E.M., Zanjani, H., Hurst, R.S., Efstratiadis, A., Zeitlin, S., & Chesselet, M.F. (1999). Enhanced sensitivity to N-methyl-D-aspartate receptor activation in transgenic and knock-in mouse models of Huntington's disease. *Journal of Neuroscience Research*, Vol. 58, (November 1999), pp. 515-532, ISSN 0360-4012

Leyva, M.J., Degiacomo, F., Kaltenbach, L.S., Holcomb, J., Zhang, N., Gafni, J., Park, H., Lo, D.C., Salvesen, G.S., Ellerby, L.M., & Ellman, J.A. (2010). Identification and evaluation of small molecule pan-caspase inhibitors in Huntington's disease models. *Chemistry & Biology*, Vol. 17, No. 11, (November 2010), pp. 1189-1200, ISSN 1074-5521

Li, L., Murphy, T.H., Hayden, M.R., & Raymond, L.A. (2004). Enhanced striatal NR2B containing N-methyl-D-aspartate receptor-mediated synaptic currents in a mouse model of Huntington disease. *Journal of Neurophysiology*, Vol. 92, (November 2004), pp. 2738–2746, ISSN 0022-3077

Lievens, J.C., Woodman, B., Mahal, A., Spasic-Boscovic, O., Samuel, D., Kerkerian-Le Goff, L., & Bates, G.P. (2001). Impaired glutamate uptake in the R6 Huntington's disease transgenic mice. *Neurobiology of Disease*, Vol. 8, No. 5, (October 2001), pp. 807-821, ISSN 0969-9961

Lin, C.H., Tallaksen-Greene, S., Chien, W.M., Cearley, J.A., Jackson, W.S., Crouse, A.B., Ren, S., Li, X.J., Albin, R.L., & Detloff, P.J. (2001). Neurological abnormalities in a knock-in mouse model of Huntington's disease. *Human Molecular Genetics*, Vol. 10, No. 2, (January 2001), pp. 137-144, ISSN 0964-6906

Lovinger, D.M. (2010). Neurotransmitter roles in synaptic modulation, plasticity, and learning in the dorsal striatum. *Neuropharmacology*, Vol. 58, (June 2010), pp. 951-961, ISSN 0964-6906

Massey, P.V., & Bashir, Z.I. (2007). Long-term depression: multiple forms and implications for brain function. *Trends in Neurosciences*, Vol. 30, No. 4, (April 2007), pp. 176-184, ISSN 0166-2236

Matamales, M., Bertran-Gonzalez, J., Salomon, L., Degos, B., Deniau, J.M., Valjent, E., Herve, D., & Girault, J.A. (2009). Striatal medium-sized spiny neurons: identification by nuclear staining and study of neuronal subpopulations in BAC transgenic mice. *PloS One*, Vol. 4, No. 3, (March 2009), pp. e4770, ISSN 1932-6203

Matell, M.S., & Meck, W.H. (2004). Cortico-striatal circuits and interval timing: coincidence detection of oscillatory responses. *Cognitive Brain Research*, Vol. 21, (October 2004), pp. 139-170, ISSN 0926-6410

Mattson, M.P. (2000). Apoptosis in neurodegenerative disorders. *Nature Reviews Molecular Cell Biology*, Vol. 1, (November, 2000), pp. 120-129, ISSN 1471-0072

Mattson, M.P., & Furukawa K. (1996). Programmed cell life: anti-apoptotic signaling and therapeutic strategies for neurodegenerative disorders. *Restorative Neurology and Neuroscience*, Vol. 9, No. 4, (January 1996), pp. 191-205, ISSN 0922-6028

Matyas, F., Yanovski, Y., Mackie, K., Kelsch, W., Misgeld, U., & Freund, T.F. (2006). Subcellular localization of type 1 cannabinoid receptors in the rat basal ganglia. *Neuroscience*, Vol. 137, No. 1, (January 2006), pp. 337-361, ISSN 0306-4522

McBain, C.J., & Kauer, J.A. (2009). Presynaptic plasticity: targeted control of inhibitory networks. *Current Opinion in Neurobiology*, Vol. 19, No. 3, (June 2009), pp. 254-262, ISSN 0959-4388

Middleton, F.A., & Strick, P.L. (2000). Basal ganglia and cerebellar loops: motor and cognitive circuits. *Brain Research Bulletin*, Vol. 31, (March 2000) pp. 236-250, ISSN 0361-9230

Miller, B.R., Dorner, J.L., Shou, M., Sari, Y., Barton, S.J., Sengelaub, D.R., Kennedy, R.T., & Rebec, G.V. (2008a). Up-regulation of GLT1 expression increases glutamate uptake and attenuates the Huntington's disease phenotype in the R6/2 mouse. *Neuroscience*, Vol. 153, No. 1, (April 2008), pp. 329-337, ISSN 0306-4522

Miller, B.R., Walker, A.G., Barton, S.J., & Rebec, G.V. (2011). Dysregulated Neuronal Activity Patterns Implicate Corticostriatal Circuit Dysfunction in Multiple Rodent Models of Huntington's Disease. *Frontiers in Systems Neuroscience*, Vol. 5, (May 2011), pp. 26, ISSN 1662-5137

Miller, B.R., Walker, A.G., Fowler, S.C., von Horsten, S., Riess, O., Johnson, M.A., & Rebec, G.V. (2010). Dysregulation of coordinated neuronal firing patterns in striatum of freely behaving transgenic rats that model Huntington's disease. *Neurobiology of Disease*, Vol. 37, No. 1, (January 2010) pp. 106-113, ISSN 0969-9961

Miller, B.R., Walker, A.G., Shah, A.S., Barton, S.J., & Rebec, G.V. (2008b). Dysregulated information processing by medium spiny neurons in striatum of freely behaving mouse models of Huntington's disease. *Journal of Neurophysiology*, Vol. 100, No. 4, (October 2008), pp. 2205-2216, ISSN 0022-3077

Milner, P. (2003). A brief history of the Hebbian learning rule. *Canadian Psychology*, Vol. 44, (February 2003), pp. 5-9, ISSN 0708-5591

Milnerwood, A.J., Gladding, C.M., Pouladi, M.A., Kaufman, A.M., Hines, R.M., Boyd, J.D., Ko, R.W.Y., Vasuta, O.C., Graham, R.K., Hayden, M.R., Murphy, T., & Raymond, L.A. (2010). Early increase in extrasynaptic NMDA receptor signaling and expression contributes to phenotype onset in Huntington's disease mice. *Neuron*, Vol. 65, (January 2010), pp. 178-190, ISSN 0896-6273

Montoya, A., Price, B.H., Menear, M., & Lepage, M. (2006). Brain imaging and cognitive dysfunctions in Huntington's disease. *Journal of Psychiatry & Neuroscience*, Vol. 31, No. 1, (January 2006), pp. 21-29, ISSN 1180-4882

Niatsetskaya, Z., Basso, M., Speer, R.E., McConoughey, S.J., Coppola, G., Ma, T.C., & Ratan, R.R. (2010). HIF prolyl hydroxylase inhibitors prevent neuronal death induced by mitochondrial toxins: Therapeutic implications for Huntington's disease and Alzheimer's disease. *Antioxidants & Redox Signaling*, Vol. 12, No. 4, (April 2010), pp. 435-443, ISSN 1523-0864

Nicniocaill, B., Haraldsson, B., Hansson, O., O'Connor, W.T., & Brundin, P. (2001). Altered striatal amino acid neurotransmitter release monitored using microdialysis in R6/1 Huntington transgenic mice. *European Journal of Neuroscience*, Vol. 13, No. 1, (January 2001), pp. 206-210, ISSN 0953-816X

Nieoullon, A. (2002). Dopamine and the regulation of cognition and attention. *Progress in Neurobiology*, Vol. 67, (May 2002), pp. 53-83, ISSN 0301-0082

Okamoto, S., Pauladi, M.A., Talantova, M., Yao, D., Xia, P., Ehrnhoefer, D.E., Zaldi, R., Clemente, A., Kaul, M., Grayham, R.K., Zhang, D., Vincent Chen, H.S., Tong, G., Hayden, M.R., & Lipton, S.A. (2009). Balance between synaptic versus extrasynaptic NMDA receptor activity influences inclusions and neurotoxicity of mutant huntingtin. *Nature Medicine*, Vol. 15, No. 12, (December 2009), pp. 1407-1413, ISSN 1078-8956

Ortiz, A.N., Kurth, B.J., Osterhaus, G.L., & Johnson, M.A. (2010). Dysregulation of intracellular dopamine stores revealed in the R6/2 mouse striatum. *Journal of Neurochemistry*, Vol. 112, No. 3, (February 2010), pp. 755-761, ISSN 0022-3042

Ortiz, A.N., Kurth, B.J., Osterhaus, G.L., & Johnson, M.A. (2011). Impaired dopamine release and uptake in R6/1 Huntington's disease model mice. *Neuroscience Letters*, Vol. 492, No. 1, (March 2011), pp. 11-14, ISSN 0304-3940

Owen, A.M., James, M., Leigh, P.N., Summers, B.A., Marsden, C.D., Quinn, N.P., Lange, K.W., & Robbins, T.W. (1992). Fronto-striatal cognitive deficits at different stages of Parkinson's disease. *Brain*, Vol. 115, (December 1992), pp. 1727-1751, ISSN 0006-8950

Pajski, M.L., & Venton, B.J. (2010). Adenosine release evoked by short electrical stimulations in striatal brain slices is primarily activity dependent. *ACS Chemical Neuroscience*, Vol. 1, (October 2010), pp. 775-787, ISSN 1948-7193

Paoletti, P., Vila, I., Rifé, M., Lizcano, J.M., Alberch, J., & Ginés, S. (2008). Dopaminergic and glutamatergic signaling crosstalk in Huntington's disease neurodegeneration: The role of p25/cyclin-dependent kinase 5. *Journal of Neuroscience*, Vol. 28, No. 40, (October 2008), pp. 10090-10101, ISSN 0270-6474

Pascual, O., Casper, K.B., Kubera, C., Zhang, J., Revilla-Sanchez, R., Sul, J.-Y., Takano, H., Moss, S.J., McCarthy, K., & Haydon, P.G. (2005). Astrocytic purinergic signaling coordinates synaptic networks. *Science*, 310, (October 2005), pp. 113-116, 0036-8075

Paulsen, J.S., (2010). Early detection of Huntington's disease: Review. *Future Neurology*, Vol. 5, No. 1, pp. 85-104, ISSN 1479-6708

Paulsen, J.S., & Conybeare, R.A. (2005). Cognitive changes in Huntington's disease. *Advances in Neurology*, Vol. 96, pp. 209-225, ISSN 0091-3952

Paulsen, J.S., Langbehn, D.R., Stout, J.C., Aylward, E., Ross, C.A., Nance, M., Guttman, M., Johnson, S., MacDonald, M., Beglinger, L.J., Duff, K., Kayson, E., Biglan, K., Shoulson, I., Oakes, D., Hayden, M., & The Predict-HD Investigators and Coordinators of the Huntington Study Group. (2008). Detection of Huntington's disease decades before diagnosis: the Predict-HD study. *Journal of Neurology, Neurosurgery, and Psychiatry*, Vol. 79, (August 2008), pp. 874-880, ISSN 0022-3050

Paulsen, J.S., Zimbelman, J.L., Hinton, S.C., Langbehn, D.R., Leveroni, C.L., Benjamin, M.L., Reynolds, N.C., & Rao, S.M. (2004). fMRI biomarker of early neuronal dysfunction in presymptomatic Huntington's disease. *American Journal of Neuroradiology*, Vol. 25, No. 10, (November 2010), pp. 1715-1721, ISSN 0195-6108

Pavese, N., Andrews, T.C., Brooks, D.J., Ho, A.K., Rosser, A.E., Barker, R.A., Robbins, T.W., Sahakian, B.J., Dunnett, S.B., & Piccini, P. (2003). Progressive striatal and cortical dopamine receptor dysfunction in Huntington's disease: a PET study. *Brain*, Vol. 126, (May 2003), pp. 1127-1135, ISSN 0006-8950

Pavese, N., Politis, M., Tai, Y.F., Barker, R.A., Tabrizi, S.J., Mason, S.L., Brooks, D.J., & Piccini, P. (2010). Cortical dopamine dysfunction in symptomatic and premanifest Huntington's disease gene carriers. *Neurobiology of Disease*, Vol. 37, No. 2, (February 2010), pp. 356-361, ISSN 0969-9961

Peyser, C.E., Folstein, M., Chase, G.A., Starkstein, S., Brandt, J., Cockrell, J.R., Bylsma, F., Coyle, J.T., McHugh, P.R., & Folstein, S.E. (1995). Trial of d-alpha-tocopherol in Huntington's disease. *American Journal of Psychiatry*, Vol. 152, No. 12, (December 1995), pp. 1171-1175, ISSN 0002-953X

Pierce, R.C., & Rebec, G.V. (1995). Iontophoresis in the neostriatum of awake, unrestrained rats: differential effects of dopamine, glutamate and ascorbate on motor- and nonmotor-related neurons. *Neuroscience*, Vol. 67, No. 2, (July 1995), pp. 313-324, ISSN 0306-4522

Qian, Y., Guan, T., Tang, X., Huang, M., Li, Y., Sun, H., Yu, R., & Zhang, F. (2011). Astrocytic glutamate transporter-dependent neuroprotection against glutamate toxicity: an in vitro study of maslinic acid. *European Journal of Pharmacology*, Vol. 651, No. 1-3, (January 2011), pp. 59-65, ISSN 0014-2999

Quarrell, O. (2008). *Huntington's Disease: The Facts*. Second Ed., Oxford University Press, ISBN 0199212015, New York.

Rebec, G.V., Conroy, S.K., & Barton, S.J. (2006). Hyperactive striatal neurons in symptomatic Huntington R6/2 mice: variations with behavioral state and repeated ascorbate treatment. *Neuroscience*, Vol. 137, No. 1, pp. 327-336, ISSN 0306-4522

Rosas, H.D., Hevelone, N.D., Zaleta, A.K., Greve, D.N., Salat, D.H., & Fischl, B. (2005). Regional cortical thinning in preclinical Huntington disease and its relationship to cognition. *Neurology*, Vol. 65, No. 5, (September 2005), pp. 745-747, ISSN 0028-3878

Rowe, K.C., Paulsen, J.S., Langbehn, D.R., Duff, K., Beglinger, L.J., Wang, C., O'Rourke, J.J., Stout, J.C., & Moser, D.J. (2010). Self-paced timing detects and tracks change in prodromal Huntington disease. *Neuropsychology*, Vol. 24, (July 2010), pp. 435-442, ISSN 0894-4105

Salgado-Pineda, P., Delaveau, P., Blin, O., & Nieoullon, A. (2005). Dopaminergic contribution to the regulation of emotional perception. *Clinical Neuropharmacology*, Vol. 28, No. 5, (September-October 2005), pp. 228-237, ISSN 0362-5664

Sandstrom, M.I., & Rebec, G.V. (2003). Characterization of striatal activity in conscious rats: Contribution of NMDA and AMPA/kainate receptors to both spontaneous and glutamate-driven firing. *Synapse*, Vol. 47, No. 2, (February 2003), pp. 91-100, ISSN 0887-4476

Sari, Y., Prieto, A.L., Barton, S.J., Miller, B.R., & Rebec, G.V. (2010). Ceftriaxone-induced up-regulation of cortical and striatal GLT1 in the R6/2 model of Huntington's disease. *Journal of Biomedical Science*, Vol. 17, (July 2010), pp. 62, ISSN 1021-7770

Sax, D.S., Powsner, R., Kim, A., Tilak, S., Bhatia, R., Cupples, L.A., & Myers, R.H. (1996). Evidence of cortical metabolic dysfunction in early Huntington's disease by single-photon-emission computed tomography. *Movement Disorders*, Vol. 11, No. 6, (November 1996), pp. 671-677, ISSN 0885-3185

Snowden, J.S., Craufurd, D., Griffiths, H., Thompson, J., & Neary, D. (2001). Longitudinal evaluation of cognitive disorder in Huntington's disease. *Journal of the International Neuropsychological Society: JINS*, Vol. 7, No. 1, (January 2001), pp. 33-44, ISSN 1355-6177

Snowden, J.S., Craufurd, D., Thompson, J., & Neary, D. (2002). Psychomotor, executive, and memory function in preclinical Huntington's disease. *Journal of Clinical and Experimental Neuropsychology*, 24(2), (April 2002), pp. 133-145, 0168-8634

Sprengelmeyer, R., Schroeder, U., Young, A.W., & Epplen, J.T. (2006). Disgust in preclinical Huntington's disease: A longitudinal study. *Neuropsychologia*,Vol. 44, pp. 518-533, ISSN 0028-3932

Starling, A.J., André, V.M., Cepada, C., de Lima, M., Chandler, S.H., & Levine, M.S. (2005). Alterations in N-methyl-D-aspartate receptor sensitivity and magnesium blockade occur early in development in the R6/2 mouse model of Huntington's disease. *Journal of Neuroscience Research*, Vol. 82, (November 2005), pp. 377-386, ISSN 0360-4012

Stern, E.A., Jaeger, D., & Wilson, C.J. (1998). Membrane potential synchrony of simultaneously recorded striatal spiny neurons in vivo. *Nature*, Vol. 394, (July 1998), pp. 475-478, ISSN 0028-0836

Tang, T.S., Chen, X., Liu, J., & Bezprozvanny, I. (2007). Dopaminergic signaling and striatal neurodegeneration in Huntington's disease. *Journal of Neuroscience*, Vol. 27. No. 30, (July 2007), pp. 7899-7910, ISSN 0270-6474

Thomas, C.G., Tian, H., & Diamond, J.S. (2011a). The relative roles of diffusion and uptake in clearing synaptically released glutamate change during early postnatal development. *Journal of Neuroscience*, Vol. 3, No. 12, (March 2011), pp. 4743-4754, ISSN 0270-6474

Thomas, E.A., Coppola, G., Tang, B., Kuhn, A., Kim, S., Geschwind, D.H., Brown, T.B., Luthi-Carter, R., & Ehrlich, M.E. (2011b). In vivo cell-autonomous transcriptional abnormalities revealed in mice expressing mutant huntingtin in striatal but not cortical neurons. *Human Molecular Genetics*, Vol. 20, No. 6, (March 2011), pp. 1049-1060, ISSN 0964-6906

Tossman, U., Jonsson, G., & Ungerstedt, U. (1986). Regional distribution and extracellular levels of amino acids in rat central nervous system. *ACTA Physiologica Scandinavica*, Vol. 127, No. 4, (August 1986), pp. 533-545, ISSN 0001-6772

Trueman, R.C., Brooks, S.P., Jones, L., & Dunnett, S.B. (2007). The operant serial implicit learning task reveals early onset motor learning deficits in the Hdh^{Q92} knock-in mouse model of Huntington's disease. *European Journal of Neuroscience*, Vol. 25, (January 2007), pp. 551-558, ISSN 0953-816X

Trueman, R.C., Brooks, S.P., Jones, L., & Dunnett, S.B. (2008). Time course of choice reaction time deficits in the Hdh^{Q92} knock-in mouse model of Huntington's disease in the operant serial implicit learning task (SILT). *Behavioural Brain Research*, Vol. 189, (June 2008), pp. 317-324, ISSN 0166-4328

Uchigashima, M., Narushima, M., Fukaya, M., Katona, I., Kano, M., & Watanabe, M. (2007). Subcellular arrangement of molecules for 2-arachidonoyl-glycerol-mediated retrograde signaling and its physiological contribution to synaptic modulation in the striatum. *Journal of Neuroscience*, Vol. 27, No. 14, (April 2007), pp. 3663-3676, ISSN 0270-6474

Valjent, E., Bertran-Gonzalez-J., Herve, D., Fisone, G., & Girault, J.-A. (2009). Looking BAC at striatal signaling: cell-specific analysis in new transgenic mice. *Trends in Neurosciences*, Vol. 32, No. 10, (October 2009), pp. 538-547, ISSN 0166-2236

van Oostrom, J.C., Dekker, M., Willemsen, A.T., de Jong, B.M., Roos, R.A., & Leenders, K.L. (2009). Changes in striatal dopamine D2 receptor binding in pre-clinical Huntington's disease. *European Journal of Neurology*, Vol. 16, No. 2, (February 2009), pp. 226-231, ISSN 1351-5101

van Raamsdonk, J.M., Pearson, J., Bailey, C.D., Rogers, D.A., Johnson, G.V., Hayden, M.R., & Leavitt, B.R. (2005a). Cystamine treatment is neuroprotective in the YAC128 mouse model of Huntington disease. *Journal of Neurochemistry*, Vol. 95, No. 1, (October 2005), pp. 210-220, ISSN 0022-3042

van Raamsdonk, J.M., Pearson, J., Slow, E.J., Hossain, S.M., Leavitt, B.R., & Hayden, M.R. (2005b). Cognitive dysfunction precedes neuropathology and motor abnormalities in the YAC128 mouse model of Huntington's disease. *Journal of Neuroscience*, Vol. 25, No. 16, (April 2005), pp. 4169-4180, ISSN 0270-6474

Verbessem, P., Lemiere, J., Eijnde, B.O., Swinnen, S., Vanhees, L., Van Leemputte, M., Hespel, P. & Dom, R. (2003). Creatin supplementation in Huntington's disease. *Neurology*, Vol. 61, No. 7, (October 2003), pp. 925-930, ISSN 0028-3878

Waelti, P., Dickinson, A., & Schultz, W. (2001). Dopamine responses comply with basic assumptions of formal learning theory. *Nature*, Vol. 412, (July 2001), pp. 43-48, ISSN 0028-0836

Walker, A.G., Miller, B.R., Fritsch, J.N., Barton, S.J., & Rebec, G.V. (2008). Altered information processing in the prefrontal cortex of Huntington's disease mouse models. *Journal of Neuroscience*, Vol. 28, No. 36, (September 2008), pp. 8973-8982, ISSN 0270-6474

Wang, H., Chen, X., Li, Y., Tang, T.S., & Bezprozvanny, I. (2010). Tetrabenazine is neuroprotective in Huntington's disease mice. *Molecular Neurodegeneration*, Vol. 5, (April 2010), pp. 18, ISSN 1750-1326

Wang, W., Duan, W., Igarashi, S., Morita, H., Nakamura, M., & Ross, C.A. (2005). Compounds blocking mutant huntingtin toxicity identified using a Huntington's disease neuronal cell model. *Neurobiology of Disease*, Vol. 20, (November 2005), pp. 500-508, ISSN 0969-9961

Wheeler, V.C., White, J.K,. Gutekunst, C.A., Vrbanac, V., Weaver, M., Li, X.J., Li, S.H., Yi, H., Vonsattel, J.P., Gusella, J.F., Hersch, S., Auerbach, W., Joyner, A.L., & MacDonald, M.E. (2000). Long glutamine tracts cause nuclear localization of a novel form of huntingtin in medium spiny striatal neurons in HdhQ92 and HdhQ111 knock-in mice. *Human Molecular Genetics*, Vol. 9, No. 4, (March 2000), pp. 503-513, ISSN 0964-6906

Wilson, C.J. (1993). The generation of natural firing patterns in neostriatal neurons. *Progress in Brain Research*, Vol. 99, pp. 277-297, ISSN 0079-6123

Wilson, C.J. (2004). Basal Ganglia. In: *The Synaptic Organization of the Brain*. Shepherd, G.M. (Ed.), pp. 361-414, Oxford University Press, ISBN 0-19-511824-3, Oxford.

Wilson, C.J., Chang, H.T., & Kitai, S.T. (1990). Firing patterns and synaptic potentials of identified giant aspiny interneurons in the rat neostriatum. *Journal of Neuroscience*, Vol. 10, (February 1990), pp. 508-519, 0270-6474

Wilson, C.J., & Kawaguchi, Y. (1996). The origins of two-state spontaneous membrane potential fluctuations of neostriatal spiny neurons. *Journal of Neuroscience*, Vol. 16, (April 1996), pp. 2397-2410, ISSN 0270-6474

Winter, S., Deikmann, M., & Schwabe, K. (2009). Dopamine in the prefrontal cortex regulates rats behavioral flexibility to changing reward value. *Behavioural Brain Research*, Vol. 198, (March 2009), pp. 206-213, ISSN 0166-4328

Wolf, R.C., Sambataro, F., Vasic, N., Schönfeldt-Lecuona, C., Ecker, D. & Landwehrmeyer, G.B. (2008). Altered frontostriatal coupling in pre-manifest Huntington's disease: effects of increasing cognitive load. *European Journal of Neurology*, Vol. 15, (November 2008), pp. 1180-1190, ISSN 1351-5101

Wolf, R.C., Vasic, N., Schönfeldt-Lecuona, C., Landwehrmeyer, G.B., & Ecker, D. (2007). Dorsolateral prefrontal cortex dysfunction in presymptomatic Huntington's disease: evidence from event-related fMRI. *Brain*, Vol. 130, (November 2007), pp. 2845-2857, ISSN 0006-8950

Wu, Y., Richard, S., & Parent, A. (2000). The organization of the striatal output system: a single-cell juxtacellular labeling study in the rat. *Neuroscience Research*, Vol. 38, (November 2000), pp. 49-62, ISSN 0168-0102

Yohrling, G.J. 4th, Jiang, G.C., DeJohn, M.M., Miller, D.W., Young, A.B., Vrana, K.E., & Cha, J.H. (2003). Analysis of cellular, transgenic and human models of Huntington's disease reveals tyrosine hydroxylase alterations and substantia nigra neuropathology. *Molecular Brain Research*, Vol. 119, No. 1, (November 2003), pp. 28-36, ISSN 0169-328X

Young, A.B., Penney, J.B., Starosta-Rubinstein, S., Markel, D.S., Berent, S., Giordani, B., Ehrenkaufer, R., Jewett, D., & Hichwa, R. (1986). PET scan investigations of Huntington's disease: Cerebral metabolic correlates of neurological features and functional decline, *Annals of Neurology*, Vol. 20, (September 1986), pp. 296-303, ISSN 0364-5134

Zeron, M.M., Hansson, O., Chen, N., Wellington, C.L., Leavitt, B.R., Brundin, P., Hayden, M.R., & Raymond, L.A. (2002). Increased sensitivity to N-methyl-D-aspartate receptor-mediated excitotoxicity in a mouse model of Huntington's disease. *Neuron*, Vol. 33, (March 2002), pp. 849–860, ISSN 0896-6273

9

Endogenous Attention in Normal Elderly, Presymptomatic Huntington's Disease and Huntington's Disease Subjects

Charles-Siegfried Peretti[1,3], Charles Peretti[1],
Virginie-Anne Chouinard[2,3] and Guy Chouinard[3,4]
[1]Service de Psychiatrie, Hôpital Saint-Antoine, Université Pierre et Marie Curie, Paris,
[2]Massachusetts General Hospital/McLean Adult Psychiatry,
Residency Training Program, Harvard Medical School, Boston, Massachusetts,
[3]Clinical Psychopharmacology Unit, McGill University, Montreal,
[4]Fernand Seguin Research Center, University of Montreal, Montreal,
[1]France
[2]USA
[3,4]Canada

1. Introduction

1.1 Definition and fragmentation of attention

Attention is a cognitive process sensitive to the effects of aging and neurodegenerative diseases, and its decline contributes to a decrease in cognitive performances during aging[1-4]. Attention is viewed as a set of different components derived from the Posner's model rather than a global unitary model[5]. Visuospatial attention is divided into explicit or implicit attention depending on the presence or absence of awareness[6,7]. Attention is also classified as exogenous and endogenous attention. Exogenous or automatic attention is directed by external stimuli, whereas endogenous or voluntary attention is directed by voluntary acts. Furthermore, shifts of visuospatial attention involve separate processes, such as shift between objects, as opposed to shift within objects[8,9]. The distinction between data-driven attention which is sensitive to aging, and memory-driven attention which is hardly sensitive to aging, also corresponds to the distinction between fluid and crystallized intelligence[10]. These distinctions have been established through findings of impairment caused by aging on subtests of the Wechsler Adult Intelligence Scale (WAIS) measuring data-driven attention (Digit Span, Digit Symbol and Vocabulary subtests), while other WAIS subtests remained unimpaired by aging[11,12].

Visual perception depends on the occipito-temporal or the "what" pathway for object vision and on the occipital parietal or the "where" pathway for spatial vision[13]. Milner and Goodale[14] proposed that the "where" pathway or dorsal stream contributes to action control by first selecting the location of an object, and then the ventral stream or the "what" pathway recognizes and analyzes the spatially defined part of the scene[15]. Different models of attention describe mechanisms of how visual features (e.g. colour) guide spatial attention and maximum activation[16-18].

Posner postulated that there are three different types of operations involved in visual attention: disengaging, shifting and engaging[19]. These have been studied in Alzheimer's disease (AD), Huntington's disease (HD) and Parkinson's disease (PD)[20-22]. In AD, decreased attention has been attributed to a process that accelerates aging deficits, thus altering all cognitive components (memory, language, executive functions, etc.). However, this model does not explain deficits of attention found in normal aging and other neurodegenerative diseases[22,23]. Patients with AD and PD were found to have distinct attention deficits when required to shift attention to targets contained within the same visual stimulus[20,21]. PD patients also had impaired ability in shifting attention as shown by the number of perceptual errors made in identifying target stimuli[20]. The ability of HD patients to shift attention has been less investigated. Nonetheless, cognitive deficits associated with PD and HD tend to be more similar than different, and diseases that affect subcortical structures were found to produce similar patterns (or a similar pattern) of attention deficits[22].

1.2 Normal aging of attention

Studies investigating normal aging of attention have consistently shown that attention declines with age[24,25]. The exogenous component of attention is less impaired by aging as can been seen in simple detection tests when the target is explicitly cued[24,25]. In contrast, the endogenous component of attention is more sensitive to aging in discrimination tests when the target is not explicitly cued. From the age of 75, the effects of age and neurodegenerative diseases on attention are similar[25], which is largely due to misleading cues increasing reaction time (RT). These results suggest that the decline in attention associated with aging comes from a decreased ability to shift attention[3,23,25]. In AD patients, impaired ability to shift attention was found in both spatial- and object-based attention[26].

Response latency is shorter when the target is surrounded by congruent flankers (arrows pointing in the correct direction to target location), than when the target is surrounded by arrows pointing in the direction opposite to target location. Target detection latency depends on surrounding stimuli (flankers), which could be congruent or incongruent[27]. Thus, the conflicting nature of the stimulation allows the study of voluntary components of endogenous attention[28]. Furthermore, the anterior region of the Cingulate Gyrus, part of the exogenous attention network, was also found to be activated during perceptual conflicting situations[29].

1.3 Attentional neuronal networks

Exogenous and endogenous attention correspond to separate neuronal networks[30]. Two attentional neuronal networks have been identified: a posterior automatic network which includes the posterior parietal cortex, the thalamus pulvinar nuclei and the superior colliculus[31], and an anterior executive network which consists of the dorsolateral and ventromedial prefrontal cortex, the anterior Cingulate Gyrus and striatum.

The discovery of posterior and anterior attentional networks has led to a better understanding of normal and pathological aging of attention. If attentional networks do not undergo alterations other than those expected to occur during normal aging, one would expect a relative preservation of both endogenous and exogenous attention. In HD with dementia, neurodegeneration occurs in the striatum, putamen and caudate, and is often associated with neuronal death in the prefrontal cortex. In contrast to normal aging, damage to sub-cortical frontal circuits would lead to an alteration of the endogenous attention anterior network, and preservation of the exogenous attention posterior network. In AD, progressive degeneration occurs simultaneously in multimodal association cortex (parietal and prefrontal regions), with impairment of posterior and anterior attentional networks, affecting both endogenous and exogenous attention[22].

The present study was designed to evaluate these two components of attention in normal aging, presymptomatic Huntington's Disease (presymptomatic HD) and HD.

2. Methods

2.1 Subjects

2.1.1 HD patients, presymptomatic HD patients and controls

The HD group included 10 symptomatic subjects (3 women and 7 men) with hyperkinetic rather than hypokinetic symptoms. The age of HD patients ranged from 35 to 47 years (mean=42 years [SD=4.2]), and Mattis total score ranged from 110 to 125 (mean=120 [SD=5.8]). Seven presymptomatic HD (4 women and 3 men) were included with the transmittable mutation of the gene, IT15, on the short arm of chromosome 4[32]. DNA amplification technique[33] makes it possible to detect if an allele of the Huntington encoding gene displays repeats of the CAG sequences exceeding the threshold of 37 repeats and thus, detect carriers. The age of the presymptomatic HD group ranged from 35 to 46 years (mean=39 years [SD=3.6]), and Mattis total score ranged from 130 to 145 (mean=137 [SD=5.6]). The control group included 18 patients suffering from various medical conditions (8 women and 10 men) who were tested while awaiting clinical investigation at the Maison-Blanche Hospital of Reims, and for whom the main inclusion criterion was the integrity of frontal and striatal regions. Their mean age was 38 years [SD=4.7] and the mean Mattis total score was 140 [SD=4.1].

The study was funded by a Champagne-Ardenne Regional Grant and approved by the Ethic Committee of the Champagne-Ardenne Region. Participants were informed of the purpose of the research protocols and gave informed written consent. All subjects were tested during the same time period, 2002-2003.

2.2 Procedures

2.2.1 Visuospatial attention protocol

Protocols derived from Posner's visuospatial orientation model allow the study of the different components of attention[34,6,19,30]. In these protocols, the subject's attention is focused first on the centre of the screen and then to the left or right, using a cue, which is either peripheral (brightness of the place where the target will appear) or central (with an arrow pointing to left or right). The target (a letter to be detected, a side to be localized or a target to be identified) is then presented either to the left or the right. If the target appears on the same side as the preparatory cue, the cue is considered "valid". If it appears on the other side, it is misleading or "invalid". When both sides are simultaneously signalled, the cue is considered neutral since it does not permit a preparatory process. In addition to the peripheral or central nature of the cue, other experimental factors may be used: time interval between the cue and target, called cue-target Stimulus-Onset Asynchrony (SOA), which may range from 50 ms to several seconds, and % proportion of valid to invalid preparatory cue, which may range from 80%/20%, 50%/50%, to 20%/80%.

Results obtained from visuospatial protocols show that RT is shorter for valid compared to neutral cues, and shorter for neutral than for invalid cues[35]. These measurements make it possible to calculate an attention "benefit" by subtracting Neutral RTs minus Valid RTs and an attention "cost" by subtracting Invalid RTs minus Neutral RTs. Subtracting Invalid RTs minus Valid RTs bypasses bias associated with neutral cues[36], and allows a more global measurement, called the RT difference score[37,38].

The RT difference score is generally thought to be a relatively pure measure of attention. However, RT difference score may be influenced by eye movements, which we controlled by using a SOA of less than 200 ms (less than the saccadic eye movement latency). An auditory version of the orientation task can also be used to control this potential bias.

Exogenous attention defined as automatic, involuntary and unaffected by memory load, was studied with 1) a peripheral cue, 2) a brief SOA (< 200 ms), and 3) an equal proportion of valid and invalid cues (50%/50%). Endogenous executive attention characterised as voluntary, controlled, effortful and affected by memory load was studied with 1) a central cue which permits the subject to decode the symbol presented in the middle of the screen, 2) a long SOA> 200 ms, which allows strategic display of attention, and 3) a higher proportional frequency of valid to invalid cues (80%/20%), which allows the subject to display anticipatory attention.

2.2.2 Audiovisual congruence

To evaluate endogenous attention, an additional procedure was used. Subjects were asked to watch a stimuli on a screen and heard the stimuli through earphones. Targets were designed to combine both auditory and visual modalities in "congruent" and "incongruent" situations. These compound stimuli consisted of everyday objects or actions (a dog barking, a drum roll, a liquid being poured into a glass, etc.). Two such compound stimuli were presented simultaneously. In congruent situations, corresponding auditory and visual stimuli were presented on the same side (e.g. an image of a dog barking to the left, and the sound of the barking in the left ear, together with the image of a drum to the right and the sound of the drum roll in the right ear). In incongruent situations, corresponding stimuli are dissociated (image of the dog to the left, and the sound of barking in the right ear, image of the drum to the right, and the sound of drum in the left ear). The preparatory cue in such tests, occurring 350 ms before the target, was a pointer either to right or left, or in neutral condition favouring neither side. These pointers could be a pair of eyes looking right, left or straight ahead, or a dog "pricking up its ears" to right or left, or with lowered ears. The subject's task was to identify as quickly as possible the side of one of the target stimuli. Thus, prior to each trial, and throughout the trial, there would be a question on-screen such as: "On which side do you see the dog?" or "In which ear do you hear the drum roll?". Given the configuration of question (cue and target), the cue could be a valid one or an invalid (misleading) one. When the cue is valid, the target is presented on the cued side of the target presentation. When it is invalid, the target is presented on the opposite side. All four combinations of congruent/incongruent and valid/invalid were assessed. Congruent and incongruent conditions occurred randomly, each in 50% of the trials. Valid and invalid trials occurred randomly in 75% and 25% of the trials respectively. The procedure has been described in detail by Camus and Gely-Nargeot[24].

2.2.3 Statistical methods

Dependent variables were the recorded RTs[1] and RT difference scores[37]. To reduce the effect of extreme values, we used the median RT per experimental condition (usual procedure with this type of data). We calculated the median RT difference score, defined as the difference between median Invalid RT minus median Valid RT for each experimental condition. We also calculated an Index by dividing the RT difference score by the overall RT mean for matching valid and invalid trials. This adjustment, recommended by Faust and Balota[1], consists of calculating the Proportional Cue Effect expressed as the proportion of Invalid RT minus Valid RT divided by the overall mean RT.

Statistical significance was defined at an alpha level less than 0.05 and all tests were two-tailed tests.

Results were submitted to factorial analysis of variance (ANOVA) using age groups, cue validity (valid versus invalid RT), congruence (congruent versus incongruent RT) and modality (auditory versus visual) as factors. For each analysis of variance (ANOVA) the following orthogonal comparisons were made: 1) main effect of age, 2) main effect of cue validity, 3) main effect of congruence, 4) main effect of modality, and 5) their interactions. As a test of statistical significance, each comparison was compared with the error mean square by means of an F test. An F test was made for individual "a priori" comparisons. In addition, to evaluate the differences between HD, presymptomatic HD and controls, independent Student t-tests were performed to compare the means between groups.

3. Results

3.1 HD patients

ANOVA analyses of RTs for endogenous attention revealed a statistically significant main effect of group ($F=19.61$, $df=2,34$, $p<0.001$), a significant main effect of cue validity ($F=18.34$, $df=1,34$, $p<0.001$) and a significant main congruence effect ($F=11.89$, $df=1,34$, $p<0.01$). The interaction of group and cue validity was significant ($F=6.35$, $df=2,34$, $p<0.01$) due to the greater effects of cue validity in HD and presymptomatic HD patients compared to controls. However, we did not observe significant interactions between group and congruence ($F=2.27$, $df=2,34$, $p=0.49$) and cue validity and congruence ($F=0.73$, $df=1,34$, $p=0.39$).

Analyses of RT difference score revealed that the RT difference score of HD patients (mean=810 ms [SD=863]) were significantly ($t=3.54$, $df=26$, $p<0.002$) different from those of the control group (mean=93 ms [SD=92]). Presymptomatic HD RT difference score (mean=339 ms [SD=397]) did not differ significantly ($t=-1.34$, $df=15$, $p=0.20$) from HD RT difference score (mean=810 ms [SD=863]), but were also significantly ($t=2.54$, $df=23$, $p<0.02$) different from controls (mean=93 ms [SD=92]) (**Figure 1**).

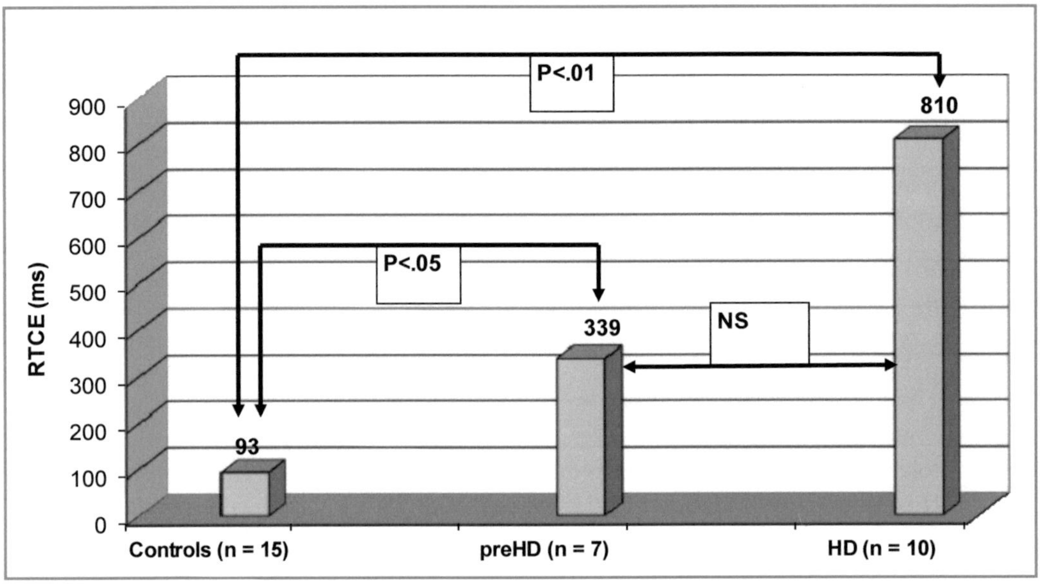

Fig. 1. Mean RTCE (Valid minus Invalid cue RT) in Huntington's Disease (HD), presymptomatic HD patients and matched healthy controls in the Audiovisual cued target condition.

4. Comment

4.1 Attention abnormalities in presymptomatic HD and HD

We found attention deficits in the endogenous component of attention in patients with HD. While patients with HD did not exhibit attention deficits in exogenous attention, they showed a significant increase in RT difference scores (RT difference between invalid and valid cues) compared to controls. These results for endogenous attention are similar to those obtained in normal elderly subjects. However, the results adjusted by Proportional Cue Effect of mean RT are no longer abnormal in the normal elderly group, while they remain abnormal in patients with HD who have a proportional cue effect of 0.37 compared to controls with a proportional cue effect of 0.05. This suggests that abnormalities observed in normal aging are limited to selective situations, whereas those observed in HD are the manifestations of a pathological deficit. HD patients have shown deficits in engaging attention[22] and our findings in HD patients are in agreement with these results[22].

Attention impairment has been consistently reported in patients with HD[39]. However, it is unclear if HD attention deficits are independent or associated with other memory deficits (episodic memory), or language difficulties (verbal fluency). Our study showed deficits in both patients with HD and presymptomatic HD, thus their attention disorders would not depend on the progression of dementia. CAG trinucleotides exceeding the threshold of 37 repeats express abnormally high number of glutamates in the huntingtin protein[33] and are present in both HD and presymptomatic HD patients; which can explain our similar results obtained with both HD and presymptomatic HD patients.

The most common forms of HD are characterized by the appearance of involuntary choreic movements in sub-cortico-frontal dementia. Many of the prominent hyperkinetic symptoms of early HD can be understood as an inability to suppress dominant response tendencies such as ballism and coprolalia. Patients with HD are impaired on the antisaccade paradigm which tests subjects' ability to suppress automatic saccadic eye movement. One theory to explain attention disorders in HD proposes that the caudate-putamen degeneration impairs functioning of the fronto-subcortical loops, resulting in disruption of the anterior attention network. Within the framework of Posner's model, there would be a dissociation between preserved exogenous and impaired endogenous attention components.

In summary, normal aging was characterized by impaired attention in situations of endogenous or voluntary attention, particularly in perceptual conflict situations. However, this impairment is no longer significant when the data are proportionally cue adjusted. In presymptomatic HD and HD, the endogenous component was also found to be impaired in situations of perceptual conflict, but remains markedly impaired after the data are proportionally cue adjusted.

Theories proposing that attention deficits are the result of a slowing of cognitive processes or a decrease of attentional resources are not supported by our findings. In the present study, cue effects were estimated by calculating RT differences between invalid and valid cues, thus correcting for for the confounding effects of motor-sensory and other non-attention components on RT values. During audio-visual congruent situations, endogenous voluntary attention was preserved in normal aging as well as in HD, and attention abnormalities appeared in situations of perceptual conflict, indicating that some endogenous components were preserved while others were impaired. The theory of attention deficit through depletion of attention resources does not allow for these distinctions to be made.

The proposed model of attention includes components that might respond differently to the effects of age and to cortical/sub cortical neurodegeneration. Such a model has the following advantages: 1) defining cognitive attention by updating former approaches; 2) providing a methodology to investigate and to measure attention abnormalities; and 3) permitting integration of data obtained from brain neuroimaging and cognitive neurosciences. The model proposed allows the separation of attention pathologies and the distinction of pathological categories, thus improving clinical evaluations of the patients.

5. References

[1] Faust ME, Balota DA. Inhibition of return and visuospatial attention in healthy older adults and individuals with dementia of the Alzheimer type. Neuropsychology 1997;11:13-29.

[2] Festa-Martino E, Ott BR, Heindel WC. Interactions between phasic alerting and spatial orienting: effects of normal aging and Alzheimer's disease. Neuropsychology 2004;18:258-68.

[3] Greenwood PM, Parasuraman R, Alexander GE. Controlling the focus of spatial attention during visual search: effects of advanced aging and Alzheimer disease. Neuropsychology 1997;11:3-12.

[4] Langley LK, Overmier JB, Knopman DS, Prod'Homme MM. Inhibition and habituation: preserved mechanisms of attentional selection in aging and Alzheimer's disease. Neuropsychology 1998;12:353-66.
[5] Camus J: Neuropsychologie de l'attention. L'apport des réseaux attentionnels neurocérébraux. Revue de Neuropsychologie 1998;8:25-51.
[6] Posner MI, Cohen Y. Components of visual Orienting. In Attention & Performance, Vol X. Edited by Bouma H, Bowhuis D. Hillsdale, NJ: Erlbaum, 1984,pp 551-556.
[7] Posner MI, Snyder C. Attention and Cognitive Control. In Information, processing and cognition: The Loyola Symposium.Edited by Solso R. Hillsdale, NJ: Erlbaum, 1975,pp 55-85.
[8] Driver J, Baylis G. Attention and visual object segmentation. In The Attentive Brain. Edited by Parasuraman R. Cambridge, MA: MIT Press, 1998,pp 299-325.
[9] Humphreys GW, Riddoch M. Attention to within object and between object spatial representations: multiple sites for visual selection. Cognitive Neuropsychology 1994;11:207-241.
[10] Rabbitt P. Crystal Quest: a search for the basis of maintenance of practiced skills into old age. In Attention: selection, awareness and control. Edited by Baddeley A, Weiskrantz L. Oxford: Clarendon Press, 1993.
[11] Stankov L. Aging, attention, and intelligence. Psychol Aging 1988;3:59-74.
[12] Van Zomeren A, Brouwer W. Assessment of Attention. In A Handbook of neuropsychological assessment. Edited by Crawford J, Parker D, McKinley W. Hove, UK, Erlbaum, 1994,pp 241-266.
[13] Mishkin M, Ungerleider L, Macko K. Object vision and spatial vision: Two cortical pathways. Trends in Neurosciences 1983;6:414-417.
[14] Milner AD, Goodale MA. Visual pathways to perception and action. Prog Brain Res 1993;95:317-37.
[15] Koch C, Ullman S. Shifts in selective visual attention: towards the underlying neural circuitry. Hum Neurobiol 1985;4:219-27.
[16] Hamker FH. A dynamic model of how feature cues guide spatial attention. Vision Res 2004;44:501-21.
[17] Treisman A, Sato S. Conjunction search revisited. J Exp Psychol Hum Percept Perform 1990;16:459-78.
[18] Wolfe JM, Cave KR, Franzel SL. Guided search: an alternative to the feature integration model for visual search. J Exp Psychol Hum Percept Perform 1989;15:419-33.
[19] Posner MI, Petersen SE. The attention system of the human brain. Annu Rev Neurosci 1990;13:25-42.
[20] Filoteo J, Delis D, Demadura T, Salmon D, Roman M, Shults C. Abnormally rapid disengagement of covert attention to global and local stimulus levels may underlie the visual-perceptual impairment in patients with Parkinson's disease. Neuropsychology 1994;8:218-226.
[21] Filoteo JV, Delis DC, Massman PJ, Demadura T, Butters N, Salmon DP. Directed and divided attention in Alzheimer's disease: impairment in shifting of attention to global and local stimuli. J Clin Exp Neuropsychol 1992;14:871-83.

[22] Filoteo JV, Delis DC, Roman MJ, Demadura T, Ford E, Butters N, Salmon DP, Paulsen J, Shults CW, Swenson M, Swerdlow N. Visual attention and perception in patients with Huntington's disease: comparisons with other subcortical and cortical dementias. J Clin Exp Neuropsychol 1995;17:654-67.
[23] Parasuraman R, Greenwood P. Selective attention in aging and dementia. In The Attentive Brain. Edited by Parasuraman R. Cambridge, MA, MIT Press, 1998,pp 461-488.
[24] Camus J, Gely-Nargeot M. Existe-t-il un vieillissement de l'attention? In Le vieillissement cognitif normal: vers un modèle explicatif du vieillissement. Edited by Brouillet D, Syssau A. Bruxelles, DeBoeck Université, 2000,pp 53-74.
[25] Greenwood P, Parasuraman R. Attention disengagement deficit in nondemented elderly over 75 years of age. Aging and Cognition 1994;1:188-202.
[26] Buck BH, Black SE, Behrmann M, Caldwell C, Bronskill MJ. Spatial- and object-based attentional deficits in Alzheimer's disease. Relationship to HMPAO-SPECT measures of parietal perfusion. Brain 1997;120 (Pt 7):1229-44.
[27] Eriksen C. The flanker's task and response competition: A useful tool for investigating a variety of cognitive problems. Visual Cognition 1995;2:101-118.
[28] Fan J, McCandliss BD, Sommer T, Raz A, Posner MI. Testing the efficiency and independence of attentional networks. J Cogn Neurosci 2002;14:340-7.
[29] Fan J, Flombaum JI, McCandliss BD, Thomas KM, Posner MI. Cognitive and brain consequences of conflict. Neuroimage 2003;18:42-57.
[30] Posner MI, Raichle M. Images of Mind. New York: Freeman and Company 1994,pp 1-257.
[31] Balkenius C. Attention, Habituation and Conditioning: Toward a Computational Model. Cognitive Science Quarterly 2000;1:1-29.
[32] Gusella JF, Wexler NS, Conneally PM, Naylor SL, Anderson MA, Tanzi RE, Watkins PC, Ottina K, Wallace MR, Sakaguchi AY,Young AB, Shoulson I, Bonilla E, Martin JB. A polymorphic DNA marker genetically linked to Huntington's disease. Nature 1983;306:234-8.
[33] Riess O, Noerremoelle A, Soerensen SA, Epplen JT. Improved PCR conditions for the stretch of (CAG)n repeats causing Huntington's disease. Hum Mol Genet 1993;2:1523.
[34] Posner MI. Orienting of attention. Q J Exp Psychol 1980;32:3-25.
[35] Parasuraman R, Greenwood PM, Haxby JV, Grady CL. Visuospatial attention in dementia of the Alzheimer type. Brain 1992;115 (Pt 3):711-33.
[36] Jonides J, Mack R. On the cost and benefit of cost and benefit. Psychological Bulletin 1984;96:29-44.
[37] Chapman LJ, Chapman JP, Curran T, Miller MB. Do children and the elderly show heightened semantic priming ? How to answer the question. Developmental Review 1994;14:159-185.
[38] Curran T, Hills A, Patterson MB, Strauss ME. Effects of aging on visuospatial attention:an ERP study. Neuropsychologia 2001;39:288-301.

[39] Lawrence AD, Hodges JR, Rosser AE, Kershaw A, French-Constant C, Rubinsztein DC, Robbins TW, Sahakian BJ. Evidence for specific cognitive deficits in preclinical Huntington's disease. Brain 1998;121 (Pt 7):1329-41.

10

Computational Investigations of Cognitive Impairment in Huntington's Disease

Eddy J. Davelaar
Birkbeck College, University of London
United Kingdom

1. Introduction

Huntington's Disease (HD) is a genetic disorder involving progressive loss of the neostriatal cells. The most prominent symptom in early clinical HD is chorea, which is assumed to reflect a dysfunctional error-feedback system. Cognitive impairments are also observable in early stages of the disease in mild form at first, but more severe in later stages. Cognitive decline is present years before the clinical manifestation of HD. In a recent report, the term "mild cognitive impairment" in preclinical HD (pHD) was advocated with an amnestic and nonamnestic variant (Duff, et al., 2010). Considerable effort is invested in identifying which combinations of neuropsychological tests have high clinical utility.

Currently, there is no treatment for HD and much research addresses possible interventions, (see for extensive review, Zuccato, et al., 2010). In this chapter, the focus is on the profile of cognitive impairment during the period before clinical symptoms are present and is investigated via computational methods. A computational model of the basal ganglia is used to simulate cognitive performance. The cognitive tasks that have been shown to be sensitive to HD pathology cover such domains as memory (e.g., verbal learning), executive functioning (e.g., random sequence generation), and attention (e.g., flanker task). The HD pathology is simulated in the model and the pattern of cognitive decline is compared with published reports, when present.

This chapter is structured as follows. Section 2 reviews the functional neuroanatomy of the basal ganglia and summarizes the status of computational modeling of the basal ganglia. In section 3, the neuropathology and cognitive impairments in HD are reviewed and a new classification scheme of cognitive tasks is presented. To preview the scheme, tasks cluster in accordance with their reliance on the internal dynamics of the basal ganglia. A computational study will be presented in section 4 that addresses compensatory mechanisms in HD as well as specific deficits that differentiate early pHD from late pHD and amnestic pHD from nonamnestic pHD. The model sheds light on why certain tasks detect deficits at clinical stages (memory span), while other tasks detects deficits both at preclinical and clinical stages (episodic memory). The chapter concludes with a reflection on the utility of computational models in HD research.

2. The basal ganglia

The basal ganglia are a group of functionally related subcortical nuclei that have predominantly been described as being involved in movement control. In the last decades, it has been shown that the basal ganglia are involved in many cognitive domains, such as learning, memory, and planning. The functionality of the basal ganglia can be understood by the interconnections among the nuclei and the various neurotransmitters used by the structures (for a recent update see, Delong & Wichmann, 2009). Computational studies at various levels of biological realism have enhanced our understanding of the complex neurodynamics of the basal ganglia and its role in cognitive performance.

2.1 Neuroanatomy of the basal ganglia

The basal ganglia consist of the putamen, the caudate nucleus, the globus pallidus, the substantia nigra, and the subthalamic nucleus (see Fig. 1). The caudate nucleus and the putamen constitute the striatum (STR) and receive cortical input. The striatum projects to the globus pallidus and substantia nigra. The globus pallidus is divided into the internal (GPi) and external (GPe) segments. The substantia nigra also contains functionally separate parts: the pars compacta (SNc) and the reticulata (SNr). The GPi and the SNr project to nuclei in the thalamus and form the output of the basal ganglia. The thalamus is reciprocally connected with frontal cortical areas. This cascade of projections, i.e., cortex → BG → Thal → cortex, is referred to as the cortico-basal ganglia loop of which there are several, each with its specific functional role. The loops are thought to implement a selection mechanism through which only the most appropriate actions (or thoughts) are selected (Doya, 2007).

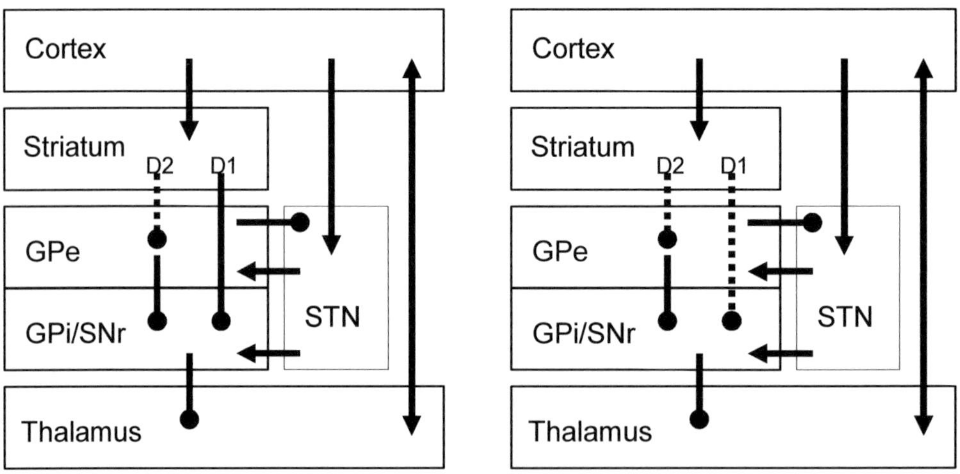

Fig. 1. Architecture of the basal ganglia. Not all connections are presented. Left: simplified situation in pHD with the degraded D2-pathway. Right: simplified situation in HD with both D1 and D2 pathways degraded.

There are several pathways within the basal ganglia in which the cortex projects to the basal ganglia and receive, via the thalamus, the output of the basal ganglia (see Fig. 1). In the direct pathway, projections go from the cortex to STR, then to GPi/SNr and then the thalamus. In the indirect pathway, projections go from the cortex to STR, then to GPe, then to the subthalamic nucleus (STN), then to GPi/SNr and then to the thalamus. A third pathway called the hyperdirect pathway (Nambu, et al., 2000, 2002) involves direct projections from the neocortex to the STN, which then influences the activation in the GPi/SNr.

The projections from the cortex to the striatum are topographically organized both in the anterior-posterior and in the lateral-medial directions (e.g., Nambu, 2011). This even extends to the level of specific body parts, such as the face, arm, and leg (Alexander & Crutcher, 1990). Although this finding is consistent with a view that the various cortico-basal ganglia loops are separate parallel circuits, there is evidence showing that the loops share information (Graybiel, 1995) and that the amount of information sharing is modulated by dopamine (Bergman, et al., 1998).

The neurochemistry of the basal ganglia is complex with inhibitory and excitatory neurotransmitters and receptors distributed in a precise architecture. The striatum consists mainly of medium spiny neurons (MSN) that are predominantly GABA-ergic and contain either dopamine D1- or D2-receptor. MSNs with both D1- and D2-receptors have also been found (see for a review, Perreault, et al., 2011). The neural space of the striatum is made up of a large area called the matrix that is rich in acetylcholinesterase and smaller islands of acetylcholinesterase-poor neurons called striosomes (Cichetti, et al., 2000). Striosomal MSNs have a different input/output connectivity and receptor expression than matrix MSNs (Eblen & Graybiel, 1995; Joyce, et al., 1986; Lévesque & Parent, 2005), which has been interpreted to show a functional differentiation between the two compartments (see for review, Cichetti, et al., 2000).

The prevailing view is that the MSNs in the striosomes receive input from parts of the limbic system and project directly or indirectly to dopaminergic cells in the SNc (but see, Lévesque & Parent, 2005). The MSNs in the matrix that contain D1-receptors project to the GPe/SNr using the neurotransmitters GABA and substance P and form the striatal-pallidal leg of the direct pathway. The GPe/SNr-neurons inhibit the thalamus and are tonically active. This makes the direct pathway one that disinhibits the reciprocal cortico-thalamic loop. It is important to stress that disinhibition is different from excitation in that excitation leads to thalamic activation in the absence of cortico-thalamic input, whereas disinhibition essentially lowers the threshold to allow already present cortico-thalamic input to activate the thalamic neurons (Chevalier & Deniau, 1990). The MSNs in the matrix that express D2-receptors contain GABA and enkephalin and project to the GPi, forming the striatal-pallidal leg of the indirect pathway. The GPe inhibits the STN, which has glutamatergic projections to GPe and GPi/SNr. The indirect and hyperdirect pathways thus increase the GPi/SNr activity, leading to more inhibition of the thalamus. Functionally, these pathways are preventing thalamic activation. The nigrostriatal dopaminergic projections increase the firing rate of MSNs of the direct pathway that contain D1-receptors and decrease the firing rate of MSNs of the indirect pathway that contain D2-receptors.

The striatal neurons are surrounded by various tonically active interneurons. The afferent and efferent connections of the interneurons respect the striosomal-matrix boundaries, apart from those of the cholinergic interneurons, which are located around the striosomal-matrix boundaries and are thought to be critical in allowing crosstalk between MSNs in both striatal areas (Cichetti, et al., 2000).

The number of participating neurons decreases from cortex to striatum to globus pallidus. This convergence was interpreted as evidence that the basal ganglia play a role in evaluating contextual information for generating appropriate motor responses (Graybiel, 1991; Houk & Wise, 1995; Joel & Weiner, 1994). A complementary view is that the convergence of information together with the inhibitory interconnections is the neurobiological equivalent of dimensionality reduction through principal component analysis (Bar-Gad, et al., 2003). This process allows the cortex to focus on and learn the underlying statistical structure of a large amount of activity patterns.

2.2 Computational models of the basal ganglia

The connections among the basal ganglia structures form complex feedback loops that modulate each other at different time-scales. This continuous interaction makes it very hard to intuit how a single manipulation will affect the cortico-subcortical dynamics. This is even further complicated by the different timecourses of neurodegeneration of the two pathways in HD. Building and testing computational models will help in understanding the functional and impaired dynamics of the basal ganglia.

Many detailed computational models of the basal ganglia exist (for reviews on cognitive models see e.g., Bullock, 2004; Bullock, et al., 2009; Cohen & Frank, 2009). Most models are concerned with the motor deficits seen in diseases such as Parkinson's Disease, but some have been developed to further understand the functional roles of the connections among subcortical nuclei (e.g., Gurney, et al., 2001) or the roles of the entire collection of basal ganglia in cognitive performance.

Computational models that investigate the functional roles of the basal ganglia vary in the level of biological realism of the neurons and in the level of detail regarding the interconnections among the basal ganglia. Models may employ (1) simplified rate-coding neurons that are used as a proxy for groups of individual firing neurons (e.g., Berns & Sejnowski, 1998; Frank & Claus, 2006; Frank, et al., 2001; Gurney, et al., 2001; Monchi, et al., 2000) or (2) spiking neurons with various levels of intracellular detail (e.g., J. Brown, et al., 1999; Guthrie, et al., 2009; Humphries, et al., 2006).

Many of these models address the role of the basal ganglia in specific cognitive domains, such as working memory (e.g., Berns & Sejnowski, 1998; Frank, et al., 2001; Monchi, et al., 2000), decision making and action selection (e.g., Frank & Claus, 2006; Gurney, et al., 2001; Guthrie, et al., 2009), and reinforcement learning (e.g., J. Brown, et al., 1999; Frank & Claus, 2006). Although there is a bias towards addressing dopaminergic influences, which explains the large volume of computational models that address cognitive performance in Parkinson's Disease and schizophrenia, associative learning of cortico-striatal connections will not be addressed in this chapter. An extensive literature exists on the role of dopamine in reward-based or reinforcement learning and is closely related to the mechanisms of long-term potentiation and long-term depression (for recent review see, Manninen et al., 2010).

The various computational models that deal with cognitive phenomena are based on the original idea of the direct and indirect pathways (Albin, et al., 1989). However, recent modeling work has focused on the role of the STN (Frank, 2006; Gurney, et al., 2001). In particular, the hyperdirect pathway (Nambu, et al., 2000, 2002) has been attributed the function of relaying stopping decisions from the cortex (Frank, 2006). Cortical activation of the STN causes a global increase in GPe activation, which in turn prevents any channel from disinhibiting the thalamus. Gurney et al., like Nambu, et al. (2000, 2002), view the STN as an input structure complementing the striatum. The dual-input design creates an off-center/on-surround in GPe, which leads to the selection of the relevant channel while inhibiting closely related channels. In both scenarios, the STN plays a more central role than in previous models (e.g., Albin, et al., 1989).

3. Huntington's Disease

Huntington's Disease is an autosomal-dominant progressive neurodegenerative disorder that presents with motor disturbances, psychiatric symptoms, and cognitive decline. The onset of clinical symptoms is in middle-age, but the disorder can manifest at any time between infancy and old age. In early HD, hyperkinesia is observed, whereas in later stages, hypokinesia dominates.

3.1 Neuropathology of Huntington's Disease

Huntington's Disease results from a gene mutation on chromosome 4, leading to an increase in CAG repeats. The gene codes for the protein huntingtin (htt) and the mutant variant thus has an extended number of glutamine repeats, varying from 11 to 34 in normal individuals and 36 to 121 units in HD patients (The Huntington's Disease Collaborative Research Group, 1993). The role of huntingtin is yet unclear, but evidence suggests that it is involved in neurodevelopmental processes (e.g., Hebb, et al., 2004). The pathophysiological mechanisms of Huntington's Disease are poorly understood, but research with transgenic animal models of the disorder is providing insights into the causative factors and potential treatments.

3.1.1 Huntington's Disease involves targeted cell death

The disease triggers striatal neurons to go into apoptosis (commit suicide) and the neuropathology spreads in all directions. In the early stage of the disease, there is up to 60% loss of GABA/enkephalin striatopallidal neurons that project to GPe. These neurons are part of the indirect pathway and express D2-receptors. The amount of D2-receptor binding sites correlates with the estimated years of disease onset (Feigin, et al., 2007). In the intermediate stage, up to 50% of GABA/substance P striatonigral neurons that project to SNr are lost. These neurons are part of the direct pathway and express D1-receptors. Finally, at the last stage, there is loss of GABA/substance P striatopallidal neurons that project to GPi that are also part of the D1-receptor expressing direct pathway (Glass, et al., 2000).

Animal studies with knock-in mice have verified the two-stage neurodegenerative process that starts with hyperkinesia and continues with hypokinesia, mirroring the progression of cell death from predominantly D2-receptor MSNs associated with the indirect pathway followed by the D1-receptor MSNs that are associated with the direct pathway (Menalled et al., 2002).

Although some studies report that neurodegeneration starts in the striatal matrix and continues to the striosomal neurons, others have reported the reverse progression (e.g., Hedreen & Folstein, 1995). Tippett et al. (2007) observed that both directions of progression occur and that the heterogenic pattern is associated with the heterogeneity in clinical symptoms. Interestingly, the balance between striosomal and matrix loss was associated with the number of CAG repeats.

3.1.2 The mutant huntingtin protein destabilizes cells

Several studies have shown that the mutant huntingtin protein is neurotoxic. For example, phosphorylation of amino acids on the huntingtin protein reduces its toxicity (Gu, et al., 2007). The precise pathway of toxicity is yet unknown, but evidence suggests that the mutant huntingtin protein interacts with DNA transcription factors, such as CREB (c-AMP response element-binding protein) and dysregulates DNA transcription processes. In particular, there is evidence showing that the mutant huntingtin interferes selectively with transcription factors in the medium spiny neurons (Gomez, et al., 2006). In transgenic mice, Hebb et al. (2004) observed decreased levels of PDE10A mRNA before motor symptoms appear and Gomez et al. (2006) observed decreased levels of DARPP-32 mRNA in MSNs, but not in other DARPP-32 mRNA expressing tissue, such as the kidneys.

Several interacting intracellular pathways have been identified that are negatively affected by the mutant huntingtin protein (see for reviews, Cha, 2007; Luthi-Carter & Cha, 2003). Apart from dysregulation of DNA transcription, mutant huntingtin has also been found to be associated with abnormal protein-protein interactions and with energy dysregulation in the mitochondria. The affected cells eventually enter a cascade resulting in apoptosis. It has been found that BDNF, which is needed for expression of DARPP32 (Ivkovic & Ehrlich, 1999), and other neurotrophins are not only important in regulating the phenotype of striatal projection neurons (see for reviews, Pérez-Navarro, et al., 2000; Zuccato & Cattaneo, 2007; Zuccato, et al., 2010), but also protected striatal neurons from the accelerated degeneration induced by huntingtin (Pérez-Navarro, et al., 2000).

3.1.3 Compensatory mechanisms

Substantial striatal atrophy precedes clinical motor symptoms (Aylward, 2006). In addition, cortical white matter decreases as the estimated years to disease onset decreases (Paulsen, et al., 2006; Stoffers, et al., 2010). However, cortical gray matter is above normal in pHD with large estimated years to onset and decreases to below normal the nearer the estimated onset time (Paulsen, et al., 2006; also observable in Stoffers et al., 2010). This increased gray matter volume is consistent with the increased functional activation in medial frontal areas (Beste, et al., 2007; Paulsen, et al., 2004), which is interpreted as reflecting a compensatory neurodevelopmental mechanism.

An important aspect to take into consideration with Huntington's Disease is that the individual has the CAG repeats from birth, but that the clinical symptoms are manifest at a much later stage in life. It is not inconceivable that during child development, the underresponsive indirect pathway is compensated for by the hyperdirect pathway. Using the hyperdirect pathway as a compensatory mechanism will affect cognitive tasks that depend on stopping an ongoing response or cognitive operation. This will be further addressed in sections 3.2.1 and 4.2.

3.1.4 Predicting disease onset

Predicting the onset of HD is of great importance to the affected individuals and their families. As discussed by Aylward (2006), although detection of the mutated gene can be done at any time, treatment will not begin until the very earliest signs of change. The literature is somewhat biased with regard to what constitutes an early sign. Langbehn et al. (2004) obtained clinical data from 2913 individuals from 40 centers involved in the PREDICT-HD study. Using survival analyses which include the number of CAG repeats in symptomatic and presymptomatic individuals they derived a statistical model that predicts the disease onset given the number of CAG repeats and current age of the individual. Critical for conducting such an analysis is the need for an agreed upon endpoint. Langbehn et al. (2004) chose as the end point "the first-time neurological signs representing a permanent change from the normal state" (p. 268). This assumes that physician's accuracy in determining the clinical status of an individual is infallible. As Aylward (2006) discusses, this is seldom the case and much research remains needed to find objective biomarkers that help pinpoint the disease onset (Weir, et al., 2011). One such objective measure is striatal volume loss (Aylward, 2006). The loss of striatal neurons is a gradual process that starts around 10 years before the clinical onset and is associated with cognitive decline. In other words, one can discern a second earlier endpoint beyond which there is an accelerated loss of striatal volume and an increase in cognitive impairment. During the 10 year period, "presymptomatic" individuals are more prone to commit cognitive errors due to difficulty in concentration or inability to operate multiple complex tasks simultaneously. Although the acknowledgement of cognitive decline in presymptomatic individuals has attracted more cross-disciplinary research, it also comes with the fear that the presymptomatic individuals are at risk of committing preventable (and potentially fatal) accidents at the workplace. It is therefore critical to understand exactly what types of cognitive deficits coincide with striatal neuron loss and predict the onset of the classical symptoms by which HD is diagnosed.

There exists a strong regularity among trinucleotide disorders with the age of disease onset being an exponential function of number of trinucleotide repeats (Kaplan, et al., 2007; Walker, et al., 2007). Kaplan et al. (2007) proposed a model based on three assumptions. First, above a lower disease-related threshold, the trinucleotide repeats lead to damage that requires DNA repair processes. Second, these repair processes have a tendency to become error prone with more repeats, leaving the repaired DNA with even longer repeats. This repair-error-repair continues at an increasingly faster rate. Third, when the number of trinucleotide repeats reaches an upper critical threshold in a certain number of cells, the clinical symptoms of the disease become manifest. This simple model abstracts away from the various detailed molecular processes mentioned in the previous section. Nevertheless, it captures the exponential onset curve seen across trinucleotide repeat disorders and provides a statistical account of the correlations between disease progression and disease onset (the earlier the onset the faster the disease progression).

3.2 Cognitive decline in Huntington's Disease

The basal ganglia are involved in various cognitive processes (Aglioti, 1997; Grahn, et al., 2009; Heyder, et al., 2004) and therefore it is expected that neurodegeneration of striatal MSNs will have concomitant effects on cognitive processing, even in the absence of motor deficits. There is an abundance of studies investigating the cognitive decline in pHD and early stage HD employing various tasks and methods (e.g., Ho, et al., 2002, 2003; Lawrence, et al., 1996; Montoya, et al., 2008; Peretti, et al., 2008; Solomon, et al., 2007; Stout, et al., 2011).

Understanding the pattern of cognitive decline beyond the standard neuropsychological tests is vital. For example, many test measures do not differentiate between patient populations and most measures do not address compensatory mechanisms. A clinical solution is to use batteries of tests and combine the scores to improve overall diagnostic utility. A complementary solution, advocated here, is to combine tests that are sensitive to the specific cognitive mechanism(s) that is/are affected in the disorder. To do this, a very detailed understanding of the tests is required. For several tasks this is indeed the situation and much of the research in mathematical psychology and cognitive science is dedicated to understanding the cognitive operations that take place in a particular task. The measures that are accounted for are the usual primary measures, such as accuracy, response times, and higher moments of response time distributions, but also secondary measures, such as speed-accuracy tradeoff functions, conditional probabilities, and clustering. Some of those tasks have also been used extensively in conjunction with brain imaging and electrophysiological recordings, enriching our understanding of which brain structures process the information and when. Together with neuropsychological investigations of these tasks in HD and pHD, a detailed picture emerges of the progression of cognitive decline that could be combined with objective biomarkers to not only answer the question "When will I get HD?", but also answer the question "Where am I in my neurocognitive journey?"

Space does not allow a thorough overview of the various tests used HD research and the reader is invited to consult the relevant articles for tests that are not discussed in the following sections and for further information about the tests that will be discussed.

3.2.1 Flanker task: Evidence for compensatory mechanisms

In this task, participants are required to respond to a target character that is presented on a computer screen. For example, if the target character is a right- or left-pointing arrow, the right of left response key needs to be depressed, respectively. The target character is flanked by distractor arrows on either side that are either pointing in the same (congruent condition) or opposite (incongruent condition) direction as the target arrow. The response time in the incongruent condition is slower than in the congruent condition, which is explained by the interference caused by the distractors. Huntington's patients are generally slower in this task, but do not show an abnormal interference effect (Beste, et al., 2008).

Apart from behavioral measures, Beste et al. (2006, 2007, 2008) recorded electrophysiological responses and observed that compared to controls, HD showed reduced stimulus-related potentials (i.e., N1), slowed motor-related components (i.e., lateralized readiness potentials), and reduced error-related processing (i.e., Ne/ERN). Interestingly, the CAG-index of HD patients correlated with the size of the Ne/ERN response (Beste, et al., 2006). Compared to HD, pHD showed stronger brain responses, but not more than controls, apart from error-related processing (Beste, et al., 2007, 2008). It has been suggested that pHD may exhibit compensatory activation in other brain regions (Feigin, et al., 2006; Paulsen, et al., 2004). The enhanced Ne/ERN in pHD might indicate an upregulated hyperdirect pathway.

3.2.2 Memory: Seeing the trees through the forest

There are many processes involved in memory and it is still uncertain which processes are differentially affected in HD. There is a myriad of theories about the role of the basal ganglia in learning and memory. In the interest of brevity, the reader is referred to the review articles that address research in animals and computational studies on long-term potentiation, depression and reinforcement learning (e.g., Da Cunha, et al., 2009; White, 1997). In a nutshell, the basal ganglia build stimulus-response mappings that provide the largest amount of reward. These types of learning can be considered procedural to the extent that no deliberate act of learning takes place other than valuating the stimulus-response contingency. Here, the role of the basal ganglia in deliberate memory encoding and retrieval will be addressed.

With regard to memory encoding, the basal ganglia have been compared to a gatekeeper that allows entry of memoranda into working memory (McNab & Klingberg, 2008). This gateway function is implemented by the disinhibition of the cortico-thalamic loop. Davelaar et al. (2005) showed how this gateway mechanism influences working memory updating and it is expected that this function is impaired in HD. At memory retrieval, Rohrer et al. (1999) showed that HD exhibit slow episodic retrieval, which in turn would lead to lowered total recall under speeded conditions. In a simulation model, Davelaar (2007) showed how degradation of the same gatekeeper pathway, the direct pathway, captures the slowed, but accurate memory performance.

3.2.3 Random number generation

For a task to be clinically useful, performance on the task should be stable and vary only in relation to the disease progression. That means that the ideal task does not show learning effects and is novel every time it is administered. Of course, from a clinician's point of view a task that requires electrophysiological recording or analyses of response time distributions is far from ideal. A task that requires two minutes and ticks the boxes is the random number generation task. In this task, the participant is required to produce one hundred digits from one to ten in a random fashion at a rate of one digit per second. This is relatively fast, but faster and slower rates have been used in the cognitive literature to understand the influence of production rate on executive functioning (Jahanshahi, et al., 2006; Towse, 1998). People are seldom truly random and the deviation from randomness can be captured with a variety of measures (Jahanshahi, et al., 2006; Towse & Neil, 1998). People have a tendency to produce a sequence of digits that exhibits single-digit increments or decrements (e.g., "1, 2, 3", "7, 6, 5"). This tendency is captured quantitatively in the adjacency score and is calculated as the number of adjacent pairs divided by the total number of response pairs (see Towse & Neil, 1998). In HD, counting in ones shows the greatest between group differences, with HD patients differing from pHD and controls, while there is no difference between the last two groups (Ho, et al., 2004).

3.3 A classification scheme for tasks used in HD research

Most of the cognitive deficits can be understood as a general inability to engage in or disengage from cognitive processing or where processing does occur it is slowed down. Based on considering a number of tasks and the results of simulations, the following classification scheme is proposed in which tasks are grouped in accordance with the required basal ganglia pathways.

Extra-basal ganglia (EBG) tasks - these are tasks that do not require the basal ganglia for much of the processing, other than as an output system. These are tasks such as simple and choice response time and are likely to be affected in later stages of HD. Despite the central role of working memory updating, it is still a form of output, albeit at the cognitive instead of motor level. Therefore, a task that only requires the gatekeeper pathway is an EBG task.

Intra-basal ganglia (IBG) tasks - these are tasks that are sensitive to the release of a recently chosen channel. Tasks that require a disengagement of selected information are random number generation and the use of retrieval cues in episodic memory tasks.

The direction of neuropathology, from striosomal to matrix or the reverse, was shown to be correlated with mood changes (Tippett, et al., 2007). It is assumed that the dysfunctional striosomal STR-SNc loop underlies this pattern and involves the reinforcement learning system. Thus depending on the precise pathology, category learning may be affected in pHD[1]. As responses in category learning end the respective trial, category learning tasks are also EBG tasks. This adds another dimension, namely the presence/absence of the necessity of reinforcement learning that can be crossed with the extra/intra-BG dimension, producing four types of tasks: EBG^-, EBG^+, IBG^-, and IBG^+. It should be stressed that classification of a task as an EBG^- task, does not mean that no learning takes place. In fact, the flanker task would be classified as an EBG^- task, but the observed sequential effects are signatures of associative learning (Davelaar & Stevens, 2009). Therefore, great care should be taken to refer to the precise dependent measure that is being used. No example, of an IBG^+ task is given here, as no such task has been employed in HD. The simulations in section 4 address only EBG^- and IBG^- tasks (and task measures).

4. Modeling Huntington's Disease

In this section, results of simulation studies are presented that address cognitive decline before the clinical onset of HD. Details of most the models are available in the relevant publications, but critical features of existing models or novel details are presented where possible.

4.1 Differential consequences of loss of D1- or D2-pathway

Before presenting simulations of cognitive tasks, the influence of the direct and indirect pathways on the thalamic output needs to be clarified, as results are crucial in understanding the successes and failures of different tasks and task measures in detecting cognitive decline in pHD. Consider the simplified network in Fig. 2. When a single signal is presented to the network (see Fig. 3), the direct pathway releases the thalamic inhibition, whereas the indirect pathway stops the release. This results in a pulse as the thalamic output. When the integrity of the D2-pathway decreases, as in pHD, the thalamus remains disinhibited for a longer time. When the D1-pathway is compromised, as in HD, the thalamic output is slowed.

[1]Stout et al., (2011) presented results showing no impairment on category learning in pHD. However, as task performance was not correlated with neuroanatomical data, the hypothesis of impaired reinforcement learning remains to be tested.

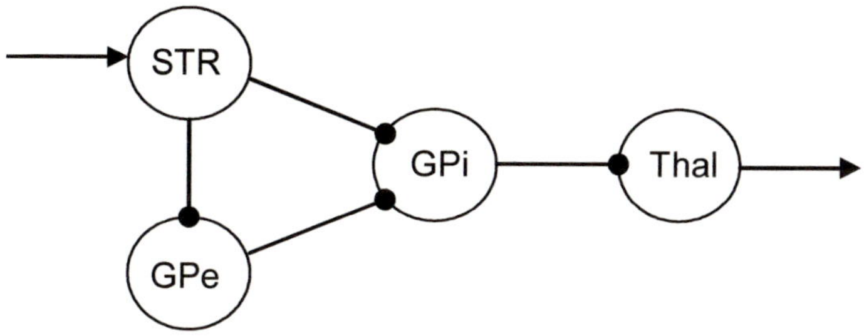

Fig. 2. Simplified basal ganglia network to address the consequences of loss of the D1- and D2-pathways.

When a sequence of signals is presented to the network, individual pulses can be discerned in the thalamic output. D2-pathway loss leads to interference in the ability to release the chosen channel, as seen in the middle column of Fig. 3. Thus, the continued disinhibition of the previous response interferes with the onset of the response to the next stimulus. This can be alleviated by compensation via the hyperdirect pathway. Loss of D1-receptor MSNs results in overall slower response onsets without interference.

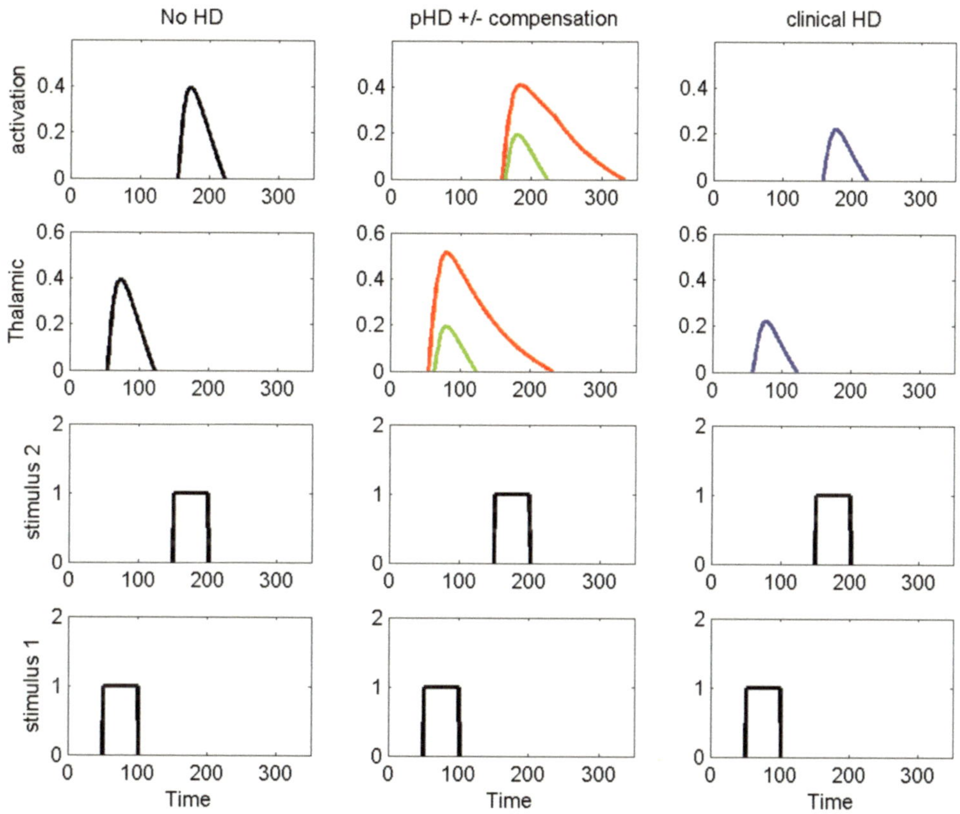

Fig. 3. The effect of loss of the D1- and D2-pathways in HD on thalamic output. First column: control situation. Middle column: pHD without (red lines) and with (green lines) compensation via the hyperdirect pathway. Last column: thalamic output in clinical HD.

4.2 Neural compensation for loss of D2-pathway: Evidence from flanker task

As described in section 3.2.1, in the flanker task, stimuli made up of arrows (e.g., <<<<<, <<><<) need to be categorized based on the central target character. In a series of papers, Beste et al. (2006, 2007, 2008) presented various analyses of conflict processing in presymptomatic and symptomatic HD. Although the usual slow down in response times was observed for HD, this was absent in pHD. The critical data that will be focused on here are the N2 and the Ne/ERN components, both of which have been argued to reflect error-related controlled processing (Yeung & Cohen, 2006). Both ERP components are calculated with respect to the response time, i.e., the individual trials are aligned to the time of response in each individual trial. The N2 is calculated as the difference between the ERP for correct incongruent trials and correct congruent trials[2]. The Ne/ERN is calculated as the difference between the erroneous incongruent trials and the correct incongruent trials.

4.2.1 Method

For this simulation, the model by Davelaar (2008) was used which implements the arrow flanker task. The model was developed to address the role of conflict monitoring in attentional control. In particular, the conflict was assumed to be monitored by medial frontal areas and this conflict influenced the amount of attention paid to filter out the distracting flanker characters. The model provided a good account of complete response time distributions, ERP latencies and fMRI BOLD responses. The details of the model are in Davelaar (2008). Here, one addition was made. The part of the model that represents the medial frontal areas which monitors the conflict signal was modulated by dopamine, while the input gradually decreases. Loss of D2-receptor MSNs is assumed to lead to increased sensitivity of cortical neurons. The resulting activation response functions with the nonmonotonic pattern of compensation are presented in Fig. 4 (left panel). This provides a dynamical control system by which the hyperdirect pathway takes over from the indirect pathway, which is one interpretation of the increased cingulate activation (Paulsen, et al., 2004).

4.2.2 Results and discussion

Table 1 shows the mean response times (with standard deviations in brackets) and accuracy for the two conditions for each simulated group. No effect of group is found on the behavioral measures. The ERP components, however, do differ among the groups. Compared to the no HD group the two pHD groups show larger N2 and Ne/ERN amplitudes, whereas the HD group shows a marked decreased amplitude (see Fig. 4). This pattern mirrors that of Beste et al. (2006, 2007, 2008) for the Ne/ERN and is consistent with the putative compensatory mechanism of the cingulate cortex (Paulsen, et al. 2004).

---

[2]Beste et al. (2008) computed an N2 within incongruent trials. This is a different measure than that used in studies on cognitive control. Nevertheless, the simulation results with the standard N2 calculation is used and can be treated as a prediction by the model.

	No HD	"pHD far"	"pHD near"	"HD onset"
RT - Congruent	193 (16)	192 (18)	193 (17)	192 (17)
RT - Incongruent	267 (26)	167 (24)	266 (24)	267 (25)
Accuracy - Congruent	100%	100%	100%	100%
Accuracy - Incongruent	89%	87%	88%	88%
N2-peak	0.68	1.47	2.15	0.17
Ne/ERN-peak	0.47	0.96	1.15	0.06

Table 1. Results of flanker simulation with compensation via hyperdirect pathway. The stages of HD pathology are tentative, but the sequence follows the progressive loss of the D2-pathway.

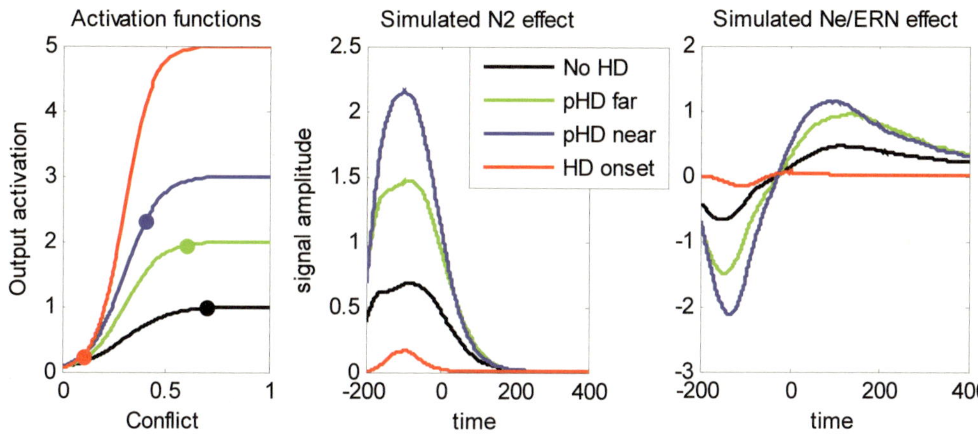

Fig. 4. Simulation results of the flanker task with compensation via the hyperdirect pathway. Left panel: illustration of how increased sensitivity of medial frontal areas together with decreased input leads to a nonmonotonic level of activation. Middle panel: simulation results for the N2-component. Right panel: simulation results for the Ne/ERN-component.

4.3 Working memory capacity: The role of D1-pathway

As mentioned in 3.2.2, the basal ganglia have been attributed a gatekeeper function (McNab & Klingberg, 2008). Davelaar et al. (2005) showed how this gatekeeper function affects updating the contents of working memory. Yet, in neuropsychological studies, the memory span task is used to measure working memory capacity. In the following simulation, it is shown how capacity is affected by the D1-pathway. Note that the D2-pathway will not influence performance in this simulation as the cognitive process stops after an item is gated into working memory.

4.3.1 Method

The working memory model by Davelaar (Davelaar, 2007; Davelaar, et al., 2005) was used to simulate a memory span task. In this task, items are presented sequentially and need to be reported in the correct serial order. The capacity of the model is the maximum number of items that can be presented for which all items are reported in the correct serial order. To implement pHD, a 40% loss of the D2-pathway was chosen, while a further 50% loss of the D1-pathway was chosen to implement HD (cf. Glass, et al., 2000).

4.3.2 Results and discussion

The results are as was expected with simulations of memory spans for controls, pHD, and HD of 3.88, 3.76, and 2.54, respectively. The gatekeeper pathway involves the D1-pathway and thus deficits in memory span are observable after onset of HD.

4.4 Random number generation: Information sharing in D1-pathway

The random number generation task has been shown to be sensitive to basal ganglia pathology and an ideal test for investigating executive functioning in neuropsychological patients. As mentioned in section 3.2.3, HD patients show more stereotyped behavior than pHD and controls, whereas pHD and controls do not differ. The following simulation provides an answer for this difference.

4.4.1 Method

The model is a working version of a network model proposed by R. G. Brown et al. (2000, see also Jahanshahi, et al., 1998) and is based on the memory model of sequential retrieval by Davelaar (2007) that is augmented with a semantic memory system containing the digits in an associative fashion. That is, the digit 3 was associated strongly with the digits 2 and 4 and less strongly with digits 1 and 5. This gradient was used for all digits. The actual simulation uses an abstracted version to speed up the computer time, but the model makes use of two cortico-basal loops (see Fig. 5). The first is the loop that selects an individual digit, while the second loop selects the motor program associated with this digit. From Fig. 5 it can be seen that the associative structure corresponds to each digit being associated with multiple motor programs. Neuropsychological investigations on frontal patients using a speeded naming task are suggestive of such architecture (Thompson-Schill & Botvinick, 2006).

Initially, all digits are activated and when the digit is selected it receives inhibition, leading to a lower likelihood of selecting the same digit in succession. This implements the ubiquitous repetition avoidance (Towse, 1998). The remaining activations may be selected when they have not yet decayed to baseline. The reason for separating the selection of the digit and the selection of the motor program, is that a single loop has been shown to (incorrectly) predict a decrease in stereotyped behavior with increasing production rate (Davelaar, 2004) instead of an increase (Towse, 1998). This process continues until a hundred digits have been reported.

Loss of the indirect pathway was modeled by decreasing the rate at which selection suppression and recovery takes place. Loss of the direct pathway was modeled by increasing the threshold for entry in the set of selection candidates. This loss was implemented in both loops simultaneously.

4.4.2 Results and discussion

Fig. 6 (left panel) shows the distribution of first-order differences at three levels of the HD progression. The distributions show clear repetition avoidance and a tendency to report adjacent digits. Importantly, this counting behavior is larger for the HD simulation than for the pHD and control simulations. The right panel in Fig. 6 shows the adjacency score as standardized score in which a zero score means perfect random generation (see Towse, 1998). In the model, the adjacency score is not influenced by loss of the indirect pathway, but is sensitive to loss of neurons in the direct pathway.

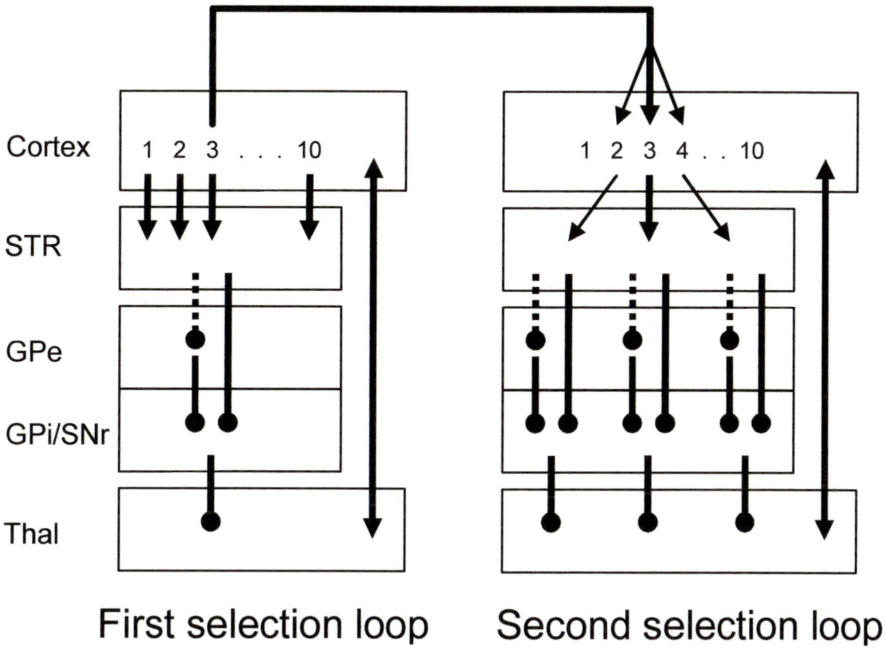

Fig. 5. Basal ganglia model of the random number generation task involving two selection loops. The first loop selects the digits from memory and the second loop selects the associated motor response. The associative network resides in the cortico-cortico connections. For simplicity the STN and other within basal ganglia connections have been omitted.

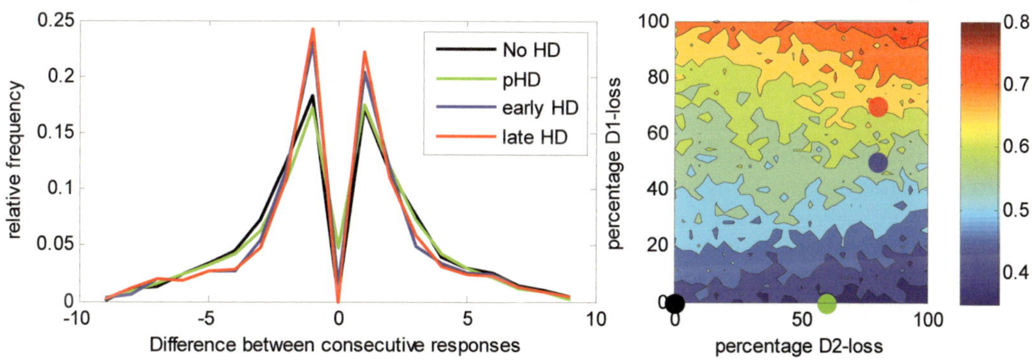

Fig. 6. Results of the simulation on random number generation. Left: distribution of first-order differences. Distributions of the highlighted combinations of D1/D2-loss in the right panel are shown. Right: standardized adjacency scores at various levels of loss in D1- and D2-pathways. Percentage loss is approximately corresponding to reports by Glass et al. (2000).

4.5 Memory strategies: A window into D2-pathway loss

The simulations showed that HD pathology influence memory only via the direct pathway which would mean that pHD would not involve memory deficits. Yet, several studies have shown various memory deficits in pHD (Solomon, et al., 2007; Stout, et al., 2011). Critically, Stout et al. (2011) showed that performance on the Hopkins Verbal Learning Test (HVLT) still predicted the probability of disease onset in 5 years after controlling for the standard Unified Huntington's Disease Rating Scale (UHDRS) Motor Score. How to reconcile these opposing views?

The model of random number generation is an example of a system that first selects a cue and then uses this cue to select candidates for output. This is closely related to the use of cues in memory tasks and the HVLT in particular. In this task, participants are presented with items for immediate or delayed free recall. The sequence of items consists of four words from three categories. The strategic use of cues, such as category labels improves overall recall performance. Therefore it is better to first select a cue and then use the cue to select items from memory than to directly try to retrieve the items from memory. This extra step would involve the first selection loop in the random generation model. With the HVLT, the selection is not random, but involves choosing which of the three category labels to use. Once the category label is chosen, words that belong to that category and have episodic contextual list information will be selected as candidates for retrieval.

This two-stage retrieval process requires the indirect pathway to let go of the chosen cue when it is not needed anymore. It is expected that pHD would have difficulty in performing this cognitive operation. Even though loss of the indirect pathway did not influence the adjacency score in random number generation, none of the measures used in the literature on random number generation is specifically designed to capture the operation of the first selection loop. In the verbal learning test, not being able to disengage from a cue will limit the total recall performance and thus pHD should show selective memory impairment.

Duff et al. (2010) discuss the possibility of dividing pHD into an amnestic and a nonamnestic group. The measure used to distinguish this is the total recall on the HVLT together with scores from other tasks, with subnormal performance on any non-memory resulting in a nonamnestic categorization. Assuming that an amnestic and nonamnestic variant of pHD exist, how does this fit with the modeling framework? To answer this question, it needs to be realized that none of the verbal learning tests control for strategy use. That is, some individuals may use memory strategies, such as semantic retrieval cues, whereas others do not. Only when cues are being utilized will the inability to disengage from cues be observed. Thus, individuals that are classified as amnestic pHD might be those who tend to use optimal memory strategies. In longitudinal studies, this will be observed as a high performing affected individual declining faster than low performing individuals. The final simulation is designed to demonstrate this pattern.

4.5.1 Method

A computational model of chunking is used that accounts for idiosyncratic chunking behavior and is a combination of the random generation model and the search of associative memory (SAM; Raaijmakers & Shiffrin, 1981). To model the longitudinal trajectory of pHD the probability of disengaging a cue was decreased over the course of pHD. To increase the performance for the simulation of nonamnestic pHD, the short-term capacity was increased. This is justified under the assumption that people choose to either focus more during encoding or work harder during retrieval. Note that in this simulation the selection threshold stays fixed. In the model for the random number generation and in a simulation study on recall latencies in HD (Davelaar, 2007), the threshold increase relates to the stage after clinical onset. In simulation studies that focus on the longitudinal trajectory of memory impairment after disease onset this parameter would need to be increased as the disease progresses.

4.5.2 Results and discussion

The results are shown in Fig. 7. As discussed above, total recall is better when memory cues are being used. When no strategy is used involving sequential cue retrieval, the memory performance stays relatively constant and fluctuations are within normal ranges. When the retrieval strategy is used, there is an advantage at the beginning of the pHD, which turns into a disadvantage the nearer the disease onset time. These results partly explain why memory tests are good short-term predictors of disease onset (Stout, et al., 2011).

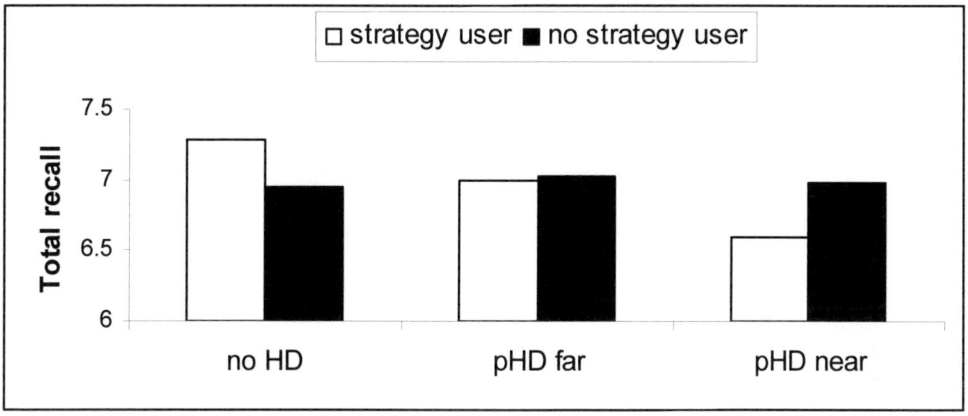

Fig. 7. Simulation results of a verbal learning experiment.

At each stage, until just prior to disease onset, the total recall performance leads to misclassification of those individuals who use memory cues to aid their retrieval. The choice of using memory cues is voluntary and thus an individual (e.g., no strategy user) may be classified as amnestic at an early stage and nonamnestic at a later time. This is possible explanation for the erratic pattern presented in Duff, et al. (2010).

5. General discussion

This chapter has presented several simulation results regarding the cognitive decline in symptomatic and presymptomatic HD. The architectures of the computational models respect the known functional neuroanatomy and neuropathology in HD. The results showed how compensation via the hyperdirect pathway influences performance in the flanker task (section 4.2) and why memory span (section 4.3) and random number generation (section 4.4) only show a decline when the D1-pathway is affected. Finally, with regard to mild cognitive impairment in pHD with amnestic and nonamnestic variants, simulations verify the MCI variants and provide an intuition on how to maximize the differential classification (section 4.5).

As can be seen in Fig. 3, loss of D2-receptor MSNs result in increased times to disengage the chosen response. Only when D1-receptor MSNs are lost, does the response time increase. This has important implications for longitudinal studies that aim to assess the cognitive decline in pHD. With the assumption that pHD is mainly associated with loss of the indirect pathway (cf. Glass, et al., 2000), it is expected that tasks that depend on disengagement after a choice are ideal candidates for screening. However, compensatory mechanisms may mask the degeneration of the indirect pathway, making it difficult to observe deficits in cognitive and motor stopping behavior until near the onset of HD.

Yet, failure to disengage after a choice can be observed in situations where normally a maintained focus would be beneficial. In section 4.5, memory retrieval was addressed, which involves a two-stage process, with selection of a cue followed by selection of an item. Failure to disengage from the cue impacted on total recall performance. The strategy of using retrieval cues is under voluntary control, leading to variability among individuals in the likelihood of using memory strategies. It is likely that some pHD individuals will attempt using retrieval cues and some do not. For pHD individuals who do attempt cued retrieval, a memory enhancement is expected when they are far removed from the disease onset, but a memory deficit is expected when they are closer to the disease onset.

The division of mild cognitive impairment in pHD into an amnestic and a nonamnestic variant Duff et al. (2010) would map onto those who do and those who do not attempt memory strategies. In addition, the choice of using a memory strategy is unlikely to involve any components that are measured using the UHDRS Motor Score. This would explain why the HVLT has predictive power after controlling for UHDRS Motor Score (Stout, et al., 2011). Currently, these are still preliminary predictions that need to be confirmed in empirical studies.

Apart from the choice of tasks and task measures used in studies and the classification of presymptomatic HD, there is a rich understanding of the neuropathology and its consequences for motor behavior. At this level, detailed computational models of the basal ganglia can be employed to extrapolate back in time to see how the deficit in motor behavior unfolds from early pHD to HD. With the added assumption that the cognitive loops undergo the same changes, albeit at a different sensitivity, the lifetime approach can be used to identify tasks that are sensitive to pHD status. These types of models therefore bridge the detailed findings about the progression of the neuropathology with the overall cognitive outcome.

Computational models are in essence complex statistical models. Conventional statistical tests may take simplifying assumptions, such as simplified functional forms, normally-distributed variance, equality of variance, and many more. Computational models are not developed within the framework of statistical theory, but can be developed to correspond to functional neuroanatomical and neuropathological reality. In this way, neurally plausible computational models can be used to extrapolate back in time and are arguably the best reality-grounded statistical models we have. Verification of the model predictions is still needed. Studies such as PREDICT-HD (e.g., Stout, et al., 2011) employed tasks for which computational models exist and therefore the required database is present to provide the final impetus to include computational modeling as an additional methodology in HD research.

Throughout this chapter, the critical data that grounded the models was the observation that disease onset coincides with loss of D1-receptor MSNs. The proposed classification of tasks is based on the tasks' reliance of the D1- and D2-pathways and associative learning (section 3.3). The power of the tasks to predict disease onset depends on the precise profile of D1-MSN loss. The computational approach requires reliable and objective biomarkers. Once these have been validated, new HD-specific cognitive batteries can be developed that predict the probability of disease onset even after controlling for UHDRS Motor Score. This will further improve the clinician's ability to grade a person's cognitive decline and support the affected individual and families.

6. Conclusion

This chapter presented computer simulations of tasks that have been used in the cognitive research on presymptomatic and symptomatic HD. The models were developed to be close to the underlying neurobiological factors that drive the disease. The models are still premature, but have the potential to contribute to intervention strategies. To bring forth tangible results, interdisciplinary research is vital and will benefit those with HD, those who are aware they will get HD, and the society who needs to prepare and provide care for those affected. Such a research program remains for the future. This chapter presented a first stage, demonstrating the ability of neuroanatomically grounded computational models to capture cognitive performance and decline in presymptomatic Huntington's Disease.

7. References

Aglioti, S. (1997). The role of the thalamus and basal ganglia in human cognition. *Journal of Neurolinguistics, 10*, 255-265, ISSN 0911-6044

Albin, R. L., Young, A. B., & Penney, J. B. (1989). The functional anatomy of basal ganglia disorders. *Trends in Neurosciences, 12*, 366-375, ISSN 0166-2236

Alexander, G. E., & Crutcher, M. D. (1990). Functional architecture of basal ganglia circuits: neural substrates of parallel processing. *Trends in Neurosciences, 13*, 266-271, ISSN 0166-2236

Aylward, E. H. (2006). Change in MRI striatal volumes as a biomarker in preclinical Huntington's disease. *Brain Research Bulletin, 72*, 152-158, ISSN 0361-9230

Bar-Gad, I., Morris, G., & Bergman, H. (2003). Information processing, dimensionality reduction and reinforcement learning in the basal ganglia. *Progress in Neurobiology, 71*, 439–473, ISSN 0301-0082

Bergman, H., Feingold, A., Nini, A., Raz, A., Slovin, H., Abeles, M., & Vaadia, W. (1998). Physiological aspects of information processing in the basal ganglia of normal and parkinsonian primates. *Trends in Neurosciences, 21*, 32-38, ISSN 0166-2236

Berns, G. S., & Sejnowski, T. J. (1998). A computational model of how the basal ganglia produce sequences. *Journal of Cognitive Neuroscience, 10*, 108-121, ISSN 0898-929X

Beste, C., Saft, C., Andrich, J., Gold, R., & Falkenstein, M. (2006). Error processing in Huntington's disease. *PLoS ONE, 1*, e86. doi: 10.1371/journal.pone.0000086, ISSN 1932-6203

Beste, C., Saft, C., Andrich, J., Gold, R., & Falkenstein, M. (2008). Stimulus-response compatibility in Huntington's disease: a cognitive-neurophysiological analysis. *Journal of Neurophysiology, 99*, 1213-1223, ISSN 0022-3077

Beste, C., Saft, C., Yordanova, J., Andrich, J., Gold, R., Falkenstein, M., & Kolev, V. (2007). Functional compensation or pathology in cortico-subcortical interactions in preclinical Huntington's disease? *Neuropsychologia, 45,* 2922-2930, ISSN 0028-3932

Brown, J., Bullock, D., & Grossberg, S. (1999). How the basal ganglia use parallel excitatory and inhibitory learning pathways to selectively respond to unexpected rewarded cues. *Journal of Neuroscience, 19,* 10502-10511, ISSN 0270-6474

Brown, R. G., Soliveri, P., & Jahanshahi, M. (1998). Random response generation in Parkinson's disease and normal controls. Executive processes and the role of the prefrontal cortex. *Neuropsychologia, 36,* 1355–1362, ISSN 0028-3932

Bullock, D. (2004). Adaptive neural models of queuing and timing in fluent action. *Trends in Cognitive Sciences, 8,* 426-433, ISSN 1364-6613

Bullock, D., Tan, C. O., & John, Y. J. (2009). Computational perspectives on forebrain microcircuits implicated in reinforcement learning, action selection, and cognitive control. *Neural Networks, 22,* 757-765, ISSN 0893-6080

Cha, J. J. (2007). Transcriptional signatures in Huntington's disease. *Progress in Neurobiology, 83,* 228-248, ISSN 0301-0082

Chevalier, G., & Deniau, J. M. (1990). Disinhibition as a basic process in the expression of striatal functions. *Trends in Neurosciences, 13,* 277-280, ISSN 0166-2236

Cicchetti, F., Prensa, L., Wu, Y., & Parent, A. (2000). Chemical anatomy of striatal interneurons in normal individuals and in patients with Huntington's disease. *Brain Research Reviews, 34,* 80-101, ISSN 0165-0173

Cohen, M. X., & Frank, M. J. (2009). Neurocomputational models of basal ganglia function in learning, memory and choice. *Behavioural Brain Research, 199,* 141-156, ISSN 0166-4328

Da Cunha, C., Wietzikoski, E. C., Dombrowski, P., Bortolanza, M., Santos, L. M., Boschen, S. L., & Miyoshi, E. (2009). Learning processing in the basal ganglia: a mosaic of broken mirrors. *Behavioural Brain Research, 199,* 157-170, ISSN 0166-4328

Davelaar, E. J., Goshen-Gottstein, Y., Ashkenazi, A., Haarmann, H. J., & Usher, M. (2005). The demise of short-term memory revisited: empirical and computational investigations of recency effects. *Psychological Review, 112,* 3-42, ISSN 0033-295X

Davelaar, E. J. (2004). *Random generation of items from memory: empirical and computational explorations.* Poster presented at the 2nd International Conference on Working Memory, Kyoto, Japan.

Davelaar, E. J. (2007). Sequential retrieval and inhibition of parallel (re)activated representations: a neurocomputational comparison of competitive queuing and resampling models. *Adaptive Behavior, 15,* 51-71, ISSN 1059-7123

Davelaar, E. J. (2008). A computational study of conflict-monitoring at two levels of processing: reaction time distributional analyses and hemodynamic responses. *Brain Research, 1202,* 109-119, ISSN 0006-8993

Davelaar, E. J., & Stevens, J. (2009). Sequential dependencies in the Eriksen flanker task: a direct comparison of two competing accounts. *Psychonomic Bulletin & Review, 16,* 121-126, ISSN 1069-9384

DeLong, M., & Wichmann, T. (2009). Update on models of basal ganglia function and dysfunction. *Parkinsonism and Related Disorders, 1553,* S237-S240, ISSN 1353-8020

Doya, K. (2007). Reinforcement learning: Computational theory and biological mechanisms. *HFSP Journal, 1,* 30-40.

Duff, K., Paulsen, J., Mills, J., Beglinger, L. J., Moser, D. J., Smith, M. M., Langbehn, D., Stout, J., Queller, S., & Harrington, D. L. (2010). Mild cognitive impairment in prediagnosed Huntington disease. *Neurology, 75,* 500-507, ISSN 0028-3878

Eblen, F., & Graybiel, A. M. (1995). Highly restricted origin of prefrontal cortical inputs to striosomes in the macaque monkey. *Journal of Neuroscience, 15*, 5999-6013, ISSN 0270-6474

Feigin, A., Tang, C., Ma, Y., Mattis, P., Zgaljardic, D., Guttman, M., Paulsen, J. S., Dhawan, V., Eidelberg, D. (2007). Thalamic metabolism and symptom onset in preclinical Huntington's disease. *Brain, 130*, 2858-2867, ISSN 0006-8950

Frank, M. J. (2006). Hold your horses: a dynamic computational role for the subthalamic nucleus in decision making. *Neural Networks, 19*, 1120-1136, ISSN 0893-6080

Frank, M. J., Loughry, B., & O'Reilly, R. C. (2001). Interactions between the frontal cortex and basal ganglia in working memory: a computational model. *Cognitive, Affective, and Behavioral Neuroscience, 1*, 137-160, ISSN 1530-7026

Frank, M. J., & Claus, E. D. (2006). Anatomy of a decision: striato-orbitofrontal interactions in reinforcement learning, decision making, and reversal. *Psychological Review, 113*, 300-326, ISSN 0033-295X

Glass, M., Dragunow, M., & Faull, R. L. M. (2000). The pattern of neurodegeneration in Huntington's disease: a comparative study of cannabinoid, dopamine, adenosine and GABAA receptor alterations in the human basal ganglia in Huntington's disease. *Neuroscience, 97*, 505-519, ISSN 0306-4522

Gomez, G. T., Hu, H., McCaw, E. A., & Denovan-Wright, E. M. (2006). Brain-specific factors in combination with mutant huntingtin induce gene-specific transcriptional dysregulation. *Molecular and Cellular Neuroscience, 31*, 661-675, ISSN 1044-7431

Grahn, J. A., Parkinson, J. A., & Owen, A. M. (2009). The role of the basal ganglia in learning and memory: neuropsychological studies. *Behavioural Brain Research, 199*, 53-60, ISSN 0166-4328

Graybiel, A. M. (1991). Basal ganglia - input, neural activity, and relation to the cortex. *Current Opinion in Neurobiology, 1*, 644-651, ISSN 0959-4388

Graybiel, A. M. (1995). Building action repertoires: memory and learning functions of the basal ganglia. *Current Opinion in Neurobiology, 5*, 733-741, ISSN 0959-4388

Graybiel, A. M. (2005). The basal ganglia: learning new tricks and loving it. *Current Opinion in Neurobiology, 15*, 638-644, ISSN 0959-4388

Gu, X., Greiner, E. R., Mishra, R., Kodali, R., Osmand, A., Finkbeiner, S., Steffan, J. S., Thompson, L. M., Wetzel, R., & Yang, X. W. (2009). Serines 13 and 16 are critical determinants of full-length human mutant huntingtin induced disease pathogenesis in HD mice. *Neuron, 64*, 828-840, ISSN 0896-6273

Gurney, K., Prescott, T. J., & Redgrave, P. (2001). A computational model of action selection in the basal ganglia II: analysis and simulation of behaviour. *Biological Cybernetics, 84*, 411-423, ISSN 0340-1200

Guthrie, M., Myers, C. E., & Gluck, M. A. (2009). A neurocomputational model of tonic and phasic dopamine in action selection: a comparison with cognitive deficits in Parkinson's disease. *Behavioural Brain Research, 200*, 48-59, ISSN 0166-4328

Hebb, A. L. O., Robertson, H. A., & Denovan-Wright, E. M. (2004). Striatal phosphadiesterase mRNA and protein levels are reduced in Huntington's disease transgenic mice prior to the onset of motor symptoms. *Neuroscience, 123*, 967-981, ISSN 0306-4522

Hedreen, J. C., & Folstein, S. E. (1995). Early loss of neostriatal striosome neurons in Huntington's disease. *Journal of Neuropathology and Experimental Neurology, 54*,105-120, ISSN 0022-3069

Heyder, K., Suchan, B., & Daum, I. (2004). Cortico-subcortical contributions to executive control. *Acta Psychologica, 115,* 271-289, ISSN 0001-6918

Ho, A. K., Sahakian, B. J., Brown, R. G., Barker, R. A., Hodges, J. R., Ané, M., Snowden, J., Thompson, J., Esmonde, T., Gentry, R., Moore, J. W., & Bodner, T. (2003). Profile of cognitive progression in early Huntington's disease. *Neurology, 61,* 1702-1706, 0028-3878

Ho, A. K., Sahakian, B. J., Robbins, T. W., & Barker, R. A. (2004). Random number generation in patients with symptomatic and presymptomatic Huntington's disease. *Cognitive and Behavioral Neurology, 17,* 208-212, ISSN 1543-3633

Ho, A. K., Sahakian, B. J., Robbins, T. W., Barker, R. A., Rosser, A. E., & Hodges, J. R. (2002). Verbal fluency in Huntington's disease: a longitudinal analysis of phonemic and semantic clustering and switching. *Neuropsychologia, 40,* 1277-1284, ISSN 0028-3932

Houk, J.C., & Wise, S. P. (1995). Distributed modular architectures linking basal ganglia, cerebellum, and cerebral cortex: their role in planning and controlling action. *Cerebral Cortex, 5,* 95-110, ISSN 1047-3211

Humphries, M. D., Stewart, R. D., & Gurney, K. N. (2006). A physiologically plausible model of action selection and oscillatory activity in the basal ganglia. *Journal of Neuroscience, 26,* 12921-12942, ISSN 0270-6474

The Huntington's Disease Collaborative Research Group (1993). A novel gene containing a trinucleotide repeat that is expanded and unstable on Huntington's disease chromosomes. *Cell, 72,* 971-983, ISSN 0092-8674

Ivkovic, S., & Ehrlich, M. E. (1999). Expression of the striatal DARPP-32/ARPP-21 phenotype in GABAergic neurons requires neurotrophins in vivo and in vitro. *Journal of Neuroscience, 19,* 5409-5419, ISSN 0270-6474

Jahanshahi, M., Saleem, T., Jo, A. K., Dirnberger, G., & Fuller, R. (2006). Random number generation as an index of controlled processing. *Neuropsychology, 20,* 391-399, ISSN 0894-4105

Jahanshahi, M., Profice, P., Brown, R. G., Ridding, M. C., Dirnberger, G., & Rothwell, J. C. (1998). The effects of transcranial magnetic stimulation over the dorsolateral prefrontal cortex on suppression of habitual counting during random number generation. *Brain, 121,* 1533-1544, ISSN 0006-8950

Joel, D., & Weiner, I. (1994). The organization of the basal ganglia-thalamocortical circuits: open interconnected rather than closed segregated. *Neuroscience, 63,* 363-379, ISSN 0306-4522

Joyce, J. N., Sapp, D. W., Marshall, J. F. (1986). Human striatal dopamine receptors are organized in compartments. *Proceedings of the National Academy of Sciences, 83,* 8002-8006, ISSN 1091-6490

Kaplan, S., Itzkovitz, S., & Shapiro, E. (2007). A universal mechanism ties genotype to phenotype in trinucleotide diseases. *PLoS Computational Biology, 3,* e235. doi: 10.1371/journal.pcbi.0030235, ISSN 1553-734X

Langbehn, D. R., Brinkman, R. R., Gaulsh, D., Paulsen, J. S., & Hayden, M. R. (2004). A new model for prediction of the age of onset and penetrance for Huntington's disease based on CAG length. *Clinical Genetics, 65,* 267-277, ISSN 0009-9163

Lawrence, A. D., Sahakian, B. J., Hodges, J. R., Rosser, A. E., Lange, K. W., & Robbins, T. W. (1996). Executive and mnemonic functions in early Huntington's disease. *Brain, 119,* 1633-1645, ISSN 0006-8950

Lévesque, M., & Parent, A. (2005). The striatofugal fiber system in primates: a reevaluation of its organization based on single-axon tracing studies. *Proceedings of the National Academy of Sciences, 102,* 11888–11893, ISSN 1091-6490

Luthi-Carter, R., & Cha, J. J. (2003). Mechanisms of transcriptional dysregulation in Huntington 's disease. *Clinical Neuroscience Research, 3,* 165-177, ISSN 1566-2772

Manninen, T., Hituri, K., Kotaleski, J. H., Blackwell, K. T., & Linne, M. (2010). Postsynaptic signal transduction models of long-term potentiation and depression. *Frontiers in Computational Neuroscience, 4,* 152. doi: 10.3389/fncom.2010.00152, ISSN 1662-5188

McNab, F., & Klingberg, T. (2008). Prefrontal cortex and basal ganglia control access to working memory. *Nature Neuroscience, 11,* 103-107, ISSN 1097-6256

Menalled, L. B., Sison, J. D., Wu, Y., Olivieri, M., Li, X., Li, H., Zeitlin, S., & Chesselet, M. (2002). Early motor dysfunction and striosomal distribution of huntingtin microaggregates in Huntington's disease knock-in mice. *Journal of Neuroscience, 22,* 8266-8276, ISSN 0270-6474

Monchi, O., Taylor, J. G., & Dagher, A. (2000). A neural model of working memory processes in normal subjects, Parkinson's disease and schizophrenia for fMRI design and predictions. *Neural Networks, 13,* 953-973, ISSN 0893-6080

Montoya, A., Price, B. H., Menear, M., & Lepage, M. (2006). Brain imaging and cognitive dysfunctions in Huntington's disease. *Journal of Psychiatry and Neuroscience, 31,* 21-29, ISSN 1180-4882

Nambu, A. (2011). Somatotopic organization of the primate basal ganglia. *Frontiers in Neuroanatomy 5,* 26. doi: 10.3389/fnana.2011.00026, ISSN 1662-5129

Nambu, A., Tokuno, H., & Takada, M. (2002). Functional significance of the cortico-subthalamo-pallidal 'hyperdirect' pathway. *Neuroscience Research, 43,* 111-117, ISSN 0168-0102

Nambu, A., Tokuno, H., Hamada, I., Kita, H., Imanishi, M., Akazawa, T., Ikeuchi, Y., & Hasegawa, N. (2000). Excitatory cortical inputs to pallidal neurons via the subthalamic nucleus in the monkey. *Journal of Neurophysiology, 84,* 289-300, ISSN 0022-3077

Paulsen, J. S., Magnotta, V. A., Mikos, A. E., Paulson, H. L., Penziner, E., Andreasen, N. C., & Nopoulos, P. C. (2006). Brain structure in preclinical Huntington's disease. *Biological Psychiatry, 59,* 57-63, ISSN 0006-3223

Paulsen, J. S., Zimbelman, J. L., Hinton, S. C., Langbehn, D. R., Leveroni, C. L., Benjamin, M. L., Reynolds, N. C., Rao, S. M. (2004). fMRI biomarker of early neuronal dysfunction in presymptomatic Huntington's disease. *American Journal of Neuroradiology, 25,* 1715-1721, ISSN 0195-6108

Peretti, C., Ferreri, F., Blanchard, F., Bakchine, S., Peretti, C. R., Dobrescu, A., Chouinard, V., & Chouinard, G. (2008). Normal and pathological aging of attention in presymptomatic Huntington's, Huntington's and Alzheimer's disease, and nondemented elderly subjects. *Psychotherapy and Psychosomatics, 77,* 139-146, ISSN 0033-3190

Pérez-Navarro, E., Canudas, A. M., Åkerud, P., Alberech, J., & Arenas, E. (2000). Brain-derived neurotrophic factor, neurotrophin-3, and neurotrophin-4/5 prevent the death of striatal projection neurons in a rodent model of Huntington's disease. *Journal of Neurochemistry, 75,* 2190-2199, ISSN 1471-4159

Perreault, M. L., Hasbi, A., O'Dowd, B. F., & George, S. R. (2011). The dopamine D1-D2 receptor heteromer in striatal medium spiny neurons: evidence for a third distinct neuronal pathway in basal ganglia. *Frontiers in Neuroanatomy, 5,* 31 doi:10.3389/fnana.2011.00031, ISSN 1662-5129

Rohrer, D., Salmon, D. P., Wixted, J. T., & Paulsen, J. S. (1999). The disparate effects of Alzheimer's disease and Huntington's disease on semantic memory. *Neuropsychology, 13,* 381–388, ISSN 0894-4105

Raaijmakers, J. G. W., & Shiffrin, R. M. (1981). Search of associative memory. *Psychological Review, 88,* 93-134, ISSN 0033-295X

Solomon, A. C., Stout, J. C., Johnson, S. A., Langbehn, D. R., Aylward, E. H., Brandt, J., Ross, C. A., Beglinger, L., Hayden, M. R., Kieburtz, K., Kayson, E., Julian-Baros, E., Duff, K., Guttman, M., Nance, M., Oakes, D., Shoulson, I., Penziner, E., & Paulsen, J. S. (2007). Verbal episodic memory declines prior to diagnosis in Huntington's disease. *Neuropsychologia, 45,* 1767-1776, ISSN 0028-3932

Stoffers, D., Sheldon, S., Kuperman, J. M., Goldstein, J., Corey-Bloom, J., Aron, A. R. (2010). Contrasting gray and white matter changes in preclinical Huntington's disease. *Neurology, 74,* 1208-1216, ISSN 0028-3878

Stout, J. C., Paulsen, J. S., Queller, S., Solomon, A. C., Whitlock, K. B., Campbell, J. C., Carlozzi, N., Duff, K., Beglinger, Langbehn, D. R., Johnson, S. A., Biglan, K. M., Aylward, E. H. (2011). Neurocognitive signs in prodromal Huntington disease. *Neuropsychology, 25,* 1-14, ISSN 0894-4105

Thompson-Schill, S. L., & Botvinick, M. M. (2006). Resolving conflict: A response to Martin and Cheng (2006). *Psychonomic Bulletin & Review, 13,* 402-408, ISSN 1069-9384

Towse, J. N. (1998). On random generation and the central executive of working memory. *British Journal of Psychology, 89,* 77-101, ISSN 2044-8295

Towse, J. N., & Neil, D. (1998). Analyzing human random generation behavior: A review of methods used and a computer program for describing performance. *Behavior Research Methods, Instruments & Computers, 30,* 583-591, ISSN 0743-3808

Tippett, L. J., Waldvogel, H. J., Thomas, S. J., Hogg, V. M., van Roon-Mom, W., Synek, B. J., Graybiel, A. M., & Faull, R. L. M. (2007). Striosomes and mood dysfunction in Huntington's disease. *Brain, 130,* 206-221, ISSN 0006-8950

Walker, F. O. (2007). Huntington's disease. *Lancet, 369,* 218-228, ISSN 0140-6736

Weir, D. W., Sturrock, A., & Leavitt, B. R. (2011). Development of biomarkers for Huntington's disease. *Lancet, 10,* 573-590, ISSN 0140-6736

White, N. M. (1997). Mnemonic functions of the basal ganglia. *Current Biology, 7,* 164-169, ISSN 0960-9822

Yeung, N., & Cohen, J. D. (2006).The impact of cognitive deficits on conflict monitoring: Predictable dissociations between the ERN and N2. *Psychological Science, 17,* 164-171, ISSN 1467-9280

Zuccato, C., & Cattaneo, E. (2007). Role of brain-derived neurotrophic factor in Huntington's disease. *Progress in Neurobiology, 81,* 294-330, ISSN 0301-0082

Zuccato, C., Valenza, M., & Cattaneo, E. (2010). Molecular mechanisms and potential therapeutical targets in Huntington's disease. *Physiological Reviews, 90,* 905-981, ISSN 0031-9333

Part 4

Transcriptional and Post-Transcriptional Dysregulation in Huntington's Disease

11

Targeting Transcriptional Dysregulation in Huntington's Disease: Description of Therapeutic Approaches

Manuela Basso
Burke Medical Research Institute, Weill Medical College, Cornell University, USA

1. Introduction

Huntington's Disease (HD) is a dominantly inherited neurodegenerative disease affecting cognitive, emotional and motor systems. While alterations in the huntingtin gene (HTT) have been identified as causative for nearly two decades, an effective treatment has yet to be developed. Prior studies have shown that mutant huntingtin (mHTT), via its polyglutamine-expanded repeats, can affect cellular function in many ways, such as alteration of gene transcription, one of the best-characterized pathobiological events leading to HD. Microarray studies in mouse models of HD and in postmortem brain samples from HD patients report a decrease in transcriptional levels of hundreds of genes, most of them selectively expressed in the striatum, the affected brain region in HD. mHTT has been shown to inhibit the interactions of several transcription factors and to repress the transcription of genes necessary for neuronal function and survival, such as Brain Derived Neurotrophin (BDNF) or the co-activator Peroxisome proliferator-activated receptor gamma coactivator 1-alpha (PGC1-alpha).

The main question that arises is how the changes in transcriptional expression are triggered. Several studies from multiple laboratories focus only on one transcription factor as causative of the disease, but a comprehensive view of all the described events is missing and drug treatments able to correct the transcriptional dysregulation in this incurable disease are warranted. Global transcriptional modulators, like Histone deacetylase (HDAC) inhibitors, have been seen as a potential therapy for this disease. On the other hand, transcription can be regulated modulating the activity of histone demethylases, histone acetyl transferases, microRNAs and new approaches have been developed recently. An alternative way to modulate transcription in HD resides in the inhibition of transglutaminase 2 (TGase 2). The multifunctional enzyme TGase 2 is hyperactivated in several neurodegenerative diseases and acute injuries leading to neuronal death and its pharmacological or genetic deletion leads to partial rescue in mouse models of HD. Our study (McConoughey et al., 2010), along with more recent publications (Munsie et al., 2011), unravels the important role of nuclear TGase 2 in HD and defines that in the presence of mHTT, TGase 2 is recruited to chromatin, where it binds to histone H3 and participates in transcriptional silencing of genes that control mitochondrial biogenesis, chromatin structure, protein folding and DNA repair. In our results TGase 2 inhibition regulates the gene expression of PGC1-alpha, a transcriptional coactivator, and cytochrome c, a transcription factor, both important in mitochondrial biogenesis. TGase 2 inhibition can normalize 40% of the dysregulated gene expression in a HD cell model and for this reason TGase 2 may act as a broader transcriptional modulator. TGase 2 might negatively modulate transcription of neuroprotective genes, inhibiting the interaction between transcription factors and their co-activators and thereby repressing gene expression designed to compensate, for instance, for mitochondrial dysfunction in HD. Specific TGase 2 inhibitors, along with other therapies targeting transcriptional dysregulation, may offer a beneficial effect to this incurable disease.

2. Genes dysregulated in HD

Transcriptional profiles of several *in vivo* and *in vitro* models of HD revealed a notable dysregulation of coding and non-coding RNAs expression (Tang et al., 2011). The cause of this impairment is linked to an alteration (loss or gain) of mHTT functions. mHTT is susceptible to protein cleavage by caspase-6 and its N-terminal fragments shuttle prevalently into the nuclear compartments where they form inclusions. Several transcription factors and enzymes involved in chromatin regulation were shown to interact with mHTT or to be present in intranuclear aggregates. The loss of these proteins contributes to global transcriptional dysregulation, typical of this neurodegenerative disease (Zhai et al., 2005). A series of very elegant papers published at the beginning of the millennium described the dysregulation of transcription factors and co-activators or co-repressors and their most well characterized downstream genes in HD, such as: the transcription factor **CREB** (cAMP Responsive Element-Binding), the co-activator **CBP** (CREB-Binding Protein), the co-repressor **NREST** (Neuronal Specific Responsive Element 1 (RE1) Silencing Transcription factor) and the DNA binding Specific Protein 1 (**Sp1**).

2.1 CREB

CREB is a transcription factor known to mediate stimulus-dependent expression of genes critical for plasticity, growth, and survival of neurons (Lonze &Ginty, 2002). The earliest observation that CREB signalling is compromised in HD came from Ross and collaborators in 2001 where the expression of different lengths of mHTT in N2A cells induced aggregation of the co-activator CBP and downregulation of CRE-mediated signalling (Nucifora et al., 2001). In the same year, Wyttenbach et al. confirmed this important observation in PC12 cells, where inducible mHTT expression impairs, primarily, the cAMP-regulated response (Wyttenbach et al., 2001). Subsequent works on the same line demonstrated the early CREB-signalling dysregulation in immortalized striatal cell lines (Gines et al., 2003) and in R6/2 mice (Sugars et al., 2004). Its reduced signalling became a promising target for therapeutic intervention; from a pharmacological point, specific phosphodiesterases inhibitors, like rolipam and TP10, were tested to maintain CREB in its active form (phosphorylated) and preserved neuronal viability (DeMarch et al., 2007; Giampa et al., 2006; Giampa et al., 2009). As a genetic approach, CREB overexpression was sufficient to rescue polyglutamine-dependent lethality in *Drosophila* (Iijima-Ando et al., 2005).

CREB regulates many genes and controls the transcription of the coactivator PGC1-alpha. Recent data from our group and others indicate that PGC1-alpha is necessary and sufficient to overcome mitochondrial toxicity in rodent models of HD and in other neurodegenerative diseases (Cui et al., 2006; Lin et al., 2004; McConoughey et al., 2010; St-Pierre et al., 2006; Weydt et al., 2006). PGC1-alpha can be regulated by and interact with transcription factors such as CREB, NRF-1, FOXO, MEF-2 and PPARγ to recruit the basal transcriptional machinery to genes involved in mitochondrial biogenesis, mitochondrial function and antioxidant defence (Figure 1). Additional functions of PGC1-alpha have been recently described, such as its role in cholesterol biosynthesis and myelination (Xiang et al., 2011), essential for neuronal functionality.

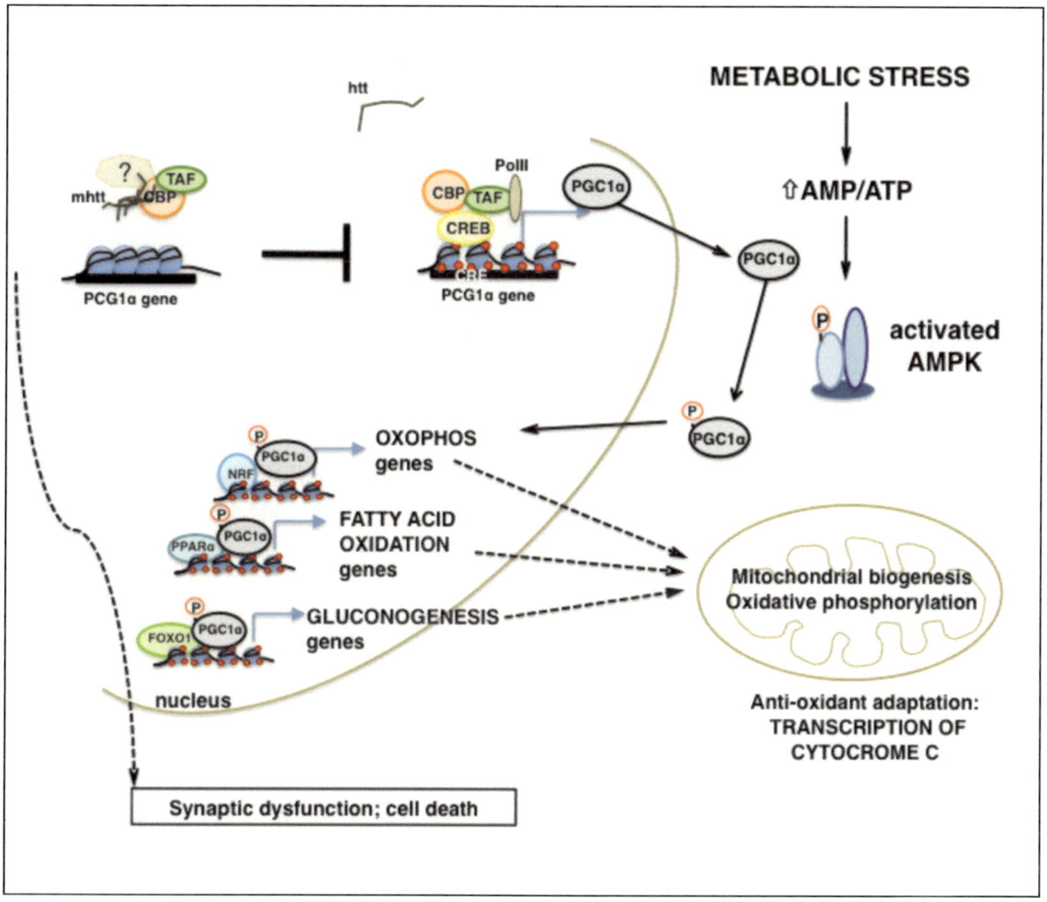

Fig. 1. The transcription of PGC1-alpha is regulated by metabolic stress. When PGC1-alpha is expressed and phosphorylated by AMPK, translocates to the nucleus and regulates the transcription of several genes involved in mitochondrial biogenesis and oxidative phosphorylation. These events lead to the activation of mitochondrial anti-oxidant adaptation and the increased transcription of several genes such as cytochrome c. mHTT has been shown to block the transcription of PGC1-alpha gene, recruiting CBP in intranuclear aggregates and blocking PolII activation.

2.2 CBP

CBP, best know as CREB co-activator, modulates the activation of many transcription factors (Goldman et al., 1997) by facilitating the recruitment of the transcriptional machinery. CBP has a key role in the nervous system; its mutations or deletions are associated to the Rubinstein-Taybi syndrome. In 2001 Steffan and colleagues showed that CBP and p300/CBP-associated factor (P/CAF) interact directly with mHTT blocking their acetyltransferase function (Figure 1). Additionally, CBP activity is reduced by its presence in polyglutamine aggregates (Nucifora et al., 2001) or by its increased proteasomal degradation (Cong et al., 2005; Jiang et al., 2003; Sadri-Vakili et al., 2007). Of note, CBP regulates the transcription of genes involved in the urea cycle, compromised in the liver of HD patients (Chiang et al., 2007) and this dysfunction contributes to the development of the disease.

2.3 REST/NREST

The Brain-Derived Neurotrophic Factor (BDNF) is an essential neurothrophin for the Central Nervous System. Its decreased levels have been well documented in HD human tissues and in mouse models. Its transcriptional regulation has been thoroughly described by Cattaneo and colleagues and it offers a different example of how mHTT can accomplish its detrimental effects. BDNF transcription can be switch off by a corepressor called REST. Usually REST interacts with *wild type* huntingtin and resides in the cytosol. mHTT fails to bind REST, which translocates to the nucleus and binds the Repressor-Element 1 (RE1) blocking BDNF gene transcription (Zuccato et al., 2001; Zuccato et al., 2003). Strategies to limit the repressive REST/NREST complex with pharmacological modulators, such as 2-aminothiazole derivatives (Leone et al., 2008) or decoys (Soldati et al., 2011) are now under investigation. Furthermore, REST modulates many microRNAs (miRs) and long non-coding RNAs, important in neuronal functions and dysregulated in HD (Bithell et al., 2009; Buckley et al., 2010; Johnson &Buckley, 2009; Johnson et al., 2008). One of them, miR-9, is downregulated by mHTT and fails to repress REST itself, contributing to the enhancement of its repressive activity (Packer et al., 2008).

2.4 Sp1

Sp1 is a member of an extended family of DNA-binding proteins that has three zinc finger motifs and binds to GC-rich DNA (Bouwman &Philipsen, 2002). Although classically thought to regulate the constitutive expression of numerous housekeeping genes, Sp1 transcriptional activities have been found to change in association with differentiation and proliferation and to regulate gene expression in association with these as well as other functions. In HD, the evidence that Sp1 dependent transcription is inhibited is extensive. mHTT interacts specifically with glutamine rich activation domains in Sp1 (Dunah et al., 2002) and blocks its direct binding to DNA. This aberrant interaction nullifies the ability of Sp1 to induce transcription of important genes including those encoding neurotransmitter receptors, downregulated in HD patients and rodents models (Cha et al., 1998). Sp1 overexpression (Dunah et al., 2002) or Sp1 acetylation (Ryu et al., 2003a) provide protection in HD. Interestingly, two anthracycline antibiotics, mithramycin and chromomycin, were shown to bind DNA inhibiting Sp1 activity and they provided the higher rate of survival reported to date in R6/2 mice (Ferrante et al., 2004; Stack et al., 2007). Unfortunately, the clinical trial on mithramycin was interrupted for low tolerability in humans. A recent paper from our group described promising analogs and showed the ability of these antibiotics to induce a promoter-specific displacement of Sp1, favouring the pro-survival effects of this transcription factor and inhibiting its pro-death activities (Sleiman et al., 2011).

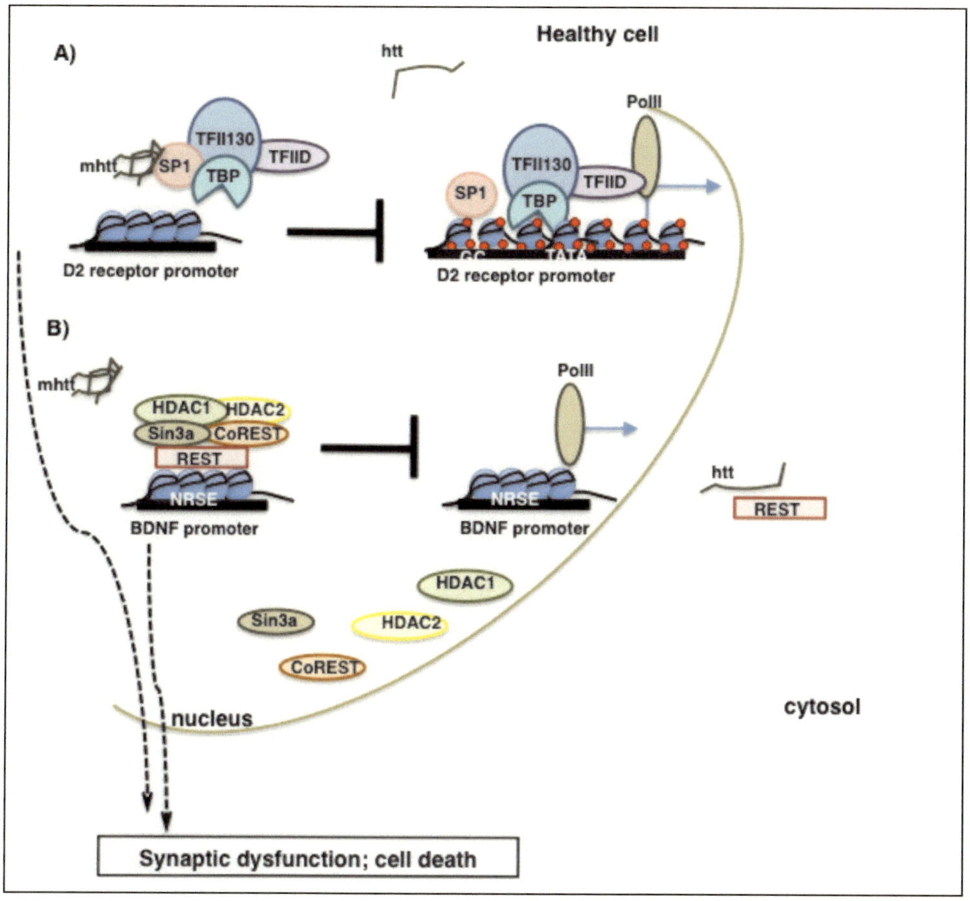

Fig. 2. mHTT recruits Sp1 and the transcription machinery in intranuclear inclusions, downregulating the expression of Sp1-dependent genes (A). At the same time, mHTT fails to interact and inhibit NREST repressive activity in the nucleus, leading to an aberrant inhibition of BDNF transcription (B).

3. Global histone modifications and transcriptional modulation

Within the eukaryotic nucleus, DNA is packaged into chromatin domain. The basic subunit of chromatin is the nucleosome, which is composed of DNA coiled around an octamer of histone proteins, two molecules each of histone H2A, H2B, H3 and H4. Histone H1 associates with chromatin outside the nucleosome. The amino-terminal tail of each histone is evolutionarily conserved and it is the target of numerous post-translational modifications (PTM). PTM of histones are major players in transcriptional control. These modifications include acetylation, methylation, phosphorylation, ADP-ribosylation, mono-ubiquitylation, citrullination, sumoylation and polyamination. The specific pattern of histone modification, identified as histone code, is used by proteins involved in chromatin organization to establish a transcriptionally silent or active state.

mHTT impacts transcription not only trough the direct binding on DNA (Benn et al., 2008) or transcription factors (e.g. CREB, FOXO) (Zhai et al., 2005) but also inducing a global modification of histone proteins. On one side, mHTT recruits histone acetyl transferases (HATs), such as CBP, in intranuclear aggregates and reduces their ability to acetylate histones; on the other side, mHTT facilitates polycomb repressive complex 2 (PRC2), which methylates histone H3 in lysine 27 and mediates transcriptional repression (Seong et al., 2010).

3.1 Histone acetylation and HDACs

Among the myriad of modifications that are normally occurring at the histone tails, acetylation is the most common. Histone acetylation and deacetylation are regulated by a delicate interplay between Histone Acetyl Transferases (HATs) and Deacetylases (HDACs). In a simplistic view, histone acetylation is usually associated with increase in gene transcription; conversely, histone deacetylation represses transcription. Several works described a global inhibition of acetylation in HD mouse models, human samples and cell lines, due to the propensity of mHTT to recruit HATs such as CBP (Steffan et al., 2000) in intracellular inclusions. HAT activity and global histone acetylation were significantly decreased in several models of HD (Igarashi et al., 2003; Sadri-Vakili et al., 2007). Difficulties in upregulating the acetyl transferase activity moved the attention on the other enzymes involved in the acetylation homeostasis: HDACs. HDAC inhibitors have been tested in various HD models to restore transcription, although their expression and activity are not altered by mHTT (Hockly et al., 2003) (Table 1). The first evidence that HDAC inhibitors would have been promising therapeutic agents in HD came from Leslie Thompson and collaborators in 2001, where butyrate and suberoylanilide hydroxamic acid (SAHA) reduced lethality in two *Drosophila* models of polyglutamine disease (Steffan et al., 2001). Sodium butyrate ameliorated HD symptoms in R6/2 mice and increased histones and Sp1 acetylation (Ferrante et al., 2003). Phenylbutyrate increased the lifespan of N171-82Q mice (Gardian et al., 2005) and it has been reported as safe and tolerable in humans (Hogarth et al., 2007). Other protective HDAC inhibitors are: SAHA, tested in R6/2 mice (Hockly et al., 2003); trichostatin A (TSA) is effective in immortalized cell lines (Dompierre et al., 2007; Oliveira et al., 2006); the inhibitor 4b effective in R6/2(300Q) transgenic mice (Thomas et al., 2008); valproate alone or in combination with lithium in N171-82Q mice (Zadori et al., 2009; Chiu et al., 2011). Clinical trials for valproate showed some beneficial effects (Saft et al., 2006; Grove et al., 2000). Finally, a role for the NAD+-dependent HDACs is emerging (Pallos et al., 2008; Hathorn et al., 2011) in relation to cholesterol synthesis in the HD brain (Luthi-Carter et al., 2010). Trials to assess the safety, tolerability and pharmacokinetics of sirtuins inhibitors are on going (SEN0014196) (Gray, 2010).

There is an emerging believe that global HDAC inhibition may exert partial toxicity due to the suppression of pro-survival isoforms. Genetic deletion of single isoforms have been performed revealing that HDAC4 may be the only causative in HD. Specific HDAC4 inhibitors are now under investigation (Munoz-Sanjuan &Bates, 2011).

3.1.1 Protein acetylation in HD

Acetylation is important not only on histone tails but on several proteins and transcription factors to recruit specific transcriptional regulatory complexes (Xu et al., 2007) or to mediate signalling. Sp1 acetylation, for instance, is necessary to activate the adaptive response to oxidative stress *in vitro* and *in vivo* (Ryu et al., 2003b) and alpha-tubulin acetylation increases BDNF trafficking and release in neurons (Dompierre et al., 2007). It has been recently reported that ribosomal DNA transcription is also impaired in HD due to decreased acetylation of the upstream binding factor-1 (UBF-1) (Lee et al., 2011); similarly, decreased levels of acetylation in p53 (lysine 382) correlate with the accumulation of DNA damage in HD (Illuzzi et al., 2011). Nevertheless, HTT itself is usually acetylated and degraded by autophagy; mHTT conformation impedes acetylation at lysine 444 and mediates its accumulation in intracellular inclusions (Jeong et al., 2009).

HDAC inhibitor	HD Model	References
SAHA	*Drosophila*	Steffan, 2001
Sodium butyrate	Fibroblast from HD patients	Kegel, Meloni et al. 2002
Sodium butyrate	R6/2 HD mouse model	Ferrante, Kubilus et al. 2003
SAHA	R6/2 HD mouse model	Hockly, Richon et al. 2003
Phenylbutyrate	N171-82Q HD mouse model	Gardian, Browne et al. 2005
HDAC3 shRNA	Caenorhabditis elegans expressing a human huntingtin fragment with an expanded polyglutamine tract (Htn-Q150)	Bates, Victor et al. 2006
Trichostatin A (TSA)/ Sodium butyrate	STHdh cell line	Oliveira, Chen et al. 2006
TSA and HDAC6 shRNA	Primary neurons	Dompierre, Godin et al. 2007
Phenyl butyrate and sodium butyrate	STHdh cell line and R6/2 mouse model	Sadri-Vakili, Bouzou et al. 2007
Phenylbutyrate	Humans/Clinical Trial	Hogarth, Lovrecic et al. 2007
HDAC1 and Sirt2 knock down	*Drosophila* (UAS-Httex1p Q93 flies)	Pallos, Bodai et al. 2008
Pimelic diphenylamide HDAC inhibitor, HDACi 4b	R6/2 mouse model	Thomas, Coppola et al. 2008
Nicodinamide to block Sirtuins	R6/1 mouse model	Hathorn, Snyder-Keller et al. 2011
SIRT2	*Drosophila* (UAS-Httex1p Q93 flies) and primary cultures trasduced with mHTT	Luthi-Carter, Taylor et al. 2010

Table 1. HDAC inhibitors tested in different models of HD.

3.2 Beyond acetylation: Methylation, ubiquitylation, polyamination

Decreased acetylation is associated usually with an increase of histone methylation at specific arginine and lysine residues (e.g. H3K9me, H3K27me). Histone methylation, in fact, has a similar dynamic regulation than histone acetylation and it is controlled by histone demethylases and histone methyltransferases. Levels of trimethylated histone H3 Lysine 9 are upregulated in HD human and mouse tissues by the dysregulated transcription of a Lysine methyl transferase, ESET (Ryu et al., 2006). Accordingly, partial deletion of CBP induces ESET transcription (Lee et al., 2008), suggesting that it is important to preserve the homeostatic equilibrium of the enzymes that regulate chromatin. The decrease of CBP involves reduced acetylation and shifts the equilibrium towards methylation.

Despite the simplistic concept of transcriptional repression mediated by a decrease of acetyl transferases activity and a consequent increase of global histone methylation, other histone modifications can lead to the same repressive result. Due to a disrupted interaction between mHTT and Bmi-1, part of the ubiquitin ligase complex, histone H2A monoubiquitylation is aberrantly increased in genes downregulated in HD. Consequently, monoubiquitylation of histone H2A promotes methylation in histone H3, lysine 9, a repressive mark (Kim et al., 2008). Conversely, the genes that are not altered by mHTT present normal levels of monoubiquitylated H2A and increased levels of monoubiquitylated H2B that induces methylation in histone H3 lysine 4, an active mark. In light of these important results, it is plausible to hypothesizes that new therapeutic avenues will be embraced by the HD scientific community in order to understand better how to modulate histone methylation in relation to dysregulation.

An emerging field in epigenetic modulation involves small cationic metabolites called polyamines. Polyamines are organic compounds with two or more primary amino groups able to regulated gene expression. They interact with DNA, RNA and control cell proliferation and growth. Their avidity for DNA on a charge base makes them ideally suited to regulate its conformation. Attaching them to proteins provides an elegant way to manipulate charge concentrations locally and alter DNA binding affinity (highly negatively charged due to phosphate backbone) to assume a compact (silenced) conformation. Recent papers showed that polyamines or polyamines analogs inhibit Lysine Specific Demethylase 1 (LSD1), a FAD-dependent histone demethylases, able to demethylate mono and dimethyl lysine 4 of histone H3, active marks of transcription (Huang et al., 2007; Shi et al., 2004) and they can block HDACs activity sitting in their catalytic pocket (Varghese et al., 2005). In a number of in vitro studies, polyamines can be crosslinked to glutamine tails of histones by transglutaminase 2 (TGase 2). Indeed, Ballestar identified polyamination of histone H3 in glutamine 5 and 19 and polyamination of histone H2B in glutamine 22 and correlated these modification with a change in the nucleosome structure (Ballestar et al., 1996; Ballestar et al., 2001).

3.2.1 Transglutaminase 2 and HD: Protein crosslinking or protein polyamination?

Transcriptional proteins that are inhibited in HD contain glutamine rich activation domains (Sp1, CBP, TAF4). Glutamines in proteins are substrates for a class of enzymes called transglutaminases (TGase 2) (Jeon et al., 2003). In humans, eight distinct TGases, encoded by different genes and referred to as TGase 1-7 and coagulation factor XIIIa have been previously identified. All members of the class have common catalytic activity and protein structure. The activity of each of these enzymes leads either to the formation of covalent bonds within or between polypeptide chains (γ-glutamyl-lysine; GGEL; Figure 3A) or the incorporation of polyamines into substrate proteins. This generates one of two possible types of products of TGase 2-polyamination: the N-(γ-glutamyl)polyamine and bis-(γ-glutamyl)polyamine (Figure 3B). In a recent study (Jeitner et al., 2008), increased levels of (γ-glutamyl)polyamines were seen in the CSF of HD patients suggesting a link between TGase 2 activity and polyamination in HD.

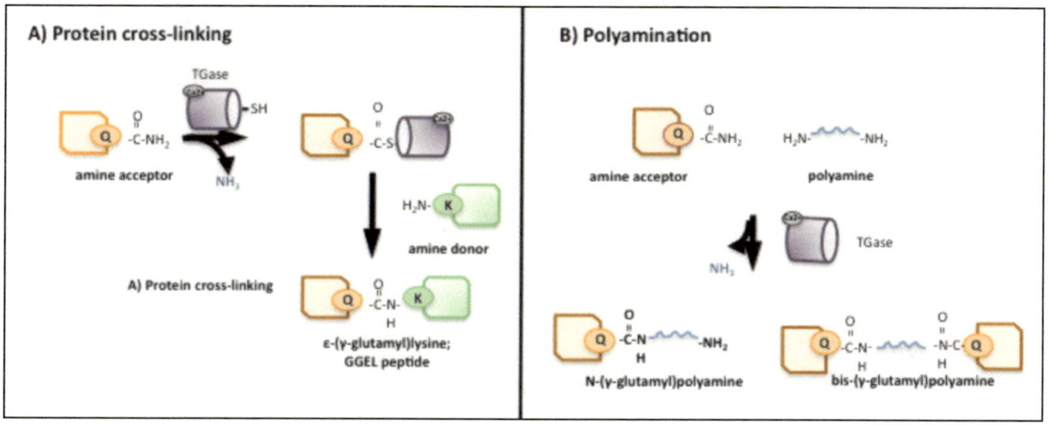

Fig. 3. TGase 2 catalyzes cross-links between glutamine and lysines in proteins leading to gamma-glutamyl-lysine covalent bonds (A) or the incorporation of polyamines into substrate proteins (B).

Investigations of TGase 2 in HD date back to 1993. Since then, a number of studies have documented increases in TGase 2 activity in a host of tissues, including in nuclei of human HD brains (Karpuj et al., 1999; Lesort et al., 1999). In the 80s, transglutaminase was first suspected to participate in HD pathogenesis via its ability to promote aggregates of polyglutamine (PolyQ) peptides and polyQ-huntingtin. Subsequently, Finkbeiner and colleagues suggested that aggregates were beneficial rather than pathogenic in HD (Arrasate et al., 2004). These findings suggested that TGase 2 inhibition prevented HD pathology by mechanisms independent of huntingtin aggregation. In the last ten years, several studies described the effect of TGase 2 inhibition in HD. Cystamine, a broad TGase 2 inhibitor, has been shown to be protective in R6/2 mice (Dedeoglu et al., 2002; Karpuj et al., 2002; Wang et al., 2005) and in YAC128 mice (Van Raamsdonk et al., 2005), both established models of the disease. Karpuj et al. in 2002 correlated the beneficial effects of TGase 2 inhibition with the transcriptional upregulation of a DNAJ-type heat shock protein, but did not offer any specific data on how TGase 2 might regulate DNAJ message levels in HD. The general model garnered support through a subsequent study by Borrel-Pages (Borrell-Pages et al., 2006) that showed that the levels of the DNAJ-containing protein HSJ1B are reduced in HD samples and that pharmacological inhibition of TGase 2 could restore message and protein levels in this context. The findings showed that TGase 2-mediated reduction in HSJ1B is critical for HD pathogenesis via its ability to delay brain-derived neurotrophic factor BDNF trafficking and release. Again, the findings were consistent with an effect of TGase 2 on message and protein levels, but did not offer a model of how TGase 2 might exert these effects. The crossbreeding between the TGase 2-/- and R6/1 or R6/2 mice resulted in reduced neuronal death, improved motor performance and increased survival (Mastroberardino et al., 2002, Bailey &Johnson, 2006). These positive results were not as encouraging as the HD community expected but it is important to consider that TGase 2 is ubiquitously expressed and among its several functions, it also has a role in normal development (Bailey et al., 2004). Deletion of TGase 2 induces compensation by the other seven transglutaminases that probably masked the real beneficial effect of TGase 2 inhibition.

We have proposed a novel TGase 2 function and demonstrated that TGase 2 inhibition normalized transcription in HD (McConoughey et al., 2010). In cells expressing mHTT, TGase 2 is recruited at the promoters or genomic regions of repressed genes. Microarray analysis indicates that TGase 2 inhibition via a selective inhibitor corrects transcriptional dysregulation in HD more efficiently than canonical TGase 2 inhibitors (cystamine) or HDAC inhibitors (TSA). However, TGase 2 inhibition does not affect histone acetylation (H4), suggesting a parallel and additive mechanism for histone regulation by HDAC inhibitors and TGase 2 inhibitors. Our results suggest that TGase 2 inhibition is a significant driver of transcriptional dysregulation in HD and should further stimulate efforts to understand how it exerts this function.

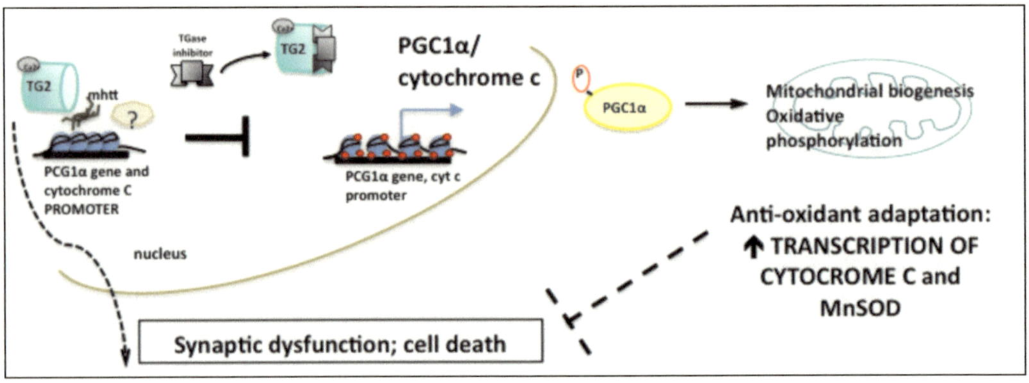

Fig. 4. Proposed mechanism of action for TGase 2 in HD. In the presence of mHTT, TGase2 is hyperactived and it can bind to the promoter of genes such as cytochrome c and PGC1-alpha repressing transcription. The use of specific TGase 2 inhibitors displace TGase 2 from these promoters and block synaptic dysfunction and consequent cell death.

4. Conclusion

Targeting transcriptional dysregulation is one of the most promising avenues for this untreatable disease. The continuous understanding of how transcriptional regulation occurs in vivo along with the development of more specific modulators of chromatin remodelling enzymes will lead hopefully to a cure for HD in the early future. In the last ten years, since the involvement of transcriptional dysfunction has been reported in the field, huge efforts have been invested by researchers, founding agencies, private foundations and patients, all over the world. Broad HDAC inhibitors, specific HDAC inhibitors, CREB activators, SP1 modulators, TGase 2 inhibitors have been tested so far in mouse models and clinical trials. Unfortunately, the results in humans are not as promising as observed in mouse models, suggesting that a deeper understanding of the molecular mechanisms leading to neurodegeneration and the design of combined therapies are still required.

5. Acknowledgment

A special thank to Dr. Ratan for his support, Dr. Sama Sleiman for discussions on transcription and neurodegeneration, Dr. Sivaramakrishnan Muthuswamy for critical revisions of this chapter and Sergio Robbiati for suggestions on the manuscript.

6. References

Arrasate, M., Mitra, S., Schweitzer, E. S., Segal, M. R. & Finkbeiner, S. (2004). Inclusion body formation reduces levels of mutant huntingtin and the risk of neuronal death. *Nature*, 431, 7010, (Oct 14, 2004), pp. (805-810), ISSN 1476-4687

Bailey, C. D., Graham, R. M., Nanda, N., Davies, P. J. & Johnson, G. V. (2004). Validity of mouse models for the study of tissue transglutaminase in neurodegenerative diseases. *Mol Cell Neurosci*, 25, 3, (Mar, 2004), pp. (493-503), ISSN 1044-7431

Bailey, C. D. & Johnson, G. V. (2006). The protective effects of cystamine in the R6/2 Huntington's disease mouse involve mechanisms other than the inhibition of tissue transglutaminase. *Neurobiol Aging*, 27, 6, (Jun, 2006), pp. (871-879), 0197-4580 (Print) 0197-4580 (Linking)

Ballestar, E., Abad, C. & Franco, L. (1996). Core histones are glutaminyl substrates for tissue transglutaminase. *J Biol Chem*, 271, 31, (Aug 2, 1996), pp. (18817-18824), ISSN 0021-9258

Ballestar, E., Boix-Chornet, M. & Franco, L. (2001). Conformational changes in the nucleosome followed by the selective accessibility of histone glutamines in the transglutaminase reaction: effects of ionic strength. *Biochemistry*, 40, 7, (Feb 20, 2001), pp. (1922-1929), ISSN 0006-2960

Bates, E. A., Victor, M., Jones, A. K., Shi, Y. & Hart, A. C. (2006). Differential contributions of Caenorhabditis elegans histone deacetylases to huntingtin polyglutamine toxicity. *J Neurosci*, 26, 10, (Mar 8, 2006), pp. (2830-2838), ISSN 1529-2401

Benn, C. L., Sun, T., Sadri-Vakili, G., McFarland, K. N., DiRocco, D. P., Yohrling, G. J., Clark, T. W., Bouzou, B. & Cha, J. H. (2008). Huntingtin modulates transcription, occupies gene promoters in vivo, and binds directly to DNA in a polyglutamine-dependent manner. *J Neurosci*, 28, 42, (Oct 15, 2008), pp. (10720-10733), ISSN 1529-2401

Bithell, A., Johnson, R. & Buckley, N. J. (2009). Transcriptional dysregulation of coding and non-coding genes in cellular models of Huntington's disease. *Biochem Soc Trans*, 37, Pt 6, (Dec, 2009), pp. (1270-1275), ISSN 1470-8752

Borrell-Pages, M., Canals, J. M., Cordelieres, F. P., Parker, J. A., Pineda, J. R., Grange, G., Bryson, E. A., Guillermier, M., Hirsch, E., Hantraye, P., Cheetham, M. E., Neri, C., Alberch, J., Brouillet, E., Saudou, F. & Humbert, S. (2006). Cystamine and cysteamine increase brain levels of BDNF in Huntington disease via HSJ1b and transglutaminase. *J Clin Invest*, 116, 5, (May, 2006), pp. (1410-1424), ISSN 0021-9738

Bouwman, P. & Philipsen, S. (2002). Regulation of the activity of Sp1-related transcription factors. *Mol Cell Endocrinol*, 195, 1-2, (Sep 30, 2002), pp. (27-38), ISSN 0303-7207

Buckley, N. J., Johnson, R., Zuccato, C., Bithell, A. & Cattaneo, E. (2010). The role of REST in transcriptional and epigenetic dysregulation in Huntington's disease. *Neurobiol Dis*, 39, 1, (Jul, 2010), pp. (28-39), ISSN 1095-953X

Cha, J. H., Kosinski, C. M., Kerner, J. A., Alsdorf, S. A., Mangiarini, L., Davies, S. W., Penney, J. B., Bates, G. P. & Young, A. B. (1998). Altered brain neurotransmitter receptors in transgenic mice expressing a portion of an abnormal human huntington disease gene. *Proc Natl Acad Sci U S A*, 95, 11, (May 26, 1998), pp. (6480-6485), ISSN 0027-8424

Chiang, M. C., Chen, H. M., Lee, Y. H., Chang, H. H., Wu, Y. C., Soong, B. W., Chen, C. M., Wu, Y. R., Liu, C. S., Niu, D. M., Wu, J. Y., Chen, Y. T. & Chern, Y. (2007). Dysregulation of C/EBPalpha by mutant Huntingtin causes the urea cycle deficiency in Huntington's disease. *Hum Mol Genet*, 16, 5, (Mar 1, 2007), pp. (483-498), ISSN 0964-6906

Chiu, C. T., Liu, G., Leeds, P. & Chuang, D. M. (2011). Combined Treatment with the Mood Stabilizers Lithium and Valproate Produces Multiple Beneficial Effects in Transgenic Mouse Models of Huntington's Disease. *Neuropsychopharmacology*, (Jul 27, 2011), pp. ISSN 1740-634X

Cong, S. Y., Pepers, B. A., Evert, B. O., Rubinsztein, D. C., Roos, R. A., van Ommen, G. J. & Dorsman, J. C. (2005). Mutant huntingtin represses CBP, but not p300, by binding and protein degradation. *Mol Cell Neurosci*, 30, 4, (Dec, 2005), pp. (560-571), ISSN 1044-7431

Cui, L., Jeong, H., Borovecki, F., Parkhurst, C. N., Tanese, N. & Krainc, D. (2006). Transcriptional repression of PGC-1alpha by mutant huntingtin leads to mitochondrial dysfunction and neurodegeneration. *Cell*, 127, 1, (Oct 6, 2006), pp. (59-69), ISSN 0092-8674

Dedeoglu, A., Kubilus, J. K., Jeitner, T. M., Matson, S. A., Bogdanov, M., Kowall, N. W., Matson, W. R., Cooper, A. J., Ratan, R. R., Beal, M. F., Hersch, S. M. & Ferrante, R. J. (2002). Therapeutic effects of cystamine in a murine model of Huntington's disease. *J Neurosci*, 22, 20, (Oct 15, 2002), pp. (8942-8950), ISSN 1529-2401

DeMarch, Z., Giampa, C., Patassini, S., Martorana, A., Bernardi, G. & Fusco, F. R. (2007). Beneficial effects of rolipram in a quinolinic acid model of striatal excitotoxicity. *Neurobiol Dis*, 25, 2, (Feb, 2007), pp. (266-273), ISSN 0969-9961

Dompierre, J. P., Godin, J. D., Charrin, B. C., Cordelieres, F. P., King, S. J., Humbert, S. & Saudou, F. (2007). Histone deacetylase 6 inhibition compensates for the transport deficit in Huntington's disease by increasing tubulin acetylation. *J Neurosci*, 27, 13, (Mar 28, 2007), pp. (3571-3583), ISSN 1529-2401

Dunah, A. W., Jeong, H., Griffin, A., Kim, Y. M., Standaert, D. G., Hersch, S. M., Mouradian, M. M., Young, A. B., Tanese, N. & Krainc, D. (2002). Sp1 and TAFII130 transcriptional activity disrupted in early Huntington's disease. *Science*, 296, 5576, (Jun 21, 2002), pp. (2238-2243), ISSN 1095-9203

Ferrante, R. J., Kubilus, J. K., Lee, J., Ryu, H., Beesen, A., Zucker, B., Smith, K., Kowall, N. W., Ratan, R. R., Luthi-Carter, R. & Hersch, S. M. (2003). Histone deacetylase inhibition by sodium butyrate chemotherapy ameliorates the neurodegenerative phenotype in Huntington's disease mice. *J Neurosci*, 23, 28, (Oct 15, 2003), pp. (9418-9427), ISSN 1529-2401

Ferrante, R. J., Ryu, H., Kubilus, J. K., D'Mello, S., Sugars, K. L., Lee, J., Lu, P., Smith, K., Browne, S., Beal, M. F., Kristal, B. S., Stavrovskaya, I. G., Hewett, S., Rubinsztein, D. C., Langley, B. & Ratan, R. R. (2004). Chemotherapy for the brain: the antitumor antibiotic mithramycin prolongs survival in a mouse model of Huntington's disease. *J Neurosci*, 24, 46, (Nov 17, 2004), pp. (10335-10342), ISSN 1529-2401

Gardian, G., Browne, S. E., Choi, D. K., Klivenyi, P., Gregorio, J., Kubilus, J. K., Ryu, H., Langley, B., Ratan, R. R., Ferrante, R. J. & Beal, M. F. (2005). Neuroprotective effects of phenylbutyrate in the N171-82Q transgenic mouse model of Huntington's disease. *J Biol Chem*, 280, 1, (Jan 7, 2005), pp. (556-563), ISSN 0021-9258

Giampa, C., DeMarch, Z., D'Angelo, V., Morello, M., Martorana, A., Sancesario, G., Bernardi, G. & Fusco, F. R. (2006). Striatal modulation of cAMP-response-element-binding protein (CREB) after excitotoxic lesions: implications with neuronal vulnerability in Huntington's disease. *Eur J Neurosci*, 23, 1, (Jan, 2006), pp. (11-20), ISSN 0953-816X

Giampa, C., Patassini, S., Borreca, A., Laurenti, D., Marullo, F., Bernardi, G., Menniti, F. S. & Fusco, F. R. (2009). Phosphodiesterase 10 inhibition reduces striatal excitotoxicity in the quinolinic acid model of Huntington's disease. *Neurobiol Dis*, 34, 3, (Jun, 2009), pp. (450-456), ISSN 1095-953X

Gines, S., Seong, I. S., Fossale, E., Ivanova, E., Trettel, F., Gusella, J. F., Wheeler, V. C., Persichetti, F. & MacDonald, M. E. (2003). Specific progressive cAMP reduction implicates energy deficit in presymptomatic Huntington's disease knock-in mice. *Hum Mol Genet*, 12, 5, (Mar 1, 2003), pp. (497-508), ISSN 0964-6906

Goldman, P. S., Tran, V. K. & Goodman, R. H. (1997). The multifunctional role of the co-activator CBP in transcriptional regulation. *Recent Prog Horm Res*, 52, 1997), pp. (103-119; discussion 119-120), ISSN 0079-9963

Gray, S. G. (2010). Targeting histone deacetylases for the treatment of Huntington's disease. *CNS Neurosci Ther*, 16, 6, (Dec, 2010), pp. (348-361), ISSN 1755-5949

Grove, V. E., Jr., Quintanilla, J. & DeVaney, G. T. (2000). Improvement of Huntington's disease with olanzapine and valproate. *N Engl J Med*, 343, 13, (Sep 28, 2000), pp. (973-974), ISSN 0028-4793

Hathorn, T., Snyder-Keller, A. & Messer, A. (2011). Nicotinamide improves motor deficits and upregulates PGC-1alpha and BDNF gene expression in a mouse model of Huntington's disease. *Neurobiol Dis*, 41, 1, (Jan, 2011), pp. (43-50), ISSN 1095-953X

Hockly, E., Richon, V. M., Woodman, B., Smith, D. L., Zhou, X., Rosa, E., Sathasivam, K., Ghazi-Noori, S., Mahal, A., Lowden, P. A., Steffan, J. S., Marsh, J. L., Thompson, L. M., Lewis, C. M., Marks, P. A. & Bates, G. P. (2003). Suberoylanilide hydroxamic acid, a histone deacetylase inhibitor, ameliorates motor deficits in a mouse model of Huntington's disease. *Proc Natl Acad Sci U S A*, 100, 4, (Feb 18, 2003), pp. (2041-2046), ISSN 0027-8424

Hogarth, P., Lovrecic, L. & Krainc, D. (2007). Sodium phenylbutyrate in Huntington's disease: a dose-finding study. *Mov Disord*, 22, 13, (Oct 15, 2007), pp. (1962-1964), ISSN 0885-3185

Huang, Y., Greene, E., Murray Stewart, T., Goodwin, A. C., Baylin, S. B., Woster, P. M. & Casero, R. A., Jr. (2007). Inhibition of lysine-specific demethylase 1 by polyamine analogues results in reexpression of aberrantly silenced genes. *Proc Natl Acad Sci U S A*, 104, 19, (May 8, 2007), pp. (8023-8028), ISSN 0027-8424

Igarashi, S., Morita, H., Bennett, K. M., Tanaka, Y., Engelender, S., Peters, M. F., Cooper, J. K., Wood, J. D., Sawa, A. & Ross, C. A. (2003). Inducible PC12 cell model of Huntington's disease shows toxicity and decreased histone acetylation. *Neuroreport*, 14, 4, (Mar 24, 2003), pp. (565-568), ISSN 0959-4965

Iijima-Ando, K., Wu, P., Drier, E. A., Iijima, K. & Yin, J. C. (2005). cAMP-response element-binding protein and heat-shock protein 70 additively suppress polyglutamine-mediated toxicity in Drosophila. *Proc Natl Acad Sci U S A*, 102, 29, (Jul 19, 2005), pp. (10261-10266), ISSN 0027-8424

Illuzzi, J. L., Vickers, C. A. & Kmiec, E. B. (2011). Modifications of p53 and the DNA Damage Response in Cells Expressing Mutant Form of the Protein Huntingtin. *J Mol Neurosci*, (Apr 5, 2011), pp. ISSN 1559-1166

Jeitner, T. M., Matson, W. R., Folk, J. E., Blass, J. P. & Cooper, A. J. (2008). Increased levels of gamma-glutamylamines in Huntington disease CSF. *J Neurochem*, 106, 1, (Jul, 2008), pp. (37-44), ISSN 1471-4159

Jeon, J. H., Choi, K. H., Cho, S. Y., Kim, C. W., Shin, D. M., Kwon, J. C., Song, K. Y., Park, S. C. & Kim, I. G. (2003). Transglutaminase 2 inhibits Rb binding of human papillomavirus E7 by incorporating polyamine. *Embo J*, 22, 19, (Oct 1, 2003), pp. (5273-5282), ISSN 0261-4189

Jeong, H., Then, F., Melia, T. J., Jr., Mazzulli, J. R., Cui, L., Savas, J. N., Voisine, C., Paganetti, P., Tanese, N., Hart, A. C., Yamamoto, A. & Krainc, D. (2009). Acetylation targets mutant huntingtin to autophagosomes for degradation. *Cell*, 137, 1, (Apr 3, 2009), pp. (60-72), ISSN 1097-4172

Jiang, H., Nucifora, F. C., Jr., Ross, C. A. & DeFranco, D. B. (2003). Cell death triggered by polyglutamine-expanded huntingtin in a neuronal cell line is associated with degradation of CREB-binding protein. *Hum Mol Genet*, 12, 1, (Jan 1, 2003), pp. (1-12), ISSN 0964-6906

Johnson, R. & Buckley, N. J. (2009). Gene dysregulation in Huntington's disease: REST, microRNAs and beyond. *Neuromolecular Med*, 11, 3, 2009), pp. (183-199), ISSN 1559-1174

Johnson, R., Zuccato, C., Belyaev, N. D., Guest, D. J., Cattaneo, E. & Buckley, N. J. (2008). A microRNA-based gene dysregulation pathway in Huntington's disease. *Neurobiol Dis*, 29, 3, (Mar, 2008), pp. (438-445), ISSN 1095-953X

Karpuj, M. V., Becher, M. W. & Steinman, L. (2002). Evidence for a role for transglutaminase in Huntington's disease and the potential therapeutic implications. *Neurochem Int*, 40, 1, (Jan, 2002), pp. (31-36), ISSN 0197-0186

Karpuj, M. V., Garren, H., Slunt, H., Price, D. L., Gusella, J., Becher, M. W. & Steinman, L. (1999). Transglutaminase aggregates huntingtin into nonamyloidogenic polymers, and its enzymatic activity increases in Huntington's disease brain nuclei. *Proc Natl Acad Sci U S A*, 96, 13, (Jun 22, 1999), pp. (7388-7393), ISSN 0027-8424

Kegel, K. B., Meloni, A. R., Yi, Y., Kim, Y. J., Doyle, E., Cuiffo, B. G., Sapp, E., Wang, Y., Qin, Z. H., Chen, J. D., Nevins, J. R., Aronin, N. & DiFiglia, M. (2002). Huntingtin is present in the nucleus, interacts with the transcriptional corepressor C-terminal binding protein, and represses transcription. *J Biol Chem*, 277, 9, (Mar 1, 2002), pp. (7466-7476), ISSN 0021-9258

Kim, M. O., Chawla, P., Overland, R. P., Xia, E., Sadri-Vakili, G. & Cha, J. H. (2008). Altered histone monoubiquitylation mediated by mutant huntingtin induces transcriptional dysregulation. *J Neurosci*, 28, 15, (Apr 9, 2008), pp. (3947-3957), ISSN 1529-2401

Lee, J., Hagerty, S., Cormier, K. A., Kim, J., Kung, A. L., Ferrante, R. J. & Ryu, H. (2008). Monoallele deletion of CBP leads to pericentromeric heterochromatin condensation through ESET expression and histone H3 (K9) methylation. *Hum Mol Genet*, 17, 12, (Jun 15, 2008), pp. (1774-1782), ISSN 1460-2083

Lee, J., Hwang, Y. J., Boo, J. H., Han, D., Kwon, O. K., Todorova, K., Kowall, N. W., Kim, Y. & Ryu, H. (2011). Dysregulation of upstream binding factor-1 acetylation at K352 is linked to impaired ribosomal DNA transcription in Huntington's disease. *Cell Death Differ*, (May 6, 2011), pp. ISSN 1476-5403

Leone, S., Mutti, C., Kazantsev, A., Sturlese, M., Moro, S., Cattaneo, E., Rigamonti, D. & Contini, A. (2008). SAR and QSAR study on 2-aminothiazole derivatives, modulators of transcriptional repression in Huntington's disease. *Bioorg Med Chem*, 16, 10, (May 15, 2008), pp. (5695-5703), ISSN 1464-3391

Lesort, M., Chun, W., Johnson, G. V. & Ferrante, R. J. (1999). Tissue transglutaminase is increased in Huntington's disease brain. *J Neurochem*, 73, 5, (Nov, 1999), pp. (2018-2027), ISSN 0022-3042

Lin, J., Wu, P. H., Tarr, P. T., Lindenberg, K. S., St-Pierre, J., Zhang, C. Y., Mootha, V. K., Jager, S., Vianna, C. R., Reznick, R. M., Cui, L., Manieri, M., Donovan, M. X., Wu, Z., Cooper, M. P., Fan, M. C., Rohas, L. M., Zavacki, A. M., Cinti, S., Shulman, G. I., Lowell, B. B., Krainc, D. & Spiegelman, B. M. (2004). Defects in adaptive energy metabolism with CNS-linked hyperactivity in PGC-1alpha null mice. *Cell*, 119, 1, (Oct 1, 2004), pp. (121-135), ISSN 0092-8674

Lonze, B. E. & Ginty, D. D. (2002). Function and regulation of CREB family transcription factors in the nervous system. *Neuron*, 35, 4, (Aug 15, 2002), pp. (605-623), ISSN 0896-6273

Luthi-Carter, R., Taylor, D. M., Pallos, J., Lambert, E., Amore, A., Parker, A., Moffitt, H., Smith, D. L., Runne, H., Gokce, O., Kuhn, A., Xiang, Z., Maxwell, M. M., Reeves, S. A., Bates, G. P., Neri, C., Thompson, L. M., Marsh, J. L. & Kazantsev, A. G. (2010). SIRT2 inhibition achieves neuroprotection by decreasing sterol biosynthesis. *Proc Natl Acad Sci U S A*, 107, 17, (Apr 27, 2010), pp. (7927-7932), ISSN 1091-6490

Mastroberardino, P. G., Iannicola, C., Nardacci, R., Bernassola, F., De Laurenzi, V., Melino, G., Moreno, S., Pavone, F., Oliverio, S., Fesus, L. & Piacentini, M. (2002). 'Tissue' transglutaminase ablation reduces neuronal death and prolongs survival in a mouse model of Huntington's disease. *Cell Death Differ*, 9, 9, (Sep, 2002), pp. (873-880), ISSN 1350-9047

McConoughey, S. J., Basso, M., Niatsetskaya, Z. V., Sleiman, S. F., Smirnova, N. A., Langley, B. C., Mahishi, L., Cooper, A. J., Antonyak, M. A., Cerione, R. A., Li, B., Starkov, A., Chaturvedi, R. K., Beal, M. F., Coppola, G., Geschwind, D. H., Ryu, H., Xia, L., Iismaa, S. E., Pallos, J., Pasternack, R., Hils, M., Fan, J., Raymond, L. A., Marsh, J. L., Thompson, L. M. & Ratan, R. R. (2010). Inhibition of transglutaminase 2 mitigates transcriptional dysregulation in models of Huntington disease. *EMBO Mol Med*, 2, 9, (Sep, 2010), pp. (349-370), ISSN 1757-4684

Munoz-Sanjuan, I. & Bates, G. P. (2011). The importance of integrating basic and clinical research toward the development of new therapies for Huntington disease. *J Clin Invest*, 121, 2, (Feb 1, 2011), pp. (476-483), ISSN 1558-8238

Munsie, L., Caron, N., Atwal, R. S., Marsden, I., Wild, E. J., Bamburg, J. R., Tabrizi, S. J. & Truant, R. (2011). Mutant huntingtin causes defective actin remodeling during stress: defining a new role for transglutaminase 2 in neurodegenerative disease. *Hum Mol Genet*, 20, 10, (May 15, 2011), pp. (1937-1951), ISSN 1460-2083

Nucifora, F. C., Jr., Sasaki, M., Peters, M. F., Huang, H., Cooper, J. K., Yamada, M., Takahashi, H., Tsuji, S., Troncoso, J., Dawson, V. L., Dawson, T. M. & Ross, C. A. (2001). Interference by huntingtin and atrophin-1 with cbp-mediated transcription leading to cellular toxicity. *Science*, 291, 5512, (Mar 23, 2001), pp. (2423-2428), ISSN 0036-8075

Oliveira, J. M., Chen, S., Almeida, S., Riley, R., Goncalves, J., Oliveira, C. R., Hayden, M. R., Nicholls, D. G., Ellerby, L. M. & Rego, A. C. (2006). Mitochondrial-dependent Ca2+ handling in Huntington's disease striatal cells: effect of histone deacetylase inhibitors. *J Neurosci*, 26, 43, (Oct 25, 2006), pp. (11174-11186), ISSN 1529-24010270-6474 (Linking)

Packer, A. N., Xing, Y., Harper, S. Q., Jones, L. & Davidson, B. L. (2008). The bifunctional microRNA miR-9/miR-9* regulates REST and CoREST and is downregulated in Huntington's disease. *J Neurosci*, 28, 53, (Dec 31, 2008), pp. (14341-14346), ISSN 1529-2401

Pallos, J., Bodai, L., Lukacsovich, T., Purcell, J. M., Steffan, J. S., Thompson, L. M. & Marsh, J. L. (2008). Inhibition of specific HDACs and sirtuins suppresses pathogenesis in a Drosophila model of Huntington's disease. *Hum Mol Genet*, 17, 23, (Dec 1, 2008), pp. (3767-3775), ISSN 1460-2083

Ryu, H., Lee, J., Hagerty, S. W., Soh, B. Y., McAlpin, S. E., Cormier, K. A., Smith, K. M. & Ferrante, R. J. (2006). ESET/SETDB1 gene expression and histone H3 (K9) trimethylation in Huntington's disease. *Proc Natl Acad Sci U S A*, 103, 50, (Dec 12, 2006), pp. (19176-19181), ISSN 0027-8424

Ryu, H., Lee, J., Olofsson, B. A., Mwidau, A., Dedeoglu, A., Escudero, M., Flemington, E., Azizkhan-Clifford, J., Ferrante, R. J. & Ratan, R. R. (2003a). Histone deacetylase inhibitors prevent oxidative neuronal death independent of expanded polyglutamine repeats via an Sp1-dependent pathway. *Proc Natl Acad Sci U S A*, 100, 7, (Apr 1, 2003a), pp. (4281-4286), ISSN 0027-8424

Ryu, H., Lee, J., Zaman, K., Kubilis, J., Ferrante, R. J., Ross, B. D., Neve, R. & Ratan, R. R. (2003b). Sp1 and Sp3 are oxidative stress-inducible, antideath transcription factors in cortical neurons. *J Neurosci*, 23, 9, (May 1, 2003b), pp. (3597-3606), ISSN 1529-2401

Sadri-Vakili, G., Bouzou, B., Benn, C. L., Kim, M. O., Chawla, P., Overland, R. P., Glajch, K. E., Xia, E., Qiu, Z., Hersch, S. M., Clark, T. W., Yohrling, G. J. & Cha, J. H. (2007). Histones associated with downregulated genes are hypo-acetylated in Huntington's disease models. *Hum Mol Genet*, 16, 11, (Jun 1, 2007), pp. (1293-1306), ISSN 0964-6906

Saft, C., Lauter, T., Kraus, P. H., Przuntek, H. & Andrich, J. E. (2006). Dose-dependent improvement of myoclonic hyperkinesia due to Valproic acid in eight Huntington's Disease patients: a case series. *BMC Neurol*, 6, 2006), pp. (11), ISSN 1471-2377

Seong, I. S., Woda, J. M., Song, J. J., Lloret, A., Abeyrathne, P. D., Woo, C. J., Gregory, G., Lee, J. M., Wheeler, V. C., Walz, T., Kingston, R. E., Gusella, J. F., Conlon, R. A. & MacDonald, M. E. (2010). Huntingtin facilitates polycomb repressive complex 2. *Hum Mol Genet*, 19, 4, (Feb 15, 2010), pp. (573-583), ISSN 1460-2083

Shi, Y., Lan, F., Matson, C., Mulligan, P., Whetstine, J. R., Cole, P. A. & Casero, R. A. (2004). Histone demethylation mediated by the nuclear amine oxidase homolog LSD1. *Cell*, 119, 7, (Dec 29, 2004), pp. (941-953), ISSN 0092-8674

Sleiman, S. F., Langley, B. C., Basso, M., Berlin, J., Xia, L., Payappilly, J. B., Kharel, M. K., Guo, H., Marsh, J. L., Thompson, L. M., Mahishi, L., Ahuja, P., Maclellan, W. R., Geschwind, D. H., Coppola, G., Rohr, J. & Ratan, R. R. (2011). Mithramycin Is a Gene-Selective Sp1 Inhibitor That Identifies a Biological Intersection between Cancer and Neurodegeneration. *J Neurosci*, 31, 18, (May 4, 2011), pp. (6858-6870), ISSN 1529-2401

Soldati, C., Bithell, A., Conforti, P., Cattaneo, E. & Buckley, N. J. (2011). Rescue of gene expression by modified REST decoy oligonucleotides in a cellular model of Huntington's disease. *J Neurochem*, 116, 3, (Feb, 2011), pp. (415-425), ISSN 1471-4159

St-Pierre, J., Drori, S., Uldry, M., Silvaggi, J. M., Rhee, J., Jager, S., Handschin, C., Zheng, K., Lin, J., Yang, W., Simon, D. K., Bachoo, R. & Spiegelman, B. M. (2006). Suppression of reactive oxygen species and neurodegeneration by the PGC-1 transcriptional coactivators. *Cell*, 127, 2, (Oct 20, 2006), pp. (397-408), ISSN 0092-8674

Stack, E. C., Del Signore, S. J., Luthi-Carter, R., Soh, B. Y., Goldstein, D. R., Matson, S., Goodrich, S., Markey, A. L., Cormier, K., Hagerty, S. W., Smith, K., Ryu, H. & Ferrante, R. J. (2007). Modulation of nucleosome dynamics in Huntington's disease. *Hum Mol Genet*, 16, 10, (May 15, 2007), pp. (1164-1175), ISSN 0964-6906

Steffan, J. S., Bodai, L., Pallos, J., Poelman, M., McCampbell, A., Apostol, B. L., Kazantsev, A., Schmidt, E., Zhu, Y. Z., Greenwald, M., Kurokawa, R., Housman, D. E., Jackson, G. R., Marsh, J. L. & Thompson, L. M. (2001). Histone deacetylase inhibitors arrest polyglutamine-dependent neurodegeneration in Drosophila. *Nature*, 413, 6857, (Oct 18, 2001), pp. (739-743), ISSN 0028-0836

Steffan, J. S., Kazantsev, A., Spasic-Boskovic, O., Greenwald, M., Zhu, Y. Z., Gohler, H., Wanker, E. E., Bates, G. P., Housman, D. E. & Thompson, L. M. (2000). The Huntington's disease protein interacts with p53 and CREB-binding protein and represses transcription. *Proc Natl Acad Sci U S A*, 97, 12, (Jun 6, 2000), pp. (6763-6768), ISSN 0027-8424

Sugars, K. L., Brown, R., Cook, L. J., Swartz, J. & Rubinsztein, D. C. (2004). Decreased cAMP response element-mediated transcription: an early event in exon 1 and full-length cell models of Huntington's disease that contributes to polyglutamine pathogenesis. *J Biol Chem*, 279, 6, (Feb 6, 2004), pp. (4988-4999), ISSN 0021-9258

Tang, B., Seredenina, T., Coppola, G., Kuhn, A., Geschwind, D. H., Luthi-Carter, R. & Thomas, E. A. (2011). Gene expression profiling of R6/2 transgenic mice with different CAG repeat lengths reveals genes associated with disease onset and progression in Huntington's disease. *Neurobiol Dis*, 42, 3, (Jun, 2011), pp. (459-467), ISSN 1095-953X

Thomas, E. A., Coppola, G., Desplats, P. A., Tang, B., Soragni, E., Burnett, R., Gao, F., Fitzgerald, K. M., Borok, J. F., Herman, D., Geschwind, D. H. & Gottesfeld, J. M. (2008). The HDAC inhibitor 4b ameliorates the disease phenotype and transcriptional abnormalities in Huntington's disease transgenic mice. *Proc Natl Acad Sci U S A*, 105, 40, (Oct 7, 2008), pp. (15564-15569), ISSN 1091-6490

Van Raamsdonk, J. M., Pearson, J., Bailey, C. D., Rogers, D. A., Johnson, G. V., Hayden, M. R. & Leavitt, B. R. (2005). Cystamine treatment is neuroprotective in the YAC128 mouse model of Huntington disease. *J Neurochem*, 95, 1, (Oct, 2005), pp. (210-220), ISSN 0022-3042

Varghese, S., Gupta, D., Baran, T., Jiemjit, A., Gore, S. D., Casero, R. A., Jr. & Woster, P. M. (2005). Alkyl-substituted polyaminohydroxamic acids: a novel class of targeted histone deacetylase inhibitors. *J Med Chem*, 48, 20, (Oct 6, 2005), pp. (6350-6365), ISSN 0022-2623

Wang, X., Sarkar, A., Cicchetti, F., Yu, M., Zhu, A., Jokivarsi, K., Saint-Pierre, M. & Brownell, A. L. (2005). Cerebral PET imaging and histological evidence of transglutaminase inhibitor cystamine induced neuroprotection in transgenic R6/2 mouse model of Huntington's disease. *J Neurol Sci*, 231, 1-2, (Apr 15, 2005), pp. (57-66), ISSN 0022-510X

Weydt, P., Pineda, V. V., Torrence, A. E., Libby, R. T., Satterfield, T. F., Lazarowski, E. R., Gilbert, M. L., Morton, G. J., Bammler, T. K., Strand, A. D., Cui, L., Beyer, R. P., Easley, C. N., Smith, A. C., Krainc, D., Luquet, S., Sweet, I. R., Schwartz, M. W. & La Spada, A. R. (2006). Thermoregulatory and metabolic defects in Huntington's disease transgenic mice implicate PGC-1alpha in Huntington's disease neurodegeneration. *Cell Metab*, 4, 5, (Nov, 2006), pp. (349-362), ISSN 1550-4131

Wyttenbach, A., Swartz, J., Kita, H., Thykjaer, T., Carmichael, J., Bradley, J., Brown, R., Maxwell, M., Schapira, A., Orntoft, T. F., Kato, K. & Rubinsztein, D. C. (2001). Polyglutamine expansions cause decreased CRE-mediated transcription and early gene expression changes prior to cell death in an inducible cell model of Huntington's disease. *Hum Mol Genet*, 10, 17, (Aug 15, 2001), pp. (1829-1845), ISSN 0964-6906

Xiang, Z., Valenza, M., Cui, L., Leoni, V., Jeong, H. K., Brilli, E., Zhang, J., Peng, Q., Duan, W., Reeves, S. A., Cattaneo, E. & Krainc, D. (2011). Peroxisome-proliferator-activated receptor gamma coactivator 1 alpha contributes to dysmyelination in experimental models of Huntington's disease. *J Neurosci*, 31, 26, (Jun 29, 2011), pp. (9544-9553), ISSN 1529-2401

Xu, W. S., Parmigiani, R. B. & Marks, P. A. (2007). Histone deacetylase inhibitors: molecular mechanisms of action. *Oncogene*, 26, 37, (Aug 13, 2007), pp. (5541-5552), ISSN 0950-9232

Zadori, D., Geisz, A., Vamos, E., Vecsei, L. & Klivenyi, P. (2009). Valproate ameliorates the survival and the motor performance in a transgenic mouse model of Huntington's disease. *Pharmacol Biochem Behav*, 94, 1, (Nov, 2009), pp. (148-153), ISSN 1873-5177

Zhai, W., Jeong, H., Cui, L., Krainc, D. & Tjian, R. (2005). In vitro analysis of huntingtin-mediated transcriptional repression reveals multiple transcription factor targets. *Cell*, 123, 7, (Dec 29, 2005), pp. (1241-1253), ISSN 0092-8674

Zuccato, C., Ciammola, A., Rigamonti, D., Leavitt, B. R., Goffredo, D., Conti, L., MacDonald, M. E., Friedlander, R. M., Silani, V., Hayden, M. R., Timmusk, T., Sipione, S. & Cattaneo, E. (2001). Loss of huntingtin-mediated BDNF gene transcription in Huntington's disease. *Science*, 293, 5529, (Jul 20, 2001), pp. (493-498), ISSN 0036-8075

Zuccato, C., Tartari, M., Crotti, A., Goffredo, D., Valenza, M., Conti, L., Cataudella, T., Leavitt, B. R., Hayden, M. R., Timmusk, T., Rigamonti, D. & Cattaneo, E. (2003). Huntingtin interacts with REST/NRSF to modulate the transcription of NRSE-controlled neuronal genes. *Nat Genet*, 35, 1, (Sep, 2003), pp. (76-83), ISSN 1061-4036

12

ZNF395 (HDBP2 /PBF) is a Target Gene of Hif-1α

Darko Jordanovski, Christine Herwartz and Gertrud Steger
Institute of Virology, University of Cologne, Cologne, Germany

1. Introduction

The extention of the polyglutamin (polyQ) repeats within the N-terminus of the Huntingtin (Htt) protein causes Huntington´s disease (D) associated with aging and the accumulation of mutant (mt) Htt in the diseased neurons. The level of the mtHtt proteins and the length of the polyQ repeat determine the severity and progression of the disease. The intracellular aggregation of mtHtt to form insoluble inclusion bodies may be crucial to the development of the disease. One of the most important mechanisms by which the mutant Htt leads to cell cytotoxicity comprises transcriptional dysregulation (Cha, 2007). A variety of mechanisms have been attributed to result in large changes in the expression of coding and non-coding RNAs. In addition to the cytoplasm, the proteolytic fragments comprising the N-terminus of mtHtt aggregate in the nucleus where they sequester several co-factors and transcription factors such as CBP, p300, mSIN3a and Sp1. This might limit their access to DNA and decrease their normal transcriptional activity (Buckley et al., 2010).

Genes repressed by mtHtt include those controlling adaption to low mitochondrial energy charge. This may cause mitochondrial dysfunction resulting in aberrant energy metabolism which is one of the primary defects in Huntington´s D (Cui et al., 2006). A way to compensate for mitochondrial energy deficit is to shift the cell´s energy production from oxidative phosphorylation towards aerobic glycolysis, as observed upon adaption to hypoxia. The transcriptional upregulation of glycolytic enzymes in the presence of low O_2 tension occurs primarily through the hypoxia inducible transcription factor-1α (Hif-1α) that functions as a global regulator of O_2 homeostasis and adaption to low energy (reviewed in (Denko, 2008; Majmundar et al., 2010)). Under normoxia Hif-1α is an unstable protein. In the presence of O_2, prolylhydroxylases (PHD) act as oxygen sensors and hydroxylate a prolin in Hif-1α, which is a signal to initiate the proteasome mediated degradation of Hif-1α. Upon hypoxia, PHD are inactive, resulting in the stabilization of Hif-1α, which can then dimerize with its interaction partner Hif-1β and bind to its specific recognition sequence, the hypoxia response element (HRE), present in the control regions of its targets genes. Hif-1α activates more than 100 genes associated with the adaption to hypoxic stress, including genes involved in angiogenesis, cell survival and aerobic glycolysis. Hif-1α stimulates glycolytic energy production by transactivating genes involved in extracellular glucose import (such as GLUT1) and coding for enzymes responsible for the breakdown of intracellular glucose. Small molecule inhibitors of PHD have been shown to protect neurons from ischemic or oxidative injury. PHD inhibitors, resulting in activation of Hif-1α, are able to prevent neuronal death induced by mitochondrial toxins and have therapeutic implications in the treatment for Huntington´s D and Alzheimer D (Niatsetskaya et al., 2010). Inhibition of PHD may also prevent mitochondrial toxicity in glioma cells. Thus, PHD inhibitors are regarded as promising candidates for preventing cell death in Huntington´s D as well as other neurodegenerative Ds associated with metabolic stress (Harten et al., 2010). On the other side, Hif-1α was among the genes whose expression was significantly upregulated in brain from post-mortem Huntington´s D patients as well as in blood samples from symptomatic patients in contrast to non symptomatic patients and healthy individuals, suggesting a role of these factors in disease development. Elevated level of these factors including Hif-1α was also correlated with disease progression and response to treatment (Borovecki et al., 2005; Lovrecic et al., 2009) indicating that Hif-1α activation may not only be beneficial for the disease outcome. In order to consider PHD inhibitors as neurological therapeutics, it is necessary to characterize their effect in the cells at the molecular level. We describe here that Hif-1α activates the expression of Huntington´s D binding protein 2 (HDPB2), a protein binding to a DNA segment within the Htt promoter that mediates neuronal cell specific activation of Htt expression (Tanaka et al., 2004) and discuss its potential implication for Huntington´s D.

2. ZNF395 is identical to HDBP2, binding to a neuronal specific regulatory element of the Htt promoter

HDBP1 and HDBP2 are two closely related proteins that were identified as transcription factors binding to a 7bp GC rich sequence which resides in triplicate at intervals of 13bp within and proximal to the -20bp direct repeat sequences of the Htt promoter (for overview see Fig. 1A). Two years earlier, we have identified the cellular factor HDBP2 by its ability to bind to regulatory regions in papillomaviruses (PV) and subsequently called the protein papillomavirus binding factor (PBF) (Boeckle et al., 2002). The official gene name is ZNF395, which we will use here. HDBP1 is identical to GLUT4-Enhancer factor (GEF) which activates the gene expression of GLUT4, a glucose transporter. HDBP1/GEF and ZNF395 are closely related to the mouse glucocorticoid induced gene 1 (GIG1, human ZNF704). It has been suggested that these three proteins that are conserved from drosophila to vertebrates build up a new family of transcription factors. They share three conserved regions CR1, CR2 and CR3, a domain rich in serines and prolines and have the potential to form a zinc-finger structure (see Fig. 1C for overview). The C-terminal CR3 is responsible for DNA-binding (Sichtig et al., 2007a; Tanaka et al., 2004). This region is highly similar to the 30-amino acid auxiliary DNA interaction motif present in "E" variants of TCF transcription factors (Atcha et al., 2007). Although the recognition motif of ZNF395 has not yet been determined, it recognizes GC rich sequences which is supported by the finding that ZNF395 was "very strongly" excluded from DNA after CpG methylation in a genome-wide screen (Bartke et al., 2010). Moreover, it binds to GCCGGCG in the Huntington´s D gene promoter (Tanaka et al., 2004) and a CCGG in HPV8 (Boeckle et al., 2002) while GEF binds ACCGG within GLUT4 (Knight et al., 2003; Oshel et al., 2000). While in Drosophila GEF was found to be required for normal wing-positioning (Yazdani et al., 2008), a physiological role of these factors in vertebrates is unknown. ZNF395 was characterized as a nucleo-cytoplasmic shuttling protein (Tanaka et al., 2004). We could show that its subcellular localization seems to be regulated by growth factors, since recombinant ZNF395 entered the nucleus upon withdrawal of growth factors from the cell culture medium. The binding to 14-3-3β contributes to the control of the subcellular localization of ZNF395. Moreover, over-expression of ZNF395 resulted in inhibition of cell growth (Sichtig et al., 2007b). ZNF395-mediated growth inhibition of osteosarcoma cell lines was shown to rely on apoptosis (Tsukahara et al., 2008).

2.1 ZNF395 is a repressor of PV gene expression

In the case of PV, mutations abolishing the DNA binding of ZNF395 reduced the promoter activity, from which we concluded that ZNF395 is a transcriptional activator (Boeckle et al., 2002). Surprisingly, the over-expression of ZNF395 resulted in repression of transcription from the PV promoters. This repression was dependent on the recruitment of the mSIN3A/HDAC1/2 complex via a direct interaction of ZNF395 with Sin3A associated protein of 30kDa (SAP30), a component of this complex. Moreover, transcriptional repression required the intact CR3, indicating that ZNF395 has to bind to DNA (Sichtig et al., 2007a).

2.2 Recombinant ZNF395 is a repressor of the Htt promoter

Similar to the situation observed with PV promoters, mutations within the 7bp motif that abolished binding of GEF/HDBP1 and ZNF395/HDBP2, reduced the Htt promoter activity in a neuronal cell line, while there was no effect in HeLa cells (Tanaka et al., 2004), indicating that these two factors are involved in neuronal specific gene expression of Htt. However, neither the direct involvement of these two factors in the control of Htt expression nor their specific activity has been analyzed. In order to investigate the role of ZNF395 on the expression of Htt we performed transient transfections with two different reporter constructs. The first construct contained the Htt promoter and 1032bp of its upstream regulatory region (-1032-Htt-Luc) while the second had 324bp of the upstream region of the Htt promoter (-324-Htt-Luc). Both constructs that were kindly provided by Coles et al. (Coles et al., 1998) contained the 21 base pair repeat flanked by three copies of the 7bp GC rich sequence, the putative DNA segment bound by HDBP1/GEF and HDBP2/ZNF395 (Tanaka et al., 2004; see Fig. 1A). We used an immortalized keratinocyte cell line, since we have initially isolated the ORF for ZNF395 from these cells. As shown in Fig. 1, transfecting increasing amounts of an expression vector for ZNF395 resulted in a dose dependent repression of the promoter up to 90%. The level of repression was similar with both constructs. Thus, consistent with the situation in PV, heterologous ZNF395 acts as repressor of Htt. In order to exclude that a cell specific factor is required for ZNF395 to activate we used U87 MG cells, a human glioblastoma cell line. Again, over-expression of ZNF395 induced 80% repression of the promoter with the -324-Htt-Luc reporter. In order to further address the mechanism of repression we tested a set of ZNF395 mutants. Most of the repression was relieved when over-expressing ZNF395mtCR3, devoid of DNA-binding, indicating that ZNF395 has to bind to DNA to act as transcriptional repressor. ZNF395Δ280-312 also revealed a reduced repression although it was still able to decrease luc activity by 60%. Thus, recruitment of the mSIN3A/HDAC1 complex via interaction of amino acids 280-312 with SAP30 may be involved, although regions outside contribute to the repression. The co-transfection of an expression vector for HDBP1/GEF did not stimulate the activity of the HD promoter as well. In contrast to ZNF395, GEF only slightly repressed the HD promoter in this assay (Fig. 1C), even when increasing amounts of expression vectors have been transfected (data not shown), indicating that both proteins might affect Huntington´s D gene expression differentially.

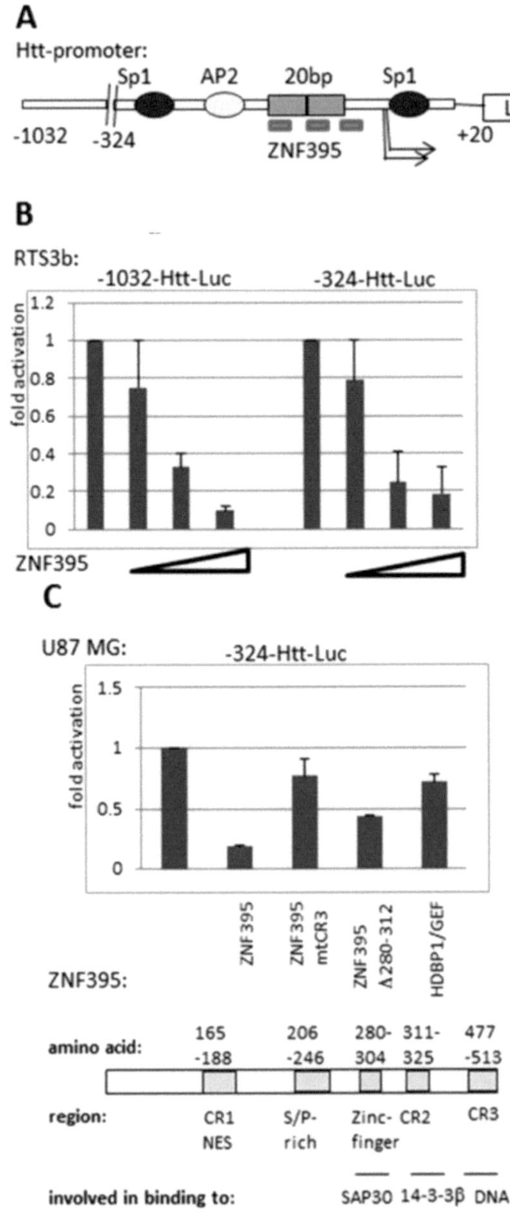

Fig. 1. ZNF395 represses the Huntington´s D gene promoter. RTS3b cells (B) and U87 MG cells (C) were transiently transfected with luciferase reporter constructs containing the Htt promoter including its upstream region up to -324 or -1032, respectively, together with an expression vector for ZNF395 (5, 10ng and 20ng in B and 10ng in C) or for HDBP1/GEF (described in Knight et al., 2003) as indicated. The cells were transfected by the FuGene reagent (Roche diagnostics) and 48h later luciferase activity was determined. The results represent the means of two (in C) and three (in B) independent experiments and the standard deviations are shown. The structure of the Htt promoter is given in (A) and the structure of ZNF395 with its domains that are described in the text in (C).

2.3 ZNF395 is over-expressed in cancers and is a target gene of Hif-1α

Data obtained from transcriptional profiling implied that ZNF395 is a target gene of Hif-1α. For instance, ZNF395 was among the genes activated by hypoxia, by over-expression of Hif-1α, in the absence of von Hippel Lindau (VHL) proteins or by treating the cells with a chemical inducer of Hif-1α, DMOG (dimethyloxalyl-glycine) (Jiang et al., 2003; Lal et al., 2001). Consistent with these reports, ZNF395 was among hypoxia inducible genes that represent a hypoxic signature in neuroblastoma cell lines and neuroblastomas as well as in glioblastomas (Fardin et al., 2010; Murat et al., 2009). These reports imply that ZNF395 over-expression may have a functional role in cancer progression and in Hif-1α regulated pathways.

2.3.1 Hif-1α activates the expression of ZNF395

In order to investigate a role of Hif-1α in the regulation of expression of ZNF395, we treated RTS3b cells with DMOG for 24h prior harvesting. The Western Blot shown in Fig. 2 demonstrates that the level of Hif-1α protein increased, which is in line with the stabilization of Hif-1α due to the inhibition of PHD. Only from extracts of cells that have been treated with DMOG, we were able to precipitate ZNF395 by a specific antibody, indicating that Hif-1α mediated activation of ZNF395 expression is also reflected by increased protein level. In order to address the role of Hif-1α in regulation of ZNF395 expression in more detail, we cloned a cellular DNA fragment harboring the putative promoter of ZNF395 by PCR. A fragment spanning 1190 bases upstream of the initiation site (-1190) to 51 bases downstream (+51) of the mRNA for ZNF395 was amplified from total genomic DNA of RTS3b cells and cloned into a luciferase reporter gene vector. An analysis of putative transcription factor binding sites predicted a high affinity HRE at pos. -815 (Fig. 2B). The co-transfection of an expression vector for Hif-1α increased the promoter activity 1.5 fold. The deletion of the segment from -830 to –565, thus removing the HRE at pos. -815, eliminated this small activation indicating that the effect is specific and the HRE is required, although the activation is much smaller than observed in microarrays, where up to 7 fold inductions of ZNF395 specific mRNA level were described (Jiang et al., 2003; Lal et al., 2001). A second putative HRE located 2000bp further upstream and not included in the DNA segment in our reporter construct might contribute to the Hif-1α mediated regulation of the ZNF395 expression. Moreover, our preliminary results indicate that Hif-1α cooperates with other transcription factors binding to the promoter region of ZNF395 as well (own unpublished results).

3. Conclusions

Activation of the Hif-1α-pathway by inhibitors of PHDs was shown to prevent neuronal cell death in a Huntington´s D cell culture model, thus PHD inhibitors might be considered as therapeutics. Our data shown here imply that PHD inhibitors will also induce ZNF395 via Hif-1α. ZNF395 will then bind to the Htt promoter and contribute to the control of the expression of mtHtt. Our findings that ZNF395 represses the Htt-promoter implicate that ZNF395 contributes to an amelioration of the disease and/or a slowing down of disease progression achieved by PHD inhibitors. However, until now an involvement of ZNF395 in the regulation of Htt is not convincingly shown at all and has to be analyzed carefully.

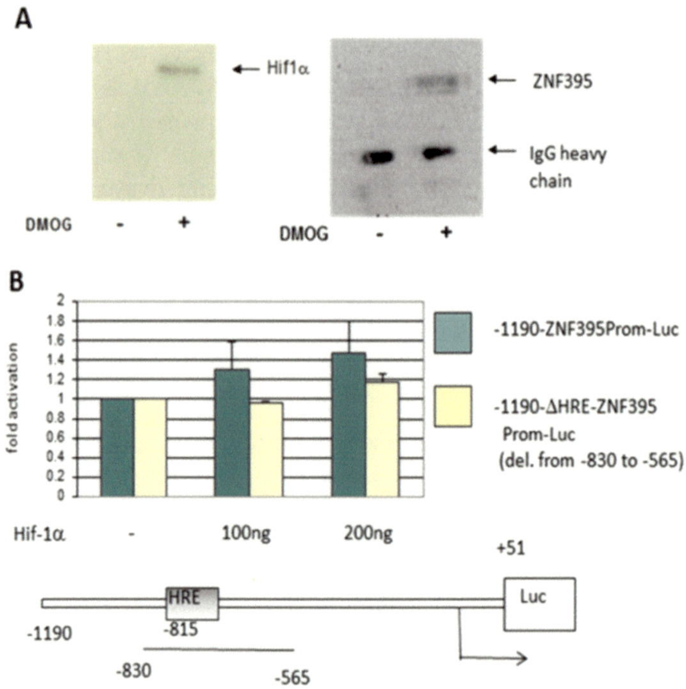

Fig. 2. Hif-1α activates ZNF395 expression. (A) RTS3b cells were treated with 1mM DMOG for 24h. Cell extracts were used in Western Blot developed with an antibody against Hif-1α (from Epitomics) (on the left) or for an immunoprecipitation with an antibody against ZNF395 followed by Western Blot, developed with the anti-ZNF395 antibody described in (Boeckle et al., 2002). The positions of Hif-1α and ZNF395 are indicated. (B) RTS3b cells were transiently transfected as described in figure 1, with a reporter construct containing a 1190 bp fragment upstream of the initiation site of the ZNF395 gene in front of the luciferase and an expression vector for Hif-1α. In the construct -1190-ΔHRE-ZNF395Prom-Luc the segment from -830 to -565 has been deleted. The structure of the ZNF395 promoter with the position of the putative HRE is shown beneath the graph. The graph represents the means of three independent experiments and the standard deviations are shown.

Tanaka et al. concluded that ZNF395 acts as activator since the mutations that reduced promoter activity also resulted in loss of binding of ZNF395 in vitro (Tanaka et al., 2004). However, it cannot not be excluded that the mutations affected the binding of another factor recognizing a similar sequence and mediating activation. A chromatin immunoprecipitation assay might be performed to reveal the presence of endogenous ZNF395 on the Htt promoter in stratial cells. A neuronal cell specific knock out of ZNF395 in Htt mouse models that have been described will provide evidence for the implication and the specific role of ZNF395 in the control of Htt expression. The effect of ZNF395 on the Htt promoter is strikingly similar to that observed for the PV promoters. Consistently, eliminating the binding of ZNF395 in vitro reduced the PV-promoter activity, but over-expression of the recombinant protein efficiently repressed the PV promoter, which required the DNA-binding domain of ZNF395 and the segment binding to SAP30 (Sichtig et al., 2007a). This may reflect that ZNF395 acts as activator or as repressor of transcription. The specific effect of ZNF395, including its stability and subcellular localization might be controlled by post-translational modifications such as phosphorylation and ubiquitination. In line with this, we found that Akt-kinase-mediated phosphorylation of ZNF395 at S447/449/451 creates an interaction motif for 14-3-3β, which contributes to the control of the subcellular localization of ZNF395 and its cell growth inhibitory function (Sichtig et al., 2007b). Elucidating these modifications, the associated pathways and their consequences for the activity of ZNF395 is a prerequisite to understand a role in Huntington´s D.

4. Acknowledgment

We thank D. Richards, D. C. Rubinsztein and L. Olson for providing plasmids and the members of the Institute of Virology for helpful discussion. This work was supported by the Deutsche Forschungsgemeinschaft (STE604/5-1) and the Köln Fortune Program of the Medical Faculty of the University of Cologne.

5. References

Atcha, F. A., Syed, A., Wu, B., Hoverter, N. P., Yokoyama, N. N., Ting, J. H., Munguia, J. E., Mangalam, H. J., Marsh, J. L. & Waterman, M. L. (2007). A unique DNA binding domain converts T-cell factors into strong Wnt effectors. *Mol Cell Biol* 27, 8352-8363.

Bartke, T., Vermeulen, M., Xhemalce, B., Robson, S. C., Mann, M. & Kouzarides, T. (2010). Nucleosome-interacting proteins regulated by DNA and histone methylation. *Cell* 143, 470-484.

Boeckle, S., Pfister, H. & Steger, G. (2002). A new cellular factor recognizes E2 binding sites which mediate transcriptional repression by E2. *Virology* 293, 103-117.

Borovecki, F., Lovrecic, L., Zhou, J., Jeong, H., Then, F., Rosas, H. D., Hersch, S. M., Hogarth, P., Bouzou, B., Jensen, R. V. & Krainc, D. (2005). Genome-wide expression profiling of human blood reveals biomarkers for Huntington's disease. *Proc Natl Acad Sci U S A* 102, 11023-11028.

Buckley, N. J., Johnson, R., Zuccato, C., Bithell, A. & Cattaneo, E. (2010). The role of REST in transcriptional and epigenetic dysregulation in Huntington's disease. *Neurobiol Dis* 39, 28-39.

Cha, J. H. (2007). Transcriptional signatures in Huntington's disease. *Prog Neurobiol* 83, 228-248.

Coles, R., Caswell, R. & Rubinsztein, D. C. (1998). Functional analysis of the Huntington's disease (HD) gene promoter. *Hum Mol Genet* 7, 791-800.

Cui, L., Jeong, H., Borovecki, F., Parkhurst, C. N., Tanese, N. & Krainc, D. (2006). Transcriptional repression of PGC-1alpha by mutant huntingtin leads to mitochondrial dysfunction and neurodegeneration. *Cell* 127, 59-69.

Denko, N. C. (2008). Hypoxia, HIF1 and glucose metabolism in the solid tumour. *Nat Rev Cancer* 8, 705-713.

Dunah, A. W., Jeong, H., Griffin, A., Kim, Y. M., Standaert, D. G., Hersch, S. M., Mouradian, M. M., Young, A. B., Tanese, N. & Krainc, D. (2002). Sp1 and TAFII130 transcriptional activity disrupted in early Huntington's disease. *Science* 296, 2238-2243.

Fardin, P., Barla, A., Mosci, S., Rosasco, L., Verri, A., Versteeg, R., Caron, H. N., Molenaar, J. J., Ora, I., Eva, A., Puppo, M. & Varesio, L. (2010). A biology-driven approach identifies the hypoxia gene signature as a predictor of the outcome of neuroblastoma patients. *Mol Cancer* 9, 185.

Harten, S. K., Ashcroft, M. & Maxwell, P. H. (2010). Prolyl hydroxylase domain inhibitors: a route to HIF activation and neuroprotection. *Antioxid Redox Signal* 12, 459-480.

Jiang, Y., Zhang, W., Kondo, K., Klco, J. M., St Martin, T. B., Dufault, M. R., Madden, S. L., Kaelin, W. G., Jr. & Nacht, M. (2003). Gene expression profiling in a renal cell carcinoma cell line: dissecting VHL and hypoxia-dependent pathways. *Mol Cancer Res* 1, 453-462.

Knight, J. B., Eyster, C. A., Griesel, B. A. & Olson, A. L. (2003). Regulation of the human GLUT4 gene promoter: interaction between a transcriptional activator and myocyte enhaner factor 2A. *Proc Natl Acad Sci U S A* 100, 14725-14730.

Lal, A., Peters, H., St Croix, B., Haroon, Z. A., Dewhirst, M. W., Strausberg, R. L., Kaanders, J. H., van der Kogel, A. J. & Riggins, G. J. (2001). Transcriptional response to hypoxia in human tumors. *J Natl Cancer Inst* 93, 1337-1343.

Li, S. H., Cheng, A. L., Zhou, H., Lam, S., Rao, M., Li, H. & Li, X. J. (2002). Interaction of Huntington disease protein with transcriptional activator Sp1. *Mol Cell Biol* 22, 1277-1287.

Lovrecic, L., Kastrin, A., Kobal, J., Pirtosek, Z., Krainc, D. & Peterlin, B. (2009). Gene expression changes in blood as a putative biomarker for Huntington's disease. *Mov Disord* 24, 2277-2281.

Majmundar, A. J., Wong, W. J. & Simon, M. C. (2010). Hypoxia-inducible factors and the response to hypoxic stress. *Mol Cell* 40, 294-309.

Murat, A., Migliavacca, E., Hussain, S. F., Heimberger, A. B., Desbaillets, I., Hamou, M. F., Ruegg, C., Stupp, R., Delorenzi, M. & Hegi, M. E. (2009). Modulation of angiogenic and inflammatory response in glioblastoma by hypoxia. *PLoS One* 4, e5947.

Niatsetskaya, Z., Basso, M., Speer, R. E., McConoughey, S. J., Coppola, G., Ma, T. C. & Ratan, R. R. (2010). HIF prolyl hydroxylase inhibitors prevent neuronal death induced by mitochondrial toxins: therapeutic implications for Huntington's disease and Alzheimer's disease. *Antioxid Redox Signal* 12, 435-443.

Oshel, K. M., Knight, J. B., Cao, K. T., Thai, M. V. & Olson, A. L. (2000). Identification of a 30-base pair regulatory element and novel DNA binding protein that regulates the human GLUT4 promoter in transgenic mice. *J Biol Chem* 275, 23666-23673.

Sichtig, N., Körfer, N. & Steger, G. (2007a). Papillomavirus binding factor represses transcription via recruitment of the Sin3/HDAC co-repressor complex. *Archives of Biochemistry and Biophysics* 467, 67-75.

Sichtig, N., Silling, S. & Steger, G. (2007b). Papillomavirus binding factor (PBF) mediated inhibition of cell growth is regulated by 14-3-3 ß. *Archives of Biochemistry and Biophysics* 464, 90-99.

Tanaka, K., Shouguchi-Miyata, J., Miyamoto, N. & Ikeda, J.-E. (2004). Novel nuclear shuttle proteins, HDBP1 and HDBP2, bind to neuronal cell-specific cis-regulatory element in the promoter for the human Huntington disease gene. *J Biol Chem* 279, 7275-7286.

Tsukahara, T., Kimura, S., Ichimiya, S., Torigoe, T., Kawaguchi, S., Wada, T., Yamashita, T. & Sato, N. (2009). Scythe/BAT3 regulates apoptotic cell death induced by papillomavirus binding factor in human osteosarcoma. *Cancer Sci.* 100, 47-53.

Yazdani, U., Huang, Z. & Terman, J. R. (2008). The glucose transporter (GLUT4) enhancer factor is required for normal wing positioning in Drosophila. *Genetics* 178, 919-929.

13

Role of Huntington's Disease Protein in Post-Transcriptional Gene Regulatory Pathways

Brady P. Culver and Naoko Tanese
NYU School of Medicine
USA

1. Introduction

This chapter will focus on the potential role that misregulation of post-transcriptional control of gene expression could have on the development or progression of Huntington's disease (HD). Every cell in our bodies possesses the same genetic material, and yet every cell is not the same. We also all know that the tremendous diversity of biology present within each individual is accomplished through unique patterns of gene expression on a cell-by-cell basis. Of course the timing and amounts of gene expression also contribute to this diversity of phenotype and function. The complexity, however, goes even deeper. Within individual genes there is information to produce multiple different messenger RNAs and often multiple different proteins, each with different functional implications. We see then, that the multi-step process of gene expression using a number of genes only modestly greater than what is found in certain species of ciliates is capable of generating a being of vastly more complicated biology. With this complexity in mind, it is therefore possible that small defects at any of the steps of gene expression could have deleterious consequences on the identity, ability to appropriately respond to environmental cues, and on the survival of cells.

In HD, certain parts of the brain are primarily affected over others. Furthermore, the Huntington's disease gene product huntingtin (Htt) is ubiquitously expressed. So how is specificity of the disease manifested when the mutant protein is present everywhere in our bodies at all times? One possible explanation for this observation is that the expression of specific genes important for the survival and function of the cells affected in HD are disrupted to a more significant extent than others. These alterations need not be drastic; rather they are more likely to be the result of an accumulation of small changes, which over time could lead to the dysfunction or death of particular types of neurons. There is already a large body of research on the role of Htt in transcriptional control, and many studies have identified reproducible alterations in gene expression patterns between mouse models of HD and postmortem human HD patient brain samples (Hodges et al. 2006; Seredenina and Luthi-Carter 2011). While we do not discount this potential mechanism, we suggest that mutant Htt may also influence gene expression at steps downstream of transcription.

This hypothesis, although unique amongst HD researchers, is either gaining traction or is already widely accepted as the basis for other neurodegenerative diseases. These diseases would affect the processing of multiple messenger RNAs and present with a broad phenotypic spectrum, reflecting the loss of function or aberrant processing of specific mRNAs. For example, amyotrophic lateral sclerosis (ALS) results from the death of cortical motor neurons and the spinal cord motor neurons on which they synapse. Although the majority of ALS cases have no genetic predisposition, a small percentage of cases have been linked to mutations in the RNA binding proteins TDP-43 and TLS (Lagier-Tourenne, Polymenidou, and Cleveland 2010). However, in both genetic (TDP-43 driven) and sporadic cases of ALS, TDP-43 forms abnormal intracellular aggregates, which are thought to influence the expression patterns of mRNAs dependent upon TDP-43 function for their normal post-transcriptional processing (Mackenzie et al. 2007). Similarly, Fragile X syndrome (FXS) and the related Fragile X tremor ataxia syndrome (FXTAS) are also known to result from mutations in the Fragile X gene (*FMR1*) whose protein product is involved in translational control of mRNAs to which the protein is bound (Willemsen, Levenga, and Oostra 2011). These examples and others argue that the dysfunction of RNA binding proteins can produce cell-type specific effects based on the RNA binding proteins affected and their associated RNAs. Here we provide an overview of the steps at which gene expression may be controlled downstream of transcription, examples in which each of these processes may be perturbed in other neurodegenerative diseases, and review the evidence implicating Htt in control of post-transcriptional gene expression. We aim to incite enthusiasm in the reader for this underappreciated hypothesis and cite the numerous parallels between HD and other neurodegenerative diseases involving the dysfunction of normal post-transcriptional RNA processing.

2. Gene expression is controlled at multiple steps downstream of transcription

Transcription produces a full-length RNA copy of the DNA sequence of a gene, which is heavily edited through removal of intronic sequences, the joining of exonic sequences, the cleavage of the mRNA at specific sites, and the addition of elements not coded for in the genome. The processed RNA must then be exported from the nucleus to be translated. Upon nuclear export, mRNAs may be stored in a translationally repressed state bound by RNA binding proteins until this translational repression is relieved, or they may be immediately translated. Translationally repressed mRNAs can be trafficked to distant sites within the cell to impart an additional level of control to gene expression. mRNAs have a finite lifespan and the levels of mRNA can be controlled through degradation in addition to rates of transcription. Each of these steps is controlled by the activity of specific proteins, which ultimately enable a tight control of protein content, amounts, and location within the cell.

2.1 Alternative patterns of RNA splicing produce multiple messages from a common gene

Alternative splicing is the process by which different exons are joined together to form different sequences from the same precursor mRNA (pre-mRNA). The different patterns of exon joining in alternatively spliced transcripts are determined by sequences contained within the introns of pre-mRNAs and the presence or absence of the proteins that recognize these sequences, although sometimes exons themselves also play a role in this process (Wahl, Will, and Luhrmann 2009). Regardless of the location of the *cis*-acting elements within pre-mRNAs, specific RNA binding proteins promote the inclusion or exclusion of particular exons. These exons may contain protein-coding sequence or untranslated regions (UTRs) if they are present outside of the main open reading frame (ORF). Failure to remove an intron or the exclusion of a particular exon(s) can produce mRNAs that are targeted for degradation before any protein can be produced from these messages (Rebbapragada and Lykke-Andersen 2009). Most human genes are composed of multiple exons (Venter et al. 2001); therefore, the splicing process must be carefully orchestrated to ensure the generation of a meaningful and high fidelity transcript. Furthermore, most human genes are alternatively spliced, which vastly increases the diversity of RNAs and proteins coded for in the entire genome.

2.1.1 Alternative splicing generates different protein isoforms

Splicing events are catalyzed by a large ribonucleoprotein complex called the spliceosome (Wahl, Will, and Luhrmann 2009). Specificity is imparted through the action of distinct splicing factors that recognize particular sequence elements within a pre-mRNA and recruit the spliceosome to these sites. Additionally, splicing factors may mask individual splice sites so they are not included, or spliced out of the resultant mRNA. The activity of these splicing factors is required for the process of alternative splicing as well as to ensure the generation of an intron-free mature mRNA. Alternative splicing can change the amino acid coding potential of an mRNA and thereby generate multiple different proteins from a single gene. Proteins are often composed of modular domains. For example, a protein may harbor a membrane targeting sequence at its N-terminus and a catalytic domain at its C-terminus. Therefore, an mRNA encompassing both of these features would produce a membrane bound protein with catalytic activity. If the membrane-targeting domain of a hypothetical protein were contained within a single exon, then omission of this exon in an alternatively spliced version of this mRNA would produce a cytoplasmically localized protein with catalytic activity (Figure 1).

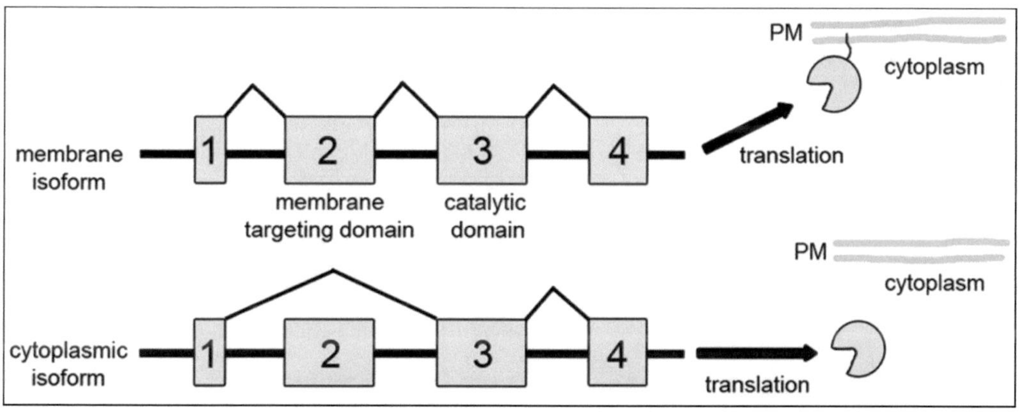

Fig. 1. Alternative splicing can generate different protein isoforms from a single gene. The cartoon illustrates a hypothetical gene with 4 exons (grey boxes numbered 1-4). The solid line represents the unspliced transcript, with the lines above indicating the splicing pattern of the hypothetical mRNA. Inclusion of exon 2 results in the translation of a protein with a plasma membrane (grey squiggly lines, PM) -targeting domain. Exclusion of exon 2 produces a cytoplasmically localized protein.

The protein tau, which is one of the main components of the neurofibrillary tangles in dementia and Alzheimer's disease, is present in as many as 30 different isoforms within neurons, all of which are generated through alternative splicing (Andreadis 2011). Tau helps organize axonal microtubules and functions as a cytoskeletal scaffold in post-synaptic densities (Pritchard et al. 2011). Alterations in the ratio of tau isoforms are the cause of familial cases of frontotemporal lobar degeneration (FTLD) (Gasparini, Terni, and Spillantini 2007). These mutations do not change the coding potential of the *tau* gene, but instead alter the frequency of alternative splicing events such that appropriate stoichiometry of tau isoforms is disrupted.

2.1.2 Defects in mRNA splicing lead to abnormal protein translation or mRNA degradation

The inclusion of an intron(s) within a mature mRNA can lead to its degradation or the translation of a protein with an unintended amino acid sequence. Splicing aberrations are normally detected by the nonsense-mediated decay (NMD) pathway. This pathway recognizes mRNAs that contain a premature termination codon and targets them for destruction. Premature termination codons are recognized by the context in which they are found, vis-à-vis the presence of protein complexes on the transcript at defined positions. Normally, mature mRNAs are marked at splice sites by proteins of the exon-exon junction complex (EJC) (Rebbapragada and Lykke-Andersen 2009). If the NMD protein machinery encounters a stop codon upstream of an EJC, then this signals that the termination codon may be premature. However, the positioning of other factors on the mRNA may prevent NMD from occurring. In short, mRNAs are bound by proteins that recognize specific elements within their sequence. The positioning of these elements provides a context in which to determine whether the message has been appropriately processed. mRNAs that escape the NMD pathway are available for translation and when translated can provoke unforeseen cellular responses. Figure 2 illustrates the consequences of errors in splicing.

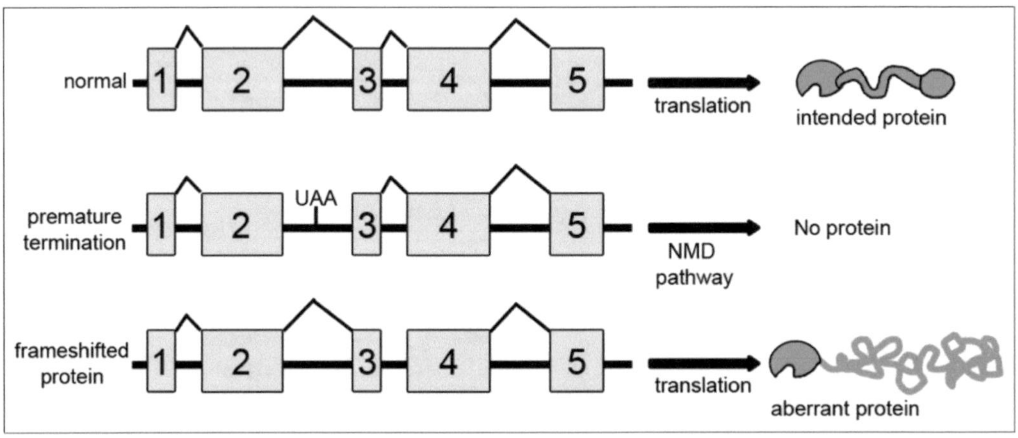

Fig. 2. Errors in splicing can lead to mRNA degradation or aberrant protein synthesis. The cartoon shows how errors in splicing can lead to the inclusion of an intron, which may contain a termination codon (UAA) and activate the NMD pathway. Alternatively, the intron may not contain a premature termination codon, but produce a frame-shifted transcript coding for an abnormal protein.

In ALS, aberrant splicing results in the inclusion of introns in the mature mRNA for the intermediate filament protein peripherin (Xiao et al. 2008). Peripherin is present within the protein aggregations observed in ALS patient tissue and overexpression of peripherin is sufficient to induce neurodegeneration in transgenic mice (Robertson et al. 2003). Furthermore, expression of the aberrant transcript in cultured motor neurons resulted in its aggregation, and was associated with the death of these cells (Robertson et al. 2003). These findings suggest that the abnormal splicing events can trigger neurodegeneration in a cell type specific manner.

2.2 Non-protein coding sequences influence stability, location, and translational potential of mRNAs

Eukaryotic mRNAs contain untranslated regions (UTRs) at their 5′ and 3′ ends that contain information used in determining the stability, translational potential, and location of an RNA. As the name suggests, untranslated regions of an mRNA are not translated into protein, but serve as *cis*-acting elements within an mRNA that provide binding surfaces for other molecules through the secondary structural elements conferred by the sequence of the UTR (Spriggs, Bushell, and Willis 2010). While exons most often contain the information required for the assembly of the encoded protein, they also include UTRs at the 5′ and 3′ end of a mature mRNA. Sometimes different 5′ and 3′ UTRs for the same mRNA arise from alternative splicing events, as has been observed for the brain derived neurotrophic factor, BDNF (Pruunsild et al. 2007). Alternative versions of 3′ UTRs may also be generated through the cleavage of precursor mRNAs in anticipation of poly-A (poly-adenosine monophosphate) tail addition (Hughes 2006). The vast majority of mRNAs include a stretch of poly-A residues at their 3′ end that are not coded for in the genome and are added post-transcriptionally. The poly-A tail protects the mRNA from degradation at the 3′ end and is involved in initiation of protein translation through its interaction with poly-A binding protein (PABP) (Lemay et al. 2010). Many pre-mRNAs may be poly-adenylated at multiple positions based on the presence of multiple poly-A signal sequences in the pre-mRNA (Tian et al. 2005). The selection of the cleavage site on the pre-mRNA can therefore directly determine the extent and content of the 3′ UTR contained within an mRNA. We provide a brief overview of the functions of 5′ and 3′ UTRs and highlight examples where perturbations in these processes may contribute to neurodegeneration.

2.2.1 The 5' UTR functions in mRNA translation

5′ UTRs typically contain sequences necessary for the initiation of translation. The 5′ end of most mammalian mRNAs is capped by a modified ribonucleotide, the m7G cap. This modified nucleotide is attached to the 5′ end of the mRNA through an atypical 5′ to 5′ linkage. The m7G cap is bound by a translation initiation factor complex (eIF4E), which is then used to circularize the transcript through binding to PABP. Circularization is thought to improve the efficiency of translation (Gingras, Raught, and Sonenberg 1999). In addition to the 5′ cap, 5′ UTRs often contain upstream open reading frames (uORFs). The presence of a uORF is usually inhibitory; a ribosome scanning along the mRNA will encounter the start codon of a uORF and begin translation at this position. This results in fewer ribosomes recognizing the main ORF start codon and decreased protein translation. uORF-encoded polypeptides can also directly inhibit translation by binding to the ribosome and preventing the translation of the major downstream ORF (Lovett and Rogers 1996).

Beta site APP-cleaving enzyme 1 (BACE1) cleaves the amyloid precursor protein (APP) producing the toxic fragment that forms the amyloid plaques that define Alzheimer's disease (Sisodia 1992). Elevated levels of BACE1 protein expression are normally kept in check by the presence of six uORFs (Zhou and Song 2006). Alzheimer's disease patients typically display elevated levels of BACE1 protein compared with unaffected individuals, but the levels of *BACE1* mRNA do not always reflect this (Mihailovich et al. 2007). This suggests that translational efficiency of BACE1 may be enhanced in Alzheimer's disease compared with unaffected cases. These uORFs are therefore thought to be important in keeping the levels of BACE1 protein low.

5′ UTRs also typically contain highly structured regions, which can be bound by proteins that recognize these structures. Protein binding to the secondary structures within the 5′ UTR can promote 5′ m7G cap-independent translation initiation (Pickering and Willis 2005). In this context, these secondary structures are known as IRES (internal ribosome entry sites). These sites are used under conditions of cellular stress when the translation of many other proteins is globally inhibited (Spriggs, Bushell, and Willis 2010). Global translational inhibition is mediated through the phosphorylation of eIF2A. eIF2A can be phosphorylated by four different kinases (PKR, PERK, GCN2, and HRI), which are activated by different cellular stresses. Phosphorylated eIF2A inhibits translation by preventing recycling of initiation complexes to translation start sites. The presence of an IRES in an mRNA can bypass this translational repression by recruiting ribosomes to start codons by an eIF4A- and m7G cap-independent mechanism (Spriggs, Bushell, and Willis 2010).

Fragile X syndrome (FXS) and fragile X tremor ataxia syndrome (FXTAS) are caused by differing degrees of a CGG repeat expansion at the 5′ non-protein-coding end of the fragile X mental retardation 1 (*FMR1*) gene. The extent of CGG repeat expansion determines the disease; longer CGG expansions (> 200 repeats) result in FXS and shorter CGG expansions (55-200) result in FXTAS. The repeat expansion lengths of FXS silence transcription of *FMR1*, effectively producing a null mutation through increased methylation of the *FMR1* gene promoter. By contrast, the shorter CGG expansions of FXTAS do not result in transcriptional silencing, but instead produce an mRNA containing a large stem-loop structure composed of CGG repeats in the 5′ UTR (Willemsen, Levenga, and Oostra 2011). Transgenic mice expressing FXTAS-correlated CGG repeat lengths, outside of a protein translation context, phenocopy symptoms of the human condition, which indicates that the RNA itself may be the toxic agent in FXTAS (Van Dam et al. 2005). The stem-loop generated by expanded CGG repeats in the 5′ UTR of *FMR1* could act as a sponge to pull the RNA binding proteins away from their normal targets and therefore affect the processing of other RNAs normally bound by these RNA binding proteins (Jin et al. 2003). This mechanism implies that the processing of many other RNAs may be affected by the single mutation in *FMR1*.

2.2.2 The 3' UTR functions in mRNA stability and defines its localization

3′ UTRs are typically much longer than 5′ UTRs and therefore contain significantly more information that can be used to affect the fate of mRNA to which they are attached. This information is decoded from the 3′ UTR sequence by the binding of proteins or other RNAs to the 3′ UTR. A combination of sequence and secondary structure allows specific proteins to bind to 3′ UTRs, where they affect the stability, localization, and translational potential of the mRNA. All of these processes are important regulators of gene expression, and defects in any one of these processes may lead to cell dysfunction and undesirable effects depending on the affected mRNAs (Andreassi and Riccio 2009).

Recently, the role of non-coding RNAs (ncRNAs) in the regulation of gene expression has received considerable attention. ncRNAs contribute to the regulation of gene expression at both the transcriptional and post-transcriptional steps. As nucleic acids, ncRNAs are able to form base-pair interactions with perfect or imperfect complementarity. This feature provides a simple mechanism to allow for the targeting of specific sequences by these ncRNAs. ncRNAs could therefore serve as an adaptor molecule to facilitate RNA-protein interactions (Mattick and Makunin 2006). In this model, a protein would have an affinity for either a sequence or a structural feature present within the ncRNA. The ncRNA in turn would have sequence elements within it that allow it to recognize and form base-pair interactions with distinct RNA(s). In this manner, an ncRNA could allow a protein to interact with a large variety of messages. No other system is as well characterized with regards to this phenomenon as the RNA interference pathway. In this pathway, small RNAs of 21-23 nucleotides in length recognize sequences in a target mRNA, and through base-paired interactions direct the assembly of a protein complex called RISC (RNA-induced silencing complex) onto the matched mRNA. Perfect complementarity between the small RNA (siRNA) and RNA target leads to cleavage and degradation of the RNA mediated by the endonucleolytic activity of the protein Argonaute 2 (Ago2) (Siomi and Siomi 2009). RNAi (RNA interference) has now become an invaluable tool in experimental molecular biology and has exciting potential therapeutic applications. Most small RNAs or microRNAs (miRNAs) present in mammalian cells, however, have imperfect sequence complementarity with their targets resulting in the translational silencing of the affected message through Ago family members Ago1, 2, 3, and 4 (Siomi and Siomi 2009), or degradation of target mRNAs . Current estimates suggest that around 60% of all human genes are regulated by miRNAs (Friedman et al. 2009). miRNAs and components of the RISC complex are found in dendrites where they repress the translation of synaptic proteins (Swanger and Bassell 2011). Although the effects of miRNAs on protein levels are subtle, the elimination of miRNAs from adult brain results in a neurodegenerative phenotype in mice (Hebert et al. 2010).

Changes in RNA stability may contribute to the pathology of ALS. The brains of ALS affected individuals possess intracellular accumulations of neurofilament proteins. Neurofilament proteins come in three different isotypes: neurofilament heavy, medium, and light, all encoded by separate genes. The stoichiometry of neurofilaments is hypothesized to be important to prevent their aggregation (Xu et al. 1993). TDP-43 stabilizes the neurofilament light chain (*NFL*) mRNA by binding to its 3'UTR (Strong et al. 2007). As discussed in the introduction, TDP-43 is present within intracellular inclusions in ALS. This sequestration of TDP-43 may abrogate its binding to the *NFL* mRNA and thus decrease the stability of the mRNA. Indeed, *NFL* mRNA levels are reduced in ALS patient brains (Volkening et al. 2009). Therefore, this reduction in mRNA could translate to a reduction in NFL protein, altered neurofilament stoichiometry, and protein aggregation.

Myotonic dystrophy type 1 (DM1) is caused by a CTG repeat expansion in the 3' UTR of the dystrophia myotonica-protein kinase (*DMPK*) gene (Mahadevan et al. 1992). This repeat expansion results in the sequestration and aggregation of CUG-expanded RNA in the nucleus. Here the protein muscleblind-like (MBNL) binds to the CUG-expanded RNA and is prevented from performing its normal role in splicing (Jiang et al. 2004), which is similar to the proposed mechanistic explanation for FXTAS. Additionally, expression of a CUG repeat-expanded RNA was sufficient to cause the formation of large RNA and protein aggregations called stress granules in cell culture (Huichalaf et al. 2010). Stress granules are sites of RNA storage, where translation is inhibited and RNAs are kept from a potentially damaging cytoplasmic environment (Buchan and Parker 2009). Stress granules contain a heterogeneous population of mRNAs, and therefore the induction of the stress granule assembly in response to CUG-expanded RNA expression could affect the translation of many different mRNAs. Could CAG-expanded RNA expression in Huntington's disease produce a similar response?

2.2.3 The poly-A tail contributes to stability and translational potential of an mRNA

Polyadenylation is a requisite step in the biogenesis of most mRNAs. This modification serves to protect mRNAs from degradation by 3′ to 5′ exonucleases (Mangus, Evans, and Jacobson 2003). This protection is imparted to the mRNA through the binding of PABP to the poly-A sequence. As a polyadenylated mature mRNA ages, the length of the poly-A tail shrinks. This poly-A shortening is analogous to the process that occurs at telomeres as a cell ages. Just as a critically short telomere signals senescence, a critically short poly-A tail is unable to stave off the exonucleases wishing to make a lunch of it, and it is degraded (Meyer, Temme, and Wahle 2004). The poly-A tail therefore helps to establish a lifespan for RNAs. Polyadenylation most often occurs in the nucleus, but can also take place in the cytoplasm. Cytoplasmic mRNAs lacking a poly-A tail or possessing a shortened poly-A tail are sequestered in ribonucleoprotein particles until they are acted upon by a cytoplasmic polyadenylation element binding protein (CPEB), which promotes poly-A tail extension and subsequent translation (Richter 2007). The addition and function of the poly-A tail therefore provide yet another step at which gene expression may be regulated.

Brain derived neurotrophic factor (BDNF) possesses two cytoplasmic polyadenylation elements in its 3′ UTR generating short and long forms of mRNAs. One of these elements is required for constitutive BDNF mRNA trafficking to dendrites, while the other is important for activity dependent trafficking (Oe and Yoneda 2010), although different conclusions were drawn from another study (An et al. 2008). BDNF plays important roles in the health and survival of neurons, and the protein is present at reduced levels in both Alzheimer's disease and HD patient brain tissue (Ferrer et al. 2000; Narisawa-Saito et al. 1996). In Alzheimer's disease, BDNF mRNA levels are reduced, and the severity of BDNF mRNA reduction correlates with the Aβ aggregation size in mouse models of AD (Peng et al. 2009).

The autosomal dominant disorder, oculopharyngeal muscular dystrophy (OPMD) is caused by a GCG repeat expansion in the coding region of poly(A) binding protein nuclear 1 (*PABPN1*). Heterozygous *PABPN1* mutant carriers display myopathic symptoms: proximal limb weakness, dysphagia, and ptosis (Davies, Berger, and Rubinsztein 2006). However, individuals with homozygous mutations often display neurological disturbances: cognitive decline, depression, and psychosis (Blumen et al. 2009). The GCG expansion is translated into an expanded poly-alanine tract, which results in the nuclear aggregation of PABPN1 (Davies, Berger, and Rubinsztein 2006). The aggregation of PABPN1 into filamentous nuclear inclusions correlates with retention of large amounts of poly-A containing mRNA (Calado et al. 2000). This suggests that multiple mRNAs may be affected by the mutation in a single gene and fits into a paradigm for diseases caused by malfunctioning of RNA binding proteins.

2.3 The multiple steps of RNA processing downstream of transcription allow for multiple regulations and potential errors

Interactions between mRNA and protein, and mRNA and RNA are responsible for executing distinct steps of post-transcriptional processing on particular RNAs. Most RNA binding proteins and ncRNAs influence the fate of multiple transcripts and therefore problems with either an RNA binding protein or an ncRNA will affect multiple messages. RNA binding proteins can interact with RNA and other proteins. An RNA binding protein has stabilizing and destabilizing effects on target RNAs, depending upon its interactions with other proteins. Figure 3 depicts a hypothetical mRNA with common features indicated and the positions/features where RNA binding proteins and ncRNAs likely bind. A cytoplasmic complex of RNA and proteins is referred to as ribonucleoprotein particle (RNP) and distinctions between types of RNPs are made based upon the protein constituents of different complexes and the fate of transcripts within an RNP. There are many different types of RNPs, but for brevity sake we will simplify the discussion to transport RNPs, P-bodies, and stress granules. Their functions and consequences of their dysfunction as they relate to neurodegenerative diseases will be described in the following sections.

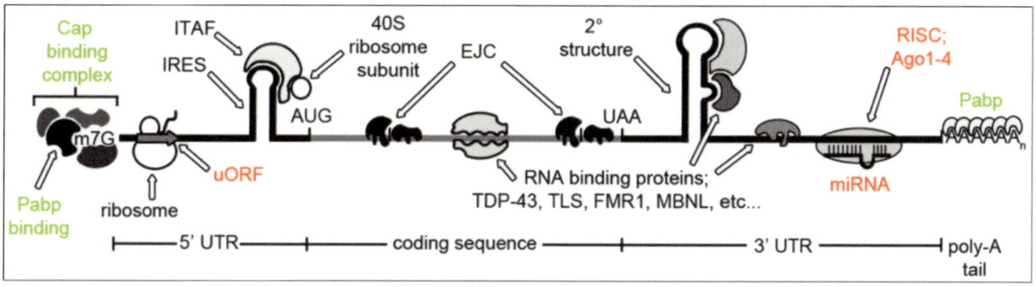

Fig. 3. mRNA processing is mediated by multiple proteins and ncRNAs. The cartoon shows a hypothetical mRNA organized by regions: 5′ UTR and 3′ UTR (black line segments) and coding sequence (gray line segment). Proteins/entities labeled in green in general have a positive effect on mRNA translation, while those in red tend to inhibit protein translation or mRNA stability. Proteins in black may have positive or negative effects on mRNA translation or stability. The abbreviations used are: m7G (7-methylguanosine), uORF (upstream open reading frame), IRES (internal ribosome entry site), ITAF (IRES-transacting factor), EJC (exon-exon junction complex), miRNA (micro RNA), and RISC (RNA-induced silencing complex).

2.3.1 Transport RNPs deliver mRNAs to discrete locations within a cell

RNA transport is an efficient means to spatially control gene expression patterns in a single cell. In transporting an mRNA a cell is able to produce multiple protein copies from a single molecule of mRNA at a discrete location, and thus reduce the energy expenditure that would be required if individual protein molecules were instead transported. This mechanism also eliminates the need to suppress the activity of a protein with properties that could be detrimental if present in an inappropriate context (Andreassi and Riccio 2009). Transported mRNAs are translationally repressed until they are delivered to their final destinations by protein- and ncRNA-dependent mechanisms (Wang, Martin, and Zukin 2010). This section will focus on transport RNPs in neurons.

RNA transport is mediated by specific proteins and/or ncRNAs that function to tether the mRNA to a motor protein complex (Wang, Martin, and Zukin 2010). Movement of transport RNPs out of the soma occurs largely on the backs of the microtubule-based motor proteins kinesin and dynein (Wang, Martin, and Zukin 2010). The microtubules present in axons are all oriented with their plus ends pointed away from the cell interior, and so only the activity of plus-end directed kinesin motors can move cargoes into axons. In contrast, dendrites have a mixed polarity of microtubules in proximal segments; therefore, minus-end directed dynein motors are capable of directing cargoes into dendrites in addition to plus-end directed kinesin motors (Kapitein et al. 2010). Actin-based myosin motors also contribute to mRNA transport and are most likely involved in moving mRNAs from larger bore dendritic chambers into the smaller diameter dendritic spines where actin filaments predominate (Hirokawa, Niwa, and Tanaka 2010).

mRNAs bound for transport are first recognized in the nucleus by trans-acting factors that recognize specific sequence elements present within the mRNA. Although 3′ UTR sequence seems a prime candidate for the placement of these location-defining elements, they may also occur within the coding sequence of an mRNA. The protein-bound mRNA is then exported from the nucleus where additional proteins can be recruited to the transcript through binding sites on the *trans*-acting factor or mRNA (Sossin and DesGroseillers 2006). These RNPs are then thought recognized by motor protein complexes, which then move the mRNA to an appropriate location in the cell.

Fragile X mental retardation protein (FMR1) is an RNA binding protein important for the transport and localization of specific RNAs. FMR1 binds to distinct RNAs and represses their translation, through either a direct influence on the processivity of bound ribosomes and/or an association with Argonaute proteins and miRNAs (De Rubeis and Bagni 2010; Muddashetty et al. 2011). Fragile X syndrome is an inherited intellectual disability that results from a loss of *FMR1* expression and is therefore predicted to affect the trafficking and translation of the RNAs normally bound by FMR1 (De Rubeis and Bagni 2010). Many of the RNAs bound by FMR1 are involved in pre- and postsynaptic functions, which suggest that loss of *FMR1* may drastically impair neuronal function (Darnell et al. 2011). Interestingly, Huntington's disease protein huntingtin was identified as an FMR1-associated RNA in this study. Staufen is another protein involved in RNA transport and was originally identified in *Drosophila*, where mutations were shown to affect the asymmetric localization of mRNAs important for embryonic axis generation (St Johnston, Beuchle, and Nusslein-Volhard 1991). A human homolog of Staufen (hStau) co-purifies and co-localizes with RNA, FMR1, kinesin, dynein, and myosin proteins in human cell (Villace, Marion, and Ortin 2004). This example illustrates the combinatorial control and potential redundancy that is utilized for RNA transport.

Once delivered to their final destinations, transported RNAs could be tethered to the cytoskeleton through binding to cytoskeletal associated proteins with affinity for particular mRNAs, or proteins present in the transport RNPs (Kim and Coulombe 2010). Cytoplasmic fractions of cultured cells are enriched for certain mitochondrial RNAs and ribosomal protein RNAs (Russo et al. 2008). Furthermore, disassembly of the microtubule cytoskeleton by nocodazole treatment results in shifting of these mRNAs into the soluble portion of the cytoplasm. A definitive identification of the proteins responsible for cytoskeletal tethering of mRNAs in neurons has not yet been made, although a tug-of-war between different polarity-directed motors could effectively produce a localized mRNA. Cytoplasmic FMR1-interacting proteins 1 and 2 (CYFIP1, 2) interact with FMR1 and are involved in actin cytoskeletal remodeling (Anitei et al. 2010). Although it has yet to be shown if CYFIP1 or 2 act to anchor FMR1-associated RNAs to the cytoskeleton, this seems like a possibility based on the proteins' affinity for FMR1 and actin remodeling proteins.

Localized transcripts are translationally silenced until the repression is relieved by the dissociation of the inhibitory factor(s). PSD-95 is a post-synaptic scaffolding protein that functions in the regulation of AMPA-type glutamate receptor endocytosis and in the maintenance of dendritic spine architecture. FMR1 along with miR125a (microRNA 125a) and Ago2 bind to the PSD-95 transcript to suppress its translation. The binding of these molecules to the PSD-95 mRNA is dependent upon the phosphorylation of FMR1. When FMR1 is dephosphorylated following group I metabotropic glutamate receptor stimulation, miR125a and Ago2 dissociate from the 3′ UTR of PSD-95 and the mRNA is translated (Muddashetty et al. 2011). This example illustrates how post-translational modifications of RNA binding proteins control their inhibitory or stimulatory effects on mRNA translation.

2.3.2 P-bodies control mRNA stability

Processing bodies (P-bodies) share components with transport RNPs, and considerable grey area exists in discriminating between the two types of particles based on protein associations. P-bodies are functionally defined as constitutively present cytoplasmic outposts of RNA degradation or storage (Buchan and Parker 2009). In some instances, mRNAs may be rescued from P-body association and be translated (Brengues, Teixeira, and Parker 2005), although the mechanism for this alternative fate is unclear. P-bodies contain the ribonucleases responsible for 5′ m7G cap removal, 5′ exoribonucleases, deadenylases, as well as components of the RISC pathway (Parker and Sheth 2007).

2.3.3 Stress granules are large ribonucleoprotein particles that assemble in response to various cellular stressors

Cellular stress caused by a variety of factors results in the phosphorylation of eIF2A and the subsequent translational silencing of mRNAs not involved in the stress response (Spriggs, Bushell, and Willis 2010). Translationally silenced mRNAs are sequestered away from large ribosomal subunits and the potentially damaging cytoplasmic environment by recruitment into large assemblies of proteins and RNAs called stress granules. Stress granules contain translation initiation factors, small ribosomal subunits, specific RNA binding proteins and their associated RNAs, and general stress granule assembly factors (Buchan and Parker 2009). Experiments in yeast demonstrated that stress granules are dependent upon P-bodies for their assembly, but P-bodies can form when stress granule assembly is inhibited by genetic means (Buchan, Muhlrad, and Parker 2008). mRNAs stored in stress granules can be transferred to P-bodies for degradation and vice-versa (Buchan and Parker 2009). The signals that promote this switch have yet to be identified.

Stress granules form in part through self-association properties, similar to what takes place during protein aggregation. Indeed, several P-body and stress granule associated proteins with glutamine- and asparagine- (Q/N) rich regions depend on these regions for the self assembly requisite for stress granule and P-body formation (Buchan and Parker

2009). TDP-43 localizes to stress granules and C-terminal cleavage products generated by caspase 3 are prone to cytoplasmic aggregation (Liu-Yesucevitz et al. 2010). In the event that TDP-43 levels are elevated and there is increased production of C-terminal caspase 3 cleavage products, then this could lead to the nucleation of constitutive stress granules. These stress granules may contain TDP-43 target mRNAs and therefore reduce the levels of the proteins encoded in these mRNAs. By another conspicuous coincidence, caspase 3 is known to cleave mutant huntingtin, which also has intrinsic aggregation properties (Wellington et al. 1998).

Mutations in TLS are associated with a subset of familial cases of ALS (Lagier-Tourenne, Polymenidou, and Cleveland 2010). When these mutant alleles of TLS are expressed in HeLa cells, they localize to stress granules in the absence of added cellular stress (Bosco et al. 2010). Endogenous TLS is also found in stress granules, but only upon experimentally induced cellular stress. These mutant TLS-induced stress granules did not recruit endogenous TLS or TDP-43 into these abnormal structures. If different RNA binding proteins are responsible for delivering their RNA targets to stress granules upon induction of the stress response, then the absence of endogenous TDP-43 from the mutant TLS-induced stress granules could mean that TDP-43 specific mRNAs are also absent from these aggregates.

FMR1 is a well-established stress granule marker and seems to be required for stress granule assembly in mouse and human cells in culture (Didiot et al. 2009). Mouse embryonic fibroblasts from *FMR1* null mice do not assemble stress granules, yet P-body assembly is unaffected. FMR1 protein levels increase in response to stress (Didiot et al. 2009). There are currently no mutants in FMR1 that can separate its role in stress granule assembly from its role in transport RNPs, and these structures may in fact be inextricably linked (Kiebler and Bassell 2006). It would be interesting to determine if Fragile X syndrome results from impairment in stress granule assembly or from a reduction in RNA transport, or a combination of the two.

3. Huntington's disease may involve detrimental changes in post-transcriptional gene expression patterns

There is mounting circumstantial and experimental evidence implicating deviations in post-transcriptional RNA processing events in the establishment and/or progression of HD. Changes in post-transcriptional processing of RNA could account for the enhanced susceptibility of certain types of cells in HD if specific RNAs important for the survival or proper functioning of these cells are affected by mutant huntingtin (Htt) expression. Here we will review the evidence that suggests Htt is normally involved in RNA processing, and mutant Htt expression may impair normal RNA processing or activate stress responses that alters gene expression.

3.1 Huntingtin association with Ago2 is important for RNA silencing pathway

Our group discovered an Htt-Ago2 association through affinity purification and mass spectrometry of a FLAG-tagged N-terminal fragments of wild-type (25 glutamines) and mutant (97 glutamines) Htt expressed in HeLa cells (Savas et al. 2008). We went on to show that endogenous Htt co-localizes with Ago2 and Dcp1 at P-bodies in U2OS cells.

Because these N-terminal Htt purifications did not contain dicer or other proteins involved in the biogenesis of miRNAs, we hypothesized that Htt was involved in the effector stage of the RNA silencing pathway. Knockdown of Htt in U2OS cells by RNAi revealed that a subset of P-bodies required Htt for their assembly. Incomplete inhibition of P-body assembly could have resulted from incomplete knockdown of Htt. Alternatively, a specific subset of RNPs could require Htt for their incorporation into functional P-bodies. Interestingly, a knock-in striatal precursor cell line expressing mutant Htt (Trettel et al. 2000) formed fewer P-bodies than a wild-type striatal precursor cell line (Savas et al. 2008). This observation suggests that mutant Htt may have a dominant negative or loss-of-function effect on the RNA silencing pathway. We further demonstrated that knockdown of Htt inhibits the RNA interference response by reporter assays, and that mutant Htt-expressing cells are less efficient in this pathway. Finally, fluorescence recovery after photobleaching (FRAP) analysis of GFP-Ago2 dynamics in wild-type (25 glutamines) and mutant (97 glutamines) Htt fragment-expressing cells demonstrated that mutant Htt inhibits the recruitment of Ago2 to P-bodies.

In a follow-up study, we focused our analysis on cultured cortical neurons and a neuroblastoma cell line. This work showed that Htt co-localizes, co-fractionates, and co-purifies with Ago2 in cultured cortical neurons (Savas et al. 2010). We also discovered that tethering Htt to a luciferase reporter mRNA through a λN element-box-B interaction represses luciferase expression. This repression was found to be at least partially dependent on the presence of Ago2 in the assay. In summary, the Ago2-Htt interaction uncovered by affinity purification has been verified by numerous assays and is functionally important in the RNA silencing pathway. Moreover, mutant Htt-expressing cells seem to be less able to utilize this pathway, possibly through a decreased recruitment of Ago2 to P-bodies. We have since found that full-length FLAG-tagged mutant Htt expressed from the endogenous locus associates with Ago2 in mouse brains (Culver, Savas, et. al. submitted). Based on these observations we suggest that Htt may contribute to the silencing of a specific subset of messages and that mutant Htt interfering with this process could give rise to inappropriate levels of particular target proteins.

3.2 Huntingtin contributes to mRNA transport in neurons

Our studies indicate that Htt co-localizes with mRNAs, and lentiviral-mediated knockdown of Htt drastically reduces the number of punctate polyadenylated RNA-containing particles detected by FISH in cultured cortical neurons. Furthermore, Htt co-localizes with Staufen and co-traffics with an MS2-tagged IP_3R1 3′ UTR (inositol 1,4,5-trisphosphate receptor 1) mRNA in cultured cortical neurons (Savas et al. 2010). Htt is known to associate with the microtubule-based motor dynein complex (Caviston et al. 2007). These data seem to suggest that Htt may be required for all directed dendritic RNA transport in cultured neurons. Although we did not demonstrate that mutant Htt expression had any effect on this process, based on the inhibitory effect that mutant Htt expression has on vesicular transport of BDNF- and APP-containing vesicles (Gauthier et al. 2004; Her and Goldstein 2008), it seems likely that mutant Htt will also inhibit RNA transport. If this is the case, then many mRNAs may be inappropriately localized and neurons could be less able to respond to stimuli through local translation.

BDNF levels are reduced in HD patient brains and exogenous delivery of BDNF rescues many of the phenotypes in a mouse model of HD (Gharami et al. 2008). Furthermore, mutant Htt expression inhibits the trafficking of BDNF-containing vesicles in cultured cortical neurons (Gauthier et al. 2004). We have recently found that Htt co-localizes with BDNF mRNA in cultured cortical neurons and in brain cortical sections (Ma et al. 2010). These observations suggest that not only Htt is important for BDNF protein trafficking, but also mRNA trafficking. It is also interesting that wild-type Htt overexpression increases BDNF mRNA levels, while mutant Htt overexpression reduces BDNF mRNA. Furthermore, BDNF mRNA levels are reduced in HD brain compared with unaffected individuals (Zuccato et al. 2001). It would therefore seem that Htt influences BDNF levels and location from transcription to mRNA localization, to delivery of the translated protein to its sites of action.

3.3 Mutant Htt aggregates may cause RNA processing defects through sequestration of RNA binding proteins

Cytoplasmic Htt-containing aggregates in HD brain tissue contain the RNA binding proteins TDP-43 and TLS (Doi et al. 2010; Schwab et al. 2008). Intriguingly, these same two RNA binding proteins also form aggregates in ALS and FTLD (a type of dementia). TDP-43 is a widely expressed RNA binding protein with roles in RNA splicing, stability, and regulation of protein translation. TDP-43 purifications from UV cross-linked sources identified thousands of mRNAs bound by TDP-43. Many of the RNAs were involved in RNA metabolic processes, and Htt was among the list of mRNAs that co-purified with cross-linked TDP-43 (Sephton et al. 2011). TDP-43 also binds to its own mRNA and leads to its degradation (Ayala et al. 2011). This finding, along with the findings discussed in section 2.2.2, suggest that TDP-43 activity is important for controlling the levels of particular transcripts. Over-expression of TDP-43 is sufficient to induce neurodegenerative phenotypes in model organisms (Ash et al. 2010; Tatom et al. 2009). Therefore, a reduced capacity of TDP-43 to regulate its own mRNA levels through sequestration of the protein in aggregates could produce increasing levels of TDP-43 available for aggregation and thus accelerate toxicity.

TLS is similar to TDP-43 in that it is involved in many different steps of RNA processing, from transcription and RNA splicing, to mRNA stability, and transport. It is also similar to TDP-43 in that it forms aggregates in ALS in cases where mutations in TLS correlate with disease presentation (Lagier-Tourenne, Polymenidou, and Cleveland 2010). TDP-43 and TLS aggregation seem to be mutually exclusive, as TDP-43 aggregations were not observed in ALS cases associated with mutations in TLS (Vance et al. 2009). This observation strongly argues that alterations in RNA processing events can cause ALS and FTLD by multiple independent means. Clinical presentations of HD are somewhat heterogeneous: some patients display more severe forms of psychiatric and intellectual disturbances, as well as varying degrees of mobility impairment. These differences in clinical presentation are not correlated with differences in CAG repeat lengths (Weigell-Weber, Schmid, and Spiegel 1996), which suggests that additional genetic or environmental factors contribute to the differences seen in HD patients. We hypothesize that these differences could be accounted for by unique combinations of alterations in the post-transcriptional processing of specific RNAs.

3.4 Htt associates with proteins involved in RNA splicing and cleavage

A yeast two-hybrid screen using the N-terminus of Htt identified three RNA binding proteins that interacted with Htt. Two of these proteins are involved in RNA splicing activity and are widely conserved in eukaryotic evolution (PRPF40A and PRPF40B). They are both general components of the core splicing machinery and as such, would be predicted to have broad effects if their activity were perturbed (Faber et al. 1998). As would be expected of a protein involved in mRNA splicing, PRPF40A is predominantly a nuclear protein. However, the protein is redirected to the cytoplasm when co-overexpressed with a mutant form of a Htt fragment (Jiang et al. 2011). Furthermore, mutant Htt fragments more strongly interact with PRPF40A than wild-type fragments. The authors suggest that mutant Htt may actively sequester PRPF40A in the cytoplasm and thereby inhibit the protein's normal splicing activity.

The other RNA binding protein identified in this yeast two-hybrid screen is known as symplekin (SYMPK) and is involved in polyadenylation of mRNAs. SYMPK is present in a large complex containing other proteins involved in mRNA cleavage and polyadenylation (Kolev and Steitz 2005). Although SYMPK was initially identified by yeast two-hybrid, the protein was also present along with other members of the cleavage and polyadenylation specificity factor (CPSF) complex in Htt purifications from HeLa cells (our unpublished observations). SYMPK is required for both nuclear and cytoplasmic polyadenylation events (Barnard et al. 2004). Therefore, any changes in SYMPK activity could have a global impact on the stability and translational potential of mRNAs. Interestingly however, SYMPK was identified in a genome-wide screen for modifiers of mitotic fidelity (Cappell et al. 2010). The influence of SYMPK on spindle positioning was shown to occur through its role in polyadenylation, as knockdown of other genes involved in this process produced a similar effect. It is therefore possible that specific cellular processes are more sensitive to a reduction in polyadenylated transcripts than others.

3.5 Mutant Htt mRNA may have detrimental effects independent of coding for mutant Htt protein

Mutant Htt mRNA can form hairpin loops through G-C base pairing amongst the CAG repeats and the downstream CGG repeats of the poly-proline encoding region (de Mezer et al. 2011). These hairpins are bound by the double-stranded RNA dependent protein kinase PKR (Peel et al. 2001). Furthermore, PKR activity is elevated in HD patient brain tissue (Bando et al. 2005). PKR is activated by viral infection and acts to repress 5′ m7G cap-dependent translation of endogenous transcripts through phosphorylation of eIF2A (Spriggs, Bushell, and Willis 2010). Activated PKR induces the stress response and preferentially allows for the translation of mRNAs containing an IRES element immediately upstream of the initiation codon. Htt contains a uORF in its 5′ UTR, which can inhibit the translation of a CAT reporter when fused upstream of the CAT ATG codon (Lee et al. 2002). This uORF is predicted to reduce the translation of Htt under normal conditions when cap-dependent translation is employed. The intervening sequence between the uORF in the Htt 5′ UTR and the initiation codon of Htt is GC-rich. Furthermore, this sequence is predicted to form a long hairpin structure by the RNAfold webserver sequence analysis program. In the future it would be interesting to determine if Htt can be translated in the absence of a 5′ m7G cap.

The ability of CAG-expanded mRNA to form extended hairpin loops leaves open the possibility that the mRNA could act as a molecular sponge for other RNA binding proteins and produce a toxic response similar to what has been proposed for myotonic dystrophy type 1 (DM1) and fragile X tremor ataxia syndrome (FXTAS). In this model, mutant Htt mRNA could actively recruit RNA binding proteins that recognize the CAG/CGG hairpin, and either prevents the RNA binding proteins from performing their normal activities through an effective reduction in their levels, or initiates an inappropriate response through a scaffolding-type effect between different RNA binding proteins such as PKR. We have noticed that mutant Htt knock-in cell lines and animals produce less mutant protein than wild-type, despite identical 5′ regions (unpublished observations). The expanded CAG region could therefore inhibit the translation of mutant Htt protein through the CAG/CGG hairpin structure.

CNG (N represents any nucleotide) expanded mRNAs were demonstrated to lead to translation initiation independent of an initiation codon (Zu et al. 2011). This ATG-independent translation occurred only when the extent of expansion was beyond a certain threshold. Strikingly, the threshold for CAG expansion necessary to produce ATG independent translation occurred at 42 CAG repeats, which is remarkably close to the threshold for HD diagnosis (Group 1993). This exciting finding raises the possibility that mutant Htt may be translated by a 5′ m7G cap-independent mechanism. Furthermore, these authors (Zu et al. 2011) showed that the CAG-expanded RNA was capable of producing protein in all three frames of translation, meaning that poly-alanine, poly-serine, and poly-glutamine proteins would be produced from this expansion. In SCA3 (spinocerebellar ataxia type 3), CAG repeat expansion also produces a poly-alanine peptide in addition to the poly-glutamine peptide (Gaspar et al. 2000). Similarly, poly-alanine and poly-serine proteins were found in HD patient brain tissues where they localized to ubiquitin-positive intranuclear inclusions (Davies and Rubinsztein 2006). Poly-alanine expansion in PABPN1 results in nuclear aggregation of the protein in oculopharyngeal muscular dystrophy (Davies, Berger, and Rubinsztein 2006). Poly-alanine or poly-serine containing peptides produced from frame-shifted translation of Htt would encounter stop codons very shortly after the homopolymeric stretch of amino acids. This would result in the production of very small proteins, which could diffuse through the nuclear pore without the aid of the import machinery or a nuclear localization element. Poly-alanine stretches are highly hydrophobic and are predicted to form stable and compact β-sheets, which are assembled into insoluble fibrils as the lowest energy conformation (Shinchuk et al. 2005).

3.6 Mutant Htt purifications from mouse brain are enriched in RNA binding proteins and protein translation machinery components

We have recently used affinity purification and mass spectrometry to identify the cellular pathways most likely affected by mutant Htt expression in mouse brains. Our as yet unpublished observations demonstrated that mutant Htt associates with vast numbers of proteins involved in translation initiation and RNA metabolic processes. These functional categories were significantly better represented in mutant purifications than wild-type, which argues that mutant Htt disproportionately affects these processes compared with wild-type. We also found that mutant Htt expression influenced the solubility of two of its newly identified interaction partners, FMR1 and PABP by affecting their sensitivity to treatment with RNAse. These observations suggest that mutant Htt expression affects the activity of at least two RNA binding proteins known to target many mRNAs.

One of the commonalities between the many RNA binding proteins and translational proteins identified in our purifications was their known involvement in stress granule assembly or function. This led us to discover that both wild-type and mutant Htt localize to stress granules (Culver, Savas, et. al, submitted). Similar to what we previously observed with regards to mutant Htt expression on P-body formation (Savas et al. 2008), we found that mutant Htt-expressing striatal precursor cell line (Trettel et al. 2000) formed fewer, but larger stress granules than a wild-type version of these cells (Culver, Savas, et. al, submitted). Htt is not required for stress granule assembly, however, as Htt-null mouse ES cells are able to form stress granules as well as wild-type cells (our unpublished observation). Based on these data, we propose that Htt may help to deliver specific mRNAs to stress granules, but is not required for their assembly.

Late stage HD patient brains exhibit reduced electron transport chain activity in mitochondrial complex II, III, and IV when assayed postmortem (Gu et al. 1996). In addition, systemic delivery of the complex II inhibitor 3-nitropropionic acid results in striatal-specific cell death in rodents and non-human primates (Brouillet et al. 2005). Inhibition of the electron transport chain can lead to accumulation of reactive oxygen species (ROS) (Chen et al. 2007). ROS species can activate the stress response, lead to the formation of stress granules, and subsequently suppress the translation of many different kinds of proteins. Since the striatum seems to be exquisitely sensitive to electron transport chain inhibition, perhaps the death of the cells within this tissue stems from a reduced ability to synthesize the proteins required for maintaining neuronal activity and viability. This hypothesis agrees with the increased levels of activated PKR seen in HD patient brain tissue as mentioned above. Indeed, ROS generation has been shown to increase PKR transcription and hence activity of the kinase (Pyo, Lee, and Choi 2008). We speculate that cellular stress, either owing to mutant Htt expression, or normally experienced by the brain, acts to activate the stress response, which results in a chronic down regulation of translation of proteins vital to the integrity of the neurons of the striatum. A similar type of process may also contribute to the neurodegeneration accompanying a stroke or the chronic head injuries experienced by athletes involved in contact sports.

Mutant Htt aggregates in HD contain RNA binding proteins (Doi et al. 2010; Schwab et al. 2008) and we have since found that expression of an aggregate-prone fragment of mutant Htt in mouse neuroblastoma cell line N2A produces mutant Htt aggregates that also contain RNA (Culver, Savas et. al, submitted). It is therefore possible that the mutant Htt aggregates present in HD also contain RNA. There are two possible interpretations for the aggregation of RNA binding proteins in neurodegenerative diseases that are not mutually exclusive. One is that that the aggregation of RNA binding proteins prevents them from acting on their normal targets, thereby altering the expression patterns of particular RNAs. It is also possible that certain RNAs are also sequestered within protein aggregates and this sequestration prevents the translation of these RNAs. One could imagine that disease-by-disease combinations of aggregated RNA binding proteins and RNAs could produce the defining symptoms and features for a particular disease. These defining features would depend on the combinations of genes affected in each disease.

3.7 Htt protein associates with its own mRNA

One of the necessary conditions for the hypothesis that mutant Htt influences post-transcriptional gene expression is that there must be specific mRNAs adversely affected over others. It seems unlikely that a global interference with this process by mutant Htt would produce the specific effects of the disease. We have attempted to identify these specific mRNAs through affinity purification and microarray profiling of Htt-associated RNAs in mouse brains. In agreement with the results of our mass spectrometry data, we have found that mutant Htt purifications contained substantially more enriched mRNAs than wild-type purifications (unpublished observations). Although, many of these enriched mRNAs have been reproduced and have direct relevance to HD, the identifications are still too preliminary to be reported here. However, we are confident that Htt protein purifications reproducibly recovered substantial amounts of Htt mRNA. Despite reduced amounts of mutant Htt protein recovery, these purifications contained substantially more Htt mRNA than wild-type purifications (Figure 4).

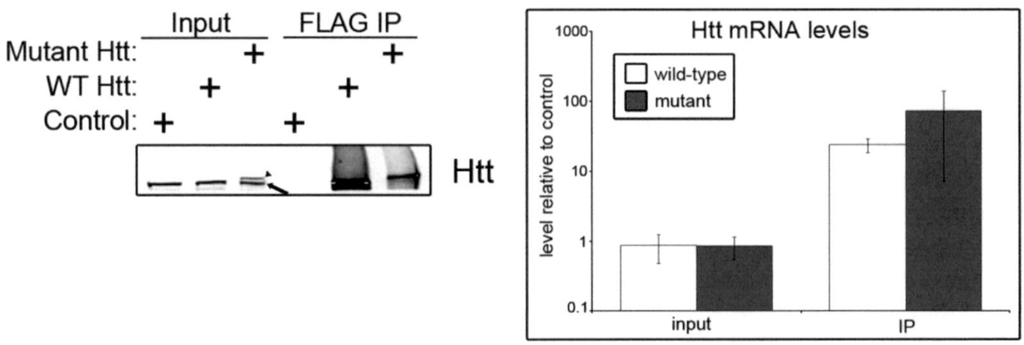

Fig. 4. Mutant Htt associates with more of its own mRNA than wild-type in mouse brains. The western blot on the left shows that wild-type Htt purifications recover more protein than mutant purifications. Htt was recovered (FLAG IP) from a cytoplasmic fraction of FLAG-tagged wild-type and mutant Htt knock-in mouse brains. Control purifications were from non-transgenic littermates. The blot is probed with an anti-Htt antibody. The bar graph on the right shows the levels of Htt mRNA in the inputs and affinity purifications (IP) from three independent experiments relative to the control.

This result has since been reproduced in cultured neurons and neuroblastoma cell lines using an Htt specific antibody (Culver, unpublished observations). This unexpected and exciting finding has interesting parallels to TDP-43 autoregulation of its own mRNA, and our lab is currently investigating if a similar phenomenon occurs with respect to Htt and its own mRNA.

4. Conclusion

There is now considerable evidence that Htt influences multiple steps of post-transcriptional gene expression. Htt interacts with proteins involved in RNA splicing, and expression of a mutant Htt fragment results in the cytoplasmic retention of a pre-mRNA processing factor

PRPF40a. Splicing occurs in the nucleus and therefore the absence of a splicing factor from here is functionally equivalent to a loss-of-function of the protein. This could have far reaching consequences in cases where redundancy mechanisms preclude splicing activity rescue for PRPF40a. The advent of highly sensitive RNA sequencing technology could be used to identify small or subtle differences in the splicing patterns of mRNAs in HD compared with an unaffected population. Potential differences in splicing patterns could be informative in characterizing how HD pathology progresses. Ideally, we would like to see a link between affected mRNAs and splicing proteins known to associate with Htt.

Htt interacts with Ago2, localizes to P-bodies, and participates in RNA silencing. Furthermore, mutant Htt-expressing cells possess fewer P-bodies and less efficiently execute the RNA silencing response. Artificial tethering of Htt to a luciferase reporter reduces its expression and this inhibition is partially mediated by Ago2. If the RNA silencing response is less effective in HD, then this could lead to increased levels of mRNAs and proteins whose levels are normally tightly controlled. This key step in regulation of gene expression is critical to ensure that potentially damaging gene products are only expressed under the right conditions. An impressive microarray study of HD patients and unaffected individuals has already pinpointed several mRNAs whose levels are elevated in HD compared with unaffected individuals (Hodges et al. 2006). In the future it will be important to determine if any of these mRNAs is regulated post-transcriptionally by Argonaute proteins and the RNA silencing pathway.

Mutant Htt mRNA can form hairpin structures composed of CAG and neighboring CGG repeats. These hairpins are recognized by the double stranded RNA binding protein kinase PKR, which acts to repress translation by inhibiting m7G cap-dependent translation through phosphorylation of eIF2A. PKR levels are elevated in HD patient brain tissue and elevated PKR activity can induce the formation of stress granules through global m7G cap-dependent translational inhibition. Expression of a CUG-expanded RNA is sufficient to induce stress granule formation in a cell culture model of myotonic dystrophy type I. This response is mediated by PKR activation (Huichalaf et al. 2010). These stress granules sequester an mRNA encoding a key DNA repair enzyme. Both wild-type and mutant Htt localize to stress granules, although Htt seems to be dispensable for their formation. Aggregation of proteins required for stress granule and P-body assembly involves glutamine-rich regions. We propose that stress granules are assembled in Htt either through activation of PKR by expanded CAG hairpins, increased ROS generation through electron transport chain impairment, or a combination of both processes. Htt localization to stress granules could lead to increased recruitment of Htt to stress granules through the expanded polyglutamine repeat sequence of mutant Htt. In this model, stress granules therefore serve as a nucleating factor in Htt aggregate assembly. The RNAs that are trapped within these stress granule-nucleated aggregates will therefore be prevented from being translated. In theory this could generate a feedback loop which leads to increased accumulation of Htt as the stress response is perpetuated. In cell culture, mutant Htt aggregates contain RNA. In the future we would like to identify the specific RNAs that may be trapped within insoluble Htt aggregates as they could be key factors in determining HD disease symptoms.

Htt is involved in trafficking RNAs in cultured neurons. Lentiviral-mediated Htt knockdown in cultured cortical neurons results in a drastic reduction in the amount of large ribonucleoprotein particles containing polyadenylated RNA present in dendrites. Htt co-localizes with proteins important for transporting RNAs and with the BDNF mRNA in rat brain slices and in cultured cortical neurons. Mutant Htt expression inhibits transport of vesicular cargoes and so it is likely that mutant Htt expression will also inhibit RNA transport by a similar mechanism. Localized protein translation is known to be important for synaptic plasticity. If mutant Htt reduces the amount of RNA transported, then this could manifest as a decreased ability of cells to quickly respond to changes in their environment. Data from our lab suggests that Htt plays a central role in RNA transport, with little specificity for particular mRNAs. Further experiments in simpler model systems (*e.g.* flies) will be required to determine if this is in fact the case. If defects in systemic RNA transport contribute to HD, then genetic studies in mouse on genes with central roles in RNA transport should produce a similar constellation of phenotypes as HD.

If HD results in part from the post-transcriptional deregulation of specific RNAs, then these transcripts will need to be identified to strengthen this hypothesis. These genes should function in a process that is known to be perturbed in HD or be important for the proper functioning or survival of striatal and cortical neurons. We are currently identifying Htt-associated RNAs by affinity purification of wild-type and mutant Htt from mouse brains. The significance of these potential interactions will be verified in animal models and human tissue. It is our hope that these experiments will further our understanding of HD and possibly contribute to treatment and a cure.

5. References

An, J. J., et al. 2008. Distinct role of long 3' UTR BDNF mRNA in spine morphology and synaptic plasticity in hippocampal neurons. *Cell* 134 (1):175-87.

Andreadis, A. 2011. Tau splicing and the intricacies of dementia. *Journal of cellular physiology*.

Andreassi, C., & A. Riccio. 2009. To localize or not to localize: mRNA fate is in 3'UTR ends. *Trends in cell biology* 19 (9):465-74.

Anitei, M., et al. 2010. Protein complexes containing CYFIP/Sra/PIR121 coordinate Arf1 and Rac1 signalling during clathrin-AP-1-coated carrier biogenesis at the TGN. *Nature cell biology* 12 (4):330-40.

Ash, P. E., et al. 2010. Neurotoxic effects of TDP-43 overexpression in C. elegans. *Human molecular genetics* 19 (16):3206-18.

Ayala, Y. M., et al. 2011. TDP-43 regulates its mRNA levels through a negative feedback loop. *The EMBO journal* 30 (2):277-88.

Bando, Y., et al. 2005. Double-strand RNA dependent protein kinase (PKR) is involved in the extrastriatal degeneration in Parkinson's disease and Huntington's disease. *Neurochemistry international* 46 (1):11-8.

Barnard, D. C., et al. 2004. Symplekin and xGLD-2 are required for CPEB-mediated cytoplasmic polyadenylation. *Cell* 119 (5):641-51.

Blumen, S. C., et al. 2009. Cognitive impairment and reduced life span of oculopharyngeal muscular dystrophy homozygotes. *Neurology* 73 (8):596-601.

Bosco, D. A., et al. 2010. Mutant FUS proteins that cause amyotrophic lateral sclerosis incorporate into stress granules. *Human molecular genetics* 19 (21):4160-75.

Brengues, M., D. Teixeira, & R. Parker. 2005. Movement of eukaryotic mRNAs between polysomes and cytoplasmic processing bodies. *Science* 310 (5747):486-9.

Brouillet, E., et al. 2005. 3-Nitropropionic acid: a mitochondrial toxin to uncover physiopathological mechanisms underlying striatal degeneration in Huntington's disease. *Journal of neurochemistry* 95 (6):1521-40.

Buchan, J. R., D. Muhlrad, & R. Parker. 2008. P bodies promote stress granule assembly in Saccharomyces cerevisiae. *The Journal of cell biology* 183 (3):441-55.

Buchan, J. R., & R. Parker. 2009. Eukaryotic stress granules: the ins and outs of translation. *Molecular cell* 36 (6):932-41.

Calado, A., et al. 2000. Nuclear inclusions in oculopharyngeal muscular dystrophy consist of poly(A) binding protein 2 aggregates which sequester poly(A) RNA. *Human molecular genetics* 9 (15):2321-8.

Cappell, K. M., et al. 2010. Symplekin specifies mitotic fidelity by supporting microtubule dynamics. *Molecular and cellular biology* 30 (21):5135-44.

Caviston, J. P., et al. 2007. Huntingtin facilitates dynein/dynactin-mediated vesicle transport. *Proceedings of the National Academy of Sciences of the United States of America* 104 (24):10045-50.

Chen, Y., et al. 2007. Mitochondrial electron-transport-chain inhibitors of complexes I and II induce autophagic cell death mediated by reactive oxygen species. *Journal of cell science* 120 (Pt 23):4155-66.

Darnell, J. C., et al. 2011. FMRP Stalls Ribosomal Translocation on mRNAs Linked to Synaptic Function and Autism. *Cell* 146 (2):247-61.

Davies, J. E., Z. Berger, & D. C. Rubinsztein. 2006. Oculopharyngeal muscular dystrophy: potential therapies for an aggregate-associated disorder. *The international journal of biochemistry & cell biology* 38 (9):1457-62.

Davies, J. E., & D. C. Rubinsztein. 2006. Polyalanine and polyserine frameshift products in Huntington's disease. *Journal of medical genetics* 43 (11):893-6.

de Mezer, M., et al. 2011. Mutant CAG repeats of Huntingtin transcript fold into hairpins, form nuclear foci and are targets for RNA interference. *Nucleic acids research* 39 (9):3852-63.

De Rubeis, S., & C. Bagni. 2010. Fragile X mental retardation protein control of neuronal mRNA metabolism: Insights into mRNA stability. *Molecular and cellular neurosciences* 43 (1):43-50.

Didiot, M. C., et al. 2009. Cells lacking the fragile X mental retardation protein (FMRP) have normal RISC activity but exhibit altered stress granule assembly. *Molecular biology of the cell* 20 (1):428-37.

Doi, H., et al. 2010. The RNA-binding protein FUS/TLS is a common aggregate-interacting protein in polyglutamine diseases. *Neuroscience research* 66 (1):131-3.

Faber, P. W., et al. 1998. Huntingtin interacts with a family of WW domain proteins. *Human molecular genetics* 7 (9):1463-74.

Ferrer, I., et al. 2000. Brain-derived neurotrophic factor in Huntington disease. *Brain research* 866 (1-2):257-61.

Friedman, R. C., et al. 2009. Most mammalian mRNAs are conserved targets of microRNAs. *Genome research* 19 (1):92-105.

Gaspar, C., et al. 2000. CAG tract of MJD-1 may be prone to frameshifts causing polyalanine accumulation. *Human molecular genetics* 9 (13):1957-66.

Gasparini, L., B. Terni, & M. G. Spillantini. 2007. Frontotemporal dementia with tau pathology. *Neuro-degenerative diseases* 4 (2-3):236-53.

Gauthier, L. R., et al. 2004. Huntingtin controls neurotrophic support and survival of neurons by enhancing BDNF vesicular transport along microtubules. *Cell* 118 (1):127-38.

Gharami, K., et al. 2008. Brain-derived neurotrophic factor over-expression in the forebrain ameliorates Huntington's disease phenotypes in mice. *Journal of neurochemistry* 105 (2):369-79.

Gingras, A. C., B. Raught, & N. Sonenberg. 1999. eIF4 initiation factors: effectors of mRNA recruitment to ribosomes and regulators of translation. *Annual review of biochemistry* 68:913-63.

Group, The Huntington's Disease Collaborative Research. 1993. A novel gene containing a trinucleotide repeat that is expanded and unstable on Huntington's disease chromosomes. *Cell* 72 (6):971-83.

Gu, M., et al. 1996. Mitochondrial defect in Huntington's disease caudate nucleus. *Annals of neurology* 39 (3):385-9.

Hebert, S. S., et al. 2010. Genetic ablation of Dicer in adult forebrain neurons results in abnormal tau hyperphosphorylation and neurodegeneration. *Human molecular genetics* 19 (20):3959-69.

Her, L. S., & L. S. Goldstein. 2008. Enhanced sensitivity of striatal neurons to axonal transport defects induced by mutant huntingtin. *The Journal of neuroscience : the official journal of the Society for Neuroscience* 28 (50):13662-72.

Hirokawa, N., S. Niwa, & Y. Tanaka. 2010. Molecular motors in neurons: transport mechanisms and roles in brain function, development, and disease. *Neuron* 68 (4):610-38.

Hodges, A., et al. 2006. Regional and cellular gene expression changes in human Huntington's disease brain. *Human molecular genetics* 15 (6):965-77.

Hughes, T. A. 2006. Regulation of gene expression by alternative untranslated regions. *Trends in genetics : TIG* 22 (3):119-22.

Huichalaf, C., et al. 2010. Expansion of CUG RNA repeats causes stress and inhibition of translation in myotonic dystrophy 1 (DM1) cells. *The FASEB journal : official publication of the Federation of American Societies for Experimental Biology* 24 (10):3706-19.

Jiang, H., et al. 2004. Myotonic dystrophy type 1 is associated with nuclear foci of mutant RNA, sequestration of muscleblind proteins and deregulated alternative splicing in neurons. *Human molecular genetics* 13 (24):3079-88.

Jiang, Y. J., et al. 2011. Interaction with Polyglutamine-expanded Huntingtin Alters Cellular Distribution and RNA Processing of Huntingtin Yeast Two-hybrid Protein A (HYPA). *The Journal of biological chemistry* 286 (28):25236-45.

Jin, P., et al. 2003. RNA-mediated neurodegeneration caused by the fragile X premutation rCGG repeats in Drosophila. *Neuron* 39 (5):739-47.

Kapitein, L. C., et al. 2010. Mixed microtubules steer dynein-driven cargo transport into dendrites. *Current biology : CB* 20 (4):290-9.

Kiebler, M. A., & G. J. Bassell. 2006. Neuronal RNA granules: movers and makers. *Neuron* 51 (6):685-90.

Kim, S., & P. A. Coulombe. 2010. Emerging role for the cytoskeleton as an organizer and regulator of translation. *Nature reviews. Molecular cell biology* 11 (1):75-81.

Kolev, N. G., & J. A. Steitz. 2005. Symplekin and multiple other polyadenylation factors participate in 3'-end maturation of histone mRNAs. *Genes & development* 19 (21):2583-92.

Lagier-Tourenne, C., M. Polymenidou, & D. W. Cleveland. 2010. TDP-43 and FUS/TLS: emerging roles in RNA processing and neurodegeneration. *Human molecular genetics* 19 (R1):R46-64.

Lee, J., et al. 2002. An upstream open reading frame impedes translation of the huntingtin gene. *Nucleic acids research* 30 (23):5110-9.

Lemay, J. F., et al. 2010. Crossing the borders: poly(A)-binding proteins working on both sides of the fence. *RNA biology* 7 (3):291-5.

Liu-Yesucevitz, L., et al. 2010. Tar DNA binding protein-43 (TDP-43) associates with stress granules: analysis of cultured cells and pathological brain tissue. *PLoS One* 5 (10):e13250.

Lovett, P. S., & E. J. Rogers. 1996. Ribosome regulation by the nascent peptide. *Microbiological reviews* 60 (2):366-85.

Ma, B., et al. 2010. Localization of BDNF mRNA with the Huntington's disease protein in rat brain. *Molecular neurodegeneration* 5:22.

Mackenzie, I. R., et al. 2007. Pathological TDP-43 distinguishes sporadic amyotrophic lateral sclerosis from amyotrophic lateral sclerosis with SOD1 mutations. *Annals of neurology* 61 (5):427-34.

Mahadevan, M., et al. 1992. Myotonic dystrophy mutation: an unstable CTG repeat in the 3' untranslated region of the gene. *Science* 255 (5049):1253-5.

Mangus, D. A., M. C. Evans, & A. Jacobson. 2003. Poly(A)-binding proteins: multifunctional scaffolds for the post-transcriptional control of gene expression. *Genome biology* 4 (7):223.

Mattick, J. S., & I. V. Makunin. 2006. Non-coding RNA. *Human molecular genetics* 15 Spec No 1:R17-29.

Meyer, S., C. Temme, & E. Wahle. 2004. Messenger RNA turnover in eukaryotes: pathways and enzymes. *Critical reviews in biochemistry and molecular biology* 39 (4):197-216.

Mihailovich, M., et al. 2007. Complex translational regulation of BACE1 involves upstream AUGs and stimulatory elements within the 5' untranslated region. *Nucleic acids research* 35 (9):2975-85.

Muddashetty, R. S., et al. 2011. Reversible inhibition of PSD-95 mRNA translation by miR-125a, FMRP phosphorylation, and mGluR signaling. *Molecular cell* 42 (5):673-88.

Narisawa-Saito, M., et al. 1996. Regional specificity of alterations in NGF, BDNF and NT-3 levels in Alzheimer's disease. *Neuroreport* 7 (18):2925-8.

Oe, S., & Y. Yoneda. 2010. Cytoplasmic polyadenylation element-like sequences are involved in dendritic targeting of BDNF mRNA in hippocampal neurons. *FEBS letters* 584 (15):3424-30.

Parker, R., & U. Sheth. 2007. P bodies and the control of mRNA translation and degradation. *Molecular cell* 25 (5):635-46.

Peel, A. L., et al. 2001. Double-stranded RNA-dependent protein kinase, PKR, binds preferentially to Huntington's disease (HD) transcripts and is activated in HD tissue. *Human molecular genetics* 10 (15):1531-8.

Peng, S., et al. 2009. Decreased brain-derived neurotrophic factor depends on amyloid aggregation state in transgenic mouse models of Alzheimer's disease. *The Journal of neuroscience : the official journal of the Society for Neuroscience* 29 (29):9321-9.

Pickering, B. M., & A. E. Willis. 2005. The implications of structured 5' untranslated regions on translation and disease. *Seminars in cell & developmental biology* 16 (1):39-47.

Pritchard, S. M., et al. 2011. The toxicity of tau in Alzheimer disease: turnover, targets and potential therapeutics. *Journal of cellular and molecular medicine* 15 (8):1621-35.

Pruunsild, P., et al. 2007. Dissecting the human BDNF locus: bidirectional transcription, complex splicing, and multiple promoters. *Genomics* 90 (3):397-406.

Pyo, C. W., S. H. Lee, & S. Y. Choi. 2008. Oxidative stress induces PKR-dependent apoptosis via IFN-gamma activation signaling in Jurkat T cells. *Biochemical and biophysical research communications* 377 (3):1001-6.

Rebbapragada, I., & J. Lykke-Andersen. 2009. Execution of nonsense-mediated mRNA decay: what defines a substrate? *Current opinion in cell biology* 21 (3):394-402.

Richter, J. D. 2007. CPEB: a life in translation. *Trends in biochemical sciences* 32 (6):279-85.

Robertson, J., et al. 2003. A neurotoxic peripherin splice variant in a mouse model of ALS. *The Journal of cell biology* 160 (6):939-49.

Russo, A., et al. 2008. cis-acting sequences and trans-acting factors in the localization of mRNA for mitochondrial ribosomal proteins. *Biochimica et biophysica acta* 1779 (12):820-9.

Savas, J. N., et al. 2010. A role for huntington disease protein in dendritic RNA granules. *The Journal of biological chemistry* 285 (17):13142-53.

Savas, J. N., et al. 2008. Huntington's disease protein contributes to RNA-mediated gene silencing through association with Argonaute and P bodies. *Proceedings of the National Academy of Sciences of the United States of America* 105 (31):10820-5.

Schwab, C., et al. 2008. Colocalization of transactivation-responsive DNA-binding protein 43 and huntingtin in inclusions of Huntington disease. *Journal of neuropathology and experimental neurology* 67 (12):1159-65.

Sephton, C. F., et al. 2011. Identification of neuronal RNA targets of TDP-43-containing ribonucleoprotein complexes. *The Journal of biological chemistry* 286 (2):1204-15.

Seredenina, T., & R. Luthi-Carter. 2011. What have we learned from gene expression profiles in Huntington's disease? *Neurobiology of disease.*

Shinchuk, L. M., et al. 2005. Poly-(L-alanine) expansions form core beta-sheets that nucleate amyloid assembly. *Proteins* 61 (3):579-89.

Siomi, H., & M. C. Siomi. 2009. On the road to reading the RNA-interference code. *Nature* 457 (7228):396-404.

Sisodia, S. S. 1992. Beta-amyloid precursor protein cleavage by a membrane-bound protease. *Proceedings of the National Academy of Sciences of the United States of America* 89 (13):6075-9.

Sossin, W. S., & L. DesGroseillers. 2006. Intracellular trafficking of RNA in neurons. *Traffic* 7 (12):1581-9.

Spriggs, K. A., M. Bushell, & A. E. Willis. 2010. Translational regulation of gene expression during conditions of cell stress. *Molecular cell* 40 (2):228-37.

St Johnston, D., D. Beuchle, & C. Nusslein-Volhard. 1991. Staufen, a gene required to localize maternal RNAs in the Drosophila egg. *Cell* 66 (1):51-63.

Strong, M. J., et al. 2007. TDP43 is a human low molecular weight neurofilament (hNFL) mRNA-binding protein. *Molecular and cellular neurosciences* 35 (2):320-7.

Swanger, S. A., & G. J. Bassell. 2011. Making and breaking synapses through local mRNA regulation. *Current opinion in genetics & development* 21 (4):414-21.

Tatom, J. B., et al. 2009. Mimicking aspects of frontotemporal lobar degeneration and Lou Gehrig's disease in rats via TDP-43 overexpression. *Molecular therapy : the journal of the American Society of Gene Therapy* 17 (4):607-13.

Tian, B., et al. 2005. A large-scale analysis of mRNA polyadenylation of human and mouse genes. *Nucleic acids research* 33 (1):201-12.

Trettel, F., et al. 2000. Dominant phenotypes produced by the HD mutation in STHdh(Q111) striatal cells. *Human molecular genetics* 9 (19):2799-809.

Van Dam, D., et al. 2005. Cognitive decline, neuromotor and behavioural disturbances in a mouse model for fragile-X-associated tremor/ataxia syndrome (FXTAS). *Behavioural brain research* 162 (2):233-9.

Vance, C., et al. 2009. Mutations in FUS, an RNA processing protein, cause familial amyotrophic lateral sclerosis type 6. *Science* 323 (5918):1208-11.

Venter, J. C., et al. 2001. The sequence of the human genome. *Science* 291 (5507):1304-51.

Villace, P., R. M. Marion, & J. Ortin. 2004. The composition of Staufen-containing RNA granules from human cells indicates their role in the regulated transport and translation of messenger RNAs. *Nucleic acids research* 32 (8):2411-20.

Volkening, K., et al. 2009. Tar DNA binding protein of 43 kDa (TDP-43), 14-3-3 proteins and copper/zinc superoxide dismutase (SOD1) interact to modulate NFL mRNA stability. Implications for altered RNA processing in amyotrophic lateral sclerosis (ALS). *Brain research* 1305:168-82.

Wahl, M. C., C. L. Will, & R. Luhrmann. 2009. The spliceosome: design principles of a dynamic RNP machine. *Cell* 136 (4):701-18.

Wang, D. O., K. C. Martin, & R. S. Zukin. 2010. Spatially restricting gene expression by local translation at synapses. *Trends in neurosciences* 33 (4):173-82.

Weigell-Weber, M., W. Schmid, & R. Spiegel. 1996. Psychiatric symptoms and CAG expansion in Huntington's disease. *American journal of medical genetics* 67 (1):53-7.

Wellington, C. L., et al. 1998. Caspase cleavage of gene products associated with triplet expansion disorders generates truncated fragments containing the polyglutamine tract. *The Journal of biological chemistry* 273 (15):9158-67.

Willemsen, R., J. Levenga, & B. Oostra. 2011. CGG repeat in the FMR1 gene: size matters. *Clinical genetics* 80 (3):214-25.

Xiao, S., S. Tjostheim, et al. 2008. An aggregate-inducing peripherin isoform generated through intron retention is upregulated in amyotrophic lateral sclerosis and associated with disease pathology. *The Journal of neuroscience : the official journal of the Society for Neuroscience* 28 (8):1833-40.

Xu, Z., et al. 1993. Increased expression of neurofilament subunit NF-L produces morphological alterations that resemble the pathology of human motor neuron disease. *Cell* 73 (1):23-33.

Zhou, W., and W. Song. 2006. Leaky scanning and reinitiation regulate BACE1 gene expression. *Molecular and cellular biology* 26 (9):3353-64.

Zu, T., et al. 2011. Non-ATG-initiated translation directed by microsatellite expansions. *Proceedings of the National Academy of Sciences of the United States of America* 108 (1):260-5.

Zuccato, C., et al. 2001. Loss of huntingtin-mediated BDNF gene transcription in Huntington's disease. *Science* 293 (5529):493-8.

Part 5

Metabolic Dysregulation in Huntington's Disease

14

Energy Metabolism in Huntington's Disease

Fabíola M. Ribeiro[1], Tomas Dobransky[3],
Eduardo A. D. Gervásio-Carvalho[1], Jader S. Cruz[1]
and Fernando A. Oliveira[2]
*[1]Department of Biochemistry and Immunology,
Universidade Federal de Minas Gerais (UFMG),
[2]Department of Biological Sciences,
Universidade Federal de São Paulo (UNIFESP),
[3]DB Biotech, Kosice,
[1,2]Brazil
[3]Slovakia*

1. Introduction

Neurodegenerative diseases are pathological processes characterized by neuronal death and morbid evolution leading to occupational injury and serious neuropsychiatric disorders. The natural course of neurodegenerative diseases does not show regression of symptoms or cure and current treatments are far from producing a real improvement in the quality of patient's life. Several studies have been conducted in an attempt to find causes of cellular disturbances focusing new pharmacological targets priming to successful therapeutic interventions. Studies have been directed to investigate possible changes in energy metabolism pathways. Indeed, some disturbances in glycolytic pathway and mitochondrial dysfunctions have been associated with Huntington's Disease and other neurodegenerative diseases and are often related to the events of cell death. In this section, an overview of the energy metabolism pathways will be presented and the particular aspects of energetic metabolism in Huntington's Disease will be discussed.

2. Glucose transport

Under ordinary conditions, the basic substrate for brain metabolism is glucose. In the resting state, adults use about 20% of whole-body glucose for brain metabolism. The brain has an exquisite dependence on glucose for energy production and as an important carbon source for biosynthesis of a variety of simple and complex molecules (Siegel et al., 1999). Such dependence is well demonstrated and a transient decline in the metabolism of glucose would cause a serious disruption of brain function (Oliveira et al., 2007). As glucose is a water-soluble substance its entry into brain from blood is greatly restricted (Vannucci et al., 1997). The major reason for that is the presence of an anatomical-physiological barrier - so called Blood-Brain-Barrier (BBB). The BBB is a specialized barrier made up of microvascular endothelial cells that are held together by tight junction complexes that effectively avoid the paracellular diffusion of solutes (Wilhelm et al., 2011). Even small molecules do not simply diffuse across this physical barrier. To manage this problem, not only for glucose but also for other metabolites important to brain metabolism, there are a large family of specific membrane transporters responsible to carry these molecules across the BBB by means of facilitated diffusion through the luminal and ablumenal endothelial membranes (Vannucci et al., 1997).

It is established that transport and uptake of glucose is performed by a super-family of glucose transporters belonging to the GLUT gene family. The GLUT family includes 12 genes encoding 12 GLUT proteins (Vannucci et al., 1997; Klepper and Voit, 2002). Various members of this membrane protein family have been detected in brain (GLUT-1, GLUT-3 and GLUT-5), but the initial glucose transport step across the BBB is mediated exclusively by the facilitative glucose transporter protein type 1 (GLUT-1) (Lund-Andersen, 1979; Pardridge et al., 1990; Klepper and Voit, 2002). Important to note is the fact that despite such facilitative transport mechanism, the actual concentration of glucose in brain is lower than it is in structures lacking a barrier, e.g. peripheral nerves. There is, then, no safety device to supplement carbohydrate reserve during hypoglycemia episodes making the regulation of GLUTs extremely relevant for brain physiology.

Following entry into the brain glucose is transported from the interstitial fluid into neurons primarily via GLUT-3 and into glia primarily via GLUT-1 transporters. GLUT-5 protein is known to be present in brain microglia, although its function in these cells remains unclear. This transport into the intracellular compartment is relatively rapid, making the BBB the rate-limiting step for glucose entry into brain cells (Pardridge et al., 1990).

Glucose, entering the neuronal cells, is phosphorylated irreversibly to glucose-6-phosphate (G-6~P) and metabolized in the pentose phosphate shunt or the Embden-Meyerhoff pathway, or converted to glycogen. The Embden-Meyerhoff metabolic pathway permits glycolytic conversion of glucose to pyruvate. Glycogen synthesis provides a source of fuel during periods of metabolic stress (3.3 mmol/kg rat). Evidently a decreased entry of glucose into the brain limits these three pathways and potentially contributes to the development of innumerous neuronal pathologies.

Brain metabolic needs are demanding and the way in which it circumvents this situation has long remained unclear. For the past 15 years authors are looking for a more universal explanation but it turns out to be rather complicated. The subject is controversial and the field is still very active.

Neurons and astrocytes, the two major types of brain cells, are largely responsible for the massive consumption of O_2 and glucose in the brain. Just to point out how important they are, under resting conditions, astrocytes release ~85% of the glucose they consume as lactate. On the contrary, neurons contribute minimally to glucose consumption by the brain. Thus, despite their shared localization, neurons and astrocytes exhibit a different preference for glucose consumption and utilization (Nehlig and Coles, 2007).

The ATP-dependent phosphorylation of glucose to G-6~P is the first step of glycolysis and it is catalyzed by hexokinase (HK) (Berg et al., 2006). The reaction is practically irreversible and has been recognized as a key point in the regulation of carbohydrate metabolism in brain. This whole process concomitantly generates local ADP which is important as a recycling mechanism. Important to note that hexokinase activity in neurons and in other cell types also participates in various essential processes including ATP production, apoptosis, controlling glutathione levels, and preventing neuronal oxidative unbalance (Saraiva et al., 2010). In the brain, HK-1 is the major expressed enzyme isoform. It localizes into cytosol or firmly attached to outer mitochondrial membrane. The bound enzyme is more active and the extent of binding is thought to be inversely related to the ATP/ADP ratio (Siegel et al., 1999). Interestingly, conditions where energy utilization is greater than the substrate supply there is a shift in the solubilization equilibrium towards the membrane-bound enzyme form which per se provokes a greater potential to ignite glycolysis to meet the energy demand.

The activity and specific subcellular localization of neuronal HK-1 are regulated by distinct mechanisms that act synergistically to fine-tuning glycolytic flux in response to changes in cellular environment. At this point it may be important to bring up the idea that mitochondrial-bound hexokinase 1 could be neuroprotective as has been discussed by different research groups in this field.

The reaction product of HK-1 is G-6~P which represents a major branch point in metabolism because it is a common substrate for enzymes involved in glycolytic, pentose-phosphate shunt, and glycogen-forming pathways. In glycolysis, G-6~P is the substrate of isomerase producing fructose-6-phosphate (F-6~P). This reaction is promptly reversible (small free energy change), however its equilibrium ratio in brain greatly favors G-6~P accumulation. F-6~P is phosphorylated by phosphofructokinase-1 (PFK-1), which is considered one of the most regulated catalysts of the glycolytic sequence, to form fructose-1,6-bisphosphate (F-1,6-Bis~P). As observed in other regulatory biochemical reactions it is also essentially irreversible.

A number of studies focusing into this particular metabolic reaction led to the observation that astrocytes' PFK-1 is about two fold more active than in neurons under baseline conditions (Herrero-Mendez et al., 2009). One possible reason for that is the concentration of fructose-2,6-bisphosphate (F-2,6-Bis~P), an powerful allosteric modulator of PFK-1 activity, is significantly greater in astrocytes. This dramatic difference could be ascribed to the near absence in neurons of phosphofructokinase-2/fructose-2,6-phosphatase (PFK-2/F-2,6-Pase) the bifunctional enzyme responsible for the F-2,6-Bis~P synthesis and degradation (Herrero-Mendez et al., 2009).

The question to be raised is how neurons actually control PFK-2/F-2,6-Pase activity? There are four isoforms of PFK-2/F-2,6-Pase, each encoded by a separate gene [Pfkfb1, Pfkfb2, Pfkfb3, and Pfkfb4] (Okar et al., 2001). It is reasonable to suppose that each isoform displays significantly different regulatory and kinetic features and this, in turn, determines the ratio of kinase/phosphatase activity, and as a consequence the concentration of F-2,6-Bis~P in different tissues (Yalcin et al., 2009). In a well conducted study Herrero-Mendez and colleagues (2009), using RT-PCR analysis of RNA extracts from rat cortical neurons and astrocytes, indicated that Pfkfb3 mRNA was present in neurons as well as astrocytes. Importantly, the authors demonstrated that Pfkfb3 was the most abundant mRNA expressed in both cell types. The authors expected that if neurons contained the same relative abundance of Pfkfb3 mRNA as astrocytes and lower levels of the mature enzyme that the final concentration of PFK-2/F-2,6-Bis~Pase in neurons might be regulated post-transcriptionally. Some studies exploring the mechanism of such proteasomal degradation revealed that Pfkfb3, contains a KEN box that starts at position 142 (Pesin and Orr-Weaver, 2008; Herrero-Mendez et al., 2009). This motif targets proteins for ubiquitylation by the anaphase-promoting complex/cyclosome when bound to its activator CDH1 (Pesin and Orr-Weaver, 2008). Interestingly, CDH1 silencing in neurons induced PFK-2/F-2,6-Bis~Pase accumulation and as expected increased the rate of glycolysis.

Fructose-1,6-Bis~P is split by brain aldolase to glyceraldehyde-3-phosphate (Gal-3~P) and dihydroxyacetone phosphate (DHAP). DHAP is the common substrate for both glyceraldehyde-3-phosphate dehydrogenase and triose phosphate isomerase. One important point to comment is the equilibrium between DHAP and Gal-3~P maintained by the action of triose phosphate isomerase. In brain, the equilibrium favors accumulation of DHAP. After this reaction step, glycolysis in the brain proceeds through the usual biochemical reactions to produce pyruvate (Berg et al., 2006).

As we have been discussing throughout the chapter neurons use glucose basically to maintain their antioxidant balance status. In order to do this, neurons have to downregulate glycolysis. Under this situation one the question is raised: Where does glucose go to be further metabolized? To answer this question we need to remember that pentose-phosphate pathway (PPP) uses G-6~P as substrate and that PPP is metabolic linked to glycolysis. One interesting, however somehow surprising, observation made by different laboratories is the fact that, in contrast to astrocytes, neurons do not display increased glycolytic rate upon mitochondrial inhibition as one could expect (Bolanos et al., 2008; Bolanos et al., 2010), but instead these neuronal cells entry into cell death program. These results lead us to assume that the increased glycolytic rate in astrocytes served to preserve cells from ATP depletion and cell death, most probably because glycolytic ATP was used to drive the reverse activity of ATP synthase to maintain the mitochondrial membrane potential (Nehlig and Coles, 2007; Bolanos et al., 2010; Cunnane et al., 2011). On the other hand, such treatment caused neuronal ATP depletion and apoptotic cell death (Nehlig and Coles, 2007). Based on all facts presented so far one can hypothesize that neurodegenerative diseases (including Alzheimer's Disease and Huntington's Disease) may present a diminished neuronal glycolytic activity (Oliveira et al., 2007). Data in support of this hypothesis will be discussed later.

Accumulated evidence suggest that in neurons a significant proportion of G-6~P is directed towards the PPP (Bolanos et al., 2008). Besides its role at supplying ribose-5-phosphate for nucleic acid biosynthesis, glucose oxidation through the PPP is a major component of the cytosolic NADPH regenerating cell machinery (Nelson and Cox, 2004; Bolanos et al., 2008). The rate-limiting step in PPP activity is catalyzed by glucose-6-phosphate dehydrogenase, which oxidizes G-6~P into 6-phosphogluconate, conserving the redox energy as NADPH. Next, 6-phosphogluconate is further oxidized by 6-phosphogluconate dehydrogenase, which also conserves redox energy in the form of NADPH (Nelson and Cox, 2004; Bolanos et al., 2008).

It is important to note: NADPH is a necessary cofactor in the regeneration of reduced glutathione and for the reductive reactions for lipid biosynthesis. This mechanism is not exclusive for neurons it also operates in astrocytes which have high concentrations of glutathione due to their high activity of γ-glutamyl cysteine synthetase. This enzyme catalyzes the rate-limiting step in glutathione synthesis providing a metabolic scenario to build a robust antioxidant system (Heales and Bolanos, 2002). As neurons have rather low concentrations of glutathione, and low activity of γ-glutamyl cysteine synthetase there are compelling evidence that glucose entry in the PPP is important to regenerate glutathione and provide an effective defense mechanism against oxidative stress.

The energy output and oxygen consumption in adult brain are associated with high levels of enzyme activity in the tricarboxylic acid cycle (TCA). The TCA is organized into a supramolecular complex that interacts with mitochondrial membranes and the electron transport chain (Berg et al., 2006). Therefore, mitochondria have a central role for the energetic metabolism, their main function is oxidation of acetyl-coenzyme A derived from carbohydrates, amino-acids and fatty acids to produce ATP (Nelson and Cox, 2004). These organelles provide energy for a plethora of cellular processes and the highest number of mitochondria is present in organs demanding the most of energy, such as brain, liver and muscles.

Actually, the processes responsible for energy production are recognized as oxidative phosphorylation which is coupled to the electron transport chain (ETC). The ETC is a set of five protein complexes sitting on the inner mitochondrial membrane. Three protein complexes (complex I, III and IV) work as a proton pump transferring protons through the membrane into the intermembrane space. Chemiosmotic theory predicts that most of the ATP synthesis comes from the electrochemical gradient across the inner membranes of mitochondria by ATP synthase. Energy saved in ATP is used in synaptic ion homeostasis and phosphorylation reactions. ATP is essential for the excitability and survival of neurons, oxidative phosphorylation is involved in synaptic signaling and is related to changes of neuronal structure and function.

The major role is given to complex I (NADH dehydrogenase [ubiquinone]) in controlling mitochondrial oxidative phosphorylation; its malfunctioning can result in mitochondrial dysfunction.(Davey et al., 1998; Hroudova and Fisar, 2011). Thus, many mitochondrial diseases originate from complex I deficiencies.

In adult brain, the enzyme succinate dehydrogenase (SDH), which catalysis the oxidation of succinate to fumarate, is tightly bound to mitochondrial inner membrane. In brain, SDH may also have a regulatory role when its steady state is disturbed. Important to note that the levels of succinate and isocitrate in brain tissue are little affected by changes in the flux of the TCA, as long as proper glucose supply is available. The highly unfavorable free-energy change of the malate dehydrogenase reaction is bypassed by the very rapid removal of oxaloacetate, which is maintained at low concentrations under steady-state conditions through the condensation reaction with acetyl-coenzyme A (Berg et al., 2006); Nelson and Cox, 2004).

3. Energetic metabolism deficit in Huntington's Disease

Huntington´s Disease patients face pronounced weight loss, despite sustained caloric intake, which was a first indication that alterations in energetic metabolism could play a role in Huntington´s Disease pathogenesis (O'Brien et al., 1990). In agreement with this hypothesis, Huntington´s Disease patients exhibit alterations in cerebral glucose consume, lactate levels, and mitochondrial enzymes activity involved in glucose metabolism. Moreover, ATP depletion was directly demonstrated in Huntington´s Disease brain tissue (Mochel et al., 2010a). Thus far, various mechanisms underlying energy deficit in Huntington´s Disease brain have been identified, including impaired oxidative phosphorylation (Milakovic and Johnson, 2005), altered oxidative stress (Tabrizi et al., 1999), impaired mitochondrial calcium handling (Lim et al., 2008), abnormal mitochondria trafficking (Li et al., 2010), and deregulation of the transcriptional coactivator PPARγ coactivator-1α (PGC-1α), which is a crucial factor of mitochondrial biogenesis (Cui et al., 2006), and decreased glycolysis (Powers et al., 2007).

3.1 Glucose levels are reduced in Huntington's Disease patient's brain

Positron emission tomography (PET) studies revealed that glucose metabolism in the basal ganglia and cerebral cortex is markedly reduced in Huntington´s Disease patients (Kuwert et al., 1990; Andrews and Brooks, 1998). Moreover, the decrease in glucose metabolism is specific to cortical areas, caudate and putamen, and starts in the asymptomatic phase of the disease (Kuhl et al., 1985). Regardless of the severity of symptoms and despite apparent shrinkage of brain tissue, glucose utilization appears normal throughout the rest of the brain of Huntington´s Disease patients (Kuhl et al., 1985).

Furthermore, studies performed in presymptomatic Huntington´s Disease gene carriers revealed a pattern of cerebral metabolism characterized by relative increases in thalamic, occipital, and cerebellar glucose metabolism, despite reduced caudate and putamen metabolism. Following Huntington´s Disease symptoms appearance, this pattern was altered as thalamic metabolism, which was previously elevated, was reduced (Feigin et al., 2007). These data highlights the importance of the region specific alterations in glucose metabolism for Huntington´s Disease pathology.

Interestingly, a recent report supports the idea that the hypothalamus, but not the basal ganglia, is the brain region responsible for the metabolic abnormalities that take place in Huntington´s Disease (Hult et al., 2011). Selective hypothalamic expression of a short fragment of mutant huntingtin was sufficient to recapitulate the glucose metabolic disturbances that occur in Huntington´s Disease patients. In addition, selective hypothalamic inactivation of the mutant huntingtin gene prevented the development of the metabolic phenotype in a Huntington´s Disease mouse model, BACHD mice (Hult et al., 2011). Further studies will be important to point all the regions involved in Huntington´s Disease metabolic alterations.

In addition to its role as an energetic molecule, glucose also plays a role as a signaling molecule. It has been demonstrated that increased intracellular glucose levels decreases aggregate formation and is neuroprotective in cultured cells transfected with a mutant huntingtin construct (Ravikumar et al., 2003). Glucose metabolism appears altered in Huntington´s Disease, as huntingtin transfected PC12 cells exhibit disturbed expression levels of four genes involved in glucose metabolism (Glut1, Pfkm, Aldolase A, and Enolase), as well as a reduction in cell death following over-expression of Glut1 and Pfkm (another key regulatory protein for glycolysis) (Kita et al., 2002). Glucose reduces phosphorylation of mTOR, which is a negative regulator of authophagy, and its downstream effector S6K1 (Ravikumar et al., 2003). Thus, glucose-mediated negative regulation of mTOR could induce autophagy and clearance of the mutated huntingtin protein, as well as influence other mTOR mediated activities involving cell survival, growth, and translation of protein transcripts. Furthermore, glucose can regulate Akt and GSK3, which influence cell growth and survival (Clodfelder-Miller et al., 2005). Importantly, Akt activation can protect against neuronal death (Datta et al., 1999; Kandel and Hay, 1999). Akt can also promote phosphorylation of mutated Htt protein, which functions to reduce Htt aggregate formation and neuronal cell death, providing a protective pathway in Huntington´s Disease (Humbert et al., 2002; Warby et al., 2009). Highlighting the importance of Akt in Huntington´s Disease pathology, both NMDA and metabotropic glutamate receptor 5 receptors can increase Akt activation in striatal neurons from Huntington´s Disease mouse models (Gines et al., 2003a). These observations highlight the role of glucose as an important molecule not only for its energetic properties but also for its capacity to activate key molecules involved in cell survival and huntingtin clearance.

3.2 Enzymes involved in energetic metabolism are altered in Huntington´s Disease

Mutated huntingtin protein alters the function and/or expression of a number of enzymes involved in energetic metabolism and many of these alterations can have important implications in Huntington´s Disease pathology. Alteration of enzyme expression by mutated huntingtin may occur due to huntingtin-mediated regulation of transcriptional factors. For example, mutant huntingtin inhibits expression of PGC-1α, which is a transcriptional coactivator that regulates several metabolic processes, including mitochondrial biogenesis and respiration (Cui et al., 2006). Mutated huntingtin may also alter enzyme function by incorporation and sequestration of transcriptional factors and enzymes into mutated huntingtin aggregates (Yamanaka et al., 2008). The enzyme alterations caused by mutated huntingtin can have important deleterious consequence, including the metabolic deficit observed in Huntington´s Disease patients.

Glyceraldehyde-3-phosphate dehydrogenase (GAPDH) and the α-ketoglutarate dehydrogenase complex can be inactivated by long polyglutamine domains, which may cause a deficit in cerebral energy metabolism (Cooper et al., 1997). Further studies have shown that the mutated huntingtin protein alters the subcellular localization of GAPDH, increasing its nuclear localization in both human fibroblasts and in neurons from a transgenic mouse model (Mazzola and Sirover, 2001, 2002; Senatorov et al., 2003). The appearance of an abnormal high molecular weight form of GAPDH in fibroblast nuclei has also been associated with decreased glycolytic activity (Mazzola and Sirover, 2001).

Pyruvate dehydrogenase activity is decreased in basal ganglia and this deficit was significantly augmented with increasing duration of illness, possibly due to a progressive loss of neurons in Huntington's Disease caudate nucleus (Butterworth et al., 1985). Activities of the complexes II, III and IV of the electron transport chain were reduced in both Huntington´s Disease caudate and putamen of advanced grade (3 and 4) Huntington´s Disease patients (Gu et al., 1996; Browne et al., 1997). It has been shown that lactate concentration is increased in the basal ganglia and the occipital cortex of Huntington´s Disease patients (Jenkins et al., 1993; Jenkins et al., 1998). The lactate itself is not thought to be a neurotoxic metabolite, but may represent a marker for energetic changes such as reduced ATP production and excitotoxicity, which may have a direct effect on neuronal function and survival in Huntington´s Disease.

A number of studies have shown that the activity of transglutaminase 2, an enzyme primarily known for cross-linking proteins, is increased in Huntington´s Disease affected brain areas and that transglutaminase 2 causes an increase in huntingtin aggregation (Karpuj et al., 1999; Lesort et al., 1999; Karpuj et al., 2002). The most compelling evidence for a role of transglutaminase 2 in Huntington´s Disease is provided by the work of

Mastroberardino et al. (2002). These studies reported a reduction in neuronal cell death, improved behavior and prolonged survival in R6/1 X transglutaminase $2^{-/-}$, as compared to R6/1 X transglutaminase $2^{+/+}$. In addition to affecting huntingtin aggregation, increased transglutaminase 2 activity in Huntington´s Disease caudate may contribute to mitochondrial dysfunction by incorporating aconitase into inactive polymers and dramatically decreasing aconitase activity (Kim et al., 2005).

3.3 Mitochondrial deficit in Huntington´s Disease

Studies published so far suggest that mitochondrial defects play a major role in Huntington´s Disease etiology, underlined by decreased mitochondrial biogenesis, oxidative stress, ATP deficit, increased apoptosis, and, ultimately, a central and peripheral energy deficit (Browne and Beal, 2004).

Degenerated mitochondria have been detected in the striatum of symptomatic Huntington´s Disease mice (R6/2, R6/1, N171-82Q, and Hdh150CAG mice) and could be detected before other neuronal pathological changes and concomitant with symptom onset (Yu et al., 2003). These degenerating mitochondria exhibit swelling, disruption of the cristae and mitochondrial membranes, and eventual condensation and lysosomal engulfment. Interestingly, this study shows that neuronal cell death aspects due to motichondrial alterations varied among different Huntington´s Disease mouse models (Hickey and Chesselet, 2003; Yu et al., 2003).

Mutated huntingtin protein destabilizes mitochondrial Ca^{2+} regulation (Panov et al., 2002; Choo et al., 2004). Mitochondrial Ca^{2+} abnormalities occur early in Huntington´s Disease pathogenesis and appear to be caused by a direct effect of mutant huntingtin, as incubation of normal human lymphoblast mitochondria with a fusion protein containing a long polyglutamine repeat recapitulates the mitochondrial calcium defect observed in Huntington´s Disease (Panov et al., 2002). Further studies have also demonstrated that the huntingtin protein binds to the outer membrane of mitochondria from human neuroblastoma cells and from cultured striatal cells from WT and transgenic mice (Choo et al., 2004). Moreover, binding of mutated huntingtin protein, but not of wild type, increases sensitivity to calcium-induced opening of the mitochondrial permeability transition (MPT), leading to the release of cytochrome c in normal liver mitochondria (Choo et al., 2004).

Mitochondria play an important role in buffering cytoplasmic calcium and increased neuronal calcium modifies mitochondrial ATP production by uncoupling oxidative phosphorylation (Nicholls, 2009). Calcium overload may result in discharge of the mitochondrial membrane potential, opening of the MPT pore, release of cytochrome c, and activation of cell death pathways (Nicholls, 2009). Mutated huntingtin protein causes sensitization of both the NMDA receptor and the inositol-1,4,5-triphosphate (IP3) receptor, increasing entrance of extracellular Ca^{2+} and the release of Ca^{2+} from intracellular stores, respectively (Chen et al., 1999; Sun et al., 2001; Tang et al., 2005). The final result is an increase in intracellular Ca^{2+} levels. The role of NMDA receptors on mitochondrial biogenesis has been further characterized, as the reduced mitochondrial ATP levels and decreased ATP/ADP ratio found in mutant Htt-containing striatal cells is normalized by blocking NMDA receptor-mediated calcium influx (Seong et al., 2005). Moreover, mitochondria isolated from both lymphoblasts of Huntington´s Disease patients and brains of transgenic mice have a reduced membrane potential and depolarize at lower Ca^{2+} concentrations than control mitochondria (Panov et al., 2002).

Thus, data obtained so far points to a close relationship between mitochondria deficit and NMDA-mediated excitotoxicity. The glutamatergic system plays a substantial role in neuronal cell death and there are consistent data implicating NMDA receptor activation with the excitotoxic neuronal loss that takes place in Huntington´s Disease (Zeron et al., 2002; Schiefer et al., 2004). Selective depletion of NMDA receptors has been found in Huntington´s Disease striatum, suggesting that neurons expressing NMDA receptors are preferentially vulnerable to degeneration (Dure et al., 1991). Prior to the identification of the genetic mutation responsible for Huntington´s Disease, a Huntington´s Disease mouse model was developed by the introduction of quinolinic acid, which is an NMDA receptor agonist that produces excitotoxic striatal lesions that closely resemble those seen in Huntington´s Disease brain (Beal et al., 1986). Moreover, some studies suggest that the sensitization of the NMDA receptor containing the subunit NR1/NR2B by the mutated Htt protein is responsible for causing the selective cell death of the medium sized spiny neurons present in the striatum, since these neurons express high level of this NMDA receptor subtype (Chen et al., 1999; Zeron et al., 2001).

Mitochondrial toxins that deplete ATP production can also mediate excitotoxic processes (Schulz et al., 1996; Browne and Beal, 2002). Systemic administration of 3-nitropropionic acid, which is a mitochondrial toxicant that inhibits succinate dehydrogenase, results in striatum lesions similar to those observed in Huntington´s Disease (Wullner et al., 1994). Nevertheless, 3-nitropropionic acid and malonate lesions can be prevented by NMDA antagonists, such as MK-801 and memantine (Wullner et al., 1994). Taken together these observations suggest that mitochondrial-mediated excitotoxicity is promoted by secondary mechanisms involving glutamate receptors. It has been shown that omission of glucose, exclusion of oxygen, or inclusion of inhibitors of oxidative phosphorylation or of the sodium/potassium pump, enables glutamate to express its neurotoxic effects via NMDAR (Novelli et al., 1988; Henneberry et al., 1989; Zeevalk and Nicklas, 1991). Thus, in a context of reduced intracellular energy levels an otherwise harmless amount of glutamate becomes toxic.

Studies performed with Huntington´s Disease transgenic models have implicated decreased transcription of genes regulated by cyclic adenosine 3',5'-monophosphate (cAMP) responsive element (CRE) binding protein (CREB) to Huntington´s Disease pathology (Luthi-Carter et al., 2000; Shimohata et al., 2000; Steffan et al., 2000; Nucifora et al., 2001; Wyttenbach et al., 2001). These genes include brain derived neurotrophic factor (BDNF) (Zuccato et al., 2001) and a host of others involved in diverse processes ranging from neurotransmission (Bibb et al., 2000; Luthi-Carter et al., 2000) to cholesterol metabolism (Sipione et al., 2002).

Reduced CREB dependent transcription of BDNF is a robust feature of Huntington´s Disease pathophysiology. By grades II and III of the disease, BDNF protein and mRNA levels in frontoparietal cortex are halved, and this effect can be mimicked by expressing full-length human mutant huntingtin in a rat CNS parental cell line (Ferrer et al., 2000; Zuccato et al., 2001). Reduced levels of cortical and striatal BDNF have been demonstrated in multiple mouse models of Huntington´s Disease expressing mutant Huntingtin (including

R6/2, N171-82Q, Hdh, and YAC-72 lines) (Luthi-Carter et al., 2000; Zuccato et al., 2001; Luthi-Carter et al., 2002; Gines et al., 2003b). Importantly, the diminished CREB-mediated gene transcription appears to be linked to energy impairment and deficient cAMP, which has been shown to be decreased in the cerebral spinal fluid of symptomatic Huntington´s Disease patients (Sawa et al., 1999). Furthermore, PC12 cells stimulated with forskolin, which activates adenylyl cyclases to produce cAMP from ATP, exhibit ameliorated mutant huntingtin-fragment induced phenotypes, further supporting the hypothesis that low levels of cAMP might be implicated in Huntington´s Disease pathology (Wyttenbach et al., 2001). Levels of both cAMP and CRE-signaling are decreased prior to Huntington´s Disease symptoms in ***Hdh***^{Q111} mice (Gines et al., 2003b). These data suggest that mutant huntingtin might lead to an early metabolic deficit that amplifies the disease cascade by altering cAMP-dependent processes, including CRE-mediated gene transcription (Gines et al., 2003b).

4. Treatment options for the metabolic deficit

The presence of the mutated huntingtin gene can be detected early in life which makes substrates capable of slowing disease progression an attracting therapeutic tool. A number of energy-related therapeutic approaches have been used in preclinical models and/or Huntington´s Disease patients, such as coenzyme Q_{10}, creatine, antioxidant therapies, anaplerotic therapies, and PGC-1α agonists.

Creatine is an important energy molecule in the brain (O'Gorman et al., 1996). Creatine administration increases brain concentrations of phosphocreatine and inhibits activation of the MPT, both of which may exert neuroprotective effects (Hemmer and Wallimann, 1993; O'Gorman et al., 1996). Moreover, creatine appears to be neuroprotective in a rodent mitochondrial toxin model via enhancing cerebral energy metabolism (Koroshetz et al., 1997; Matthews et al., 1998). The R6/2 mice exhibit lower levels of creatine and ATP in the brain (Dedeoglu et al., 2003). In addition, pre-symptomatic dietary creatine supplementation extends survival in the R6/2 and N171–82Q transgenic Huntington´s Disease mice while significantly improving the clinical and neuropathological phenotype (Ferrante et al., 2000; Andreassen et al., 2001). Creatine supplementation in symptomatic R6/2 mice also has clinical benefits (Dedeoglu et al., 2003). However, so far, clinical trials have demonstrated no substantial benefit for creatine administration to Huntington´s Disease patients (Verbessem et al., 2003; Tabrizi et al., 2005). One year of creatine intake, at a rate that can improve muscle functional capacity in healthy subjects and patients with neuromuscular disease, did not improve functional, neuromuscular, and cognitive status in patients with stage I to III Huntington´s Disease (Verbessem et al., 2003). Even the low levels of cerebral creatine and phosphocreatine observed in these previous studies have been disputed, as more recent studies in which in vivo concentrations of brain metabolites were preserved found increased brain levels of creatine and phosphocreatine in the same mouse model of Huntington´s Disease used in previous studies (Tkac et al., 2007; Mochel et al., 2010a). Thus, it is still unclear whether creatine has a clinical benefit to Huntington´s Disease patients.

Q_{10} is an antioxidant and promoter of respiratory chain function that has also been tested as a treatment for Huntington´s Disease. Oral administration of Q_{10} ameliorates elevated lactate levels seen in the cortex of Huntington´s Disease patients, an effect that is reversible on withdrawal of the agent (Koroshetz et al., 1997). In addition, combination of Coenzyme

Q_{10} and creatine produces additive neuroprotective effects in reducing striatal lesion volumes produced by chronic subcutaneous administration of 3-NP to rats, improves motor performance, and extends survival in the transgenic R6/2 Huntington´s Disease mice (Yang et al., 2009). However, one large-scale study assessing the potential neuroprotective effects of coenzyme Q_{10} revealed that, at the tested dosages, Q_{10} produced no significant slowing in functional decline in early Huntington´s Disease (Huntington Study Group, 2001).

Other antioxidants, such as ascorbate and BN82451, have been shown to improve motor performance and survival of R6 mice (Klivenyi et al., 2003; Rebec et al., 2003). The level of ascorbate is significantly diminished in the striatum of Huntington´s Disease mouse models, which highlights the importance of studying the effect of ascorbate supplementation to treat Huntington´s Disease (Rebec et al., 2002; Dorner et al., 2007). However further studies will be necessary to determine whether either ascorbate or other antioxidant is capable of slowing Huntington´s Disease progression in patients.

A decrease in branched-chain amino acid (BCAA) levels has been observed in the plasma of Huntington´s Disease patients (Mochel et al., 2007). Decreased BCAA levels might occur to compensate the energetic deficit observed in Huntington´s Disease, which is caused by impaired glycolysis, citric acid cycle and/or oxidative phosphorylation, as earlier described in this chapter (Tabrizi et al., 1999; Browne and Beal, 2004; Milakovic and Johnson, 2005). Based on this hypothesis, a short-term therapeutic clinical trial was performed using triheptanoin, a triglyceride containing seven carbon fatty acids that is metabolized to acetyl-CoA and propionyl-CoA, which is an anaplerotic compound that is a precursor of the citric acid cycle intermediate, succinate (Mochel et al., 2010b). This study shows that triheptanoin therapy can improve peripheral energy metabolism in Huntington´s Disease patients, and in particular oxidative phosphorylation in skeletal muscle (Mochel et al., 2010b). However, the benefit of anaplerotic approaches to the brain energy metabolism remains to be established.

Peroxisome proliferator-activated receptor (PPAR)γ is a member of the nuclear hormone receptor family of ligand-activated transcription factors (Rosen and Spiegelman, 2001). PPARγ is the target of the insulin-sensitizing thiazolidinediones (TZDs) drugs used to treat type II diabetes and recent studies suggest that treatment of insulin resistance with a PPARγ agonist retards the development of Alzheimer's Disease (Watson and Craft, 2003; Watson et al., 2005). There is evidence suggesting that PPARγ agonists are neuroprotective and increase mitochondrial function (Schutz et al., 2005; Hunter et al., 2007). Moreover, oral treatment with rosiglitazone, which is a thiazolidinedione drug, induces mitochondrial biogenesis in mouse brain (Strum et al., 2007). Interestingly, a significant defect in the PPARγ signaling pathway has been found in mutant huntingtin-expressing cells, as compared to cells expressing wild-type huntingtin protein (Quintanilla et al., 2008). In addition, pretreatment of mutant huntingtin-expressing cells with rosiglitazone avoids the loss of mitochondrial potential, mitochondrial calcium deregulation, and oxidative stress overproduction in response to intracellular calcium overload (Quintanilla et al., 2008). Rosiglitazone also increases mitochondrial mass levels, suggesting a role for the PPARγ pathway in mitochondrial function in striatal cells (Quintanilla et al., 2008). PPARγ protein levels are decreased in the brain and peripheral tissue of R6/2 mice and in lymphocytes of Huntington´s Disease patients, probably due to a decrease in transcription as well as recruitment of PPARγ protein to huntingtin aggregates (Chiang et al., 2010). R6/2 mice treatment with TZD results in beneficial effects on energy deficiency and on several major

Huntington´s Disease phenotypes, decreasing weight loss, lessening motor deterioration, reducing mutant huntingtin aggregate formation, improving the reduced levels of two neuroprotective factor, Bcl-2 e BDNF, and increasing mouse life span (Chiang et al., 2010). Moreover, the protective effects described above appear to have been exerted, at least partially, via direct activation of PPARγ in the brain (Chiang et al., 2010).

Peroxisome proliferator-activated receptor-γ coactivator (PGC)-1α, which is a potent co-activator of PPARγ transcriptional coactivator, is a member of a family of transcription coactivators that plays a central role in the regulation of cellular energy metabolism and stimulates mitochondrial biogenesis, participating in the regulation of both carbohydrate and lipid metabolism (Liang and Ward, 2006). PGC-1α knockout mice exhibit striatum lesions resembling Huntington´s Disease, which was first evidence that this molecule could be involved in Huntington´s Disease pathology (Lin et al., 2004). Further studies have demonstrated that downregulation of PGC-1α in Huntington´s Disease striatum affects mitochondrial energy metabolism, possibly by impairing oxidative phosphorylation (Cui et al., 2006). Moreover, over-expression of exogenous PGC-1α in Huntington´s Disease striatal neurons was protective against 3-NP treatment (Weydt et al., 2006). Decreasing levels of PGC-1α were shown to parallel markers of mitochondrial dysfunction with disease progression in Huntington´s Disease patients (Kim et al., 2010). Of note, PGC-1α polymorphisms in Huntington´s Disease patients may modify Huntington´s Disease onset age (Taherzadeh-Fard et al., 2009). Interestingly, it has been shown that mutated huntingtin protein can promote transcripition repression of PGC-1alpha (Cui et al., 2006). Mutant huntingtin represses PGC-1α gene transcription by associating with the promoter and interfering with the CREB/TAF4-dependent transcriptional pathway (Cui et al., 2006). These data support a link between transcriptional deregulation and mitochondrial dysfunction in Huntington´s Disease.

Resveratrol is a polyphenol that increases the activity of SIRT1, which is an activator factor capable of increasing PGC-1α activity and mitochondrial biogenesis, as evidenced by increased oxidative-type muscle fibers, enhanced resistance to muscle fatigue, and increased tolerance to cold observed in mice treated with resveratrol (Lagouge et al., 2006). Repeated treatment with resveratrol for a period of 8 days beginning 4 days prior to 3-nitropropionic acid administration, which induces symptoms similar to Huntington´s Disease, significantly improves the 3-nitropropionic acid-induced motor and cognitive impairment (Kumar et al., 2006). When tested in the context of a transgenic mouse model of Huntington´s Disease, resveratrol increased PGC-1α mRNA levels and had protective effects in peripheral tissues, by reducing vacuolation in the brown adipose tissue and decreasing elevated blood glucose levels (Ho et al., 2010). However, there was no improvement of motor performance, weight loss, striatal atrophy and survival in Huntington´s Disease transgenic mice treated with resveratrol, which was consistent with no increase in PGC-1α mRNA levels in the striatum (Ho et al., 2010). Thus, resveratrol appears to protect against the peripheral energetic deficit in Huntington´s Disease, but it is not effective to alleviate CNS Huntington´s Disease pathology.

As stated previously, transglutaminase 2 activity appears to be increased in Huntington´s Disease, leading to huntingtin aggregation and mitochondrial aconitase inhibition (Karpuj et al., 1999; Lesort et al., 1999; Kim et al., 2005). Based on these studies, cystamine, which is a drug capable of inhibiting transglutaminase, was tested as a therapeutic tool to treat Huntington´s Disease. Cystamine treatment prolongs the lifespan and reduces associated tremor and abnormal movements in a Huntington´s Disease transgenic mice, possibly in part due to inhibition of transglutaminase 2 activity (Dedeoglu et al., 2002; Karpuj et al., 2002). However, cystamine does not inhibits transglutaminase 2 specifically, which might invalidate its therapeutic use (Jeitner et al., 2005). In addition to cystamine, there are a number of other transglutaminase inhibitors that have been tested in Huntington´s Disease, such as a set of irreversible peptidic inhibitors, the allosteric reversible small-molecule hydrazides, and inhibitors that bind to the guanosine triphosphate (GTP) binding site (Duval et al., 2005; Lai et al., 2008). However, each of these compounds was found to be inadequate for *in vivo* testing because of a general lack of selectivity or poor cellular potency (Schaertl et al., 2010).

Huntingtin-related proteomic studies represent an area of intense research. Proteins interacting with huntingtin pathological form exhibit altered patterns and metabolism. Mitochondrial metabolism alterations are observed in mouse models expressing various types of the mutated huntingtin (Browne, 2008). Moreover, knock-in mice models with pathogenic CAG repeats inserted into the murine homolog Hdh (e.g. Hdh^{Q111}, CAG 140, CAG 150) develop similar cerebral pathologies, including reduction of striatal-related dopamine receptor (Menalled, 2005). Protein-protein interaction studies resulted in the identification of many proteins interacting with huntingtin, characterizing it rather as a scaffolding, membrane-associated protein, involved in axonal trafficking of mitochondria and vesicles (Truant et al., 2006). From the large group of huntingtin-interacting proteins, the most research interest is focused on proteins that have a direct implication on its biological functions. Most of these proteins bind to the amino terminus of huntingtin near the polyglutamine domain (Holbert et al., 2001; Ferrier, 2002; McPherson, 2002). However, huntingtin-associated protein 40 (HAP40) which is affecting Rab5-mediated endosomal motility in complex with huntingtin, interacts with huntingtin through its carboxy-terminal domain (Pal et al., 2006).

Also, the first conserved 17 amino acids in the amino-terminal huntingtin represent possible membrane association signal, which may influence the polyglutamine expansion (Ross, 1997). It has been shown that specific cleavage (at the residue Arg^{167}) and related presence of defined truncation at the N-terminus of huntingtin, mediate mutant huntingtin toxicity in Huntington's Disease (Ratovitski et al., 2009). Association of huntingtin with the production of brain-derived neurotrophic factors in cortical cells, a pro-survival factor for striatal neurons, just completes the complexity of its functions (Zuccato et al., 2001). Thus, detailed studies of huntingtin-related proteome, including the role of associated proteins in regulation of basic signaling pathways of wild type protein and mutants/polyglutamine forms in Huntington´s Disease, play an important role in research for discrepancies in energetic metabolism in Huntington's Disease and other neurodegenerative disorders, alternatively. Monospecific antibodies mapping targeting well defined epitopes on huntingtin and associated proteins may play a crucial role not only in basic research and clinical diagnostics, but also in the development of an efficient treatment for Huntington's Disease.

5. Conclusion

Despite all the efforts to obtain a drug that could overcome the energetic deficit that takes place in Huntington´s Disease, no such treatment has so far been successful to treat

Huntington´s Disease patients. However, as huntingtin protein has multiple functions in cell metabolism, it is possible that combined therapeutic approaches could improve the mutated huntingtin-mediated energy debit and slow down Huntington´s Disease course.

6. Acknowledgment

CNPq - Conselho Nacional de Desenvolvimento Científico e Tecnológico; Capes - Coordenação de Aperfeiçoamento de Pessoal de Nível Superior; J.S. Cruz is a CNPq fellow researcher; F.A. Oliveira was under a Capes fellowship (postdoctoral REUNI) in the beginning of this work. E. A. D. Gervásio-Carvalho held a scholarship from FAPEMIG-SANTANDER program.

7. References

Andreassen, O.A., Jenkins, B.G., Dedeoglu, A., Ferrante, K.L., Bogdanov, M.B., Kaddurah-Daouk, R., Beal, M.F. (2001) Increases in cortical glutamate concentrations in transgenic amyotrophic lateral sclerosis mice are attenuated by creatine supplementation. *Journal of neurochemistry*, 77, 383-390; ISSN 0022-3042.

Andrews, T.C., Brooks, D.J. (1998) Advances in the understanding of early Huntington's disease using the functional imaging techniques of PET and SPET. *Molecular medicine today*, 4, 532-539; ISSN 1357-4310.

Beal, M.F., Kowall, N.W., Ellison, D.W., Mazurek, M.F., Swartz, K.J., Martin, J.B. (1986) Replication of the neurochemical characteristics of Huntington's disease by quinolinic acid. *Nature*, 321, 168-171; ISSN 0028-0836.

Berg, J.M., Tymoczko, J.L., Stryer, L., eds (2006) *Biochemistry*, 6th Edition. New York: W H Freeman; 10: 0-7167-3051-0.

Bibb, J.A., Yan, Z., Svenningsson, P., Snyder, G.L., Pieribone, V.A., Horiuchi, A., Nairn, A.C., Messer, A., Greengard, P. (2000) Severe deficiencies in dopamine signaling in presymptomatic Huntington's disease mice. *Proceedings of the National Academy of Sciences of the United States of America*, 97, 6809-6814; ISSN 0027-8424.

Bolanos, J.P., Almeida, A., Moncada, S. (2010) Glycolysis: a bioenergetic or a survival pathway? *Trends in biochemical sciences*, 35, 145-149; ISSN 0968-0004.

Bolanos, J.P., Delgado-Esteban, M., Herrero-Mendez, A., Fernandez-Fernandez, S., Almeida, A. (2008) Regulation of glycolysis and pentose-phosphate pathway by nitric oxide: impact on neuronal survival. *Biochimica et biophysica acta*, 1777, 789-793; ISSN 0006-3002.

Browne, S.E. (2008) Mitochondria and Huntington's disease pathogenesis: insight from genetic and chemical models. *Annals of the New York Academy of Sciences*, 1147, 358-382; ISSN 1749-6632.

Browne, S.E., Beal, M.F. (2002) Toxin-induced mitochondrial dysfunction. *International review of neurobiology*, 53, 243-279; ISSN 0074-7742.

Browne, S.E., Beal, M.F. (2004) The energetics of Huntington's disease. *Neurochemical research*, 29, 531-546; ISSN 0364-3190.

Browne, S.E., Bowling, A.C., MacGarvey, U., Baik, M.J., Berger, S.C., Muqit, M.M., Bird, E.D., Beal, M.F. (1997) Oxidative damage and metabolic dysfunction in Huntington's disease: selective vulnerability of the basal ganglia. *Annals of neurology*, 41, 646-653; ISSN 0364-5134.

Butterworth, J., Yates, C.M., Reynolds, G.P. (1985) Distribution of phosphate-activated glutaminase, succinic dehydrogenase, pyruvate dehydrogenase and gamma-glutamyl transpeptidase in post-mortem brain from Huntington's disease and agonal cases. *Journal of the neurological sciences*, 67, 161-171; ISSN 0022-510X.

Chen, N., Luo, T., Wellington, C., Metzler, M., McCutcheon, K., Hayden, M.R., Raymond, L.A. (1999) Subtype-specific enhancement of NMDA receptor currents by mutant huntingtin. *Journal of neurochemistry*, 72, 1890-1898; ISSN 0022-3042.

Chiang, M.C., Chen, C.M., Lee, M.R., Chen, H.W., Chen, H.M., Wu, Y.S., Hung, C.H., Kang, J.J., Chang, C.P., Chang, C., Wu, Y.R., Tsai, Y.S., Chern, Y. (2010) Modulation of energy deficiency in Huntington's disease via activation of the peroxisome proliferator-activated receptor gamma. *Human molecular genetics*, 19, 4043-4058; ISSN 0964-6906.

Choo, Y.S., Johnson, G.V., MacDonald, M., Detloff, P.J., Lesort, M. (2004) Mutant huntingtin directly increases susceptibility of mitochondria to the calcium-induced permeability transition and cytochrome c release. *Human molecular genetics*, 13, 1407-1420; ISSN 0964-6906.

Clodfelder-Miller, B., De Sarno, P., Zmijewska, A.A., Song, L., Jope, R.S. (2005) Physiological and pathological changes in glucose regulate brain Akt and glycogen synthase kinase-3. *The Journal of biological chemistry*, 280, 39723-39731; ISSN 0021-9258.

Cooper, A.J., Sheu, K.R., Burke, J.R., Onodera, O., Strittmatter, W.J., Roses, A.D., Blass, J.P. (1997) Transglutaminase-catalyzed inactivation of glyceraldehyde 3-phosphate dehydrogenase and alpha-ketoglutarate dehydrogenase complex by polyglutamine domains of pathological length. *Proceedings of the National Academy of Sciences of the United States of America*, 94, 12604-12609; ISSN 0027-8424.

Cui, L., Jeong, H., Borovecki, F., Parkhurst, C.N., Tanese, N., Krainc, D. (2006) Transcriptional repression of PGC-1alpha by mutant huntingtin leads to mitochondrial dysfunction and neurodegeneration. *Cell*, 127, 59-69; ISSN 0092-8674.

Cunnane, S., Nugent, S., Roy, M., Courchesne-Loyer, A., Croteau, E., Tremblay, S., Castellano, A., Pifferi, F., Bocti, C., Paquet, N., Begdouri, H., Bentourkia, M., Turcotte, E., Allard, M., Barberger-Gateau, P., Fulop, T., Rapoport, S.I. (2011) Brain fuel metabolism, aging, and Alzheimer's disease. *Nutrition*, 27, 3-20; ISSN 0899-9007.

Datta, S.R., Brunet, A., Greenberg, M.E. (1999) Cellular survival: a play in three Akts. *Genes & development*, 13, 2905-2927; ISSN 0890-9369.

Davey, G.P., Peuchen, S., Clark, J.B. (1998) Energy thresholds in brain mitochondria. Potential involvement in neurodegeneration. *The Journal of biological chemistry*, 273, 12753-12757; ISSN 0021-9258.

Dedeoglu, A., Kubilus, J.K., Jeitner, T.M., Matson, S.A., Bogdanov, M., Kowall, N.W., Matson, W.R., Cooper, A.J., Ratan, R.R., Beal, M.F., Hersch, S.M., Ferrante, R.J. (2002) Therapeutic effects of cystamine in a murine model of Huntington's disease. *The Journal of neuroscience : the official journal of the Society for Neuroscience*, 22, 8942-8950; ISSN 0270-6474.

Dedeoglu, A., Kubilus, J.K., Yang, L., Ferrante, K.L., Hersch, S.M., Beal, M.F., Ferrante, R.J. (2003) Creatine therapy provides neuroprotection after onset of clinical symptoms in Huntington's disease transgenic mice. *Journal of neurochemistry*, 85, 1359-1367; ISSN 0022-3042.

Dorner, J.L., Miller, B.R., Barton, S.J., Brock, T.J., Rebec, G.V. (2007) Sex differences in behavior and striatal ascorbate release in the 140 CAG knock-in mouse model of Huntington's disease. *Behavioural brain research*, 178, 90-97; ISSN 0166-4328.

Dure, L.S.t., Young, A.B., Penney, J.B. (1991) Excitatory amino acid binding sites in the caudate nucleus and frontal cortex of Huntington's disease. *Annals of neurology*, 30, 785-793; ISSN 0364-5134.

Duval, E., Case, A., Stein, R.L., Cuny, G.D. (2005) Structure-activity relationship study of novel tissue transglutaminase inhibitors. *Bioorganic & medicinal chemistry letters*, 15, 1885-1889; ISSN 0960-894X.

Feigin, A., Tang, C., Ma, Y., Mattis, P., Zgaljardic, D., Guttman, M., Paulsen, J.S., Dhawan, V., Eidelberg, D. (2007) Thalamic metabolism and symptom onset in preclinical Huntington's disease. *Brain : a journal of neurology*, 130, 2858-2867; ISSN 0006-8950.

Ferrante, R.J., Andreassen, O.A., Jenkins, B.G., Dedeoglu, A., Kuemmerle, S., Kubilus, J.K., Kaddurah-Daouk, R., Hersch, S.M., Beal, M.F. (2000) Neuroprotective effects of creatine in a transgenic mouse model of Huntington's disease. *The Journal of neuroscience : the official journal of the Society for Neuroscience*, 20, 4389-4397; ISSN 0270-6474.

Ferrer, I., Goutan, E., Marin, C., Rey, M.J., Ribalta, T. (2000) Brain-derived neurotrophic factor in Huntington disease. *Brain research*, 866, 257-261; ISSN 0006-8993.

Ferrier, V. (2002) Hip, hip, hippi! *Nature cell biology*, 4, E30; ISSN 1465-7392.

Gines, S., Ivanova, E., Seong, I.S., Saura, C.A., MacDonald, M.E. (2003a) Enhanced Akt signaling is an early pro-survival response that reflects N-methyl-D-aspartate receptor activation in Huntington's disease knock-in striatal cells. *The Journal of biological chemistry*, 278, 50514-50522; ISSN 0021-9258.

Gines, S., Seong, I.S., Fossale, E., Ivanova, E., Trettel, F., Gusella, J.F., Wheeler, V.C., Persichetti, F., MacDonald, M.E. (2003b) Specific progressive cAMP reduction implicates energy deficit in presymptomatic Huntington's disease knock-in mice. *Human molecular genetics*, 12, 497-508; ISSN 0964-6906.

Group, H.S. (2001) A randomized, placebo-controlled trial of coenzyme Q10 and remacemide in Huntington's disease. *Neurology*, 57, 397-404; ISSN 0028-3878.

Gu, M., Gash, M.T., Mann, V.M., Javoy-Agid, F., Cooper, J.M., Schapira, A.H. (1996) Mitochondrial defect in Huntington's disease caudate nucleus. *Annals of neurology*, 39, 385-389; ISSN 0364-5134.

Heales, S.J., Bolanos, J.P. (2002) Impairment of brain mitochondrial function by reactive nitrogen species: the role of glutathione in dictating susceptibility. *Neurochemistry international*, 40, 469-474; ISSN 0197-0186.

Hemmer, W., Wallimann, T. (1993) Functional aspects of creatine kinase in brain. *Developmental neuroscience*, 15, 249-260; ISSN 0378-5866.

Henneberry, R.C., Novelli, A., Cox, J.A., Lysko, P.G. (1989) Neurotoxicity at the N-methyl-D-aspartate receptor in energy-compromised neurons. An hypothesis for cell death in aging and disease. *Annals of the New York Academy of Sciences*, 568, 225-233; ISSN 0077-8923.

Herrero-Mendez, A., Almeida, A., Fernandez, E., Maestre, C., Moncada, S., Bolanos, J.P. (2009) The bioenergetic and antioxidant status of neurons is controlled by continuous degradation of a key glycolytic enzyme by APC/C-Cdh1. *Nature cell biology*, 11, 747-752; ISSN 1465-7392.

Hickey, M.A., Chesselet, M.F. (2003) Apoptosis in Huntington's disease. *Progress in neuro-psychopharmacology & biological psychiatry*, 27, 255-265; ISSN 0278-5846.

Ho, D.J., Calingasan, N.Y., Wille, E., Dumont, M., Beal, M.F. (2010) Resveratrol protects against peripheral deficits in a mouse model of Huntington's disease. *Experimental neurology*, 225, 74-84; ISSN 0014-4886.

Holbert, S., Denghien, I., Kiechle, T., Rosenblatt, A., Wellington, C., Hayden, M.R., Margolis, R.L., Ross, C.A., Dausset, J., Ferrante, R.J., Neri, C. (2001) The Gln-Ala repeat transcriptional activator CA150 interacts with huntingtin: neuropathologic and genetic evidence for a role in Huntington's disease pathogenesis. *Proceedings of the National Academy of Sciences of the United States of America*, 98, 1811-1816; ISSN 0027-8424.

Hroudova, J., Fisar, Z. (2011) Connectivity between mitochondrial functions and psychiatric disorders. *Psychiatry and clinical neurosciences*, 65, 130-141; ISSN 1323-1316.

Hult, S., Soylu, R., Bjorklund, T., Belgardt, B.F., Mauer, J., Bruning, J.C., Kirik, D., Petersen, A. (2011) Mutant huntingtin causes metabolic imbalance by disruption of hypothalamic neurocircuits. *Cell metabolism*, 13, 428-439; ISSN 1550-4131.

Humbert, S., Bryson, E.A., Cordelieres, F.P., Connors, N.C., Datta, S.R., Finkbeiner, S., Greenberg, M.E., Saudou, F. (2002) The IGF-1/Akt pathway is neuroprotective in Huntington's disease and involves Huntingtin phosphorylation by Akt. *Developmental cell*, 2, 831-837; ISSN 1534-5807.

Hunter, R.L., Dragicevic, N., Seifert, K., Choi, D.Y., Liu, M., Kim, H.C., Cass, W.A., Sullivan, P.G., Bing, G. (2007) Inflammation induces mitochondrial dysfunction and dopaminergic neurodegeneration in the nigrostriatal system. *Journal of neurochemistry*, 100, 1375-1386; ISSN 0022-3042.

Jeitner, T.M., Delikatny, E.J., Ahlqvist, J., Capper, H., Cooper, A.J. (2005) Mechanism for the inhibition of transglutaminase 2 by cystamine. *Biochemical pharmacology*, 69, 961-970; ISSN 0006-2952.

Jenkins, B.G., Koroshetz, W.J., Beal, M.F., Rosen, B.R. (1993) Evidence for impairment of energy metabolism in vivo in Huntington's disease using localized 1H NMR spectroscopy. *Neurology*, 43, 2689-2695; ISSN 0028-3878.

Jenkins, B.G., Rosas, H.D., Chen, Y.C., Makabe, T., Myers, R., MacDonald, M., Rosen, B.R., Beal, M.F., Koroshetz, W.J. (1998) 1H NMR spectroscopy studies of Huntington's disease: correlations with CAG repeat numbers. *Neurology*, 50, 1357-1365; ISSN 0028-3878.

Kandel, E.S., Hay, N. (1999) The regulation and activities of the multifunctional serine/threonine kinase Akt/PKB. *Experimental cell research*, 253, 210-229; ISSN 0014-4827.

Karpuj, M.V., Becher, M.W., Springer, J.E., Chabas, D., Youssef, S., Pedotti, R., Mitchell, D., Steinman, L. (2002) Prolonged survival and decreased abnormal movements in transgenic model of Huntington disease, with administration of the transglutaminase inhibitor cystamine. *Nature medicine*, 8, 143-149; ISSN 1078-8956.

Karpuj, M.V., Garren, H., Slunt, H., Price, D.L., Gusella, J., Becher, M.W., Steinman, L. (1999) Transglutaminase aggregates huntingtin into nonamyloidogenic polymers, and its enzymatic activity increases in Huntington's disease brain nuclei. *Proceedings of the National Academy of Sciences of the United States of America*, 96, 7388-7393; ISSN 0027-8424.

Kim, J., Moody, J.P., Edgerly, C.K., Bordiuk, O.L., Cormier, K., Smith, K., Beal, M.F., Ferrante, R.J. (2010) Mitochondrial loss, dysfunction and altered dynamics in Huntington's disease. *Human molecular genetics*, 19, 3919-3935; ISSN 0964-6906.

Kim, S.Y., Marekov, L., Bubber, P., Browne, S.E., Stavrovskaya, I., Lee, J., Steinert, P.M., Blass, J.P., Beal, M.F., Gibson, G.E., Cooper, A.J. (2005) Mitochondrial aconitase is a transglutaminase 2 substrate: transglutamination is a probable mechanism contributing to high-molecular-weight aggregates of aconitase and loss of aconitase activity in Huntington disease brain. *Neurochemical research*, 30, 1245-1255; ISSN 0364-3190.

Kita, H., Carmichael, J., Swartz, J., Muro, S., Wyttenbach, A., Matsubara, K., Rubinsztein, D.C., Kato, K. (2002) Modulation of polyglutamine-induced cell death by genes identified by expression profiling. *Human molecular genetics*, 11, 2279-2287; ISSN 0964-6906.

Klepper, J., Voit, T. (2002) Facilitated glucose transporter protein type 1 (GLUT1) deficiency syndrome: impaired glucose transport into brain-- a review. *European journal of pediatrics*, 161, 295-304; ISSN 0340-6199.

Klivenyi, P., Ferrante, R.J., Gardian, G., Browne, S., Chabrier, P.E., Beal, M.F. (2003) Increased survival and neuroprotective effects of BN82451 in a transgenic mouse model of Huntington's disease. *Journal of neurochemistry*, 86, 267-272; ISSN 0022-3042.

Koroshetz, W.J., Jenkins, B.G., Rosen, B.R., Beal, M.F. (1997) Energy metabolism defects in Huntington's disease and effects of coenzyme Q10. *Annals of neurology*, 41, 160-165; ISSN 0364-5134.

Kuhl, D.E., Markham, C.H., Metter, E.J., Riege, W.H., Phelps, M.E., Mazziotta, J.C. (1985) Local cerebral glucose utilization in symptomatic and presymptomatic Huntington's disease. *Research publications - Association for Research in Nervous and Mental Disease*, 63, 199-209; ISSN 0091-7443.

Kumar, P., Padi, S.S., Naidu, P.S., Kumar, A. (2006) Effect of resveratrol on 3-nitropropionic acid-induced biochemical and behavioural changes: possible neuroprotective mechanisms. *Behavioural pharmacology*, 17, 485-492; ISSN 0955-8810.

Kuwert, T., Lange, H.W., Langen, K.J., Herzog, H., Aulich, A., Feinendegen, L.E. (1990) Cortical and subcortical glucose consumption measured by PET in patients with Huntington's disease. *Brain : a journal of neurology*, 113 (Pt 5), 1405-1423; ISSN 0006-8950.

Lagouge, M., Argmann, C., Gerhart-Hines, Z., Meziane, H., Lerin, C., Daussin, F., Messadeq, N., Milne, J., Lambert, P., Elliott, P., Geny, B., Laakso, M., Puigserver, P., Auwerx, J. (2006) Resveratrol improves mitochondrial function and protects against metabolic disease by activating SIRT1 and PGC-1alpha. *Cell*, 127, 1109-1122; ISSN 0092-8674.

Lai, T.S., Liu, Y., Tucker, T., Daniel, K.R., Sane, D.C., Toone, E., Burke, J.R., Strittmatter, W.J., Greenberg, C.S. (2008) Identification of chemical inhibitors to human tissue transglutaminase by screening existing drug libraries. *Chemistry & biology*, 15, 969-978; ISSN 1074-5521.

Lesort, M., Chun, W., Johnson, G.V., Ferrante, R.J. (1999) Tissue transglutaminase is increased in Huntington's disease brain. *Journal of neurochemistry*, 73, 2018-2027; ISSN 0022-3042.

Li, X.J., Orr, A.L., Li, S. (2010) Impaired mitochondrial trafficking in Huntington's disease. *Biochimica et biophysica acta*, 1802, 62-65; ISSN 0006-3002.

Liang, H., Ward, W.F. (2006) PGC-1alpha: a key regulator of energy metabolism. *Amerian Journal of Physiology*, 30, 145-151; ISSN 1043-4046.

Lim, D., Fedrizzi, L., Tartari, M., Zuccato, C., Cattaneo, E., Brini, M., Carafoli, E. (2008) Calcium homeostasis and mitochondrial dysfunction in striatal neurons of Huntington disease. *The Journal of biological chemistry*, 283, 5780-5789; ISSN 0021-9258.

Lin, J. et al. (2004) Defects in adaptive energy metabolism with CNS-linked hyperactivity in PGC-1alpha null mice. *Cell*, 119, 121-135; ISSN 0092-8674.

Lund-Andersen, H. (1979) Transport of glucose from blood to brain. *Physiological reviews*, 59, 305-352; ISSN 0031-9333.

Luthi-Carter, R., Hanson, S.A., Strand, A.D., Bergstrom, D.A., Chun, W., Peters, N.L., Woods, A.M., Chan, E.Y., Kooperberg, C., Krainc, D., Young, A.B., Tapscott, S.J., Olson, J.M. (2002) Dysregulation of gene expression in the R6/2 model of polyglutamine disease: parallel changes in muscle and brain. *Human molecular genetics*, 11, 1911-1926; ISSN 0964-6906.

Luthi-Carter, R., Strand, A., Peters, N.L., Solano, S.M., Hollingsworth, Z.R., Menon, A.S., Frey, A.S., Spektor, B.S., Penney, E.B., Schilling, G., Ross, C.A., Borchelt, D.R., Tapscott, S.J., Young, A.B., Cha, J.H., Olson, J.M. (2000) Decreased expression of striatal signaling genes in a mouse model of Huntington's disease. *Human molecular genetics*, 9, 1259-1271; ISSN 0964-6906.

Mastroberardino, P.G., Iannicola, C., Nardacci, R., Bernassola, F., De Laurenzi, V., Melino, G., Moreno, S., Pavone, F., Oliverio, S., Fesus, L., Piacentini, M. (2002) 'Tissue' transglutaminase ablation reduces neuronal death and prolongs survival in a mouse model of Huntington's disease. *Cell death and differentiation*, 9, 873-880; ISSN 1350-9047.

Matthews, R.T., Yang, L., Jenkins, B.G., Ferrante, R.J., Rosen, B.R., Kaddurah-Daouk, R., Beal, M.F. (1998) Neuroprotective effects of creatine and cyclocreatine in animal models of Huntington's disease. *The Journal of neuroscience : the official journal of the Society for Neuroscience*, 18, 156-163; ISSN 0270-6474.

Mazzola, J.L., Sirover, M.A. (2001) Reduction of glyceraldehyde-3-phosphate dehydrogenase activity in Alzheimer's disease and in Huntington's disease fibroblasts. *Journal of neurochemistry*, 76, 442-449; ISSN 0022-3042.

Mazzola, J.L., Sirover, M.A. (2002) Alteration of nuclear glyceraldehyde-3-phosphate dehydrogenase structure in Huntington's disease fibroblasts. *Brain research Molecular brain research*, 100, 95-101; ISSN 0169-328X.

McPherson, P.S. (2002) The endocytic machinery at an interface with the actin cytoskeleton: a dynamic, hip intersection. *Trends in cell biology*, 12, 312-315; ISSN 0962-8924.

Menalled, L.B. (2005) Knock-in mouse models of Huntington's disease. *NeuroRx : the journal of the American Society for Experimental NeuroTherapeutics*, 2, 465-470; ISSN 1545-5343.

Milakovic, T., Johnson, G.V. (2005) Mitochondrial respiration and ATP production are significantly impaired in striatal cells expressing mutant huntingtin. *The Journal of biological chemistry*, 280, 30773-30782; ISSN 0021-9258.

Mochel, F., Charles, P., Seguin, F., Barritault, J., Coussieu, C., Perin, L., Le Bouc, Y., Gervais, C., Carcelain, G., Vassault, A., Feingold, J., Rabier, D., Durr, A. (2007) Early energy deficit in Huntington disease: identification of a plasma biomarker traceable during disease progression. *PloS one*, 2, e647; ISSN 1932-6203.

Mochel, F., Durant, B., Schiffmann, R., Durr, A. (2010a) Characterization of the locoregional brain energy profile in wild-type mice and identification of an energy deficit in a neurodegenerative model. *Journal of inherited metabolic disease*, 33, S181; ISSN 0141-8955.

Mochel, F., Duteil, S., Marelli, C., Jauffret, C., Barles, A., Holm, J., Sweetman, L., Benoist, J.F., Rabier, D., Carlier, P.G., Durr, A. (2010b) Dietary anaplerotic therapy improves peripheral tissue energy metabolism in patients with Huntington's disease. *European journal of human genetics*, 18, 1057-1060; ISSN 1018-4813.

Nehlig, A., Coles, J.A. (2007) Cellular pathways of energy metabolism in the brain: is glucose used by neurons or astrocytes? *Glia*, 55, 1238-1250; ISSN 0894-1491.

Nelson, D.L., Cox, M.M., eds (2004) *Lehninger Principles of Biochemistry*, 4th Edition: W. H. Freeman; 10: 0716743396

Nicholls, D.G. (2009) Mitochondrial calcium function and dysfunction in the central nervous system. *Biochimica et biophysica acta*, 1787, 1416-1424; ISSN 0006-3002.

Novelli, A., Reilly, J.A., Lysko, P.G., Henneberry, R.C. (1988) Glutamate becomes neurotoxic via the N-methyl-D-aspartate receptor when intracellular energy levels are reduced. *Brain research*, 451, 205-212; ISSN 0006-8993.

Nucifora, F.C., Jr., Sasaki, M., Peters, M.F., Huang, H., Cooper, J.K., Yamada, M., Takahashi, H., Tsuji, S., Troncoso, J., Dawson, V.L., Dawson, T.M., Ross, C.A. (2001) Interference by huntingtin and atrophin-1 with cbp-mediated transcription leading to cellular toxicity. *Science*, 291, 2423-2428; ISSN 0036-8075.

O'Gorman, E., Beutner, G., Wallimann, T., Brdiczka, D. (1996) Differential effects of creatine depletion on the regulation of enzyme activities and on creatine-stimulated mitochondrial respiration in skeletal muscle, heart, and brain. *Biochimica et biophysica acta*, 1276, 161-170; ISSN 0006-3002.

O'Brien, C.F., Miller, C., Goldblatt, D., Welle, S., Forbes, G., Lipinski, B., Panzik, J., Peck, R., Plumb, S., Oakes, D., Kurlan, R., Shoulson, I. (1990) Extraneural metabolism in early Huntington's disease. Ann. Neurol. 28:300–301. *Annals of neurology*, 28, 300-301; ISSN 0364-5134.

Okar, D.A., Manzano, A., Navarro-Sabate, A., Riera, L., Bartrons, R., Lange, A.J. (2001) PFK-2/FBPase-2: maker and breaker of the essential biofactor fructose-2,6-bisphosphate. *Trends in biochemical sciences*, 26, 30-35; ISSN 0968-0004.

Oliveira, F.A., Galan, D.T., Ribeiro, A.M., Santos Cruz, J. (2007) Thiamine deficiency during pregnancy leads to cerebellar neuronal death in rat offspring: role of voltage-dependent K+ channels. *Brain research*, 1134, 79-86; ISSN 0006-8993.

Pal, A., Severin, F., Lommer, B., Shevchenko, A., Zerial, M. (2006) Huntingtin-HAP40 complex is a novel Rab5 effector that regulates early endosome motility and is up-regulated in Huntington's disease. *The Journal of cell biology*, 172, 605-618; ISSN 0021-9525.

Panov, A.V., Gutekunst, C.A., Leavitt, B.R., Hayden, M.R., Burke, J.R., Strittmatter, W.J., Greenamyre, J.T. (2002) Early mitochondrial calcium defects in Huntington's disease are a direct effect of polyglutamines. *Nature neuroscience*, 5, 731-736; ISSN 1097-6256.

Pardridge, W.M., Boado, R.J., Farrell, C.R. (1990) Brain-type glucose transporter (GLUT-1) is selectively localized to the blood-brain barrier. Studies with quantitative western blotting and in situ hybridization. *The Journal of biological chemistry*, 265, 18035-18040; ISSN 0021-9258.

Pesin, J.A., Orr-Weaver, T.L. (2008) Regulation of APC/C activators in mitosis and meiosis. *Annual review of cell and developmental biology*, 24, 475-499; ISSN 1081-0706.

Powers, W.J., Videen, T.O., Markham, J., McGee-Minnich, L., Antenor-Dorsey, J.V., Hershey, T., Perlmutter, J.S. (2007) Selective defect of in vivo glycolysis in early Huntington's disease striatum. *Proceedings of the National Academy of Sciences of the United States of America*, 104, 2945-2949; ISSN 0027-8424.

Quintanilla, R.A., Jin, Y.N., Fuenzalida, K., Bronfman, M., Johnson, G.V. (2008) Rosiglitazone treatment prevents mitochondrial dysfunction in mutant huntingtin-expressing cells: possible role of peroxisome proliferator-activated receptor-gamma (PPARgamma) in the pathogenesis of Huntington disease. *The Journal of biological chemistry*, 283, 25628-25637; ISSN 0021-9258.

Ratovitski, T., Gucek, M., Jiang, H., Chighladze, E., Waldron, E., D'Ambola, J., Hou, Z., Liang, Y., Poirier, M.A., Hirschhorn, R.R., Graham, R., Hayden, M.R., Cole, R.N., Ross, C.A. (2009) Mutant huntingtin N-terminal fragments of specific size mediate aggregation and toxicity in neuronal cells. *The Journal of biological chemistry*, 284, 10855-10867; ISSN 0021-9258.

Ravikumar, B., Stewart, A., Kita, H., Kato, K., Duden, R., Rubinsztein, D.C. (2003) Raised intracellular glucose concentrations reduce aggregation and cell death caused by mutant huntingtin exon 1 by decreasing mTOR phosphorylation and inducing autophagy. *Human molecular genetics*, 12, 985-994; ISSN 0964-6906.

Rebec, G.V., Barton, S.J., Ennis, M.D. (2002) Dysregulation of ascorbate release in the striatum of behaving mice expressing the Huntington's disease gene. *The Journal of neuroscience: the official journal of the Society for Neuroscience*, 22, RC202; ISSN 0270-6474.

Rebec, G.V., Barton, S.J., Marseilles, A.M., Collins, K. (2003) Ascorbate treatment attenuates the Huntington behavioral phenotype in mice. *Neuroreport*, 14, 1263-1265; ISSN 0959-4965.

Rosen, E.D., Spiegelman, B.M. (2001) PPARgamma : a nuclear regulator of metabolism, differentiation, and cell growth. *The Journal of biological chemistry*, 276, 37731-37734; ISSN 0021-9258.

Ross, C.A. (1997) Intranuclear neuronal inclusions: a common pathogenic mechanism for glutamine-repeat neurodegenerative diseases? *Neuron*, 19, 1147-1150; ISSN 0896-6273.

Saraiva, L.M., Seixas da Silva, G.S., Galina, A., da-Silva, W.S., Klein, W.L., Ferreira, S.T., De Felice, F.G. (2010) Amyloid-β Triggers the Release of Neuronal Hexokinase 1 from Mitochondria. *PloS one*, 5, e15230; ISSN 1932-6203.

Sawa, A., Wiegand, G.W., Cooper, J., Margolis, R.L., Sharp, A.H., Lawler, J.F., Jr., Greenamyre, J.T., Snyder, S.H., Ross, C.A. (1999) Increased apoptosis of Huntington disease lymphoblasts associated with repeat length-dependent mitochondrial depolarization. *Nature medicine*, 5, 1194-1198; ISSN 1078-8956.

Schaertl, S., Prime, M., Wityak, J., Dominguez, C., Munoz-Sanjuan, I., Pacifici, R.E., Courtney, S., Scheel, A., Macdonald, D. (2010) A profiling platform for the characterization of transglutaminase 2 (TG2) inhibitors. *Journal of biomolecular screening*, 15, 478-487; ISSN 1087-0571.

Schiefer, J., Sprunken, A., Puls, C., Luesse, H.G., Milkereit, A., Milkereit, E., Johann, V., Kosinski, C.M. (2004) The metabotropic glutamate receptor 5 antagonist MPEP and the mGluR2 agonist LY379268 modify disease progression in a transgenic mouse model of Huntington's disease. *Brain research*, 1019, 246-254; ISSN 0006-8993.

Schulz, J.B., Matthews, R.T., Henshaw, D.R., Beal, M.F. (1996) Neuroprotective strategies for treatment of lesions produced by mitochondrial toxins: implications for neurodegenerative diseases. *Neuroscience*, 71, 1043-1048; ISSN 0306-4522.

Schutz, B., Reimann, J., Dumitrescu-Ozimek, L., Kappes-Horn, K., Landreth, G.E., Schurmann, B., Zimmer, A., Heneka, M.T. (2005) The oral antidiabetic pioglitazone protects from neurodegeneration and amyotrophic lateral sclerosis-like symptoms in superoxide dismutase-G93A transgenic mice. *The Journal of neuroscience : the official journal of the Society for Neuroscience*, 25, 7805-7812; ISSN 0270-6474.

Senatorov, V.V., Charles, V., Reddy, P.H., Tagle, D.A., Chuang, D.M. (2003) Overexpression and nuclear accumulation of glyceraldehyde-3-phosphate dehydrogenase in a transgenic mouse model of Huntington's disease. *Molecular and cellular neurosciences*, 22, 285-297; ISSN 1044-7431.

Seong, I.S., Ivanova, E., Lee, J.M., Choo, Y.S., Fossale, E., Anderson, M., Gusella, J.F., Laramie, J.M., Myers, R.H., Lesort, M., MacDonald, M.E. (2005) HD CAG repeat implicates a dominant property of huntingtin in mitochondrial energy metabolism. *Human molecular genetics*, 14, 2871-2880; ISSN 0964-6906.

Shimohata, T., Nakajima, T., Yamada, M., Uchida, C., Onodera, O., Naruse, S., Kimura, T., Koide, R., Nozaki, K., Sano, Y., Ishiguro, H., Sakoe, K., Ooshima, T., Sato, A., Ikeuchi, T., Oyake, M., Sato, T., Aoyagi, Y., Hozumi, I., Nagatsu, T., Takiyama, Y. , Nishizawa, M., Goto, J., Kanazawa I., Davidson, I., Tanese, N., Takahashi, H., Tsuji, S. (2000) Expanded polyglutamine stretches interact with TAFII130, interfering with CREB-dependent transcription. *Nature genetics*, 26, 29-36; ISSN 1061-4036.

Siegel, G.J., Agranoff, B.W., Albers, W., Fisher, S.K., Uhler, M.D., eds (1999) *Basic Neurochemistry - Molecular, Cellular and Medical Aspects*. Philadelphia: Lippincott-Raven; ISBN 10: 0-397-51820-X.

Sipione, S., Rigamonti, D., Valenza, M., Zuccato, C., Conti, L., Pritchard, J., Kooperberg, C., Olson, J.M., Cattaneo, E. (2002) Early transcriptional profiles in huntingtin-inducible striatal cells by microarray analyses. *Human molecular genetics*, 11, 1953-1965; ISSN 0964-6906.

Steffan, J.S., Kazantsev, A., Spasic-Boskovic, O., Greenwald, M., Zhu, Y.Z., Gohler, H., Wanker, E.E., Bates, G.P., Housman, D.E., Thompson, L.M. (2000) The Huntington's disease protein interacts with p53 and CREB-binding protein and represses transcription. *Proceedings of the National Academy of Sciences of the United States of America*, 97, 6763-6768; ISSN 0027-8424.

Strum, J.C., Shehee, R., Virley, D., Richardson, J., Mattie, M., Selley, P., Ghosh, S., Nock, C., Saunders, A., Roses, A. (2007) Rosiglitazone induces mitochondrial biogenesis in mouse brain. *Journal of Alzheimer's disease*, 11, 45-51; ISSN 1387-2877.

Sun, Y., Savanenin, A., Reddy, P.H., Liu, Y.F. (2001) Polyglutamine-expanded huntingtin promotes sensitization of N-methyl-D-aspartate receptors via post-synaptic density 95. *The Journal of biological chemistry*, 276, 24713-24718; ISSN 0021-9258.

Tabrizi, S.J., Blamire, A.M., Manners, D.N., Rajagopalan, B., Styles, P., Schapira, A.H., Warner, T.T. (2005) High-dose creatine therapy for Huntington disease: a 2-year clinical and MRS study. *Neurology*, 64, 1655-1656; ISSN 0028-3878.

Tabrizi, S.J., Cleeter, M.W., Xuereb, J., Taanman, J.W., Cooper, J.M., Schapira, A.H. (1999) Biochemical abnormalities and excitotoxicity in Huntington's disease brain. *Annals of neurology*, 45, 25-32; ISSN 0364-5134.

Taherzadeh-Fard, E., Saft, C., Andrich, J., Wieczorek, S., Arning, L. (2009) PGC-1alpha as modifier of onset age in Huntington disease. *Molecular neurodegeneration*, 4, 10; ISSN 1750-1326.

Tang, T.S., Slow, E., Lupu, V., Stavrovskaya, I.G., Sugimori, M., Llinas, R., Kristal, B.S., Hayden, M.R., Bezprozvanny, I. (2005) Disturbed Ca2+ signaling and apoptosis of medium spiny neurons in Huntington's disease. *Proceedings of the National Academy of Sciences of the United States of America*, 102, 2602-2607; ISSN 0027-8424.

Tkac, I., Dubinsky, J.M., Keene, C.D., Gruetter, R., Low, W.C. (2007) Neurochemical changes in Huntington R6/2 mouse striatum detected by in vivo 1H NMR spectroscopy. *Journal of neurochemistry*, 100, 1397-1406; ISSN 0022-3042.

Truant, R., Atwal, R., Burtnik, A. (2006) Hypothesis: Huntingtin may function in membrane association and vesicular trafficking. *Biochemistry and cell biology = Biochimie et biologie cellulaire*, 84, 912-917; ISSN 0829-8211.

Vannucci, S.J., Maher, F., Simpson, I.A. (1997) Glucose transporter proteins in brain: delivery of glucose to neurons and glia. *Glia*, 21, 2-21; ISSN 0894-1491.

Verbessem, P., Lemiere, J., Eijnde, B.O., Swinnen, S., Vanhees, L., Van Leemputte, M., Hespel, P., Dom, R. (2003) Creatine supplementation in Huntington's disease: a placebo-controlled pilot trial. *Neurology*, 61, 925-930; ISSN 0028-3878.

Warby, S.C., Doty, C.N., Graham, R.K., Shively, J., Singaraja, R.R., Hayden, M.R. (2009) Phosphorylation of huntingtin reduces the accumulation of its nuclear fragments. *Molecular and cellular neurosciences*, 40, 121-127; ISSN 1044-7431.

Watson, G.S., Cholerton, B.A., Reger, M.A., Baker, L.D., Plymate, S.R., Asthana, S., Fishel, M.A., Kulstad, J.J., Green, P.S., Cook, D.G., Kahn, S.E., Keeling, M.L., Craft, S. (2005) Preserved cognition in patients with early Alzheimer disease and amnestic mild cognitive impairment during treatment with rosiglitazone: a preliminary study. *The American journal of geriatric psychiatry : official journal of the American Association for Geriatric Psychiatry*, 13, 950-958; ISSN 1064-7481.

Watson, G.S., Craft, S. (2003) The role of insulin resistance in the pathogenesis of Alzheimer's disease: implications for treatment. *CNS drugs*, 17, 27-45; ISSN 1172-7047.

Weydt, P., Pineda, V.V., Torrence, A.E., Libby, R.T., Satterfield, T.F., Lazarowski, E.R., Gilbert, M.L., Morton, G.J., Bammler, T.K., Strand, A.D., Cui, L., Beyer, R.P., Easley, C.N., Smith, A.C., Krainc, D., Luquet, S., Sweet, I.R., Schwartz, M.W., La Spada, A.R. (2006) Thermoregulatory and metabolic defects in Huntington's disease transgenic mice implicate PGC-1alpha in Huntington's disease neurodegeneration. *Cell metabolism*, 4, 349-362; ISSN 1550-4131.

Wilhelm, I., Fazakas, C., Krizbai, I.A. (2011) In vitro models of the blood-brain barrier. *Acta neurobiologiae experimentalis*, 71, 113-128; ISSN 0065-1400.

Wullner, U., Young, A.B., Penney, J.B., Beal, M.F. (1994) 3-Nitropropionic acid toxicity in the striatum. *Journal of neurochemistry*, 63, 1772-1781; ISSN 0022-3042.

Wyttenbach, A., Swartz, J., Kita, H., Thykjaer, T., Carmichael, J., Bradley, J., Brown, R., Maxwell, M., Schapira, A., Orntoft, T.F., Kato, K., Rubinsztein, D.C. (2001) Polyglutamine expansions cause decreased CRE-mediated transcription and early gene expression changes prior to cell death in an inducible cell model of Huntington's disease. *Human molecular genetics*, 10, 1829-1845; ISSN 0964-6906.

Yalcin, A., Telang, S., Clem, B., Chesney, J. (2009) Regulation of glucose metabolism by 6-phosphofructo-2-kinase/fructose-2,6-bisphosphatases in cancer. *Experimental and molecular pathology*, 86, 174-179; ISSN 0014-4800.

Yamanaka, T., Miyazaki, H., Oyama, F., Kurosawa, M., Washizu, C., Doi, H., Nukina, N. (2008) Mutant Huntingtin reduces HSP70 expression through the sequestration of NF-Y transcription factor. *The EMBO journal*, 27, 827-839; ISSN 0261-4189.

Yang, L., Calingasan, N.Y., Wille, E.J., Cormier, K., Smith, K., Ferrante, R.J., Beal, M.F. (2009) Combination therapy with coenzyme Q10 and creatine produces additive neuroprotective effects in models of Parkinson's and Huntington's diseases. *Journal of neurochemistry*, 109, 1427-1439; ISSN 0022-3042.

Yu, Z.X., Li, S.H., Evans, J., Pillarisetti, A., Li, H., Li, X.J. (2003) Mutant huntingtin causes context-dependent neurodegeneration in mice with Huntington's disease. *The Journal of neuroscience : the official journal of the Society for Neuroscience*, 23, 2193-2202; ISSN 0270-6474ISSN.

Zeevalk, G.D., Nicklas, W.J. (1991) Mechanisms underlying initiation of excitotoxicity associated with metabolic inhibition. *The Journal of pharmacology and experimental therapeutics*, 257, 870-878; ISSN 0022-3565.

Zeron, M.M., Chen, N., Moshaver, A., Lee, A.T., Wellington, C.L., Hayden, M.R., Raymond, L.A. (2001) Mutant huntingtin enhances excitotoxic cell death. *Molecular and cellular neurosciences*, 17, 41-53; ISSN 1044-7431.

Zeron, M.M., Hansson, O., Chen, N., Wellington, C.L., Leavitt, B.R., Brundin, P., Hayden, M.R., Raymond, L.A. (2002) Increased sensitivity to N-methyl-D-aspartate receptor-mediated excitotoxicity in a mouse model of Huntington's disease. *Neuron*, 33, 849-860; ISSN 0896-6273.

Zuccato, C., Ciammola, A., Rigamonti, D., Leavitt, B.R., Goffredo, D., Conti, L., MacDonald, M.E., Friedlander, R.M., Silani, V., Hayden, M.R., Timmusk, T., Sipione, S., Cattaneo, E. (2001) Loss of huntingtin-mediated BDNF gene transcription in Huntington's disease. *Science*, 293, 493-498; ISSN 0036-8075.

15

The Use of the Mitochondrial Toxin 3-NP to Uncover Cellular Dysfunction in Huntington's Disease

Elizabeth Hernández-Echeagaray, Gabriela De la Rosa-López and Ernesto Mendoza-Duarte
Laboratorio de Neurofisiología del Desarrollo y la Neurodegeneración, Unidad de Biomedicina, FES-Iztacala, Universidad Nacional Autónoma de México, México

1. Introduction

Degenerative diseases that affect the nervous system are characterized by a progressive alteration of specific neuronal populations and normally end in cell death. Some neurodegenerative disorders exhibit a clear genetic origin, such as in the case of Huntington's Disease (HD); however, most neurodegenerative diseases do not have a genetic cause, suggesting that other mechanisms cause these alterations.

Even though each neurodegenerative disease exhibits specific features, many similarities in the degenerative process are also shown; the study of these similarities could provide ideas for the therapeutic management of such diseases. For example, Parkinson's, Alzheimer's, and Huntington's Diseases exhibit atypical protein assemblies, excitotoxicity, metabolic alterations, oxidative stress and mitochondrial failure (Shoffner et al., 1991; Sims et al., 1996; Kapogiannis and Mattson, 2011). All of these cellular alterations can trigger one or more forms of cell death, namely apoptosis, necrosis and/or autophagy. Normally, excitotoxicity and inflammatory activations are associated with necrosis as a type of neuronal death (Artal-Sanz and Tavernarakis, 2005); the apoptotic type of cell death is often associated with the activation of cysteine protease caspase-3 (Kroemer et al., 1998). However, in post-mortem tissue from HD patients, no clear histological or pathological data are available in support of apoptotic cell death (Vis et al., 2005), although DNA damage in the progression of degeneration was shown to be evident (Brouillet et al., 1999). Also, in animal models of neurodegeneration, oxidative stress promotes apoptotic damage, triggered by the activation of caspases (Burke et al., 1996; Krantic et al., 2005). In neurodegenration, it is important to understand which type of cell death mechanism is involved however, it is also important to be aware that the speed of cell death process in "sick" neurons is slow (Kanazawa, 2001).

2. Mitochondrial dysfunction

Studies over several decades have documented experimental evidence to support the fact that mitochondrial dysfunction and oxidative stress are part of the cellular mechanisms underlying neurodegeneration. Mitochondrial dysfunction can occur early on in the pathogenesis of several diseases, including HD (Koroshetz et al., 1997; Jenkins et al., 1993, 1998; Panov et al., 2002; Lin and Beal, 2006).

During mitochondrial respiration, reactive oxygen species (ROS) are produced as by products of respiratory chain activity; their overproduction generates oxidative stress, and is a hallmark of neurodegenerative disorders. The respiratory chain is composed of five complexes: complexes I and II collect electrons from the catabolism of fats, proteins and carbohydrates and transfer them to co-enzyme Q10, complex III and complex IV. Importantly, complexes I, III and IV utilize the energy produced by the electron gradient generated by pumping protons across the inner mitochondrial membrane; this proton gradient is used by complex V to condense Adenosine diphosphate (ADP) and inorganic phosphate into Adenosine-5'-triphosphate (ATP). Any alteration in the mitochondrial complex is expected to produce ATP deficiency (Fridovich, 1999), release of cytochrome *c* which activates the intrinsic pathways of neuronal death and the activation of caspases related to apoptotic damage (Maciel et al., 2004).

3. Huntington Disease

Huntington's Disease, also called Huntington's chorea because of the presence of rapid and incessant choreic movements accompanied by cognitive and psychiatric alterations, is an inherited neurodegenerative disease that affects cell projections in a specific region of the brain known as the nucleus striatum. This neurodegenerative illness was described in 1872 by George Huntington, although some features that resemble HD had already been described earlier (Walker, 2007).

The symptoms manifest in the third or fourth decade of life in most cases, and progression of the illness is slow (15 to 20 years). In comparison to other neurodegenerative diseases, the development of HD is associated with the mutation of a single gene called *Interesting Transcrip 15* (IT15), or the *Hd* gene. The mutation originates the expansion of the CAG nucleotide repeats in a single protein called huntingtin; wich is called mutant huntingtin (mhtt), when the CAG expansion is present (Taylor et al., 2002). This single mutation produces diverse cellular, physiological and anatomical changes; however, the cellular mechanisms underlying HD neurodegeneration are not yet fully understood.

A number of studies have shown that mitochondria from HD patients and animal models are damaged (Jenkings et al., 1993), suggesting that disturbances in the cellular metabolism of HD patients originate via mitochondrial dysfunction (Beal, 2005). Deficits in energy metabolism become manifest in the pre-symptomatic and symptomatic HD brain and peripheral tissues (Kuhl et al., 1982; Mazziotta et al., 1987; Grafton et al., 1990; Koroshetz et al., 1997; Lodi et al., 2000). In particular, factors related to mitochondrial functioning seem to underlie the selective vulnerability of striatal cells (Saft et al., 2005; Seong et al., 2005); for example, the reduction of the chaperone, protease and intramembrane mitochondrial molecule Omi/HTr2 (Inagaki et al., 2008), and alterations in oxidative phosphorylation functioning in general (Pickrell et al., 2011). Post-mortem studies on HD brain tissue showed decreased activity in complexes II, III and IV of the mitochondrial respiratory chain (Gu et al., 1996). Also, animal models of HD showed deficits in mitochondrial respiration (Browne et al., 1997); for example; the systemic administration of mitochondrial toxins or inhibitors generated striatal pathology and movement disorders such as chorea and dystonia, which resemble HD (Ludolph et al., 1991; Browne, 2008). In fact, accidental ingestion of the irreversible mitochondrial inhibitor 3-nitropropionic acid (3-NP) was found to cause striatal degeneration and the HD phenotype. Moreover, the systemic administration of 3-NP caused striatal cell loss and movement alterations in rats (Beal et al., 1993) and primates (Palfi et al., 1996; Dautry et al., 2000). However, an analysis of unbiased gene expression showed that changes in energy metabolism in mhtt of transgenic mice compared to 3-NP treated animals were different: while 3-NP affected mitochondrial pathway gene expression, the effects of mhtt on metabolism were extramitochondrial (Lee et al., 2007), which suggested that mitochondrial toxins such as 3-NP do not quite cause HD pathology (Olivera, 2010). However, mhtt generates mitochondrial dysfunction in HD (Grunewald & Beal 1999) and reductions in ATP generation (Seong et al., 2005), and it also alters Ca^{2+} buffering (Reddy et al., 2009) and mitochondrial trafficking (Li et al., 2010). Therefore, irrespective of whether mhtt impacts mitochondria directly or secondarily, the repercussions of mitochondrial dysfunction are devastating to cells and may underlie the disruptions in numerous cellular processes, resulting in HD pathogenesis. As a result, the identification of respiratory chain changes in complex II of respiratory chain in HD post-mortem brains led to the use of mitochondrial complex II inhibitors to generate toxicity models that replicate aspects of HD striatal pathology *in vivo*. Thus, studies on mitochondrial toxins are relevant and important for understanding defects in cellular metabolism and the energetic pathogenesis of Huntington's Disease.

4. 3-Nitropropionic acid (3-NP)

3-NP is a highly specific, time dependent and irreversible inhibitor of succinate dehydrogenase (SDH) and the Krebs cycle (Alston et al., 1977). The levels of inhibition of this enzyme by 3-NP correlate well with the levels of inhibition of the tricarboxylic acid cycle (Henry et al., 2002).

4.1 Changes in the central nervous system

3-NP treatment was found to cause striatal degeneration in rodents (Gould and Gustine 1982; Gould et al., 1985; Beal et al., 1992; Broulliet et al., 2005). In non-human primates, 3-NP produced cognitive deficits similar to those displayed by frontal-type and abnormal choreiform movements, followed by evident striatal degeneration (Palfi et al., 1998; Brouillet et al., 1999; Dautry et al., 2000).

It is known that 3-NP imitates the symptoms of dystonia, glutaric aciduria, Leber's disease and HD (Novotny et al., 1986; Janavs and Aminoff 1998; Strauss and Morton, 2003), but after discovering that the accidental ingestion of 3-NP (He et al., 1995; Ming, 1995) caused damage that was concentrated in the striatum, 3-NP was used in experimental models to study the cellular mechanisms underlying striatal neural degeneration (Alexi et al., 1998).

The initial studies suggested that excitotoxicity plays a central role in the physiological and cellular effects of 3-NP on striatal degeneration (Hamilton and Gould, 1987; Novelli et al., 1988; Zeevalk and Nicklas, 1990) because of the presence of massive glutamatergic afferents in the nuclei (Di Figlia et al., 1990); also, metabolic insults were suggested as playing a part in the mechanistic damage (Browne et al., 1997; Brouillet et al., 1999). Chronic treatment with 3-NP in rats produced astrogliosis and selective degeneration of medium-sized spiny neurons, similar to the neurochemical and histological pathology observed in post-mortem HD tissue. Interestingly, 3-NP treatment was found to retain terminals from large cholinergic interneurons and NADPH-diaphorase-positive aspiny interneurons (Beal et al., 1993; Brouillet et al., 1999).

The type of death triggered by 3-NP treatment depends on how the toxin was administered: intraparenchymal applications induce ischaemic injury features while intraperitoneal applications induce striatal degeneration, which shows more of an HD phenotype (Borlongan et al., 1997a). Another concern about the 3-NP pharmacological model of HD is related to the variations in cellular damage, which depend on whether a study is carried out *in vitro* or *in vivo*, whether rats or mice are used, whether 3-NP is administered intrastriatally or intraperitoneally, whether the treatment is acute or chronic, and whether low, sub-toxic or toxic concentrations are provided (Brouillet et al., 1999, 2005).

3-NP administered by subcutaneous (s.c.) or intraperitoneal (i.p.) injections is more toxic in rats than in mice. In rodents, the toxicity of 3-NP depends on the strain (Brouillet et al., 2005). Fisher rats, for example, are more susceptible to 3-NP toxicity than Sprague-Dawley, Wistar and Lewis strains (Ouary et al., 2000), and C57BL/6 and Balb/c mice are more resistant to 3-NP toxicity than 129SVEMS and FVB/n mice (Gabrielson et al., 2001). The strain-dependent differences observed following 3-NP intoxication are probably related to differences in elimination/detoxification of the compound (Ouary et al., 2000). Vulnerability to different agents is not restricted to animal strains; it occurs in all human groups and is due to genetic variations that give rise to different responses to drugs (Weinshilboum et al., 2003; Weinshilboum and Wang, 2006), so we need to be cautious when generalizing about the results obtained among different strains.

Other important factors are the age and gender of the animals used in experimental protocols (Brouillet et al., 1993). Interestingly, female rats are less sensitive to 3-NP than males, which suggests that oestrogen protection can affect the degree of sensitivity (Nishino et al., 1998; Mogami et al., 2002).

The method used for 3-NP delivery also influences the physiological effect; acute treatments of a single i.p. dose of 3-NP were found to lead to striatal degeneration within 6-12 h after injection (Alexi et al., 1998; Brouillet et al. 1999). Sub-chronic treatments consisting of daily repeated i.p. injections led to striatal degeneration over a few days (Beal et al., 1993; Schulz et al., 1996; Guyot et al., 1997). Chronic treatments (of more than 5 days up to 4 weeks) with the continuous systemic administration of 3-NP using subcutaneously implanted osmotic minipumps also produced striatal degeneration. Besides the mode of delivery, the treatment dose concentration also has an impact on 3-NP toxicity.

Depending on the time period over which 3-NP is administered, and the dose administered, rodents treated with 3-NP exhibit HD-like motor disorders with hyperkinetic and hypokinetic symptoms (Borlogan et al., 1997), and rats were shown to be more sensitive to the effects of 3-NP than mice (Brouillet, 2005). In rats, the administration of 3-NP (10 mg/kg i.p.) over several days was found to induce the onset of hypokinetic symptoms (Guyot et al., 1997), while its administration in two individual doses caused hyperkinetic symptoms (Borlongan et al., 1997b). However, besides the mode of delivery, the treatment dose also has an impact on 3-NP toxicity, where a 3-NP concentration of ~20 mg/kg was found to induce the expression of an HD behavioural phenotype after two injections. Nevertheless, these animals did not display extra-striatal lesions, which are frequently observed in the initial stages of HD (Beal et al., 1993; Guyot et al., 1997; Borlogan et al., 1997b). The chronic administration of low doses of 3-NP (~10 mg/kg, per day) for more than 3 weeks was found to induce sustained metabolic alterations and some other cellular features exhibited in HD patients, but did it not cause clear dyskinetic movements resembling chorea (Borlogan et al., 1997a, b; Brouillet et al., 1999).

Since there are differences in the response to 3-NP treatment among different animal strains (Ouary et al., 2000), we designed an administration plan of low concentration doses of 3-NP (15 mg/kg, i.p.) over a sub-chronic period (5 days) in C57BL/6 mice, which are known to be more resistant to 3-NP toxicity; this advantage enabled us to observe histopathological changes that mimic those found in the initial steps of the illness (Rodriguez et al., 2010), plus motor alterations such as orofacial dyskinesias and clasping behaviour (Hernández-Echeagaray et al., 2011), and spontaneous behaviours that resemblance the HD phenotype. Figure 1, shows that mice treated with low dose concentration of 3-NP displayed motor hyperactivity, as evaluated in open field tests. Hyperkinetic symptoms are exhibited in the initial steps of striatal damage in animals treated with 3-NP, whereas the hypokinetic phenotype develops later during striatal deterioration (Borlongan et al., 1997b).

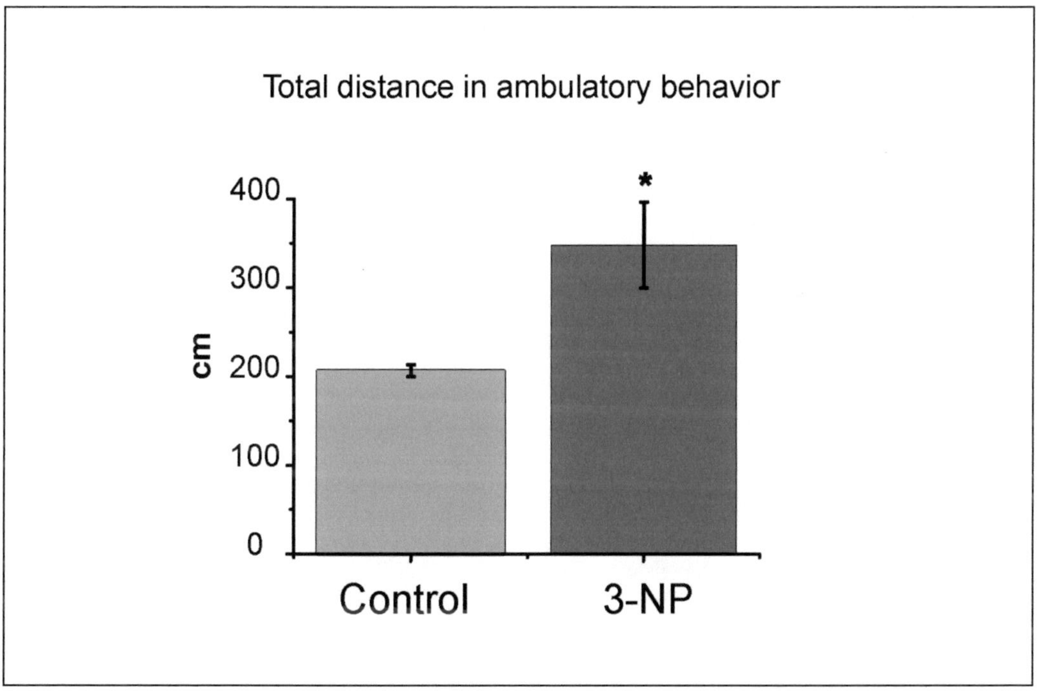

The graph shows the effects of 3-NP (15 mg / kg, per 5 days) in the spontaneous ambulatory motor behaviour of mice, scored as the total distance during 10 minutes period, evaluated in the open field test (Versadata, 3.02-1E7E software, Accuscan Instruments, INC.). There was a significant increase in locomotor activity in the 3-NP treated group (t-test, *p=0.0213). Arithmetic means and standard errors are plotted.

Fig. 1. 3-NP increases the ambulatory behavior evaluated in open field test.

We are especially interested in discerning the cellular events that take place during the beginning of the neurodegenerative process; the understanding of early dysfunctions will help in the planning of therapeutic strategies to reduce or delay cellular damage. The neuronal damage caused by low doses of 3-NP administration is restricted to striatal calbindin-positive cells (Fig. 2), leaving sparing parvalbumin and calretinin positive interneurons as previously suggested (Ferrante et al., 1987a, 1987b, Kowall et al., 1987¸ Massouh et al., 2008). Cellular alteration can also be initiated by caspase 3-dependent apoptosis, although necrotic cell death is also present (Rodriguez et al., 2010).

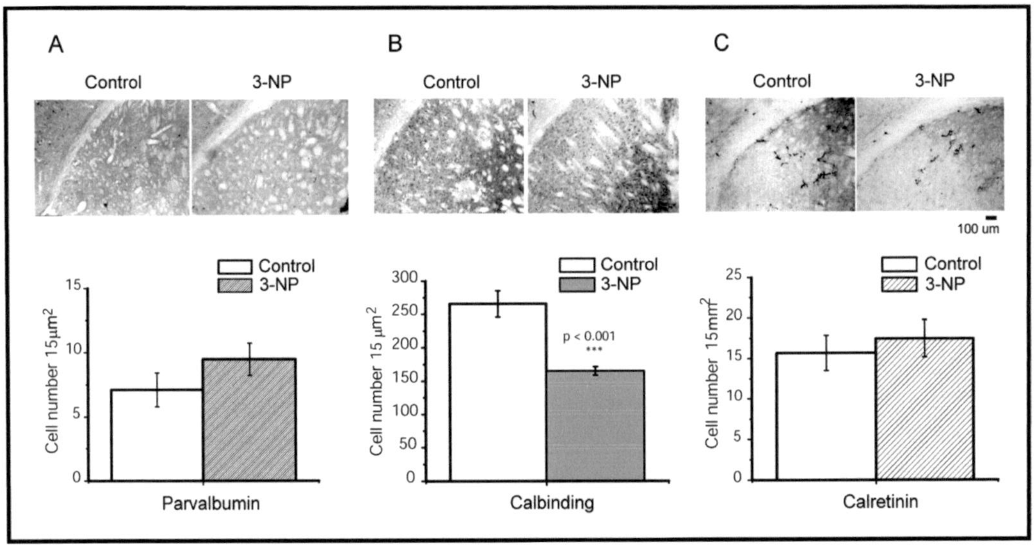

Top: light micrographs of calcium binding proteins immune localization in the neostriatal tissue. A, illustrates cells immune positive to parvalbumin. Cells were homogenously distributed in the striatum in both groups. The staining of immune positive cells was located in the soma. Graph in the bottom shows that there were no differences between groups in the number of cells expressing parvalbumin (t35 =-1.519, p=0.138). B illustrates striatal cells that were immune positive to calbinding. Calbinding positive cells were homogenously distributed a long the striatum in both groups and its localization was concentrated in the cell soma. Graph in the bottom shows that in tissue of 3-NP treated group, immune positive cells were significant reduced in comparison to control group (T99=2073.5, p<0.001). C illustrates calretinin immune positive cells. These cells were localized in dorsal striatum and staining was observed in cell body, dendrites and axonal ramifications. Interestingly tissue from 3-NP treated mice exhibited a decrease in cell ramifications but as shown in the bottom graph, the immune stained cell number did not change between groups (T82= 1686.5, p=0.523). Scale bar is 100 μm.

Fig. 2. 3-NP significantly decreases medium spiny neurons identified with the calcium binding protein calbinding.

3-NP treatment induces the abnormal production of reactive oxygen species (ROS), as well as highly reactive molecules derived from the formation of nitric oxide (NO). Succinate dehydrogenase inhibition interferes with the electron transport cascade and oxidative phosphorylation, which results in a decrease in ATP production and a cellular energy deficit (Jana et al., 2001; Lunkes et al., 2002). In studies by our group, the systemic administration of low sub-chronic doses of 3-NP did not produce a significant augmentation of NO in the brain (Rodriguez et al., 2010); however, NO and lipid peroxidation (LPO) increased in skeletal muscle (Hernández-Echeagaray et al., 2011). Previous reports looking at the involvement of NO in the toxicity of 3-NP did not draw any clear conclusions, but it has been suggested that 3-NP acts as an NO donor, increasing the levels of nitro anions (Jana et al., 2001; Lunkes et al., 2002).

Increases in NO and LPO along with signs of necrosis that manifest at the ultra-structural level in 3-NP-treated animals, like cellular oedema, are suggestive of the inflammatory process. Gene expression of the anti-inflammatory cytokine Interleukin-10 (IL-10) was found to significantly decrease whereas expression of the inflammatory cytokine Interferon-gamma (IFNγ) increased, indicating that, in low doses, 3-NP reduces the activation of anti-inflammatory cytokines (Fig. 3). This may have negative effects to prevent cellular damage.

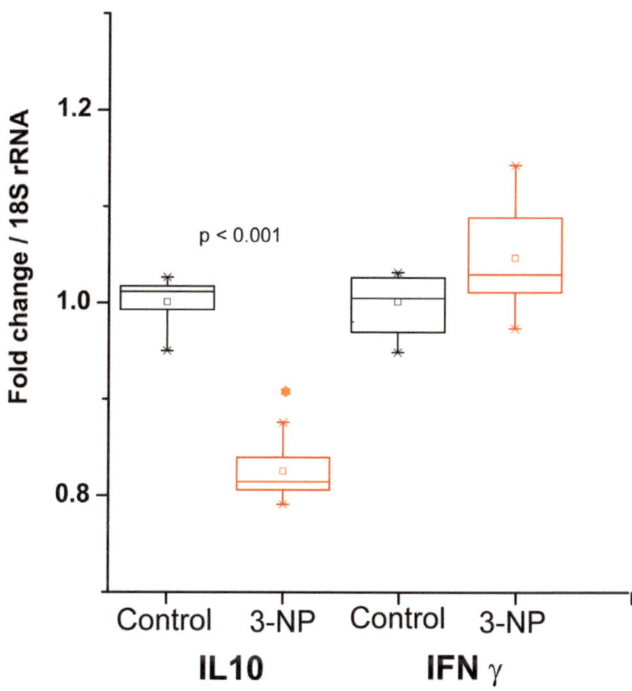

PCR products of IL-10 and INFγ in striatum of control and 3-NP treated groups are displayed. Data were analyzed by measuring continuously gene-specific PCR products and differences were assessed with the 2_ΔΔCT method. Data are presented as the fold increase in gene transcripts normalized to the 18S rRNA expression and relative to the control. IL-10 expression was significantly reduced in the 3-NP group ($t13 = 12.904$, $p<0.001$). INFγ exhibited a no significant increase in the 3-NP treated group ($t16= 0.232$, $p= 0.819$).

Fig. 3. Expression of striatal IL-10 and INFγ mRNA by real time RT-PCR.

Cells obtain energy from oxidative phosphorylation and from glycolysis; the glycolytic enzyme GAPDH has been implicated in neuronal degeneration (Taylor et al., 2002; Huntington's Disease Collaborative Research Group, 1993) and a reduction in GAPDH activity was demonstrated in HD patients and in a transgenic mouse model of HD (Burke et al., 1996; Matthews et al., 1997). Some studies have suggested that mitochondrial failure is secondary to striatal damage in HD (Lee et al., 2007), and then another energy source that may fail and generate damage is glycolysis. Both oxidative phosphorylation and glycolysis can be involved in the pathogenesis of neural degeneration in the striatum. However, inhibition of the glycolytic enzyme GAPDH was found to cause apoptotic damage, which is independent of the activation of caspase 3 (Rodriguez et al., 2010). Hence, alterations in glycolysis may be a critical point in neuronal death, but its inhibition activates different cellular signals than oxidative phosphorylation.

4.2 Changes in the periphery

Any deficit in energy metabolism can damage the entire physiology of an organism. Muscles metabolism is essential for locomotion and heat production (Zierath and Hawley, 2004). Patients afflicted by HD experience many deficits (Bradshaw et al., 1992; Aron et al., 2003; Abbruzzese et al., 2003), and weight loss is a characteristic feature (Robbins et al., 2006; Ciammola et al., 2011; Mochel and Haller, 2011). Transgenic mice model in an HD exhibited protein inclusions (Sathasivam et al., 1999) and mhtt aggregation in skeletal muscle (Ribchester et al., 2004). The motor impairments displayed by patients and animal models could result from central neurodegeneration, but also from alterations in muscle metabolism, reflecting peripheral disturbances as a general metabolic failure. Where metabolic collapse occurs in an organism, this could partly clarify the muscle alterations and body weight loss documented in HD patients (Stoy and McKay, 2000; Robbins et al., 2006; Ciammola et al., 2011) and transgenic HD mice (Sathasivam et al., 1999; Ribchester et al., 2004; She et al., 2011) and the alterations caused by metabolic alterations (Simoneau et al., 1995; Petersen et al., 2004).

Initially, it was thought that 3-NP did not produce major peripheral effects (Hamilton and Gould 1987). Until now, most of the studies that were carried out using 3-NP as a tool to investigate neurodegeneration focused on evaluating central damage (Beal et al., 1993; Borlongan et al., 1997; Brouillet, et al., 1999, 2005), even though in a number of them 3-NP was systemically administered. However, Gabrielson (et al., 2001) showed that 3-NP induced modifications in cardiac muscle physiology.

In addressing the possible changes that 3-NP might generate outside of the brain, our group documented modifications in skeletal muscle after uncoupling oxidative metabolism with 3-NP in low sub-chronic doses. Our hypothesis was that abnormal mitochondrial functioning during first stages of neurodegenerative disease is responsible for body weight loss in patients afflicted with HD or illness where there is a metabolic dysfunction (Stoy and McKay, 2000; Robbins et al., 2006). The systemic administration of low doses of 3-NP altered enzymatic activity in muscle, as well as the organization of the sarcomere (Hernández-Echeagaray et al., 2011), suggesting that energy failure exacerbates metabolic activity in the whole organism, producing high metabolic demands to counteract the failure of bioenergetics. When comparing muscle modifications in animals where glycolysis was inhibited, we found that iodoacetate IOA also alters the ultrastructure of the gastrocnemius muscle; this disorganization was more pronounced in animals that were treated with 3-NP (Fig. 4).

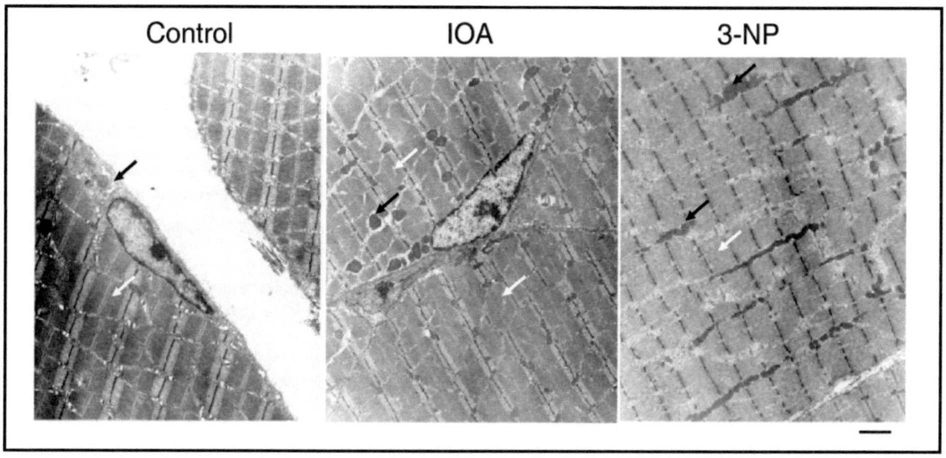

Electron microscopy of gastrocnemius muscle from control, IOA, and 3-NP treated animals. Control muscle shows normal sarcomere organization (white arrow) and mitochondria morphology (black arrow). However, sarcomere of IOA and 3-NP treated groups was altered; also the number of mithoxhondria and mitochondria morphology were modified. In particular muscles from 3-NP treated group were most affected than those from the IOA group. Micrographics magnification is 7000X. Scale bar is 500 nm.

Fig. 4. Metabolic uncoupling induces ultra structural modification in the mouse gastrocnemius muscle.

Oxidative phosphorylation or metabolic failure may generate an increase in energy consumed by muscles, as has been suggested in several degenerative disorders where alterations in oxidative phosphorylation and metabolism are compromised (Ristow, 2004). It is important to mention that in models of food restriction or malnutrition, the size and body weight of animals were affected before the nervous system became involved in the behavioural phenotype (Woodall et al., 1996; Clapham, 2004). Being aware of the peripheral and central damage due to metabolic dysfunctions might help in the clinical management of the early stages of the disease. For example, treatments designed to improve energy metabolism might modify the course of the illness and delay the progression of the disease.

5. Conclusions

The goal of this chapter was to illustrate the fact that 3-NP in low sub-chronic doses induces cellular alterations that emulate the damage exhibited in the striatum and muscle of patients afflicted by HD in the early stages of the illness. Even though 3-NP does not reproduce all signs and symptoms displayed in HD patients, it is true that it has helped in the understanding of the cellular physiology of animal models where a general failure in bioenergetics has been presumed.

6. Acknowledgements

The authors wish to thank A Ruelas for carrying out the open field test evaluation. This work was partially supported by grants from CONACyT, DGAPA-PAPIIT and Institutional support from FES-I, UNAM.

7. References

Abbruzzese G, Berardelli A (2003). Sensorimotor integration in movement disorders. Mov Disord 18; 231-40

Alston TA, Mela L, Bright HJ (1977). 3-Nitropropionate, the toxic substance of Indigofera, is a suicide inactivator of succinate dehydrogenase. Proc Natl Acad Sci USA 74; 3767–3771.

Alexi T, Hughes P E, Knusel B, Tobin A J (1998). Metabolic compromise with systemic 3-nitropropionic acid produces striatal apoptosis in Sprague–Dawley rats but not in BALB/c ByJ mice. Exp Neurol 153; 74–93.

Aron AR, Watkins L, Sahakian BJ, Monsell S, Barker RA, Robbins TW (2003). Task-set switching deficits in early-stage Huntington's disease: implications for basal ganglia function. J Cogn Neurosci 15; 629-42

Artal-Sanz M and Tavernarakis N (2005). Proteolytic mechanisms in necrotic cell death and neurodegeneration FEBS Letter 579; 3287–3296

Beal MF, Brouillet E, Jenkins BG, Ferrante RJ, Kowall NW, Miller JM, Storey E, Srivastava R, Rosen BR, Hyman BT (1993). Neurochemical and histologic characterization of striatal excitotoxic lesions produced by the mitochondrial toxin 3-nitropropionic acid. J Neurosci 13; 4181–4192

Beal MF (2005). Mitochondria take center stage in aging and neurodegeneration. Ann Neurol. 58; 495-505

Borlongan CV, Nishino H, Sanberg PR (1997a). Systemic but not intraparenquimal administration of 3-nitropropionic acid mimics the neuropathology of Huntington`s disease: a speculative explanation. Neurosci Res 28; 185-189

Borlongan CV, Koutouzis TK, Freeman TB, Hauser RA, Cahill DW, Sanberg R (1997b). Hyperactivity and hypo activity in a rat model of Huntington's Disease: The systemic 3- nitropropionic acid model. Brain Res Protoc 1; 253–257.

Bradshaw JL, Phillips JG, Dennis C, Mattingley JB, Andrewes D, Chiu E, Pierson JM, Bradshaw JA (1992). Initiation and execution of movement sequences in those suffering from and at-risk of developing Huntington's disease. J Clin Exp Neuropsychol 14;179-92.

Brouillet E, Jenkins BG, Hyman BT, Ferrante RJ, Kowall NW, Srivastava R, Roy DS, Rosen BR, Beal MF (1993). Age dependent vulnerability of the striatum to the mitochondrial toxin 3-nitropropionic acid. J Neurochem 60; 356–359.

Brouillet E, Conde F, Beal MF, Hantraye P (1999). Replicating Huntington's disease phenotype in experimental animals Prog Neurobiol 59; 427.

Brouillet E, Jacquard C, Bizat N, Blum D (2005). 3-Nitropropionic acid: A mitochondrial toxin to uncover physiopathological mechanisms underlying striatal degeneration in Huntington's disease. J Neurochem 95; 1521–1540.

Browne SE, Bowling AC, MacGarvey U, Baik MJ, Berger SC, Muqit MM, Bird ED, Beal MF (1997). Oxidative damage and metabolic dysfunction in Huntington's Disease: selective vulnerability of the basal ganglia. Ann Neurol 41; 646–653.

Browne, SE (2008). Mitochondrial and Huntington's Disease pathogenesis: Insight from genetic and chemical models. Ann NY Acad 1147; 358–382.

Burke JR, Enghild JJ, Martin ME, Jou YS, Myers RM, Roses AD, Vance JM, Strittmatter WJ (1996). Huntingtin and DRPLA proteins selectively interact with the enzyme GAPDH. Nat Med 2; 347–350

Ciammola A, Sassone J, Sciacco M, Mencacci NE, Ripolone M, Bizzi C, Colciago C, Moggio M, Parati G, Silani V, Malfatto G (2011). Low Anaerobic Threshold and Increased Skeletal Muscle Lactate Production in Subjects with Huntington's Disease. Movement Disorders, 26; 493-499.

Clapham JC (2004). Treating obesity: pharmacology of energy expenditure. Cur Drug Targets 5; 309-323.

Dautry C, Vaufrey F, Brouillet E, Bizat N, Henry PG, Conde F, Bloch G, Hantraye P (2000). Early N-acetylaspartate depletion is a marker of neuronal dysfunction in rats and primates chronically treated with the mitochondrial toxin 3-nitropropionic acid. J Cereb Blood Flow Metab 20; 789–799.

DiFiglia M (1990). Excitotoxic injury of the neostriatum: a model for Huntington's disease. TINS 13; 286-289.

Ferrante RJ, Beal MF, Kowall NW, Richardson EPJr, Martin JB (1987a). Sparing of acetylcholinesterase-containing striatal neurons in Huntington's Disease. Brain Res 411; 162-166.

Ferrante RJ, Kowall NW, Beal MF, Martin JB, Bird ED, Richardson EP Jr (1987b). Morphologic and histochemical characteristics of a spared subset of striatal neurons in Huntington's Disease. J Neuropathol Exp Neurol 46; 12-27.

Fridovich I (1999). Fundamental aspects of reactive oxygen species, or what's the matter with oxygen? Ann NY Acad Sci 893; 13–28.

Gabrielson KL, Hogue BA, Bohr VA, Cardounel AJ, Nakajima W, Kofler J, Zweier JL, Rodriguez ER, Martin LJ, de Souza Pinto NC, Bressier J (2001). Mitochondrial toxin 3-Nitropropionic acid induces cardiac and neurotoxicity differentially in mice. Am J Pathol 159; 1507-1520.

Gould DH, Gustine DL (1982). Basal ganglia degeneration,myelin alterations, and enzyme inhibition induced in mice by the plant toxin 3-nitropropanoic acid. Neuropathol Appl Neurobiol 8; 377–393.

Gould D H, Wilson MP, Hamar DW (1985). Brain enzyme and clinical alterations induced in rats and mice by nitroaliphatic toxicants. Toxicol Lett 27; 83–89.

Grafton ST, Mazziotta JC, Pahl JJ, St George-Hyslop P, Haines JL, Gusella J, Hoffman J M MD, Baxter LR, Phelps ME (1990). A comparison of neurological, metabolic, structural, and genetic evaluations in persons at risk for Huntington's disease. Ann Neurol 28; 614–621.

Grünewald T, BEAL M F (1999). Bioenergetics in Huntington's Disease. Ann NY Acad Sci 893; 203–213.

Gu M, Gash MT, Mann VM, Javoy-Agid F, Cooper JM, Schapira AH (1996). Mitochondrial defect in Huntington's disease caudate nucleus. Ann Neurol 39; 385–389.

Guyot MC, Palfi S, Stutzmann JM, Maziere M, Hantraye P, Brouillet E (1997). Riluzole protects from motor deficits and striatal degeneration produced by systemic 3-nitropropionic acid intoxication in rats. Neurosci 81; 141–149.

Hamilton BF, Gould DH (1987). Nature and distribution of brain lesions in rats intoxicated with 3-nitropropionic acid: a type of hypoxic (energy deficient) brain damage. Acta Neuropathol 72; 286–297.

He F, Zhang S, Qian F, Zhang C (1995). Delayed dystonia with striatal CT lucencies induced by a mycotoxin (3-nitropropionic acid). Neurol 45; 2178–2183.

Henry PG, Lebon V, Vaufrey F, Brouillet E, Hantraye P, Bloch G (2002). Decreased TCA cycle rate in the rat brain after acute 3-NP treatment measured by in vivo 1H-[13C] NMR spectroscopy. J Neurochem 82; 857–866.

Hernández-Echeagaray E, González N, Ruelas A, Mendoza E, Rodríguez-Martínez E, Antuna-Bizarro R (2011). Low doses of 3-nitropropionic acid in vivo induce damage in mouse skeletal muscle. Neurol Sci 32; 241-54.

Huntington's Disease Collaborative Research Group (1993). Cell; 72, 971.

Inagaki R, Tagawa K, Qi M-L, Enokido Y, Ito H, Tamura T, Shimizu S, Oyanagi K, Arai N, Kanazawa I, Wanker E E, Okazawa H (2008). Omi / Htr2 is relevant to the selective vulnerability of striatal neurons in Huntinton's disease Eur J Neurosci 28; 30-40.

Janavs J L, Aminoff M J (1998). Dystonia and chorea in acquired systemic disorders. J Neurol Neurosurg Psychiatry 65; 436–445.

Jana NR, Zemskov EA, Wang G, Nukina N (2001). Altered proteasomal function due to the expression of polyglutamine-expanded truncated N-terminal huntingtin induces apoptosis by caspase activation through mitochondrial cytochrome c release. Hum Molec Genet 10; 1049.

Jenkins BG, Koroshetz WJ, Beal MF, Rosen BR (1993). Evidence for impairment of energy metabolism in vivo in Huntington's disease using localized 1H NMR spectroscopy Neurol 43; 2689–2695.

Jenkins BG, Rosas HD, Chen YC, Makabe T, Myers R, MacDonald M, Rosen BR, Beal MF, Koroshetz WJ (1998). 1H NMR spectroscopy studies of Huntington's disease: Correlations with CAG repeat numbers. Neurology 50: 1357–1365.

Kanazawa I (2001). How do neurons die in neurodegenerative diseases? Trends in Molecular Medic 7; 339-344.

Kapogiannis D, Mattson MP (2011). Disrupted energy metabolism and neuronal circuit dysfunction in cognitive impairment and Alzheimer's disease. Lancet Neurobiol 10; 187-198.

Koroshetz WJ, Jenkins BG, Rosen BR, Beal MF (1997). Energy metabolism defects in Huntington's disease and effects of coenzyme Q10. Ann Neurol 41; 160–165.

Kowall NW, Ferrante RJ, Martin JB (1987). Patterns of cell loss in Huntington's Disease. TINS 10; 24-29.

Krantic S, Mechawar N, Reix S, Quirion R (2005). Molecular basis of programmed cell death involved in neurodegeneration. TINS, 28; 670-676.

Kroemer G, Dallaporta B, Resche-Rigon M (1998). The mitochondrial death/life regulator in apoptosis and necrosis. Annu Rev Physiol 60; 619–642.

Kuhl DE, Phelps ME, Markham CH, Metter EJ, Riege WH, Winter J (1982). Cerebral metabolism and atrophy in Huntington's disease determined by 18FDG and computed tomographic scan. Ann Neurol 12; 425–434.

Lee J M, Ivanova E V, Seong I S, Cashorali T, Kohane I, Gusella JF, MacDonald ME (2007). Unbiased gene expression analysis implicates the huntingtin polyglutamine tract in extramitochondrial energy metabolism. PLoS Genet 3; e135.

Li XJ, Orr AL, Li S (2010). Impaired mitochondrial trafficking in Huntington's disease. Biochem Biophys Acta 1802; 62–65.

Lin MT, Beal MF (2006). Mitochondrial dysfunction and oxidative stress in neurodegenerative diseases. Nat 443; 787–795.

Lodi R, Schapira AH, Manners D, Styles P, Wood NW, Taylor DJ, Warner TT (2000). Abnormal in vivo skeletal muscle energy metabolism in Huntington's disease and dentate rubropallidoluysian atrophy. Ann Neurol 48: 72–76.

Ludolph AC,He F, Spencer PS, Hammerstad J, Sabri M (1991). 3-Nitropropionic acid-exogenous animal neurotoxin and possible human striatal toxin. Can J Neuro. Sci 18; 492–498.

Lunkes A, Lindenberg KS, Ben-Haiem L, Weber C, Devys D, Landwehrmeyer GB, Mandel JL, Trottier Y (2002). Proteases acting on mutant huntingtin generate cleaved products that differentially build up cytoplasmic and nuclear inclusions. Mol Cell 10; 259-265.

Mazziotta JC, Phelps ME, Pahl JJ, Huang SC, Baxter LR, Riege WH, Hoffman JM, Kuhl DE, Lanto AB,. Wapenski JA, Markham CH (1987). Reduced cerebral glucose metabolism in asymptomatic subjects at risk for Huntington's disease. N Engl J Med 316: 357-362.

Maciel EN, Kowaltowski A J, Schwalm FD, Rodrigues J M, Souza DO, Vercesi AE, Wajner M, Castilho RF (2004). Mitochondrial permeability transition in neuronal damage promoted by Ca2+ and respiratory chain complex II inhibition. J Neurochem 90; 1025-1035.

Massouh M, Wallman MJ, Pourcher E, Parent A (2008). The fate of the large striatal interneurons expressing calretinin in Huntington's Disease. Neurosci Res 62; 216-224.

Matthews RT, Ferrante RJ, Jenkins BG, Browne SE, Goetz K, Berger S, Chen YC, Beal MF (1997). Iodoacetate produces striatal excitotoxic lesions. J Neurochem 69; 285-289.

Ming L (1995). Moldy sugarcane poisoning-a case report with a brief review. J Toxicol Clin Toxicol 33; 363-367.

Mochel F, Haller RG (2011). Energy deficit in Huntington disease: why it matters. J Clin Invest 121; 493-499.

Mogami M, Hida H, Hayashi Y, Kohri K, Kodama Y, Gyun JC, Nishino H (2002). Estrogen blocks 3-nitropropionic acid induced Ca2+ increase and cell damage in cultured rat cerebral endothelial cells. Brain Res 956; 116-125.

Nishino H, Nakajima K, Kumazaki M, Fukuda A, Muramatsu K, Deshpande SB, Inubushi T, Morikawa S, Borlongan CV, Sanberg PR (1998). Estrogen protects against while testosteroneexacerbates vulnerability of the lateral striatal artery to chemicalhypoxia by 3-nitropropionic acid. Neurosci Res 30; 303-312.

Novelli A, Reilly JA, Lysko PG, Henneberry RC (1988). Glutamate becomes neurotoxic via the N-methyl-D-aspartate receptor when intracellular energy levels are reduced. Brain Res 451; 205-212.

Novotny E JJr, Singh G, Wallace DC, Dorfman LJ, Louis A, Sogg RL, Steinman L (1986). Leber's disease and dystonia: a mitochondrial disease. Neurol 36; 1053-1060.

Oliveira JM (2010). Nature and cause of mitochondrial dysfunction in Huntington's disease: focusing on huntingtin and the striatum. J Neurochem 114; 1-12.

Ouary S, Bizat N, Altairac S, Menetrat H, Mittoux V, Conde F, Hantraye P, Brouillet E (2000). Major strain differences in response to chronic systemic administration of the mitochondrial toxin 3-nitropropionic acid in rats: implications for neuroprotection studies. Neurosci 97; 521-530.

Palfi S, Ferrante RJ, Brouillet E, Beal MF, Dolan R, Guyot MC, Peschanski M, Hantraye P (1996). Chronic 3-nitropropionic acid treatment in baboons replicates the cognitive and motor deficits of Huntington's Disease. J Neurosci 16; 3019- 3025.

Panov AV, Gutekunst CA, Leavitt BR, Hayden MR, Burke JR, Strittmatter WJ, Greenamyre JT (2002). Early mitochondrial calcium defects in Huntington's Disease are a direct effect of polyglutamines. Nat Neurosci 5; 731-736.

Petersen KF, Durour S, Befroy D, García R, Shulman GI (2004). Impaired mitochondrial activity in the insuline resistant offspring of patients with type 2 diabetes. N Engl J Med 350: 664-671.

Pickrell AM, Fukui H, Wang X, Pinto M, Moraes CT (2011). The striatum is highly susceptible to mitochondrial oxidative phosphorylation dysfunctions. J Neurosci, 31: 9895-9904.

Reddy PH, Mao P, Manczak M (2009). Mitochondrial structural and functional dynamics in Huntington's disease. Brain Res Rev 61: 33-48.

Ribchester RR, Thomson D, Wood NI, Hinks T, Gillingwater TH, Wishart TM, Court FA, Morton AJ (2004). Progressive abnormalities in skeletal muscle and neuromuscular junctions of transgenic mice expressing the Huntington's disease mutation. Eur J Neurosci 20: 3092-3114.

Ristow M (2004). Neurodegenerative disorders associated with diabetes mellitus. J Mol Med 82; 510-529.

Robbins AO, Ho AK, Barker RA (2006). Weight changes in Huntington's disease. Eur J Neurol 13; e7.

Rodríguez E, Rivera I, Astroga S, Medoza E, García F, Hernández-Echeagaray E (2010). Uncoupling Oxidative/Energy Metabolism with Low sub Chronic Doses of 3-Nitropropionic acid or Iodoacetate in vivo Produces Striatal Cell Damage. Int J Biol Sci 6; 199-212.

Saft C, Zange J, Andrich J, Muller K, Lindenberg K, Landwehrmeyer B, Vorgerd M, Kraus PH, Przuntek H, Schöls L (2005). Mitochondrial impairment in patients and asymptomatic mutation carriers of Huntington's disease. Mov Disord 20; 674–679.

Sathasivam K, Hobbs C, Turmaine M, Mangiarini L, Mahal A, Bertaux F, Wanker EE, Doherty P, Davies SW, Bates G (1999). Formation of polyglutamine inclusions in non-CNS tissue. Hum Mol Gen 8; 813-822.

Schulz JB, Henshaw DR, MacGarvey U, Beal MF (1996). Involvement of oxidative stress in 3-nitropropionic acid neurotoxicity. Neurochem Int 29; 167–171.

Seong IS, Ivanova E, Lee JM, Choo YS, Fossale E, Anderson M, Gusella JF, Laramie JM, Myers RH, Lesort M, MacDonald ME (2005). HD CAG repeat implicates a dominant property of huntingtin in mitochondrial energy metabolism. Hum Mol Genet 14; 2871–2880.

Shoffner JM, Watts RL, Juncos JL, Torroni A, Wallace DC (1991). Mitochondrial oxidative phosphorylation defects in parkinson's disease Ann Neurol 30; 332–339.

Simoneau JA, Colberg SR, Thaete FL, Kelley DE (1995). Skeletal muscle glycolytic and oxidative enzyme capacities are determinants of insulin sensitivity and muscle composition in obese women FASEB J 9; 273-278.

Sims NR (1996). Energy Metabolism, Oxidative Stress and Neuronal Degeneration in Alzheimer's Disease Neurodegeneration. Volume 5; 435-440.

Stoy N, McKay E (2000). Weight loss in Huntington's disease. Ann Neurol 48; 130-131.

Strauss KA, Morton DH (2003). Type I glutaric aciduria, part 2: a model of acute striatal necrosis. Am J Med Genet 121, 53–70.

Taylor PJ, Hardy J, Fischbeck KH (2002). Toxic proteins in neurodegenerative disease. Science 296; 1991-1995.

Vis JC, Schipper E, Boer-van Huizen RT, Verbeek MM, de Waal RMW, Wesseling P, Donkelaar H J Kremer B (2005). Expression pattern of apoptosis-related markers in Huntington's Disease Acta Neuropathol 109; 321-328.

Walker OW (2007). Huntington's disease. Lancet 369; 218-225.

Weinshilboum R (2003). Inheritance and drug response. N Engl J Med 348; 529–37.

Weinshilboum RM, Wang L (2006). Pharmacogenetics and pharmacogenomics: development, science, and translation. Annu Rev Genomics Hum Genet 7; 223-245.

Woodall SM, Breier BH, Johnston BM, Gluckman PD (1996). A model of intrauterine growth retardation caused by chronic maternal undernutrition in the rat: effects on the somatotrophic axis and postnatal growth. J Endocrinol 150; 231–242.

Zeevalk GD, Nicklas WJ (1990). Chemically induced hypoglycemia and anoxia: relationship to glutamate receptor-mediated btoxicity in retina. J Pharmacol Exp Ther 253; 1285–1292.

Zierath JR, Hawley JA (2004). Skeletal muscle fiber type: influence on contractile and metabolic properties. PLOS Biol 2; 337-348.

16

Consequences of Mitochondrial Dysfunction in Huntington's Disease and Protection via Phosphorylation Pathways

Teresa Cunha-Oliveira[1]*, Ildete Luísa Ferreira[1]* and A. Cristina Rego[1,2]
[1]CNC-Center for Neuroscience and Cell Biology, University of Coimbra,
[2]Faculty of Medicine, University of Coimbra,
Portugal

1. Introduction

Huntington's Disease (HD) is an autosomal dominant neurodegenerative disorder clinically characterized by psychiatric disturbances, progressive cognitive impairment and choreiform movements. These symptoms are associated with the selective atrophy and neuronal loss in the striatum, cortex and hypothalamus. The disease is caused by a mutation at the 5′ terminal of the huntingtin (*HTT*) gene involving the expansion of CAG triplet, which encodes for glutamine. Mutant huntingtin (mHtt) may be cleaved by proteases originating neurotoxic fragments, and also undergoes conformational changes that lead to the formation of protein aggregates (Gil and Rego 2008, for review). Among several mechanisms of neurodegeneration, mHtt is related to mitochondrial dysfunction and relevant changes in energy metabolism in both central and peripheral cells, which may underlie cell death (Gil and Rego 2008, for review).

In this review chapter we emphasize the role of mitochondrial dysfunction in neurodegeneration in HD, particularly centering on loss of mitochondrial activity and the regulation of intrinsic apoptosis in central and peripheral HD human tissue or cells, and in animal models of HD. We focus on the changes in energy metabolism, oxidative stress, the link to transcriptional dysfunction and the regulation of intrinsic apoptosis. We further explore the therapeutic role of promoting phosphorylation pathways through selective inhibition of phosphatases (e.g. with FK506) and/or activation of kinase signaling cascades mediated by neurotrophins, namely brain-derived neurotrophic factor (BDNF) and nerve growth factor (NGF).

2. Mitochondrial dysfunction and apoptosis in HD

2.1 Mitochondrial dysfunction

The mechanisms by which neurons die in HD are uncertain, however, mitochondrial dysfunction and apoptosis have been implicated. Mitochondria are important organelles

---

*These authors contributed equally

that regulate the life and death of cells and neurons are particularly dependent on these organelles due to their high energy requirements.

Mitochondrial dysfunction is considered a common feature in the pathogenesis of neurodegenerative disorders like HD (Kim *et al.* 2010;Oliveira 2010;Parker, Jr. *et al.* 1990), and constitutes a cellular hallmark for neurodegeneration, occurring as a consequence of defective mitochondrial composition, trafficking to synapses, calcium handling, ATP production, transcription abnormalities and/or electron transport chain (ETC) impairment (Rosenstock *et al.* 2010, for review). Moreover, cell and animal models of HD exhibit mitochondrial impairment and metabolic deficits similar to those found in HD patients (reviewed in Damiano *et al.* 2010;Quintanilla and Johnson 2009). mHtt may cause mitochondrial dysfunction by directly interacting with the organelle (Panov et al. 2002) by evoking defects in mitochondrial dynamics, organelle trafficking and fission and fusion, which, in turn, may result in bioenergetic failure, or indirectly by perturbing transcription of nuclear-encoded mitochondrial proteins (Bossy-Wetzel *et al.* 2008, for review).

The hypothesis that mitochondrial dysfunction contributes to the pathogenesis of HD was first tested pharmacologically by using 3-nitropropionic acid (3-NP) and malonate, irreversible and reversible inhibitors of succinate dehydrogenase (a component of both the tricarboxylic acid cycle and the complex II of the ETC), respectively. Administration of these inhibitors to animals results in pathological characteristics of HD, such as marked increases in striatal lactate concentration, striatal lesions and motor disturbances (Beal *et al.* 1993;Brouillet *et al.* 1993;Frim *et al.* 1993), involving an immediate ATP drop and secondary increase in reactive oxygen species (ROS), which is correlated with profound mitochondrial fragmentation (Brouillet *et al.* 1999). Selective striatal neurodegeneration induced by 3-NP appears to be related to the early expression and activation of matrix metalloproteinase-9 by ROS which can digest the endothelial basal lamina, leading to the disruption of the blood–brain barrier and to progressive striatal damage (Kim *et al.* 2003). Concordant with 3-NP mimicking the disease, in 1974 a defect in succinate dehydrogenase was reported in the caudate and, to a lesser extent, in the cortex of postmortem HD brains (Stahl and Swanson 1974). Moreover, yeast expressing mHtt showed a significant reduction in oxidative phosphorylation due to a decrease in complexes II and III activities (Solans *et al.* 2006).

Furthermore, early studies of cortical biopsies obtained from patients with either juvenile or adult onset HD showed abnormal mitochondria morphology and function (Goebel *et al.* 1978;Tellez-Nagel *et al.* 1974). Functional changes in mitochondrial ETC were also observed in HD, namely decreased mitochondrial complexes II/III activity and succinate oxidation in striatal tissue from HD patients (Stahl and Swanson 1974;Gu *et al.* 1996;Browne *et al.* 1997;Benchoua *et al.* 2006). Moreover, a decrease in complex IV activity was found in HD striatum (Browne *et al.* 1997;Gu *et al.* 1996).

In skeletal muscle, mHtt was reported to affect the activity of mitochondrial complex I (Arenas *et al.* 1998) and also complexes II/III (Ciammola *et al.* 2006;Turner *et al.* 2007), along with mitochondrial depolarization, cytochrome c release and caspases activation (Ciammola *et al.* 2006;Turner *et al.* 2007). In platelets from HD patients, some authors also found a decrease in complex I activity (Parker, Jr. *et al.* 1990), whereas others reported no changes in the activity of mitochondrial complexes (Gu *et al.* 1996;Powers *et al.* 2007a). A decrease in mitochondrial complex II/III activity was also found in lymphoblasts of HD patients (Sawa *et al.* 1999). No significant differences were observed in complexes I and IV but a correlation was found between complex II/III activity and disease duration and progress and inclusion formation in muscle (Turner *et al.* 2007).

Cybrids, an *ex-vivo* human peripheral cell model in which the contribution of mitochondrial defects from patients may be isolated, are an interesting approach to study mitochondrial dysfunction (King and Attardi 1989). Results from our laboratory showed that HD cybrids, prepared from the fusion of HD human platelets with NT2 rho0 cells, depleted of mitochondrial DNA, did not exhibit significant modifications in the activity of ETC complexes I–IV or specific mitochondrial DNA (mtDNA) sequence variations, suggestive of a primary role in mitochondrial susceptibility in the subpopulation of HD carriers studied (Ferreira et al. 2010). In accordance, Swerdlow and collaborators (1999) showed that HD cybrids did not present changes in ETC activity, oxidative stress or calcium homeostasis. Despite unchanged activity of mitochondrial complexes, this cell model presented evidences of mitochondrial dysfunction based on significant changes on mitochondrial membrane potential and increased ROS generation (Ferreira et al. 2010). The presence of mtDNA variations, including an 8656A N G variant in one patient, was previously shown in a screening study for mutations in the tRNA(leu/lys) and MTATP6 genes of 20 patients with HD (Kasraie *et al.* 2008). However, the nucleotides 8915-9207 of the same gene did not present any sequence variation in our HD cybrids (Ferreira et al. 2010). One of our HD cybrid lines carried the 3394T N C mutation with status "unclear" (Ferreira et al. 2010), previously described in cases suffering from Leber Hereditary Optic Neuropathy (LHON), which was shown to be related with HD features (Morimoto *et al.* 2004). In addition, a decrease in mitochondrial DNA content was found in cerebral cortex of HD patients (Horton *et al.* 1995).

It is accepted that mHtt not only impairs mitochondrial function, but also compromises cytosolic and mitochondrial calcium homeostasis, which contributes to neuronal dysfunction and death in HD (Damiano *et al.* 2010;Quintanilla and Johnson 2009, for review). Multiple changes in mitochondrial calcium handling (Panov *et al.* 2002;Oliveira *et al.* 2007), metabolism (Damiano *et al.* 2010), and susceptibility to apoptosis (Sawa *et al.* 1999) were suggested to be related with mitochondrial localization of mHtt (Orr *et al.* 2008). Indeed, mHtt interaction with neuronal mitochondria of YAC72 transgenic mice (Panov et al. 2002) was directly linked to mitochondrial calcium abnormalities (Choo et al. 2004;Panov et al. 2002). In this respect our group has also demonstrated changes in calcium handling linked to mitochondrial dysfunction in striatal neurons from YAC128 HD mice and cells derived from knock-in mice (Oliveira *et al.* 2006). Interestingly, increased vulnerability of striatal mitochondria to calcium loads was found to be present in both intact neurons and astrocytes, when compared with their cortical counterparts. Moreover, a lower mitochondrial calcium buffering capacity in intact striatal *versus* cortical astrocytes, associated with increased cyclosporin A-dependent permeability transition, suggested that the striatum is at higher risk for disturbed interactions between neurons and astrocytes (Oliveira and Goncalves 2009).

Various mitochondrial abnormalities observed in human patient samples, postmortem HD brains, cellular, invertebrate and vertebrate models of the disease, cooperate with mitochondrial ETC dysfunction in the genesis of HD (Pandey *et al.* 2010, for review). These include imbalance of calcium buffering capacity and oxidative stress, impaired axonal transport and abnormal fission and fusion of mitochondria, which are further described in this Chapter.

2.2 Altered mitochondrial trafficking and dynamics

Mitochondrial shape and structure are maintained by mitochondrial fission and fusion and disruption of mitochondrial dynamics was shown to be involved in HD (Chen and Chan 2009, for review). Fission is controlled by dynamin-related protein 1 (Drp1), mostly localized in the cytoplasm and in the mitochondrial outer membrane (MOM), and fission 1 (Fis1), localized to the MOM. On the other hand, mitochondrial fusion is ruled by mitofusin 1 (Mfn1) and mitofusin 2 (Mfn2), localized in the MOM, and optic atrophy-1 (Opa1), localized in the mitochondrial inner membrane (MIM) (Chen and Chan 2009, for review). In a healthy neuron, fission and fusion mechanisms balance equally and mitochondria alter their shape and size to move from cell body to the axons, dendrites, and synapses, and back to the cell body through mitochondrial trafficking. Recently, a role for abnormal mitochondrial networking in HD pathogenesis was described, involving mitochondrial fragmentation and cristae alterations, in different cellular models of HD (lymphoblasts from HD patients, striatal progenitor cell lines isolated from knock-in HdhQ111 mouse embryos and in YAC128 primary striatal neurons), explaining their increased susceptibility to apoptosis (Costa *et al.* 2010). Thus, increased cytotoxicity induced by overexpression of Htt proteins containing expanded polyglutamine (polyQ) tracts is likely mediated, at least in part, by an alteration in normal mitochondrial dynamics, which results in increased mitochondrial fragmentation (Wang *et al.* 2009). In striatal neurons from moderate-to-severe grade HD patients, both mitochondrial loss and altered mitochondrial morphogenesis have been described, with increased mitochondrial fission and reduced fusion (Kim *et al.* 2010). Indeed, mHtt was recently shown to bind the mitochondrial fission Drp-1 and increase its enzymatic activity (Song *et al.* 2011). Furthermore, overexpression of proteins that stimulate mitochondrial fusion attenuates the toxicity of Htt proteins containing expanded polyQ tracts in both HeLa cells and *C. elegans* (Wang *et al.* 2009).

Efficient mitochondrial trafficking is especially important in neurons with long axons and dendrites, to ensure high metabolic energy requirements for neuronal signaling, plasticity and neurotransmitter release. mHtt impairs axonal transport of mitochondria, decreases mitochondrial function and damages neurons in affected regions of HD patients' brains (Shirendeb *et al.* 2011). In particular, specific N-terminal fragments of mHtt (produced before aggregate formation) were shown to preferentially associate with mitochondria *in vivo*, in an age-dependent way, directly affecting the mitochondrial traffic in an HD-knock-in mouse model (Orr *et al.* 2008). In rat cortical neurons expressing full-length mHtt, an early event in HD pathophysiology is the aberrant mobility and trafficking of mitochondria caused by cytosolic Htt aggregates (Chang *et al.* 2006). Sequestration of mitochondrial proteins along with defective trafficking might lead to failure of ATP synthesis, energy depletion, and ultimately cell death in striatal neurons isolated from transgenic mice expressing mHtt with 72 glutamines (Trushina et al. 2004). Thus, disruption of mitochondrial trafficking in neurodegenerative diseases and abnormal mitochondrial dynamics, due to the perturbation of balance between fission and fusion, may mediate and amplify mitochondrial dysfunction in HD, compromising the supply of energy for normal neuronal function (Bossy-Wetzel *et al.* 2008, for review).

2.3 Changes in energy metabolism

Neurons are largely dependent on ATP to perform their functions and, thus, a decrease in mitochondrial energy metabolism may highly contribute to neurodegeneration. Moreover, mitochondria in striatal neurons, especially in the GABAergic medium-sized spiny neurons (MSNs), seem to be selectively vulnerable to metabolic stress, which may contribute to the selective loss of these neurons in HD (Jin and Johnson 2010, for review). Evidences of altered energy metabolism in HD include a decrease in glucose metabolism, observed in the caudate, putamen and cortex of symptomatic and pre-symptomatic HD patients (Kuhl *et al.* 1982;Kuwert *et al.* 1990). Modified glycolytic energy metabolism, in particular, has been described in HD patients, both in central and in peripheral tissues. This includes elevated levels of lactate in the striatum (Jenkins *et al.* 1993) and in the cortex (Jenkins *et al.* 1993;Koroshetz *et al.* 1997), and increased lactate/pyruvate ratio in the CSF (Koroshetz *et al.* 1997). However, decreased astrocytic glucose metabolism, with preserved oxygen metabolism, was described in the striatum of early symptomatic HD patients (Powers *et al.* 2007b). A significant decrease in phosphocreatine/inorganic phosphate ratio was found in resting muscle (Koroshetz *et al.* 1997) of HD patients, evidencing bioenergetic changes in HD peripheral tissues. Previous studies showed low levels of phosphocreatine/inorganic phosphate ratio in muscle of HD patients, compared to control subjects (Lodi *et al.* 2000), and a delayed recovery of phosphocreatine levels in HD patients in response to exercise (Saft *et al.* 2005). Moreover, reduced ATP production was observed in muscle of both presymptomatic and symptomatic HD patients (Lodi *et al.* 2000). In fact, the onset of energy-related manifestations at the presymptomatic stages of the disease, such as alterations in brain and muscle metabolism and weight loss, suggest that the energy deficit is likely to be an early phenomenon in the cascade of events leading to HD pathogenesis (Mochel and Haller 2011). Conversely, in HD N171-82Q mice model, increased glucose metabolism and ATP levels were found in brain tissue, suggesting that the neuronal damage in HD tissue may be associated with increased energy metabolism at the tissue level, leading to modified levels of various intermediary metabolites (Olah *et al.* 2008). Interestingly, we observed that HD cybrid lines exhibited increased glycolytic ATP levels compared to control cybrids, which were correlated with increased lactate/pyruvate levels (Ferreira *et al.* 2011). In these cybrids, the activity of G6PD, a key enzyme of the pentose phosphate pathway, was decreased (Ferreira *et al.* 2011), suggesting that glucose metabolization occurs primarily through the glycolytic pathway. Furthermore, mitochondrial NADH/NADt ratio was decreased (Ferreira *et al.* 2011), which was further correlated with a large decrease in the activity and protein levels of pyruvate dehydrogenase (PDH) (Ferreira *et al.* 2011). Nevertheless, the activity of alpha-ketoglutarate dehydrogenase (KGDH), another NADH producer in the tricarboxylic acid cycle, was increased, suggesting a compensatory mechanism to counterbalance the decrease in NADH production through the PDH. Decreased PDH activity was also previously observed in the caudate and putamen of HD patients (Sorbi *et al.* 1983), which was correlated with increasing duration of the illness (Butterworth *et al.* 1985). Moreover, PDH expression was shown to decrease with age in the striatum of R6/2 transgenic mice (Perluigi *et al.* 2005). A decrease in mitochondrial alanine and an increase in mitochondrial glutamate levels observed in these cybrids may be interpreted as an attempt to recover ketoglutarate levels and thus mitochondrial NADH (Ferreira *et al.* 2011). Alanine levels were also found to be decreased in the CSF of HD patients, along with decreased pyruvate levels and increased lactate/pyruvate ratio (Koroshetz *et al.* 1997). Our results demonstrated that HD cybrid lines possess inherent bioenergetically dysfunctional mitochondria derived from HD patients' platelets in the presence of a functional nuclear background (Ferreira et al. 2011). Mitochondrial dysfunction at the level of PDH, upstream the oxidative phosphorylation, affected amino acid metabolic fluxes and the cellular bioenergetics through glycolysis stimulation, which assumed a greater importance in promoting ATP production (Ferreira et al. 2011).

2.4 Oxidative stress

Oxidative phosphorylation at the level of mitochondrial ETC is a major source of ROS, such as superoxide anion (the radical formed from the direct reduction of oxygen due to electron leakage at the ETC), hydrogen peroxide and hydroxyl radical (the most reactive and unstable radical). In the absence of effective antioxidants, ROS generated by dysfunctional mitochondria may attack mitochondrial components, promoting intracellular oxidative stress and leading to protein, lipid and DNA oxidation, further contributing to mitochondrial dysfunction.

Oxidative damage was shown to play an important role in the pathogenesis and progression of HD in the R6/2 transgenic mouse model (Perluigi *et al.* 2005) and also in post-mortem samples obtained from the striatum and cortex of human HD brain (Sorolla *et al.* 2010). An increase in DCF fluorescence, indicative of an increase in hydroperoxide levels, was also described in the striatum of R6/1 mice 11-35 weeks (Perez-Severiano *et al.* 2004). In accordance, we demonstrated that, under basal conditions, HD cybrids were endowed with a significant higher production of hydroperoxides when compared to control cybrids (Ferreira et al., 2010). These data differ from a previous study showing no evidence of ROS generation in untreated HD cybrids (Swerdlow *et al.* 1999); however, these authors did not exclude a subtle mitochondrial pathology in these cells. In agreement, we showed that HD cybrids are more vulnerable than control cybrids to produce superoxide upon exposure to 3-NP or staurosporine (STS), whereas increased hydroperoxide production was mainly evoked by STS, suggesting that the presence of higher amounts of hydroperoxides in untreated HD cybrids masks the effect caused by 3-NP-induced mitochondrial inhibition (Ferreira *et al.* 2010).

Several biomarkers of oxidative stress, such as oxidized macromolecules, were found in HD patients and in HD models. Oxidized DNA was found in the caudate of HD patients (Browne *et al.* 1997), whereas oxidized mtDNA was reported in the parietal cortex of late stage (grade 3-4) HD patients (Polidori *et al.* 1999). 8-Hydroxy-deoxyguanosine was also found in peripheral blood of HD patients (Chen *et al.* 2007;Hersch *et al.* 2006). Moreover, oxidized DNA markers were also found in forebrain, striatum (Tabrizi *et al.* 2000;Bogdanov *et al.* 2001), urine, plasma and striatal dialysates of R6/2 mice at 12 and 14 weeks of age (Bogdanov *et al.* 2001). An increase in lipid peroxidation markers was also found in HD human blood (Chen *et al.* 2007;Stoy *et al.* 2005) or brain (Browne *et al.* 1999) and in R6/2 mouse brain (Tabrizi *et al.* 2000;Perez-Severiano *et al.* 2000). Protein oxidation markers, such as carbonyl levels, were also found to be increased in mitochondrial enzymes, resulting in decreased mitochondrial activity in the striatum of Tet/HD94 conditional HD mice (Sorolla *et al.* 2010).

Decreased activities of the antioxidant enzymes Cu–Zn-superoxide dismutase and glutathione peroxidase in erythrocytes (Chen *et al.* 2007), and decreased catalase activity were found in skin fibroblasts from HD patients (del Hoyo *et al.* 2006). A decrease in the antioxidant enzyme Cu/Zn-superoxide dismutase was also observed in R6/1 mice at 35 weeks (Santamaria *et al.* 2001). Moreover, the antioxidant agents lipoic acid and BN-82451 are neuroprotective in HD mice (R6/2 and N171-82Q lines), increasing survival and delaying striatal atrophy in these genetic models of HD (Andreassen *et al.* 2001;Klivenyi *et al.* 2003), further evidencing participation of oxidative damage in the process of neurodegeneration in HD. However, 3-NP *in vivo* exposure induced antioxidant response element (ARE)-dependent gene expression in cultured astrocytes through the transcription factor nuclear factor (erythroid-derived 2)-like 2 (Nrf2), leading to gene expression of antioxidant and detoxification genes (Shih *et al.* 2005).

2.5 Transcriptional deregulation

Nuclear localization of mHtt was shown to play a role in toxicity (Saudou *et al.* 1998), possibly due to interference of the mutant protein with nuclear transcription factors and co-factors (Benn *et al.* 2008;Zhai *et al.* 2005). Moreover, mitochondrial dysfunction in HD has been related to transcriptional deregulation.

Mitochondrial gene expression is regulated in the nucleus by the transcriptional co-activator peroxisome proliferative activated receptor gamma coactivator 1 alpha (PGC-1alpha) (Lin *et al.* 2004;Lin *et al.* 2005), and in the mitochondria, by the nuclear-encoded mitochondrial transcription factor A (Tfam) (Kaufman *et al.* 2007), which also regulate mitochondrial function and biogenesis.

Abnormal PGC-1alpha function was shown to result in significant mitochondrial impairment (Kim *et al.* 2010). The levels of PGC-1alpha and Tfam were found to be reduced in HD (Cui et al. 2006;Chaturvedi et al. 2009). Moreover, both proteins have been reported to be significantly reduced in brain lysates from HD patients, which was correlated with HD progression (Kim et al. 2010). A significant decrease in PGC-1alpha mRNA was found in the caudate nucleus in asymptomatic HD patients, accompanied by reduced expression of genes involved in energy metabolism (Cui *et al.* 2006). Interestingly, decreased expression of PGC-1alpha was observed in MSNs (largely affected in HD), whereas striatal interneurons showed increased mRNA levels for PGC-1alpha (Cui *et al.* 2006) which could, at least partially, explain the different vulnerability of these striatal neuronal populations. PGC-1alpha and Tfam were also reduced in muscle biopsies and myoblast cultures from HD subjects (Chaturvedi *et al.* 2009). Transcriptional repression of PGC-1alpha by mHtt leads not only to mitochondrial dysfunction, but also to neurodegeneration, suggesting a key role for PGC-1alpha in the control of energy metabolism in the early stages of HD pathogenesis (Cui *et al.* 2006). Thermoregulatory and metabolic defects in HD transgenic mice also implicate PGC-1alpha in HD neurodegeneration (Weydt *et al.* 2006), and polymorphisms at the PGC-1alpha gene modify the age at onset in HD (Weydt *et al.* 2009). In accordance, activation of PGC-1alpha/peroxisome proliferator-activated receptor gamma (PPARgamma) seems to protect against neurodegeneration (St-Pierre *et al.* 2006).

PGC-1alpha controls many aspects of oxidative metabolism, including respiration and mitochondrial biogenesis by co-activating and enhancing the expression and activity of several transcription factors, including the nuclear respiratory factors (NRF)-1 and NRF-2 (also known as GA-binding protein, GABP), PPARgamma and the estrogen related receptor alpha (ERRalpha) (Scarpulla 2002;Scarpulla 2011). It was recently shown that PGC-1alpha downstream transcription factors NRF-1 and Tfam are genetic modifiers of HD

(Taherzadeh-Fard *et al.* 2011). PGC-1alpha is indirectly involved in regulating the expression of mtDNA transcription *via* increased expression of Tfam, which is co-activated by NRF-1 (Scarpulla 2002;Kelly and Scarpulla 2004). Moreover, mitochondrial-dependent generation of ROS in HD seems to be due, at least in part, to suppression of PGC-1alpha in the presence of mHtt, as this transcription coactivator is required for the induction of ROS-detoxifying enzymes, namely Mn-superoxide dismutase and glutathione peroxidase (St-Pierre *et al.* 2006), implicating PGC-1alpha as an important protector against oxidative damage in HD. Importantly, activation of PPARgamma was recently shown to rescue mitochondrial dysfunction in HD (Chiang *et al.* 2011).

An important and key event in the signaling cascade that regulates PGC-1alpha expression is related with mitogen- and stress-activated protein kinase 1 (MSK-1) activation (Martin *et al.* 2011). MSK1 induces neuroprotection in HD, involving chromatin remodeling at the PGC-1 alpha promoter (Martin *et al.* 2011).

cAMP response element-binding (CREB) is a major transcription factor for PGC-1alpha (Cui *et al.* 2006). CREB is widely expressed and has a well-established role in neuronal protection (Lee *et al.* 2005). mHtt was shown to interfere with CREB transcriptional processes, through direct interaction with CREB-binding protein (CBP) (Steffan *et al.* 2000) and with TATA box-binding protein (TBP)-associated factor TAF4/TAFII130 (Dunah *et al.* 2002;Shimohata *et al.* 2000), leading to an increase in mHtt-induced cytotoxicity (Steffan *et al.* 2001). TAFII130 is a co-factor for CREB-dependent transcriptional activation that binds to polyQ, strongly suppressing CREB-mediated transcription (Shimohata *et al.* 2000). Reduction in cAMP levels in HD mice and HD patients likely contributes to the significant reduction in CREB activation (Gines *et al.* 2003). Moreover, CBP co-localizes with mHtt (Nucifora, Jr. *et al.* 2001), being found in nuclear inclusions in HD mice (Nucifora, Jr. *et al.* 2001;Steffan *et al.* 2001) and human brain (Nucifora, Jr. *et al.* 2001). In accordance, CRE-response genes such as corticotrophin-releasing hormone, proenkephalin, substance P were found to be reduced in brain tissue in HD patients (Augood *et al.* 1996;De Souza 1995) and R6/2 mice (Luthi-Carter *et al.* 2002).

Our group has previously shown that dysregulation of CREB activation and histone acetylation occurs in 3-NP-treated cortical neurons (Almeida *et al.* 2010), an *in vitro* model of mitochondrial complex II inhibition in HD. The phosphorylation status of CREB is critical for its activity and several protein kinases, such as calcium/calmodulin-dependent kinase II and IV, protein kinase C, PI3K, Akt, MAPK, and Rsk2, have been reported to promote the activation of CREB (Yamamoto *et al.* 1988;Matthews *et al.* 1994;Du and Montminy 1998;Bito *et al.* 1996;Impey *et al.* 1998;Perkinton *et al.* 2002). Phosphorylation on Ser133 leads to CREB activation and promotes the transcription of a large number of genes, through interaction with its nuclear partner CBP (Mayr and Montminy 2001). Results from our laboratory showed that 3-NP treatment of cortical neurons decreased both CREB phosphorylation on Ser133 and CBP levels (Almeida *et al.* 2010), strongly suggesting reduced CREB-dependent gene expression/activation. The decrease in CREB phosphorylation was possibly due to the activation of phosphatases in response to 3-NP exposure. Several studies have shown that calcineurin, whose expression is regulated by 3-NP (Napolitano *et al.* 2004), also regulate the duration of CREB phosphorylation (Bito *et al.* 1996). However, the concentration of 3-NP used in our study did not significantly alter calcineurin (Almeida *et al.* 2004). The decrease in total CBP levels after 3-NP exposure could be explained by an independent mechanism, related with caspase-3 (Almeida *et al.* 2010), but not calpain activation (Almeida *et al.* 2004). CBP has previously been reported to be specifically targeted for cleavage by caspases (and also by calpains) at the onset of neuronal apoptosis (Rouaux *et al.* 2003). A decrease in CBP was correlated with reduced acetylation of histones H3 and H4 and with a reduction in CBP/p300 HAT activity, even while total HAT activity remained unchanged (Rouaux *et al.* 2003). Similarly, we showed that 3-NP did not alter total HAT activity, but significantly decreased overall HDAC activity, likely explaining why we did not observe a reduction in H3 or H4 acetylation (Almeida *et al.* 2010). Instead, we observed an increase in both H3 and H4 acetylation in cortical neurons upon exposure to 3-NP. Because 3-NP induces caspase-3 activation (Almeida *et al.* 2004), we hypothesized that caspase-3 plays a role in inactivating HDACs. On the other hand, inhibition of HDAC may constitute a mechanism of protection of cells exposed to mild metabolic stress. Indeed, neuroprotection induced by HDAC inhibitors in HD striatal cells involves more efficient calcium handling, thus improving the neuronal ability to cope with excitotoxic stimuli (Oliveira *et al.* 2006).

mHtt was previously reported to bind p53 and upregulate its expression and transcriptional activity (Bae *et al.* 2005). It was demonstrated that some of the alterations induced by mHtt in mitochondrial homeostasis and cell death were dependent on p53 (Bae *et al.* 2005). Recently, mHtt expression was correlated with an increase in phosphorylated p53 at Ser15, a decrease in acetylation at Lys382, altered ubiquitination pattern, and oligomerization activity. The lack of a proper p53-mediated signaling cascade or its alteration in the presence of DNA damage may contribute to the slow progression of cellular dysfunction which is a hallmark of HD pathology (Illuzzi *et al.* 2011).

Specific protein-1 (Sp1) is another transcription factor that was found to bind mHtt, resulting in inhibition of Sp1-mediated transcription of genes in post-mortem brain tissue of pre-symptomatic and symptomatic HD patients (Dunah *et al.* 2002), such as NGF receptor (Li *et al.* 2002). Sp1 is a regulatory protein that binds to guanine-cytosine boxes and mediates transcription through its glutamine-rich activation domains which target components of the basal transcriptional complex, such as TAFII130 (Sugars and Rubinsztein 2003,for review). Furthermore, it has also been shown that, despite normal protein levels and nuclear binding activity, the binding of Sp1 to specific promoters of susceptible genes is significantly decreased in transgenic HD mouse brains, striatal HD cells and human HD brains, suggesting that mHtt dissociates Sp1 from target promoters, inhibiting the transcription of specific genes (Chen-Plotkin *et al.* 2006). Sequestration of Sp1 and TAFII130 into nuclear inclusions leads to the inhibition of Sp1-mediated transcription (Dunah *et al.* 2002;Li *et al.* 2002). Moreover, shorter N-terminal Htt fragments, which are more prone to misfold and aggregate, are more competent to bind and inhibit Sp1 (Cornett *et al.* 2006). Interestingly, this effect was reversed *in vitro* by HSP40, a molecular chaperone that reduces mHtt misfolding (Cornett *et al.* 2006).

mHtt may also lose the ability to bind and interact with other transcription factors regulated by wild-type huntingtin (Htt), as is the case of the neuron-restrictive silencer element (NRSE)-binding transcription factors, in which the failure of mHtt to interact with transcriptional factor complex repressor-element-1 transcription factor (REST)/neuron-restrictive silencer factor (NRSF) in the cytoplasm leads to its nuclear accumulation. There, it binds to NRSE sequences and promotes histone deacetylation, leading to the remodeling of the chromatin into a closed structure, resulting in the suppression of NRSE-containing genes, including the bdnf gene (Zuccato *et al.* 2003). In this case, the loss of the normal Htt function may have profound effects, leading to decreased levels of BDNF, an important survival factor for striatal neurons (section 2.2). Indeed, BDNF-knockout models were shown to largely recapitulate the expression profile of human HD (Strand *et al.* 2007), suggesting that striatal MSNs suffer similar insults in HD and BDNF-deprived environments.

2.6 Regulation of mitochondrial-driven apoptosis

Neurodegeneration in HD has been associated with increased cell death by apoptosis, particularly by the intrinsic pathway, highly regulated by mitochondria. Previous studies demonstrated the presence of caspases cleavage sites in Htt, a mechanism that may also contribute to apoptotic death by generating truncated toxic fragments of this protein (Wellington *et al.* 1998), although the CAG length does not seem to modulate the susceptibility for cleavage. mHtt is a substrate for several caspases and calpains (Kim *et al.* 2001) and the polyglutamine fragments of Htt may present enhanced toxicity, promoting caspases activation by interfering with mitochondrial function, thus amplifying the generation of toxic truncated mHtt (Graham *et al.* 2010). Moreover, sequestration of pro-caspases in the aggregates is thought to promote their activation, triggering an intracellular cascade of proteolytic events (Gil and Rego 2008, for review). Interestingly, wild-type Htt was found to have antiapoptotic properties against a variety of apoptotic stimuli, including serum withdrawal, death receptors, and proapoptotic Bcl-2 homologs (Rigamonti *et al.* 2000), namely through inhibition of cytochrome c-dependent procaspase-9 processing and activity (Rigamonti *et al.* 2001). Furthermore, calpain (Gafni and Ellerby 2002), caspase-1 (Ona *et al.* 1999) and caspase-8 (Sanchez *et al.* 1999) activities are increased in HD human brains, suggesting that an apoptotic mechanism is responsible for HD neuronal loss (Gil and Rego 2008, for review). Moreover, cultured blood cells from patients homozygous for CAG repeat mutations and heterozygous with high size mutations causing juvenile onset presented significantly increased caspases -2, -3, -6, -8 and -9 activities, decreased cell viability and pronounced mitochondria morphological abnormalities, compared with cells from HD patients carrying low mutation size and controls (Squitieri *et al.* 2011).

Cell death by necrosis and apoptosis, along with energy deficiency, were previously described in striatal, cortical and hippocampal cells exposed to 3-NP (Behrens *et al.* 1995;Pang and Geddes 1997;Almeida *et al.* 2004;Almeida *et al.* 2006;Brouillet *et al.* 2005), and both processes of cell damage have been proven to involve mitochondria (Kroemer and Reed 2000). Concordantly with a higher role of intrinsic apoptosis in HD, Ferrer and collaborators (2000) found a reduction in Fas and FasL expression levels in the caudate and putamen of HD patients. Mitochondria has been largely recognized to play a critical role in cell death by releasing apoptogenic factors, such as cytochrome c and apoptosis-inducing factor (AIF), from the intermembrane space into the cytoplasm.

As described before, by directly interacting with the mitochondria (Panov *et al.* 2002), mHtt may cause mitochondrial abnormalities in HD, leading to cytochrome c release (Panov *et al.* 2002), and a decrease in mitochondrial membrane potential (Sawa *et al.* 1999). Release of cytochrome *c* along with the activation of caspases -1, -8, and -9 have been demonstrated in HD (Ona *et al.* 1999;Sanchez *et al.* 1999;Kiechle *et al.* 2002), and increased Bcl-2 and Bax were also reported in HD patients' brain, especially in the most severely affected (Vis *et al.* 2005).

Overexpression of mHtt, but not the normal protein, increases oxidative stress-induced mitochondrial fragmentation in HeLa cells, which correlates with increased caspase-3 activation and cell death (Wang *et al.* 2009). Results from our laboratory highly suggested that 3-NP induces both caspase-dependent and -independent cell death (Almeida *et al.* 2006). Our group also showed that exposure of HD cybrid cell lines to 3-NP or STS caused DNA fragmentation and moderate caspase-3 activation, evidencing an increased susceptibility of HD cybrids to apoptosis (Ferreira *et al.* 2010). In contrast, 3-NP-treated control cybrids died predominantly by necrosis, not involving caspase-3 activation (Ferreira *et al.* 2010), suggesting that HD mitochondria are endowed with pro-apoptotic machinery and thus more susceptible to this type of cell death. Moreover, preserved ATP in HD cybrids compared to control cybrids (Ferreira et al. 2011) may facilitate apoptotic cell death. Mitochondrial-dependent apoptosis in HD cybrids subjected to 3-NP was correlated with increased release of mitochondrial cytochrome c, AIF, Bax translocation, caspase-3 activation and ROS formation (Ferreira *et al.* 2010). Increased mitochondrial Bim and Bak levels, and a slight release of cytochrome c in untreated HD cybrids further explained their moderate susceptibility to mitochondrial-dependent apoptosis under basal conditions (Ferreira et al. 2010). These data appear to be consistent with possible subtle effects of mHtt in the mitochondria of HD cybrids. 3-NP has been also shown to collapse mitochondrial membrane potential and to downregulate striatal Bcl-2 levels (Zhang *et al.* 2009b), promoting cytochrome c release from mitochondria, transient caspase-9 processing, activation of calpains and subsequent striatal apoptosis (Bizat *et al.* 2003;Zhang *et al.* 2009b). 3-NP-induced decrement in Bcl-2 may also play a role in mitochondrial-dependent autophagy activation (through the release of Beclin 1 from hVps34 complex), which was also involved in striatal neuronal apoptosis (Zhang *et al.* 2009a).

Our group has also reported that 3-NP causes mitochondrial-dependent apoptotic neuronal death through the release of cytochrome c and consequent activation of caspases, or the release of AIF in cortical neurons, depending on the concentration of 3-NP (Almeida *et al.* 2004;Almeida *et al.* 2006;Almeida *et al.* 2009). Enhanced mitochondrial-dependent apoptosis was also observed in 3-NP-treated cortical neurons as a result of decreased Bim turnover (Almeida *et al.* 2004). mHtt fragments were previously shown to directly induce the opening of the mitochondrial permeability transition pore (PTP) in isolated mouse liver mitochondria, with the consequent release of cytochrome c (Choo *et al.* 2004), which evokes caspase cascade activation. Choo and collaborators (2004) also described that mitochondria from liver of knock-in mouse model of HD and from homozygous ST*Hdh*^{Q111} cells were more sensitive to calcium-induced cytochrome c release, swelling at lower calcium loads. An increased striatal mitochondrial susceptibility to the induction of permeability transition (Brustovetsky *et al.* 2003) may be responsible to the striatal selectivity for energy deficit associated with mHtt. An age- and polyQ-dependent decrease in the amount of calcium necessary to induce permeability transition in striatal mitochondria was observed in severe (R6/2 mice) and in mild (*HdhQ92* knock-in mice) HD mouse models (Brustovetsky *et al.* 2003). Moreover, increased mitochondrial calcium loading capacity, previously shown in isolated mitochondria from 12-13 week-old R6/2 and 12 month-old YAC mice brain (Oliveira *et al.* 2007) could constitute a compensatory mechanism, to extend neuronal function and survival or, alternatively, it could simply reflect an artifact resulting from mitochondria isolation, as it was not observed in neuronal *in situ* experiments following exposure to excitotoxic stimuli (Oliveira *et al.* 2007).

Myoblasts obtained from presymptomatic and symptomatic HD subjects also showed mitochondrial depolarization, cytochrome c release and increased activities of caspases -3, -8 and -9 (Ciammola *et al.* 2006). In addition, peripheral blood cells, in particularly B lymphocytes from HD patients, may reflect changes observed in HD brain. Our group previously found increased Bax expression in B and T lymphocytes, and monocytes from HD patients, with no alterations in Bcl-2 expression levels, and decreased mitochondrial membrane potential in B lymphocytes (Almeida *et al.* 2008), further suggesting that an adverse effect of mHtt is not limited to neurons. Moreover, mitochondria from lymphoblasts of HD patients have been shown to present increased susceptibility to apoptotic stimuli due to an abnormal mitochondrial transmembrane potential (Sawa et al. 1999). Lymphoblasts derived from HD patients also showed increased stress-induced apoptotic cell death associated with caspase-3 activation, abnormal calcium homeostasis and mitochondrial dysfunction (Panov *et al.* 2002;Sawa *et al.* 1999).

3. Protective effects involving modulation of phosphorylation pathways – The case of FK506 and the neurotrophins BDNF and NGF

Even though HD has a single genetic cause, the multiplicity of pathogenic mechanisms involved suggests that several different targets must be taken into account in order to slow down HD progression. Despite important advances in elucidating the molecular pathways involved in HD neurodegeneration, there is currently no therapy that delays the onset of the disease. In this respect, stimulation of phosphorylation pathways by neurotrophins or calcineurin inhibitors (such as FK506) may be a promising strategy.

3.1 FK506 – An inhibitor of calcineurin

It is well accepted that mHtt is associated with calcium handling abnormalities (Quintanilla and Johnson 2009, for review). Calcineurin can be activated by abnormal calcium levels occurring in HD. Classically, calcineurin (or protein phosphatase 3, formerly known as protein phosphatase 2B) can promote apoptosis through dephosphorylation of Bad at Ser112 and Ser136 (Wang *et al.* 1999), a proapoptotic member of the Bcl-2 family. Dephosphorylated Bad translocates from the cytosol to the mitochondria, where it inhibits antiapoptotic activity of Bcl-2 and Bcl-xL, ultimately leading to cell death. Calcineurin couples intracellular calcium to the dephosphorylation of other selected substrates, which include transcription factors [nuclear factor of activated T-cells (NFAT)], ion channels (inositol-1,4,5 triphosphate receptor), proteins involved in vesicular trafficking (amphyphysin, dynamin), scaffold proteins (AKAP79), and phosphatase inhibitors (DARPP-32 inhibitor-1) (Aramburu *et al.* 2000).

Calcineurin was recently shown to be involved in the dephosphorylation of Drp1, thus increasing Drp1 association with mitochondria and promoting fission, cristae disruption, cytochrome c release and apoptosis (Costa *et al.* 2010;Cereghetti *et al.* 2010). Concordantly, the calcineurin inhibitor PPD1 blocked Drp1 translocation to mitochondria and fragmentation of the organelle, delaying intrinsic apoptosis by preventing fragmentation and release of cytochrome *c*, suggesting an important function of calcineurin in mitochondrial fragmentation and in the amplification of cell death (Cereghetti *et al.* 2010).

FK506, also known as tacrolimus, is a selective inhibitor of calcineurin (Griffith *et al.* 1995) that has shown to exert neuroprotective effects in several cellular and animal models of HD. Kumar and Kumar (2009) showed that systemic treatment with FK506 significantly improved cognitive function in a 3-NP rodent model. In the 3-NP neuronal model, we have previously shown that FK506 precludes cytochrome c release, activation of caspase-3 and DNA fragmentation in cultured cortical neurons (Almeida *et al.* 2004). FK506 neuroprotection against 3-NP-induced apoptosis was associated with the redistribution of Bcl-2 and Bax in the mitochondrial membrane of cortical neurons (Almeida *et al.* 2004). Moreover, FK506 significantly attenuated oxidative stress as evidenced by restoring glutathione levels and acetylcholinesterase activity in 3-NP treated animals (Kumar and Kumar 2009), highlighting the therapeutic potential of this compound. In a recent study from our laboratory FK506 has shown neuroprotective effects against apoptosis and necrosis under mild cell death stimulus, in the presence of full-length mHtt, in 3-NP-treated primary striatal neurons and immortalized striatal cells derived from HD knock-in mice (ST*Hdh*$^{Q111/Q111}$ mutant cells) (Rosenstock *et al.* 2011).

In the context of mHtt expression, intraperitoneal injection of calcineurin inhibitors was shown to accelerate the neurological phenotype in R6/2 mice (Hernandez-Espinosa and Morton 2006), which are resistant to excitotoxicity (Hansson *et al.* 1999). Interestingly, reduced calcineurin protein levels and activity were observed in this HD animal model (Xifro *et al.* 2009). In contrast, calcineurin is involved in cell death induced by activation of *N*-methyl-D-aspartate receptors (NMDARs) in knock-in striatal cells expressing full-length mHtt (Xifro *et al.* 2008). Moreover, FK506 and the genetic inactivation of calcineurin protected against mHtt toxicity through increased phosphorylation of Htt (Pardo *et al.* 2006) and further ameliorated the defect in BDNF axonal transport (Pineda *et al.* 2009).

3.2 BDNF and NGF – Activators of survival pathways

Trophic support to neurons largely influences neuronal survival and function. BDNF, a pro-survival factor that is produced by cortical neurons, is necessary for the survival of striatal neurons in the brain. This is particularly relevant in HD since its transcription (Zuccato *et al.* 2001) and axonal transport (Gauthier *et al.* 2004) are decreased by the presence of mHtt, affecting the survival of both striatal and cortical neurons. Members of the neurotrophin family, namely BDNF and NGF, have been suggested as therapeutic candidates to treat neurodegenerative disorders because they promote neuronal survival in different lesion models (Connor and Dragunow 1998). Indeed, implantation of NGF-secreting fibroblasts was found to reduce the size of adjacent striatal 3-NP lesions (Frim *et al.* 1993).

Wild-type Htt was demonstrated to promote the expression of BDNF by interacting with the REST/NRSF in the cytoplasm, preventing this complex from translocating into the nucleus and binding to NRSE present in the promoter of the bdnf gene (Zuccato *et al.* 2003). Wild-type Htt also promoted the vesicular transport of BDNF along the microtubules through a mechanism involving Htt -associated protein 1 (HAP1) and the p150 subunit of dynactin (Gauthier *et al.* 2004). Thus, wild-type Htt controls neurotrophic support and survival of striatal neurons by promoting BDNF transcription and vesicular transport along microtubules (Gauthier *et al.* 2004).

In contrast, mHtt decreases transcription of BDNF, which results in decreased production of cortical BDNF in HD (Zuccato *et al.* 2001), leading to insufficient neurotrophic support for striatal neurons, which then die. Accordingly, a reduction in cortical BDNF mRNA levels was shown to correlate with the progression of the disease in a mouse model of HD (Zuccato *et al.* 2005). In addition, BDNF-knockout models were shown to largely recapitulate the expression profiling of human HD (Strand *et al.* 2007), suggesting that striatal MSNs suffer similar insults in HD and BDNF-deprived environments. Moreover, mHtt appears to be responsible for altering the wild-type Htt /HAP1/p150 complex, causing an impaired association between motor proteins and microtubules, and attenuating BDNF transport, which results in loss of neurotrophic support (Gauthier *et al.* 2004). Thus, restoring wild-type Htt activity and increasing BDNF production are promising therapeutic approaches for treating HD (Zuccato *et al.* 2001).

BDNF was previously shown to prevent the death of different populations of striatal projection neurons in a quinolinic acid model of HD (Perez-Navarro *et al.* 2000;Kells *et al.* 2004) and in striatal neurons exposed to 3-NP (Ryu *et al.* 2004). The effects of BDNF are mainly mediated by TrkB receptor-induced activation of key signaling pathways, including phosphoinositide phospholipase C (PLC-γ), rat sarcoma GTPase (Ras)/MEK/ Ras–mitogen-activated protein kinase (MAPK) and PI3K/Akt pathways (Huang and Reichardt 2003), which have been shown to regulate apoptotic cell death by increasing the transcription of neuroprotective proteins such as Bcl-2 (Pugazhenthi *et al.* 2000) and/or by posttranslational modifications of proteins such as Bad and Bim (Scheid et al. 1999;Luciano et al. 2003). Bim phosphorylation by MAPK promotes its subsequent ubiquitination and degradation (Ley *et al.* 2003), whereas serine phosphorylation of Bad is associated with protein 14-3-3 binding and inhibition of Bad-induced cell death (Masters *et al.* 2001). Data from our laboratory support an important role for BDNF in protecting cortical neurons against apoptotic cell death caused by 3-NP through the activation of PI3K and MEK1/2 intracellular signaling pathways and the regulation of Bim turnover (Almeida et al. 2009). Moreover, signaling of BDNF and NGF culminates in the transcription of neuroprotective proteins through the activation of critical transcription factors such as CREB and nuclear factor-kB (NFkB) (Huang and Reichardt 2003). As described in section 1.5, when activated by phosphorylation, CREB binds to its co-activator CBP and the complex is competent to initiate gene transcription (Mayr and Montminy 2001). Similarly, phosphorylation of IkB releases the p65:p50 NFkB heterodimers, which then translocate to the nucleus to initiate transcription. Pro-survival proteins whose expression is dependent on these transcription factors include proteins such as Bcl-2, Mn-superoxide dismutase and BDNF (Saha et al. 2006). A recent study from our laboratory also suggested that BDNF and NGF induce positive changes in the levels and activities of CREB and NFkB, and both neurotrophins counteracted 3-NP-induced chromatin-bound histone acetylation modifications. The latter finding was correlated with BDNF-induced hyperphosphorylation of HDAC2, explaining the neuroprotective role of this neurotrophin in the context of mitochondrial dysfunction (Almeida *et al.* 2010).

4. Conclusions

In summary, biochemical studies support the view that mitochondrial dysfunction, including impaired oxidative phosphorylation, tricarboxylic acid cycle dysfunction, and oxidative stress are important determinants of altered energy metabolism in HD. Bioenergetic changes in HD may be related with impaired intracellular transport and transcriptional deregulation in the disease (Mochel and Haller 2011). Impaired bioenergetics in HD likely represents downstream effects of both a mHtt toxic gain-of-function and a loss-of-function of the wild-type protein. Thus, therapeutic strategies designed to improve energy metabolism and survival pathways dependent on kinase signaling in the HD brain will possibly impact the course of the disease, delaying its onset and the rate of progression. BDNF support (which can be rescued by wild-type Htt) and FK506 may have important therapeutic effects as enhancers of phosphorylation pathways, preventing mitochondrial dysfunction caused by mHtt and mitochondrial-dependent apoptosis.

5. Acknowledgements

T.C.O. holds a postdoctoral fellowship from 'Fundação para a Ciência e a Tecnologia' (FCT), Portugal (SFRH/BPD/34711/2007). A.C.R. acknowledges financial support from FCT Grant PTDC/SAU-FCF/108056/2008.

6. References

Almeida S., Brett A. C., Gois I. N., Oliveira C. R. and Rego A. C. (2006) Caspase-dependent and -independent cell death induced by 3-nitropropionic acid in rat cortical neurons. *J. Cell Biochem.* 98, 93-101. ISSN: 0730-2312 (Print); ISSN: 0730-2312 (Linking)

Almeida S., Cunha-Oliveira T., Laco M., Oliveira C. R. and Rego A. C. (2010) Dysregulation of CREB activation and histone acetylation in 3-nitropropionic acid-treated cortical neurons: prevention by BDNF and NGF. *Neurotox. Res.* 17, 399-405. ISSN: 1476-3524 (Electronic); ISSN: 1029-8428 (Linking)

Almeida S., Domingues A., Rodrigues L., Oliveira C. R. and Rego A. C. (2004) FK506 prevents mitochondrial-dependent apoptotic cell death induced by 3-nitropropionic acid in rat primary cortical cultures. *Neurobiol. Dis.* 17, 435-444. ISSN: 0969-9961 (Print); ISSN: 0969-9961 (Linking)

Almeida S., Laco M., Cunha-Oliveira T., Oliveira C. R. and Rego A. C. (2009) BDNF regulates BIM expression levels in 3-nitropropionic acid-treated cortical neurons. *Neurobiol. Dis.* 35, 448-456. ISSN:1095-953X (Electronic); ISSN: 0969-9961 (Linking)

Almeida S., Sarmento-Ribeiro A. B., Januario C., Rego A. C. and Oliveira C. R. (2008) Evidence of apoptosis and mitochondrial abnormalities in peripheral blood cells of Huntington's disease patients. *Biochem. Biophys. Res. Commun.* 374, 599-603. ISSN: 1090-2104 (Electronic); ISSN: 0006-291X (Linking)

Andreassen O. A., Ferrante R. J., Dedeoglu A. and Beal M. F. (2001) Lipoic acid improves survival in transgenic mouse models of Huntington's disease. *Neuroreport* 12, 3371-3373. ISSN: 0959-4965 (Print); ISSN: 0959-4965 (Linking)

Aramburu J., Rao A. and Klee C. B. (2000) Calcineurin: from structure to function. *Curr. Top. Cell Regul.* 36, 237-295. ISSN: 0070-2137 (Print); ISSN: 0070-2137 (Linking)

Arenas J., Campos Y., Ribacoba R., Martin M. A., Rubio J. C., Ablanedo P. and Cabello A. (1998) Complex I defect in muscle from patients with Huntington's disease. *Ann. Neurol.* 43, 397-400. ISSN: 0364-5134 (Print); ISSN: 0364-5134 (Linking)

Augood S. J., Faull R. L., Love D. R. and Emson P. C. (1996) Reduction in enkephalin and substance P messenger RNA in the striatum of early grade Huntington's disease: a detailed cellular in situ hybridization study. *Neuroscience* 72, 1023-1036. ISSN: 0306-4522 (Print); ISSN: 0306-4522 (Linking)

Bae B. I., Xu H., Igarashi S., Fujimuro M., Agrawal N., Taya Y., Hayward S. D., Moran T. H., Montell C., Ross C. A., Snyder S. H. and Sawa A. (2005) p53 mediates cellular dysfunction and behavioral abnormalities in Huntington's disease. *Neuron* 47, 29-41. ISSN: 0896-6273 (Print); ISSN: 0896-6273 (Linking)

Beal M. F., Brouillet E., Jenkins B., Henshaw R., Rosen B. and Hyman B. T. (1993) Age-dependent striatal excitotoxic lesions produced by the endogenous mitochondrial inhibitor malonate. *J. Neurochem.* 61, 1147-1150. ISSN: 0022-3042 (Print); ISSN: 0022-3042 (Linking)

Behrens M. I., Koh J., Canzoniero L. M., Sensi S. L., Csernansky C. A. and Choi D. W. (1995) 3-Nitropropionic acid induces apoptosis in cultured striatal and cortical neurons. *Neuroreport* 6, 545-548. ISSN: 0959-4965 (Print); ISSN: 0959-4965 (Linking)

Benchoua A., Trioulier Y., Zala D., Gaillard M. C., Lefort N., Dufour N., Saudou F., Elalouf J. M., Hirsch E., Hantraye P., Deglon N. and Brouillet E. (2006) Involvement of mitochondrial complex II defects in neuronal death produced by N-terminus fragment of mutated huntingtin. *Mol. Biol. Cell* 17, 1652-1663. ISSN: 1059-1524 (Print); ISSN: 1059-1524 (Linking)

Benn C. L., Sun T., Sadri-Vakili G., McFarland K. N., DiRocco D. P., Yohrling G. J., Clark T. W., Bouzou B. and Cha J. H. (2008) Huntingtin modulates transcription, occupies gene promoters in vivo, and binds directly to DNA in a polyglutamine-dependent manner. *J. Neurosci.* 28, 10720-10733. ISSN: 1529-2401 (Electronic); ISSN: 0270-6474 (Linking)

Bito H., Deisseroth K. and Tsien R. W. (1996) CREB phosphorylation and dephosphorylation: a Ca(2+)- and stimulus duration-dependent switch for hippocampal gene expression. *Cell* 87, 1203-1214. ISSN: 0092-8674 (Print); ISSN: 0092-8674 (Linking)

Bizat N., Hermel J. M., Boyer F., Jacquard C., Creminon C., Ouary S., Escartin C., Hantraye P., Kajewski S. and Brouillet E. (2003) Calpain is a major cell death effector in selective striatal degeneration induced in vivo by 3-nitropropionate: implications for Huntington's disease. *J. Neurosci.* 23, 5020-5030. ISSN: 1529-2401 (Electronic); ISSN: 0270-6474 (Linking)

Bogdanov M. B., Andreassen O. A., Dedeoglu A., Ferrante R. J. and Beal M. F. (2001) Increased oxidative damage to DNA in a transgenic mouse model of Huntington's disease. *J. Neurochem.* 79, 1246-1249. ISSN: 0022-3042 (Print); ISSN: 0022-3042 (Linking)

Bossy-Wetzel E., Petrilli A. and Knott A. B. (2008) Mutant huntingtin and mitochondrial dysfunction. *Trends Neurosci.* 31, 609-616. ISSN: 0166-2236 (Print); ISSN: 0166-2236 (Linking)

Brouillet E., Conde F., Beal M. F. and Hantraye P. (1999) Replicating Huntington's disease phenotype in experimental animals. *Prog. Neurobiol.* 59, 427-468. ISSN: 0301-0082 (Print); ISSN: 0301-0082 (Linking)

Brouillet E., Jacquard C., Bizat N. and Blum D. (2005) 3-Nitropropionic acid: a mitochondrial toxin to uncover physiopathological mechanisms underlying striatal degeneration in Huntington's disease. *J. Neurochem.* 95, 1521-1540. ISSN: 0022-3042 (Print); ISSN: 0022-3042 (Linking)

Brouillet E., Jenkins B. G., Hyman B. T., Ferrante R. J., Kowall N. W., Srivastava R., Roy D. S., Rosen B. R. and Beal M. F. (1993) Age-dependent vulnerability of the striatum to the mitochondrial toxin 3-nitropropionic acid. *J. Neurochem.* 60, 356-359. ISSN: 0022-3042 (Print); ISSN: 0022-3042 (Linking)

Browne S. E., Bowling A. C., MacGarvey U., Baik M. J., Berger S. C., Muqit M. M., Bird E. D. and Beal M. F. (1997) Oxidative damage and metabolic dysfunction in Huntington's disease: selective vulnerability of the basal ganglia. *Ann. Neurol.* 41, 646-653. ISSN: 0364-5134 (Print); ISSN: 0364-5134 (Linking)

Browne S. E., Ferrante R. J. and Beal M. F. (1999) Oxidative stress in Huntington's disease. *Brain Pathol.* 9, 147-163. ISSN: 1015-6305 (Print): ISSN: 1015-6305 (Linking)

Brustovetsky N., Brustovetsky T., Purl K. J., Capano M., Crompton M. and Dubinsky J. M. (2003) Increased susceptibility of striatal mitochondria to calcium-induced permeability transition. *J. Neurosci.* 23, 4858-4867. ISSN: 1529-2401 (Electronic); ISSN: 0270-6474 (Linking)

Butterworth J., Yates C. M. and Reynolds G. P. (1985) Distribution of phosphate-activated glutaminase, succinic dehydrogenase, pyruvate dehydrogenase and gamma-glutamyl transpeptidase in post-mortem brain from Huntington's disease and agonal cases. *J. Neurol. Sci.* 67, 161-171. ISSN: 0022-510X (Print); ISSN: 0022-510X (Linking)

Cereghetti G. M., Costa V. and Scorrano L. (2010) Inhibition of Drp1-dependent mitochondrial fragmentation and apoptosis by a polypeptide antagonist of calcineurin. *Cell Death. Differ.* 17, 1785-1794. ISSN: 1476-5403 (Electronic); ISSN: 1350-9047 (Linking)

Chang D. T., Rintoul G. L., Pandipati S. and Reynolds I. J. (2006) Mutant huntingtin aggregates impair mitochondrial movement and trafficking in cortical neurons. *Neurobiol. Dis.* 22, 388-400. ISSN: 0969-9961 (Print); ISSN: 0969-9961 (Linking)

Chaturvedi R. K., Adhihetty P., Shukla S., Hennessy T., Calingasan N., Yang L., Starkov A., Kiaei M., Cannella M., Sassone J., Ciammola A., Squitieri F. and Beal M. F. (2009) Impaired PGC-1alpha function in muscle in Huntington's disease. *Hum. Mol. Genet.* 18, 3048-3065. ISSN: 1460-2083 (Electronic); ISSN: 0964-6906 (Linking)

Chen C. M., Wu Y. R., Cheng M. L., Liu J. L., Lee Y. M., Lee P. W., Soong B. W. and Chiu D. T. (2007) Increased oxidative damage and mitochondrial abnormalities in the peripheral blood of Huntington's disease patients. *Biochem. Biophys. Res. Commun.* 359, 335-340. ISSN: 0006-291X (Print); ISSN: 0006-291X (Linking)

Chen H. and Chan D. C. (2009) Mitochondrial dynamics--fusion, fission, movement, and mitophagy--in neurodegenerative diseases. *Hum. Mol. Genet.* 18, R169-R176. ISSN: 1460-2083 (Electronic); ISSN: 0964-6906 (Linking)

Chen-Plotkin A. S., Sadri-Vakili G., Yohrling G. J., Braveman M. W., Benn C. L., Glajch K. E., DiRocco D. P., Farrell L. A., Krainc D., Gines S., MacDonald M. E. and Cha J. H. (2006) Decreased association of the transcription factor Sp1 with genes downregulated in Huntington's disease. *Neurobiol. Dis.* 22, 233-241. ISSN: 0969-9961 (Print); ISSN: 0969-9961 (Linking)

Chiang M. C., Chern Y. and Huang R. N. (2011) PPARgamma rescue of the mitochondrial dysfunction in Huntington's disease. *Neurobiol. Dis.* ISSN: 1095-953X (Electronic); ISSN: 0969-9961 (Linking)

Choo Y. S., Johnson G. V., MacDonald M., Detloff P. J. and Lesort M. (2004) Mutant huntingtin directly increases susceptibility of mitochondria to the calcium-induced permeability transition and cytochrome c release. *Hum. Mol. Genet.* 13, 1407-1420. ISSN: 0964-6906 (Print); ISSN: 0964-6906 (Linking)

Ciammola A., Sassone J., Alberti L., Meola G., Mancinelli E., Russo M. A., Squitieri F. and Silani V. (2006) Increased apoptosis, Huntingtin inclusions and altered differentiation in muscle cell cultures from Huntington's disease subjects. *Cell Death. Differ.* 13, 2068-2078. ISSN: 1350-9047 (Print); ISSN: 1350-9047 (Linking)

Connor B. and Dragunow M. (1998) The role of neuronal growth factors in neurodegenerative disorders of the human brain. *Brain Res. Brain Res. Rev.* 27, 1-39. ISSN: 0165-0173

Cornett J., Smith L., Friedman M., Shin J. Y., Li X. J. and Li S. H. (2006) Context-dependent dysregulation of transcription by mutant huntingtin. *J. Biol. Chem.* 281, 36198-36204. ISSN: 0021-9258 (Print); ISSN: 0021-9258 (Linking)

Costa V., Giacomello M., Hudec R., Lopreiato R., Ermak G., Lim D., Malorni W., Davies K. J., Carafoli E. and Scorrano L. (2010) Mitochondrial fission and cristae disruption increase the response of cell models of Huntington's disease to apoptotic stimuli. *EMBO Mol. Med.* 2, 490-503. ISSN: 1757-4684 (Electronic); ISSN: 1757-4676 (Linking)

Cui L., Jeong H., Borovecki F., Parkhurst C. N., Tanese N. and Krainc D. (2006) Transcriptional repression of PGC-1alpha by mutant huntingtin leads to mitochondrial dysfunction and neurodegeneration. *Cell* 127, 59-69. ISSN: 0092-8674 (Print); ISSN: 0092-8674 (Linking)

Damiano M., Galvan L., Deglon N. and Brouillet E. (2010) Mitochondria in Huntington's disease. *Biochim. Biophys. Acta* 1802, 52-61. ISSN: 0006-3002 (Print); ISSN: 0006-3002 (Linking)

De Souza E. B. (1995) Corticotropin-releasing factor receptors: physiology, pharmacology, biochemistry and role in central nervous system and immune disorders. *Psychoneuroendocrinology* 20, 789-819. ISSN: 0306-4530 (Print); ISSN: 0306-4530 (Linking)

del Hoyo P., Garcia-Redondo A., de B. F., Molina J. A., Sayed Y., Alonso-Navarro H., Caballero L., Arenas J. and Jimenez-Jimenez F. J. (2006) Oxidative stress in skin fibroblasts cultures of patients with Huntington's disease. *Neurochem. Res.* 31, 1103-1109. ISSN: 0364-3190 (Print); ISSN: 0364-3190 (Linking)

Du K. and Montminy M. (1998) CREB is a regulatory target for the protein kinase Akt/PKB. *J. Biol. Chem.* 273, 32377-32379. ISSN: 0021-9258 (Print); ISSN: 0021-9258 (Linking)

Dunah A. W., Jeong H., Griffin A., Kim Y. M., Standaert D. G., Hersch S. M., Mouradian M. M., Young A. B., Tanese N. and Krainc D. (2002) Sp1 and TAFII130 transcriptional activity disrupted in early Huntington's disease. *Science* 296, 2238-2243. ISSN: 1095-9203 (Electronic); ISSN: 0036-8075 (Linking)

Ferreira I. L., Cunha-Oliveira T., Nascimento M. V., Ribeiro M., Proenca M. T., Januario C., Oliveira C. R. and Rego A. C. (2011) Bioenergetic dysfunction in Huntington's disease human cybrids. *Exp. Neurol.* 231, 127-134. ISSN: 1090-2430 (Electronic); ISSN: 0014-4886 (Linking)

Ferreira I. L., Nascimento M. V., Ribeiro M., Almeida S., Cardoso S. M., Grazina M., Pratas J., Santos M. J., Januario C., Oliveira C. R. and Rego A. C. (2010) Mitochondrial-dependent apoptosis in Huntington's disease human cybrids. *Exp. Neurol.* 222, 243-255. ISSN: 1090-2430 (Electronic); ISSN: 0014-4886 (Linking)

Ferrer I., Blanco R., Cutillas B. and Ambrosio S. (2000) Fas and Fas-L expression in Huntington's disease and Parkinson's disease. *Neuropathol. Appl. Neurobiol.* 26, 424-433. ISSN: 0305-1846 (Print); ISSN: 0305-1846 (Linking)

Frim D. M., Simpson J., Uhler T. A., Short M. P., Bossi S. R., Breakefield X. O. and Isacson O. (1993) Striatal degeneration induced by mitochondrial blockade is prevented by biologically delivered NGF. *J. Neurosci. Res.* 35, 452-458. ISSN: 0360-4012 (Print); ISSN: 0360-4012 (Linking)

Gafni J. and Ellerby L. M. (2002) Calpain activation in Huntington's disease. *J. Neurosci.* 22, 4842-4849. ISSN: 1529-2401 (Electronic); ISSN: 0270-6474 (Linking)

Gauthier L. R., Charrin B. C., Borrell-Pages M., Dompierre J. P., Rangone H., Cordelieres F. P., De M. J., MacDonald M. E., Lessmann V., Humbert S. and Saudou F. (2004) Huntingtin controls neurotrophic support and survival of neurons by enhancing BDNF vesicular transport along microtubules. *Cell* 118, 127-138. ISSN: 0092-8674 (Print); ISSN: 0092-8674 (Linking)

Gil J. M. and Rego A. C. (2008) Mechanisms of neurodegeneration in Huntington's disease. *Eur. J. Neurosci.* 27, 2803-2820. ISSN: 1460-9568 (Electronic); ISSN: 0953-816X (Linking)

Gines S., Seong I. S., Fossale E., Ivanova E., Trettel F., Gusella J. F., Wheeler V. C., Persichetti F. and MacDonald M. E. (2003) Specific progressive cAMP reduction implicates energy deficit in presymptomatic Huntington's disease knock-in mice. *Hum. Mol. Genet.* 12, 497-508. ISSN: 0964-6906 (Print); ISSN: 0964-6906 (Linking)

Goebel H. H., Heipertz R., Scholz W., Iqbal K. and Tellez-Nagel I. (1978) Juvenile Huntington chorea: clinical, ultrastructural, and biochemical studies. *Neurology* 28, 23-31. ISSN: 0028-3878 (Print); ISSN: 0028-3878 (Linking)

Graham R. K., Deng Y., Carroll J., Vaid K., Cowan C., Pouladi M. A., Metzler M., Bissada N., Wang L., Faull R. L., Gray M., Yang X. W., Raymond L. A. and Hayden M. R. (2010) Cleavage at the 586 amino acid caspase-6 site in mutant huntingtin influences caspase-6 activation in vivo. *J. Neurosci.* 30, 15019-15029. ISSN: 1529-2401 (Electronic); ISSN: 0270-6474 (Linking)

Griffith J. P., Kim J. L., Kim E. E., Sintchak M. D., Thomson J. A., Fitzgibbon M. J., Fleming M. A., Caron P. R., Hsiao K. and Navia M. A. (1995) X-ray structure of calcineurin inhibited by the immunophilin-immunosuppressant FKBP12-FK506 complex. *Cell* 82, 507-522. ISSN: 0092-8674 (Print); ISSN: 0092-8674 (Linking)

Gu M., Gash M. T., Mann V. M., Javoy-Agid F., Cooper J. M. and Schapira A. H. (1996) Mitochondrial defect in Huntington's disease caudate nucleus. *Ann. Neurol.* 39, 385-389. ISSN: 0364-5134 (Print); ISSN: 0364-5134 (Linking)

Hansson O., Petersen A., Leist M., Nicotera P., Castilho R. F. and Brundin P. (1999) Transgenic mice expressing a Huntington's disease mutation are resistant to quinolinic acid-induced striatal excitotoxicity. *Proc. Natl. Acad. Sci. U. S. A* 96, 8727-8732. ISSN: 0027-8424 (Print); ISSN: 0027-8424 (Linking)

Hernandez-Espinosa D. and Morton A. J. (2006) Calcineurin inhibitors cause an acceleration of the neurological phenotype in a mouse transgenic for the human Huntington's disease mutation. *Brain Res. Bull.* 69, 669-679. ISSN: 0361-9230 (Print); ISSN: 0361-9230 (Linking)

Hersch S. M., Gevorkian S., Marder K., Moskowitz C., Feigin A., Cox M., Como P., Zimmerman C., Lin M., Zhang L., Ulug A. M., Beal M. F., Matson W., Bogdanov M., Ebbel E., Zaleta A., Kaneko Y., Jenkins B., Hevelone N., Zhang H., Yu H., Schoenfeld D., Ferrante R. and Rosas H. D. (2006) Creatine in Huntington disease is safe, tolerable, bioavailable in brain and reduces serum 8OH2'dG. *Neurology* 66, 250-252. ISSN: 1526-632X (Electronic); ISSN: 0028-3878 (Linking)

Horton T. M., Graham B. H., Corral-Debrinski M., Shoffner J. M., Kaufman A. E., Beal M. F. and Wallace D. C. (1995) Marked increase in mitochondrial DNA deletion levels in the cerebral cortex of Huntington's disease patients. *Neurology* 45, 1879-1883. ISSN: 0028-3878 (Print); ISSN: 0028-3878 (Linking)

Huang E. J. and Reichardt L. F. (2003) Trk receptors: roles in neuronal signal transduction. *Annu. Rev. Biochem.* 72, 609-642. ISSN: 0066-4154 (Print); ISSN: 0066-4154 (Linking)

Illuzzi J. L., Vickers C. A. and Kmiec E. B. (2011) Modifications of p53 and the DNA Damage Response in Cells Expressing Mutant Form of the Protein Huntingtin. *J. Mol. Neurosci.* 45, 256-268. ISSN: 1559-1166 (Electronic); ISSN: 0895-8696 (Linking)

Impey S., Obrietan K., Wong S. T., Poser S., Yano S., Wayman G., Deloulme J. C., Chan G. and Storm D. R. (1998) Cross talk between ERK and PKA is required for Ca2+ stimulation of CREB-dependent transcription and ERK nuclear translocation. *Neuron* 21, 869-883. ISSN: 0896-6273 (Print); ISSN: 0896-6273 (Linking)

Jenkins B. G., Koroshetz W. J., Beal M. F. and Rosen B. R. (1993) Evidence for impairment of energy metabolism in vivo in Huntington's disease using localized 1H NMR spectroscopy. *Neurology* 43, 2689-2695. ISSN: 0028-3878 (Print); ISSN: 0028-3878 (Linking)

Jin Y. N. and Johnson G. V. (2010) The interrelationship between mitochondrial dysfunction and transcriptional dysregulation in Huntington disease. *J. Bioenerg. Biomembr.* 42, 199-205. ISSN: 1573-6881 (Electronic); ISSN: 0145-479X (Linking)

Kasraie S., Houshmand M., Banoei M. M., Ahari S. E., Panahi M. S., Shariati P., Bahar M. and Moin M. (2008) Investigation of tRNA(Leu/Lys) and ATPase 6 genes mutations in Huntington's disease. *Cell Mol. Neurobiol.* 28, 933-938. ISSN: 1573-6830 (Electronic); ISSN: 0272-4340 (Linking)

Kaufman B. A., Durisic N., Mativetsky J. M., Costantino S., Hancock M. A., Grutter P. and Shoubridge E. A. (2007) The mitochondrial transcription factor TFAM coordinates the assembly of multiple DNA molecules into nucleoid-like structures. *Mol. Biol. Cell* 18, 3225-3236. ISSN: 1059-1524 (Print): ISSN: 1059-1524 (Linking)

Kells A. P., Fong D. M., Dragunow M., During M. J., Young D. and Connor B. (2004) AAV-mediated gene delivery of BDNF or GDNF is neuroprotective in a model of Huntington disease. *Mol. Ther.* 9, 682-688. ISSN: 1525-0016 (Print); ISSN: 1525-0016 (Linking)

Kelly D. P. and Scarpulla R. C. (2004) Transcriptional regulatory circuits controlling mitochondrial biogenesis and function. *Genes Dev.* 18, 357-368. ISSN: 0890-9369 (Print); ISSN: 0890-9369 (Linking)

Kiechle T., Dedeoglu A., Kubilus J., Kowall N. W., Beal M. F., Friedlander R. M., Hersch S. M. and Ferrante R. J. (2002) Cytochrome C and caspase-9 expression in Huntington's disease. *Neuromolecular. Med.* 1, 183-195. ISSN: 1535-1084 (Print); ISSN: 1535-1084 (Linking)

Kim G. W., Gasche Y., Grzeschik S., Copin J. C., Maier C. M. and Chan P. H. (2003) Neurodegeneration in striatum induced by the mitochondrial toxin 3-nitropropionic acid: role of matrix metalloproteinase-9 in early blood-brain barrier disruption? *J. Neurosci.* 23, 8733-8742. ISSN: 1529-2401 (Electronic); ISSN: 0270-6474 (Linking)

Kim J., Moody J. P., Edgerly C. K., Bordiuk O. L., Cormier K., Smith K., Beal M. F. and Ferrante R. J. (2010) Mitochondrial loss, dysfunction and altered dynamics in Huntington's disease. *Hum. Mol. Genet.* 19, 3919-3935. ISSN: 1460-2083 (Electronic); ISSN: 0964-6906 (Linking)

Kim Y. J., Yi Y., Sapp E., Wang Y., Cuiffo B., Kegel K. B., Qin Z. H., Aronin N. and DiFiglia M. (2001) Caspase 3-cleaved N-terminal fragments of wild-type and mutant huntingtin are present in normal and Huntington's disease brains, associate with membranes, and undergo calpain-dependent proteolysis. *Proc. Natl. Acad. Sci. U. S. A* 98, 12784-12789. ISSN: 0027-8424 (Print); ISSN: 0027-8424 (Linking)

King M. P. and Attardi G. (1989) Human cells lacking mtDNA: repopulation with exogenous mitochondria by complementation. *Science* 246, 500-503. ISSN: 0036-8075 (Print); ISSN: 0036-8075 (Linking)

Klivenyi P., Ferrante R. J., Gardian G., Browne S., Chabrier P. E. and Beal M. F. (2003) Increased survival and neuroprotective effects of BN82451 in a transgenic mouse model of Huntington's disease. *J. Neurochem.* 86, 267-272. ISSN: 0022-3042 (Print); ISSN: 0022-3042 (Linking)

Koroshetz W. J., Jenkins B. G., Rosen B. R. and Beal M. F. (1997) Energy metabolism defects in Huntington's disease and effects of coenzyme Q10. *Ann. Neurol.* 41, 160-165. ISSN: 0364-5134 (Print); ISSN: 0364-5134 (Linking)

Kroemer G. and Reed J. C. (2000) Mitochondrial control of cell death. *Nat. Med.* 6, 513-519. ISSN: 1078-8956 (Print); ISSN: 1078-8956 (Linking)

Kuhl D. E., Phelps M. E., Markham C. H., Metter E. J., Riege W. H. and Winter J. (1982) Cerebral metabolism and atrophy in Huntington's disease determined by 18FDG and computed tomographic scan. *Ann. Neurol.* 12, 425-434. ISSN: 0364-5134 (Print); ISSN: 0364-5134 (Linking)

Kumar P. and Kumar A. (2009) Neuroprotective effect of cyclosporine and FK506 against 3-nitropropionic acid induced cognitive dysfunction and glutathione redox in rat: possible role of nitric oxide. *Neurosci. Res.* 63, 302-314. ISSN: 0168-0102 (Print); ISSN: 0168-0102 (Linking)

Kuwert T., Lange H. W., Langen K. J., Herzog H., Aulich A. and Feinendegen L. E. (1990) Cortical and subcortical glucose consumption measured by PET in patients with Huntington's disease. *Brain* 113 (Pt 5), 1405-1423. ISSN: 0006-8950 (Print); ISSN: 0006-8950 (Linking)

Lee J., Kim C. H., Simon D. K., Aminova L. R., Andreyev A. Y., Kushnareva Y. E., Murphy A. N., Lonze B. E., Kim K. S., Ginty D. D., Ferrante R. J., Ryu H. and Ratan R. R. (2005) Mitochondrial cyclic AMP response element-binding protein (CREB) mediates mitochondrial gene expression and neuronal survival. *J. Biol. Chem.* 280, 40398-40401. ISSN: 0021-9258 (Print); ISSN: 0021-9258 (Linking)

Ley R., Balmanno K., Hadfield K., Weston C. and Cook S. J. (2003) Activation of the ERK1/2 signaling pathway promotes phosphorylation and proteasome-dependent degradation of the BH3-only protein, Bim. *J. Biol. Chem.* 278, 18811-18816. ISSN: 0021-9258 (Print); ISSN: 0021-9258 (Linking)

Li S. H., Cheng A. L., Zhou H., Lam S., Rao M., Li H. and Li X. J. (2002) Interaction of Huntington disease protein with transcriptional activator Sp1. *Mol. Cell Biol.* 22, 1277-1287. ISSN: 0270-7306 (Print); ISSN: 0270-7306 (Linking)

Lin J., Wu P. H., Tarr P. T., Lindenberg K. S., St-Pierre J., Zhang C. Y., Mootha V. K., Jager S., Vianna C. R., Reznick R. M., Cui L., Manieri M., Donovan M. X., Wu Z., Cooper M. P., Fan M. C., Rohas L. M., Zavacki A. M., Cinti S., Shulman G. I., Lowell B. B., Krainc D. and Spiegelman B. M. (2004) Defects in adaptive energy metabolism with CNS-linked hyperactivity in PGC-1alpha null mice. *Cell* 119, 121-135. ISSN: 0092-8674 (Print); ISSN: 0092-8674 (Linking)

Lin J., Yang R., Tarr P. T., Wu P. H., Handschin C., Li S., Yang W., Pei L., Uldry M., Tontonoz P., Newgard C. B. and Spiegelman B. M. (2005) Hyperlipidemic effects of dietary saturated fats mediated through PGC-1beta coactivation of SREBP. *Cell* 120, 261-273. ISSN: 0092-8674 (Print); ISSN: 0092-8674 (Linking)

Lodi R., Schapira A. H., Manners D., Styles P., Wood N. W., Taylor D. J. and Warner T. T. (2000) Abnormal in vivo skeletal muscle energy metabolism in Huntington's disease and dentatorubropallidoluysian atrophy. *Ann. Neurol.* 48, 72-76. ISSN: 0364-5134 (Print); ISSN: 0364-5134 (Linking)

Luciano F., Jacquel A., Colosetti P., Herrant M., Cagnol S., Pages G. and Auberger P. (2003) Phosphorylation of Bim-EL by Erk1/2 on serine 69 promotes its degradation via the proteasome pathway and regulates its proapoptotic function. *Oncogene* 22, 6785-6793. ISSN: 0950-9232 (Print); ISSN: 0950-9232 (Linking)

Luthi-Carter R., Hanson S. A., Strand A. D., Bergstrom D. A., Chun W., Peters N. L., Woods A. M., Chan E. Y., Kooperberg C., Krainc D., Young A. B., Tapscott S. J. and Olson J. M. (2002) Dysregulation of gene expression in the R6/2 model of polyglutamine disease: parallel changes in muscle and brain. *Hum. Mol. Genet.* 11, 1911-1926. ISSN: 0964-6906 (Print); ISSN: 0964-6906 (Linking)

Martin E., Betuing S., Pages C., Cambon K., Auregan G., Deglon N., Roze E. and Caboche J. (2011) Mitogen- and stress-activated protein kinase 1-induced neuroprotection in Huntington's disease: role on chromatin remodeling at the PGC-1-alpha promoter. *Hum. Mol. Genet.* 20, 2422-2434. ISSN: 1460-2083 (Electronic); ISSN: 0964-6906 (Linking)

Masters S. C., Yang H., Datta S. R., Greenberg M. E. and Fu H. (2001) 14-3-3 inhibits Bad-induced cell death through interaction with serine-136. *Mol. Pharmacol.* 60, 1325-1331. ISSN: 0026-895X (Print); ISSN: 0026-895X (Linking)

Matthews R. P., Guthrie C. R., Wailes L. M., Zhao X., Means A. R. and McKnight G. S. (1994) Calcium/calmodulin-dependent protein kinase types II and IV differentially regulate CREB-dependent gene expression. *Mol. Cell Biol.* 14, 6107-6116. ISSN: 0270-7306 (Print); ISSN: 0270-7306 (Linking)

Mayr B. and Montminy M. (2001) Transcriptional regulation by the phosphorylation-dependent factor CREB. *Nat. Rev. Mol. Cell Biol.* 2, 599-609. ISSN: 1471-0072 (Print); ISSN: 1471-0072 (Linking)

Mochel F. and Haller R. G. (2011) Energy deficit in Huntington disease: why it matters. *J. Clin. Invest* 121, 493-499. ISSN: 1932-6203 (Electronic); ISSN: 1932-6203 (Linking)

Morimoto N., Nagano I., Deguchi K., Murakami T., Fushimi S., Shoji M. and Abe K. (2004) Leber hereditary optic neuropathy with chorea and dementia resembling Huntington disease. *Neurology* 63, 2451-2452. ISSN: 1526-632X (Electronic); ISSN: 0028-3878 (Linking)

Napolitano M., Centonze D., Gubellini P., Rossi S., Spiezia S., Bernardi G., Gulino A. and Calabresi P. (2004) Inhibition of mitochondrial complex II alters striatal expression of genes involved in glutamatergic and dopaminergic signaling: possible implications for Huntington's disease. *Neurobiol. Dis.* 15, 407-414. ISSN: 0969-9961 (Print); ISSN: 0969-9961 (Linking)

Nucifora F. C., Jr., Sasaki M., Peters M. F., Huang H., Cooper J. K., Yamada M., Takahashi H., Tsuji S., Troncoso J., Dawson V. L., Dawson T. M. and Ross C. A. (2001) Interference by huntingtin and atrophin-1 with cbp-mediated transcription leading to cellular toxicity. *Science* 291, 2423-2428. ISSN: 0036-8075 (Print); ISSN: 0036-8075 (Linking)

Olah J., Klivenyi P., Gardian G., Vecsei L., Orosz F., Kovacs G. G., Westerhoff H. V. and Ovadi J. (2008) Increased glucose metabolism and ATP level in brain tissue of Huntington's disease transgenic mice. *FEBS J.* 275, 4740-4755. ISSN: 1742-464X (Print); ISSN: 1742-464X (Linking)

Oliveira J. M. (2010) Nature and cause of mitochondrial dysfunction in Huntington's disease: focusing on huntingtin and the striatum. *J. Neurochem.* 114, 1-12. ISSN: 1471-4159 (Electronic); ISSN: 0022-3042 (Linking)

Oliveira J. M., Chen S., Almeida S., Riley R., Goncalves J., Oliveira C. R., Hayden M. R., Nicholls D. G., Ellerby L. M. and Rego A. C. (2006) Mitochondrial-dependent Ca2+ handling in Huntington's disease striatal cells: effect of histone deacetylase inhibitors. *J. Neurosci.* 26, 11174-11186. ISSN: 1529-2401 (Electronic); ISSN: 0270-6474 (Linking)

Oliveira J. M. and Goncalves J. (2009) In situ mitochondrial Ca2+ buffering differences of intact neurons and astrocytes from cortex and striatum. *J. Biol. Chem.* 284, 5010-5020. ISSN: 0021-9258 (Print); ISSN: 0021-9258 (Linking)

Oliveira J. M., Jekabsons M. B., Chen S., Lin A., Rego A. C., Goncalves J., Ellerby L. M. and Nicholls D. G. (2007) Mitochondrial dysfunction in Huntington's disease: the bioenergetics of isolated and in situ mitochondria from transgenic mice. *J. Neurochem.* 101, 241-249. ISSN: 0022-3042 (Print); ISSN: 0022-3042 (Linking)

Ona V. O., Li M., Vonsattel J. P., Andrews L. J., Khan S. Q., Chung W. M., Frey A. S., Menon A. S., Li X. J., Stieg P. E., Yuan J., Penney J. B., Young A. B., Cha J. H. and Friedlander R. M. (1999) Inhibition of caspase-1 slows disease progression in a mouse model of Huntington's disease. *Nature* 399, 263-267. ISSN: 0028-0836 (Print); ISSN: 0028-0836 (Linking)

Orr A. L., Li S., Wang C. E., Li H., Wang J., Rong J., Xu X., Mastroberardino P. G., Greenamyre J. T. and Li X. J. (2008) N-terminal mutant huntingtin associates with mitochondria and impairs mitochondrial trafficking. *J. Neurosci.* 28, 2783-2792. ISSN: 1529-2401 (Electronic); ISSN: 0270-6474 (Linking)

Pandey M., Mohanakumar K. P. and Usha R. (2010) Mitochondrial functional alterations in relation to pathophysiology of Huntington's disease. *J. Bioenerg. Biomembr.* 42, 217-226. ISSN: 1573-6881 (Electronic); ISSN: 0145-479X (Linking)

Pang Z. and Geddes J. W. (1997) Mechanisms of cell death induced by the mitochondrial toxin 3-nitropropionic acid: acute excitotoxic necrosis and delayed apoptosis. *J. Neurosci.* 17, 3064-3073. ISSN: 0270-6474 (Print); ISSN: 0270-6474 (Linking)

Panov A. V., Gutekunst C. A., Leavitt B. R., Hayden M. R., Burke J. R., Strittmatter W. J. and Greenamyre J. T. (2002) Early mitochondrial calcium defects in Huntington's disease are a direct effect of polyglutamines. *Nat. Neurosci.* 5, 731-736. ISSN: 1097-6256 (Print); ISSN: 1097-6256 (Linking)

Pardo R., Colin E., Regulier E., Aebischer P., Deglon N., Humbert S. and Saudou F. (2006) Inhibition of calcineurin by FK506 protects against polyglutamine-huntingtin toxicity through an increase of huntingtin phosphorylation at S421. *J. Neurosci.* 26, 1635-1645. ISSN: 1529-2401 (Electronic); ISSN: 0270-6474 (Linking)

Parker W. D., Jr., Boyson S. J., Luder A. S. and Parks J. K. (1990) Evidence for a defect in NADH: ubiquinone oxidoreductase (complex I) in Huntington's disease. *Neurology* 40, 1231-1234. ISSN: 0028-3878 (Print); ISSN: 0028-3878 (Linking)

Perez-Navarro E., Canudas A. M., Akerund P., Alberch J. and Arenas E. (2000) Brain-derived neurotrophic factor, neurotrophin-3, and neurotrophin-4/5 prevent the death of striatal projection neurons in a rodent model of Huntington's disease. *J. Neurochem.* 75, 2190-2199. ISSN: 0022-3042 (Print); ISSN: 0022-3042 (Linking)

Perez-Severiano F., Rios C. and Segovia J. (2000) Striatal oxidative damage parallels the expression of a neurological phenotype in mice transgenic for the mutation of Huntington's disease. *Brain Res.* 862, 234-237. ISSN: 0006-8993 (Print); ISSN: 0006-8993 (Linking)

Perez-Severiano F., Santamaria A., Pedraza-Chaverri J., Medina-Campos O. N., Rios C. and Segovia J. (2004) Increased formation of reactive oxygen species, but no changes in glutathione peroxidase activity, in striata of mice transgenic for the Huntington's disease mutation. *Neurochem. Res.* 29, 729-733. ISSN: 0364-3190 (Print); ISSN: 0364-3190 (Linking)

Perkinton M. S., Ip J. K., Wood G. L., Crossthwaite A. J. and Williams R. J. (2002) Phosphatidylinositol 3-kinase is a central mediator of NMDA receptor signalling to MAP kinase (Erk1/2), Akt/PKB and CREB in striatal neurones. *J. Neurochem.* 80, 239-254. ISSN: 0022-3042 (Print); ISSN: 0022-3042 (Linking)

Perluigi M., Poon H. F., Maragos W., Pierce W. M., Klein J. B., Calabrese V., Cini C., De M. C. and Butterfield D. A. (2005) Proteomic analysis of protein expression and oxidative modification in r6/2 transgenic mice: a model of Huntington disease. *Mol. Cell Proteomics.* 4, 1849-1861. ISSN: 1535-9476 (Print); ISSN: 1535-9476 (Linking)

Pineda J. R., Pardo R., Zala D., Yu H., Humbert S. and Saudou F. (2009) Genetic and pharmacological inhibition of calcineurin corrects the BDNF transport defect in Huntington's disease. *Mol. Brain* 2, 33. ISSN: 1756-6606 (Electronic); ISSN: 1756-6606 (Linking)

Polidori M. C., Mecocci P., Browne S. E., Senin U. and Beal M. F. (1999) Oxidative damage to mitochondrial DNA in Huntington's disease parietal cortex. *Neurosci. Lett.* 272, 53-56. ISSN: 0304-3940 (Print); ISSN: 0304-3940 (Linking)

Powers W. J., Haas R. H., Le T., Videen T. O., Hershey T., McGee-Minnich L. and Perlmutter J. S. (2007a) Normal platelet mitochondrial complex I activity in Huntington's disease. *Neurobiol. Dis.* 27, 99-101. ISSN: 0969-9961 (Print); ISSN: 0969-9961 (Linking)

Powers W. J., Videen T. O., Markham J., McGee-Minnich L., Antenor-Dorsey J. V., Hershey T. and Perlmutter J. S. (2007b) Selective defect of in vivo glycolysis in early Huntington's disease striatum. *Proc. Natl. Acad. Sci. U. S. A* 104, 2945-2949. ISSN: 0027-8424 (Print); ISSN: 0027-8424 (Linking)

Pugazhenthi S., Nesterova A., Sable C., Heidenreich K. A., Boxer L. M., Heasley L. E. and Reusch J. E. (2000) Akt/protein kinase B up-regulates Bcl-2 expression through cAMP-response element-binding protein. *J. Biol. Chem.* 275, 10761-10766. ISSN: 0021-9258 (Print); ISSN: 0021-9258 (Linking)

Quintanilla R. A. and Johnson G. V. (2009) Role of mitochondrial dysfunction in the pathogenesis of Huntington's disease. *Brain Res. Bull.* 80, 242-247. ISSN: 1873-2747 (Electronic); ISSN: 0361-9230 (Linking)

Rigamonti D., Bauer J. H., De-Fraja C., Conti L., Sipione S., Sciorati C., Clementi E., Hackam A., Hayden M. R., Li Y., Cooper J. K., Ross C. A., Govoni S., Vincenz C. and Cattaneo E. (2000) Wild-type huntingtin protects from apoptosis upstream of caspase-3. *J. Neurosci.* 20, 3705-3713. ISSN: 1529-2401 (Electronic); ISSN: 0270-6474 (Linking)

Rigamonti D., Sipione S., Goffredo D., Zuccato C., Fossale E. and Cattaneo E. (2001) Huntingtin's neuroprotective activity occurs via inhibition of procaspase-9 processing. *J. Biol. Chem.* 276, 14545-14548. ISSN: 0021-9258 (Print); ISSN: 0021-9258 (Linking)

Rosenstock T. R., de Brito O. M., Lombardi V., Louros S., Ribeiro M., Almeida S., Ferreira I. L., Oliveira C. R. and Rego A. C. (2011) FK506 ameliorates cell death features in Huntington's disease striatal cell models. *Neurochem. Int.* ISSN: 1872-9754 (Electronic); ISSN: 0197-0186 (Linking)

Rosenstock T. R., Duarte A. I. and Rego A. C. (2010) Mitochondrial-associated metabolic changes and neurodegeneration in Huntington's disease - from clinical features to the bench. *Curr. Drug Targets.* 11, 1218-1236. ISSN: 1873-5592 (Electronic); ISSN: 1389-4501 (Linking)

Rouaux C., Jokic N., Mbebi C., Boutillier S., Loeffler J. P. and Boutillier A. L. (2003) Critical loss of CBP/p300 histone acetylase activity by caspase-6 during neurodegeneration. *EMBO J.* 22, 6537-6549. ISSN: 0261-4189 (Print); ISSN: 0261-4189 (Linking)

Ryu J. K., Kim J., Cho S. J., Hatori K., Nagai A., Choi H. B., Lee M. C., McLarnon J. G. and Kim S. U. (2004) Proactive transplantation of human neural stem cells prevents degeneration of striatal neurons in a rat model of Huntington disease. *Neurobiol. Dis.* 16, 68-77. ISSN: 0969-9961 (Print); ISSN: 0969-9961 (Linking)

Saft C., Zange J., Andrich J., Muller K., Lindenberg K., Landwehrmeyer B., Vorgerd M., Kraus P. H., Przuntek H. and Schols L. (2005) Mitochondrial impairment in patients and asymptomatic mutation carriers of Huntington's disease. *Mov Disord.* 20, 674-679. ISSN: 0885-3185 (Print); ISSN: 0885-3185 (Linking)

Saha R. N., Liu X. and Pahan K. (2006) Up-regulation of BDNF in astrocytes by TNF-alpha: a case for the neuroprotective role of cytokine. *J. Neuroimmune. Pharmacol.* 1, 212-222. ISSN: 1557-1904 (Electronic); ISSN: 1557-1890 (Linking)

Sanchez I., Xu C. J., Juo P., Kakizaka A., Blenis J. and Yuan J. (1999) Caspase-8 is required for cell death induced by expanded polyglutamine repeats. *Neuron* 22, 623-633. ISSN: 0896-6273 (Print); ISSN: 0896-6273 (Linking)

Santamaria A., Perez-Severiano F., Rodriguez-Martinez E., Maldonado P. D., Pedraza-Chaverri J., Rios C. and Segovia J. (2001) Comparative analysis of superoxide dismutase activity between acute pharmacological models and a transgenic mouse model of Huntington's disease. *Neurochem. Res.* 26, 419-424. ISSN: 0364-3190 (Print); ISSN: 0364-3190 (Linking)

Saudou F., Finkbeiner S., Devys D. and Greenberg M. E. (1998) Huntingtin acts in the nucleus to induce apoptosis but death does not correlate with the formation of intranuclear inclusions. *Cell* 95, 55-66. ISSN: 0092-8674 (Print); ISSN: 0092-8674 (Linking)

Sawa A., Wiegand G. W., Cooper J., Margolis R. L., Sharp A. H., Lawler J. F., Jr., Greenamyre J. T., Snyder S. H. and Ross C. A. (1999) Increased apoptosis of Huntington disease lymphoblasts associated with repeat length-dependent mitochondrial depolarization. *Nat. Med.* 5, 1194-1198. ISSN: 1078-8956 (Print); ISSN: 1078-8956 (Linking)

Scarpulla R. C. (2002) Nuclear activators and coactivators in mammalian mitochondrial biogenesis. *Biochim. Biophys. Acta* 1576, 1-14. ISSN: 0006-3002 (Print); ISSN: 0006-3002 (Linking)

Scarpulla R. C. (2011) Metabolic control of mitochondrial biogenesis through the PGC-1 family regulatory network. *Biochim. Biophys. Acta* 1813, 1269-1278. ISSN: 0006-3002 (Print); ISSN: 0006-3002 (Linking)

Scheid M. P., Schubert K. M. and Duronio V. (1999) Regulation of bad phosphorylation and association with Bcl-x(L) by the MAPK/Erk kinase. *J. Biol. Chem.* 274, 31108-31113. ISSN: 0021-9258 (Print); ISSN: 0021-9258 (Linking)

Shih A. Y., Imbeault S., Barakauskas V., Erb H., Jiang L., Li P. and Murphy T. H. (2005) Induction of the Nrf2-driven antioxidant response confers neuroprotection during mitochondrial stress in vivo. *J. Biol. Chem.* 280, 22925-22936. ISSN: 0021-9258 (Print); ISSN: 0021-9258 (Linking)

Shimohata T., Nakajima T., Yamada M., Uchida C., Onodera O., Naruse S., Kimura T., Koide R., Nozaki K., Sano Y., Ishiguro H., Sakoe K., Ooshima T., Sato A., Ikeuchi T., Oyake M., Sato T., Aoyagi Y., Hozumi I., Nagatsu T., Takiyama Y., Nishizawa M., Goto J., Kanazawa I., Davidson I., Tanese N., Takahashi H. and Tsuji S. (2000) Expanded polyglutamine stretches interact with TAFII130, interfering with CREB-dependent transcription. *Nat. Genet.* 26, 29-36. ISSN: 1061-4036 (Print); ISSN: 1061-4036 (Linking)

Shirendeb U., Reddy A. P., Manczak M., Calkins M. J., Mao P., Tagle D. A. and Reddy P. H. (2011) Abnormal mitochondrial dynamics, mitochondrial loss and mutant huntingtin oligomers in Huntington's disease: implications for selective neuronal damage. *Hum. Mol. Genet.* 20, 1438-1455. ISSN: 1460-2083 (Electronic); ISSN: 0964-6906 (Linking)

Solans A., Zambrano A., Rodriguez M. and Barrientos A. (2006) Cytotoxicity of a mutant huntingtin fragment in yeast involves early alterations in mitochondrial OXPHOS complexes II and III. *Hum. Mol. Genet.* 15, 3063-3081. ISSN: 0964-6906 (Print); ISSN: 0964-6906 (Linking)

Song W., Chen J., Petrilli A., Liot G., Klinglmayr E., Zhou Y., Poquiz P., Tjong J., Pouladi M. A., Hayden M. R., Masliah E., Ellisman M., Rouiller I., Schwarzenbacher R., Bossy B., Perkins G. and Bossy-Wetzel E. (2011) Mutant huntingtin binds the mitochondrial fission GTPase dynamin-related protein-1 and increases its enzymatic activity. *Nat. Med.* 17, 377-382. ISSN: 1546-170X (Electronic); ISSN: 1078-8956 (Linking)

Sorbi S., Bird E. D. and Blass J. P. (1983) Decreased pyruvate dehydrogenase complex activity in Huntington and Alzheimer brain. *Ann. Neurol.* 13, 72-78. ISSN: 0364-5134 (Print); ISSN: 0364-5134 (Linking)

Sorolla M. A., Rodriguez-Colman M. J., Tamarit J., Ortega Z., Lucas J. J., Ferrer I., Ros J. and Cabiscol E. (2010) Protein oxidation in Huntington disease affects energy production and vitamin B6 metabolism. *Free Radic. Biol. Med.* 49, 612-621. ISSN: 1873-4596 (Electronic); ISSN: 0891-5849 (Linking)

Squitieri F., Maglione V., Orobello S. and Fornai F. (2011) Genotype-, aging-dependent abnormal caspase activity in Huntington disease blood cells. *J. Neural Transm.* ISSN: 1435-1463 (Electronic); ISSN: 0300-9564 (Linking)

St-Pierre J., Drori S., Uldry M., Silvaggi J. M., Rhee J., Jager S., Handschin C., Zheng K., Lin J., Yang W., Simon D. K., Bachoo R. and Spiegelman B. M. (2006) Suppression of reactive oxygen species and neurodegeneration by the PGC-1 transcriptional coactivators. *Cell* 127, 397-408. ISSN: 0092-8674 (Print); ISSN: 0092-8674 (Linking)

Stahl W. L. and Swanson P. D. (1974) Biochemical abnormalities in Huntington's chorea brains. *Neurology* 24, 813-819. ISSN: 0028-3878 (Print); ISSN: 0028-3878 (Linking)

Steffan J. S., Bodai L., Pallos J., Poelman M., McCampbell A., Apostol B. L., Kazantsev A., Schmidt E., Zhu Y. Z., Greenwald M., Kurokawa R., Housman D. E., Jackson G. R., Marsh J. L. and Thompson L. M. (2001) Histone deacetylase inhibitors arrest polyglutamine-dependent neurodegeneration in Drosophila. *Nature* 413, 739-743. ISSN: 0028-0836 (Print); ISSN: 0028-0836 (Linking)

Steffan J. S., Kazantsev A., Spasic-Boskovic O., Greenwald M., Zhu Y. Z., Gohler H., Wanker E. E., Bates G. P., Housman D. E. and Thompson L. M. (2000) The Huntington's disease protein interacts with p53 and CREB-binding protein and represses transcription. *Proc. Natl. Acad. Sci. U. S. A* 97, 6763-6768. ISSN: 0027-8424 (Print); ISSN: 0027-8424 (Linking)

Stoy N., Mackay G. M., Forrest C. M., Christofides J., Egerton M., Stone T. W. and Darlington L. G. (2005) Tryptophan metabolism and oxidative stress in patients with Huntington's disease. *J. Neurochem.* 93, 611-623. ISSN: 0022-3042 (Print); ISSN: 0022-3042 (Linking)

Strand A. D., Baquet Z. C., Aragaki A. K., Holmans P., Yang L., Cleren C., Beal M. F., Jones L., Kooperberg C., Olson J. M. and Jones K. R. (2007) Expression profiling of Huntington's disease models suggests that brain-derived neurotrophic factor depletion plays a major role in striatal degeneration. *J. Neurosci.* 27, 11758-11768. ISSN: 1529-2401 (Electronic); ISSN: 0270-6474 (Linking)

Sugars K. L. and Rubinsztein D. C. (2003) Transcriptional abnormalities in Huntington disease. *Trends Genet.* 19, 233-238. ISSN: 0168-9525 (Print); ISSN: 0168-9525 (Linking)

Swerdlow R. H., Parks J. K., Cassarino D. S., Shilling A. T., Bennett J. P., Jr., Harrison M. B. and Parker W. D., Jr. (1999) Characterization of cybrid cell lines containing mtDNA from Huntington's disease patients. *Biochem. Biophys. Res. Commun.* 261, 701-704. ISSN: 0006-291X (Print); ISSN: 0006-291X (Linking)

Tabrizi S. J., Workman J., Hart P. E., Mangiarini L., Mahal A., Bates G., Cooper J. M. and Schapira A. H. (2000) Mitochondrial dysfunction and free radical damage in the Huntington R6/2 transgenic mouse. *Ann. Neurol.* 47, 80-86. ISSN: 0364-5134 (Print); ISSN: 0364-5134 (Linking)

Taherzadeh-Fard E., Saft C., Akkad D. A., Wieczorek S., Haghikia A., Chan A., Epplen J. T. and Arning L. (2011) PGC-1alpha downstream transcription factors NRF-1 and TFAM are genetic modifiers of Huntington disease. *Mol. Neurodegener.* 6, 32. ISSN: 1750-1326 (Electronic); ISSN: 1750-1326 (Linking)

Tellez-Nagel I., Johnson A. B. and Terry R. D. (1974) Studies on brain biopsies of patients with Huntington's chorea. *J. Neuropathol. Exp. Neurol.* 33, 308-332. ISSN: 0022-3069 (Print); ISSN: 0022-3069 (Linking)

Trushina E., Dyer R. B., Badger J. D., Ure D., Eide L., Tran D. D., Vrieze B. T., Legendre-Guillemin V., McPherson P. S., Mandavilli B. S., Van H. B., Zeitlin S., McNiven M., Aebersold R., Hayden M., Parisi J. E., Seeberg E., Dragatsis I., Doyle K., Bender A., Chacko C. and McMurray C. T. (2004) Mutant huntingtin impairs axonal trafficking in mammalian neurons in vivo and in vitro. *Mol. Cell Biol.* 24, 8195-8209. ISSN: 0270-7306 (Print); ISSN: 0270-7306 (Linking)

Turner C., Cooper J. M. and Schapira A. H. (2007) Clinical correlates of mitochondrial function in Huntington's disease muscle. *Mov Disord.* 22, 1715-1721. ISSN: 0885-3185 (Print); ISSN: 0885-3185 (Linking)

Vis J. C., Schipper E., de Boer-van Huizen RT, Verbeek M. M., de Waal R. M., Wesseling P., ten Donkelaar H. J. and Kremer B. (2005) Expression pattern of apoptosis-related markers in Huntington's disease. *Acta Neuropathol.* 109, 321-328. ISSN: 0001-6322 (Print); ISSN: 0001-6322 (Linking)

Wang H., Lim P. J., Karbowski M. and Monteiro M. J. (2009) Effects of overexpression of huntingtin proteins on mitochondrial integrity. *Hum. Mol. Genet.* 18, 737-752. ISSN: 1460-2083 (Electronic); ISSN: 0964-6906 (Linking)

Wang H. G., Pathan N., Ethell I. M., Krajewski S., Yamaguchi Y., Shibasaki F., McKeon F., Bobo T., Franke T. F. and Reed J. C. (1999) Ca2+-induced apoptosis through calcineurin dephosphorylation of BAD. *Science* 284, 339-343. ISSN: 0036-8075 (Print); ISSN: 0036-8075 (Linking)

Wellington C. L., Ellerby L. M., Hackam A. S., Margolis R. L., Trifiro M. A., Singaraja R., McCutcheon K., Salvesen G. S., Propp S. S., Bromm M., Rowland K. J., Zhang T., Rasper D., Roy S., Thornberry N., Pinsky L., Kakizuka A., Ross C. A., Nicholson D. W., Bredesen D. E. and Hayden M. R. (1998) Caspase cleavage of gene products associated with triplet expansion disorders generates truncated fragments containing the polyglutamine tract. *J. Biol. Chem.* 273, 9158-9167. ISSN: 0021-9258 (Print); ISSN: 0021-9258 (Linking)

Weydt P., Pineda V. V., Torrence A. E., Libby R. T., Satterfield T. F., Lazarowski E. R., Gilbert M. L., Morton G. J., Bammler T. K., Strand A. D., Cui L., Beyer R. P., Easley C. N., Smith A. C., Krainc D., Luquet S., Sweet I. R., Schwartz M. W. and La Spada A. R. (2006) Thermoregulatory and metabolic defects in Huntington's disease transgenic mice implicate PGC-1alpha in Huntington's disease neurodegeneration. *Cell Metab* 4, 349-362. ISSN: 1550-4131 (Print); ISSN: 1550-4131 (Linking)

Weydt P., Soyal S. M., Gellera C., Didonato S., Weidinger C., Oberkofler H., Landwehrmeyer G. B. and Patsch W. (2009) The gene coding for PGC-1alpha modifies age at onset in Huntington's Disease. *Mol. Neurodegener.* 4, 3.ISSN: 1750-1326 (Electronic); ISSN: 1750-1326 (Linking)

Xifro X., Garcia-Martinez J. M., Del T. D., Alberch J. and Perez-Navarro E. (2008) Calcineurin is involved in the early activation of NMDA-mediated cell death in mutant huntingtin knock-in striatal cells. *J. Neurochem.* 105, 1596-1612. ISSN: 1471-4159 (Electronic); ISSN: 0022-3042 (Linking)

Xifro X., Giralt A., Saavedra A., Garcia-Martinez J. M., Diaz-Hernandez M., Lucas J. J., Alberch J. and Perez-Navarro E. (2009) Reduced calcineurin protein levels and activity in exon-1 mouse models of Huntington's disease: role in excitotoxicity. *Neurobiol. Dis.* 36, 461-469. ISSN: 1095-953X (Electronic); ISSN: 0969-9961 (Linking)

Yamamoto K. K., Gonzalez G. A., Biggs W. H., III and Montminy M. R. (1988) Phosphorylation-induced binding and transcriptional efficacy of nuclear factor CREB. *Nature* 334, 494-498. ISSN: 0028-0836 (Print); ISSN: 0028-0836 (Linking)

Zhai W., Jeong H., Cui L., Krainc D. and Tjian R. (2005) In vitro analysis of huntingtin-mediated transcriptional repression reveals multiple transcription factor targets. *Cell* 123, 1241-1253. ISSN: 0092-8674 (Print); ISSN: 0092-8674 (Linking)

Zhang X. D., Wang Y., Wang Y., Zhang X., Han R., Wu J. C., Liang Z. Q., Gu Z. L., Han F., Fukunaga K. and Qin Z. H. (2009a) p53 mediates mitochondria dysfunction-triggered autophagy activation and cell death in rat striatum. *Autophagy.* 5, 339-350. ISSN: 1554-8635 (Electronic); ISSN: 1554-8627 (Linking)

Zhang X. D., Wang Y., Wu J. C., Lin F., Han R., Han F., Fukunaga K. and Qin Z. H. (2009b) Down-regulation of Bcl-2 enhances autophagy activation and cell death induced by mitochondrial dysfunction in rat striatum. *J. Neurosci. Res.* 87, 3600-3610. ISSN: 1097-4547 (Electronic); ISSN: 0360-4012 (Linking)

Zuccato C., Ciammola A., Rigamonti D., Leavitt B. R., Goffredo D., Conti L., MacDonald M. E., Friedlander R. M., Silani V., Hayden M. R., Timmusk T., Sipione S. and Cattaneo E. (2001) Loss of huntingtin-mediated BDNF gene transcription in Huntington's disease. *Science* 293, 493-498. ISSN: 0036-8075 (Print); ISSN: 0036-8075 (Linking)

Zuccato C., Liber D., Ramos C., Tarditi A., Rigamonti D., Tartari M., Valenza M. and Cattaneo E. (2005) Progressive loss of BDNF in a mouse model of Huntington's disease and rescue by BDNF delivery. *Pharmacol. Res.* 52, 133-139. ISSN: 1043-6618 (Print); ISSN: 1043-6618 (Linking)

Zuccato C., Tartari M., Crotti A., Goffredo D., Valenza M., Conti L., Cataudella T., Leavitt B. R., Hayden M. R., Timmusk T., Rigamonti D. and Cattaneo E. (2003) Huntingtin interacts with REST/NRSF to modulate the transcription of NRSE-controlled neuronal genes. *Nat. Genet.* 35, 76-83. ISSN: 1061-4036 (Print); ISSN: 1061-4036 (Linking)

17

Cholesterol Metabolism in Huntington's Disease

Valerio Leoni[1], Claudio Caccia[1] and Ingemar Björkhem[2]
[1]*IRCCS National Institute of Neurology "C. Besta", Milano,*
[2]*Karolinska Institute, Stockholm,*
[1]*Italy*
[2]*Sweden*

1. Introduction

Cholesterol is present in all vertebrate cells where has several important functions. Being a structural element in plasma membranes it supports the structure and function of lipid bilayers and is regarded to be the most important "fluidity buffer" of the membrane. It is also a precursor of bile acids, steroid hormones and vitamin D. While cholesterol is involved in many cellular processes, a strong indicator of its importance is that it is the only major lipid in mammals not used in energy generation.

De novo synthesis and uptake from circulating lipoproteins cover the cholesterol needs of the cells. Almost all the mammalian cells are able to synthesize cholesterol and express the sophisticated and energy demanding enzymatic machinery required for the de novo synthesis.

In general, the cells in the body are able to release and take up cholesterol to maintain their cholesterol homeostasis: some are able to produce an excess to provide other cells, some others need exogenous cholesterol because of limited synthetic capacity.

Excess cholesterol may be toxic for the cells and a number of strategies have evolved either to export it or to store it in an esterified form. The exogenous cell supply is covered via the Low Density Lipoproteins (LDL) cycle and most of the excess is exported by the High Density Lipoprotein (HDL) mechanism (reverse cholesterol transport), mediated by members of the ATP-Binding Cassette (ABC)-transporter family.

Under normal conditions, about the 60% of the body's cholesterol is synthesized (about 700 mg/day) and the remaining is provided by the diet. The liver accounts for approximately 10% of total synthesis in humans, as does the small intestines.

The biosynthesis of cholesterol may be divided into five stages: 1) synthesis of mevalonate from acetyl-coenzymeA (CoA); 2) synthesis of isoprenoid units from mevalonate by loss of CO2; 3) condensation of six isoprenoids units to form squalene; 4) cyclization of squalene to give the parental steroid, lanosterol; 5) formation of cholesterol by rearranging the lanosterol molecule (Fig. 1).

The most important rate limiting step is the conversion of the 3α-hydroxy-3-methylglutarylcoenzyme A (HMG-CoA) into mevalonate, catalysed by the microsomal HMG-CoA reductase (Rodwell et al., 1976). The activity of the enzyme is regulated by a negative feedback mechanism both at the protein level and at the transcriptional level. To some extent the latter effects may be mediated by oxysterols and bile acids.

Lanosterol is the first sterol formed during cholesterol biosynthesis by conversion of squalene, while lathosterol is a further precursor synthesised in later steps (Kandutsch-Russel pathway). In humans, lanosterol and lathosterol are regarded to be suitable plasma surrogate marker for whole body cholesterol synthesis (Kempen et al., 1988; Bloch et al., 1957; Matthan et al., 2000).

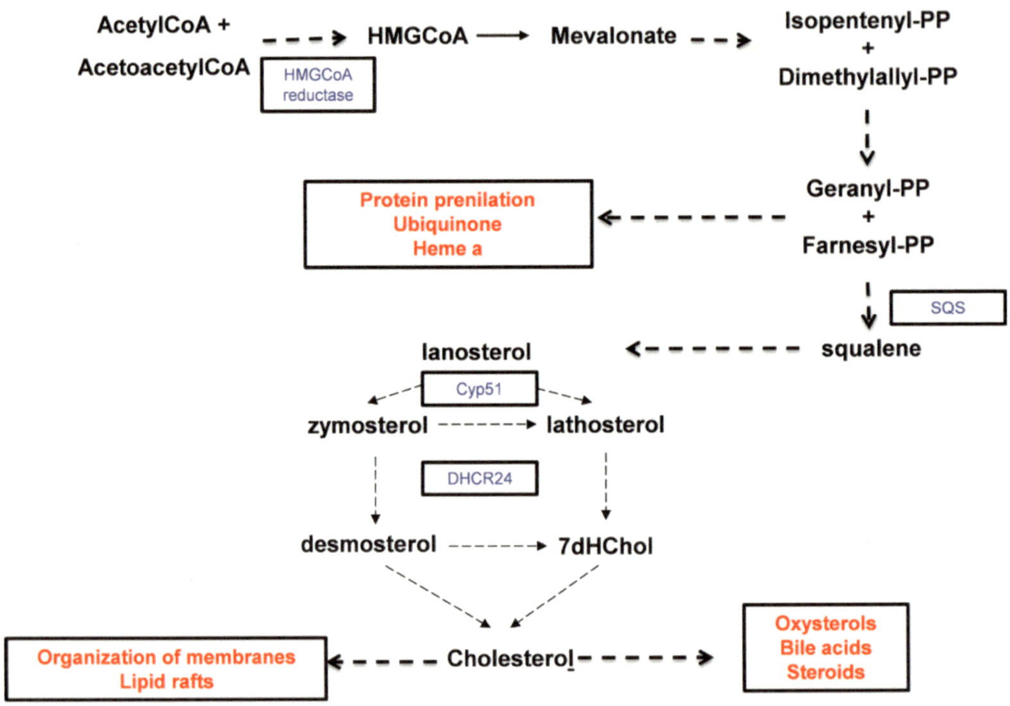

Fig. 1. Simplified diagram of cholesterol metabolism in the cells. Filled arrows mean direct enzymatic reaction, dot arrows mean metabolic reactions not presented in the figure.

Cholesterol synthesis begins with the transport of acetyl-CoA from the mitochondrion to the cytosol. Rate limiting step occurs at the 3-hydroxy-3-methylglutaryl-CoA (HMG-CoA) reducatase followed by mevalonate formation. Phosphorylation is required to solubilize the isoprenoid intermediates in the pathway (the PP abbreviation stands for pyrophosphate). Intermediates in the pathway are used for the synthesis of prenylated proteins, dolichol, coenzyme Q and the side chain of heme a. Pyrophosphated isoprenoids are condensed and ciclised by squalene synthetase (SQS) then the first sterol, lanosterol is formed. Two althernative pathways (Block and Kandush-Russel) lead to cholesterol formation. Precursor sterols can be converted by 3β-hydroxycholesterol Δ^{24} reductase (DHCR24). Cholesterol is involved in structure, organisation and function of cellular membranes and is precursor of oxysterols, bile acids and steroids.

Cholesterol is insoluble in water and is transported in the circulation associated with lipoproteins. Cholesterol is absorbed from the intestinal lumen and transported to the liver via chylomicrons. The cholesterol in these particles can be esterified, converted into bile acids or secreted into bile or collected in Very Low Density Lipoproteins (VLDL) to be transported to the extrahepatic tissues. VLDL can be remodelled by the action of lipoprotein lipase that removes triacyl-glycerol, transferring Apolipoprotein A (ApoA) and Apolipoprotein C (ApoC) from VLDL to HDL. The product of these remodelling is LDL which supplies peripheral tissues with cholesterol. The intake of cellular LDL is strictly regulated via the LDL receptors (LDLR) and Apolipoprotein B (ApoB). The influx of cholesterol inhibits HMG-CoA reductase and cholesterol synthesis and stimulates the cholesterol esterification by acylCoA:cholesterol acyltransferase (ACAT).

The reverse cholesterol transport, whereby cells from different organs eliminate excess cholesterol through the liver, is mediated by HDL. The HDL particles contain ApoA1 and acquire cholesterol directly from the plasma membrane. This transfer is mediated by members of the ABC-transporter family.

About 1 g of cholesterol is eliminated from the body every day. Approximately half of this is excreted into the faeces after conversion into bile acids; the remainder is excreted as non-metabolized cholesterol or the bacterial metabolite coprostanol. The bile acids formed have an important role in the solubilisation and absorption of fats, cholesterol, vitamins and drugs. Approximately 95% of the bile acids are reabsorbed from the intestine and reach the liver via the portal vein (entero-hepatic cycling).

There are two different major pathways in bile acid synthesis. The neutral pathway is initiated by the rate-limiting enzyme cholesterol 7α-hydroxylase which is mainly expressed in hepatocytes. Under normal conditions the neutral pathway dominates in healthy adult humans (Russel, 2003). In contrast to the acid pathway the neutral pathway is under strict metabolic control.

In many cells and organs cholesterol is eliminated by side chain oxidation as an alternative to the classical HDL-mediated reversed cholesterol transport. Thus, almost all cells in the body contain the enzyme sterol 27-hydroxylase (CYP27A1) located in the inner membrane of the mitochondria. This enzyme is particularly expressed by macrophages. At high levels of CYP27A1, 27OHC may be further oxidized by CYP27A1 into 3β-hydroxy-5-cholestenoic acid. The latter acid may be further converted into 7α-hydroxy-3-oxo-cholesten-4-cholestenoic acid and then proceed in the acidic pathway for bile acid synthesis in the liver. The latter pathway is responsible for formation of about 10 % of the daily production of bile acids in humans (Duane & Javitt, 1999; Brown & Jessup, 2009).

2. Brain cholesterol metabolism

The content of cholesterol in the brain is about 10-fold higher than in any other organ and about the 25% of the total body cholesterol is located there (Dietschy & Turley, 2004). Synthesis and storage of such a large amount of cholesterol indicates a close link between the evolution of the nervous system and a specific role for cholesterol. Within the brain about 70% of cholesterol is present in myelin. It is likely that the requirement for efficient signalling despite a small transverse diameter of axons was a key selective pressure driving the accretion of cholesterol in the mammalian brain (Dietschy & Turley, 2004; Snipe & Suter, 1997; Björkhem & Meaney, 2004). The importance of such structural role is also suggested by the long half-life of brain cholesterol: overall brain cholesterol turns over some 250-300 times slower than that in the circulation (Björkhem & Meaney, 2004).

Myelin sheath is formed by sections of plasma membrane repeatedly wrapped around an axon, with the extrusion of virtually all of the cytoplasm. Myelin is formed by two very specialized cells: the oligodendrocyte in CNS and the Schwann cell in the peripheral nervous system. As an individual axon may be ensheathed by myelin from several oligodendrocytes, periodic gaps are present in the sheath. These are called the "nodes of Ranvier" and are the site of propagation of the action potential. Myelin can thus be regarded as a discontinuous insulation that enables the saltatory conduction of the action potential (Dietschy & Turley, 2004; Snipe & Suter, 1997). In addition to a large lipid component myelin also contains many specific proteins such as proteo-lipid protein and myelin basic protein.

Recently evidence has been presented that cholesterol can regulate the correct targeting of one of the major membrane proteins of the periphery nervous system and thereby myelin compaction. These data extend the role of cholesterol in myelin from an essential structural component to a regulator of overall myelin structure (Saher et al., 2009).

Oligodendrocytes differentiate postnatally and the process of myelination both in rodents and humans occurs during the first weeks (or months) postnatally with a coordinated accumulation of cholesterol and myelin basic protein (Dietschy and Turley, 2004). During brain maturation there is a progressive accumulation of cholesterol which ends in the adulthood when the myelin formation is completed. Interestingly the rate of cholesterol synthesis is higher in the early stage of myelination and in the regions with myelin (and white matter, such as mid brain and spinal cord compared to cortex). Together with cholesterol, myelin basic protein is one of the major proteins of CNS myelin and it represents about the 30% of the brain total protein. Severe alterations to the myelin were described in case of the shiverer mutant mouse with deletion of the myelin basic protein gene (Baumann & Pham-Dinh, 2001). The remaining 30% of brain cholesterol is divided between glial cells (20%) and neurons (10%), mainly located in cellular membranes (Maxfield & Tabas, 2005). Cholesterol is organized in microdomains called lipid rafts which are involved in the maintenance of the properties of membrane proteins such as receptors and ion channels (Allen et al., 2007). In addition to its structural role, cholesterol is involved in synapthogenesis, turnover, maintenance, stabilisation and restore of synapses (Koudinov & Koudinova, 2001). In addition it is a limiting factor for outgrowth of neurites and involved in vesicle transport and exocytosis at synaptic levels (for review see also Pfrieger, 2003 a and b; Pfrieger, 2011).

According to various *in vitro* studies with cultured cells, astrocytes synthesize at least 5-10 fold more cholesterol than neurons, while oligodendrocytes have an even higher capacity for cholesterol synthesis, at least during periods of active myelination (Pitas et al., 1987; Björkhem & Meaney, 2004). According to the "outsorcing" hypothesis it was suggested that neurons down-regulate their cholesterol synthesis and rely at least in part on delivery of cholesterol from astrocytes which differentiate postnatally and release cholesterol rich lipoproteins (Pfrieger, 2011) (Fig. 2). This strategy may allow neurons to focus on generation of electrical activity rather than dispense energy on costly cholesterol synthesis. This may be of particular importance in presynaptic terminals and dendritic spines, which are distant from the soma (Dietschy & Turley, 2004; Pfrieger, 2003a; Pfrieger, 2003b; Pfrieger, 2011). ApoE is the main lipid carrier protein in the Central Nervous System (CNS) and is released by astrocytes in order to supply neurons and synaptogenesis with lipids and cholesterol (Bu, 2009; Posse de Chaves & Narayanaswam, 2008; Björkhem et al., 2010).

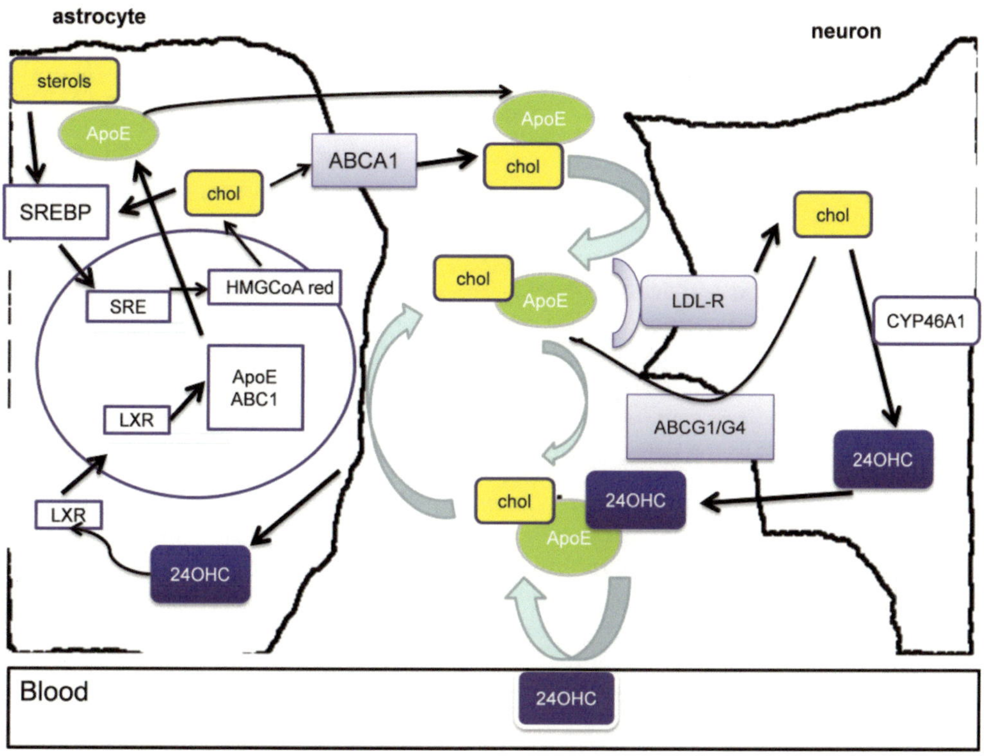

Fig. 2. Proposed model for cholesterol turnover in neurons and astrocytes. In mature brain neurons down regulate their cholesterol synthesis and relay on astrocytes delivery of cholesterol via ApoE lipoproteins. Cholesterol metabolism is a complex multisteps pathway involving endoplasmic reticulum or peroxisomes. Expression of hydroxy-methyl-glutaryl-Coenzyme A reductase (HMGCR), the rate limiting enzyme of the pathway, is regulated by feedback inhibition via the sterol-regulated element binding protein (SREBP) that binds to the sterol-regulated element (SRE-1) in the HMGCR gene. Cholesterol is loaded by astrocytes on ApoE involving also in the transport the ATP binding cassette (ABC) transporter A1 (ABCA1). The apoE-cholesterol complex is internalized via low-density lipoprotein receptors (LDLR). Excess of cholesterol is converted by neurons into the more polar 24S-hydroxycholesterol (24OHC) via the cholesterol 24-hydroxylase, CYP46. 24OHC and other oxysterols are important ligands of the liver X-activated receptor (LXR), which translocate to the nucleus (as circle in the figure) and induces expression of both the APOE and the ABCA1 genes in astrocytes. Cholesterol and 24OHC are excreted from neurons via ABC G1/G4 to ApoE particles or to CSF. 24OHC through the blood–brain barrier can be delivered into plasma for further elimination. The above proposed model is supported by a number of *in vitro* experiments with isolated neurons and astrocytes (Pfrieger, 2003a; Pfrieger, 2011; Abildayeva et al., 2006). The relation between 24OHC and ApoE is consistent with the finding of a correlation between these two factors in CSF from patients with neurodegeneration (Shafaati et al., 2007; Leoni et al., 2010). Such a relation was not found in CSF from healthy volunteers, however. The proposed model predicts a relation between levels of 24OHC and expression of LXR-target genes such as ApoE, ABCA1 and ABCG1. Such a relation was not found, however, in the brain of mice with elevated levels of 24OHC due to overexpression of CYP46A1 (Shafaati et al., 2011).

Neurons appear to produce a sufficient amounts of cholesterol to survive, to differentiate axons and dendrites and to form few and inefficient synapses. The massive formation of synapses, however, requires additional cholesterol delivered by astrocytes via ApoE-containing lipoproteins (Pfrieger 2003a; Pfrieger 2003 b; Pfrieger 2011).

Due to the efficient blood-brain barrier there is no passage of lipoprotein-bound cholesterol from the circulation into the brain (Dietschy & Turley, 2004; Snipe & Suter, 1997; Björkhem & Meaney, 2004). The blood-brain barrier thus prevents diffusion of large molecules at the level of tight junctional attachments between adjacent capillary endothelial cells. In addition to this, there is also no transvesicular movement of solution across the capillaries. It is possible that one or more members of the ATP binding cassette transporter superfamily may be involved in the exclusion of circulating cholesterol from the brain. All the cholesterol present in the brain (and in the peripheral nervous system) is thus formed by de novo synthesis. Except for active phases of specific pathological conditions, almost all (at least 99%) of the cholesterol in the nervous system is unesterified.

Cholesterol synthesis appears to be regulated by similar mechanisms both outside and inside the brain with hydroxy-methyl-glutaryl CoenzymeA reductase (HMGCR) being the most important regulatory enzyme (Snipe & Suter, 1997). However, in the brain, cholesterol synthesis via the 7-dehydrodesmosterol pathway seems to be preferred over the 7-dehydrocholesterol pathway and disruption of the gene coding for the delta 24-reductase (DHCR24) results in the accumulation of desmosterol without any accumulation of 7-dehydrodesmosterol (Wechsler et al., 2003).

In the adult brain most of the synthesis of cholesterol is balanced by formation of a hydroxylated metabolite, 24S-hydroxycholesterol (24OHC), which is able to pass across the blood-brain barrier and enter the circulation (Lütjohann et al., 1996; Björkhem & Meaney, 2004). About 6-8 mg/24h of cholesterol are released as 24OHC by the brain into the circulation (Lütjohann et al., 1996). In addition to this there is a small efflux of cholesterol from the brain in the form of ApoE containing lipoproteins via the cerebrospinal fluid (Xie et al., 2003).

Under normal conditions cholesterol 24-hydroxylase (CYP46A1), the enzyme system responsible for formation of 24S-OHC is only present in neuronal cells, mainly in cerebral cortex, hippocampus, dentate gyrus, amygdala, putamen and thalamus, i.e. associated with grey matter (Lund et al., 1999). The uptake of cholesterol by these cells may thus be balanced by the secretion of 24S-OHC. The 24S-OHC secreted from the neuronal cells may be of importance for regulation of cholesterol synthesis and secretion of this cholesterol in APOE-bound form from astroglia.

In the liver, the conversion of cholesterol into bile acids is regulated by highly sophisticated mechanisms (Russel, 2003). In the brain the expression of CYP46A1 appears to be resistant to regulatory axes known to regulate cholesterol homeostasis and bile acid synthesis. The promoter region of cholesterol 24-hydroxylase presents a high GC content, a feature often found in genes considered to have a largely housekeeping function (Ohyama et al., 2006). Oxidative stress was the only factor found to significantly affect its transcriptional activity. Cholesterol 24S-hydroxylase is localized in the neuronal cells and since these cells may depend on a flux of cholesterol from glial cells, it seems likely that substrate availability is an important regulatory factor for the enzyme under *in vivo* conditions (Björkhem, 2006).

24OHC is an endogenous regulator of the nuclear receptor Liver X Receptor (LXR). Under *in vitro* conditions 24OHC is able to regulate the expression, synthesis and secretion of ApoE (Abildayeva et al., 2006). Furthermore LXR-activation would be expected to increase expression of the sterol transporters ATP-binding cassette A1 (ABC-A1), G1 (ABC-G1) and G4 (ABC-G4) on astrocyte membranes, involved in the transport of cholesterol from glia to ApoE particles. Recently the importance of this mechanism *in vivo* was challenged by results obtained in a study on mice overexpressing CYP46A1 (Shafaati et al., 2011). Despite increased levels of 24OHC in the brain and in the circulation there was little or no increase in the expression of the different LXR target genes. In mice with combined Abcg1 and Abcg4 knockout results were obtained consistent with a role of these transporters in the efflux of cholesterol from neurons and glia in the CNS (Wang et al., 2008).

Cyp46a1 knock-out mice showed a modest reduction of hydroxy-methly-glutaryl-CoA-reductase activity and cholesterol synthesis rate while the total brain cholesterol levels were unaffected (Xie et al., 2003). Cyp46a1 (-/-) mice presented severe deficiencies in spatial, associative and motor learning associated with a delay of long lasting potential (Kotti et al., 2006). Also alterations in synaptic maturation were described. Treatment of hippocampal slices of wild type animals with an inhibitor of cholesterol synthesis essentially recapitulated the effects observed in the Cyp46a1 (-/-) mice (Russell et al., 2009).

Finally under *in vitro* conditions 24OHC is an efficient inhibitor of the formation of Aβ counteracting the positive effect of cholesterol on Amyloid Precursor Protein cleavage by β-secretase (BACE1) resulting in formation of the amyloidogenic Aβ1-42 fragment (Prasanthi et al., 2009; Bu, 2009).

In view of the fact that almost all 24OHC present in human circulation is of cerebral origin, (Björkhem et al., 2008) its plasma level is likely to be affected by the cholesterol homeostasis in the brain. It has been shown that the plasma levels are dependent both on the rate of secretion from the brain and the rate of hepatic metabolism (Bretillon et al., 2000a). Newborns have a size of the brain that is about three-fold that of the liver, whereas the size of the two organs is more or less similar in adults. As a consequence plasma levels of 24OHC are increased in children, infants and teenagers (Bretillon et al., 2000a) but are rather constant between the third to the seventh decades of life. According to one report the plasma concentrations of 24OHC are higher in males than in females (Vega et al., 2003). A similar observation was not reported by other more extensive studies. (Bretillion et al., 2000b; Leoni et al., 2008; Leoni et al., 2011; Leoni et al., 2002; Solomon et al., 2009b; van den Kommer et al., 2009; Burckhard et al., 2007).

In line with the fact that the number of metabolically active neuronal cells are decreased in the brain of patients with neurodegenerative diseases, the plasma levels of 24OHC have been reported to be decreased in Alzheimer's Disease (AD), Multiple Sclerosis (MS) and Huntington´s Disease (HD) (Papassotiropoulos et al., 2000; Kolsch et al., 2004; Solomon et al., 2009b; Koschack et al., 2009; Bretillion et al., 2000a; Bretillon et al., 2000b; Leoni et al., 2002; Teunissen et al., 2003; Danylaité Karrenbauer et al., 2006; Leoni et al., 2008; Leoni et al., 2011). Plasma levels of 24OHC may thus be regarded as a surrogate marker for the number of metabolically active neurons located in the grey matter of the brain (Björkhem, 2006).

In addition to 24OHC also 27OHC is able to pass the blood-brain barrier (Leoni et al., 2003). A continuous flux of 27OHC from the circulation into the mammalian brain has thus been demonstrated and this flux is of similar magnitude as the flux of 24OHC in the opposite direction (Heverin et al., 2005). In spite of the high influx of 27OHC into the brain the levels of this oxysterol is low, because of the rapid metabolism. It has been shown that most of the 27OHC present in CSF is derived from extracerebral 27OHC and that the levels are dependent upon the integrity of the blood-brain barrier. A damage of this barrier thus results in a higher flux of 27OHC from the circulation into the brain (Leoni et al., 2003; Heverin et al., 2005; Leoni et al., 2004). In view of the neurotoxic effect of 27OHC demonstrated in different *in vitro* experiments, the possibility has been discussed that the flux of 27OHC from the circulation into the brain could be a pathogenetic factor in the development of neurodegenerative diseases (Björkhem et al., 2009).

3. Cholesterol and neurodegenerative diseases other than Huntington´s Disease

The importance of cholesterol synthesis in the function, development and maturation of the central nervous system is very well illustrated by the consequence of genetic disorders affecting cholesterol synthesis or metabolism with prominent neurologic manifestations such as malformations, mental retardation, cognitive impairment and ataxia (Benarroch, 2008; Porter & Herman, 2011).

The most important neurodegenerative diseases in which a disturbance in cholesterol synthesis or metabolism is the primary pathogenetic factor is Smith-Lemli-Opitz syndrome and Niemann Pick Disease Type C. For details concerning these two diseases the reader should refer to the excellent review by Porter and Herman 2011.

Smith-Lemi-Opitz syndrome (SLOs) is an autosomal recessive malformation syndrome due to a mutation in the DHCR7 gene encoding 7-dehydroxycholesterol (7-DHC) reductase (Porter, 2008). Both the accumulation of 7-DHC and the reduced cholesterol synthesis participate to the SLOs phenotype, which is extremely broad, including CNS malformations such as holoprosencephaly and agenesis of the corpus callosum, mental retardation and motorial defects (Benarroch, 2008; Porter, 2008).

Niemann Pick Disease Type C (NPC) is a rare autosomal recessive neurovisceral lipid storage disease (Vanier & Millat, 2003). Mutations in the Niemann-Pick Disease, type C1 (NPC1) and Niemann-Pick Disease, type C2 (NPC2) genes have been identified as the genetic cause of the disease. NPC1 is a large membrane anchored protein with homology to HMGCR, SREBP cleavage activating protein (SCAP) and patched (PTCH1), a gene involved in Hedgehog signalling (Davies & Ioannou, 2000). In contrast, NPC2 is a small soluble glycoprotein (Storch & Xu, 2009). NPC has extreme clinical heterogeneity (Patterson, 2003; Benarroch, 2008) ranging from a rapidly fatal disorder in neonates to a neurodegenerative disorder in adults. The most common manifestations of adult NPC were cerebellar ataxia, vertical supranuclear ophthalmoplegia, dysarthria, and cognitive disturbances, followed by movement disorders (Patterson, 2003; Benarroch, 2008).

Important neurodegenerative diseases in which cholesterol metabolism is disturbed but not likely to be the primary pathogenetic factor are Alzheimer´s Disease and Parkinson´s Disease. For a detailed review see Björkhem et al. 2010. The most obvious link between Alzheimer´s Disease and cholesterol metabolism is the fact that presence of the E4 isoform of the cholesterol transporter ApoE as well as hypercholesterolemia are important risk factors for the disease.

4. Brain cholesterol in Huntington's Disease

Huntington's Disease (HD) is an inherited dominant neurodegenerative disorder characterised by a glutamine expansion within the N-terminus of the huntingtin protein (HTT) (Walker, 2007). The CAG trinucleotide repeats are located within the coding region of exon 1 of the HTT gene. HTT is widely expressed throughout the body and has been ascribed numerous roles in various intracellular functions including protein trafficking, vesicle transport, endocytosis, postsynaptic signalling, transcriptional regulation and an anti-apoptotic function (Gil & Rego, 2008). Gradual atrophy of the striatum (caudate nucleus and putamen) together with astrogliosis (Vonsattel et al., 1985) is a pathological characteristic of the disease. According to MRI investigations there is also a severe cortical atrophy combined with striatal degeneration (Aylward, 2007).

Cholesterol metabolism is affected in HD (Valenza & Cattaneo, 2011). The expression of some genes involved in the cholesterol biosynthetic pathway: hydroxy-methyl-glutaryl-CoA reductase , sterol 14-alpha demethylase (CYP51) and 7-dehydrocholesterol 7-reductase (DHCR7), were found to be reduced in inducible mutant HTT cell lines as well as in striatum and cortex of transgenic R6/2 HTT-fragment mice (Valenza et al., 2005; Valenza et al., 2007a)

Impairment of cholesterol metabolism in HD was confirmed in other additional studies.

The brain amount of the cholesterol precursors lanosterol and lathosterol, considered as markers for cholesterol synthesis (Xie et al., 2003) were found to be reduced. Also the levels of cholesterol were found to be significantly reduced in the brain of different rodent models for Huntington´s Disease such as the R6/2 mice (Valenza et al., 2007a), the yeast artificial chromosome (YAC) mice, the (HdhQ111/111) Hdh knock-in mice and others (Valenza et al., 2007b; Valenza et al., 2010).

Reduced levels of 24OHC were found in whole brain, striatum and cortex in the rodent models of Huntington's Disease, suggesting an impairment of cholesterol elimination by the metabolically active neuronal cells in the brain (Valenza et al., 2010).

The reductions of cholesterol synthesis, accumulation and turnover were found to be more marked with increasing length of the CAG repeats. In addition to the length of the repeats, the impairment of cholesterol synthesis was affected by the amount of mutated huntingtin. Finally, there was also an age-dependent effect. Thus the levels of cholesterol and cholesterol precursors were only slightly reduced in young animals during the process of maturation and much more reduced in older animals (Valenza et al., 2010).

A possible explanation for the molecular mechanism involved in the impairment of cholesterol metabolism is a mutant HTT-dependent decrease in the amount of active SREBP. The role of this factor is to translocate from the cytosolic compartment to the cell nucleus where, in presence of low cholesterol levels, it activates the transcription of SRE-controlled genes. Reduced SREBP translocation was thus found in cellular models of HD and in brain striatum collected from R6/2 mice (Valenza et al., 2005). A reduced entry of SREBP into the nucleus would be expected to lead to decreased cholesterol synthesis.

As referred to above cholesterol is critical for neurite outgrowth (Pfrieger, 2011). Neurite loss is an early manifestation of various neurodegenerative disorders, including HD, in which morphological abnormalities of the brain and defects in synaptic activity have been documented (Li et al., 2003; Levine et al., 2004; Schulz et al., 2004).

Wild type HTT is able to bind to some nuclear receptors involved in lipid metabolism: Liver-X-Receptor (LXR), PPARγ and vitamin D receptor (Futter et al., 2009). Overexpression of HTT was shown to activate LXR while a lack of HTT led to an inhibition of LXR-mediated transcription. The possibility must be considered that the mutated form of HTT is less able to up-regulate LXR and LXR-targeted genes, including SREBP. Such a mechanism could be a possible link between the HTT-mutation and the disturbances in cholesterol metabolism. Further work is needed, however, to establish this.

The mRNA levels of genes involved in cholesterol biosynthesis (hydroxy-methyl-glutaryl-CoA reductase, sterol 14-alpha demethylase, 7-dehydrocholesterol 7-reductase) and in cholesterol efflux (abca1 and abcg1) were found to be significantly reduced in primary astrocytes from both R6/2 and YAC 128 mice as compared to wild type controls or YAC18. Thus, astrocytes bearing a HTT mutation synthesized and secreted less ApoE than control cells. In accordance with this, the levels of HDL-like ApoE-lipoproteins present in CSF collected from YAC128 mice were reduced as compared to CSF from wt mice (Valenza et al., 2010). The results are consistent with a reduced ApoE mediated cholesterol transport.

In theory, the impairment of astrocyte cholesterol metabolism might be due to a combination of reduced activity of LXRs as a consequence of the reduced levels of 24OHC (Valenza et al., 2010) and a reduced SREBP activation. According to the study by Shafaati et al., however, the levels of 24OHC may be less important for LXR activation under *in vivo* conditions (Shafaati et al., 2011). It seems likely that there are other yet uncovered HTT-sensitive mechanisms that are of importance for synthesis, transport and delivery of cholesterol from astrocytes to neurons .

Both MRI and pathological investigations demonstrated abnormalities in oligodendrocytes and white matter in HD brains (Myers et al., 1991; Gomez-Tortosa et al., 2001; Fennema-Notestine et al., 2004; Paulsen et al., 2008; Tabrizi et al., 2009; Nopoulos et al., 2011; Rosas et al., 2010) even in pre manifesting subjects (Gomez-Tortosa et al., 2001; Bartzokis et al., 2007; Tabrizi et al., 2009). Pathological alteration of white matter may represent an early event in HD pathogenesis.

In primary oligodendrocytes, mutant HTT was found to inhibit the regulatory effect of Peroxisome-proliferator-activated receptor-gamma co-activator 1 alpha (PGC1α) on HmgCoA synthetase and HmgCoA reductase, expression of myelin basic protein and cholesterol metabolism (Xjiang et al., 2011). Brains from R6/2 and BACHAD mice had abnormal myelination, reduced expression of myelin basic protein and PGC1α (Xjiang et al., 2011). In a PGC1α knock out mice model defective myelination, reduced expression of myelin basic protein and reduction of cholesterol synthesis and accumulation has been demonstrated. The expression of HMGCoA reductase and HMGCoA synthetase and myelin basic protein was found to be reduced in this model (Xjiang et al., 2011). Peroxisome-proliferator-activated receptor gamma co-activator 1 alpha (PGC1α) plays a role in the transcriptional regulation of energy metabolism and has been implicated in several neurodegenerative disorders (Finck & Kelly, 2006), including HD (Cui et al., 2006; Weydt et al., 2006; Chaturvedi et al., 2009; Chaturvedi et al., 2010; McConoughey et al., 2010). PGC1α knockout mice exhibited vacuolar abnormalities in the CNS that were primarily associated with the white matter (Lin et al., 2006; Leone et al., 2005). It is likely that PGC1α is involved in regulation of cholesterol synthesis by direct or indirect interaction with SREBP and LXR affecting, thus, myelination.

5. Study of peripheral and cerebral cholesterol metabolism in neurodegenerative disorders

5.1 Huntington's Disease

Plasma concentration of 24OHC was found reduced in HD patients compared to healthy subjects. Both in two populations (an Italian and an English cohort), as well as in the combined two cohorts, 24OHC levels were significantly reduced at any disease stage (Leoni et al., 2008). The reduction of plasma 24OHC was found to parallel the degree of caudate atrophy (measured as reduction of caudate volume at MRI). A significant positive correlation was found between 24OHC levels and degree of caudate atrophy as measured by morphometric MRI (Leoni et al., 2008). These results support that reduction of plasma 24S-hydroxycholesterol is related to the loss of metabolically active neuronal cells in the brain and thus to the degree of brain athrophy (see also Leoni & Caccia, 2011).

Total plasma cholesterol was found to be reduced in pre-manifesting subjects and in HD patients compared to controls (Markianos et al., 2008). In other studies (Leoni et al., 2008; Leoni et al., 2011) a slight reduction was found with the progress of the disease stage. A significant reduction of cholesterol levels were however found only in the mast advanced cases (stage 3-5).

In a more detailed study on cholesterol homeostasis in HD it was reported that the cholesterol precursors lanosterol and lathosterol were reduced in plasma collected from HD patients at any disease stage. Also the level of the bile acid precursor 27-hydroxycholesterol was significantly reduced. Thus both whole-body and brain cholesterol homeostasis appear to be impaired in HD (Leoni et al., 2011).

HD gene positive carriers (named pre-manifest individuals) have been shown to present significant cognitive and neuropsychiatric dysfunction in parallel with changes in whole-brain volume, regional grey and white matter, at a stage prior to motor onset of disease. (Paulsen et al., 2008; Tabrizi et al., 2009). The plasma levels of 24OHC in pre-manifest subjects were similar to those of controls and higher than those of HD patients. However the gene positive pre-manifest subjects were heterogeneous: some subjects were very close to the motor onset with advanced neurodegeneration, others were far from onset. In subjects close to motor onset, 24OHC levels were found to be lower compared to those far from onset, and similar to the levels observed in manifest HD patients (Leoni et al., 2008). Interestingly, the markers of cholesterol synthesis lathosterol and lanosterol, and the marker of cholesterol elimination (27-hydroxycholesterol) were found to be reduced in pre-manifest subjects while the levels of 24S-hydroxycholesterol were reduced in patients proportionally to the degree of brain atrophy observed at MRI.

Presence of huntingtin mutations appears the be associated with a general global effect on cholesterol synthesis. Thus it is tempting to suggest that the huntingtin protein has a regulatory role in the normal cerebral as well as extracerebral biosynthesis of cholesterol (Valenza & Cattaneo, 2011). The production of 24OHC by the neuronal cells is likely to be dependent both on the numbers of such cells and on availability of substrate cholesterol. Both these factors are likely to be affected by HTT mutations.

5.2 Plasma sterols and oxysterols in neurodegenerative disease

Impairment of cholesterol metabolism were described also in animal models and patients affected by Multiple Sclerosis and Alzheimer's Disease.

Multiple Sclerosis (MS) is the most common autoimmune and demyelinating disorder of the CNS. Axonal damage and neurodegeneration is commonly found in the brains of patients with MS in both lesions and in normal-appearing white matter (Miller et al., 2002). Substantial neuronal loss and volume loss were demonstrated in grey matter, resulting in brain atrophy measured at Magnetic Resonance Imaging (MRI) (Cifelli et al., 2002). Plasma 24OHC was significantly reduced in relapsing-remitting and in primary progressive MS patients with a long story of disease (Leoni et al., 2002; Teunissen et al., 2003). The reduction of plasma 24OHC may reflect the total spatiotemporal burden of disease (i.e. the cumulative effects of its dissemination in space and its duration in time) since a significant correlation between plasma 24OHC and the volume of T2-weighted hyperintense lesions in relapsing-remitting and in primary progressive patients (Danylaité Karrenbauer et al., 2006). A significant direct correlation was observed between the plasma 24S-hydroxycholesterol and the Grey Matter Fraction (MRI marker of brain atrophy) of MS patients (Leoni & Caccia, 2011) . Lathosterol was found reduced in patients affected by MS (Teunissen, 2003) as well in animal model of MS (Teunissen et al., 2007). Also 27-hydroxycholesterol was found reduced in plasma collected from patients (Leoni et al., 2002; Teunissen et al., 2003) suggesting that whole body cholesterol metabolism may be altered in MS.

In AD the annual rate of global brain atrophy is 2-3% as compared with 0.2-0.5% in healthy controls (Fox et al., 1999; Jack et al., 2010). There is a prominent early involvement of medial temporal lobe structures, especially the enthorineal cortex and hippocampus (Jack et al., 1998). The progressive extensive atrophy is associated to a progressive reduction of the brain (Heverin et al., 2004) and plasma levels of 24OHC, and the latter is negatively correlated to the Mini Mental Score (Papassotiropoulos et al., 2000; Solomon et al., 2009b). A significant correlation of 24OHC with the hippocampal volume (Koschack et al., 2009) or the direct or fractional volumes of grey matter was found in mid-age or aged individuals (Solomon et al., 2009b). Such correlation was missed in case of Mild Cognitive Impairment (MCI) or AD patients: a possible explanation could be the abnormal expression of the CYP46 enzyme in glial cells that was shown in the brain of patients affected by AD (Brown et al., 2004; Bogdanovic et al., 2001), which occurs as a compensatory mechanism in neuronal degeneration.

Finally it was found that the reduction of plasma 24OHC correlated with the severity of dementia or the degree of brain atrophy (Papassotiropoulos et al, 2000; Solomon et al., 2009b).

Epidemiological studies showed an association between elevated total cholesterol at midlife and increased risk of AD (Kivipelto & Salomon, 2006). Long-term studies reported that a decline in plasma total cholesterol levels from midlife to late-life is associated with early stages in dementia development. It is likely that while high midlife cholesterol is a risk factor for AD, decreased cholesterol later in life may instead reflect an going pathological processes in the brain and should be considered as a frailty marker, predictive of worse cognitive functioning (Stewart et al., 2007; Mielke et al., 2005; Solomon et al., 2009a; Solomon et al., 2009b). A large 21- year follow-up study presented an association between serum total cholesterol changes from midlife to late-life and late-life cognitive status: a moderate decrease is associated with increased risk of a more impaired late-life cognitive status after adjusting for major confounders (Solomon et al., 2007).

No correlation between serum total cholesterol or LDL-C and CSF biomarkers was reported (Solomon et al., 2009a). No significant differences about total or LDL-cholesterol were found between aging individuals, MCI and AD patients but significant reductions of cholesterol precursors lathosterol and lanosterol and 27-hydroxycholesterol were instead observed in AD patients compared to MCI and aging individuals. As expected Aβ1-42 changed in the same way while tau and P-tau in the opposite one. Thus, the CSF biomarkers signature in aging population with cognitive decline was found associated with reduction of whole body cholesterol metabolism (Solomon et al., 2009a). In AD patients (but not in case of MCI or control individuals) lower plasma total cholesterol and LDL-C were found related to lower brain volumes/higher CSF volumes (Solomon et al., 2009a). In contrast, in the control group lower levels of the cholesterol precursors lanosterol and lathosterol (considered as marker of a lower rate of endogenous cholesterol synthesis) were related to higher brain volumes/lower CSF volumes. The positive correlations between lanosterol, lathosterol, total cholesterol and LDL-C with brain volumes in patients with AD compared to MCI and controls are consistent with the hypothesis of a central nervous system (CNS)-induced depressing effect of neurodegeneration on extracerebral cholesterol metabolism (Solomon et al., 2009).

Very recent studies on patients with Parkinson's Disease have reported a markedly decreased level of 24OHC in the plasma, which is consistent with the finding of correlation between brain atrophy, CNS neuronal mass and its plasma levels (Björkhem et al., 2009).

In addition to the above diseases, brain tumours and some severe central nervous system infections also have reduced levels of 24OHC in the circulation (Bretillon et al., 2000b).

6. Conclusion

A clear link has been established between the glutamine expansion in the huntingtin gene and cholesterol metabolism. The mechanism behind this is still unknown. Since the effect on cholesterol synthesis is global it seems likely that the huntingtin gene is of regulatory importance for cholesterol synthesis also under normal conditions. It should be noted that unexplained global effect on cholesterol homeostasis has been observed also in other neurodegenerative diseases such as Alzheimer´s Disease.

Liver integrity and clearance, presence of CNS pathology, therapies, cholesterol recommended levels, body mass index, diet were found to affect significantly the whole body cholesterol metabolism and plasma levels of 24OHC (Bretilion et al., 2000; Björkhem et al., 2009; Björkhem, 2006; Brown & Jessup, 2009; Leoni & Caccia, 2011). The criteria of selection of the control population, the pre-analytical factors of sample collection and handling, the methodology used for the study of sterols and oxysterols may affect the final findings. The use of sterols and side-chain oxidised cholesterol as biomarker for the diagnosis of neurodegenerative diseases seems to be still limited. However, the plasma level of a neuronal metabolite of cholesterol, 24S-hydroxycholesterol, appears to be a valuable biomarker for the progression of Huntington´s Disease.

7. Acknowledgments

The authors wish to gratefully acknowledge the collaboration along the years of Dr. A. Salomon, Prof. M. Kivipelto, Dr. T. Mastermann, Dr. M. Shafaati at Karolinska Insitutet, Stockholm, Sweden; Dr. C. Mariotti and Dr. S. Di Donato, IRCCS Istituo Neurologico "C. Besta", Milano, Italy; Dr. M. Valenza at University of Milano, Italy.

Financial support: Italian Mininster of Health, Fondi per giovani Ricercatori 2008, to V. Leoni; Swedish Science Council and the Swedish Brain Power to I. Björkhem.

8. References

Abildayeva, K., Jansen, P. J., Hirsch-Reinshagen, V., Bloks, V. W., Bakker, A. H., Ramaekers, J. de Vente, F. C., Groen, A. K., Wellington, C. L. , Kuipers, F., & Mulder, M. (2006). 24(S)-hydroxycholesterol participates in a liver X receptor-controlled pathway in astrocytes that regulates apolipoprotein E-mediated cholesterol efflux. *J. Biol. Chem.*, Vol.281, No.18, (May 2006), pp. 12799-12808, ISSN 0021-9258

Allen, J.A., Halverson-Tamboli, R.A., & Rasenick, M.M. (2007). Lipid raft microdomains and neurotransmitter signalling. *Nat. Rev. Neurosci.*, Vol.8, No.2, (February 2007), pp. 128–140, ISSN 1471-003X

Aylward, E.H. (2007). Change in MRI striatal volumes as a biomarker in preclinical Huntington's disease. *Brain Res. Bull.*, Vol.72, No.2-3, (April 2007), pp. 152-158, ISSN 0361-9230

Bartzokis, G., Lu, P.H., Tishler, T.A., Fong, S.M., Oluwadara, B., Finn, J.P., Huang, D., Bordelon, Y., Mintz, J., & Perlman, S. (2007). Myelin breakdown and iron changes in Huntington's disease: pathogenesis and treatment implications. *Neurochem. Res.*, Vol.32, No.10, (October 2007), pp. 1655-1664, ISSN 0364-3190

Baumann. N., & Pham-Dinh, D. (2001). Biology of oligodendrocyte and myelin in the mammalian central nervous system. *Physiol. Rev.*, Vol.81, No.2, (April 2001), pp. 871-927, ISSN 0031-9333

Benarroch, E.E. (2008). Brain cholesterol metabolism and neurologic disease. *Neurology*, Vol.71, No.17, (October 2008), pp. 1368-1373, ISSN 0028-3878

Björkhem, I. (2006). Crossing the barrier, oxysterols as cholesterol transporters and metabolic modulators in the brain. *J. Intern. Med.*, Vol.260, No.6, (November 2006), pp. 493-508, ISSN 0954-6820

Bjorkhem, I., & Meaney, S. (2004). Brain cholesterol: long secret life behind a barrier. *Arterioscler. Thromb. Vasc. Biol.*, Vol.24, No.5, (May 2004), pp. 806-815, ISSN 1079-5642

Björkhem, I., Cedazo-Minguez, A., Leoni, V., & Meaney, S. (2009). Oxysterols and neurodegenerative diseases. *Mol. Aspects Med.*, Vol.30, No.3, (June 2009), pp. 171-179, ISSN 0098-2997

Björkhem, I., Leoni, V., & Meaney, S. (2010). Genetic connections between neurological disorders and cholesterol metabolism. *J. Lipid Res.*, Vol.51, No.9, (September 2010), pp. 2489-2503, ISSN 0022-2275

Bloch, K., Clayton, R.B., & Schneider, P.B. (1957) Synthesis of lanosterol in vivo. *J. Biol. Chem.*, Vol.224, No1, (January1957), pp. 175–183, ISSN 0021-9258

Bogdanovic, N., Bretillon, L., Lund, E.G., Diczfalusy, U., Lannfelt, L., Winblad, B., Russell, D.W., & Björkhem, I. (2001). On the turnover of brain cholesterol in patients with Alzheimer's disease. Abnormal induction of the cholesterol-catabolic enzyme CYP46 in glial cells. *Neurosci. Lett.* Vol 314, No1-2 (November 2001), pp. 45-48. ISSN 0304-3940

Bretillon, L., Lütjohann, D., Stahle, L., Widhe, T., Bindl, L., Eggersten, G., Diczfalusy, U., & Bjorkhem, I., (2000a). Plasma levels of 24S-hydroxycholesterol reflect the balance between cerebral production and hepatic metabolism and are inversely related to body surface. *J. Lipid Res.*, Vol.41, No.5, (May 2000), pp. 840-845, ISSN 0022-2275

Bretillon, L., Siden, Å., Wahlund, L.O., Lütjohann, D., Minthon, L., Crisby, M., Hillert, J., Groth, G.C., Diczfalusy, U., & Björkhem, I. (2000b). Plasma levels of 24S-hydroxycholesterol in patients with neurological diseases. *Neurosci. Lett.*, Vol.293, No.2, (October 2000), pp. 87-90, ISSN 0304-3940

Brown, A.J., & Jessup, W. (2009). Oxysterols, Sources, cellular storage and metabolism, and new insights into their roles in cholesterol homeostasis. *Mol. Aspects Med.*, Vol.30, No.3, (June 2009), pp. 111-122, ISSN 0098-2997

Brown, J. 3rd, Theisler, C., Silberman, S., Magnuson, D., Gottardi-Littell, N., Lee, J.M., Yager, D., Crowley, J., Sambamurti, K., Rahman, M.M., Reiss, A.B., Eckman, C.B., & Wolozin, B. (2004). Differential expression of cholesterol hydroxylases in Alzheimer's disease. *J. Biol. Chem.* Vol 279, No 33 (August 2004), pp. 34674-34681. ISSN 0021-9258

Bu, G. (2009). Apolipoprotein E and its receptors in Alzheimer's disease, pathways, pathogenesis and therapy. *Nat. Rev. Neurosci.*, Vol.10, No.5, (May 2009), pp. 333-344, ISSN 1471-003X

Burkard, I., von Eckardstein, A., Waeber, G., Vollenweider, P., & Rentsch, K.M. (2007). Lipoprotein distribution and biological variation of 24S- and 27-hydroxycholesterol in healthy volunteers. *Atherosclerosis*. Vol.194, No.1 (September 2007), pp. 71-78. ISSN 0021-9150

Chaturvedi, R.K., Adhihetty, P., Shukla, S., Hennessy, T., Calingasan, N., Yang, L., Starkov, A., Kiaei, M., Cannella, M., Sassone, J., Ciammola, A., Squitieri, F., & Beal, M.F. (2009). Impaired PGC-1alpha function in muscle in Huntington's disease. *Hum. Mol. Genet.*, Vol.18, No.16, (August 2009), pp. 3048-3065, ISSN 0964-6906

Chaturvedi, R.K., Calingasan, N.Y., Yang, L., Hennessey, T., Johri, A., & Beal, M.F. (2010). Impairment of PGC-1alpha expression, neuropathology and hepatic steatosis in a transgenic mouse model of Huntington's disease following chronic energy deprivation. *Hum. Mol. Genet.*, Vol.19, No.16, (August 2010), pp. 3190-3205, ISSN 0964-6906

Cifelli, A., Arridge, M., Jezzard, P., Esiri, M.M., Palace, J., & Matthews, P.M. (2002). Thalamic neurodegeneration in multiple sclerosis. *Ann. Neurol.* Vol.52, No.5, (November 2002), pp. 650-653. ISSN 0364-5134

Cui, L., Jeong, H., Borovecki, F., Parkhurst, C.N., Tanese, N., & Krainc, D. (2006). Transcriptional repression of PGC-1alpha by mutant huntingtin leads to mitochondrial dysfunction and neurodegeneration. *Cell*, Vol.127, No.1, (October 2006), pp. 59-69, ISSN 0092-8674

Danylaité Karrenbauer, V., Leoni, V., Lim, E.T., Giovannoni, G., Ingle, G.T., Sastre-Garriga, J., Thompson, A.J., Rashid, W., Davies, G., Hillert, J., Miller, D.H., Björkhem, I., & Masterman, T. (2006). Plasma cerebrosterol and magnetic resonance imaging measures in multiple sclerosis. *Clin. Neurol. Neurosurg.*, Vol.108, No.5, (July 2006), pp. 456-460, ISSN 0303-8467

Davies, J. P., & Ioannou, Y. A. (2000). Topological analysis of Niemann-Pick C1 protein reveals that the membrane orientation of the putative sterol-sensing domain is identical to those of 3-hydroxy-3-methylglutaryl-CoA reductase and sterol regulatory element binding protein cleavage-activating protein. *J. Biol. Chem.*, Vol.275, No.32, (August 2000), pp. 24367-24374, ISSN 0021-9258

Dietschy, J.M., & Turley, S.D. (2004). Thematic review series, brain lipids. Cholesterol metabolism in the central nervous system during early development and in the mature animal. *J. Lipid Res.*, Vol.45, No.8, (August 2004), pp. 1375–1397, ISSN 0022-2275

Duane, W.C., & Javitt, N.B. (1999). 27-hydroxycholesterol: production rates in normal human subjects. *J. Lipid Res.*, Vol.40, No.7, (July 1999), pp. 1194-1199, ISSN 0022-2275

Fennema-Notestine, C., Archibald, S.L., Jacobson, M.W., Corey-Bloom, J., Paulsen, J.S., Peavy, G.M., Gamst, A.C., Hamilton, J.M., Salmon, D.P., & Jernigan, T.L. (2004). In vivo evidence of cerebellar atrophy and cerebral white matter loss in Huntington disease. *Neurology*, Vol.63, No.6, (September 2004), pp. 989-995, ISSN 0028-3878

Finck, B.N., & Kelly, D.P. (2006). PGC-1 coactivators: inducible regulators of energy metabolism in health and disease. *J. Clin. Invest.*, Vol.116, No.3, (March 2006), pp. 615-622, ISSN 0021-9738

Fox, N.C., Scahill, R.I., Crum, W.R., & Rossor, M.N. (1999). Correlation between rates of brain atrophy and cognitive decline in AD. *Neurology*, Vol.52, No.8, (May 1999), pp. 1687-1689, ISSN 0893-0341

Futter, M., Schoenmakers, H., Sadiq, O., Chatterjee, K., & Rubinsztein, D.C. (2009). Wild-type but not mutant huntingtin modulates the transcriptional activity of liver X erceptors. J. Med. Genetic., Vol.46, No.7, (July 2009), pp. 438-446, ISSN 0022-2593

Gil, J.M., & Rego A.C. (2008). Mechanisms of neurodegeneration in Huntington's disease. *Eur. J. Neurosci.*, Vol.27, No.11, (June 2008), pp. 2803-2820, ISSN 0953-816X

Gomez-Tortosa, E., MacDonald, M.E., Friend, J.C., Taylor, S.A., Weiler, L.J., Cupples, L.A., Srinidhi, J., Gusella, J.F., Bird, E.D., Vonsattel, J.P., & Myers, R.H. (2001). Quantitative neuropathological changes in presymptomatic Huntington's disease. *Ann. Neurol.*, Vol.49, No.1, (January 2001), pp. 29-34, ISSN 0364-5134

Heverin, M., Bogdanovic, N., Lütjohann, D., Bayer, T., Pikuleva, I., Bretillon, L., Diczfalusy, U., Winblad, B., & Björkhem, I. (2004). Changes in the levels of cerebral and extracerebral sterols in the brain of patients with Alzheimer's disease. *J. Lipid Res.*, Vol.45, No.1, (January 2004), pp. 186–193, ISSN 0022-2275

Heverin, M., Meaney, S., Lütjohann, D., Diczfalusy, U., Wahren, J., & Björkhem, I. (2005). Crossing the barrier, net flux of 27-hydroxycholesterol into the human brain. *J. Lipid Res.*, Vol.46, No.5, (May 2005), pp. 1047-1052, ISSN 0022-2275

Jack, C.R. Jr., Knopman, D.S., Jagust, W.J., Shaw, L.M., Aisen, P.S., Weiner, M.W., Petersen, R.C., & Trojanowski, J.Q. (2010). Hypothetical model of dynamic biomarkers of the Alzheimer's pathological cascade. *Lancet Neurol.*, Vol.9, No.1, (January 2010) pp. 119-128, ISSN 1474-4422

Jack, C.R.Jr., Petersen, R.C., Xu, Y., O'Brien, P.C., Smith, G.E., Ivnik, R.J., Tangalos, E.G., & Kokmen, E. (1998). Rate of medial temporal lobe atrophy in typical aging and Alzheimer's disease. *Neurology*, Vol.51, No.4, (October 1998), pp. 993-999, ISSN 0028-3878

Kempen, H.J., Glatz, J.F., Gevers Leuven, J.A., van der Voort, H.A., & Katan, M.B. (1988). Serum lathosterol concentration is an indicator of whole-body cholesterol synthesis in humans. *J. Lipid Res.*, Vol.29, No.9, (September 1988), pp. 1149–1155, ISSN 0022-2275

Kivipelto, M., & Solomon, A. (2006). Cholesterol as a risk factor for Alzheimer's disease - epidemiological evidence. *Acta Neurol. Scand. Suppl.*, Vol.114, No.s185, (August 2006), pp. 50-57, ISSN 0065-1427

Kolsch, H., Heun, R., Kerksiek, A., Bergmann, K.V., Maier, W., & Lutjohann, D. (2004). Altered levels of plasma 24S- and 27-hydroxycholesterol in demented patients. *Neurosci. Lett.*, Vol.368, No.3, (September 2004), pp. 303-308, ISSN 0304-3940

Koschack, J., Lütjohann, D., Schmidt-Samoa, C., & Irle, E. (2009). Serum 24S-hydroxycholesterol and hippocampal size in middle-aged normal individuals. *Neurobiol. Aging.*, Vol.30, No.6, (June 2009), pp. 898-902, ISSN 0197-4580

Kotti, T.J., Ramirez, D.M., Pfeiffer, B.E., Huber, K.M., & Russell, D.W. (2006). Brain cholesterol turnover required for geranylgeraniol production and learning in mice. *Proc. Natl. Acad. Sci. U.S.A.*, Vol.103, No.10, (March 2006), pp. 3869–3874, ISSN 0027-8424

Koudinov, A.R., & Koudinova, N.V.(2001). Essential role for cholesterol in synaptic plasticity and neuronal degeneration. *FASEB J.*, Vol.15, No.10, (August 2001), pp. 1858 -1860, ISSN 0892-6638

Leone, T.C., Lehman, J.J., Finck, B.N., Schaeffer, P.J., Wende, A.R., Boudina, S., Courtois, M., Wozniak, D.F., Sambandam, N., Bernal-Mizrachi, C., Chen, Z., Holloszy, J.O., Medeiros, D.M., Schmidt, R.E., Saffitz, J.E., Abel, E.D., Semenkovich, C.F., & Kelly, D.P. (2005). PGC-1alpha deficiency causes multi-system energy metabolic derangements: muscle dysfunction, abnormal weight control and hepatic steatosis. *PloS. Biol.*, Vol.3, No.4, (April 2005), e101, ISSN 1545-7885

Leoni, V., & Caccia, C. (2011). Oxysterols as biomarkers in neurodegenerative diseases. *Chem. Phys. Lipids*, Vol.164, No.6, (September 2011), pp. 515-524, ISSN 0009-3084

Leoni, V., Mariotti, C., Nanetti, L., Salvatore, E., Squitieri, F., Bentivoglio, A.R., Bandettini Del Poggio, M., Piacentini, S., Monza, D., Valenza, M., Cattaneo, E., & Di Donato, S. (2011). Whole body cholesterol metabolism is impaired in Huntington's disease. *Neurosci. Lett.*, Vol.494, No.3, (May 2011), pp. 245-249, ISSN 0304-3940

Leoni, V., Mariotti, C., Tabrizi, S.J., Valenza, M., Wild, E.J., Henley, S.M., Hobbs, N.Z., Mandelli, M.L., Grisoli, M., Björkhem, I., Cattaneo, E., & Di Donato, S. (2008). Plasma 24S-hydroxycholesterol and caudate MRI in pre-manifest and early Huntington's disease. *Brain*, Vol.131, No.11, (November 2008), pp. 2851-2859, ISSN 0006-8950

Leoni, V., Masterman, T., Diczfalusy, U., De Luca, G., Hillert, J., & Björkhem, I., (2002). Changes in human plasma levels of 24S-hydroxycholesterol during progression of multiple sclerosis. *Neurosci. Lett.*, Vol.331, No.3, (October 2002), pp. 163-166, ISSN 0304-3940

Leoni, V., Masterman, T., Patel, P., Meaney, S., Diczfalusy, U., & Björkhem, I. (2003). Side-chain oxidised oxysterols in cerebrospinal fluid and integrity of blood-brain barrier. *J. Lipid Res.*, Vol.44, No.4, (April 2003), pp. 793-799, ISSN 0022-2275

Leoni, V., Mastermann, T., Mousavi, F.S., Wretlind, B., Wahlund, L.O., Diczfalusy, U., Hillert, J., & Björkhem, I. (2004). Diagnostic use of cerebral and extracerebral oxysterols. *Clin. Chem. Lab. Med.*, Vol.42, No.2, (February 2004), pp.186-191, ISSN 1434-6621

Leoni, V., Solomon, A., & Kivipelto, M. (2010). Links between ApoE, brain cholesterol metabolism, tau and amyloid beta-peptide in patients with cognitive impairment. *Biochem. Soc. Trans.*, Vol.38, No.4, (August 2010), pp. 1021-1025, ISSN 0300-5127

Levine, M.S., Cepeda, C., Hickey, M.A., Fleming, S.M., & Chesselet, M.F. (2004). Genetic mouse models of Huntington's and Parkinson's diseases: illuminating but imperfect. *Trends Neurosci.*, Vol.27, No.11, (November 2004), pp. 691-697, ISSN 0166-2236

Li, J.Y., Plomann, M., & Brundin, P. (2003). Huntington's disease: a synaptopathy? *Trends Mol. Med.*, Vol.9, No.10, (October 2003), pp. 414-420, ISSN 1471-4914

Lin, T., Xiang, Z., Cui, L., Stallcup, W., & Reeves, S.A. (2006). New mouse oligodendrocyte precursor (mOP) cells for studies on oligodendrocyte maturation and function. *J. Neurosci. Methods.*, Vol.157, No.2, (October 2006), pp. 187-194, ISSN 0165-0270

Lund, E.G., Guileyardo, J.M., & Russell, D.W. (1999). cDNA cloning of cholesterol 24-hydroxylase, a mediator of cholesterol homeostasis in the brain. *Proc. Natl. Acad. Sci. U.S.A.*, Vol.6, No.13, (June 1999), pp. 7238–7243, ISSN 0027-8424

Lütjohann, D., Breuer, O., Ahlborg, G., Nennesmo, I., Siden, Å., Diczfalusy, U., & Björkhem, I. (1996). Cholesterol homeostasis in human brain, evidence for an age-dependent flux of 24S-hydroxycholesterol from the brain into the circulation. *Proc. Natl. Acad. Sci. U.S.A.*, Vol. 93, No.18, (September 1996), pp. 9799–9804, ISSN 0027-8424

Markianos, M., Panas, M., Kalfakis, N., & Vassilopoulos, D. (2008). Low plasma total cholesterol in patients with Huntington's disease and first-degree relatives. *Mol. Genet. Metab.*, Vol. 93, No. 3, (March 2008), pp. 341-346, ISSN 1096-7192

Matthan, N.R., Raeini-Sarjaz, M., Lichtenstein, A.H., Ausman, L.M., & Jones, P.J. (2000). Deuterium uptake and plasma cholesterol precursor levels correspond as methods for measurement of endogenous cholesterol synthesis in hypercholesterolemic women. *Lipids*, Vol.35, No.9, (September2000), 1037-1044, ISSN 0024-4201

Maxfield, F.R., & Tabas, I. (2005). Role of cholesterol and lipid organization in disease. *Nature*, Vol.438, No.7068, (December 2005), pp. 612–621, ISSN 0028-0836

McConoughey, S.J., Basso, M., Niatsetskaya, Z.V., Sleiman, S.F., Smirnova, N.A., Langley, B.C., Mahishi, L., Cooper, A.J., Antonyak, M.A., Cerione, R.A., Li, B., Starkov, A., Chaturvedi, R.K., Beal, M.F., Coppola, G., Geschwind, D.H., Ryu, H., Xia, L., Iismaa, S.E., Pallos, J., Pasternack, R., Hils, M., Fan, J., Raymond, L.A., Marsh, J.L., Thompson, L.M., & Ratan, R.R. (2010). Inhibition of transglutaminase 2 mitigates transcriptional dysregulation in models of Huntington disease. *EMBO Mol. Med.*, Vol.2, No.9, (September 2010), pp. 349-370, ISSN 1757-4676

Mielke, M.M., Zandi, P.P., Sjogren, M., Gustafson, D., Ostling, S., Steen, B., & Skoog, I. (2005). High total cholesterol levels in late-life associated with a reduced risk of dementia. *Neurology*, Vol.64, No.10, (May 2005), pp. 1689–1695, ISSN 0028-3878

Miller, D.H. (2002). MRI monitoring of MS in clinical trials. *Clin. Neurol. Neurosurg.*, Vol.104, No.3, (July 2002), pp. 236-243, ISSN 0303-8467

Myers, R.H., Vonsattel, J.P., Paskevich, P.A., Kiely, D.K., Stevens, T.J., Cupples, L.A., Richardson, E.P.Jr., & Bird E.D. (1991). Decreased neuronal and increased oligodendroglial densities in Huntington's disease caudate nucleus. *J. Neuropathol. Exp. Neurol.*, Vol.50, No.6, (November 1991), pp. 729-742, ISSN 0022-3060

Nopoulos, P.C., Aylward, E.H., Ross, C.A., Mills, J.A., Langbehn, D.R., Johnson, H.J., Magnotta, V.A., Pierson, R.K., Beglinger, L.J., Nance, M.A., Barker, R.A., & Paulsen, J.S. (2011). Smaller intracranial volume in prodromal Huntington's disease: evidence for abnormal neurodevelopment. *Brain*, Vol.134, No.1, (January 2011), pp. 137-142, ISSN 0006-8950

Ohyama, Y., Meaney, S., Heverin, M., Ekstrom, L., Brafman, A., Shafir, M., Andersson, U., Olin, M., Eggertsen, G., Diczfalusy, U., Feinstein, E., & Bjorkhem, I. (2006). Studies on the transcriptional regulation of cholesterol 24-hydroxylase (CYP46A1): marked insensitivity toward different regulatory axes. *J. Biol. Chem.*, Vol.281, No.7, (February 2006), pp. 3810-3820, ISSN 0021-9258

Papassotiropoulos, A., Lütjohann, D., Bagli, M., Locatelli, S., Jessen, F., Rao, M.L., Maier, W., Björkhem, I., von Bergmann, K., & Heun, R. (2000). Plasma 24S-hydroxycholesterol, a peripheral indicator of neuronal degeneration and potential state marker for Alzheimer's disease. *Neuroreport*, Vol.11, No.9, (June 2000), pp. 1959-1962, ISSN 0959-4965

Patterson, M.C. (2003). A riddle wrapped in a mystery: understanding Niemann-Pick disease, type C. *Neurologist*, Vol.9, No.6, (November 2003), pp. 301-310, ISSN 1074-7931

Paulsen, J.S., Langbehn, D.R., Stout, J.C., Aylward, E., Ross, C.A., Nance, M., Guttman, M., Johnson, S., MacDonald, M., Beglinger, L.J., Duff, K., Kayson, E., Biglan, K., Shoulson, I., Oakes, D., & Hayden, M. (2008). Detection of Huntington's disease decades before diagnosis: the Predict-HD study. *J. Neurol. Neurosurg. Psychiatry.*, Vol.79, No.8, (August 2008), pp. 874-880, ISSN 0022-3050

Pfrieger F.W. (2003a). Cholesterol homeostasis and function in neurons of the central nervous system. *Cell. Mol. Life Sci.*, Vol.60, No.6, (June 2003), pp. 1158-1171, ISSN 1420-682X

Pfrieger, F.W. (2003b). Outsourcing in the brain, do neurons depend on cholesterol delivery by astrocytes? *Bioessays*, Vol.25, No.1, (January 2003), pp. 72–78, ISSN 0265-9247

Pfrieger, F.W., & Ungerer, N. (2011). Cholesterol metabolism in neurons and astrocytes. Prog. Lipid. Res., Vol.50, No.4, (October 2011), pp. 357-371, ISSN 0163-7827

Pitas, R. E., Boyles, J. K., Lee, S. H., Foss, D., & Mahley, R. W. (1987). Astrocytes synthesize apolipoprotein E and metabolize apolipoprotein E-containing lipoproteins. *Biochim. Biophys. Acta*, Vol.917, No.1, (January 1987), pp. 148-161, ISSN 0006-3002

Porter, F.D. & Herman, G.E. (2011). Malformation syndromes caused by disorders of cholesterol synthesis. *J. Lipid Res.* Vol.52, No.1, (January 2011), pp. 6-34, ISSN 0022-2275

Porter, F.D. (2008). Smith-Lemli-Opitz syndrome: pathogenesis, diagnosis and management. *Eur. J. Hum. Genet.*, Vol.16, No.5, (May 2008), pp. 535-541, ISSN 1018-4813

Posse de Chaves, E., & Narayanaswami, V. (2008). Apolipoprotein E and cholesterol in aging and disease in the brain. *Future Lipidol.*, Vol.3, No.5, (October 2008), pp. 505–530, ISSN 1746-0875

Prasanthi, J.R., Huls, A., Thomasson, S., Thompson, A., Schommer, E., & Ghribi, O. (2009). Differential effects of 24-hydroxycholesterol and 27-hydroxycholesterol on beta-amyloid precursor protein levels and processing in human neuroblastoma SH-SY5Y cells. *Mol. Neurodegener.*, Vol.4, No.1, (January 2009), ISSN 1750-1326

Rodwell, V.W., Nordstrom, J.J., & Mitschelen, J.J. (1976). Regulation of HMG-CoA reductase. *Adv. Lipid Res.*, Vol.14, (1976), pp. 1-74, ISSN 0065-2849

Rona-Voros, K., & Weydt, P. (2010). The role of PGC-1alpha in the pathogenesis of neurodegenerative disorders. *Curr. Drug. Targets.*, Vol.11, No.10, (October 2010), pp. 1262-1269, ISSN 1389-4501

Rosas, H.D., Lee, S.Y., Bender, A.C., Zaleta, A.K., Vangel, M., Yu, P., Fischl, B., Pappu, V., Onorato, C., Cha, J.H., Salat, D.H., & Hersch, S.M. (2010). Altered white matter microstructure in the corpus callosum in Huntington's disease: implications for cortical "disconnection". *Neuroimage*, Vol.49, No.4, (February 2010), pp. 2995-3004, ISSN 1053-8119

Russell, D. W., Halford, R. W., Ramirez, D. M., Shah, R., & Kotti, T. (2009). Cholesterol 24-hydroxylase: an enzyme of cholesterol turnover in the brain. *Annu. Rev. Biochem.*, Vol.78, (July 2009), pp. 1017-1040, ISSN 0066-4154

Russell, D.W. (2003). The enzymes, regulation, and genetics of bile acid synthesis. *Annu. Rev. Biochem.*, Vol.72, (July 2003), pp. 137-174, ISSN 0066-4154

Saher, G., Quintes, S., Mobius, W., Wehr, M. C., Kramer-Albers, E. M., Brugger, B., & Nave, K. A. (2009). Cholesterol regulates the endoplasmic reticulum exit of the major membrane protein P0 required for peripheral myelin compaction. *J. Neurosci.*, Vol.29, No.19, (May 2009), pp. 6094-6104, ISSN 0270-6474

Schulz, J.G., Bosel, J., Stoeckel, M., Megow, D., Dirnagl, U., & Endres, M. (2004). HMG-CoA reductase inhibition causes neurite loss by interfering with geranylgeranylpyrophosphate synthesis. *J. Neurochem.*, Vol.89, No.1, (April 2004), pp. 24-32, ISSN 0022-3042

Shafaati, M., Olin, M., Båvner, A., Pettersson, H., Rozell, B., Meaney, S., Parini, P., & Björkhem, I. (2011). Enhanced production of 24S-hydroxycholesterol is not sufficient to drive liver X receptor target genes in vivo. *J. Intern. Med.*, (April 2011), [Epub ahead of print], ISSN 1365-2796

Snipe, G., & Suter, U. (1997). Cholesterol and myelin.In : Bitmaan R.ed. *Cholesterol*. New York:Plenum Press; 1998.

Solomon, A., Kåreholt, I., Ngandu, T., Winblad, B., Nissinen, A., Tuomilehto, J., Soininen, H., & Kivipelto, M. (2007). Serum cholesterol changes after midlife and late-life cognition: twenty-one-year follow-up study. *Neurology*, Vol.68, No.10, (November 2007), pp. 751-756, ISSN 0893-0341

Solomon, A., Kivipelto, M., Wolozin, B., Zhou, J., & Whitmer, R.A. (2009 b). Midlife serum cholesterol and increased risk of Alzheimer's and vascular dementia three decades later. Dement. *Geriatr. Cogn. Disord.*, Vol.28, No.1, (August 2009), pp. 75-80, ISSN 1420-8008

Solomon, A., Leoni, V., Kivipelto, M., Besga, A., Oksengård, A.R., Julin, P., Svensson, L., Wahlund, L.O., Andreasen, N., Winblad, B., Soininen, H., & Björkhem, I. (2009a). Plasma levels of 24S-hydroxycholesterol reflect brain volumes in patients without objective cognitive impairment but not in those with Alzheimer's disease. *Neurosci. Lett.*, Vol.462, No.1, (September 2009), pp. 89-93, ISSN 0304-3940

Stewart, R., White, L.R., Xue, Q.L., & Launer, L.J. (2007). Twenty-six-year change in total cholesterol levels and incident dementia: the Honolulu-Asia Aging Study. *Arch. Neurol.*, Vol.64, No.1, (January 2007), pp. 103–107, ISSN 0003-9942

Storch, J., & Xu, Z. (2009). Niemann-Pick C2 (NPC2) and intracellular cholesterol trafficking. *Biochim. Biophys. Acta*, Vol.1791, No.7, (July 2009), pp. 671-678, ISSN 0006-3002

Tabrizi, S.J., Langbehn, D.R., Leavitt, B.R., Roos, R.A., Durr, A., Craufurd, D., Kennard, C., Hicks, S.L., Fox, N.C., Scahill, R.I., Borowsky, B., Tobin, A.J., Rosas, H.D., Johnson, H., Reilmann, R., Landwehrmeyer, B., Stout, J.C., TRACK-HD investigators (2009). Biological and clinical manifestations of Huntington's disease in the longitudinal TRACK-HD study, cross-sectional analysis of baseline data. *Lancet Neurol.*, Vol.8, No.8, (September 2009), pp. 791-801, ISSN 1474-4422

Teunissen, C.E., Dijkstra, C.D., Polman, C.H., Hoogervorst, E.L., von Bergmann, K., & Lütjohann, D. (2003). Decreased levels of the brain specific 24S-hydroxycholesterol and cholesterol precursors in serum of multiple sclerosis patients. *Neurosci. Lett.*, Vol.347, No.3, (August 2003), pp. 159-162, ISSN 0304-3940

Teunissen, C.E., Floris, S., Sonke, M., Dijkstra, C.D., De Vries, H.E., & Lütjohann, D. (2007). 24S-hydroxycholesterol in relation to disease manifestations of acute experimental autoimmune encephalomyelitis. *J. Neurosci. Res.*, Vol.85, No.7, (May 2007), pp. 1499-1505, ISSN 1097-4547

Valenza, M., & Cattaneo, E. (2011). Emerging roles for cholesterol in Huntington's disease. *Trends Neurosci.*, (July 2011), [Epub ahead of print], ISSN 1878-108X

Valenza, M., Carroll, J.B., Leoni, V., Bertram, L.N., Bjorkhem, I., Singaraja, R.R., Di Donato, S., Lutjohann, D., Hayden, M.R., & Cattaneo, E. (2007b). Cholesterol biosynthesis pathway is disturbed in YAC128 mice and is modulated by huntingtin mutation. *Hum. Mol. Genet.*, Vol.16, No.18, (September 2007), pp. 2187-2198, ISSN 0964-6906

Valenza, M., Leoni, V., Karasinska, J.M., Petricca, L., Fan, J., Carroll, J., Pouladi, M.A., Fossale, E., Nguyen, H.P., Riess, O., MacDonald, M., Wellington, C., DiDonato, S., Hayden, M., & Cattaneo, E. (2010). Cholesterol defect is marked across multiple rodent models of Huntington's disease and is manifest in astrocytes. *J. Neurosci.*, Vol.30, No.32, (August 2010), pp. 10844-10850, ISSN 0270-6474

Valenza, M., Leoni, V., Tarditi, A., Mariotti, C., Bjorkhem, I., Di Donato, S., & Cattaneo, E. (2007a). Progressive dysfunction of the cholesterol biosynthesis pathway in the R6/2 mouse model of Huntington's disease. *Neurobiol. Dis.*, Vol.28, No.1, (October 2007), pp. 133-142, ISSN 0969-9961

Valenza, M., Rigamonti, D., Goffredo, D., Zuccato, C., Fenu, S., Jamot, L., Strand, A., Tarditi, A., Woodman, B., Racchi, M., Mariotti, C., Di Donato, S., Corsini, A., Bates, G., Pruss, R., Olson, J.M., Sipione, S., Tartari, M., & Cattaneo, E. (2005). Dysfunction of the cholesterol biosynthetic pathway in Huntington's disease. *J. Neurosci.*, Vol.25, No.43, (October 2005), pp. 9932-9939, ISSN 0270-6474

van den Kommer, T.N., Dik, M.G., Comijs, H.C., Fassbender, K., Lütjohann, D., & Jonker, C. (2009).Total cholesterol and oxysterols: early markers for cognitive decline in elderly? *Neurobiol. Aging*, Vol.30, No.4, (april 2009), pp. 534-545, ISSN 0197-4580

Vanier, M. T., & Millat, G. (2003). Niemann-Pick disease type C. *Clin. Genet.*, Vol.64, No.4, (October 2003), pp. 269-281, ISSN 0009-9163

Vega, G.L., Weiner, M.F., Lipton, A.M., Von Bergmann, K., Lutjohann, D., Moore, C., & Svetlik, D. (2003). Reduction in levels of 24S-hydroxycholesterol by statin treatment in patients with Alzheimer disease. *Arch. Neurol.*, Vol.60, No.4, (April 2003), pp. 510-515, ISSN 0003-9942

Vonsattel, J.P., Myers, R.H., Stevens, T.J., Ferrante, R.J., Bird, E.D., & Richardson, E.P.Jr. (1985). Neuropathological classification of Huntington's disease. *J. Neuropathol. Exp. Neurol.*, Vol.44, No.6, (November 1985), pp. 559-577, ISSN 0022-3069

Walker, F.O. (2007). Huntington´s disease. *Lancet.*, Vol.369, No.9557, (January 2007), pp. 218-228, ISSN 0140-6736

Wang, N., Yvan-Charvet, L., Lütjohann, D., Mulder, M., Vabmierlo, T., Kim, T.W., & Tall, A.R. (2008). ATP-binding cassette transporters G1 and G4 mediate cholesterol and desmosterol efflux to HDL and regulate sterol accumulation in the brain. *FASEB J.*, Vol.22, No.4, (April 2008), pp. 1073–1082, ISSN 0892-6638

Wechsler, A., Brafman, A., Shafir, M., Heverin, M., Gottlieb, H., Damari, G., Gozlan-Kelner, S., Spivak, I., Moshkin, O., Fridman, E., Becker, Y., Skaliter, R., Einat, P., Faerman, A., Bjorkhem, I., & Feinstein, E. (2003). Generation of viable cholesterol-free mice. *Science* , Vol.302, No.5653, (December 2003), pp. 2087, ISSN 0036-8075

Weydt, P., Pineda, V.V., Torrence, A.E., Libby, R.T., Satterfield, T.F., Lazarowski, E.R., Gilbert, M.L., Morton, G.J., Bammler, T.K., Strand, A.D., Cui, L., Beyer, R.P., Easley, C.N., Smith, A.C., Krainc, D., Luquet, S., Sweet, I.R., Schwartz, M.W., & La Spada, A.R. (2006). Thermoregulatory and metabolic defects in Huntington's disease transgenic mice implicate PGC-1alpha in Huntington's disease neurodegeneration. *Cell Metab.*, Vol.4, No.5, (November 2006), pp. 349-362, ISSN 1550-4131

Xiang, Z., Valenza, M., Cui, L., Leoni, V., Jeong, H.K., Brilli, E., Zhang, J., Peng, Q., Duan, W., Reeves, S.A., Cattaneo, E., & Krainc, D. (2011). Peroxisome-proliferator-activated receptor gamma coactivator 1 α contributes to dysmyelination in experimental models of Huntington's disease. *J. Neurosci.*, Vol.31, No.26 (June 2011), pp. 9544-9553, ISSN 0270-6474

Xie, C., Lund, E.G., Turley, S.D., Russell, D.W., & Dietschy, J.M. (2003). Quantification of two pathways for cholesterol excretion from the brain in normal mice and mice with neurodegeneration. *J. Lipid Res.*, Vol.44, No.9, (September 2003), pp. 1780-1789, ISSN 0022-227

Part 6

Therapeutic Targets in Huntington's Disease

18

Cellular Therapies for Huntington's Disease

C. M. Kelly and A. E. Rosser
Brain Repair Group, School of Biosciences, Cardiff
UK

1. Introduction

Huntington's Disease is an autosomal dominant neurodegenerative disorder with an incidence of 5 to 10 per 100,000 in the Caucasian community. The clinical symptoms of HD are chorea, parkinsonism, dystonia, intellectual impairment, emotional and psychiatric disturbances as well as dysphasia, dysarthria, rigidity and gait disturbances. The depression that is associated with HD is thought to be secondary to the motor abnormalities, given that it develops prior to the appearance of any other symptoms (Folstein, 1989). To date symptomatic treatments are the only available treatments for HD as there are no disease-modifying therapies.

The pathology of HD is characterised by a loss of medium spiny projection neurons in the head of the caudate and putamen of the striatum (Ross and Margolis, 2001) which form part of a complex circuitry comprising parallel feedback loops involving discrete areas of cortex and subcortical structures. As a result of the neuronal loss there is eventually significant atrophy of these structures, with a compensatory expansion of the lateral ventricles. With disease progression the overall brain weight decreases by 25-30%, which reflects additional atrophy of other brain areas such as the cerebral cortex. Gliosis is also seen in the pathology alongside the marked neuronal loss. Neuronal loss in the cortex is found to be layer specific with the greatest loss seen in layer VI and significant amounts of loss seen in layers III and V (Reddy et al., 1999, Ross and Margolis, 2001). The relatively focal loss of medium spiny GABAergic projection neurons in the striatum presents an opportunity to explore neural transplantation as a strategy for cell replacement and circuit reconstruction. For neural transplantation to be successful it is dependant on the cells surviving the transplantation procedure and being able to integrate into the host brain (striatum) and become physiologically active (Lindvall and Hagell, 2002).

2. Clinical trials for neural transplantation

Much of the ground breaking clinical research on neural transplants was done for Parkinson's Disease (PD), beginning in the late 1980s. These trials used primary human foetal mesencephalic tissue as the donor tissue and transplanted it into the host striatum, which is the normal target area of these cells. The mesencephalic tissue contains fate-committed dopaminergic neuroblasts, which have the capacity to differentiate into fully mature dopaminergic neurons following transplantation. For this to be successful certain criteria need to be adhered to, these include harvesting tissue between specific gestational ages and the optimisation of tissue preparation methodologies. If one considers the PD trials in which these principles are taken into account and which use good longitudinal assessment, then results to date in the PD trials have demonstrated improvements in a range of motor skills and many, but not all, of the patients have been able to reduce or even eliminate their daily intake of L-dopa (Hagell et al., 2002, Mendez et al., 2008, Olanow et al., 1996). However, there is variability in the success of this approach, which may be a direct result of variations in transplant methodology as well as differences in patient selection criteria (Freed et al., 2001, Freeman et al., 2000, Kordower et al., 2008, Li et al., 2008, Lindvall et al., 1990). Some of these trials have also highlighted the possibility of dyskinetic side effects in a proportion of patients (Freed et al., 2001), and the reasons for these is currently a topic of active investigation (Hagell et al., 2002, Carta et al., 2010, Lane et al., 2009a, Lane et al., 2009b, Politis et al., 2011, Politis et al., 2010, Steece-Collier et al., 2009). Following several years of round table discussions about how to move forward, a new multicentre European trial has been initiated and it is expected that patients will begin to be transplanted in 2012. This trial has taken all of the new data and the experiences learned from the past to ensure that the best possible protocols are adhered to in all centres, with the aim that the patients receive the best possible tissue transplant.

Parallel clinical trials of neural transplantation in HD are at a much earlier stage than the PD trials and are currently underway in a small number of centres around the world (Bachoud-Levi et al., 2000a, Freeman et al., 2000, Hauser et al., 2002, Kopyov et al., 1998a, Kopyov et al., 1998b, Rosser et al., 2002, Reuter et al., 2008, Philpott et al., 1997, Cicchetti et al., 2009). The French trial, based in Créteil, was the first to provide efficacy data, based on systematic long-term evaluation of their patients. Three of the five patients, having received bilateral striatal implants, were reported to show substantial improvement over several years (Bachoud-Levi et al., 2006, Bachoud-Levi et al., 2000c). More recently there has been an expansion of the French trial to include other French-speaking regions in Europe and a total of 40 patients will eventually receive transplants and will undergo follow-up, although no efficacy data is available as yet. In another study in Florida, 6 of 7 patients appeared to show improvement but one declined significantly, so that the overall group changes were not significant (Hauser et al., 2002). One patient died after 18 months due to cardiovascular disease and post mortem analysis of this patient's brain showed surviving graft tissue that was not affected by the underlying disease progression, at least at this time point (Freeman et al., 2000). The graft tissue was positive for striatal markers such as acetylcholinesterase, calbindin, calretinin, dopamine and tyrosine hydroxylase. Moreover, there was no sign of immune rejection in the graft region (Freeman et al., 2000). In the same study 3 patients developed subdural haemorrhages and 2 required surgical drainage (Hauser et al., 2002). These events may have been related to the stage of disease, which was rather more advanced than for the patients in the French or UK studies, in that more advanced cases of HD tend to have more cerebral atrophy with an accompanying increased risk of intracranial bleeding peri-operatively. Small numbers of patients have received grafts in several other centres with reports of safety (Kopyov et al., 1998a, Rosser et al., 2002), and although efficacy studies are underway in these centres, systematic reports have not yet been published. More recently it has been reported following post mortem analysis 10 years after transplantation in the Florida study, that the grafted cells had themselves been subjected to the disease process (Cicchetti et al., 2009). This raises issues about the long-term viability of transplantation as a therapy for HD, however, as this is only a report of 3 patients caution has to be taken not to misrepresent the field as a whole.

The UK trial (Rosser et al., 2002) is currently on hold due to EU regulations on tissue handling for transplantation, which now requires that all tissue be treated under good manufacturing practice (GMP). The French trial is not limited by such regulations due to their use of tissue pieces grafts as apposed to the cell suspension grafts used in the UK. There is little clear evidence to date stating that one method is better than the other (Watts et al., 2000) however, from the data that is available there is a suggestion that tissue pieces induce a more intense immune response post transplantation (Cooper et al., 2009). It is the belief in the UK that cell suspensions allow for greater integration of the cells into the host brain.

The initial studies of cell transplantation in HD are providing accumulating evidence of the conditions for safety, and preliminary evidence for efficacy. However, the limited availability of foetal tissue and the difficulty in ensuring the high degree of standardisation and quality control when a continuous source of fresh donor tissue is required from elective surgical abortion limits the widespread use of neural transplantation as a practical therapy. It has recently been shown that foetal tissue obtained from medical terminations of pregnancy is a viable source of tissue for transplant studies (Kelly et al., 2011). The use of medically sourced tissue will circumvent some of the logistical issues that were envisaged using surgical tissue due to the limited supply. Despite this new source of fresh foetal tissue there is still ethical and legislative concerns about abortion and the large number of donors required to support each operation, that restrict the number of patients that can receive grafts to a few specialist centres in a restricted number of countries. These issues have stimulated the search for alternative sources of donor cells or tissue that circumvent the problems associated with primary foetal tissue collection.

3. Alternative cell sources

The ability to generate a large stable population of cells to circumvent the supply issue and also to allow regular characterisation to ensure stability of the quality and character of the tissue, without the need for separate characterisation of each and every collection as is the case with primary tissue transplants is the ideal characteristic of an alternative cell source. Also, tissue storage methods need to be refined and validated so that the cells can be delivered on demand, to advance optimal clinical management of the recipient, rather than the surgeon and patient being constrained to surgery around an erratic schedule of tissue availability. The trials using primary foetal tissue thus provide a 'proof of concept' of the cell transplantation strategy as the basis for developing a practical therapy using a standardised, quality-controlled source of cells available to any appropriately equipped neurosurgical facility on demand. Several options are now being investigated as potential sources of donor tissue.

Stem cells are a possible source of cells for neural transplantation in HD and have attracted much attention in the last decade. Stem cells undergo self-renewal by symmetric division and can also undergo asymmetric division to produce another stem cell and a more differentiated progeny (Morrison et al., 1997, Watt and Hogan, 2000). Some multipotential cells may persist into adulthood, either by remaining quiescent in specific regions of the CNS parenchyma or by continued self-renewal (Morrison et al., 1997). Such cells are now referred to as "tissue specific stem cells" (Fuchs and Segre, 2000, Watt and Hogan, 2000). Embryonic stem (ES) cells have the potential to differentiate into all cell types under the correct conditions. Stem cells from a range of sources have potential as donor cells for neural transplantation. However, whatever the source, therapeutic application will require that cells can be directed to differentiate into the precise phenotype required to replace the cells lost to the disease process, and specifically medium spiny neurons for HD. We describe here stem cell sources under consideration as potential donor cells in this context, and the extent to which directed differentiation has been achieved. This list is not exhaustive but covers at least the main categories of stem cells that are currently being explored as alternative cell sources for neural transplantation in HD as well as a number of other neurodegenerative disorders.

3.1 Adult neural stem cells (ANSCs)

ANSCs are tissue-specific stem cell and are derived from the mature brain. Altman and colleagues provided the first clear evidence, using 3H-thymidine autoradiography, that a low level of neurogenesis is ongoing in the dentate gyrus of adult rats (Altman and Das, 1965). ANSCs have since been confirmed in two main regions of the CNS: the sub granular layers of the dentate gyrus, from where the newly-formed neurons repopulate the dentate gyrus (Gage et al., 1995); and the subventricular zone (SVZ) of the lateral ventricles (Alvarez-Buylla et al., 2002), from where the newly formed neurons migrate via the rostral migratory stream to the olfactory bulb (Lois and Alvarez-Buylla, 1994). It has also been reported that neural stem cells may also reside in other regions of the brain, albeit at an even lower concentration, including cortex (Gould et al., 1999, Rietze et al., 2000) and the medial-rostral part of the substantia nigra pars compacta in the lining of the cerebroventricular system of the midbrain (Zhao et al., 2003), although these reports remain controversial (Frielingsdorf et al., 2004).

The attraction of ANSCs as a donor supply for neural transplantation would be the possibility of autologous transplants, thus bypassing the immunological issues of graft rejection, which can be severe in the case of xenografts and not entirely benign even for allografts. Furthermore, it may eventually be possible to recruit such cells for endogenous repair without a requirement for their isolation and re-implantation. That is, it might be possible to stimulate the resident population of ANSCs to migrate to the site of degeneration. However, adult neural stem cells remain difficult to isolate and grow in culture and the factors that would be required to enhance the proliferation of these cells and their differentiation into the particular phenotypes relevant to the site of degeneration remains unknown. Therefore, these cells are less likely to be of beneficial clinical use for transplantation in Huntington's Disease patients.

3.2 Neural stem (NS) cells

Neural stem (NS) cells are those cells that are derived from the developing or adult brain but which are already committed to a neural fate. These cells can be expanded in culture where they undergo asymmetrical division and have been shown to have the potential to differentiate into all cell types of the nervous system, neurons, astrocytes and oligodendrocytes. NS cells can be isolated from the developing brain and in the presence of specific growth factors such as EGF and FGF-2 these cells will form free floating spheres of cells "neurospheres". Animal experiments using these cells have been carried out using tissue from E14 (embryonic day14) mouse striatal tissue and it was verified that the spheres formed from these cells are multipotential (Reynolds and Weiss, 1996). Clonal analysis in the presence of FGF-2 has shown it to be mitogenic for NS cells (Drago et al., 1991, Gensburger et al., 1987, Ray and Gage, 1994, Ray et al., 1993, Richards et al., 1992, Vicario-Abejon et al., 1995).

Several growth factors have the potential to enhance the neuronal differentiation of these cells down particular lineages, including nerve growth factor (NGF), insulin-like growth factor (IGF) and tumour necrosis factor (TNFα) (Arsenijevic et al., 2001, Cattaneo and McKay, 1990, Santa-Olla and Covarrubias, 1995, Tropepe et al., 1997). Identifying an appropriate growth factor cocktail appropriate to the phenotype associated with each particular application may be a necessary prelude to using these cells for transplantation.

It has been found with molecular characterisation of foetal NS cells in vitro that they retain a degree of their site-specific identity when environmental cues are absent but when co-cultured with cells of different origin they can adopt a new fate (Fricker et al., 1999, Parmar et al., 2002). IsletI and Er81 are genes associated with striatal development and their expression is found to be maintained over time in culture, but with neuronal differentiation, expression of striatal specific neuronal markers such as DARPP-32 and Islet1 are lost, although they do express homeobox transcription factors Dlx and MEIS2, which are associated with ventral forebrain development (Parmar et al., 2002, Skogh et al., 2003). Thus, it appears that expansion of NS cells in culture may restrict the differentiation potential of the cells. Further evidence for this has been demonstrated in disease models where these NS cells can survive post-transplantation following a short period of expansion; but that this is compromised by longer expansion times (Zietlow et al., 2005). One interpretation of these findings is that positional information is lost with continued expansion so that when long-term expanded cells are placed in an environment such as the adult CNS, they are not exposed to the developmental signals that they would see in the developing brain and are thus unable to differentiate into neurons appropriate to the site from which they were derived (for example medium spiny neurons from striatally-derived NS cells). However, when grafted to the neonatal brain, similar cells appear to respond to developmental signals and regional determinants by differentiating in a site-specific manner (Englund et al., 2002a, Englund et al., 2002b, Rosser et al., 2000) suggesting that they retain the capacity to respond to developmental signals if they are present.

From these and many other studies it is clear that NS cells may have the potential for neural transplantation. However, for this to become successful it is imperative that we first optimise the conditions in which these cells are expanded so as to increase the frequency at which these cells differentiate into the appropriate phenotype. It may be that we are now at a point where we can take what has been learnt from the directed differentiation of stem cells (as described below) and developmental biology and apply it to NS cells. Such factors as sonic hedgehog and dikkopf to mention but a few, which have been identified as important in striatal development and for directing the differentiation of stem cells, may be of benefit to maintaining the positional identity of the NS cells as well as directing the undifferentiated cells within this culture system that have not yet gone through their terminal differentiation, given the heterogeneic nature of the cells in question (El-Akabawy et al., 2011). The immunogenicity of these cells is another factor that needs to be taken into consideration for neural transplantation. Several studies have looked at the immunogenicity of these cells (Akesson et al., 2009, Al Nimer et al., 2004, Hori et al., 2007, Odeberg et al.,

2005, Ubiali et al., 2007, Laguna Goya et al., 2011) and it has been shown that these cells when expanded in culture become more immunogenic, and so patients receiving these cells would need to be immunosuppressed to prevent graft rejection taking place.

3.3 ES cells

Embryonic stem (ES) cells are generated from the inner cell mass (ICM) of the blastocyst passing the first step of cell differentiation and giving rise to trophectoderm (Evans and Kaufman, 1981, Martin, 1981). ES cells have the ability to extensively proliferate and self-renew whilst maintaining their pluripotentcy. ES cells are able to differentiate into all cell types of the three germ layers—ectoderm, mesoderm, and endoderm, and when transplanted are capable of germline transmission to generate chimeric animals which is in contrast to embryonic carcinoma cells (EC) cells (Bradley et al., 1984). Thus, ES cells can be used to generate models of disease and to understand developmental pathways by introducing modifications into the mouse germline. Human ES (hES) cells were first derived in 1998 (Thomson et al., 1998) and also have the potential to differentiate into cells of all three germ layers (Amit et al., 2000). However, the conditions used for culturing mouse and hES cells are different in that hES cells do not survive in leukemia inhibitory factor (LIF) containing media, a prerequisite for mouse ES cell culture (Daheron et al, 2004; Humphrey et al., 2004). They are pluripotent and can be propagated in culture for long periods of time in an undifferentiated state (Blau et al., 2001, Odorico et al., 2001, Schuldiner et al., 2001).

The ability of these cells to divide in culture over long periods of time highlights their potential for cell transplantation in that they would alleviate the logistical issues associated with primary tissue transplants. However, despite this, the important step of directing the differentiation of these cells into specific cell types is proving difficult. Whilst the default differentiation pathway appears to be that of a neural lineage, the specificity of the neuronal differentiation is limited. In the Parkinson's Disease model where large numbers of dopamine (DA) neurons are required there has been significant developments in directing the differentiation of ES cells into the required phenotype using factors such as retinoic acid and sonic hedghog (shh). In the case of HD where a population of medium spiny DARPP-32 positive neurons are required there is a very limited literature. Bouton and Kato (Bouhon et al., 2006, Kato et al., 2004) have reported on their ability to direct the differentiation of mouse ES cells into a ventral lineage where the cells maintained the expression of ventral markers that are typical of striatal neurons over a short time in culture. In the case of human ES cells, to date there is only one report of DARPP-32 positive neurons (Aubry et al., 2008). In this study the authors report that the DARPP-32 positive cells generated in culture had the ability to differentiate into such mature neurons in vivo following xenotransplantation into the lesioned rat brain model of HD. However, this is the only study reporting such data and to date there is no further evidence of a working protocol generating large populations of these neurons from stem cells. As part of a European consortium to address this issue several groups have come together to facilitate this issue and it is anticipated that new data will come to the fore in the not too distant future.

There have been significant ethical disputes associated with the derivation and use of human ES cells, including concerns over the use of human embryos, and fears related to their potential for human cloning (McHugh, 2004, Sandel, 2004). As a result of these ethical issues many countries have restricted or banned human ES cell research. Nevertheless, other countries have actively supported the development of human ES cell research because of the perceived potential for therapeutic benefit in a wide range of diseases. Some, including the UK, allow cloning of human embryos for therapeutic purposes, while imposing tight regulations to preclude their use for reproductive cloning.

However, despite the great potential of these cells one major caveat is their potential to form tumours. It is crucial that a method be developed for eliminating any possible ES cells from the differentiated population as one single ES cell could have devastating affects if transplanted into the host brain.

3.4 Induced pluripotent stem (iPS) cells

Induced pluripotent stem cells were first described in 2006 (Takahashi and Yamanaka, 2006). Introducing four exogenous transcription factors to differentiated cells and nurturing those cells in an embryonic environment the authors were able to directly reprogramme the cells as pluripotent like cells. Yamanaka and Takahashi in 2006 after much work and many different combinations of factors concluded that the four factors- Oct4, Sox2, c-Myc, and Klf4, which are present at high levels in ES cells are sufficient to transform mouse fibroblasts into cells that mimic ES cells (Figure 1). Subsequently, there have been many reports strengthening this finding (Brambrink et al., 2008, Kaji et al., 2009, Maherali et al., 2007, Meissner et al., 2007, Nakagawa et al., 2008, Stadtfeld et al., 2008, Takahashi et al., 2007, Wernig et al., 2007, Yamanaka, 2008, Yu et al., 2007, Okita et al., 2007, Wernig et al., 2008). iPS cells morphologically look identical to ES cells and as a result more detailed epigenetic characterisation is required to confirm the presence of iPS cell colonies in culture (Figure 2).

Following from this initial work, there have been many refinements to the initial viral based protocol using various methods from a non viral approach to small molecules to generate similar cells and using different factors (Haase et al., 2009, Kim et al., 2009, Sidhu, 2011, Soldner et al., 2009, Yu et al., 2009, Kaji et al., 2009, Okita et al., 2008, Park et al., 2008b, Stadtfeld et al., 2008, Woltjen et al., 2009) with each reporting positive generation of iPS cells with different levels of efficiency. In 2008 the first human iPS cells were described (Park et al., 2008a, Park et al., 2008b) and as with mouse iPS cells there has since been several reports of iPS cell generation from human tissues including; human fetal tissue, adult neural cells, adult fibroblasts, foreskin and disease specific sources, thus highlighting the potential usefulness of these cells both from a scientific and clinical perspective.

The molecular mechanisms of somatic cells reprogramming are still unclear. The wide range of time points to recognise the iPS colonies were reported from numerous studies; the epigenetic events during the reprogramming process and the up or down regulation of involving pluripotent genes are still a mystery. It is not known whether the reprogramming of somatic cells to pluripotent cells is a timed sequential process or if it's a random process, but what is important is that more studies are needed in order to understand the molecular function and use these cells before they are to be used in any clinical research.

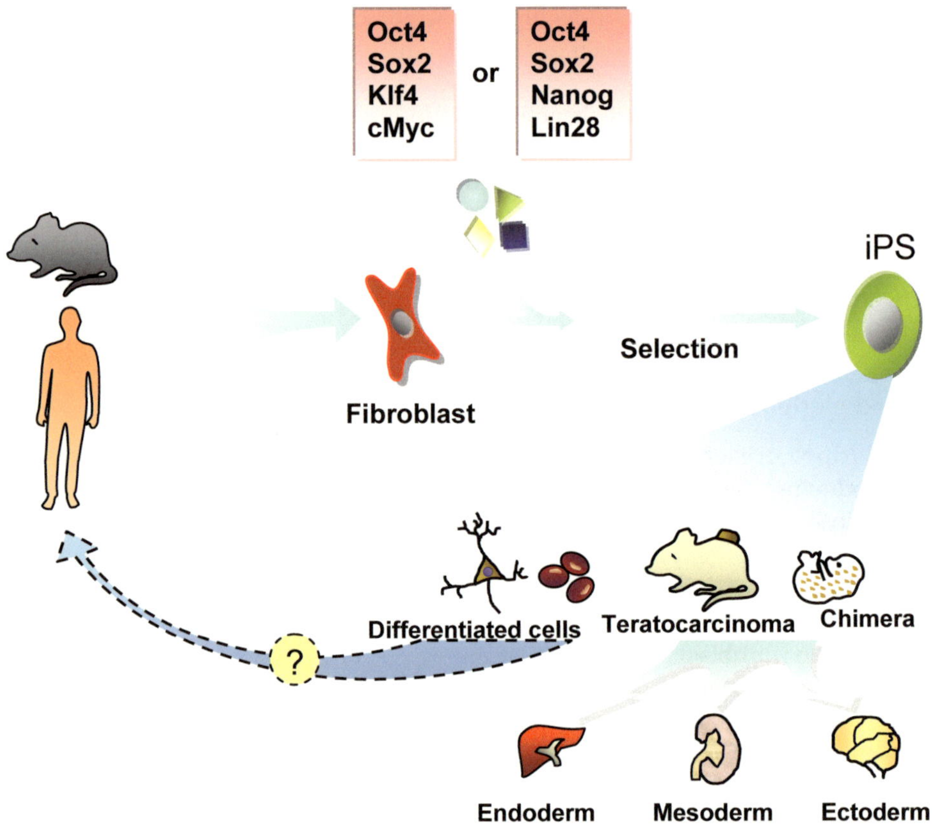

Fig. 1. Schematic drawing demonstrates direct reprogramming mouse and human fibroblasts to iPS cells with four defined factors Oct4, Sox2, Klf4, c-Myc, and Lin28. After selection, iPS cells are derived. These cells could differentiate into all cell types within three germ layers as assayed by teratoma formation and developmental contribution. (Adapted from (Welstead et al., 2008)

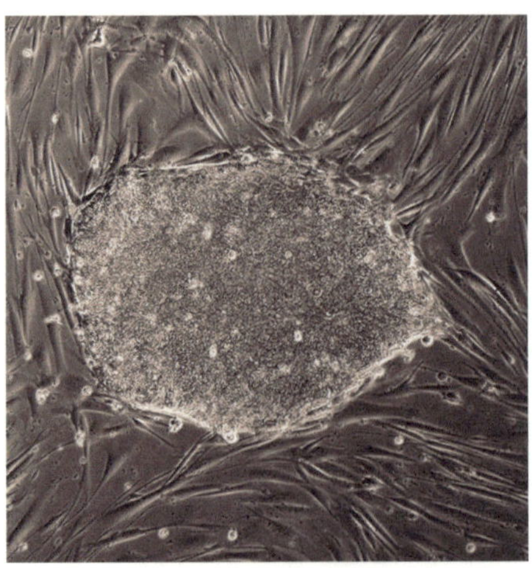

Fig. 2. A colony of pluripotent stem cells derived under non viral conditions.

3.5 Therapeutic potential of iPS cells

The ultimate goal of iPS cell derivation for degenerative diseases is to provide transplantable patient-specific cell sources that overcome ethical and immune rejection difficulties as found with other donor cell sources. Hematopoietic progenitors (HPs) derived from iPS cells were the first to be reported to successfully rescue and improve all pathological conditions in a mouse sickle cell anemia model (Hanna et al., 2007). This early study is a promising step forward in the quest to generate a suitable cell source for transplantation studies.

Interestingly, one year later, Wernig reported the functional benefits and behavioral improvements using iPS cell derived NSCs transplanted into a PD rat model (Wernig et al., 2008). Wernig claimed that in vitro NSCs generated from an iPS cell line were able to differentiate into all neural cell types (β-III tubulin, GFAP, O4 positive staining) including dopaminergic neurons after patterning factors had been added to the culture (FGF8 and sonic hedgehog) (Wernig et al., 2008). In addition, migration of NSCs from iPS cells was observed in various brain regions such as the striatum, midbrain, and hypothalamus after injection to the lateral ventricle of mouse embryos (E13.5-14.5). Furthermore, functional restoration and behavioral improvement were confirmed by action potential initiation, synaptophysin expression, and bias movement reduction from reprogramming somatic cells derived neurons (Wernig et al., 2008).

These studies provide evidence that iPS cells have high potential as an alternative donor cell source for cell replacement in regenerative medicine. To date there is no literature pertaining to the differentiation of DARPP-32 positive medium spiny neurons from iPS cells however, given the fast pace of this ever evolving field it is anticipated that such data will become available in the not too distant future.

However, as with ES cells and NSCs there is an associated risk of tumour formation. In comparison to the other cell sources, iPS cells have the potential to be generated on a patient specific basis and so may overcome this issue to some extent. The need to go through the ES like state is problematic as it renders the cells tumorigenic in nature and so makes them less attractive than initially envisaged.

Recently there have been reports about the possibility of generating functional neurons from fibroblast cells without having to go through the stem cell state (Parmar and Jakobsson, 2011, Pfisterer et al., 2011). As with iPS cells this is a viral based protocol using a combination of three different factors, namely; Ascl1, Brn2 and Mytl1 (Forsberg et al., 2010, Ieda et al., 2010). This exciting development circumvents the issue of tumorgenicity that is associated with current ES and iPS protocols but is limited in that large numbers of cells cannot be generated. Therefore whilst this approach currently is optimistic, there are issues that will need to be overcome if this is to become a viable option for cell therapy.

3.6 Trans-differentiation of other tissue-specific stem cell populations

Another approach is to attain trans-differentiation of a non-neural tissue-specific stem cell population, the classic one being bone-marrow-derived stem cells. This population have the advantage of being more easily harvested than either foetal or adult neural stem cells, but the disadvantage that they do not by default produce neurally differentiated cells.

There is some evidence that trans-differentiation can be achieved, although this remains an area of dispute. Mesenchymal stem cells (MSCs) which are derived from the bone marrow and under normal conditions give rise to chondrocytes, adipocytes and cells of the blood lineage have been reported to trans-differentiate to ectodermal and endodermal cell fates (Zhao et al., 2002). In vitro, MSCs have also been shown to differentiate to form neurons and astrocytes. MSCs transplanted into the rat brain survive and express markers of neuroectodermal cells as well as having a functional effect (Deng et al., 2006, Bertani et al., 2005). This ability to trans-differentiate is not unique to MSCs, as neural stem cells have also been shown to have this ability, where they were seen to differentiate into muscle (Galli et al., 2000). However, evidence suggests that this plasticity may be a result of cell fusion based on studies that have looked at the potential of MSCs to differentiate into hepatocytes (Vassilopoulos et al., 2003, Wang et al., 2003), the increasing body of evidence for MSC trans-differentiation would suggest that this may in fact be true. This issue will need to be clarified for these cells to be serious contenders for neural transplantation.

The best characterised of the tissue-specific stem cells are the Hematopoietic stem cells (HSCs), which are also derived from the bone marrow, and reconstitute the blood. Two classes of HSC have been identified in mouse, those that survive for around 2 months, (the short term, ST-HSC), and those that survive for greater than 6 months, (the long term, LT-HSC) (Blau et al., 2001). Fluorescence-activated cell sorting (FACS) has been used to positively select cells based on the expression of specific cell surface markers. HSCs can be highly enriched up to 10,000 fold and then transplanted into the bone marrow of patients (Lagasse et al., 2001) for the treatment of oncogenic blood diseases. In an animal model of spinal cord injury, HSCs have been shown to survive for 5 weeks after transplantation, differentiate into astrocytes, oligodendrocytes and neuronal precursors and show improvement in functional behaviour using hindlimb motor function (Koshizuka et al., 2004), although no mature neurons were identified.

Human umbilical cord blood is easily retrieved following labour without the risk of harm to the mother or child and has been reported to contain multipotential progenitor cells that apparently have the ability to trans-differentiate into neuronal and glial cells (Sanchez-Ramos et al., 2001). Transplantation of these cells into the neonatal and adult brain have shown potential to survive and differentiate into neurons and glia (Li et al., 2004, Nan et al., 2005, Sanberg et al., 2005, Willing et al., 2003, Zigova et al., 2002). It may be that intravenous delivery rather than neural transplantation will be a more advantageous method of administering these cells for therapeutic benefit, based on a study by Willing et al (2003) where there was significant improvement in certain behavioural tasks when compared to animals receiving neural transplants of cells directly to the striatum. However, further studies are necessary to validate the potential of these cells and again, the issue of cell fusion needs to be addressed in this context.

3.7 Xenogenic tissue

Xenotransplantation offers the opportunity of breeding animals for foetal striatal tissue donation under conditions where the supply can be regulated according to demand; where the breeding stock is inbred, well characterised and controlled for pathogens; and where tissue collection and preparation can be undertaken under standardised sterile good manufacturing practice (GMP) conditions. The most likely donor candidate is porcine tissue, the advantages being: the extensive experience of animal husbandry within this farm species; the reliability of breeding; the large size of the litters; the possibility of sterile collection under standardised conditions; the comparable size and time course of development of the pig and human brain; and the potential application of transgenic technology to porcine tissue, which would open up the possibility of genetic manipulation, for example to modify the immunogenicity of transplanted tissue.

Transplantation of xenogeneic tissues into the immunosuppressed host CNS has been performed using a number of species, for example human to rat, pig to rat, rat to mouse and vice versa (Armstrong et al., 2002, Deacon et al., 1999, Galpern et al., 1996, Garcia et al., 1995, Isacson et al., 2001, Svendsen et al., 1997). Both primary and expanded tissue graft experiments have been reported using xenogenic tissue. The grafted tissue has been found to survive transplantation, axonal and glial fibre projections from the grafts, and make synapses with the host brain.

Clinical studies of CNS xenotransplantation are limited. Primary porcine embryonic striatal tissue has been transplanted into the caudate and putamen of 12 immunosuppressed PD patients with some clinical improvements reported, although there was little convincing evidence of graft survival (Isacson et al., 2001). The immune response from these grafts was more vigorous than that seen in human to rodent models. One patient died 7 months post-operatively for reasons unrelated to the graft, and was found to have very small numbers of surviving neurons in the graft region, raising the possibility that the majority had been rejected. In the same series, 12 HD patients received porcine striatal grafts but, again, there was little evidence of graft survival or functional effect. Twelve months of post-operative analysis of these patients demonstrated no change in the mean total functional capacity score (Fink et al., 2000).

Two key issues need to be resolved for xenografts to progress to practical therapeutic trials. The first relates to the fact, as illustrated by the first pilot clinical trial reported above, that xenografted tissue is largely rejected in the absence of effective immune protection. Two alternative strategies were adopted in the Diacrin trial - daily treatment with CsA or treatment with an antibody against major histocompatability complex 1 (MHC 1) to block the host T cell response (Fink et al., 2000). There is no clear evidence that either strategy proved effective for yielding good cell survival in patients, and it is surprising that the study had progressed on the basis that preliminary reports of the same strategies in primates were equally ineffective. Combination immunoprotection strategies to promote xenograft survival are an area of active research (Armstrong et al., 2001, Harrower and Barker, 2004). We described a new method that would allow long term xenograft survival of human fetal tissue in the rat brain (Kelly et al., 2009). In this method we took advantage of the naïve state of the neonatal rat pup immune system and induced what we describe as desensitisation to xenogenic tissue. Subsequent neural transplants into the adult rat brain resulted in good graft survival up to 40 weeks post transplantation without any immunosuppression. This method now allows long-term evaluation of xenogenic cells both anatomically and more importantly functionally in animal models of disease. Another approach that has been described to block the immune response to ES and iPS derived cells is to use an antibody response against the co-stimulatory molecules involved in the T cell response (Pearl et al., 2011). This method has only been reported for short-term graft survival and as yet no long-term efficacy data is available for functional evaluation of xenogenic cells using this approach.

The second key issue that requires resolution relates to safety of xenografted tissues. In the light of the recent spread of bovine spongiform encephalopathy to man in the form of new variant Creutzfeld-Jacob Disease, and the difficulty in controlling the spread of animal pathogens, as exemplified by the recent UK foot-and-mouth epidemic, there is widespread concern world-wide about the difficulties of eliminating the possibility of transmitting animal diseases to man. This may be particularly risky in the context of transplantation of tissues directly into the immunosuppressed CNS. The concern is not just for the recipient but, in the case of porcine endogenous retrovirus (PERVs), whether direct transfer into the brain might provide a route of transmission that allows virus mutation into new forms of viruses that give rise to unpredicted new diseases in man, even giving rise to de novo epidemics. Although the chances of such mutation are recognised to be very low, the cost of occurrence could be devastatingly high. Moreover, the risk of generating a new disease by an unknown mechanism is one that it is impossible to absolutely exclude by any known safety screen. The regulatory climate is consequently such that any novel xenograft approach is unlikely to gain approval for trial in the foreseeable future, at least in Europe. In the absence of having suffered the same major BSE, CJD and FMD epidemics, US regulations, although strict, are somewhat more permissive, with the result that most academic and commercial research of developing xenotransplantation as a therapeutic strategy for the CNS has moved westwards across the Atlantic over the last 10 years.

3.8 Genetically engineered cells

A variety of cells may be engineered in vitro either for the purpose of producing molecules of potential importance for CNS release (for example, in the form of polymer encapsulated cells, as below), or to alter the properties of a cell to render it potentially useful for circuit reconstruction. Of course, these strategies are not necessarily mutually exclusive - trophic factor support may be crucial for transplanted cells to survive and integrate in the host brain, and genetically engineering cells to release trophic factors in the graft region is one potential method for optimising graft survival.

The herpes simplex viral vector was the first virus to be tested as a method of introducing genes into the adult CNS (During et al., 1994, Fraefel et al., 1996, Song et al., 1997). More recently, other viral vectors have been introduced, including adenovirus, the recombinant adeno-associated virus (rAAV), lentivirus and pseudotyped vectors. The rAAV vector is more efficient than the HSV in that it is possible to achieve much higher levels of expression. The use of such vectors has allowed genes to be transferred to a specific group of cells in the CNS (Janson et al., 2001), and has provided support for the efficacy of factors such as GDNF for PD (Eslamboli et al., 2003, Kirik et al., 2000, Mandel et al., 1999, Mandel et al., 1997) and CNTF for HD (Emerich, 2004, Kahn et al., 1996, Mittoux et al., 2002, Regulier et al., 2002).

Polymer capsules have been considered as a system for trophic factor delivery to the CNS as they have the advantages of being relatively cheap to produce and can also be removed from the CNS as required, but the major drawback is that the effect is not long lasting

(Emerich et al., 1994). Where a limited amount of a protein is required for relatively short periods of time, polymer microspheres are an attractive alternative as they are biodegradable and subsequent surgical procedures are not required for retrieval (Date et al., 2001). However, improvements in the duration of release have been obtained by the use of encapsulated cells engineered to produce the desired molecules (Emerich, 1999, Emerich et al., 1997). Here, cells engineered to secrete specific substances such as neurotrophic factors are protected from the host immune system by a semi-permeable selective biocompatible outer membrane (Emerich, 1999, Emerich, 2004, Emerich et al., 1998, Emerich et al., 1997, Emerich et al., 1996). The outer membrane allows the entry of nutrients to the cells whilst also allowing the exit of neuroactive molecules. The advantage of this strategy is that it allows for the implantation of xenogeneic cells, which may be much easier to obtain or engineer than human cells. This approach has been used for delivery of factors such as GDNF in animal models of PD (Date et al., 2001, Sautter et al., 1998) and CNTF in animal models of HD (Emerich et al., 1997).

In the case of HD there have been several studies using polymer encapsulated cells for the delivery of CNTF. Baby hamster fibroblasts have been genetically modified to produce hCNTF and incorporated into polymer capsules (Anderson et al., 1996, Emerich et al., 1996). Both rodent and primate studies have been carried out incorporating this method (Anderson et al., 1996; Emerich et al., 1996; Emerich et al., 1997; Emerich and Winn, 2004; Kordower et al., 2000; Mittoux et al., 2000). These animal studies suggested that CNTF can protect striatal neurons against subsequent damage from an excitotoxic lesion. As well as protecting specific populations of striatal neurons from lesion-induced cell death, behavioural improvement was observed on skilled motor and cognitive tasks when compared to control animals. Encapsulated CNTF released by BHK cells were used in a clinical trial in France (Bachoud-Levi et al., 2000b), however, it was subsequently found that on removal of the capsule the cells failed to release sufficient amount of CNTF. The trial has reported safety and feasibility of this approach but further work is required to optimise the capsule for release of CNTF (Bloch et al., 2004). Nevertheless, the use of encapsulated cells for the delivery of growth factors and neurotrophic factors is an attractive alternative and may be required in combination with neural transplantation as a means of providing trophic support to the grafted cells.

Another potential cell source is immortalised cell lines, the neurally committed lines, such as the Ntera2 cell line, RN33B and Hib5. Functional benefit has been reported using these cells in various animal models (Catapano et al., 1999, Lundberg et al., 1996, Miyazono et al., 1995, Saporta et al., 2001). The Ntera2 cell line has been the most widely used. These cells are derived from human embryonal carcinomas and are terminally differentiated in vitro with retinoic acid. They have been found to respond to environmental cues when transplanted into the excitotoxically lesioned striatum (Saporta et al., 2001; Miyazono et al., 1995), sending out target-specific projections as well as expressing a site-specific phenotype. Grafting Ntera2 cells into the excitotoxic lesioned striatum resulted in neuronal differentiation, and a preliminary study reported rather dramatic functional effects (Hurlbert et al., 1999). However, on more detailed analysis the cells did not express any striatal-specific markers and there was no sustained improvement on skilled paw reaching and cylinder placing (Fricker-Gates et al., 2004). Transplantation of the RN33B cell line to the lesioned and non-lesioned striatum of rats has demonstrated their potential to differentiate into neurons in a site-specific way and form connections with target areas such as the globus pallidus (Lundberg et al., 1996), although only a proportion of the cells showed this differentiation potential. A major disadvantage of using such cell lines is the genotypic variability that arises from the immortalization process (Renfranz et al., 1991), and the risk that cells continue to proliferate to form tumours after transplantation.

4. Good manufacturing practice (GMP)

One of the major stumbling blocks for clinical trials of cell transplantation in Europe has been the introduction of the human tissue directive. This directive stipulates that all tissues used for human patients must be handled under clean room conditions. As a result many trials have been stopped to allow time to implement these new conditions. Each member state of the EU has taken their own interpretation of the directive and in the UK this is regulated by the human tissue authority (HTA). It is a requirement that any facility used to manipulate the cells/tissue used for transplantation into patients be licenced and governed by a strict set of guidelines. As a result in the case of the UK trial (NEST-UK) no patients have been transplanted in the last decade.

5. Conclusion

For HD, Cell transplantation is a promising therapy based on the current data from clinical trials. However, it is limited by the availability of a reliable source of cells that can replace the lost cells and reform the connections required for functional benefit. The proof of principle data from human fetal tissue studies highlights the effectiveness of the approach and the need for an alternative cell source. This chapter has highlighted some of the possible alternatives available and the potential of each one. Whilst the focus is very much on stem cells it is important that other cells also be considered as each has its own caveats. Which cell source is likely to make it to the clinical is not certain at this time and it is clear that much work is required before this can happen. Whilst much is known about cell sources and their potential for PD it is clear that the HD field is a long way behind. It is important that we as scientist stay focused on the goal and work together to move the field forward. As well as the issues of differentiation, tumour formation and cell number we also need to be mindful of the regulatory issues when devising such protocols and how they will adapt to GMP conditions. There is increasing emphasis on the use of GMP grade products for such studies, which are widely available. HD is a devastating disease and we are driven by the need for a therapy that works to make the lives of these patients less distressful.

6. References

Akesson, E., Wolmer-Solberg, N., Cederarv, M., Falci, S. & Odeberg, J. 2009. Human neural stem cells and astrocytes, but not neurons, suppress an allogeneic lymphocyte response. *Stem Cell Res,* **2,** 56-67.

Al Nimer, F., Wennersten, A., Holmin, S., Meijer, X., Wahlberg, L. & Mathiesen, T. 2004. MHC expression after human neural stem cell transplantation to brain contused rats. *NeuroReport,* **15,** 1871-5.

Altman, J. & Das, G. D. 1965. Autoradiographic and histological evidence of postnatal hippocampal neurogenesis in rats. *J Comp Neurol.,* **124,** 319-335.

Alvarez-Buylla, A., Seri, B. & Doetsch, F. 2002. Identification of neural stem cells in the adult vertebrate brain. *Brain Research Bulletin,* 57, 751-758.

Amit, M., Carpenter, M. K., Inokuma, M. S., Chiu, C. P., Harris, C. P., Waknitz, M. A., Itskovitz-Eldor, J. & Thomson, J. A. 2000. Clonally derived human embryonic stem cell lines maintain pluripotency and proliferative potential for prolonged periods of culture. *Developmental Biology,* 227, 271-8.

Anderson, K. D., Panayotatos, N., Corcoran, T. L., Lindsay, R. M. & Wiegand, S. J. 1996. Ciliary neurotrophic factor protects striatal output neurons in an animal model of Huntington's Disease. *Proceedings of the National Academy of Sciences,* 93, 7346-7351.

Armstrong, R. J., Harrower, T. P., Hurelbrink, C. B., Mclaughin, M., Ratcliffe, E. L., Tyers, P., Richards, A., Dunnett, S. B., Rosser, A. E. & Barker, R. A. 2001. Porcine neural xenografts in the immunocompetent rat: immune response following grafting of expanded neural precursor cells. *Neuroscience,* 106, 201-216.

Armstrong, R. J., Hurelbrink, C. B., Tyers, P., Ratcliffe, E. L., Richards, A., Dunnett, S. B., Rosser, A. E. & Barker, R. A. 2002. The potential for circuit reconstruction by expanded neural precursor cells explored through porcine xenografts in a rat model of Parkinson's Disease. *Exp Neurol,* 175, 98-111.

Arsenijevic, Y., Weiss, S., Schneider, B. & Aebischer, P. 2001. Insulin-Like Growth Factor-1 Is Necessary for Neural Stem Cell Prolferation and Demonstartes Distinct Actions of Epidermal Growth Factor and Fibroblast Growth Factor-2. *Journal of Neuroscience,* 27(18), 7194-7202.

Aubry, L., Bugi, A., Lefort, N., Rousseau, F., Peschanski, M. & Perrier, A. L. 2008. Striatal progenitors derived from human ES cells mature into DARPP32 neurons in vitro and in quinolinic acid-lesioned rats. *Proc Natl Acad Sci U S A,* 105, 16707-12.

Bachoud-Levi, A., Bourdet, C., Brugieres, P., Nguyen, J. P., Grandmougin, T., Haddad, B., Jeny, R., Bartolomeo, P., Boisse, M. F., Barba, G. D., Degos, J. D., Ergis, A. M., Lefaucheur, J. P., Lisovoski, F., Pailhous, E., Remy, P., Palfi, S., Defer, G. L., Cesaro, P., Hantraye, P. & Peschanski, M. 2000a. Safety and tolerability assessment of intrastriatal neural allografts in five patients with Huntington's Disease. *Exp Neurol,* 161, 194-202.

Bachoud-Levi, A. C., Deglon, N., Nguyen, J. P., Bloch, J., Bourdet, C., Winkel, L., Remy, P., Goddard, M., Lefaucheur, J. P., Brugieres, P., Baudic, S., Cesaro, P., Peschanski, M. & Aebischer, P. 2000b. Neuroprotective gene therapy for Huntington's Disease using a polymer encapsulated BHK cell line engineered to secrete human CNTF. *Hum.Gene Ther.,* 11, 1723-1729.

Bachoud-Levi, A. C., Gaura, V., Brugieres, P., Lefaucheur, J. P., Boisse, M. F., Maison, P., Baudic, S., Ribeiro, M. J., Bourdet, C., Remy, P., Cesaro, P., Hantraye, P. & Peschanski, M. 2006. Effect of fetal neural transplants in patients with Huntington's Disease 6 years after surgery: a long-term follow-up study. *Lancet Neurol,* 5, 303-9.

Bachoud-Levi, A. C., Remy, P., Nguyen, J. P., Brugieres, P., Lefaucheur, J. P., Bourdet, C., Baudic, S., Gaura, V., Maison, P., Haddad, B., Boisse, M. F., Grandmougin, T., Jeny, R., Bartolomeo, P., Dalla Barba, G., Degos, J. D., Lisovoski, F., Ergis, A. M., Pailhous, E., Cesaro, P., Hantraye, P. & Peschanski, M. 2000c. Motor and cognitive improvements in patients with Huntington's Disease after neural transplantation. *Lancet,* 356, 1975-9.

Bertani, N., Malatesta, P., Volpi, G., Sonego, P. & Perris, R. 2005. Neurogenic potential of human mesenchymal stem cells revisited: analysis by immunostaining, time-lapse video and microarray. *Journal of cell science,* 118, 3925-36.

Blau, H. M., Brazelton, T. R. & Weimann, J. M. 2001. The evolving concept of a stem cell: Entity or function? *Cell,* 105, 829-841.

Bloch, J., Bachoud-Levi, A. C., Deglon, N., Lefaucheur, J. P., Winkel, L., Palfi, S., Nguyen, J. P., Bourdet, C., Gaura, V., Remy, P., Brugieres, P., Boisse, M. F., Baudic, S., Cesaro, P., Hantraye, P., Aebischer, P. & Peschanski, M. 2004. Neuroprotective gene therapy for Huntington's Disease, using polymer-encapsulated cells engineered to secrete human ciliary neurotrophic factor: results of a phase I study. *Hum.Gene Ther.,* 15, 968-975.

Bouhon, I. A., Joannides, A., Kato, H., Chandran, S. & Allen, N. D. 2006. Embryonic stem cell-derived neural progenitors display temporal restriction to neural patterning. *Stem Cells,* 24, 1908-13.

Bradley, A., Evans, M., Kaufman, M. H. & Robertson, E. 1984. Formation of germ-line chimaeras from embryo-derived teratocarcinoma cell lines. *Nature,* 309, 255-6.

Brambrink, T., Foreman, R., Welstead, G. G., Lengner, C. J., Wernig, M., Suh, H. & Jaenisch, R. 2008. Sequential expression of pluripotency markers during direct reprogramming of mouse somatic cells. *Cell Stem Cell,* 2, 151-9.

Carta, M., Carlsson, T., Munoz, A., Kirik, D. & Bjorklund, A. 2010. Role of serotonin neurons in the induction of levodopa- and graft-induced dyskinesias in Parkinson's Disease. *Mov Disord,* 25 Suppl 1, S174-9.

Catapano, L. A., Sheen, V. L., Leavitt, B. R. & Macklis, J. D. 1999. Differentiation of transplanted neural precursors varies regionally in adults striatum. *NeuroReport,* 10, 3971-3977.

Cattaneo, E. & Mckay, R. 1990. Proliferation and differentiation of neuronal stem cells regulated by nerve growth factor. *Nature,* 347, 762-765.

Cicchetti, F., Saporta, S., Hauser, R. A., Parent, M., Saint-Pierre, M., Sanberg, P. R., Li, X. J., Parker, J. R., Chu, Y., Mufson, E. J., Kordower, J. H. & Freeman, T. B. 2009. Neural transplants in patients with Huntington's Disease undergo disease-like neuronal degeneration. *Proceedings of the National Academy of Sciences of the United States of America,* 106, 12483-8.

Cooper, O., Astradsson, A., Hallett, P., Robertson, H., Mendez, I. & Isacson, O. 2009. Lack of functional relevance of isolated cell damage in transplants of Parkinson's Disease patients. *J Neurol,* 256 Suppl 3, 310-6.

Date, I., Shingo, T., Yoshida, H., Fujiwara, K., Kobayashi, K., Takeuchi, A. & Ohmoto, T. 2001. Grafting of encapsulated genetically modified cells secreting GDNF into the striatum of parkinsonian model rats. *Cell Transplantation,* 10, 397-401.

Deacon, T., Whatley, B., Leblanc, C., Lin, L. & Isacson, O. 1999. Pig Fetal Septal Neurons Implanted Into the Hippocampus of Aged or Cholinergic Deafferented Rats Grow Axons and Form Cross-Species Synapses in Appropriate Target Regions. *Cell Transplantation,* 8, 111-129.

Deng, J., Petersen, B. E., Steindler, D. A., Jorgensen, M. L. & Laywell, E. D. 2006. Mesenchymal stem cells spontaneously express neural proteins in culture and are neurogenic after transplantation. *Stem Cells,* 24, 1054-64.

Drago, J., Murphy, M., Carroll, S. M., Harvey, P. P. & Bartlett, P. F. 1991. Fibroblast growth factor-mediated proliferation of central nervous system precursors depends on endogenous production of insulin-like growth factor 1. *Proc.Natl.Acad.Sci.USA,* 88, 2199-2203.

During, M. J., Naegele, J. R., O'malley, K. L. & Geller, A. I. 1994. Long-term behavioral recovery in parkinsonian rats by an HSV vector expressing tyrosine hydroxylase. *Science,* 266, 1399-1403.

El-Akabawy, G., Medina, L. M., Jeffries, A., Price, J. & Modo, M. 2011. Purmorphamine Increases DARPP-32 Differentiation in Human Striatal Neural Stem Cells Through the Hedgehog Pathway. *Stem Cells Dev.*

Emerich, D. F. 1999. Encapsulated CNTF-producing cells for Huntington's Disease. *Cell Transplant.,* 8, 581-582.

Emerich, D. F. 2004. Sertoli cell grafts for Huntington's disease. An opinion. *Neurotox Res,* 5, 567.

Emerich, D. F., Bruhn, S., Chu, Y. & Kordower, J. H. 1998. Cellular delivery of CNTF but not NT-4/5 prevents degeneration of striatal neurons in a rodent model of Huntington's Disease. *Cell Transplant.,* 7, 213-225.

Emerich, D. F., Cain, C. K., Greco, C., Saydoff, J. A., Hu, Z. Y., Liu, H. J. & Lindner, M. D. 1997. Cellular delivery of human CNTF prevents motor and cognitive dysfunction in a rodent model of Huntington's disease. *Cell Transplantation,* 6, 249-266.

Emerich, D. F., Hammang, J. P., Baetge, E. E. & Winn, S. R. 1994. Implantation of polymer-encapsulated human nerve growth factor-secreting fibroblasts attenuates the behavioral and neuropathological consequences of quinolinic acid injections into rodent striatum. *Exp.Neurol.,* 130, 141-150.

Emerich, D. F., Lindner, M. D., Winn, S. R., Chen, E. Y., Frydel, B. R. & Kordower, J. H. 1996. Implants of encapsulated human CNTF-producing fibroblasts prevent behavioral deficits and striatal degeneration in a rodent model of Huntington's Disease. *Journal of Neuroscience,* 16, 5168-5181.

Englund, U., Bjorklund, A., Wictorin, K., Lindvall, O. & Kokaia, M. 2002a. Grafted neural stem cells develop into functional pyramidal neurons and integrate into host cortical circuitry. *Proceedings of the National Academy of Sciences,* 99, 17089.

Englund, U., Fricker-Gates, R. A., Lundberg, C., Bjorklund, A. & Wictorin, K. 2002b. Transplantation of human neural progenitor cells into the neonatal rat brain: Extensive migration and differentiation with long-distance axonal projections. *Experimental Neurology,* 173, 1-21.

Eslamboli, A., Cummings, R. M., Ridley, R. M., Baker, H. F., Muzyczka, N., Burger, C., Mandel, R. J., Kirik, D. & Annett, L. E. 2003. Recombinant adeno-associated viral vector (rAAV) delivery of GDNF provides protection against 6-OHDA lesion in the common marmoset monkey (Callithrix jacchus). *Exp.Neurol.,* 184, 536-548.

Evans, M. J. & Kaufman, M. H. 1981. Establishment in culture of pluripotential cells from mouse embryos. *Nature,* 292, 154-6.

Fink, J. S., Schumacher, J. M., Ellias, S. L., Palmer, E. P., Saint-Hilaire, M., Shannon, K., Penn, R., Starr, P., Vanhorne, C., Kott, H. S., Dempsey, P. K., Fischman, A. J., Raineri, R., Manhart, C., Dinsmore, J. & Isacson, O. 2000. Porcine xenografts in Parkinson's Disease and Huntington's Disease patients: preliminary results. *Cell Transplant.,* 9, 273-278.

Folstein, S. E. 1989. The Psychopathology of Huntingtons-Disease. *Journal of Nervous and Mental Disease,* 177, 645-645.

Forsberg, M., Carlen, M., Meletis, K., Yeung, M. S., Barnabe-Heider, F., Persson, M. A., Aarum, J. & Frisen, J. 2010. Efficient reprogramming of adult neural stem cells to monocytes by ectopic expression of a single gene. *Proceedings of the National Academy of Sciences of the United States of America,* 107, 14657-61.

Fraefel, C., Song, S., Lim, F., Lang, P., Yu, L., Wang, Y., Wild, P. & Geller, A. I. 1996. Helper virus-free transfer of herpes simplex virus type 1 plasmid vectors into neural cells. *J.Virol.,* 70, 7190-7197.

Freed, C. R., Greene, P. E., Breeze, R. E., Tsai, W. Y., Dumouchiel, W., Kao, R., Dillon, S., Winfield, H., Culver, S., Trojanowski, J. Q., Eidelberg, D. & Fahn, S. 2001. Transplantation of embryonic dopamine neurons for severe Parkinson's Disease. *The New England Journal of Medicine,* 344, 710-719.

Freeman, T. B., Cicchetti, F., Hauser, R. A., Deacon, T. W., Li, X. J., Hersch, S. M., Nauert, G. M., Sanberg, P. R., Kordower, J. H., Saporta, S. & Isacson, O. 2000. Transplanted fetal striatum in Huntington's Disease: Phenotypic development and lack of pathology. *Proceedings of the National Academy of Sciences,* 97, 13877-13882.

Fricker, R. A., Carpenter, M. K., Winkler, C., Greco, C., Gates, M. & Bjorklund, A. 1999. Site-Specific Migration and Neuroanl Differentition of Human Neural Progenitor Cells after Transplantation in the Adult Rat Brain. *The Jouranl of Neuroscience,* 19, 5990-6005.

Fricker-Gates, R. A., White, A., Gates, M. A. & Dunnett, S. B. 2004. Striatal neurons in striatal grafts are derived from both post-mitotic cells and dividing progenitors. *European Journal of Neuroscience,* 19, 513-520.

Frielingsdorf, H., Schwarz, K., Brundin, P. & Mohapel, P. 2004. No evidence for new dopaminergic neurons in the adult mammalian substantia nigra. *Proc.Natl.Acad.Sci.U.S.A,* 101, 10177-10182.

Fuchs, E. & Segre, J. A. 2000. Stem cells: a new lease on life. *Cell,* 100, 143-155.

Gage, F. H., Ray, J. & Fisher, L. J. 1995. Isolation, characterization, and use of stem cells from the CNS. *Annual Reiew of Neuroscience,* 18, 159-192.

Galli, R., Borello, U., Gritti, A., Giulia Minasi, M., Bjornson, C., Coletta, M., Mora, M., De Angelis, M. G. C., Fiocco, R., Cossu, G. & Vescovi, A. L. 2000. Skeletal myogenic potential of human and mouse neural stem cells. *Nature,* 3, 986-991.

Galpern, W. R., Burns, L. H., Deacon, T. W., Dismore, J. & Isacson, O. 1996. Xenotransplantation of Porcine Fetal Ventral Mesencephalon in a Rat Model of Parknison's Disease: Functional Recovery and Graft Morphology. *Experimental Neurology,* 140, 1-13.

Garcia, A. R., Deacon, T. W., Dinsmore, J. & Isacson, O. 1995. Extensive Axonal and Glial Fiber Growth from Fetal Porcine Cortical Xenografts in the Adult Rat Cortex. *Cell Transplantation,* 4, 515-527.

Gensburger, C., Labourdette, G. & Sensenbrenner, M. 1987. Brain basic fibroblast growth factor stimulates the proliferation of rat neuronal precursor cells in vitro. *FEBS Letters,* 217, 1-5.

Gould, E., Reeves, A. J., Graziano, M. S. A. & Gross, C. G. 1999. Neurogenesis in the neocortex of adult primates. *Science,* 286, 548-552.

Haase, A., Olmer, R., Schwanke, K., Wunderlich, S., Merkert, S., Hess, C., Zweigerdt, R., Gruh, I., Meyer, J., Wagner, S., Maier, L. S., Han, D. W., Glage, S., Miller, K., Fischer, P., Scholer, H. R. & Martin, U. 2009. Generation of induced pluripotent stem cells from human cord blood. *Cell Stem Cell,* 5, 434-41.

Hagell, P., Piccine, P., Bjâ≠◊Rklund, A., Brundin, P., Rehncrona, S., Widner, H., Crabb, L., Pavese, N., Oertel, W. H., Quinn, N., Brooks, D. J. & Lindvall, O. 2002. Dyskinesias following neural transplantation in Parkinson's Disease. *Nature Neuroscience,* 5, 627-628.

Harrower, T. P. & Barker, R. A. 2004. The emerging technologies of neural xenografting and stem cell transplantation for treating neurodegenerative disorders. *Drugs Today (Barc),* 40, 171-89.

Hauser, R. A., Furtado, S., Cimino, C. R., Delgado, H., Eichler, S., Schwartz, S., Scott, D., Nauert, G. M., Soety, E., Sossi, V., Holt, D. A., Sanberg, P. R., Stoessl, A. J. & Freeman, T. B. 2002. Bilateral human fetal striatal transplantation in Huntington's Disease. *Neurology,* 58, 687-95.

Hori, J., Ng, T. F., Shatos, M., Klassen, H., Streilein, J. W. & Young, M. J. 2007. Neural progenitor cells lack immunogenicity and resist destruction as allografts. 2003. *Ocul Immunol Inflamm,* 15, 261-73.

Humphrey, R. K., Beattie, G. M., Lopez, A. D., Bucay, N., King, C. C., Firpo, M. T., Rose-John, S. & Hayek, A. 2004. Maintenance of pluripotency in human embryonic stem cells is STAT3 independent. *Stem Cells,* 22, 522-30.

Hurlbert, M. S., Gianani, R. I., Hutt, C., Freed, C. R. & Kaddis, F. G. 1999. Neural transplantation of hNT neurons for Huntington's Disease. *Cell Transplant.,* 8, 143-151.

Ieda, M., Fu, J. D., Delgado-Olguin, P., Vedantham, V., Hayashi, Y., Bruneau, B. G. & Srivastava, D. 2010. Direct reprogramming of fibroblasts into functional cardiomyocytes by defined factors. *Cell,* 142, 375-86.

Isacson, O., Costantini, J. M., Cicchetti, F., Chung, S. & Kim, K. S. 2001. Cell implantation therapies for Parkinson's Disease using neural stem, transgenic or xenogenic donor, cells. *Parkinsonism and Related Disorders,* 7, 205-212.

Janson, C. G., Mcphee, S. W. J., Leone, P., Freese, A. & During, M. J. 2001. Viral-based gene transfer to the mammalian CNS for functional genome studies. *Trends in Neuroscience,* 24, 706-712.

Kahn, A., Haase, G., Akli, S. & Guidotti, J. E. 1996. [Gene therapy of neurological diseases]. *C.R.Seances Soc.Biol.Fil.,* 190, 9-11.

Kaji, K., Norrby, K., Paca, A., Mileikovsky, M., Mohseni, P. & Woltjen, K. 2009. Virus-free induction of pluripotency and subsequent excision of reprogramming factors. *Nature,* 458, 771-5.

Kato, H., Bouhon, I. A., Chandran, S. & Allen, N. D. 2004. Critical factors influencing fate determination and developmental plasticity of embryonic stem cells derived neural precursor cells. *Submitted.*

Kelly, C. M., Precious, S. V., Scherf, C., Penketh, R., Amso, N. N., Battersby, A., Allen, N. D., Dunnett, S. B. & Rosser, A. E. 2009. Neonatal desensitization allows long-term survival of neural xenotransplants without immunosuppression. *Nat Methods,* 6, 271-3.

Kelly, C. M., Precious, S. V., Torres, E. M., Harrison, A., Williams, D., Scherf, C., Weyrauch, U. M., Lane, E. L., Allen, N. D., Penketh, R., Amso, N., Kemp, P., Dunnett, S. B. & Rosser, A. E. 2011. Medical terminations of pregnancy: a viable source of tissue for cell replacement therapy for neurodegenerative disorders. *Cell Transplantation.*

Kim, D., Kim, C. H., Moon, J. I., Chung, Y. G., Chang, M. Y., Han, B. S., Ko, S., Yang, E., Cha, K. Y., Lanza, R. & Kim, K. S. 2009. Generation of human induced pluripotent stem cells by direct delivery of reprogramming proteins. *Cell Stem Cell,* 4, 472-6.

Kirik, D., Rosenblad, C., Bjorklund, A. & Mandel, R. J. 2000. Long-term rAAV-mediated gene transfer of GDNF in the rat Parkinson's model: intrastriatal but not intranigral transduction promotes functional regeneration in the lesioned nigrostriatal system. *Journal of Neuroscience,* 20, 4686-4700.

Kopyov, O. V., Jacques, S. & Eagle, K. S. 1998a. Fetal transplantation for the treatment of neurodegenerative diseases - Current status and future potential. *Cns Drugs,* 9, 77-83.

Kopyov, O. V., Jacques, S., Lieberman, A., Duma, C. M. & Eagle, K. S. 1998b. Safety of intrastriatal neurotransplantation for Huntington's Disease patients. *Experimental Neurology,* 149, 97-108.

Kordower, J. H., Chu, Y., Hauser, R. A., Freeman, T. B. & Olanow, C. W. 2008. Lewy body-like pathology in long-term embryonic nigral transplants in Parkinson's Disease. *Nat Med,* 14, 504-6.

Koshizuka, S., Okada, S., Okawa, A., Koda, M., Murasawa, M., Hashimoto, M., Kamada, T., Yoshinaga, K., Murakami, M., Moriya, H. & Yamazaki, M. 2004. Transplanted hematopoietic stem cells from bone marrow differentiate into neural lineage cells and promote functional recovery after spinal cord injury in mice. *J Neuropathol.Exp.Neurol.,* 63, 64-72.

Lagasse, E., Shizuru, J. A., Uchida, N., Tsukamoto, A. & Weissman, I. L. 2001. Toward regenerative medicine. *Immunity,* 14, 425-436.

Laguna Goya, R., Busch, R., Mathur, R., Coles, A. J. & Barker, R. A. 2011. Human fetal neural precursor cells can up-regulate MHC class I and class II expression and elicit CD4 and CD8 T cell proliferation. *Neurobiol Dis,* 41, 407-14.

Lane, E. L., Brundin, P. & Cenci, M. A. 2009a. Amphetamine-induced abnormal movements occur independently of both transplant- and host-derived serotonin innervation following neural grafting in a rat model of Parkinson's Disease. *Neurobiol Dis,* 35, 42-51.

Lane, E. L., Vercammen, L., Cenci, M. A. & Brundin, P. 2009b. Priming for L-DOPA-induced abnormal involuntary movements increases the severity of amphetamine-induced dyskinesia in grafted rats. *Exp Neurol,* 219, 355-8.

Li, H. J., Liu, H. Y., Zhao, Z. M., Lu, S. H., Yang, R. C., Zhu, H. F., Cai, Y. L., Zhang, Q. J. & Han, Z. C. 2004. [Transplantation of human umbilical cord stem cells improves neurological function recovery after spinal cord injury in rats]. *Zhongguo Yi.Xue.Ke.Xue.Yuan Xue.Bao.,* 26 38-42.

Li, J. Y., Englund, E., Holton, J. L., Soulet, D., Hagell, P., Lees, A. J., Lashley, T., Quinn, N. P., Rehncrona, S., Bjorklund, A., Widner, H., Revesz, T., Lindvall, O. & Brundin, P. 2008. Lewy bodies in grafted neurons in subjects with Parkinson's Disease suggest host-to-graft disease propagation. *Nat Med,* 14, 501-3.

Lindvall, O., Brundin, P., Widner, H., Rehncrona, S., Gustavii, B., Frackowiak, R., Leenders, K. L., Sawle, G., Rothwell, J. C., Marsden, D. & Bjâ≠◊Rklund, A. 1990. Grafts of fetal dopamine neurons survive and improve motor function in Parkinson's Disease. *Science,* 247, 574-577.

Lindvall, O. & Hagell, P. 2002. Cell replacement therapy in human neurodegenerative disorders. *Clinical Neuroscience Research,* 2 86-92.

Lois, C. & Alvarez-Buylla, A. 1994. Long-distance neuronal migration in the adult mammalian brain. *Science,* 264, 1145-1148.

Lundberg, C., Winkler, C., Whittemore, S. R. & Bjorklund, A. 1996. Conditionally immortalized neural progenitor cells grafted to the striatum exhibit site-specific neuronal differentiation and establish connections with the host globus pallidus. *Neurobiol.Dis.,* 3, 33-50.

Maherali, N., Sridharan, R., Xie, W., Utikal, J., Eminli, S., Arnold, K., Stadtfeld, M., Yachechko, R., Tchieu, J., Jaenisch, R., Plath, K. & Hochedlinger, K. 2007. Directly reprogrammed fibroblasts show global epigenetic remodeling and widespread tissue contribution. *Cell Stem Cell,* 1, 55-70.

Mandel, R. J., Snyder, R. O. & Leff, S. E. 1999. Recombinant adeno-associated viral vector-mediated glial cell line-derived neurotrophic factor gene transfer protects nigral dopamine neurons after onset of progressive degeneration in a rat model of Parkinson's Disease. *Exp.Neurol.,* 160, 205-214.

Mandel, R. J., Spratt, S. K., Snyder, R. O. & Leff, S. E. 1997. Midbrain injection of recombinant adeno-associated virus encoding rat glial cell line-derived neurotrophic factor protects nigral neurons in a progressive 6-hydroxydopamine-induced degeneration model of Parkinson's Disease in rats. *Proc.Natl.Acad.Sci.U.S.A,* 94, 14083-14088.

Martin, G. R. 1981. Isolation of a pluripotent cell line from early mouse embryos cultured in medium conditioned by teratocarcinoma stem cells. *Proceedings of the National Academy of Sciences of the United States of America,* 78, 7634-8.

Mchugh, P. R. 2004. Zygote and "clonote"--the ethical use of embryonic stem cells. *N.Engl.J.Med.,* 351, 209-211.

Meissner, A., Wernig, M. & Jaenisch, R. 2007. Direct reprogramming of genetically unmodified fibroblasts into pluripotent stem cells. *Nat Biotechnol,* 25, 1177-81.

Mendez, I., Vinuela, A., Astradsson, A., Mukhida, K., Hallett, P., Robertson, H., Tierney, T., Holness, R., Dagher, A., Trojanowski, J. Q. & Isacson, O. 2008. Dopamine neurons implanted into people with Parkinson's Disease survive without pathology for 14 years. *Nat Med,* 14, 507-9.

Mittoux, V., Ouary, S., Monville, C., Lisovoski, F., Poyot, T., Conde, F., Escartin, C., Robichon, R., Brouillet, E., Peschanski, M. & Hantraye, P. 2002. Corticostriatopallidal Neuroprotection by Adenovirus-Mediated Ciliary Neurotrophic Factor Gene Transfer in a Rat Model of Progressive Striatal Degeneration. *Journal of Neuroscience,* 22, 4478-4486.

Miyazono, M., Lee, V. M. & Trojanowski, J. Q. 1995. Proliferation, cell death, and neuronal differentiation in transplanted human embryonal carcinoma (NTera2) cells depend on the graft site in nude and severe combined immunodeficient mice. *Lab Invest,* 73, 273-283.

Morrison, S. J., Ahah, N. M. & Anderson, D. J. 1997. Regulatory mechanisms in stem cell biology. *Cell,* 88, 287-298.

Nakagawa, M., Koyanagi, M., Tanabe, K., Takahashi, K., Ichisaka, T., Aoi, T., Okita, K., Mochiduki, Y., Takizawa, N. & Yamanaka, S. 2008. Generation of induced pluripotent stem cells without Myc from mouse and human fibroblasts. *Nat Biotechnol,* 26, 101-6.

Nan, Z., Grande, A., Sanberg, C. D., Sanberg, P. R. & Low, W. C. 2005. Infusion of human umbilical cord blood ameliorates neurologic deficits in rats with hemorrhagic brain injury. *Ann.N.Y.Acad.Sci.,* 1049, 84-96.

Odeberg, J., Piao, J. H., Samuelsson, E. B., Falci, S. & Akesson, E. 2005. Low immunogenicity of in vitro-expanded human neural cells despite high MHC expression. *J Neuroimmunol,* 161, 1-11.

Odorico, J. S., Kaufman, D. S. & Thomson, J. A. 2001. Multilineage differentiation from Human Embryonic Stem Cell Lines. *Stem Cells,* 19, 193-204.

Okita, K., Ichisaka, T. & Yamanaka, S. 2007. Generation of germline-competent induced pluripotent stem cells. *Nature,* 448, 313-7.

Okita, K., Nakagawa, M., Hyenjong, H., Ichisaka, T. & Yamanaka, S. 2008. Generation of mouse induced pluripotent stem cells without viral vectors. *Science,* 322, 949-53.

Olanow, C. W., Kordower, J. H. & Freeman, T. B. 1996. Fetal nigral transplantation as a therapy for Parkinson's Disease. *Trends in Neurosciences,* 19, 102-109.

Park, I. H., Lerou, P. H., Zhao, R., Huo, H. & Daley, G. Q. 2008a. Generation of human-induced pluripotent stem cells. *Nat Protoc,* 3, 1180-6.

Park, I. H., Zhao, R., West, J. A., Yabuuchi, A., Huo, H., Ince, T. A., Lerou, P. H., Lensch, M. W. & Daley, G. Q. 2008b. Reprogramming of human somatic cells to pluripotency with defined factors. *Nature,* 451, 141-6.

Parmar, M. & Jakobsson, J. 2011. Turning skin into dopamine neurons. *Cell Res.*

Parmar, M., Skogh, C., Bjorklund, A. & Campbell, K. 2002. Regional specification of neurosphere cultures derived from subregions of the embryonic telencephalon. *Mol.Cell Neurosci,* 21 645-656.

Pearl, J. I., Lee, A. S., Leveson-Gower, D. B., Sun, N., Ghosh, Z., Lan, F., Ransohoff, J., Negrin, R. S., Davis, M. M. & Wu, J. C. 2011. Short-term immunosuppression promotes engraftment of embryonic and induced pluripotent stem cells. *Cell Stem Cell,* 8, 309-17.

Pfisterer, U., Kirkeby, A., Torper, O., Wood, J., Nelander, J., Dufour, A., Bjorklund, A., Lindvall, O., Jakobsson, J. & Parmar, M. 2011. Direct conversion of human fibroblasts to dopaminergic neurons. *Proc Natl Acad Sci U S A,* 108, 10343-8.

Philpott, L. M., Kopyov, O. V., Lee, A. J., Jacques, S., Duma, C. M., Caine, S., Yang, M. & Eagle, K. S. 1997. Neuropsychological functioning following fetal striatal transplantation in Huntington's chorea: three case presentations. *Cell Transplantation,* 6, 203-12.

Politis, M., Oertel, W. H., Wu, K., Quinn, N. P., Pogarell, O., Brooks, D. J., Bjorklund, A., Lindvall, O. & Piccini, P. 2011. Graft-induced dyskinesias in Parkinson's Disease: High striatal serotonin/dopamine transporter ratio. *Mov Disord.*

Politis, M., Wu, K., Loane, C., Quinn, N. P., Brooks, D. J., Rehncrona, S., Bjorklund, A., Lindvall, O. & Piccini, P. 2010. Serotonergic neurons mediate dyskinesia side effects in Parkinson's patients with neural transplants. *Sci Transl Med,* 2, 38ra46.

Ray, J. & Gage, F. H. 1994. Spinal Cord Neuroblasts Proliferate in Response to Basic Fibroblast Growth Factor. *The Jouranl of Neuroscience,* **14,** 3548-3564.

Ray, J., Peterson, M., Schinstine, M. & Gage, F. H. 1993. Proliferation,differentiation, and long-term culture of primary hippocampal neurons. *Neurobiology,* **90,** 3602-3606.

Reddy, P. H., Williams, M. & Tagle, D. A. 1999. Recent advances in understanding the pathogenesis of Huntington's Disease. *Trends in Neuroscience,* **22,** 248-254.

Regulier, E., Pereira, D. A., Sommer, B., Aebischer, P. & Deglon, N. 2002. Dose-dependent neuroprotective effect of ciliary neurotrophic factor delivered via tetracycline-regulated lentiviral vectors in the quinolinic acid rat model of Huntington's Disease. *Hum.Gene Ther.,* 13 1981-1990.

Renfranz, P. J., Cunningham, M. G. & Mckay, R. D. 1991. Region-specific differentiation of the hippocampal stem cell line HiB5 upon implantation into the developing mammalian brain. *Cell,* 66 713-729.

Reuter, I., Tai, Y. F., Pavese, N., Chaudhuri, K. R., Mason, S., Polkey, C. E., Clough, C., Brooks, D. J., Barker, R. A. & Piccini, P. 2008. Long-term clinical and positron emission tomography outcome of fetal striatal transplantation in Huntington's Disease. *Journal of neurology, neurosurgery, and psychiatry,* **79,** 948-51.

Reynolds, B. A. & Weiss, S. 1996. Clonal and Population Analyses Demonstrate That an EGF-Responsive Mammalian Embryonic CNS Precursor Is a Stem Cell. *Developmental Biology,* **175,** 1-13.

Richards, L. J., Kilpatrick, T. J. & Bartlett, P. F. 1992. De novo generation of neuroanl cells from the adult mouse brain. *Proc.Natl.Acad.Sci.USA,* **89,** 8591-8595.

Rietze, R., Poulin, P. & Weiss, S. 2000. Mitotically active cells that generate neurons and astrocytes are present in multiple regions of the adult mouse hippocampus. *Journal of Comparative Neurology,* **424,** 397-408.

Ross, C. A. & Margolis, R. L. 2001. Huntington's Disease. *Clinical Neuroscience Research,* **1,** 142-152.

Rosser, A. E., Barker, R. A., Harrower, T., Watts, C., Farrington, M., Ho, A. K., Burnstein, R. M., Menon, D. K., Gillard, J. H., Pickard, J. & Dunnett, S. B. 2002. Unilateral transplantation of human primary fetal tissue in four patients with Huntington's Disease: NEST-UK safety report ISRCTN no 36485475. *J Neurol Neurosurg Psychiatry,* **73,** 678-85.

Rosser, A. E., Tyers, P. & Dunnett, S. B. 2000. The morphological development of neurons derived from EGF- and FGF-2-driven human CNS precursors depends on their site of integration in the neonatal rat brain. *Eur J Neurosci,* **12,** 2405-13.

Sanberg, P. R., Willing, A. E., Garbuzova-Davis, S., Saporta, S., Liu, G., Sanberg, C. D., Bickford, P. C., Klasko, S. K. & El Badri, N. S. 2005. Umbilical cord blood-derived stem cells and brain repair. *Ann.N.Y.Acad.Sci.,* 1049 67-83.

Sanchez-Ramos, J., Song, S., Kamath, S. G., Zigova, T., Willing, A., Cardozo-Pelaez, F., Stedford, T., Chopp, M. & Sanberg, P. R. 2001. Expression of neural markers in human umbilical cord blood. *Experimental Neurology,* **171,** 109-115.

Sandel, M. J. 2004. Embryo ethics--the moral logic of stem-cell research. *N.Engl.J.Med.,* **351,** 207-209.

Santa-Olla, J. & Covarrubias, L. 1995. Epidermal Growth Factor (EGF), Transforming Growth Factor-Ã (T GF-Ã), and Basic Fibroblast Growth Factor (bFGF) Differentially Influence Neural Precursor Cells of Mouse Embryonic Mesencephalon. *Journal of Neuroscience Research,* **42,** 172-183.

Saporta, S., Willing, A. E., Zigova, T., Daadi, M. M. & Sanberg, P. R. 2001. Comparison of calcium-binding proteins expressed in cultured hNT neurons and hNT neurons transplanted into the rat striatum. *Exp.Neurol.,* 167, 252-259.

Sautter, J., Tseng, J. L., Braguglia, D., Aebischer, P., Spenger, C., Seiler, R. W., Widmer, H. R. & Zurn, A. D. 1998. Implants of polymer-encapsulated genetically modified cells releasing glial cell line-derived neurotrophic factor improve survival, growth, and function of fetal dopaminergic grafts. *Experimental Neurology,* 149 230-236.

Schuldiner, M., Eiges, R., Eden, A., Yanuka, O., Itskovitz-Eldor, J., Goldstein, R. S. & Benvenisty, N. 2001. Induced neuronal differentiation of human embryonic stem cells. *Brain Research,* 913, 201-205.

Sidhu, K. S. 2011. New approaches for the generation of induced pluripotent stem cells. *Expert Opin Biol Ther,* 11, 569-79.

Skogh, C., Parmar, M. & Campbell, K. 2003. The differentiation potential of precursor cells from the mouse lateral ganglionic eminence is restricted by in vitro expansion. *Neuroscience,* 120, 379-385.

Soldner, F., Hockemeyer, D., Beard, C., Gao, Q., Bell, G. W., Cook, E. G., Hargus, G., Blak, A., Cooper, O., Mitalipova, M., Isacson, O. & Jaenisch, R. 2009. Parkinson's Disease patient-derived induced pluripotent stem cells free of viral reprogramming factors. *Cell,* 136, 964-77.

Song, S., Wang, Y., Bak, S. Y., Lang, P., Ullrey, D., Neve, R. L., O'malley, K. L. & Geller, A. I. 1997. An HSV-1 vector containing the rat tyrosine hydroxylase promoter enhances both long-term and cell type-specific expression in the midbrain. *J.Neurochem.,* 68 1792-1803.

Stadtfeld, M., Nagaya, M., Utikal, J., Weir, G. & Hochedlinger, K. 2008. Induced pluripotent stem cells generated without viral integration. *Science,* 322, 945-9.

Steece-Collier, K., Soderstrom, K. E., Collier, T. J., Sortwell, C. E. & Maries-Lad, E. 2009. Effect of levodopa priming on dopamine neuron transplant efficacy and induction of abnormal involuntary movements in parkinsonian rats. *J Comp Neurol,* 515, 15-30.

Svendsen, C. N., Caldwell, M. A., Shen, J., Ter Borg, M. G., Rosser, A. E., Tyres, P., Karmiol, S. & Dunnett, S. B. 1997. Long-Term Survival of Human Central Nervous System Progenitor Cells Transplanted into a Rat Model of Parkinson's Disease. *Experimental Neurology,* 148, 135-146.

Takahashi, K., Tanabe, K., Ohnuki, M., Narita, M., Ichisaka, T., Tomoda, K. & Yamanaka, S. 2007. Induction of pluripotent stem cells from adult human fibroblasts by defined factors. *Cell,* 131, 861-72.

Takahashi, K. & Yamanaka, S. 2006. Induction of pluripotent stem cells from mouse embryonic and adult fibroblast cultures by defined factors. *Cell,* 126, 663-76.

Thomson, J. A., Itskovitz-Eldor, J., Shapiro, S. S., Waknitz, M. A., Swiergiel, J. J., Marshall, V. S. & Jones, J. M. 1998. Embryonic stem cell lines derived from human blastocysts. *Science,* 282, 1145-7.

Tropepe, V., Craig, C. G., Morshead, C. M. & Van Der Kooy, D. 1997. Transforming Growth Factor-a Null and Senescent Mice Show Decreased Neural Progenitor Cell Proliferation in the Forebrain Subependyma. *The Jouranl of Neuroscience,* 17, 7850-7859.

Ubiali, F., Nava, S., Nessi, V., Frigerio, S., Parati, E., Bernasconi, P., Mantegazza, R. & Baggi, F. 2007. Allorecognition of human neural stem cells by peripheral blood lymphocytes despite low expression of MHC molecules: role of TGF-beta in modulating proliferation. *Int Immunol,* 19, 1063-74.

Vassilopoulos, G., Wang, P. R. & Russell, D. W. 2003. Transplanted bone marrow regenerates liver by cell fusion. *Nature,* 422, 901-904.

Vicario-Abejon, C., Johe, K. K., Hazel, T. G., Collazo, D. & Mckay, R. D. 1995. Functions of basic fibroblast growth factor and neurotrophins in the differentiation of hippocampal neurons. *Neuron,* 15, 105-114.

Wang, X., Willenbring, H., Akkari, Y., Torimaru, Y., Foster, M., Al Dhalimy, M., Lagasse, E., Finegold, M., Olson, S. & Grompe, M. 2003. Cell fusion is the principal source of bone-marrow-derived hepatocytes. *Nature,* 422, 897-901.

Watt, F. M. & Hogan, B. L. M. 2000. Out of eden:Stem cells and their niches. *Science,* 287, 1427-1430.

Watts, C., Brasted, P. J. & Dunnett, S. B. 2000. The Morphology, Integration, and Functional Efficacy of Striatal Grafts Differ Between Cell Suspensions and Tissue Pieces. *Cell Transplantation,* 9 395-407.

Welstead, G. G., Schorderet, P. & Boyer, L. A. 2008. The reprogramming language of pluripotency. *Curr Opin Genet Dev,* 18, 123-9.

Wernig, M., Lengner, C. J., Hanna, J., Lodato, M. A., Steine, E., Foreman, R., Staerk, J., Markoulaki, S. & Jaenisch, R. 2008. A drug-inducible transgenic system for direct reprogramming of multiple somatic cell types. *Nat Biotechnol,* 26, 916-24.

Wernig, M., Meissner, A., Foreman, R., Brambrink, T., Ku, M., Hochedlinger, K., Bernstein, B. E. & Jaenisch, R. 2007. In vitro reprogramming of fibroblasts into a pluripotent ES-cell-like state. *Nature,* 448, 318-24.

Willing, A. E., Lixian, J., Milliken, M., Poulos, S., Zigova, T., Song, S., Hart, C., Sanchez-Ramos, J. & Sanberg, P. R. 2003. Intravenous versus intrastriatal cord blood administration in a rodent model of stroke. *J Neurosci Res.,* 73, 296-307.

Woltjen, K., Michael, I. P., Mohseni, P., Desai, R., Mileikovsky, M., Hamalainen, R., Cowling, R., Wang, W., Liu, P., Gertsenstein, M., Kaji, K., Sung, H. K. & Nagy, A. 2009. piggyBac transposition reprograms fibroblasts to induced pluripotent stem cells. *Nature,* 458, 766-70.

Yamanaka, S. 2008. Induction of pluripotent stem cells from mouse fibroblasts by four transcription factors. *Cell Prolif,* 41 Suppl 1, 51-6.

Yu, J., Hu, K., Smuga-Otto, K., Tian, S., Stewart, R., Slukvin, Ii & Thomson, J. A. 2009. Human induced pluripotent stem cells free of vector and transgene sequences. *Science,* 324, 797-801.

Yu, J., Vodyanik, M. A., Smuga-Otto, K., Antosiewicz-Bourget, J., Frane, J. L., Tian, S., Nie, J., Jonsdottir, G. A., Ruotti, V., Stewart, R., Slukvin, Ii & Thomson, J. A. 2007. Induced pluripotent stem cell lines derived from human somatic cells. *Science,* 318, 1917-20.

Zhao, L.-R., Duan, W.-M., Reyes, M., Keene, C. D., Verfaillie, C. M. & Low, W. C. 2002. Human Bone Marrow Stem Cells Exhibit Neural Phenotypes and Ameliorate Neurological Deficits after Grafting into the Ischemic Brain of Rats. *Experimental Neurology,* 174, 11-20.

Zhao, M., Momma, S., Delfani, K., Carlen, M., Cassidy, R. M., Johansson, C. B., Brismar, H., Shupliakov, O., Frisen, J. & Janson, A. M. 2003. Evidence for neurogenesis in the adult mammalian substantia nigra. *Proceedings of the National Academy of Sciences,* 100, 7925.

Zietlow, R., Pekarik, V., Armstrong, R. J., Tyers, P., Dunnett, S. B. & Rosser, A. E. 2005. The survival of neural precursor cell grafts is influenced by in vitro expansion. *Journal of anatomy,* 207, 227-40.

Zigova, T., Song, S., Willing, A. E., Hudson, J. E., Newman, M. B., Saporta, S., Sanchez-Ramos, J. & Sanberg, P. R. 2002. Human umbilical cord blood cells express neural antigens after transplantation into the developing rat brain. *Cell Transplant.,* 11, 265-274.

19

Ameliorating Huntington's Disease by Targeting Huntingtin mRNA

Melvin M. Evers[1], Rinkse Vlamings[2,3],
Yasin Temel[2,3] and Willeke M. C. van Roon-Mom[1]
*[1]Center for Human and Clinical Genetics,
Leiden University Medical Center, Leiden,
[2]Departments of Neuroscience and Neurosurgery,
Maastricht University Medical Center, Maastricht,
[3]European Graduate School of Neuroscience (EURON),
The Netherlands*

1. Introduction

To date there are 9 known neurological diseases caused by an expanded polyglutamine (polyQ) repeat, with the most prevalent being Huntington's Disease (HD) (Cummings & Zoghbi, 2000). HD is a progressive autosomal dominant disorder. It is caused by a CAG repeat expansion in the *HTT* gene, which results in an expansion of a polyQ stretch at the N-terminal end of the huntingtin (htt) protein. This polyQ expansion plays a central role in the disease and results in the accumulation of cytoplasmic and nuclear aggregates. In this chapter we will discuss wild-type htt function and the gain of toxic function of mutant htt in HD. Currently no treatment is available to delay onset or slow disease progression. However, recently developed RNA modulating therapies have great potential to lower mutant htt levels in HD. Already promising results in animal and human studies for other neurodegenerative disorders have been obtained, from which HD research can learn.

2. Huntington's Disease

HD is an autosomal dominantly inherited neurodegenerative disorder. HD is rare, but more common in Western countries. The prevalence of HD in America is approximately 5 in 100,000 (Shoulson & Young, 2011) and in Europe, the prevalence of HD may be even higher with estimates in England and Wales as high as 12 in 100,000 individuals (Rawlins, 2010).

Post-mortem studies show that there is a 10-20 percent weight reduction in HD brains (Vonsattel et al., 1985). Neurodegeneration occurs throughout the forebrain with the GABAergic medium spiny neurons of the striatum as its first prominent victim, and to a lesser extent neurons in the cerebral cortex (Levesque et al., 2003). Severe cell loss in the striatal complex, the caudate nucleus and putamen results in striatal atrophy. This is accompanied by an enlargement of the lateral ventricles. The medium spiny projection neurons, containing enkephalin, are more susceptive to degeneration than substance P containing neurons while interneurons seem to be spared (Walker, 2007). With disease progression, degeneration expands throughout the HD brain and other structures become affected (Vonsattel et al., 1985). Cortical atrophy is characterized by thinning of the cerebral cortex and the underlying white matter. Neuronal loss is abundant in cortical layers III, V and VI (Rosas et al., 2008) but is also prominent in the CA1 region of the hippocampus, with a reduction of about 9 percent (Rosas et al., 2003).

Disease onset usually occurs around midlife and is clinically characterized by a combination of symptoms: cognitive impairments, movement abnormalities, and emotional disturbances. Motor symptoms of HD include chorea and occasionally bradykinesia and dystonia (Tabrizi et al., 2009). Choreic movements, recognized as involuntary and unwanted movements, start in the distal extremities. During the course of HD these movements become more profound and eventually all other muscles of the body are affected. These symptoms can initially appear as lack of concentration or nervousness and unsteady gait (Kremer et al., 1992). Psychiatric symptoms often precede the onset of motor symptoms. Irritability is commonly one of the first signs and occurs throughout the course of the disease. Other psychiatric symptoms involve anxiety, obsessive and compulsive behavior while apathy and psychosis can appear in advanced stages. However, the most frequent psychiatric symptom is depression (Craufurd et al., 2001). Like psychiatric symptoms, cognitive symptoms can be present prior to the onset of the motor symptoms. The cognitive symptoms comprise mainly impairment in executive functions, including abstract thinking, problem solving, and attention (Snowden et al., 2002). Furthermore, the ability to learn new skills is affected (Paulsen et al., 2001). Altogether these symptoms substantially impede social and professional functioning. Eventually patients are incapable to adequately perform daily activities finally leading to progressive disability, requiring full-time care, followed by death (Simpson, 2007). Death generally occurs 15 to 20 years post diagnosis due to complications such as pneumonia, falls, dysphagia, heart disease or suicide.

The disease is caused by a CAG trinucleotide repeat expansion within the coding region of the *HTT* gene. The *HTT* gene was the first autosomal disease locus to be mapped by genetic linkage analysis in 1983 (Gusella et al., 1983) on the short arm of chromosome 4 (4p16.3). The huntingtin protein (htt) was found to be ubiquitously expressed throughout the body, with highest expression in testis and brain (Strong et al., 1993), however, cells in the brain are specifically vulnerable to the toxic function of mutant htt. The CAG repeat expansion in the *HTT* gene results in an expanded polyQ repeat in the htt protein (The Huntington's Disease Collaborative Research Group, 1993). When the number of CAG repeats exceeds 39, the gene encodes a mutated form of the htt protein that is prone to aggregation. Alleles ranging 36 to 39 repeats, lead to an incomplete and variable penetrance of the disease or to a very late onset (McNeil et al., 1997). Repeat numbers exceeding 55-60 result in clinical manifestation of the disease before the age of 20, known as Juvenile Huntington's Disease (JHD) (Andresen et al., 2007) and both sexes are affected with the same frequency (Walker, 2007). Intergenerational CAG changes are extremely rare on normal chromosomes but on expanded chromosomes changes in CAG size take place in approximately 70 percent of meioses and expansion is more likely via the paternal line (Kremer et al., 1995).

There is a strong inverse correlation between repeat numbers and the age of onset of the disease. The repeat length accounts for approximately 70 percent of the variance in age of onset (Roos, 2010). However, no correlation with repeat size is apparent for the progression and duration of the disease. Furthermore, neuropathological changes, such as atrophy and inclusion load are clearly correlated with the CAG repeat number.

For patients, only symptomatic treatment is available and a treatment to slow down the progression or delay the onset of the disease remains elusive.

2.1 Huntingtin protein

When the *HTT* gene was discovered in 1993, the htt protein had an unknown function. Since then, enormous research efforts have revealed many functions of the wild-type protein (discussed in the present paragraph) and many toxic gain of functions of the mutant protein (discussed in the next paragraph).

Wild-type htt is mainly localized in the cytoplasm, although a small proportion is present in the nucleus (de Rooij et al., 1996; Kegel et al., 2002). The protein is known to be associated with microtubules, the plasma membrane, Golgi complex, the endoplasmic reticulum, and mitochondria. Furthermore htt is associated with vesicular structures, such as clathrin-coated and non-coated vesicles, autophagic vesicles, endosomal compartments or caveolae (Kegel et al., 2005; Strehlow et al., 2007; Rockabrand et al., 2007; Atwal et al., 2007; Caviston et al., 2011).

Three of the first 17 amino acids at the amino terminus of htt are lysines, which are targets for post translational modifications that regulate htt half-life and are proposed to be involved in targeting htt to various intracellular membrane-associated organelles (Kalchman et al., 1996; Steffan et al., 2004; Kegel et al., 2005; Atwal et al., 2007; Rockabrand et al., 2007). The first 17 amino acids of htt have also been suggested to act as nuclear export signal (NES) by interaction with the nuclear pore protein translocated promoter region (Tpr) that then transports N-terminal htt fragments out of the nucleus (Cornett et al., 2005). The polyQ repeat starts at the 18th amino acid and is thought to form a polar zipper structure, which has been implicated in the interaction between different polyQ-containing transcription factors (Perutz et al., 1994; Harjes & Wanker, 2003). The polyQ stretch is followed by a polyproline repeat, which is thought to be involved in keeping the protein soluble (Steffan et al., 2004). Additionally, three main HEAT (htt, elongation factor 3, protein phosphatase 2A, and the yeast PI3-kinase TOR1) repeat motifs are identified which are known to form superhelical structures and are involved in protein-protein interactions (Takano & Gusella, 2002; Li et al., 2006). Htt is palmitoylated at the cysteine residue 214 by htt interacting protein (Hip) 14, which is thought to be involved in htt trafficking (Huang et al., 2004). Htt has various proteolytic cleavage motifs, with a hotspot between amino acid 500 and 600, which are recognized by various proteases, such as caspases 1, 3, 6, 7 and 8 and calpain (Gafni & Ellerby, 2002; Wellington et al., 2002; Kim et al., 2006). In contrast to mutant htt, the significance of wild-type htt cleavage is not completely clear.

2.2 Mutant htt gain of toxic function in HD

Expanded polyQ proteins are known to undergo conformational changes, which result in the hallmark of polyQ disorders, protein aggregates. The aggregates can already be found before the onset of the first symptoms (Weiss et al., 2008). Remarkably, there is growing evidence suggesting that these aggregates are not good indicators for disease onset and progression (Wanker, 2000; van Roon-Mom et al., 2006). The rate of aggregate formation is correlated to the length of the polyQ repeat (Legleiter et al., 2010). Whether accumulation of these aggregates is neurotoxic or neuroprotective is still under debate since evidence also suggests that soluble mutant htt is the main toxic component (Davies et al., 1997; Saudou et al., 1998; Arrasate et al., 2004). While the expanded polyQ repeat displays pathogenic properties it is probably not essential for normal function (Clabough & Zeitlin, 2006). Mutant htt is more disposed to proteolysis and it was shown that small N-terminal htt fragments are more toxic than full length mutant htt (Cooper et al., 1998). Proteolytic cleavage of mutant htt results in nuclear localization of toxic N-terminal mutant htt fragments. These N-terminal mutant htt fragments are important in the pathological process. Mutant htt fragments within the striatum of HD brains clearly differ from those of control brains, suggesting cleavage is disease specific (Mende-Mueller et al., 2001) and htt caspase-6 resistant HD mice did not show neuronal dysfunction (Graham et al., 2006).

Various transcription factors have been found to co-localize with htt aggregates, such as TATA box binding protein (TBP), CREB binding protein (CBP) and p53 (Steffan et al., 2000; van Roon-Mom et al., 2002). These co-aggregated proteins can no longer assert their normal function and could thereby contribute to HD pathology (Nucifora, Jr. et al., 2001)

Mutant htt is also suggested to act as pro-apoptotic factor triggering cell death. Htt is found to bind to the pro-apoptotic factor p53. Interestingly, p53 deficient HD mice displayed increased striatal inclusion body formation (Ryan et al., 2006). Expression of mutant htt in p53 deficient mice improved the lifespan probably by increased apoptosis initiated by mutant htt (Ryan & Scrable, 2008).

In HD the fusion machinery and axonal transport are impaired. Accumulated N-terminal fragments block the axonal machinery, resulting in transport defects (Gunawardena et al., 2003). Endocytosis is thought to be impaired since the synaptic vesicle protein PACSIN1 has an altered subcellular location in early stage HD patients (Modregger et al., 2002). Finally, various proteins involved in exocytosis are known to have decreased expression levels in HD patients. Proteins involved in docking and fusion of vesicles show reduced transcript expression, suggesting a defect in the neurotransmitter release machinery in HD patients (Smith et al., 2007).

N-terminal mutant htt fragments are found to be associated with the surface of mitochondria in transgenic and knock-in HD mice (Panov et al., 2002; Orr et al., 2008). The accumulation of mutant htt on mitochondria is increasing with age and correlates with disease progression. This impaired mitochondrial trafficking by N-terminal mutant htt could lead to decreased ATP supply in nerve terminals (Orr et al., 2008). Mutant htt is also suggested to be involved mitochondrial energy metabolism defects. Metabolic energy defects could be the result of mutant htt's capability to induce mitochondrial permeability transition pore opening. This leads to low mitochondrial membrane potential and high glutamate transmission, resulting in overactive glutamate NMDA receptors (excitotoxicity) (Choo et al., 2004). Abnormal mitochondrial respiratory chain function leads to reduced ATP levels and subsequent partially depolarized membrane. This voltage change leads to chronic calcium influx and activation of proteases, causing more reactive oxygen species (ROS) production. Further, increased ROS production gives rise to oxidative stress and could contribute to the vicious circle (Browne & Beal, 2006).

2.3 Loss of wild-type function in HD

As described above, the main cause of HD is a gain of toxic mutant htt function. Since various functions and post-translational modifications of htt are altered in HD, loss of wild-type htt function could also be involved. Htt expression is important for normal cellular function since knock-out of the homologous htt mouse gene was found to be early embryonic lethal (Zeitlin et al., 1995). Previous studies have shown that approximately 50% of htt protein level is required to maintain cell functionality (Dragatsis et al., 2000). Next to embryonic development, htt is also involved in regulation of apoptosis, transcription, intracellular transport and BDNF transcription (Zuccato et al., 2001; Imarisio et al., 2008).

Wild-type htt is reported to act as protector of the brain cells from apoptotic stimuli (Rigamonti et al., 2000). Reduced wild-type htt expression in transgenic HD mice resulted in worsening of the behavioural deficits and survival. In addition, no severe striatal abnormalities were visible in those HD mice, which could mean that the striatal phenotype is mainly caused by mutant htt toxicity (Zhang et al., 2003). Furthermore, overexpression of wild-type htt protected these mice against neurodegeneration. Removal of endogenous htt in a *Drosophila melanogaster (D. melanogaster)* HD model was found to exacerbate the neurodegenerative phenotype associated, suggesting that loss of normal htt function might also contribute to HD pathogenesis (Zhang et al., 2009).

3. HTT RNA

Although the main toxic component is the htt protein, recent evidence suggests that also HTT RNA could have toxic properties. There is also recent evidence for antisense transcription through the *HTT* locus. In this paragraph we will review the importance of these findings.

3.1 Htt RNA gain of function in HD

Trinucleotide expansion disorders occur either in untranslated genomic regions (UTRs) resulting in a toxic RNA gain of function or loss of gene function, or in coding regions resulting in a gain of toxic protein function (Orr & Zoghbi, 2007). Until recently, it was believed that HD is solely caused by a toxic gain of function of the polyQ protein and to a lesser extent, loss of wild-type function. However, recent evidence suggests that the mutant CAG repeats of the HTT RNA transcript could also have toxic properties (Fig. 1). This RNA toxicity is caused by the long hairpin structures of the expanded RNA that result in abnormal interactions with double stranded RNA-binding proteins.

The CAG repeat hairpin in the HTT transcript was found to be stabilized by the flanking (CCG)n repeat (de Mezer et al., 2011). The resulting double stranded CAG RNA hairpin formed intranuclear foci that co-localized with the muscleblind-like 1 (MBNL1) splicing factor (Jiang et al., 2004). Altered MBNL1 function is implicated in RNA toxicity of CUG repeat expansion disorders such as myotonic dystrophy type 1 (DM1) (Kanadia et al., 2003). DM1 is caused by a CTG repeat expansion at the 3′ UTR of the *DMPK* gene. The CTG repeats are known to form stable hairpin structures that are toxic by causing abnormal alternative splicing by MBNL1 binding and sequestering in nuclear foci (Fardaei et al., 2001; Kanadia et al., 2003). Similar to expanded CUG repeats in DM1, synthesized expanded CAG repeats also resulted in abnormal alternative splicing in both transiently transfected and patient-derived cells (Mykowska et al., 2011). The RNA toxicity modifier MBNL1 was also found to be involved in another polyQ disease, namely spinocerebellar ataxia 3 (SCA3). MBNL1 was found to be up-regulated in a *D. melanogaster* model of SCA3. The neurodegenerative disorder SCA3 is caused by a CAG repeat expansion in the *ATXN3* gene, which results in the expression of a polyQ containing ataxin-3 protein. Upregulation of the *D. melanogaster* homolog of MBNL1 (*mbl*) was found to enhance pathogenic ataxin-3 protein induced toxicity, as well as pathogenic mutant htt protein induced toxicity (Li et al., 2008).

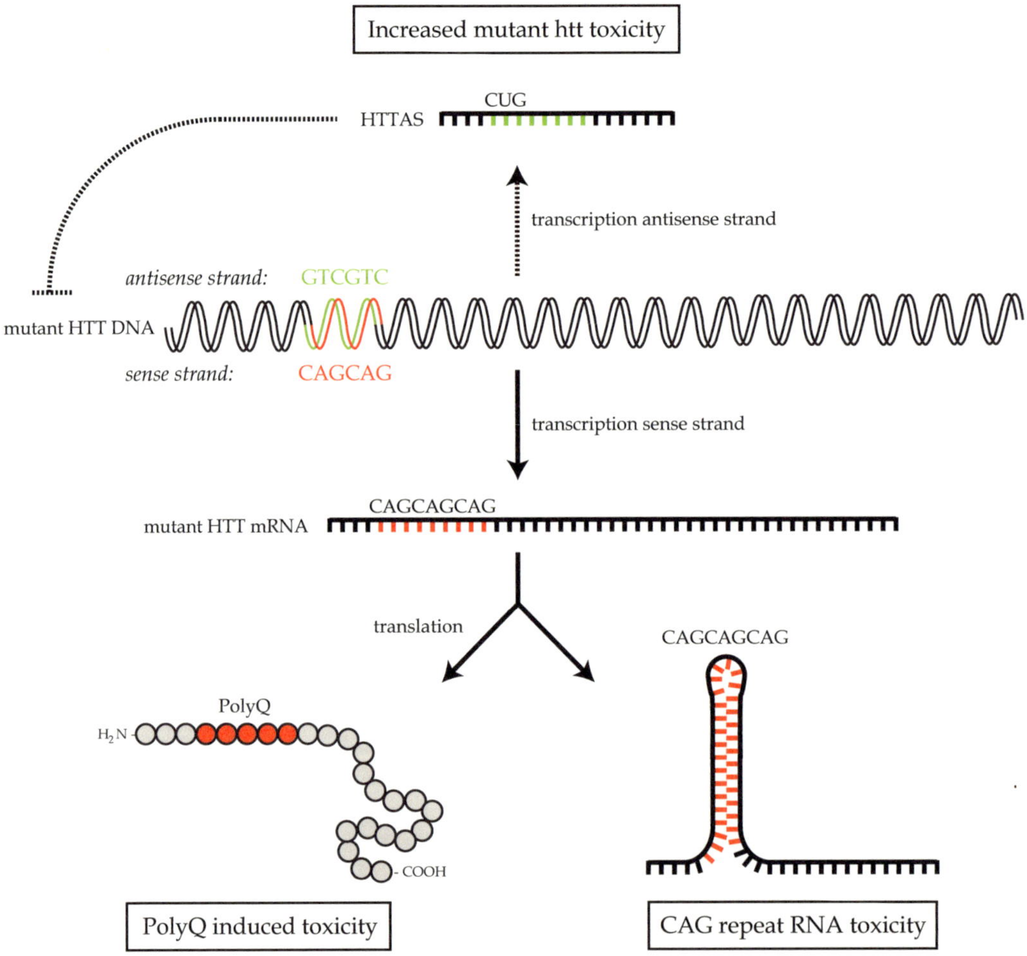

Fig. 1. Schematic representation of modes of huntingtin toxicity in HD. Mutant htt transcription results in mutant HTT mRNA which can form double stranded hairpins through the expanded CAG repeat and adjacent CCG repeat. The CAG hairpin is involved in MBNL1 induced alternative splicing and toxicity. The main pathological process in HD involves the translation of expanded CAG repeat-containing mRNA into a toxic polyQ protein. Antisense transcription through the *HTT* gene results in a HTT antisense transcript (HTTAS). This HTTAS regulates HTT sense levels. In HD there is lower expression of HTTAS, resulting in increased levels of the HTT sense transcript and increased mutant htt toxicity.

HD and SCA3 transgenes with a CAG repeat interrupted by CAA codons (expressing an identical polyQ protein as compared to a pure CAG repeat) showed only a mild phenotype, indicating the importance of the expanded pure CAG repeat for the toxic phenotype. Interestingly, both full CAG repeats and CAA interrupted CAG repeats showed similar levels of protein inclusions, indicating that the phenotype severity does not correlate with the number of inclusions (Li et al., 2008).

Recently, transgenic mice expressing a GFP construct with 200 CAGs in the 3′ UTR resulted in reduced GFP levels as compared to animals with 23 CAG repeats in their 3′ UTR of the GFP construct (Hsu et al., 2011). Furthermore, these CAG_{200} mice showed nuclear RNA foci and a reduced breeding efficiency, which supports the gain of RNA toxicity hypothesis.

Transgenic *Caenorhabditis elegans* (*C. elegans*) expressing various CAG repeat lengths in the 3′ UTR of a GFP gene showed a length-dependent toxicity. Worms with an 83 CAG repeat did not show any phenotype, whereas *C. elegans* expressing 200 CAGs died within a few days. Both 125 CUGs and 125 CAGs co-localized in nuclear foci with *C. elegans* MBNL1 homolog CeMBL and overexpression of CeMBL partly reversed the CAG 125 induced phenotype (Wang et al., 2011).

In contrast to the above studies, there is also evidence that the CAG repeat RNA is not toxic. Expression of a cDNA construct with 79 CAG repeats in the 3′ UTR did not induce cell death, whereas a construct expressing 79 CAGs in the coding region did induce cell death (Ikeda et al., 1996). This was also found in two other polyQ disorders, spinocerebellar ataxia 1 (SCA1) and spinobulbar muscular atrophy (SBMA). A SCA1 mouse model with impaired nuclear localization signal in ataxin-1 did not show nuclear inclusion bodies and did not display the disease phenotype (Klement et al., 1998). Furthermore, impairing nuclear localization of the androgen receptor (AR) in SBMA by castration showed marked improvements of disease pathology, also suggesting that the pathology is mainly caused by gain of toxic protein and not RNA (Katsuno et al., 2002). A *D. melanogaster* model of CAG toxicity expressing a repeat construct with a premature termination codon before a 93 CAG repeat, did not show any phenotype (McLeod et al., 2005). Based on these results, it was suggested that the toxicity in CAG triplet repeat disorders was exclusively the result of expanded polyQ protein gain of function.

From the above we can conclude that not only gain of toxicity by expanded polyQ protein, but also RNA toxicity from the expanded CAG repeat could be involved in HD pathology. However, the size of the CAG repeat is critical for RNA pathogenicity.

3.2 HTT antisense transcription

A large proportion of the genome can produce transcripts from both strands (Katayama et al., 2005). It has become clear that antisense transcripts are involved in triplet repeat disorders and bidirectional transcription has thus far been identified in DM1, spinocerebellar ataxia 8 (SCA8), and HD like 2 (HDL2) (Moseley et al., 2006; Wilburn et al., 2011).

In SCA8, which is caused by a CTG repeat expansion in a transcribed but not translated *ATXN8OS* gene, it was thought that the expanded CTG repeat caused RNA toxicity (Koob et al., 1999). Unexpectedly, bacterial artificial chromosome (BAC) transgenic SCA8 mice showed 1C2 positive inclusion bodies. The 1C2 antibody specifically recognizes expanded polyQ tracts, which are the hallmark of polyQ disorders. A novel transcript called ataxin-8, which encodes a polyQ protein, was expressed from the opposite strand, suggesting polyQ induced toxicity (Moseley et al., 2006). A BAC HDL2 mouse model with a pathogenic CTG repeat on the sense and expanded CAG repeat on the antisense strand at the *Junctophilin-3* locus showed both RNA toxicity caused by its expanded CUG repeat as well as protein toxicity by its polyQ translated expanded CAG repeat (Wilburn et al., 2011). These findings suggest that triplet repeat disorders can involve toxic gain of function of both protein and RNA by bidirectional transcription.

Recently, two natural HTT antisense (HTTAS) transcripts were identified at the HD locus (Chung et al., 2011). HTTAS was found to be 5′ capped, poly A-tailed and contained 3 exons. There were two different isoforms identified of which one enclosed a functional promotor and the CTG repeat. The HTTAS containing the short CTG repeat was found to be widely expressed in multiple tissues. Remarkably, expanded CTG repeat containing HTTAS was strongly reduced in HD brains. The authors state that HTTAS acts as a negative regulator for HTT transcript expression as knock-down of HTTAS resulted in higher htt levels and overexpression of HTTAS resulted in lower HTT levels (Chung et al., 2011). This negative regulating property on HTT of HTTAS could potentially have a clinical implication by overexpressing HTTAS in HD patients, thereby alleviating pathogenicity by lowering htt levels.

4. RNA modulating therapies in HD

Although the *HTT* gene was identified in 1993, there are no treatments to cure or even slow down the progression of the disease. Most therapeutic strategies under investigation are targeting one of the many altered cellular processes caused by toxic mutant htt. Targeting a single cellular process might be inadequate to be clinically beneficial.

A more effective approach would be to reduce the expression of the causative *HTT* gene and thereby inhibiting all downstream toxic effects. Recent advances to inhibit the formation of mutant polyQ proteins using RNA modulating therapies, such as RNA interference (RNAi) and antisense oligonucleotides (AONs) look promising for HD (Sah & Aronin, 2011). RNAi is an endogenous cellular process involved in transcriptional regulation and acts as cellular defense mechanism against exogenous viral components. RNAi by introducing small interfering RNA (siRNA), short hairpin RNA (shRNA), or artificial micro RNA (miRNA), is increasingly used as a potential therapeutic tool to reduce expression of target transcripts. Specific knock-down is also achieved by introducing modified single stranded AONs that can hybridize to the target RNA, which is subsequently degraded or its translation blocked.

The most frequently used htt RNA modulating strategies for HD are: Knock-down of total htt RNA levels by targeting both wild-type and mutant htt and allele-specific reduction of mutant htt RNA only (Fig. 2).

4.1 Gene therapy to lower both htt alleles in HD

Since htt has many important wild-type functions, one of the key questions that needs to be answered for htt lowering strategies to become successful is how much htt is needed for normal function, or rather, how much can htt levels be reduced before adverse effects become apparent. Below we will first describe the studies describing lowering of both wild type and mutant htt, followed by the different approaches for allele specifically lowering mutant htt only.

Various synthetic oligonucleotides with different modifications and backbones have been used in rodents to partially lower htt expression. A partial reduction of both normal and mutant htt by 25 to 35% using shRNAs was found to be well-tolerated in wild-type rats up to 9 months without signs of toxicity or striatal degeneration (Drouet et al., 2009). Total silencing using artificial miRNAs for both wild-type and mutant htt of 75% within the striatum of a transgenic HD mouse model showed reduced toxicity, extended survival, and improved motor performance, 3 months after treatment (Boudreau et al., 2009).

Striatal injection of non allele-specific artificial miRNA in wild-type mice resulted in 70% reduction of htt levels. The high murine htt transcript reduction was sustained without adverse side effects up to the end of their study, which was set at 4 months (McBride et al., 2008). Since htt lowering strategies will be most beneficial for patients when administered over many years, the long-term safety needs to be assessed. Therefore, simultaneously lowering transcript levels from both alleles can only be applied once the role of wild-type htt in the human brain is elucidated in more detail. Moreover, to date it is not known if there is equal transcription from both the mutant and wild-type htt allele. Lowering total htt transcript levels by 70% does not necessarily mean an equal reduction of both alleles by 70%.

4.2 Allele-specific reduction of mutant htt in HD

As described in previous paragraphs, endogenous htt expression is important for normal cellular function and an ideal strategy for an autosomal dominant disorder as HD would be to specifically target the mutant allele and thereby maintaining as much wild-type htt protein as possible. Suppression of 50% to 80% using siRNA specific for human mutant htt in transgenic rodent models of HD for 4 months was found to improve motor and neuropathological abnormalities and prolonged longevity in HD mice (Harper et al., 2005; Wang et al., 2005). These studies showed that lowering mutant htt without reducing wild-type htt levels, resulted in an improved pathology. These results favored an allele-specific htt lowering approach without altering the expression of endogenous wild-type htt expression. Various studies have shown that a pronounced decrease of mutant htt levels, with only minor reductions in wild-type htt is feasible using allele-specific oligonucleotides. The different approaches, their advantages and disadvantages will be discussed in the following paragraph.

4.2.1 Targeting associated SNPs in HD

Single nucleotide polymorphisms (SNPs) are DNA sequence variations that occur when a single nucleotide is different between the two alleles of a gene. One way to distinguish between the wild-type and polyQ disease-causing allele is to target such a SNP that is unique to the mutant transcript using siRNAs (Miller et al., 2003). siRNAs are known to discriminate between transcripts that differ at a single nucleotide and various studies have shown specific reduction of mutant htt mRNA using siRNAs directed against different SNPs. The first evidence of allele-specific silencing in HD using using SNP specific RNAi was obtained in human cells overexpressing htt transgenes (Schwarz et al., 2006). The first prove of principle of endogenous mutant htt silencing using SNPs in fibroblasts derived from HD patients was acquired in 2008 (van Bilsen et al., 2008). Extensive genotyping revealed a group of 22 SNPs highly associated with mutant htt alleles in a European HD cohort (Warby et al., 2009). Since then, various groups have shown that the vast majority of the HD patient population could be treated using 5 (75% of HD patients) or 7 (85% of the HD patients) different siRNAs (Lombardi et al., 2009; Pfister et al., 2009). The most promising SNP was found to be located in exon 67 of the *HTT* gene. This SNP is strongly associated with the mutant allele and 48% of the total Western HD population was heterozygous at this site (Pfister et al., 2009).

Most of the heterozygous SNPs linked to the expanded CAG repeat in exon 1, are found remote downstream from the CAG repeat in exons 25 up to 67 (Lombardi et al., 2009; Pfister et al., 2009). To determine in HD patients whether they are heterozygous and if yes, which SNP belongs to the expanded CAG repeat, a technique called SNP linkage by circularization (SLiC) was developed (Liu et al., 2008). By circulating the DNA, the CAG repeat and SNP site were brought together, making it easy to link the SNP to the expanded CAG repeat using a single PCR.

Although the selectivity obtained from above described SNP targeting siRNAs are very promising, there are some limitations. The diversity of SNPs within patient populations would make it necessary to develop multiple siRNAs. Furthermore, for HD patients that do not exhibit any of the most frequent SNPs a different treatment needs to be developed.

4.2.2 Targeting the expanded CAG repeat in mutant HTT

Another approach to achieve allele-specific silencing is based on the common denominator of all HD patients; their expanded CAG repeat. The selective silencing is either based on the hypothesis that there are structural differences between wild-type and mutant htt mRNA, or based on the larger number of CAGs in the expanded repeat and subsequent more binding possibilities. The first prove for allele discrimination by targeting the CAG repeat was achieved in HD human fibroblasts using a siRNA with 7 consecutive CUG nucleotides (Krol et al., 2007). Further studies with CAG repeat targeting siRNAs showed a low selectivity for the mutant allele, making siRNAs incompatible for CAG repeat directed allele-specific silencing (Hu et al., 2009). Other chemical modifications and oligomers show much higher specificity for expanded CAG repeat transcripts. Single stranded peptide nucleic acids (PNA), locked nucleic acids (LNA), and AONs with a 2'O methyl addition and phosphorothioate backbone targeting CAG repeats have been used to specifically reduce expanded HD transcripts *in vitro* in patient derived skin and blood cells (Hu et al., 2009; Evers et al., 2011). However, PNA selectivity was less pronounced in CAG repeat lengths (40 to 45 CAGs) that occur most frequently in the HD patient population. The allele-specific reduction after transfection of patient cells with LNAs and AONs with 7-mer CUG repeats was more pronounced in the average HD CAG repeat length. Furthermore, other endogenous CAG repeat containing transcripts with important cellular functions were unaffected by the tested CUG oligonucleotides (Hu et al., 2009; Evers et al., 2011).

The main advantages of LNAs and AONs are that they are single stranded and do not show toxicity *in vivo*. Systemic delivery of modified AONs in Duchenne muscular dystrophy (DMD) boys carrying specific deletions in the DMD gene induced the synthesis of novel, internally deleted, but likely (semi-) functional, dystrophin proteins without clinically apparent adverse event (Goemans et al., 2011).

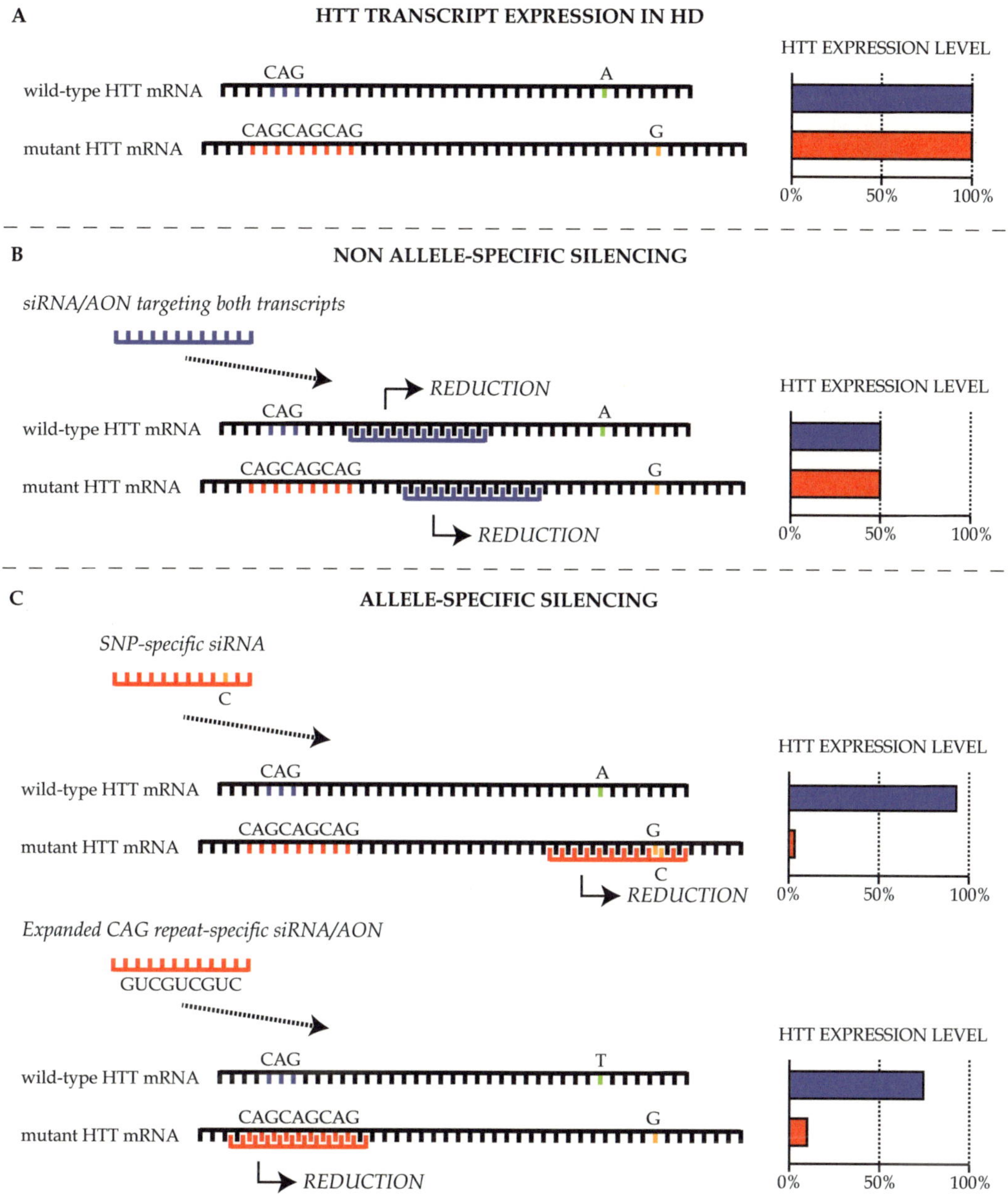

Fig. 2. RNA modulating therapeutic approaches for lowering htt. Two different HTT RNA modulating strategies used for HD are: A) Non allele-specific reduction of total HTT RNA levels by targeting a sequence that is identical in both the wild-type and mutant HTT transcript. B) Allele-specific reduction of mutant HTT RNA by targeting a unique heterozygous SNPs only present in the mutant transcript or C) Allele-specific reduction targeting the expanded CAG repeat on the mutant HTT transcripts.

Likewise, the use of only a single AON was suggested to be effective as treatment of various polyQ diseases (Hu et al., 2009; Evers et al., 2011). One expanded CAG repeat targeting AON was found to specifically reduce the expression of mutant ataxin-1 and ataxin-3 mRNA levels in SCA1 and 3, respectively, and mutant atrophin-1 in dentatorubral-pallidoluysian atrophy (DRPLA) in patient derived cells (Evers et al., 2011).

Although these results are promising, extensive research is needed to elucidate the mechanism used by those oligonucleotides to induce selective silencing and to assess specificity and safety. Likewise, the full potency of this allele-specific treatment will be revealed when the first *in vivo* results are obtained.

5. RNA modulating therapies in other neurodegenerative diseases

AONs have also been used for the treatment of neurodegenerative disorders and are found to be taken up by neurons when delivered into the cerebral lateral ventricles. Here are some examples showing therapeutic benefit in animal models and/or clinical trials.

5.1 Prevention of mutant protein translation

We will first focus on the neurodegenerative disorder amyotrophic lateral sclerosis (ALS) where RNA modulating therapeutics are used to reduce transcript levels of disease causing protein. The RNA modulating therapeutics to treat ALS are currently tested in a phase I clinical trial.

The progressive neurodegenerative muscle weakness disorder ALS is a caused by loss of motor neurons in the brain and spinal cord (Al-Chalabi & Leigh, 2000). The first mutations linked to the familial form of ALS (fALS) were found in the *superoxide dismutase 1 (SOD1)* gene. Mutated SOD1 is known to be toxic and prone to aggregation. Only approximately 1% of ALS cases is the result of mutations in the SOD1 enzyme (Bossy-Wetzel et al., 2004).

RNA modulating therapies that have been used in ALS were designed to block the translation of SOD1. In a transgenic mouse model of ALS, 2'O methoxyethyl modified AONs were used to lower mutant SOD1 levels by binding and subsequent RNase H mediated breakdown of SOD1 transcripts. Continuous ventricular infusion of the SOD1 targeting AON significantly slowed disease progression (Smith et al., 2006). The first results of a phase I study testing the safety of this SOD1 targeting AON in patients with fALS caused by mutant SOD1 are expected at the end of 2011. The outcomes of this phase I trial will be vital for future trials with RNA modulating therapies in HD.

5.2 Modulating pre-mRNA splicing

RNA modulating therapeutics are also used to modulate pre-mRNA splicing events in spinal muscular atrophy (SMA) using modified AON *in vivo*.

SMA is an autosomal recessive neuromuscular disorder caused by loss of function of the *survival motor neuron 1 (SMN1)* gene. This homozygous deletion of SMN1 results in degeneration of motor neurons in the anterior horn of the spinal cord and lower brain stem (Bowers et al., 2011). Depletion of SMN1 is not embryonic lethal because of the presence of the almost identical *SMN2* gene. However, due to a point mutation in an intron the SMN2 transcript is not correctly spliced. The majority of SMN2 transcripts are therefore lacking exon 7, which results in a truncated protein and lower expression of a functional SMN protein (Lorson et al., 2010).

Current therapeutic strategies are aimed at modulating alternative splicing of SMN2. Transfecting fibroblasts with an AON blocking intronic splicing silencers in intron 7 of SMN2 were found to result in inclusion of SMN2 exon 7 (Singh et al., 2006). Injection of differently modified AONs into the brains of SMA mouse models resulted in increased exon 7 inclusion and subsequent elevated SMN protein levels. The AON treated SMA mice displayed increased muscle size and extended survival (Williams et al., 2009; Hua et al., 2010; Passini et al., 2011).

Another modulating pre-mRNA splicing strategy involves the addition of a functional moiety to the AON to replace the missing splicing enhancer protein, thereby enhancing the inclusion of exon 7 by the splicing machinery (Cartegni & Krainer, 2003; Skordis et al., 2003). Several *in vivo* studies have shown increased SMN2 protein levels after intraventricular injection of splicing factors recruiting AONs (Dickson et al., 2008; Baughan et al., 2009).

The AONs to treat SMA show promising results *in vivo* and the progression in therapeutics will be monitored closely. Results regarding delivery of the AON to the brain in humans and how well the AON is tolerated will be very useful for the development of RNA modulating therapeutics for HD.

6. Drug delivery to the brain, how to cross the blood brain barrier?

One major challenge of AON therapies for neurodegenerative disorders is delivery of the AON to the target organ. In the following paragraph we will describe in short the blood brain barrier function and how this impairs the uptake of peripherally administered drugs. We will focus in particular on the limitations and possibilities of AON delivery to the brain and will speculate on future clinical applications.

6.1 Blood brain barrier

A unique feature of the brain is that it is separated from the blood by the blood brain barrier (BBB). This is a monolayer of endothelial cells forming tight junctions through the interaction of cell adhesion molecules (Palmer, 2010). Astrocytes with their processes surrounding the endothelial cells, pericytes located between the endothelial cells and astrocytes, macrophages, and the basement membrane, form the other structural components of the BBB. Endothelial cells of the BBB are characterized by only few fenestrae and pinocytic vesicles, limiting transport to and from the brain. In this respect, it should be noted that the BBB also separates largely the immune system from the brain. Despite this gate-controlling system, essential nutrients, such as glucose, are permitted to pass (Bernacki et al., 2008). In neurodegenerative diseases, including HD, disruption of the BBB is common (Tomkins et al., 2007; Palmer, 2010). Interestingly, in animal models, this can even lead to neurodegenerative changes itself (Tomkins et al., 2007).

The BBB has been already noticed in the work of Paul Ehrlich, Nobel Prize winning bacteriologist in the late 19th century. Injected dyes stained all organs except the brain and spinal cord. However, he did not attribute this phenomenon to the presence of a barrier but to dye characteristics. His student showed later that staining of the brain was possible when the dye was injected directly into the brain (Palmer, 2010). Subsequent studies using electron microscopy were able to directly visualize the BBB.

The BBB is a major challenge in central nervous system (CNS) drug development. When a drug is administered to the body, a fraction will be bound to proteins (e.g. serum albumin, lipoprotein etc.) and a fraction will be free. The free fraction is the pharmacologically relevant fraction, since it is available to cross the BBB (Palmer, 2010), depending on its physiochemical properties. After crossing the BBB, the drug will enter the interstitial fluid and go to the target (proteins, receptors, transporters etc.). Subsequently, the interstitial fluid drains to the cerebrospinal fluid (CSF), which is produced at a rate of 500 ml/day in humans, while the ventricle system can house only 100-150 ml. This means that there is at least 3 times CSF circulation, allowing continuous drainage of the brain's interstitial fluid.

6.2 Crossing the blood brain barrier

In the process of drug discovery, the aim is to find a substance which is potent, selective and preferably bioavailable. In addition, it needs to be able to cross the BBB, and reach the target at a sufficient concentration (Alavijeh et al., 2005). The following mechanisms are available to cross the BBB. The first one is simple diffusion. Small lipophilic substances which have a hydrogen bond are more likely to pass the BBB (Gerebtzoff & Seelig, 2006). The second mechanism is via active transport mediated by transporter molecules. The most well-known is glucose with its glucose transporter 1 (GLUT1), which is the most widely expressed among the GLUT family (13 isoforms) (Guo et al., 2005; Palmer, 2010). Other carriers are for instance lactate and amino acids. A well-known drug transported via this way is levodopa (Cotzias et al., 1967). The third mechanism to cross the BBB is via receptor-mediation. Receptor-mediated endocytosis allows macromolecules to enter the brain, such as transferrin, insulin, leptin, and insulin-like growth factor 1 (Pardridge, 2007).

Besides systemic mechanisms to cross the BBB, there are also techniques to bypass the BBB by direct infusions into the subdural space, the brain's ventricle system, or the brain parenchyma. These infusions can be single, repeated, or continuous depending on the methodology, using either simple or sophisticated pump systems. It is possible to use one probe or more probes for infusion. Using the subdural and ventricle compartments, diffuse delivery of the drug into the brain can be achieved, while using intraparenchymal delivery, a local, but well-targeted delivery can be realized

When a substance has successfully entered the brain, there are mechanisms preventing adequate functioning. One mechanism is active transport to remove the substance, also known as resistance. A superfamily of multidrug resistance proteins, belonging to the ATP-binding cassette transporters, drives substances away by an ATP-dependent process (Palmer, 2010). One of the most abundant proteins is the P-glycoprotein. This mechanism is responsible for the failure of some anticancer drugs. Another family of egress transporters is the organic anion transporting proteins.

In the field of HD, efforts are ongoing to deliver innovative drugs to the brain via the systemic route and drugs are designed to use one of the three mechanisms to cross the BBB, as explained earlier. For instance, Lee and associates described the use of a peptide nucleic acid as an antisense which was able to access endogenous transferrin transport pathways (receptor mediated endocytosis) and reach the brain in a transgenic mouse model (Lee et al., 2002). However, there are also efforts to bypass the BBB, and to deliver the drug using either the ventricle system or intraparenchymally.

7. Conclusion

To date there is no treatment to prevent or even slow down the progression of HD. Considerable research has been performed to gain more insight into HD pathology. Next to the well-known toxic gain of polyQ protein function, loss of wild-type function and a toxic gain of expanded CAG repeat RNA was also suggested recently, and needs to be examined in more detail.

Recent results using SNP specific siRNAs and CAG targeting AONs look promising both *in vitro* and *in vivo*. To develop an effective HD therapy, it is likely that a combination of different RNA modifying approaches will be optimal to lower mutant htt levels. Extensive research is required to rule out toxic off-targets effects and elucidate the exact mode of action of these RNA modulating therapeutics. Ongoing clinical trials for other neurodegenerative disorders, such as in ALS, will give us more insights in the potential of RNA modulating therapeutics.

8. Acknowledgement

MME and WvRM are supported by the Center for Biomedical Genetics (the Netherlands), AtaxiaUK (United Kingdom), Dutch ataxia charity ADCA-VN (the Netherlands), and IOP Genomics (the Netherlands). The HD related research of RV and YT received financial support from the Cure Huntington's Disease Initiative (New York, USA) and Prosensa BV (Leiden, the Netherlands).

9. References

Al-Chalabi,A. & Leigh,P.N. (2000). Recent advances in amyotrophic lateral sclerosis. *Curr. Opin. Neurol*, Vol.13, No.4, pp.397-405

Alavijeh, M.S., Chishty, M., Qaiser, M.Z. & Palmer, A.M. (2005). Drug metabolism and pharmacokinetics, the blood-brain barrier, and central nervous system drug discovery. *NeuroRx.*, Vol.2, No.4, pp.554-571

Andresen, J.M., Gayan, J., Djousse, L., Roberts, S., Brocklebank, D., Cherny, S.S., Cardon, L.R., Gusella, J.F., MacDonald, M.E., Myers, R.H., Housman, D.E. & Wexler, N.S. (2007). The relationship between CAG repeat length and age of onset differs for Huntington's disease patients with juvenile onset or adult onset. *Ann Hum. Genet*, Vol.71, No.Pt 3, pp.295-301

Arrasate, M., Mitra, S., Schweitzer, E.S., Segal, M.R. & Finkbeiner, S. (2004). Inclusion body formation reduces levels of mutant huntingtin and the risk of neuronal death. *Nature*, Vol.431, No.7010, pp.805-810

Atwal, R.S., Xia, J., Pinchev, D., Taylor, J., Epand, R.M. & Truant, R. (2007). Huntingtin has a membrane association signal that can modulate huntingtin aggregation, nuclear entry and toxicity. *Hum. Mol Genet*, Vol.16, No.21, pp.2600-2615

Baughan, T.D., Dickson, A., Osman, E.Y. & Lorson, C.L. (2009). Delivery of bifunctional RNAs that target an intronic repressor and increase SMN levels in an animal model of spinal muscular atrophy. *Hum. Mol Genet*, Vol.18, No.9, pp.1600-1611

Bernacki, J., Dobrowolska, A., Nierwinska, K. & Malecki, A. (2008). Physiology and pharmacological role of the blood-brain barrier. *Pharmacol. Rep.*, Vol.60, No.5, pp.600-622

Bossy-Wetzel, E., Schwarzenbacher, R. & Lipton, S.A. (2004). Molecular pathways to neurodegeneration. *Nat. Med.*, Vol.10 Suppl, pp.S2-S9

Boudreau, R.L., McBride, J.L., Martins, I., Shen, S., Xing, Y., Carter, B.J. & Davidson, B.L. (2009). Nonallele-specific silencing of mutant and wild-type huntingtin demonstrates therapeutic efficacy in Huntington's disease mice. *Mol Ther.*, Vol.17, No.6, pp.1053-1063

Bowers, W.J., Breakefield, X.O. & Sena-Esteves, M. (2011). Genetic therapy for the nervous system. *Hum. Mol Genet*, Vol.20, No.R1, pp.R28-R41

Browne, S.E. & Beal, M.F. (2006). Oxidative damage in Huntington's disease pathogenesis. *Antioxid. Redox. Signal.*, Vol.8, No.11-12, pp.2061-2073

Cartegni, L. & Krainer, A.R. (2003). Correction of disease-associated exon skipping by synthetic exon-specific activators. *Nat. Struct. Biol.*, Vol.10, No.2, pp.120-125

Caviston, J.P., Zajac, A.L., Tokito, M. & Holzbaur, E.L. (2011). Huntingtin coordinates the dynein-mediated dynamic positioning of endosomes and lysosomes. *Mol Biol. Cell*, Vol.22, No.4, pp.478-492

Choo, Y.S., Johnson, G.V., MacDonald, M., Detloff, P.J. & Lesort, M. (2004). Mutant huntingtin directly increases susceptibility of mitochondria to the calcium-induced permeability transition and cytochrome c release. *Hum. Mol Genet*, Vol.13, No.14, pp.1407-1420

Chung, D.W., Rudnicki, D.D., Yu, L. & Margolis, R.L. (2011). A natural antisense transcript at the Huntington's disease repeat locus regulates HTT expression. *Hum. Mol Genet*, Vol.20, No.17, pp.3467-3477

Clabough, E.B. & Zeitlin, S.O. (2006). Deletion of the triplet repeat encoding polyglutamine within the mouse Huntington's disease gene results in subtle behavioral/motor phenotypes in vivo and elevated levels of ATP with cellular senescence in vitro. *Hum. Mol Genet*, Vol.15, No.4, pp.607-623

Cooper, J.K., Schilling, G., Peters, M.F., Herring, W.J., Sharp, A.H., Kaminsky, Z., Masone, J., Khan, F.A., Delanoy, M., Borchelt, D.R., Dawson, V.L., Dawson, T.M. & Ross, C.A. (1998). Truncated N-terminal fragments of huntingtin with expanded glutamine repeats form nuclear and cytoplasmic aggregates in cell culture. *Hum. Mol Genet*, Vol.7, No.5, pp.783-790

Cornett, J., Cao, F., Wang, C.E., Ross, C.A., Bates, G.P., Li, S.H. & Li, X.J. (2005). Polyglutamine expansion of huntingtin impairs its nuclear export. *Nat. Genet*, Vol.37, No.2, pp.198-204

Cotzias, G.C., Van Woert, M.H. & Schiffer, L.M. (1967). Aromatic amino acids and modification of parkinsonism. *N Engl. J. Med.*, Vol.276, No.7, pp.374-379

Craufurd, D., Thompson, J.C. & Snowden, J.S. (2001). Behavioral changes in Huntington Disease. *Neuropsychiatry Neuropsychol. Behav. Neurol*, Vol.14, No.4, pp.219-226

Cummings, C.J. & Zoghbi, H.Y. (2000). Fourteen and counting: unraveling trinucleotide repeat diseases. *Hum. Mol Genet*, Vol.9, No.6, pp.909-916

Davies, S.W., Turmaine, M., Cozens, B.A., Difiglia, M., Sharp, A.H., Ross, C.A., Scherzinger, E., Wanker, E.E., Mangiarini, L. & Bates, G.P. (1997). Formation of neuronal intranuclear inclusions underlies the neurological dysfunction in mice transgenic for the HD mutation. *Cell*, Vol.90, No.3, pp.537-548

de Mezer M., Wojciechowska, M., Napierala, M., Sobczak, K. & Krzyzosiak, W.J. (2011). Mutant CAG repeats of Huntingtin transcript fold into hairpins, form nuclear foci and are targets for RNA interference. *Nucleic Acids Res.*, Vol.39, No.9, pp.3852-3863

de Rooij, K.E., Dorsman, J.C., Smoor, M.A., Den Dunnen, J.T. & Van Ommen, G.J. (1996). Subcellular localization of the Huntington's disease gene product in cell lines by immunofluorescence and biochemical subcellular fractionation. *Hum. Mol Genet*, Vol.5, No.8, pp.1093-1099

Dickson, A., Osman, E. & Lorson, C.L. (2008). A negatively acting bifunctional RNA increases survival motor neuron both in vitro and in vivo. *Hum. Gene Ther.*, Vol.19, No.11, pp.1307-1315

Dragatsis, I., Levine, M.S. & Zeitlin, S. (2000). Inactivation of Hdh in the brain and testis results in progressive neurodegeneration and sterility in mice. *Nat. Genet*, Vol.26, No.3, pp.300-306

Drouet, V., Perrin, V., Hassig, R., Dufour, N., Auregan, G., Alves, S., Bonvento, G., Brouillet, E., Luthi-Carter, R., Hantraye, P. & Deglon, N. (2009). Sustained effects of nonallele-specific Huntingtin silencing. *Ann Neurol*, Vol.65, No.3, pp.276-285

Evers, M.M., Pepers, B.A., van Deutekom, J.C.T., Mulders, S.A.M., Den Dunnen, J.T., Aartsma-Rus, A., Van Ommen, G.J.B. & van Roon-Mom, W.M.C. (2011). Targeting Several CAG Expansion Diseases by a Single Antisense Oligonucleotide. *PLoS. One* Vol.6, No.9,pp:e24308, Accepted

Fardaei, M., Larkin, K., Brook, J.D. & Hamshere, M.G. (2001). In vivo co-localisation of MBNL protein with DMPK expanded-repeat transcripts. *Nucleic Acids Res.*, Vol.29, No.13, pp.2766-2771

Gafni, J. & Ellerby, L.M. (2002). Calpain activation in Huntington's disease. *J. Neurosci.*, Vol.22, No.12, pp.4842-4849

Gerebtzoff, G. & Seelig, A. (2006). In silico prediction of blood-brain barrier permeation using the calculated molecular cross-sectional area as main parameter. *J. Chem. Inf. Model*, Vol.46, No.6, pp.2638-2650

Goemans, N.M., Tulinius, M., van den Akker, J.T., Burm, B.E., Ekhart, P.F., Heuvelmans, N., Holling, T., Janson, A.A., Platenburg, G.J., Sipkens, J.A., Sitsen, J.M., Aartsma-Rus, A., Van Ommen, G.J., Buyse, G., Darin, N., Verschuuren, J.J., Campion, G.V., de Kimpe, S.J. & van Deutekom, J.C. (2011). Systemic administration of PRO051 in Duchenne's muscular dystrophy. *N Engl. J. Med.*, Vol.364, No.16, pp.1513-1522

Graham, R.K., Deng, Y., Slow, E.J., Haigh, B., Bissada, N., Lu, G., Pearson, J., Shehadeh, J., Bertram, L., Murphy, Z., Warby, S.C., Doty, C.N., Roy, S., Wellington, C.L., Leavitt, B.R., Raymond, L.A., Nicholson, D.W. & Hayden, M.R. (2006). Cleavage at the caspase-6 site is required for neuronal dysfunction and degeneration due to mutant huntingtin. *Cell*, Vol.125, No.6, pp.1179-1191

Gunawardena, S., Her, L.S., Brusch, R.G., Laymon, R.A., Niesman, I.R., Gordesky-Gold, B., Sintasath, L., Bonini, N.M. & Goldstein, L.S. (2003). Disruption of axonal transport by loss of huntingtin or expression of pathogenic polyQ proteins in Drosophila. *Neuron*, Vol.40, No.1, pp.25-40

Guo, X., Geng, M. & Du, G. (2005). Glucose transporter 1, distribution in the brain and in neural disorders: its relationship with transport of neuroactive drugs through the blood-brain barrier. *Biochem. Genet*, Vol.43, No.3-4, pp.175-187

Gusella, J.F., Wexler, N.S., Conneally, P.M., Naylor, S.L., Anderson, M.A., Tanzi, R.E., Watkins, P.C., Ottina, K., Wallace, M.R., Sakaguchi, A.Y. &. (1983). A polymorphic DNA marker genetically linked to Huntington's disease. *Nature*, Vol.306, No.5940, pp.234-238

Harjes, P. & Wanker, E.E. (2003). The hunt for huntingtin function: interaction partners tell many different stories. *Trends Biochem. Sci*, Vol.28, No.8, pp.425-433

Harper, S.Q., Staber, P.D., He, X., Eliason, S.L., Martins, I.H., Mao, Q., Yang, L., Kotin, R.M., Paulson, H.L. & Davidson, B.L. (2005). RNA interference improves motor and neuropathological abnormalities in a Huntington's disease mouse model. *Proc. Natl. Acad Sci U. S. A*, Vol.102, No.16, pp.5820-5825

Hsu, R.J., Hsiao, K.M., Lin, M.J., Li, C.Y., Wang, L.C., Chen, L.K. & Pan, H. (2011). Long tract of untranslated CAG repeats is deleterious in transgenic mice. *PLoS. One.*, Vol.6, No.1, pp.e16417

Hu, J., Matsui, M., Gagnon, K.T., Schwartz, J.C., Gabillet, S., Arar, K., Wu, J., Bezprozvanny, I. & Corey, D.R. (2009). Allele-specific silencing of mutant huntingtin and ataxin-3 genes by targeting expanded CAG repeats in mRNAs. *Nat. Biotechnol.*, Vol.27, No.5, pp.478-484

Hua, Y., Sahashi, K., Hung, G., Rigo, F., Passini, M.A., Bennett, C.F. & Krainer, A.R. (2010). Antisense correction of SMN2 splicing in the CNS rescues necrosis in a type III SMA mouse model. *Genes Dev.*, Vol.24, No.15, pp.1634-1644

Huang, K., Yanai, A., Kang, R., Arstikaitis, P., Singaraja, R.R., Metzler, M., Mullard, A., Haigh, B., Gauthier-Campbell, C., Gutekunst, C.A., Hayden, M.R. & El-Husseini, A. (2004). Huntingtin-interacting protein HIP14 is a palmitoyl transferase involved in palmitoylation and trafficking of multiple neuronal proteins. *Neuron*, Vol.44, No.6, pp.977-986

Ikeda, H., Yamaguchi, M., Sugai, S., Aze, Y., Narumiya, S. & Kakizuka, A. (1996). Expanded polyglutamine in the Machado-Joseph disease protein induces cell death in vitro and in vivo. *Nat. Genet*, Vol.13, No.2, pp.196-202

Imarisio, S., Carmichael, J., Korolchuk, V., Chen, C.W., Saiki, S., Rose, C., Krishna, G., Davies, J.E., Ttofi, E., Underwood, B.R. & Rubinsztein, D.C. (2008). Huntington's disease: from pathology and genetics to potential therapies. *Biochem. J.*, Vol.412, No.2, pp.191-209

Jiang, H., Mankodi, A., Swanson, M.S., Moxley, R.T. & Thornton, C.A. (2004). Myotonic dystrophy type 1 is associated with nuclear foci of mutant RNA, sequestration of muscleblind proteins and deregulated alternative splicing in neurons. *Hum. Mol Genet*, Vol.13, No.24, pp.3079-3088

Kalchman, M.A., Graham, R.K., Xia, G., Koide, H.B., Hodgson, J.G., Graham, K.C., Goldberg, Y.P., Gietz, R.D., Pickart, C.M. & Hayden, M.R. (1996). Huntingtin is ubiquitinated and interacts with a specific ubiquitin-conjugating enzyme. *J. Biol. Chem.*, Vol.271, No.32, pp.19385-19394

Kanadia, R.N., Johnstone, K.A., Mankodi, A., Lungu, C., Thornton, C.A., Esson, D., Timmers, A.M., Hauswirth, W.W. & Swanson, M.S. (2003). A muscleblind knockout model for myotonic dystrophy. *Science*, Vol.302, No.5652, pp.1978-1980

Katayama, S., Tomaru, Y., Kasukawa, T., Waki, K., Nakanishi, M., Nakamura, M., Nishida, H., Yap, C.C., Suzuki, M., Kawai, J., Suzuki, H., Carninci, P., Hayashizaki, Y., Wells, C., Frith, M., Ravasi, T., Pang, K.C., Hallinan, J., Mattick, J., Hume, D.A., Lipovich, L., Batalov, S., Engstrom, P.G., Mizuno, Y., Faghihi, M.A., Sandelin, A., Chalk, A.M., Mottagui-Tabar, S., Liang, Z., Lenhard, B. & Wahlestedt, C. (2005). Antisense transcription in the mammalian transcriptome. *Science*, Vol.309, No.5740, pp.1564-1566

Katsuno, M., Adachi, H., Kume, A., Li, M., Nakagomi, Y., Niwa, H., Sang, C., Kobayashi, Y., Doyu, M. & Sobue, G. (2002). Testosterone reduction prevents phenotypic expression in a transgenic mouse model of spinal and bulbar muscular atrophy. *Neuron*, Vol.35, No.5, pp.843-854

Kegel, K.B., Meloni, A.R., Yi, Y., Kim, Y.J., Doyle, E., Cuiffo, B.G., Sapp, E., Wang, Y., Qin, Z.H., Chen, J.D., Nevins, J.R., Aronin, N. & Difiglia, M. (2002). Huntingtin is present in the nucleus, interacts with the transcriptional corepressor C-terminal binding protein, and represses transcription. *J. Biol. Chem.*, Vol.277, No.9, pp.7466-7476

Kegel, K.B., Sapp, E., Yoder, J., Cuiffo, B., Sobin, L., Kim, Y.J., Qin, Z.H., Hayden, M.R., Aronin, N., Scott, D.L., Isenberg, G., Goldmann, W.H. & Difiglia, M. (2005). Huntingtin associates with acidic phospholipids at the plasma membrane. *J. Biol. Chem.*, Vol.280, No.43, pp.36464-36473

Kim, Y.J., Sapp, E., Cuiffo, B.G., Sobin, L., Yoder, J., Kegel, K.B., Qin, Z.H., Detloff, P., Aronin, N. & Difiglia, M. (2006). Lysosomal proteases are involved in generation of N-terminal huntingtin fragments. *Neurobiol. Dis*, Vol.22, No.2, pp.346-356

Klement, I.A., Skinner, P.J., Kaytor, M.D., Yi, H., Hersch, S.M., Clark, H.B., Zoghbi, H.Y. & Orr, H.T. (1998). Ataxin-1 nuclear localization and aggregation: role in polyglutamine-induced disease in SCA1 transgenic mice. *Cell*, Vol.95, No.1, pp.41-53

Koob, M.D., Moseley, M.L., Schut, L.J., Benzow, K.A., Bird, T.D., Day, J.W. & Ranum, L.P. (1999). An untranslated CTG expansion causes a novel form of spinocerebellar ataxia (SCA8). *Nat. Genet*, Vol.21, No.4, pp.379-384

Kremer, B., Almqvist, E., Theilmann, J., Spence, N., Telenius, H., Goldberg, Y.P. & Hayden, M.R. (1995). Sex-dependent mechanisms for expansions and contractions of the CAG repeat on affected Huntington disease chromosomes. *Am. J. Hum. Genet*, Vol.57, No.2, pp.343-350

Kremer, B., Weber, B. & Hayden, M.R. (1992). New insights into the clinical features, pathogenesis and molecular genetics of Huntington disease. *Brain Pathol.*, Vol.2, No.4, pp.321-335

Krol, J., Fiszer, A., Mykowska, A., Sobczak, K., de, M.M. & Krzyzosiak, W.J. (2007). Ribonuclease dicer cleaves triplet repeat hairpins into shorter repeats that silence specific targets. *Mol Cell*, Vol.25, No.4, pp.575-586

Lee, H.J., Boado, R.J., Braasch, D.A., Corey, D.R. & Pardridge, W.M. (2002). Imaging gene expression in the brain in vivo in a transgenic mouse model of Huntington's disease with an antisense radiopharmaceutical and drug-targeting technology. *J. Nucl. Med.*, Vol.43, No.7, pp.948-956

Legleiter, J., Mitchell, E., Lotz, G.P., Sapp, E., Ng, C., Difiglia, M., Thompson, L.M. & Muchowski, P.J. (2010). Mutant huntingtin fragments form oligomers in a polyglutamine length-dependent manner in vitro and in vivo. *J. Biol. Chem.*, Vol.285, No.19, pp.14777-14790

Levesque, M., Bedard, A., Cossette, M. & Parent, A. (2003). Novel aspects of the chemical anatomy of the striatum and its efferents projections. *J. Chem. Neuroanat.*, Vol.26, No.4, pp.271-281

Li, L.B., Yu, Z., Teng, X. & Bonini, N.M. (2008). RNA toxicity is a component of ataxin-3 degeneration in Drosophila. *Nature*, Vol.453, No.7198, pp.1107-1111

Li, W., Serpell, L.C., Carter, W.J., Rubinsztein, D.C. & Huntington, J.A. (2006). Expression and characterization of full-length human huntingtin, an elongated HEAT repeat protein. *J. Biol. Chem.*, Vol.281, No.23, pp.15916-15922

Liu, W., Kennington, L.A., Rosas, H.D., Hersch, S., Cha, J.H., Zamore, P.D. & Aronin, N. (2008). Linking SNPs to CAG repeat length in Huntington's disease patients. *Nat. Methods*, Vol.5, No.11, pp.951-953

Lombardi, M.S., Jaspers, L., Spronkmans, C., Gellera, C., Taroni, F., Di, M.E., Donato, S.D. & Kaemmerer, W.F. (2009). A majority of Huntington's disease patients may be treatable by individualized allele-specific RNA interference. *Exp. Neurol*, Vol.217, No.2, pp.312-319

Lorson, C.L., Rindt, H. & Shababi, M. (2010). Spinal muscular atrophy: mechanisms and therapeutic strategies. *Hum. Mol Genet*, Vol.19, No.R1, pp.R111-R118

McBride, J.L., Boudreau, R.L., Harper, S.Q., Staber, P.D., Monteys, A.M., Martins, I., Gilmore, B.L., Burstein, H., Peluso, R.W., Polisky, B., Carter, B.J. & Davidson, B.L. (2008). Artificial miRNAs mitigate shRNA-mediated toxicity in the brain: implications for the therapeutic development of RNAi. *Proc. Natl. Acad Sci U. S. A*, Vol.105, No.15, pp.5868-5873

McLeod, C.J., O'Keefe, L.V. & Richards, R.I. (2005). The pathogenic agent in Drosophila models of 'polyglutamine' diseases. *Hum. Mol Genet*, Vol.14, No.8, pp.1041-1048

McNeil, S.M., Novelletto, A., Srinidhi, J., Barnes, G., Kornbluth, I., Altherr, M.R., Wasmuth, J.J., Gusella, J.F., MacDonald, M.E. & Myers, R.H. (1997). Reduced penetrance of the Huntington's disease mutation. *Hum. Mol Genet*, Vol.6, No.5, pp.775-779

Mende-Mueller, L.M., Toneff, T., Hwang, S.R., Chesselet, M.F. & Hook, V.Y. (2001). Tissue-specific proteolysis of Huntingtin (htt) in human brain: evidence of enhanced levels of N- and C-terminal htt fragments in Huntington's disease striatum. *J. Neurosci.*, Vol.21, No.6, pp.1830-1837

Miller, V.M., Xia, H., Marrs, G.L., Gouvion, C.M., Lee, G., Davidson, B.L. & Paulson, H.L. (2003). Allele-specific silencing of dominant disease genes. *Proc. Natl. Acad Sci U. S. A*, Vol.100, No.12, pp.7195-7200

Modregger, J., DiProspero, N.A., Charles, V., Tagle, D.A. & Plomann, M. (2002). PACSIN 1 interacts with huntingtin and is absent from synaptic varicosities in presymptomatic Huntington's disease brains. *Hum. Mol Genet*, Vol.11, No.21, pp.2547-2558

Morfini, G., Pigino, G., Brady, S.T. (2005). Polyglutamine expansion diseases: failing to deliver. *Trends Mol Med* 11, pp.64–70.

Moseley, M.L., Zu, T., Ikeda, Y., Gao, W., Mosemiller, A.K., Daughters, R.S., Chen, G., Weatherspoon, M.R., Clark, H.B., Ebner, T.J., Day, J.W. & Ranum, L.P. (2006). Bidirectional expression of CUG and CAG expansion transcripts and intranuclear polyglutamine inclusions in spinocerebellar ataxia type 8. *Nat. Genet*, Vol.38, No.7, pp.758-769

Mykowska, A., Sobczak, K., Wojciechowska, M., Kozlowski, P. & Krzyzosiak, W.J. (2011). CAG repeats mimic CUG repeats in the misregulation of alternative splicing. *Nucleic Acids Res.*

Nucifora, F.C., Jr., Sasaki, M., Peters, M.F., Huang, H., Cooper, J.K., Yamada, M., Takahashi, H., Tsuji, S., Troncoso, J., Dawson, V.L., Dawson, T.M. & Ross, C.A. (2001). Interference by huntingtin and atrophin-1 with cbp-mediated transcription leading to cellular toxicity. *Science*, Vol.291, No.5512, pp.2423-2428

Orr, A.L., Li, S., Wang, C.E., Li, H., Wang, J., Rong, J., Xu, X., Mastroberardino, P.G., Greenamyre, J.T. & Li, X.J. (2008). N-terminal mutant huntingtin associates with mitochondria and impairs mitochondrial trafficking. *J. Neurosci.*, Vol.28, No.11, pp.2783-2792

Orr, H.T. & Zoghbi, H.Y. (2007). Trinucleotide repeat disorders. *Annu. Rev. Neurosci.*, Vol.30, pp.575-621

Palmer, A.M. (2010). The blood-brain barrier. *Neurobiol. Dis*, Vol.37, No.1, pp.1-2

Panov, A.V., Gutekunst, C.A., Leavitt, B.R., Hayden, M.R., Burke, J.R., Strittmatter, W.J. & Greenamyre, J.T. (2002). Early mitochondrial calcium defects in Huntington's disease are a direct effect of polyglutamines. *Nat. Neurosci.*, Vol.5, No.8, pp.731-736

Pardridge, W.M. (2007). Blood-brain barrier delivery. *Drug Discov. Today*, Vol.12, No.1-2, pp.54-61

Passini, M.A., Bu, J., Richards, A.M., Kinnecom, C., Sardi, S.P., Stanek, L.M., Hua, Y., Rigo, F., Matson, J., Hung, G., Kaye, E.M., Shihabuddin, L.S., Krainer, A.R., Bennett, C.F. & Cheng, S.H. (2011). Antisense oligonucleotides delivered to the mouse CNS ameliorate symptoms of severe spinal muscular atrophy. *Sci Transl. Med.*, Vol.3, No.72, pp.72ra18

Paulsen, J.S., Ready, R.E., Hamilton, J.M., Mega, M.S. & Cummings, J.L. (2001). Neuropsychiatric aspects of Huntington's disease. *J. Neurol Neurosurg. Psychiatry*, Vol.71, No.3, pp.310-314

Perutz, M.F., Johnson, T., Suzuki, M. & Finch, J.T. (1994). Glutamine repeats as polar zippers: their possible role in inherited neurodegenerative diseases. *Proc. Natl. Acad Sci U. S. A*, Vol.91, No.12, pp.5355-5358

Pfister, E.L., Kennington, L., Straubhaar, J., Wagh, S., Liu, W., Difiglia, M., Landwehrmeyer, B., Vonsattel, J.P., Zamore, P.D. & Aronin, N. (2009). Five siRNAs targeting three SNPs may provide therapy for three-quarters of Huntington's disease patients. *Curr. Biol.*, Vol.19, No.9, pp.774-778

Rawlins, M. (2010). Huntington's disease out of the closet? *Lancet*, Vol.376, No.9750, pp.1372-1373

Rigamonti, D., Bauer, J.H., De-Fraja, C., Conti, L., Sipione, S., Sciorati, C., Clementi, E., Hackam, A., Hayden, M.R., Li, Y., Cooper, J.K., Ross, C.A., Govoni, S., Vincenz, C. & Cattaneo, E. (2000). Wild-type huntingtin protects from apoptosis upstream of caspase-3. *J. Neurosci.*, Vol.20, No.10, pp.3705-3713

Rockabrand, E., Slepko, N., Pantalone, A., Nukala, V.N., Kazantsev, A., Marsh, J.L., Sullivan, P.G., Steffan, J.S., Sensi, S.L. & Thompson, L.M. (2007). The first 17 amino acids of Huntingtin modulate its sub-cellular localization, aggregation and effects on calcium homeostasis. *Hum. Mol Genet*, Vol.16, No.1, pp.61-77

Roos, R.A. (2010). Huntington's disease: a clinical review. *Orphanet. J. Rare. Dis*, Vol.5, No.1, pp.40

Rosas, H.D., Koroshetz, W.J., Chen, Y.I., Skeuse, C., Vangel, M., Cudkowicz, M.E., Caplan, K., Marek, K., Seidman, L.J., Makris, N., Jenkins, B.G. & Goldstein, J.M. (2003). Evidence for more widespread cerebral pathology in early HD: an MRI-based morphometric analysis. *Neurology*, Vol.60, No.10, pp.1615-1620

Rosas, H.D., Salat, D.H., Lee, S.Y., Zaleta, A.K., Pappu, V., Fischl, B., Greve, D., Hevelone, N. & Hersch, S.M. (2008). Cerebral cortex and the clinical expression of Huntington's disease: complexity and heterogeneity. *Brain*, Vol.131, No.Pt 4, pp.1057-1068

Ryan, A. & Scrable, H. (2008). Mutant alleles of HD improve the life span of p53(-/-) mice. *Mech Ageing Dev.*, Vol.129, No.4, pp.238-241

Ryan, A.B., Zeitlin, S.O. & Scrable, H. (2006). Genetic interaction between expanded murine Hdh alleles and p53 reveal deleterious effects of p53 on Huntington's disease pathogenesis. *Neurobiol. Dis*, Vol.24, No.2, pp.419-427

Sah, D.W. & Aronin, N. (2011). Oligonucleotide therapeutic approaches for Huntington disease. *J. Clin. Invest*, Vol.121, No.2, pp.500-507

Saudou, F., Finkbeiner, S., Devys, D. & Greenberg, M.E. (1998). Huntingtin acts in the nucleus to induce apoptosis but death does not correlate with the formation of intranuclear inclusions. *Cell*, Vol.95, No.1, pp.55-66

Schwarz, D.S., Ding, H., Kennington, L., Moore, J.T., Schelter, J., Burchard, J., Linsley, P.S., Aronin, N., Xu, Z. & Zamore, P.D. (2006). Designing siRNA that distinguish between genes that differ by a single nucleotide. *PLoS. Genet*, Vol.2, No.9, pp.e140

Shoulson, I. & Young, A.B. (2011). Milestones in huntington disease. *Mov Disord.*, Vol.26, No.6, pp.1127-1133

Simpson, S.A. (2007). Late stage care in Huntington's disease. *Brain Res. Bull.*, Vol.72, No.2-3, pp.179-181

Singh, N.K., Singh, N.N., Androphy, E.J. & Singh, R.N. (2006). Splicing of a critical exon of human Survival Motor Neuron is regulated by a unique silencer element located in the last intron. *Mol Cell Biol.*, Vol.26, No.4, pp.1333-1346

Skordis, L.A., Dunckley, M.G., Yue, B., Eperon, I.C. & Muntoni, F. (2003). Bifunctional antisense oligonucleotides provide a trans-acting splicing enhancer that stimulates SMN2 gene expression in patient fibroblasts. *Proc. Natl. Acad Sci U. S. A*, Vol.100, No.7, pp.4114-4119

Smith, R., Klein, P., Koc-Schmitz, Y., Waldvogel, H.J., Faull, R.L., Brundin, P., Plomann, M. & Li, J.Y. (2007). Loss of SNAP-25 and rabphilin 3a in sensory-motor cortex in Huntington's disease. *J. Neurochem.*, Vol.103, No.1, pp.115-123

Smith, R.A., Miller, T.M., Yamanaka, K., Monia, B.P., Condon, T.P., Hung, G., Lobsiger, C.S., Ward, C.M., McAlonis-Downes, M., Wei, H., Wancewicz, E.V., Bennett, C.F. & Cleveland, D.W. (2006). Antisense oligonucleotide therapy for neurodegenerative disease. *J. Clin. Invest*, Vol.116, No.8, pp.2290-2296

Snowden, J.S., Craufurd, D., Thompson, J. & Neary, D. (2002). Psychomotor, executive, and memory function in preclinical Huntington's disease. *J. Clin. Exp. Neuropsychol.*, Vol.24, No.2, pp.133-145

Steffan, J.S., Agrawal, N., Pallos, J., Rockabrand, E., Trotman, L.C., Slepko, N., Illes, K., Lukacsovich, T., Zhu, Y.Z., Cattaneo, E., Pandolfi, P.P., Thompson, L.M. & Marsh, J.L. (2004). SUMO modification of Huntingtin and Huntington's disease pathology. *Science*, Vol.304, No.5667, pp.100-104

Steffan, J.S., Kazantsev, A., Spasic-Boskovic, O., Greenwald, M., Zhu, Y.Z., Gohler, H., Wanker, E.E., Bates, G.P., Housman, D.E. & Thompson, L.M. (2000). The Huntington's disease protein interacts with p53 and CREB-binding protein and represses transcription. *Proc. Natl. Acad Sci U. S. A*, Vol.97, No.12, pp.6763-6768

Strehlow, A.N., Li, J.Z. & Myers, R.M. (2007). Wild-type huntingtin participates in protein trafficking between the Golgi and the extracellular space. *Hum. Mol Genet*, Vol.16, No.4, pp.391-409

Strong, T.V., Tagle, D.A., Valdes, J.M., Elmer, L.W., Boehm, K., Swaroop, M., Kaatz, K.W., Collins, F.S. & Albin, R.L. (1993). Widespread expression of the human and rat Huntington's disease gene in brain and nonneural tissues. *Nat. Genet*, Vol.5, No.3, pp.259-265

Tabrizi, S.J., Langbehn, D.R., Leavitt, B.R., Roos, R.A., Durr, A., Craufurd, D., Kennard, C., Hicks, S.L., Fox, N.C., Scahill, R.I., Borowsky, B., Tobin, A.J., Rosas, H.D., Johnson, H., Reilmann, R., Landwehrmeyer, B. & Stout, J.C. (2009). Biological and clinical manifestations of Huntington's disease in the longitudinal TRACK-HD study: cross-sectional analysis of baseline data. *Lancet Neurol*, Vol.8, No.9, pp.791-801

Takano, H. & Gusella, J.F. (2002). The predominantly HEAT-like motif structure of huntingtin and its association and coincident nuclear entry with dorsal, an NF-kB/Rel/dorsal family transcription factor. *BMC. Neurosci.*, Vol.3, pp.15

The Huntington's Disease Collaborative Research Group (1993). A novel gene containing a trinucleotide repeat that is expanded and unstable on Huntington's disease chromosomes. *Cell*, Vol.72, No.6, pp.971-983

Tomkins, O., Friedman, O., Ivens, S., Reiffurth, C., Major, S., Dreier, J.P., Heinemann, U. & Friedman, A. (2007). Blood-brain barrier disruption results in delayed functional and structural alterations in the rat neocortex. *Neurobiol. Dis*, Vol.25, No.2, pp.367-377

van Bilsen, P.H., Jaspers, L., Lombardi, M.S., Odekerken, J.C., Burright, E.N. & Kaemmerer, W.F. (2008). Identification and allele-specific silencing of the mutant huntingtin allele in Huntington's disease patient-derived fibroblasts. *Hum. Gene Ther.*, Vol.19, No.7, pp.710-719

van Roon-Mom, W.M., Hogg, V.M., Tippett, L.J. & Faull, R.L. (2006). Aggregate distribution in frontal and motor cortex in Huntington's disease brain. *Neuroreport*, Vol.17, No.6, pp.667-670

van Roon-Mom, W.M., Reid, S.J., Jones, A.L., MacDonald, M.E., Faull, R.L. & Snell, R.G. (2002). Insoluble TATA-binding protein accumulation in Huntington's disease cortex. *Brain Res. Mol Brain Res.*, Vol.109, No.1-2, pp.1-10

Vonsattel, J.P., Myers, R.H., Stevens, T.J., Ferrante, R.J., Bird, E.D. & Richardson, E.P., Jr. (1985). Neuropathological classification of Huntington's disease. *J. Neuropathol. Exp. Neurol*, Vol.44, No.6, pp.559-577

Walker, F.O. (2007). Huntington's disease. *Lancet*, Vol.369, No.9557, pp.218-228

Wang, L.C., Chen, K.Y., Pan, H., Wu, C.C., Chen, P.H., Liao, Y.T., Li, C., Huang, M.L. & Hsiao, K.M. (2011). Muscleblind participates in RNA toxicity of expanded CAG and CUG repeats in Caenorhabditis elegans. *Cell Mol Life Sci*, Vol.68, No.7, pp.1255-1267

Wang, Y.L., Liu, W., Wada, E., Murata, M., Wada, K. & Kanazawa, I. (2005). Clinico-pathological rescue of a model mouse of Huntington's disease by siRNA. *Neurosci. Res.*, Vol.53, No.3, pp.241-249

Wanker, E.E. (2000). Protein aggregation and pathogenesis of Huntington's disease: mechanisms and correlations. *Biol. Chem.*, Vol.381, No.9-10, pp.937-942

Warby, S.C., Montpetit, A., Hayden, A.R., Carroll, J.B., Butland, S.L., Visscher, H., Collins, J.A., Semaka, A., Hudson, T.J. & Hayden, M.R. (2009). CAG expansion in the Huntington disease gene is associated with a specific and targetable predisposing haplogroup. *Am. J. Hum. Genet*, Vol.84, No.3, pp.351-366

Weiss, A., Klein, C., Woodman, B., Sathasivam, K., Bibel, M., Regulier, E., Bates, G.P. & Paganetti, P. (2008). Sensitive biochemical aggregate detection reveals aggregation onset before symptom development in cellular and murine models of Huntington's disease. *J. Neurochem.*, Vol.104, No.3, pp.846-858

Wellington, C.L., Ellerby, L.M., Gutekunst, C.A., Rogers, D., Warby, S., Graham, R.K., Loubser, O., van, R.J., Singaraja, R., Yang, Y.Z., Gafni, J., Bredesen, D., Hersch, S.M., Leavitt, B.R., Roy, S., Nicholson, D.W. & Hayden, M.R. (2002). Caspase cleavage of mutant huntingtin precedes neurodegeneration in Huntington's disease. *J. Neurosci.*, Vol.22, No.18, pp.7862-7872

Wilburn, B., Rudnicki, D.D., Zhao, J., Weitz, T.M., Cheng, Y., Gu, X., Greiner, E., Park, C.S., Wang, N., Sopher, B.L., La Spada, A.R., Osmand, A., Margolis, R.L., Sun, Y.E. & Yang, X.W. (2011). An antisense CAG repeat transcript at JPH3 locus mediates expanded polyglutamine protein toxicity in Huntington's disease-like 2 mice. *Neuron*, Vol.70, No.3, pp.427-440

Williams, J.H., Schray, R.C., Patterson, C.A., Ayitey, S.O., Tallent, M.K. & Lutz, G.J. (2009). Oligonucleotide-mediated survival of motor neuron protein expression in CNS improves phenotype in a mouse model of spinal muscular atrophy. *J. Neurosci.*, Vol.29, No.24, pp.7633-7638

Zeitlin, S., Liu, J.P., Chapman, D.L., Papaioannou, V.E. & Efstratiadis, A. (1995). Increased apoptosis and early embryonic lethality in mice nullizygous for the Huntington's disease gene homologue. *Nat. Genet*, Vol.11, No.2, pp.155-163

Zhang, S., Feany, M.B., Saraswati, S., Littleton, J.T. & Perrimon, N. (2009). Inactivation of Drosophila Huntingtin affects long-term adult functioning and the pathogenesis of a Huntington's disease model. *Dis Model Mech*, Vol.2, No.5-6, pp.247-266

Zhang, Y., Li, M., Drozda, M., Chen, M., Ren, S., Mejia Sanchez, R.O., Leavitt, B.R., Cattaneo, E., Ferrante, R.J., Hayden, M.R. & Friedlander, R.M. (2003). Depletion of wild-type huntingtin in mouse models of neurologic diseases. *J. Neurochem.*, Vol.87, No.1, pp.101-106

Zuccato, C., Ciammola, A., Rigamonti, D., Leavitt, B.R., Goffredo, D., Conti, L., MacDonald, M.E., Friedlander, R.M., Silani, V., Hayden, M.R., Timmusk, T., Sipione, S. & Cattaneo, E. (2001). Loss of huntingtin-mediated BDNF gene transcription in Huntington's disease. *Science*, Vol.293, No.5529, pp.493-498

20

Don't Take Away My P: Phosphatases as Therapeutic Targets in Huntington's Disease

Ana Saavedra[1,2,3], Jordi Alberch[1,2,3] and Esther Pérez-Navarro[1,2,3]
[1]*Departament de Biologia Cellular, Immunologia i Neurociències, Facultat de Medicina, Universitat de Barcelona, Barcelona,*
[2]*Institut d'Investigacions Biomèdiques August Pi i Sunyer (IDIBAPS), Barcelona,*
[3]*Centro de Investigación Biomédica en Red sobre, Enfermedades Neurodegenerativas (CIBERNED),*
Spain

1. Introduction

The molecular bases that account for the preferential neurodegeneration of striatal medium-sized spiny neurons (MSNs) in Huntington's Disease (HD) are still unknown, and different mechanisms have been proposed to contribute to the neurodegenerative process. These include mitochondrial dysfunction and metabolic impairment, transcriptional dysregulation, altered expression of trophic factors, dopamine toxicity, oxidative stress, and changes in autophagy, and huntingtin (htt) phosphorylation. In addition, excitotoxicity through the overactivation of N-methyl-D-aspartate (NMDA) receptors (NMDARs) has also been proposed to contribute to the preferential loss of these neurons (for review see Ehrnhoefer et al., 2011; Jin & Johnson, 2010; Perez-Navarro et al., 2006; Renna et al., 2010; Rosenstock et al., 2010; Weir et al., 2011).

Some of these mechanisms are controlled by the attachment/removal of phosphate groups through the action of protein kinases and protein phosphatases, respectively. Therefore, alterations in their levels/activity in the presence of mutant htt (mhtt) can impact on cell survival.

Htt is expressed in almost all tissues, has a widespread distribution in the brain, its expression levels are similar in control individuals and in HD patients, with no evidence of increased htt expression in the brain regions most affected in HD (reviewed by Han et al., 2010). These evidences indicate that differences in mhtt expression do not contribute to the increased vulnerability of MSNs in HD. Conversely, several cell-type specific features including morphological, biochemical, and functional characteristics might play a role in rendering MSNs more vulnerable to the toxic effects of mhtt (Han et al., 2010). In this line, it is relevant in context of the present review to mention that the phosphatases calcineurin (also known as protein phosphatase 2B - PP2B) (Goto et al., 1987) and striatal-enriched protein tyrosine phosphatase (STEP) (Lombroso et al., 1991) are enriched in MSNs, suggesting that variations in their expression levels/activity can impact seriously in the function and viability of these neurons.

Here, we will revisit the excitotoxic hypothesis in HD through the phosphatase point of view, and we will also pay attention to the importance of phosphorylation in reducing the toxicity of mhtt. We will discuss the results obtained in both exon-1 and full-length HD models, and we will integrate the potential contribution of an imbalance between the activity of phosphatases and kinases to HD pathophysiology.

1.1 Excitotoxicity

Glutamate, the major excitatory neurotransmitter in the central nervous system (CNS), is important for neural development, synaptic plasticity, and learning and memory under physiological conditions. Dysregulation of glutamate levels and/or glutamate receptor activity can result in an overstimulation of glutamate receptors leading to cell death via excitotoxicity (Olney, 1969). In HD, excitotoxicity induced by overactivation of NMDARs has been proposed to explain the preferential neurodegeneration of MSNs (reviewed by Fan & Raymond, 2007; Milnerwood & Raymond, 2010; Perez-Navarro et al., 2006). Functional NMDARs are tetrameric structures (Laube et al., 1998) composed of two NR1 and at least two NR2 subunits (Ozawa et al., 1998), and the striatum is enriched in NR2B compared with other NR2 subunits (Landwehrmeyer et al., 1995). The presence of mhtt in striatal neurons leads to a number of alterations that can explain changes in the susceptibility to excitotoxicity. These include: (1) Selective increase of the current flowing through NMDARs comprising NR1/NR2B subunits (Zeron et al., 2001, 2002); (2) Changes in NMDAR scaffolding proteins (Jarabek et al., 2004; Sun et al., 2001; Torres-Peraza et al., 2008); (3) Altered phosphorylation of NMDAR subunits (Jarabek et al., 2004; Song et al., 2003) and (4) Imbalance between synaptic and extra-synaptic NMDARs (Milnerwood et al., 2010; Okamoto, 2009). In addition to alterations at the level of NMDARs, mhtt also alters intracellular mechanisms regulated by NMDAR stimulation, such as the activity of kinases and phosphatases. Calcineurin, PP1, PP2A, and STEP are phosphatases regulated by NMDARs stimulation (Figure 1) whose levels/activity have been shown to be altered in neurons expressing exon-1 or full-length mhtt (Table 1).

1.2 Phosphorylation of htt

Htt has several known sites of phosphorylation, all of them less phosphorylated in the mutant than in the wild-type protein (reviewed by Ernhoefer et al., 2011). Among the htt phosphorylation sites identified, serine 421 (Ser421) is the most studied and thus, the best characterized. This site can be phosphorylated by Akt (Humbert et al., 2002) and serum and glucocorticoid-induced kinase (SGK) (Rangone et al., 2004), whereas calcineurin (Pardo et al., 2006; Pineda et al., 2009), PP1 and PP2A (Metzler et al., 2010) dephosphorylate it. Until now, phosphatases known to regulate htt phosphorylation at Ser421 have been shown to be altered in HD models (Table 1). Phosphorylation of Ser421 regulates htt's toxicity (Humbert et al., 2002; Pardo et al., 2006), htt's role in vesicle transport (Colin et al., 2008; Pineda et al., 2009; Zala et al., 2008), and htt cleavage by caspases (Metzler et al., 2010; Warby et al., 2009). In addition, phosphorylation of htt and mhtt at Ser421 is significantly reduced in neurons after excitotoxic stimulation of NMDARs (Metzler et al., 2010) (Figure 1). Moreover, there are other Ser and threonine (Thr) residues of htt that can be phosphorylated, and all of them regulate its toxicity. Most of the kinases that phosphorylate these sites have been identified, and include IKK (Thompson et al., 2009), cyclin-dependent kinase 5 (Cdk5) (Anne et al.,

2007; Luo et al., 2005), ERK1 (Schilling et al., 2006), and CK2 (Atwal et al., 2011). In contrast, the phosphatases acting on these residues are still unknown (reviewed by Ernhoefer et al., 2011).

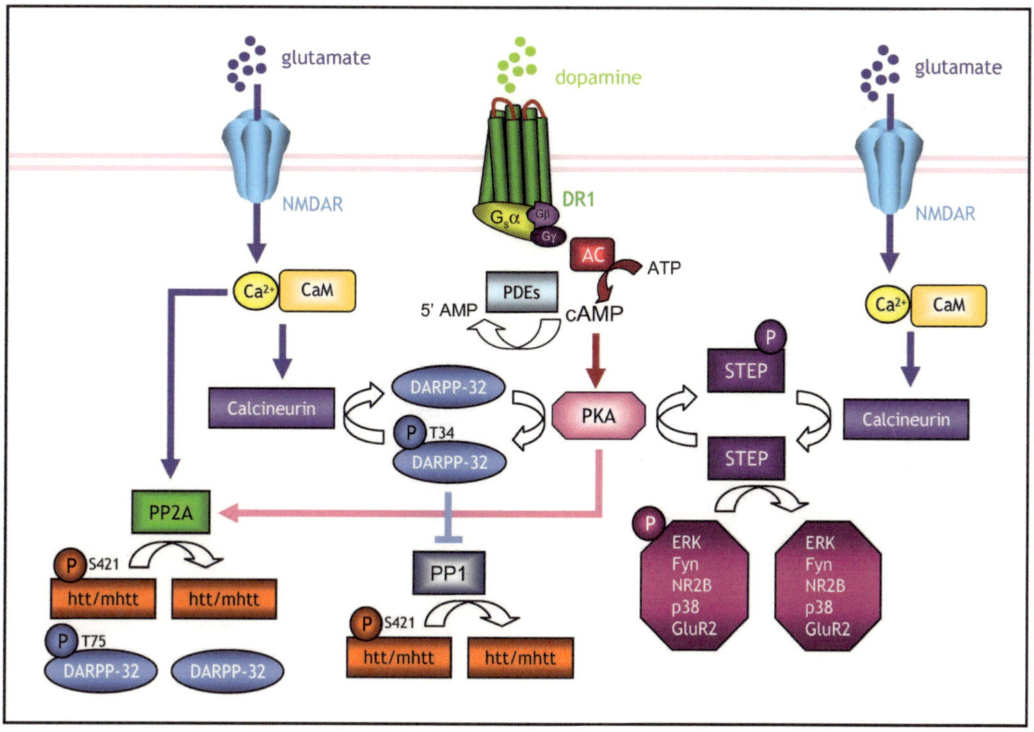

Fig. 1. Regulation of phosphatases in striatal neurons by NMDAR and dopamine D1 receptor (DR1) stimulation. Stimulation of NMDARs activates PP2A and calcineurin, which in turn will activate STEP and PP1. DR1 stimulation activates PP2A, and indirectly blocks PP1 activity. Several pathways and targets have been omitted for simplification. PDEs: Phosphodiesterases; AC: Adenylyl cyclase; PKA: cAMP-dependent protein kinase; CaM: calmodulin; DARPP-32: dopamine- and cAMP-regulated phosphoprotein of 32 kDa

2. Ser/Thr phosphatases

Ser/Thr phosphatases catalyze dephosphorylation reactions on phospho-Ser and phospho-Thr residues. They are classified into three families: protein phosphatase Mg^{2+}-activated (PPM), phosphoprotein phosphatases (PPPs) and the aspartate-based phosphatases represented by FCP/SCP (TFIIF-associating component of RNA polymerase II CTD phosphatase/small CTD phosphatase). The PPM family includes PP2C, pyruvate dehydrogenase phosphatase, and PP2C-"like" phosphatases, and the major phosphatases in the PPP family are PP1, PP2A and calcineurin (reviewed by McConnell & Wadzinski, 2009). PP1, PP2A and calcineurin are composed of catalytic and regulatory subunits, whereas PP2C exists as a monomer devoid of regulatory subunits. In the brain, the activity of these phosphatases is regulated by the regulatory subunit, interacting partners, scaffolding proteins and/or specific endogenous activators/inhibitors (reviewed by Gee & Mansuy,

2005). These phosphatases are implicated in the regulation of excitotoxicity, synaptic plasticity and cell survival, and are altered in neurodegenerative disorders such as Alzheimer's Disease (Ducruet et al., 2005; Iqbal & Grundke-Iqbal, 2007; F. Liu et al., 2006; Tian & Wang, 2002), Parkinson's Disease (Lou et al., 2010; Wera & Neyts, 1994) and HD (Metzler et al., 2010; Pineda et al., 2009; Saavedra et al., 2010; Xifro et al., 2008; 2009).

Type of phosphatase	Phosphatase	Change	HD model	Reference
Ser/Thr Phosphatase	Calcineurin	Increased	STHdh $^{Q7/Q111}$ cells and Hdh$^{Q111/Q111}$ mice	Xifro et al., 2008 Pineda et al., 2009
		Reduced	R6/1; YAC128	Xifro et al., 2009 Metzler et al., 2010
	PHLPP1	Reduced	R6/1; R6/1:BDNF; R6/2; Tet/HD94; Hdh$^{Q111/Q111}$ and STHdh $^{Q111/Q111}$ cells	Saavedra et al., 2010
	PHLPP2	Unchanged/Reduced	R6/1	Rue et al., unpublished
	PP1	Unchanged Reduced	YAC128 YAC128	Metzler et al., 2010 Ehrnhoefer et al., 2011
	PP2A	Unchanged	YAC128; R6/1	Metzler et al., 2010 Saavedra et al., 2010
		Reduced	YAC128	Ehrnhoefer et al., 2011
Tyr Phosphatase	STEP	Decreased	R6/1; R6/2; Tet/HD94; Hdh$^{Q111/Q111}$, primary striatal neurons overexpressing htt171-82Q	Saavedra et al., 2011 Runne et al., 2008
	MKP1 and MKP3	Increased	PC12 cells overexpressing exon-1 mhtt 118Q	Z. L. Wu et al., 2002
	MKP-2	Intracellular redistribution	HEK 293 cells overexpressing mhtt 138Q and NR1/NR2B	Fan et al., 2008

Calcineurin (Xifro et al., 2009) and PHLPP1 (Saavedra et al., 2010) protein levels, and STEP mRNA levels (Hodges et al., 2006) are also decreased in the caudate/putamen of HD patients. HEK: human embryonic kidney; Q: glutamine; Tet/HD94: conditional mouse model of HD

Table 1. Phosphatases altered in HD models.

2.1 Calcineurin

Calcineurin is a Ser/Thr phosphatase activated by calcium/calmodulin, highly expressed in the brain, and abundant in the cytosol, and in pre-synaptic and post-synaptic terminals (Mansuy, 2003; Shibasaki et al., 2002). It is a heterodimer composed by a calmodulin-binding catalytic subunit, calcineurin A, and an intrinsic calcium-binding regulatory subunit, calcineurin B. The dependence on calcium distinguishes calcineurin from spontaneously active PP2A and from Mg^{2+}-dependent PP2C. The binding of the calcium/calmodulin complex to calcineurin A with high affinity leads to the release of the auto-inhibitory domain from the active site and calcineurin activation. In addition to activation by calcium, calcineurin can also be activated by caspase- or calpain-mediated proteolysis, which originate a constitutively active form, insensitive to calcium/calmodulin (reviewed by A. Mukherjee & Soto, 2011).

Calcineurin is the only calcium-dependent phosphatase present in neurons, which confers it an important role in the maintenance of cellular homeostasis, and in neuronal activity (Mansuy, 2003; Shibasaki et al., 2002). Calcineurin also modulates gene expression by the regulation of transcription factors such as the cAMP responsive element binding protein (CREB) and the nuclear factor of activated T-cell (NFAT) (reviewed by A. Mukherjee & Soto, 2011).

Calcineurin is highly expressed in the striatum, and in particular in MSNs (Goto et al., 1987). The participation of calcineurin in neuronal death induced by insults that elevate intracellular calcium levels (Ankarcrona et al., 1996; Butcher et al., 1997; Dawson et al., 1993; Shamloo et al., 2005; Shibasaki & McKeon 1995; Wood & Bristow 1998; H. Y. Wu et al., 2004) suggests that this phosphatase might be a good candidate to participate in the excitotoxic events associated with HD.

The pro-apoptotic function of calcineurin has been linked to the dephosphorylation of selected substrates related to apoptosis, such as Bad (a pro-apoptotic Bcl-2 family member) (Springer et al., 2000; H. G. Wang et al., 1999), death-associated protein kinase (Shamloo et al., 2005; Xifro et al., 2008), cdk5 (Nishi et al., 2002) or transcription factors, such as NFAT (Beals et al., 1997). Importantly, calcineurin also dephosphorylates mhtt at Ser421 (Pardo et al., 2006). Consistent with the neuroprotective role of htt phosphorylation at Ser421 (Humbert et al., 2002; Rangone et al., 2004; Warby et al., 2005), inhibition of calcineurin activity in HD neuronal cells restores htt phosphorylation levels at Ser421, and prevents polyglutamine (polyQ)-mediated cell death of striatal neurons (Pardo et al., 2006). Moreover, inhibition of calcineurin by FK506 leads to sustained phosphorylation of mhtt at Ser421 and reestablishes BDNF transport in rat primary neuronal cultures expressing mhtt, and in mouse cortical neurons from $Hdh^{Q111/Q111}$ mice (Pineda et al., 2009). Recently, calcineurin has been shown to dephosphorylate the pro-fission dynamin related protein 1 (Cereghetti et al., 2008), which increases its mitochondrial translocation and activation, leading to mitochondrial fragmentation and contributing to the hypersensitivity of HD mitochondria to apoptosis (Costa et al., 2010). In fact, mitochondrial fragmentation can be prevented by genetic or pharmacological inhibition of calcineurin (Costa et al., 2010).

Studies using primary striatal cultures from YAC transgenic mice show that NMDAR stimulation produces a polyQ length-dependent increase in cell death (Shehadeh et al., 2006; Zeron et al., 2002). These observations were extended by our studies showing that

STHdh$^{Q111/Q111}$ cells are more susceptible to NMDA-mediated cell death than STHdh$^{Q7/Q7}$ cells, a phenomenon related to higher calcineurin A protein levels and calcineurin activity in mhtt knock-in striatal cells than in wild-type cells (Xifro et al., 2008). Interestingly, although calcineurin protein levels are similar in mouse brains containing wild-type and mhtt, Hdh$^{Q111/Q111}$ and Hdh$^{Q111/Q7}$ mice have significantly higher levels of calcineurin activity in the cortex than Hdh$^{Q7/Q7}$ mice (Pineda et al., 2009). In agreement with these reports showing increased calcineurin activity, the levels of the negative regulator of calcineurin RCAN1-1L are significantly down-regulated in HD brain samples (Ermak et al., 2009). Additionally, a dysregulation in the levels of cytosolic calcium, the calcineurin activator, was also reported in primary cultures from YAC128 mice (Tang et al., 2005). Calcineurin can play a toxic role in striatal cells expressing full-length mhtt at two different levels. High levels of calcineurin increase the susceptibility to excitotoxicity (Xifro et al., 2008) and, on the other hand, calcineurin can increase mhtt toxicity directly by dephosphorylation of its Ser421 (Ermak et al., 2009; Pardo et al., 2006; Pineda et al., 2009), or indirectly by regulating proteins that modulate mhtt toxicity, such as cdk5 (Luo et al., 2005) or calpain (Gafni et al., 2004).

Conversely, calcineurin A mRNA levels are decreased in human HD samples (Hodges et al., 2006). Similarly, in the striatum of R6 mouse models of HD, which express the exon-1 mhtt fragment, calcineurin levels are lower than in the wild-type mice striatum (Hernandez-Espinosa & Morton, 2006; Lievens et al., 2002; Luthi-Carter et al., 2000; Xifro et al., 2009). Interestingly, these mice are resistant to excitotoxicity (Hansson et al., 1999, 2001; Torres-Peraza et al., 2008). These findings suggest a dual regulation of calcineurin A expression during the progression of the disease, with high levels at early stages resulting in high susceptibility to excitotoxicity (Xifro et al., 2008), and low levels at end stages participating in the resistance to excitotoxic-induced cell death (Xifro et al., 2009) (Figure 2). Thus, it would be relevant to study whether this dual calcineurin regulation also occurs in full-length mouse models of HD as YAC128 mice, which were reported to be more sensitive to excitotoxicity than controls at presymptomatic stages, but resistant to intrastriatal quinolinic acid (an NMDAR agonist) injection when signs of HD are obvious (Graham et al., 2009). Consistent with resistance to excitotoxicity (Graham et al., 2009), reduced calcineurin activity has been shown in the striatum of YAC128 mice at 12 months of age (Metzler et al., 2010).

Studies performed in *in vivo* models of HD confirm the important role played by calcineurin in the excitotoxic-mediated cell death of striatal neurons. Calcineurin inhibition in wild-type mice drastically reduces quinolinic acid-induced striatal cell death (Xifro et al., 2009). Moreover, calcineurin activation induced by intrastriatal quinolinic acid injection in R6/1 mice is lower than in wild-type mice (Xifro et al., 2009), which is consistent with R6/1 animals being resistant to excitotoxicity (Hansson et al., 1999, 2001).

However, the role of calcineurin in HD remains controversial as calcineurin inhibition has been reported to have protective (Costa et al., 2010; Ermak et al., 2009; Pardo et al., 2006; Pineda et al., 2009; Xifro et al., 2008) or worsening (Hernandez-Espinosa & Morton, 2006) effects in HD models. The participation of reduced calcineurin activity caused by alteration of calcineurin A expression in the pathophysiology of HD, and in the excitotoxic resistance observed in exon-1 mouse models (Xifro et al., 2009), together with the finding that treatment with calcineurin inhibitors accelerates the progression of the disease in R6/2 mice (Hernandez-Espinosa & Morton, 2006) suggest that decreased levels of calcineurin could result in striatal neuronal dysfunction affecting the onset of motor alterations. However, since both FK506 and cyclosporine A, that does not cross the blood–brain barrier, have the same negative effect (Hernandez-Espinosa & Morton, 2006) the harmful effect of calcineurin inhibition reported in this study might be unrelated to the effect of these inhibitors in the CNS.

Taken together, these findings suggest calcineurin as an important therapeutic target for HD, by its participation in excitotoxic events, as well as by its action on phosphorylated mhtt (Ser421) to increase toxicity.

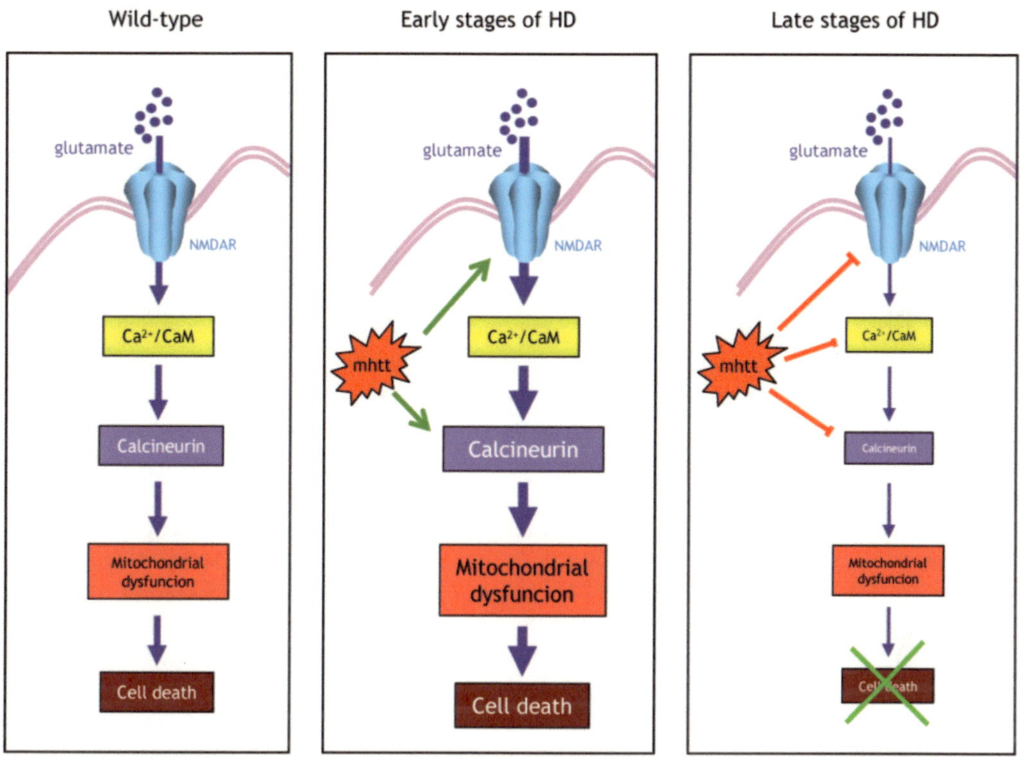

Fig. 2. Changes in striatal calcineurin levels during HD progression, and involvement in excitotoxicity. Results from Xifro et al. (2008) suggest that at early stages of HD calcineurin levels are increased and striatal neurons are more susceptible to NMDA-induced excitotoxicity. In contrast, at late stages, calcineurin levels are decreased and participate in the resistance of striatal neurons to NMDA-induced excitotoxicity (Xifro et al., 2009). CaM: calmodulin.

2.2 Pleckstrin homology (PH) domain leucine-rich repeat protein phosphatase (PHLPP)

PHLPPs constitute a subfamily within the PP2C phosphatase family. PHLPPs require Mg^{2+} and Mn^{2+} for their catalytic activity, and are not inhibited by traditional phosphatase inhibitors such as okadaic acid (Brognard et al., 2007; Gao et al., 2005). This family comprises three members: PHLPP1α, PHLPP1β and PHLPP2. PHLPP1α and PHLPP1β are splice variants from the same gene but have different sizes, whereas PHLPP2 is a different gene product and has the same domain composition of PHLPP1 (Brognard et al., 2007). PHLPP1 and PHLPP2 have an identical domain structure with a PH domain (sharing 63% amino identity) followed by a region of leucine-rich repeats, a PP2C phosphatase domain (sharing 58% amino identity) and a C terminal PDZ ligand. In addition, PHLPP1β and PHLPP2 contain a Ras-association domain preceding the PH domain (Brognard & Newton, 2008).

PHLPPs are expressed in the majority of human tissues and are localized in different cellular compartments such as cytosol, nucleus and membrane (Brognard et al., 2007; Brognard & Newton, 2008). In the CNS, PHLPP1β was the first identified as an mRNA that oscillated in a circadian rhythm-dependent manner in the suprachiasmatic nucleus (SCN) and was named SCOP (SCN circadian oscillatory protein) (Shimizu et al., 1999). PHLPP1β/SCOP is expressed in various brain regions with a relative enrichment in hippocampus and cerebellum (Shimizu et al., 1999). Its expression is highly concentrated in neurons, and is present in nuclear, mitochondrial and cytosolic fractions (Shimizu et al., 1999), as well as in membrane rafts (Shimizu et al., 2003). Recently, PHLPP1α and PHLPP2 have been shown to be also expressed in hippocampal neurons (Jackson et al., 2009; 2010) with PHLPP1α as the most abundantly expressed in the adult (Jackson et al., 2010). Although PHLPP1 and 2 can be found in the cytosolic fraction, only PHLPP1α can be localized in the nucleus of hippocampal neurons (Jackson et al., 2010). In addition, we have detected PHLPP1α in the cortex and striatum of adult mice (Saavedra et al., 2010).

So far, the known substrates for PHLPPs are the kinases Akt (also known as protein kinase B), and protein kinase C (PKC). Akt, the first identified substrate of PHLPP (Gao et al., 2005), is a key regulator of a wide range of cellular processes including growth, proliferation, metabolism, cell cycle progression, and survival. Thus, altered Akt activity has been associated with cancer and other disease conditions such as diabetes and neurodegenerative diseases (Liao & Hung, 2010). For its full catalytic activity, Akt requires phosphorylation at Thr308 in the activation loop and at Ser473 in the hydrophobic motif (Brazil & Hemmings, 2001). Its activation depends on the PI3-kinase, which produces the lipid second messenger PtdIns-3, 4, 5-P3 (PIP3) that interacts with the PH domain of Akt and recruits the kinase to the plasma membrane (Sancak et al., 2008). Subsequently, the Thr308 residue is phosphorylated by membrane-localized 3-phosphoinositide-dependent protein kinase 1 (PDK1) (Alessi et al., 1997; Calleja et al., 2007) and the Ser473 residue is phosphorylated by mTORC2 (Sarbassov et al., 2005) (Figure 3). PHLPPs specifically dephosphorylate the hydrophobic motif of Akt, resulting in a decrease of its activity (Gao et al., 2005), whereas the Thr308 site is dephosphorylated by PP2A (Bayascas & Alessi, 2005). PKC, the other PHLPPs substrate, consists in a Ser/Thr family of phosphorylating enzymes ubiquitously expressed and implicated in multiple cellular functions. There are 12 isoforms of PKC termed (1) calcium-dependent or classical PKCs, cPKCs (2) calcium-independent or novel PKCs, nPKCs, and (3) atypical PKCs, aPKCs (Amadio et al., 2006; Pearce et al., 2010). PKC isoforms, like Akt, are also activated by the phosphorylation of the activation segment and hydrophobic motif (Newton, 2003). PDK1 phosphorylates the activation segment (Dutil et al., 1998; Le Good et al., 1998), and there is increasing evidence that mTORC2 phosphorylates the hydrophobic motif of at least some isoforms (Sarbassov et al., 2004; Guertin et al., 2006). The phosphorylation of the hydrophobic motif regulates the amplitude of PKC signaling by controlling the stability of the kinase. Both PHLPP1 and PHLPP2 dephosphorylate the hydrophobic motif of conventional and novel PKC isoforms, but not atypical PKC isoforms (Gao et al., 2008). This dephosphorylation induces the degradation of PKC. Thus, depletion of PHLPP1 or PHLPP2 leads to a robust increase in PKC levels (Gao et al., 2008).

Members of the AGC kinase family like p70S6K, SGK or p90RSK, which have hydrophobic phosphorylation motifs, are other potential substrates of PHLPPs (Brognard & Newton, 2008). In addition to the dephosphorylation of Akt and PKC, PHLPP1β/SCOP negatively regulates the Ras–Raf–MEK–ERK pathway by interacting directly with Ras (Shimizu et al., 2003).

In the CNS, PHLPPs participate in the regulation of the circadian clock (Shimizu et al., 1999), learning and memory (Shimizu et al., 2007), and survival (Jackson et al., 2009; 2010; Saavedra et al., 2010). In HD, we have shown that PHLPP1α is reduced in cellular as well as in HD mouse models, and in the putamen of HD patients (Saavedra et al., 2010). STHdh$^{Q111/Q111}$ cells display decreased levels of PHLPP1α compared with STHdh$^{Q7/Q7}$ cells. Similarly, we detected reduced levels of PHLPP1α in the striatum of Hdh$^{Q111/Q111}$ mice (at 5 months of age), and also in the striatum of the exon-1 mouse models R6/1 (from 12 to 30 weeks of age), R6/1:BDNF +/- (from 12 to 30 weeks of age), R6/2 (at 12 weeks of age) and Tet/HD94 (at 22 months of age). In addition, PHLPP1α levels are also decreased in the cortex and hippocampus of R6/1 mice at 12 and 30 weeks of age. PHLPP1 expression was regulated by mhtt at the transcriptional level since we also detected decreased PHLPP1 mRNA levels in the striatum of R6/1 mice (Saavedra et al., 2010). We speculated that the down-regulation of PHLPP1 mRNA levels could be related with decreased activity of the transcription factor NF-Y, since this transcription factor is sequestered in mhtt aggregates (Yamanaka et al., 2008). It has recently been shown that the expression of PHLPP is controlled by mammalian target of rapamycin (mTOR)-dependent protein translation in colon and breast cancer cells (J. Liu et al., 2011). Interestingly, mTOR activity is reduced in HD (Ravikumar et al., 2004). Thus, it is tempting to speculate that this mechanism could also be involved in the down-regulation of PHLPP1α levels. In good correlation with decreased levels of PHLPP1α in the striatum, we observed increased phosphorylation levels of Akt (Ser473) and of its targets GSK3β (Ser9) and FoxO (Ser256). Although PHLPP1α levels were down-regulated in the cortex and hippocampus of R6/1 mice we did not observe changes in pAkt (Ser473) levels indicating that a reduction of PHLPP1α levels may not be enough to increase pAkt (Ser473) levels *in vivo* (Saavedra et al., 2010). In addition, in the striatum of Tet/HD94 mice, we observed that after shutting-down the expression of mhtt, PHLPP1α protein levels returned to wild-type levels but pAkt (Ser473) up-regulation was only partially reduced (Saavedra et al., 2010). Taken together, these results suggest that increased levels of pAkt is a specific mechanism taking place in striatal neurons expressing mhtt, which could be the sum of increased activation of kinases that phosphorylate Akt and decreased levels of PHLPP1α. Since Akt activation is one of the main mechanisms to prevent neuronal death during injury (Chong et al., 2005), and many transgenic HD mouse models show little, if any, striatal cell death (Canals et al., 2004; Diaz-Hernandez et al., 2005; Garcia-Martinez et al., 2007; Mangiarini et al., 1996; Martin-Aparicio et al., 2001), our results suggest that increased Akt activation could counteract mhtt toxicity.

In addition, we showed that decreased levels of PHLPP1α could help to maintain high levels of pAkt (Ser473) in R6/1 striatum after excitotoxicity, contributing to prevent cell death induced by NMDARs overstimulation (Saavedra et al., 2010).

Conversely, we found unchanged levels of PHLPP2 in the striatum in R6/1 mice at different stages of the disease (8, 12, 20 and 30 weeks of age), while cortical levels are decreased at 12 and 30 weeks of age.

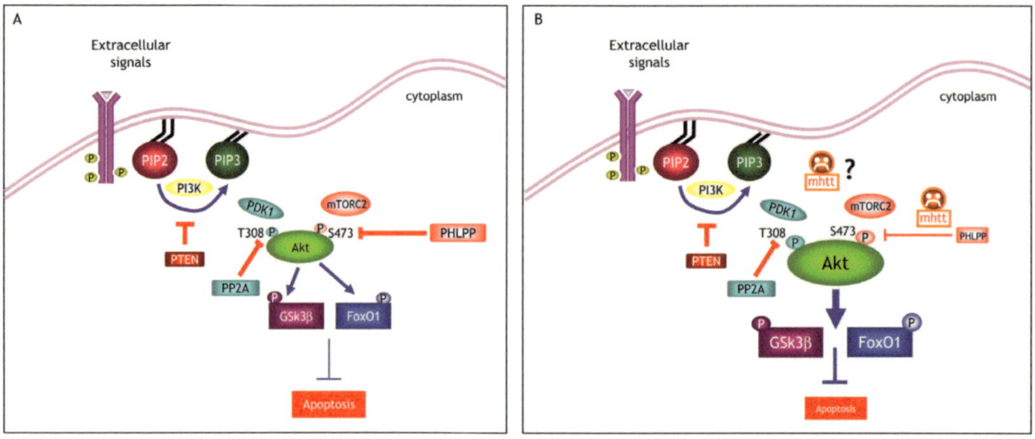

Fig. 3. PHLPP1α is down-regulated in HD striatum. (A) Scheme showing pathways that control Akt phosphorylation in wild-type cells. Akt is phosphorylated at Thr308 (T308) by PDK1 and dephosphorylated by PP2A, whereas the Ser473 residue (S473) is phosphorylated by mTORC2 and dephosphorylated by PHLPP. Once activated, Akt prevents apoptosis through the phosphorylation of several targets such as GSK3β and FoxO1. (B) Control of Akt phosphorylation in striatal cells expressing mhtt. In the presence of mhtt PHLPP1α levels are decreased and contribute to maintain high levels of Akt phosphorylated at S473 that through increased levels of phosphorylated GSK3β and FoxO1 may delay striatal cell death. Results obtained by analyzing different brain regions suggest that mhtt could also affect the activation of kinases that phosphorylate Akt in the striatum, but not in other brain regions (Saavedra et al., 2010).

2.3 PP1 and PP2A

PP1 and PP2A account for the majority of Ser/Thr phosphatase activity in mammalian cells, and are involved in diverse cellular processes such as cell growth and proliferation, development, DNA replication and repair, metabolism, neural signaling, and apoptosis. The activity of these two phosphatases can be blocked by okadaic acid and caliculin at different IC50 values (Sheppeck et al., 1997). The specific oligomeric composition of PP1 and PP2A holoenzyme is important to control their phosphatase activity. Functional PP1 enzyme consists of a catalytic subunit (PP1c) and a regulatory subunit (R subunit). The PP1c associates with more than 50 proteins that regulate substrate specificity and subcellular localization (Ceulemans & Bollen, 2004; P.T. Cohen, 2002). The interaction of PP1c with its regulatory subunit can also influence substrate specificity. In addition, its activity is regulated by endogenous inhibitory proteins like inhibitor-1 (P. Cohen & Nimmo, 1978), inhibitor-2 (Foulkes & P. Cohen, 1980), CPI-17 (Eto et al., 1997), and DARPP-32 (Walaas & Greengard, 1991), which is highly expressed in MSNs. PP2A exists in two forms: a core dimer and a heterotrimeric holoenzyme. The PP2A core dimer is composed by the scaffolding A subunit and the catalytic C subunit and associates with a regulatory B subunit to generate the heterotrimeric holoenzyme, which is the predominant form of PP2A in the cell. PP2A regulatory B subunits are divided into four different families and play a crucial role in the subcellular localization of PP2A. They can also alter the overall shape of the catalytic subunit as well as enzyme kinetics (reviewed by McConnell & Wadzinski, 2009; Shi, 2009).

In the CNS, PP1 and PP2A dephosphorylate neurotransmitter receptors and proteins localized at the post-synaptic site, thus participating in the regulation of excitatory and inhibitory transmission. PP1 dephosphorylates CaMKII when bound to post-synaptic density, whereas soluble or synaptosomal CaMKII is deposphorylated by PP2A (Shields et al., 1985; Strack et al., 1997). In addition, both phosphatases regulate NMDARs-mediated synaptic currents in an activity-dependent manner (L. Y. Wang et al., 1994; Westphal et al., 1999). PP1 dephosphorylates GABA receptor subunits (X. Wang et al., 2002) and down-regulates AMPA receptor activity and trafficking by dephosphorylation of the GluR1 subunit (reviewed by Mansuy & Shenolikar, 2006). In addition, PP1 and PP2A activity can promote apoptosis (reviewed by Garcia et al., 2003; Klumpp & Krieglstein, 2002). PP1 dephosphorylates the pro-apoptotic protein Bad with its consequent activation, and PP2A dephosphorylates the anti-apoptotic proteins Akt and Bcl-2 inactivating them. PP2A can also regulate the activity of a large number of kinases, such as ERK, PKA and p38 (reviewed by Millward et al., 1999), all of them important to neuronal survival and function.

Recently, the number of targets of PP1 and PP2A has been extended since both proteins dephosphorylate htt *in situ* and after excitotoxic stimulation of NMDARs (Metzler et al., 2010; see Figure 1). Metzler and colleagues (2010) showed that NMDARs overstimulation induces a decrease of phtt (Ser421) in primary neurons from wild-type and YAC128 transgenic mice. In addition, dephosphorylation of htt (Ser421) was also observed in YAC128 transgenic mice after quinolinic acid injection in the striatum. Dephosphorylation of htt after excitotoxicity seems to participate in the induction of cell death since blockade of PP1 and PP2A activity protects YAC128 striatal neurons from NMDA-induced cell death *in vitro*. Moreover, they showed that dopamine modulates htt phosphorylation in the striatum through the regulation of the PP1 inhibitor DARPP-32. These authors also observed a decrease in the PP1 substrate pCREB, which together with decreased levels of DARPP-32 in YAC128 striatum suggested an altered regulation of phosphatase activity in HD. However, they could not detect changes in the activity of PP1 and PP2A in YAC128 mice striatum. Although these results point to a role of htt dephosphorylation in excitotoxic-induced cell death in the striatum, it remains to be shown whether inhibition of PP1 and PP2A is also neuroprotective *in vivo*. In addition, it would be interesting to investigate whether dephosphorylation of mhtt takes place in the striatum of YAC128 mice when they are resistant to excitotoxicity. PP2A protein levels have also been analyzed in the striatum of R6/1 mice. Similarly to that observed in YAC128 mice striatum (Metzler et al., 2010), no changes in PP2A protein levels have been detected in R6/1 mouse striatum at 4, 8, 12, 16 and 30 weeks of age compared with their littermate controls (Saavedra et al., 2010).

3. Tyrosine phosphatases

Tyrosine (Tyr) phosphatases, encoded by about 107 genes in the human genome (Alonso et al., 2004; Andersen et al., 2004), have the ability to hydrolyze p-nitrophenyl phosphate, are inhibited by vanadate and are insensitive to okadaic acid. They are classified into three groups: (1) Cytoplasmic, (2) Receptor-like, and (3) Dual specificity phosphatases, which dephosphorylate Ser, Thr and Tyr residues that are in close proximity. The specificity of Tyr phosphatases is regulated by several molecular strategies such as preferential recognition of phosphopeptides, cell-type and organelle-specific expression, and assembly with other proteins (for review see S. Paul & Lombroso, 2003; Z. Y. Zhang, 2002). These phosphatases play important roles in the development and function of the CNS (Ensslen-Craig & Brady-Kalnay, 2004; S. Paul & Lombroso, 2003), and have been suggested to function as neuroprotectants. STEP, the SH2-containing Tyr phosphatases SHP1 and SHP2, and protein Tyr phosphatase alpha are among the protective candidates. However, protein Tyr phosphatase alpha and phosphatase and tensin homolog deleted from chromosome 10 (PTEN) may also induce neurotoxicity (Gee & Mansuy, 2005). Increased Tyr phosphorylation has been suggested to induce neuronal cell death in cerebral ischemia (Ohtsuki et al., 1996; R. Paul et al., 2001) and after epileptiform activity (Chun et al., 2004; Sanna et al., 2000). In addition, alterations in protein Tyr phosphatases are considered to be involved in the etiology of neural disorders such as Alzheimer's Disease (Kerr et al., 2006; Lee et al., 2004), Parkinson's Disease (Herradon & Ezquerra, 2009) and HD (Saavedra et al., 2011; Z. L. Wu et al., 2002).

3.1 STEP

STEP, encoded by the *Ptpn5* gene, is a brain-specific Tyr phosphatase involved in neuronal signal transduction. STEP plays an important role in synaptic plasticity through the opposition to synaptic strengthening (Braithwaite et al., 2006a). Additionally, STEP has been implicated in susceptibility to cell death through the modulation of ERK1/2 signaling (Choi et al., 2007; Saavedra et al., 2011), while other studies suggest that STEP can play a role in neuroprotection through the regulation of the p38 pathway (Poddar et al., 2010; Xu et al., 2009). The mechanism underlying the ability of STEP to regulate both pro-survival and pro-cell death pathways has been recently elucidated (Xu et al., 2009; see details below).

STEP is enriched in MSNs (Lombroso et al., 1991), and expressed at lower levels in the cortex, hippocampus and amygdala (Boulanger et al., 1995). STEP mRNA is alternatively spliced into several STEP isoforms (Bult et al., 1997; Sharma et al., 1995) that are differentially targeted to the post-synaptic density (Oyama et al., 1995), extra-synaptic and cytosolic compartments (Goebel-Goody et al., 2009; Xu et al., 2009). The major isoforms are $STEP_{46}$, the cytosolic isoform, and $STEP_{61}$, which is membrane-associated through the additional 172 amino acids in the N-terminus (Bult et al., 1997). Both isoforms are expressed in the striatum, whereas other brain regions only express $STEP_{61}$ (Boulanger et al., 1995).

STEP activity is regulated through phosphorylation/dephosphorylation of a Ser residue within its kinase interacting motif (KIM) domain. Stimulation of D1Rs activates PKA (Stoof & Kebabian, 1981), which phosphorylates STEP thereby inactivating it (S. Paul et al., 2000) (Figure 1). In contrast, stimulation of NMDARs results in the dephosphorylation and activation of STEP through a calcineurin/PP1 pathway (S. Paul et al., 2003; Valjent et al., 2005) (Figure 1). Additionally, STEP activity is also regulated by proteolytic cleavage (Xu et al., 2009), ubiquitin-proteasome degradation (Kurup et al., 2010; S. Mukherjee et al., 2011; Xu et al., 2009), local translation (Y. Zhang et al., 2008), and oligomerization (Deb et al., 2011).

Once activated, STEP dephosphorylates the glutamate receptor subunits NR2B (Braithwaite et al., 2006b; Pelkey et al., 2002; Snyder et al., 2005) and GluR2 (Y. Zhang et al., 2008), leading to their endocytosis, and the kinases ERK1/2, p38 and Fyn, thereby controlling the duration of their signal (Munoz et al., 2003; Nguyen et al., 2002; S. Paul et al., 2003; Pulido et al., 1998) (Figure 1).

The enrichment of STEP in MSNs, its role in the regulation of key substrates implicated in neuronal function, together with the fact that both dopaminergic and glutamatergic systems regulate STEP activity and are affected in HD patients and mouse models (Andre et al., 2010; Fan & Raymond, 2007; Jakel & Maragos, 2000) prompted us to study the possible role of STEP in the pathophysiology of HD (Saavedra et al., 2011). In fact, previous studies showed decreased mRNA levels of STEP in the caudate nucleus and cortex of HD patients (Hodges et al., 2006), in the striatum of R6 mice (Desplats et al., 2006; Luthi-Carter et al., 2000), and in primary striatal neurons overexpressing htt171-82Q (Runne et al., 2008). Our results show that R6/1 mice display reduced STEP protein levels in the striatum and cortex, and increased phosphorylation levels in the striatum, cortex and hippocampus. R6/2, Tet/HD94 and $Hdh^{Q7/Q111}$ mice striatum also displays decreased STEP protein and increased STEP phosphorylation levels (Saavedra et al., 2011). The early increase in striatal STEP phosphorylation levels correlates with a dysregulation of the PKA pathway that together with decreased calcineurin activity at later stages further contributes to an enhancement of STEP inactivation. Accordingly, the levels of phosphorylated ERK2 and p38, two targets of STEP, are increased in R6/1 mice striatum at advanced stages of the disease (Saavedra et al., 2011).

HD mouse models develop resistance to excitotoxicity (Graham et al., 2009; Hansson et al., 1999, 2001; Jarabek et al., 2004; Torres-Peraza et al., 2008), and reduced levels of calcineurin expression and activity can contribute to this phenomenon (Xifro et al., 2009). Stimulation of NMDARs activates STEP in a calcineurin-dependent manner (S. Paul et al., 2003), and disruption of STEP activity has been shown to lead to the activation of ERK1/2 signaling and to the attenuation of excitotoxic-induced cell death in the hippocampus (Choi et al., 2007). Therefore, we wondered whether STEP acts as a calcineurin target after an excitotoxic stimulus to the striatum thereby contributing to the resistance to excitotoxicity observed in HD mouse models. After intrastriatal quinolinic acid injection, we observed higher and unaltered pSTEP levels, and more sustained ERK signaling in R6/1 than in wild-type mice suggesting that STEP inactivation could mediate neuroprotection in R6/1 striatum (Saavedra et al., 2011). These findings are consistent with lower calcineurin activation which, importantly, correlates with reduced cell death in R6/1 mice striatum after quinolinic acid injection (Xifro et al., 2009). In agreement with a protective role for STEP inactivation, blockade of STEP activity with FK-506 (an inhibitor of calcineurin) allows ERK activation and confers protection to hilar interneurons of the hippocampus against excitotoxicity (Choi et al., 2007), and intrastriatal infusion of TAT-STEP, a cell-permeable form, increases quinolic acid-induced cell death in the striatum (Saavedra et al., 2011). Conversely, low striatal STEP levels and activity (increased pSTEP levels) in R6/1 mice can contribute to their reduced vulnerability to excitotoxicity (Saavedra et al., 2011).

Activation of extra-synaptic NMDARs in primary cortical neurons leads to calpain-mediated cleavage of $STEP_{61}$. This prevents STEP from binding to its substrates and contributes to the selective activation of extra-synaptically concentrated p38 (Xu et al., 2009). In contrast, synaptic NMDAR stimulation leads to the ubiquitination and degradation of $STEP_{61}$ and ERK1/2 activation (Xu et al., 2009). We did not observe $STEP_{61}$ cleavage or p38 activation which, together with ERK2 activation, suggests a preferential stimulation of synaptic NMDARs in our model (Saavedra et al., 2011). This is relevant because an imbalance between synaptic and extra-synaptic NMDARs has been shown to occur in YAC128 mice (Milnerwood et al., 2010; Okamoto et al., 2009). However, these mice develop resistance to excitotoxicity with age (Graham et al., 2009), and those studies were performed in vulnerable mice. Thus, it is likely that increased extra-synaptic NMDARs during excitotoxicity-sensitive stages might increase $STEP_{61}$ cleavage to $STEP_{33}$ enabling higher activation of p38 than in wild-type mice. In contrast, in resistant mice other mechanisms should regulate striatal cell survival in response to excitotoxicity and, according with our findings, STEP regulation of ERK activity seems to play an important role (Saavedra et al., 2011).

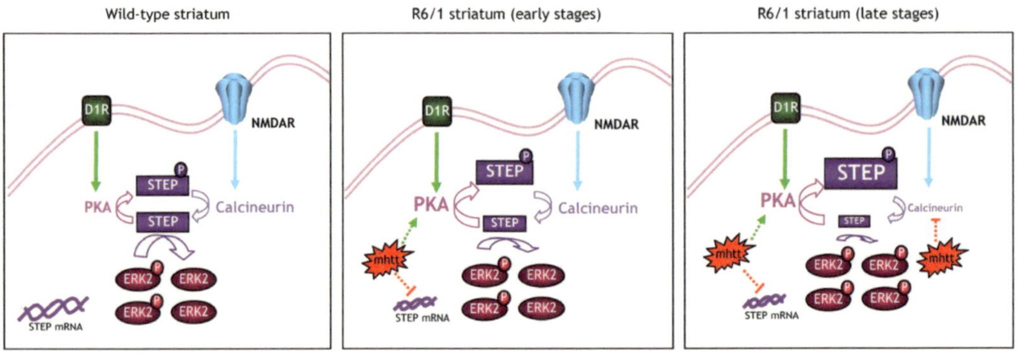

Fig. 4. Regulation of STEP levels and activity in the R6/1 mouse striatum during the progression of the disease. The presence of mhtt in the striatum alters this system at different levels: (1) At early stages mhtt induces a down-regulation of STEP mRNA and protein levels, and a dysregulation of the PKA pathway that correlates with increased STEP phosphorylation. (2) At late stages, calcineurin activity is also reduced further inactivating STEP with a consequent increase of pERK2 levels (p-p38 levels and possibly other non-analyzed STEP targets). Decreased STEP activity, through the regulation of its targets, could be involved in the development of resistance to excitotoxicity in R6/1 mice striatum. (scheme from Saavedra et al., 2011).

STEP has recently been implicated in the etiology of Alzheimer's Disease (Kurup et al., 2010; Snyder et al., 2005; Y. Zhang et al., 2010) but the alterations in the STEP pathway found in HD mouse models are specific because STEP protein levels and activity, in contrast to that observed in Alzheimer's Disease, are reduced in HD (Saavedra et al., 2011). Since the genetic reduction of STEP levels reverses cognitive and cellular deficits in Alzheimer's Disease mice (Y. Zhang et al., 2010), the modulation of STEP levels might be a good therapeutic strategy in HD. Nevertheless, the possibility of restoring STEP expression in HD is presently hampered by the lack of data about the regulation of *STEP* gene expression.

3.2 MAP kinase phosphatases (MKPs)

MKPs are intracellular dual Tyr phosphatases with an expression restricted to different subcellular compartments (S. Paul & Lombroso, 2003). Some of these MKPs, such as MKP-1, -2, -3 and -X, have been shown to be expressed in the brain with a specific distribution and different substrate preferences. MKP-1 is expressed in the cortex, thalamus, striatum and cerebellum with the following substrate specificity: p38>JNK/SAPK>>ERK (Boschert et al., 1998; Franklin & Kraft, 1997; Misra-Press et al., 1995; Takaki et al., 2001). MKP-2 is localized in the prefrontal cortex, hippocampus and cerebellum and inactivates ERK and JNK/SAPK with the same specificity, but it can also act on p38 (Chu et al., 1996; Dwivedi et al., 2001; Groom et al., 1996; Misra-Press et al., 1995). MKP-3 is detected in the cerebral cortex, striatum and hippocampus acting preferentially on ERK, but it can also inactivate JNK/SAPK and p38 with the same specificity (Boschert et al., 1998; Muda et al., 1996a,b; Takaki et al., 2001). Finally, MKP-X is expressed throughout the brain and acts preferentially on ERK, although it can also dephosphorylate p38 (Boschert et al., 1998; Dowd et al., 1998; Muda et al., 1996b; Shin et al., 1997).

Although they are expressed in the brain, their role in neuronal function is not well established. MKP-1 increases in rat brain after limbic epilepsy (Gass et al., 1996) and, together with MKP-3, upon cerebral hypoxia in neuronal nuclei of newborn piglets (Mishra & Delivoria-Papadopoulos, 2004). Moreover, both MKP-1 and -3 play important roles in neural plastic modifications after drug exposure (Takaki et al., 2001), whereas MKP-2 is increased in postmortem brains of suicide subjects with major depression (Dwivedi et al., 2001). Recently, it has been shown that MKP-1 controls axon branching of cortical neurons in response to the trophic factor BDNF (Jeanneteau et al., 2010). In addition, in PC12 cells, oxidative stress and hypoxia increase MPK-1 expression, while trophic factor treatment up-regulates both MKP-1 and -3 (Camps et al., 1998; Keyse & Emslie, 1992; Seta et al., 2001). Thus, regulation of MPKs seems to be important not only after brain injury, but also during development.

In a stable PC12 cell line expressing truncated mhtt with 118Q, Z. L. Wu and colleagues (2002) showed that MKP-1 and -3 mRNA levels, and MKP-1 protein levels, were increased at different time points after mhtt expression. In good correlation with changes in MKPs levels, they observed a substantial reduction of ERK1/2 phosphorylation. Interestingly, treatment with sodium orthovanadate and bp V (pic), two general Tyr phosphatase inhibitors, rescues cells from polyQ-induced cell death suggesting that these phosphatases are involved in mhtt-induced toxicity (Z. L. Wu et al., 2002). In HEK 293 cells transfected with NR1/NR2B and htt containing 138Q, MKP-2 has been shown to be reduced in the soluble fraction and increased in the particulate-derived fraction when compared with cells expressing htt with 15Q (Fan et al., 2008). However, the mechanism underlying this redistribution and the physiological significance of this event are presently unknown.

4. Conclusion

Understanding the pathways by which mhtt causes neuronal dysfunction and death is essential to develop efficient treatments for HD. Great progress has been made over the last years in highlighting the molecular mechanisms affected by mhtt. Here, we have reviewed the existing data about changes in the expression and regulation of phosphatases in HD models and human HD brain. From these results, it is becoming increasingly clear that alterations in phosphatases are involved in the pathogenesis of HD. So far, the phosphatases analyzed participate in the regulation of excitotoxicity and neuronal survival (through the regulation of the PI3K/Akt pathway, ERK2 and/or htt phosphorylation). In mouse models, most of them are decreased, which seems to be a compensatory mechanism induced in response to mhtt expression in order to prevent neuronal cell death. However, how this might translate to humans is still unknown as we cannot follow the disease from the beginning, and analysis of phosphatase levels and activity can be performed only at late stages of the disease. We believe that the regulation of phosphatases is a new and promising approach to treat HD. Therefore, our future challenge is to develop novel tools to treat HD based on these findings. In addition, phosphatases are also involved in the pathogenesis of other neurodegenerative disorders, and ongoing investigations of disease mechanisms in HD can also provide new therapeutic approaches to Parkinson's or Alzheimer's Diseases.

5. Acknowledgements

Research in our group is supported by Fondo de Investigaciones Sanitarias (Instituto de Salud Carlos III, PI10/01072 to E.P.-N.), Ministerio de Educación y Ciencia (Grant SAF2008-04360 to J.A.), and Generalitat de Catalunya (group of excellence; Grant 2009SGR-00326). A.S. is supported by Ministerio de Ciencia e Innovación, Juan de la Cierva subprograme, Spain (JCI-2010-08207).

6. References

Alessi D. R., James S. R., Downes C. P., Holmes A. B., Gaffney P. R., Reese C. B. & Cohen P. (1997) Characterization of a 3-phosphoinositide-dependent protein kinase which phosphorylates and activates protein kinase Balpha. *Current Biology*, Vol. 7, No. 4, pp. 261-269, ISSN 0960-9822

Alonso A., Sasin J., Bottini N., Friedberg I., Osterman A., Godzik A., Hunter T., Dixon J. & Mustelin T. (2004) Protein tyrosine phosphatases in the human genome. *Cell*, Vol. 117, No. 6, pp. 699-711, ISSN 0092-8674

Amadio M., Battaini F. & Pascale A. (2006) The different facets of protein kinases C: old and new players in neuronal signal transduction pathways. *Pharmacological Research*, Vol. 54, No. 5, pp. 317-325, ISSN 1043-6618

Andersen J. N., Jansen P. G., Echwald S. M., Mortensen O. H., Fukada T., Del Vecchio R., Tonks N. K. & Moller N. P. (2004) A genomic perspective on protein tyrosine phosphatases: gene structure, pseudogenes, and genetic disease linkage. *FASEB Journal*, Vol. 18, No. 1, pp. 8-30, ISSN 0892-6638

Andre V. M., Cepeda C. & Levine M. S. (2010) Dopamine and glutamate in Huntington's disease: A balancing act. *CNS Neuroscience & Therapeutics*, Vol. 16, No. 3, pp. 163-178, ISSN 1755-5930

Ankarcrona M., Dypbukt J. M., Orrenius S. & Nicotera P. (1996) Calcineurin and mitochondrial function in glutamate-induced neuronal cell death. *FEBS Letters*, Vol. 394, No. 3, pp. 321-324, ISSN 0014-5793

Anne S. L., Saudou F. & Humbert S. (2007) Phosphorylation of huntingtin by cyclin-dependent kinase 5 is induced by DNA damage and regulates wild-type and mutant huntingtin toxicity in neurons. *Journal of Neuroscience*, Vol. 27, No. 27, pp. 7318-7328, ISSN 0270-6474

Atwal R. S., Desmond C. R., Caron N., Maiuri T., Xia J., Sipione S. & Truant R. (2011) Kinase inhibitors modulate huntingtin cell localization and toxicity. *Nature Chemical Biology*, Vol. 7, No. 7, pp. 453-460, ISSN 1552-4450

Bayascas J. R. & Alessi D. R. (2005) Regulation of Akt/PKB Ser473 phosphorylation. *Molecular Cell*, Vol. 18, No. 2, pp. 143-145, ISSN 1097-2765

Beals C. R., Clipstone N. A., Ho S. N. & Crabtree G. R. (1997) Nuclear localization of NF-ATc by a calcineurin-dependent, cyclosporin-sensitive intramolecular interaction. *Genes & Development*, Vol. 11, No. 7, pp. 824-834, ISSN 0890-9369

Boschert U., Dickinson R., Muda M., Camps M. & Arkinstall S. (1998) Regulated expression of dual specificity protein phosphatases in rat brain. *Neuroreport*, Vol. 9, No. 18, pp. 4081-4086, ISSN 0959- 4965

Boulanger L. M., Lombroso P. J., Raghunathan A., During M. J., Wahle P. & Naegele J. R. (1995) Cellular and molecular characterization of a brain-enriched protein tyrosine phosphatase. *Journal of Neuroscience*, Vol. 15, No. 2, pp. 1532-1544, ISSN 0270-6474

Braithwaite S. P., Paul S., Nairn A. C. & Lombroso P. J. (2006a) Synaptic plasticity: one STEP at a time. *Trends in Neuroscience*, Vol. 29, No. 8, pp. 452-458, ISSN 0166-2236

Braithwaite S. P., Adkisson M., Leung J., Nava A., Masterson B., Urfer R., Oksenberg D. & Nikolich K. (2006b) Regulation of NMDA receptor trafficking and function by striatal-enriched tyrosine phosphatase (STEP). *European Journal of Neuroscience*, Vol. 23, No. 11, pp. 2847-2856, ISSN 0953-816X

Brazil D. P. & Hemmings B. A. (2001) Ten years of protein kinase B signalling: a hard Akt to follow. *Trends in Biochemical Sciences*, Vol. 26, No. 11, pp. 657-664, ISSN 0968-0004

Brognard J. & Newton A. C. (2008) PHLiPPing the switch on Akt and protein kinase C signaling. *Trends in Endocrinology and Metabolism*, Vol. 19, No. 6, pp. 223-230, ISSN 1043-2760

Brognard J., Sierecki E., Gao T. & Newton A. C. (2007) PHLPP and a second isoform, PHLPP2, differentially attenuate the amplitude of Akt signaling by regulating distinct Akt isoforms. *Molecular Cell*, Vol. 25, No. 6, pp. 917-931, ISSN 1097-2765

Bult A., Zhao F., Dirkx R., Jr., Raghunathan A., Solimena M. & Lombroso P. J. (1997) STEP: a family of brain-enriched PTPs. Alternative splicing produces transmembrane, cytosolic and truncated isoforms. *European Journal of Cell Biology*, Vol. 72, No. 4, pp. 337-344, ISSN 0171-9335

Butcher S. P., Henshall D. C., Teramura Y., Iwasaki K. & Sharkey J. (1997) Neuroprotective actions of FK506 in experimental stroke: in vivo evidence against an antiexcitotoxic mechanism. *Journal of Neuroscience*, Vol. 17, No. 18, pp. 6939-6946, ISSN 0270-6474

Calleja V., Alcor D., Laguerre M., Park J., Vojnovic B., Hemmings B. A., Downward J., Parker P. J. & Larijani B. (2007) Intramolecular and intermolecular interactions of protein kinase B define its activation in vivo. *PLoS Biology*, Vol. 5, No. 4, pp. e95, ISSN 1544-9173

Camps M., Chabert C., Muda M., Boschert U., Gillieron C. & Arkinstall S. (1998) Induction of the mitogen-activated protein kinase phosphatase MKP3 by nerve growth factor in differentiating PC12. *FEBS Letters*, Vol. 425, No. 2, pp. 271-276, ISSN 0014-5793

Canals J. M., Pineda J. R., Torres-Peraza J. F., Bosch M., Martin-Ibanez R., Munoz M. T., Mengod G., Ernfors P. & Alberch J. (2004) Brain-derived neurotrophic factor regulates the onset and severity of motor dysfunction associated with enkephalinergic neuronal degeneration in Huntington's disease. *Journal of Neuroscience*, Vol. 24, No. 35, pp. 7727-7739, ISSN 0270-6474

Cereghetti G. M., Stangherlin A., Martins d. B., Chang C. R., Blackstone C., Bernardi P. & Scorrano L. (2008) Dephosphorylation by calcineurin regulates translocation of Drp1 to mitochondria. *Proceedings of the National Academy of Sciences USA* , Vol. 105, No. 41, pp. 15803-15808, ISSN 0027-8424

Ceulemans H. & Bollen M. (2004) Functional diversity of protein phosphatase-1, a cellular economizer and reset button. *Physiological Reviews*, Vol. 84, No. 1, pp. 1-39, ISSN 0031-9333

Choi Y. S., Lin S. L., Lee B., Kurup P., Cho H. Y., Naegele J. R., Lombroso P. J. & Obrietan K. (2007) Status epilepticus-induced somatostatinergic hilar interneuron degeneration is regulated by striatal enriched protein tyrosine phosphatase. *Journal of Neuroscience*, Vol. 27, No. 11, pp. 2999-3009, ISSN 0270-6474

Chong Z. Z., Li F. & Maiese K. (2005) Activating Akt and the brain's resources to drive cellular survival and prevent inflammatory injury. *Histology and Histopathology*, Vol. 20, No. 1, pp. 299-315, ISSN 0213- 3911

Chu Y., Solski P. A., Khosravi-Far R., Der C. J. & Kelly K. (1996) The mitogen-activated protein kinase phosphatases PAC1, MKP-1, and MKP-2 have unique substrate specificities and reduced activity in vivo toward the ERK2 sevenmaker mutation. *Journal of Biological Chemistry*, Vol. 271, No. 11, pp. 6497-6501, ISSN 0021-9258

Chun J. T., Crispino M. & Tocco G. (2004) The dual response of protein kinase Fyn to neural trauma: early induction in neurons and delayed induction in reactive astrocytes. *Experimental Neurology*, Vol. 185, No. 1, pp. 109–119, ISSN 0014-4886

Cohen P. & Nimmo G. A. (1978) The purification and characterization of protein phosphatase inhibitor-1 from rabbit skeletal muscle. *Biochemical Society Transactions*, Vol. 6, No. 1, pp. 17-20, ISSN 0300-5127

Cohen P. T. (2002) Protein phosphatase 1 - targeted in many directions. *Journal of Cell Science*, Vol. 115, Pt 2, pp. 241-256, ISSN 0021-9533

Colin E., Zala D., Liot G., Rangone H., Borrell-Pages M., Li X. J., Saudou F. & Humbert S. (2008) Huntingtin phosphorylation acts as a molecular switch for anterograde/retrograde transport in neurons. *EMBO Journal*, Vol. 27, No. 15, pp. 2124-2134, ISSN 0261-4189

Costa V., Giacomello M., Hudec R., Lopreiato R., Ermak G., Lim D., Malorni W., Davies K. J., Carafoli E. & Scorrano L. (2010) Mitochondrial fission and cristae disruption increase the response of cell models of Huntington's disease to apoptotic stimuli. *EMBO Molecular Medicine*, Vol. 2, No. 12, pp. 490-503, ISSN 1757-4676

Dawson T. M., Steiner J. P., Dawson V. L., Dinerman J. L., Uhl G. R. & Snyder S. H. (1993) Immunosuppressant FK506 enhances phosphorylation of nitric oxide synthase and protects against glutamate neurotoxicity. *Proceedings of the National Academy of Sciences USA* , Vol. 90, No. 21, pp. 9808-9812, ISSN 0027-8424

Deb I., Poddar R. & Paul S. (2011) Oxidative stress-induced oligomerization inhibits the activity of the non-receptor tyrosine phosphatase STEP61. *Journal of Neurochemistry*, Vol. 116, No. 6, pp. 1097-1111, ISSN 0022-3042

Desplats P. A., Kass K. E., Gilmartin T., Stanwood G. D., Woodward E. L., Head S. R., Sutcliffe J. G. & Thomas E. A. (2006) Selective deficits in the expression of striatal-enriched mRNAs in Huntington's disease. *Journal of Neurochemistry*, Vol. 96, No. 3, pp. 743-757, ISSN 0022-3042

Diaz-Hernandez M., Torres-Peraza J., Salvatori-Abarca A., Moran M. A., Gomez-Ramos P., Alberch J. & Lucas J. J. (2005) Full motor recovery despite striatal neuron loss and formation of irreversible amyloid-like inclusions in a conditional mouse model of Huntington's disease. *Journal of Neuroscience*, Vol. 25, No. 42, pp. 9773-9781, ISSN 0270-6474

Dowd S., Sneddon A. A. & Keyse S. M. (1998) Isolation of the human genes encoding the pyst1 and Pyst2 phosphatases: characterisation of Pyst2 as a cytosolic dual-specificity MAP kinase phosphatase and its catalytic activation by both MAP and SAP kinases. *Journal of Cell Science*, Vol. 111 (Pt 22), No. pp. 3389-3399, ISSN 0021-9533

Ducruet A. P., Vogt A., Wipf P. & Lazo J. S. (2005) Dual specificity protein phosphatases: therapeutic targets for cancer and Alzheimer's disease. *Annual Review of Pharmacology and Toxicology*, Vol. 45, No. pp. 725-750, ISSN 0362-1642

Dutil E. M., Toker A. & Newton A. C. (1998) Regulation of conventional protein kinase C isozymes by phosphoinositide-dependent kinase 1 (PDK-1). *Current Biology*, Vol. 8, No. 25, pp. 1366-1375, ISSN 0960-9822

Dwivedi Y., Rizavi H. S., Roberts R. C., Conley R. C., Tamminga C. A. & Pandey G. N. (2001) Reduced activation and expression of ERK1/2 MAP kinase in the post-mortem brain of depressed suicide subjects. *Journal of Neurochemistry*, Vol. 77, No. 3, pp. 916-928, ISSN 0022-3042

Ehrnhoefer D. E., Sutton L. & Hayden M. R. (2011) Small Changes, Big Impact: Posttranslational Modifications and Function of Huntingtin in Huntington Disease. *Neuroscientist*, Vol. No. pp. ISSN 1073-8584

Ensslen-Craig S. E. & Brady-Kalnay S. M. (2004) Receptor protein tyrosine phosphatases regulate neural development and axon guidance. *Developmental Biology*, Vol. 275, No. 1, pp. 12-22, ISSN 0012-1606

Ermak G., Hench K. J., Chang K. T., Sachdev S. & Davies K. J. (2009) Regulator of calcineurin (RCAN1-1L) is deficient in Huntington disease and protective against mutant huntingtin toxicity in vitro. *Journal of Biological Chemistry*, Vol. 284, No. 18, pp. 11845-11853, ISSN 0021-9258

Eto M., Senba S., Morita F. & Yazawa M (1997) Molecular cloning of a novel phosphorylation-dependent inhibitory protein of protein phosphatase-1 (CPI17) in smooth muscle: its specific localization in smooth muscle. *FEBS Letters*, Vol. 410, No. 2-3, pp. 356-360, ISSN 0014-5793

Fan M. M. & Raymond L. A. (2007) N-methyl-D-aspartate (NMDA) receptor function and excitotoxicity in Huntington's disease. *Progress in Neurobiology*, Vol. 81, No. 5-6, pp. 272-293, ISSN 0301-0082

Fan M. M., Zhang H., Hayden M. R., Pelech S. L. & Raymond L. A. (2008) Protective up-regulation of CK2 by mutant huntingtin in cells co-expressing NMDA receptors. *Journal of Neurochemistry*, Vol. 104, No. 3, pp. 790-805, ISSN 0022-3042

Foulkes J. G. & Cohen P. (1980) The regulation of glycogen metabolism. Purification and properties of protein phosphatase inhibitor-2 from rabbit skeletal muscle. *European Journal of Biochemistry*, Vol. 105, No. 1, pp. 195-203, ISSN 0014-2956

Franklin C. C. & Kraft A.S. (1997) Conditional expression of the mitogen-activated protein kinase (MAPK) phosphatase MKP-1 preferentially inhibits p38 MAPK and stress-activated protein kinase in U937 cells. *Journal of Biological Chemistry*, Vol. 272, No. 27, pp. 16917-16923, ISSN 0021-9258

Gafni J., Hermel E., Young J. E., Wellington C. L., Hayden M. R. & Ellerby L. M. (2004) Inhibition of calpain cleavage of huntingtin reduces toxicity: accumulation of calpain/caspase fragments in the nucleus. *Journal of Biological Chemistry*, Vol. 279, No. 19, pp. 20211-20220, ISSN 0021-9258

Gao T., Furnari F. & Newton A. C. (2005) PHLPP: a phosphatase that directly dephosphorylates Akt, promotes apoptosis, and suppresses tumor growth. *Molecular Cell*, Vol. 18, No. 1, pp. 13-24, ISSN 1097-2765

Gao T., Brognard J. & Newton A. C. (2008) The phosphatase PHLPP controls the cellular levels of protein kinase C. *Journal of Biological Chemistry,* Vol. 283, No. 10, pp. 6300-6311, ISSN 0021-9258

Garcia-Martinez J. M., Perez-Navarro E., Xifro X., Canals J. M., Diaz-Hernandez M., Trioulier Y., Brouillet E., Lucas J. J. & Alberch J. (2007) BH3-only proteins Bid and Bim(EL) are differentially involved in neuronal dysfunction in mouse models of Huntington's disease. *Journal of Neuroscience Research*, Vol. 85, No. 12, pp. 2756-2769, ISSN 0360-4012

Garcia A., Cayla X., Guergnon J., Dessauge F., Hospital V., Rebollo M. P., Fleischer A. & Rebollo A. (2003) Serine/threonine protein phosphatases PP1 and PP2A are key players in apoptosis. *Biochimie*, Vol. 85, No. 8, pp. 721-726, ISSN 0300-9084

Gass P., Eckhardt A., Schroder H., Bravo R. & Herdegen T. (1996) Transient expression of the mitogen-activated protein kinase phosphatase MKP-1 (3CH134/ERP1) in the rat brain after limbic epilepsy. *Brain Research Molecular Brain Research, Vol.* 41, No. 1-2, pp. 74-80, ISSN 0169-328X

Gee C. E. & Mansuy I. M. (2005) Protein phosphatases and their potential implications in neuroprotective processes. *Cellular and Molecular Life Sciences*, Vol. 62, No. 10, pp. 1120-1130, ISSN 1420-682X

Goebel-Goody S. M., Davies K. D., Alvestad Linger R. M., Freund R. K. & Browning M. D. (2009) Phospho-regulation of synaptic and extrasynaptic N-methyl-d-aspartate receptors in adult hippocampal slices. *Neuroscience*, Vol. 158, No. 4, pp. 1446-1459, ISSN 0306-4522

Goto S., Matsukado Y., Miyamoto E. & Yamada M. (1987) Morphological characterization of the rat striatal neurons expressing calcineurin immunoreactivity. *Neuroscience*, Vol. 22, No. 1, pp. 189-201, ISSN 0306-4522

Graham R. K., Pouladi M. A., Joshi P., Lu G., Deng Y., Wu N. P., Figueroa B. E., Metzler M., Andre V. M., Slow E. J., Raymond L., Friedlander R., Levine M. S., Leavitt B. R. & Hayden M. R. (2009) Differential susceptibility to excitotoxic stress in YAC128 mouse models of Huntington disease between initiation and progression of disease. *Journal of Neuroscience*, Vol. 29, No. 7, pp. 2193-2204, ISSN 0270-6474

Groom L. A., Sneddon A. A., Alessi D. R., Dowd S. & Keyse S. M. (1996) Differential regulation of the MAP, SAP and RK/p38 kinases by Pyst1, a novel cytosolic dual-specificity phosphatase. *EMBO Journal* Vol. 15, No. 14, pp. 3621-3632, ISSN 0261-4189

Guertin D. A., Stevens D. M., Thoreen C. C., Burds A. A., Kalaany N. Y., Moffat J., Brown M., Fitzgerald K. J. & Sabatini D. M. (2006) Ablation in mice of the mTORC components raptor, rictor, or mLST8 reveals that mTORC2 is required for signaling to Akt-FOXO and PKCalpha, but not S6K1. *Developmental Cell*, Vol. 11, No. 6, pp. 859-871, ISSN 1534-5807

Han I., You Y., Kordower J. H., Brady S. T. & Morfini G. A. (2010) Differential vulnerability of neurons in Huntington's disease: the role of cell type-specific features. *Journal of Neurochemistry*, Vol. 113, No. 5, pp. 1073-1091, ISSN 0022-3042

Hansson O., Petersen A., Leist M., Nicotera P., Castilho R. F. & Brundin P. (1999) Transgenic mice expressing a Huntington's disease mutation are resistant to quinolinic acid-induced striatal excitotoxicity. *Proceedings of the National Academy of Sciences USA*, Vol. 96, No. 15, pp. 8727-8732, ISSN 0027-8424

Hansson O., Guatteo E., Mercuri N. B., Bernardi G., Li X. J., Castilho R. F. & Brundin P. (2001) Resistance to NMDA toxicity correlates with appearance of nuclear inclusions, behavioural deficits and changes in calcium homeostasis in mice transgenic for exon 1 of the huntington gene. *European Journal of Neuroscience*, Vol. 14, No. 9, pp. 1492-1504, ISSN 0953-816X

Hernandez-Espinosa D. & Morton A. J. (2006) Calcineurin inhibitors cause an acceleration of the neurological phenotype in a mouse transgenic for the human Huntington's disease mutation. *Brain Research Bulletin* Vol. 69, No. 6, pp. 669-679, ISSN 0361-9230

Herradon G. & Ezquerra L. (2009) Blocking receptor protein tyrosine phosphatase beta/zeta: a potential therapeutic strategy for Parkinson's disease. *Current Medicinal Chemistry*, Vol. 16, No. 25, pp. 3322-3329, ISSN 0929-8673

Hodges A., Strand A. D., Aragaki A. K., Kuhn A., Sengstag T., Hughes G., Elliston L. A., Hartog C., Goldstein D. R., Thu D., Hollingsworth Z. R., Collin F., Synek B., Holmans P. A., Young A. B., Wexler N. S., Delorenzi M., Kooperberg C., Augood S. J., Faull R. L., Olson J. M., Jones L. & Luthi-Carter R. (2006) Regional and cellular gene expression changes in human Huntington's disease brain. *Human Molecular Genetics*, Vol. 15, No. 6, pp. 965-977, ISSN 0964-6906

Humbert S., Bryson E. A., Cordelieres F. P., Connors N. C., Datta S. R., Finkbeiner S., Greenberg M. E., Saudou F. (2002) The IGF-1/Akt pathway is neuroprotective in Huntington's disease and involves Huntingtin phosphorylation by Akt. *Developmental Cell*, Vol. 2, No. 6, pp 831-837, ISSN 1534-5807

Iqbal K. & Grundke-Iqbal I. (2007) Developing pharmacological therapies for Alzheimer disease. *Cellular and Molecular Life Sciences*, Vol. 64, No. 17, pp. 2234-2244, ISSN 1420-682X

Jackson T. C., Rani A., Kumar A. & Foster T. C. (2009) Regional hippocampal differences in AKT survival signaling across the lifespan: implications for CA1 vulnerability with aging. *Cell Death & Differentiation*, Vol. 16, No. 3, pp. 439-448, ISSN 1350-9047

Jackson T. C., Verrier J. D., Semple-Rowland S., Kumar A. & Foster T. C. (2010) PHLPP1 splice variants differentially regulate AKT and PKCalpha signaling in hippocampal neurons: characterization of PHLPP proteins in the adult hippocampus. *Journal of Neurochemistry*, Vol. 115, No. 4, pp. 941-955, ISSN 0022-3042

Jakel R. J. & Maragos W. F. (2000) Neuronal cell death in Huntington's disease: a potential role for dopamine. *Trends in Neurosciences*, Vol. 23, No. 6, pp. 239-245, ISSN 0166-2236

Jarabek B. R., Yasuda R. P. & Wolfe B. B. (2004) Regulation of proteins affecting NMDA receptor-induced excitotoxicity in a Huntington's mouse model. *Brain*, Vol. 127, Pt 3, pp. 505-516, ISSN 0006-8950

Jeanneteau F., Deinhardt K., Miyoshi G., Bennett A. M. & Chao M. V. (2010) The MAP kinase phosphatase MKP-1 regulates BDNF-induced axon branching. *Nature Neuroscience*, Vol. 13, No. 11, pp. 1373-1379, ISSN 1097-6256

Jin Y. N. & Johnson G. V. (2010) The interrelationship between mitochondrial dysfunction and transcriptional dysregulation in Huntington disease. *Journal of Bioenergetics and Biomembranes*, Vol. 42, No. 3, pp. 199-205, ISSN 0145-479X

Kerr F., Rickle A., Nayeem N., Brandner S., Cowburn R. F. & Lovestone S. (2006) PTEN, a negative regulator of PI3 kinase signalling, alters tau phosphorylation in cells by mechanisms independent of GSK-3. *FEBS Letters*, Vol. 580, No. 13, pp. 3121-3128, ISSN 0014-5793

Keyse S. M. & Emslie E. A. (1992) Oxidative stress and heat shock induce a human gene encoding a protein-tyrosine phosphatase. *Nature*, Vol. 359, No. 6396, pp. 644-647, ISSN 0028-0836

Klumpp S. & Krieglstein J. (2002) Serine/threonine protein phosphatases in apoptosis. *Current Opinion in Pharmacology*, Vol. 2, No. 4, pp. 458-462, ISSN 1471-4892

Kurup P., Zhang Y., Xu J., Venkitaramani D. V., Haroutunian V., Greengard P., Nairn A. C. & Lombroso P. J. (2010) Abeta-mediated NMDA receptor endocytosis in Alzheimer's disease involves ubiquitination of the tyrosine phosphatase STEP61. *Journal of Neuroscience*, Vol. 30, No. 17, pp. 5948-5957, ISSN 0270-6474

Landwehrmeyer G. B., Standaert D. G., Testa C. M., Penney J. B. Jr. & Young A. B. (1995) NMDA receptor subunit mRNA expression by projection neurons and interneurons in rat striatum. *Journal of Neuroscience*, Vol. 15, No. 7, pp. 5297-5307, ISSN 0270-6474

Laube B., Kuhse J. & Betz H. (1998) Evidence for a tetrameric structure of recombinant NMDA receptors. *Journal of Neuroscience*, Vol. 18, No. 8, pp. 2954-2961, ISSN 0270-6474

Le Good J. A., Ziegler W. H., Parekh D. B., Alessi D. R., Cohen P. & Parker P. J. (1998) Protein kinase C isotypes controlled by phosphoinositide 3-kinase through the protein kinase PDK1. *Science*, Vol. 281, No. 3585, pp. 2042-2045, ISSN 0036-8075

Lee G., Thangavel R., Sharma V. M., Litersky J. M., Bhaskar K., Fang S. M., Do L. H., Andreadis A., Van Hoesen G. & Ksiezak-Reding H. (2004) Phosphorylation of tau by fyn: implications for Alzheimer's disease. *Journal of Neuroscience*, Vol. 24, No. 9, pp. 2304-2312, ISSN 0270-6474

Liao Y. & Hung M. C. (2010) Physiological regulation of Akt activity and stability. *American Journal of Translational Research*, Vol. 2, No. 1, pp. 19-42, ISSN 1943-8141

Lievens J. C., Woodman B., Mahal A. & Bates G. P. (2002) Abnormal phosphorylation of synapsin I predicts a neuronal transmission impairment in the R6/2 Huntington's disease transgenic mice. *Molecular and Cellular Neuroscience*, Vol. 20, No. 4, pp. 638-648, ISSN 1044-7431

Liu F., Liang Z. & Gong C. X. (2006) Hyperphosphorylation of tau and protein phosphatases in Alzheimer disease. *Panminerva Medica*, Vol. 48, No. 2, pp. 97-108, ISSN 0031-0808

Liu J., Stevens P. D. & Gao T. (2011) mTOR-dependent regulation of PHLPP expression controls the rapamycin sensitivity in cancer cells. *Journal of Biological Chemistry*, Vol. 286, No. 8, pp. 6510-6520, ISSN 0021-9258

Lombroso P. J., Murdoch G. & Lerner M. (1991) Molecular characterization of a protein-tyrosine-phosphatase enriched in striatum. *Proceedings of the National Academy of Sciences USA* , Vol. 88, No. 16, pp. 7242-7246, ISSN 0027-8424

Lou H., Montoya S. E., Alerte T. N., Wang J., Wu J., Peng X., Hong C. S., Friedrich E. E., Mader S. A., Pedersen C. J., Marcus B. S., McCormack A. L., Di Monte D. A., Daubner S. C. & Perez R. G. (2010) Serine 129 phosphorylation reduces the ability of alpha-synuclein to regulate tyrosine hydroxylase and protein phosphatase 2A in vitro and in vivo. *Journal of Biological Chemistry*, Vol. 285, No. 23, pp. 17648-17661, ISSN 0021-9258

Luo S., Vacher C., Davies J. E. & Rubinsztein D. C. (2005) Cdk5 phosphorylation of huntingtin reduces its cleavage by caspases: implications for mutant huntingtin toxicity. *Journal of Cell Biology*, Vol. 169, No. 4, pp. 647-656, ISSN 0021-9525

Luthi-Carter R., Strand A., Peters N. L., Solano S. M., Hollingsworth Z. R., Menon A. S., Frey A. S., Spektor B. S., Penney E. B., Schilling G., Ross C. A., Borchelt D. R., Tapscott S. J., Young A. B., Cha J. H. & Olson J. M. (2000) Decreased expression of striatal signaling genes in a mouse model of Huntington's disease. *Human Molecular Genetics*, Vol. 9, No. 9, pp. 1259-1271, ISSN 0964-6906

Mangiarini L., Sathasivam K., Seller M., Cozens B., Harper A., Hetherington C., Lawton M., Trottier Y., Lehrach H., Davies S. W. & Bates G. P. (1996) Exon 1 of the HD gene with an expanded CAG repeat is sufficient to cause a progressive neurological phenotype in transgenic mice. *Cell*, Vol. 87, No. 3, pp. 493-506, ISSN 0092-8674

Mansuy I. M. (2003) Calcineurin in memory and bidirectional plasticity. *Biochemical and Biophysical Research Communications*, Vol. 311, No. 4, pp. 1195-1208, ISSN: 0006-291X

Mansuy I. M. & Shenolikar S. (2006) Protein serine/threonine phosphatases in neuronal plasticity and disorders of learning and memory. *Trends in Neuroscience*, Vol. 29, No. 12, pp. 679-686, ISSN 0166-2236

Martin-Aparicio E., Yamamoto A., Hernandez F., Hen R., Avila J. & Lucas J. J. (2001) Proteasomal-dependent aggregate reversal and absence of cell death in a conditional mouse model of Huntington's disease. *Journal of Neuroscience*, Vol. 21, No. 22, pp. 8772-8781, ISSN 0270-6474

McConnell J. L. & Wadzinski B. E. (2009) Targeting protein serine/threonine phosphatases for drug development. *Molecular Pharmacology*, Vol. 75, No. 6, pp. 1249-1261, ISSN 0026-895X

Metzler M., Gan L., Mazarei G., Graham R. K., Liu L., Bissada N., Lu G., Leavitt B. R. & Hayden M. R. (2010) Phosphorylation of huntingtin at Ser421 in YAC128 neurons is associated with protection of YAC128 neurons from NMDA-mediated excitotoxicity and is modulated by PP1 and PP2A. *Journal of Neuroscience*, Vol. 30, No. 43, pp. 14318-14329, ISSN 0270-6474

Millward T. A., Zolnierowicz S. & Hemmings B. A. (1999) Regulation of protein kinase cascades by protein phosphatase 2A. *Trends in Biochemical Sciences*, Vol. 24, No. 5, pp. 186-191, ISSN 0968-0004

Milnerwood A. J. & Raymond L. A. (2010) Early synaptic pathophysiology in neurodegeneration: insights from Huntington's disease. *Trends in Neurosciences*, Vol. 33, No. 11, pp. 513-523, ISSN 0166-2236

Milnerwood A. J., Gladding C. M., Pouladi M. A., Kaufman A. M., Hines R. M., Boyd J. D., Ko R. W., Vasuta O. C., Graham R. K., Hayden M. R., Murphy T. H. & Raymond L. A. (2010) Early increase in extrasynaptic NMDA receptor signaling and expression contributes to phenotype onset in Huntington's disease mice. *Neuron*, Vol. 65, No. 2, pp. 178-190, ISSN 0896-6273

Mishra O. P. & Delivoria-Papadopoulos M. (2004) Effect of hypoxia on protein tyrosine kinase activity in cortical membranes of newborn piglets--the role of nitric oxide. *Neuroscience Letters*, Vol. 372, No. 1-2, pp. 114-118, ISSN 0304-3940

Misra-Press A., Rim C. S., Yao H., Roberson M. S. & Stork P. J. (1995) A novel mitogen-activated protein kinase phosphatase. Structure, expression, and regulation. *Journal of Biological Chemistry*, Vol. 270, No. 24, pp. 14587-14596, ISSN 0021-9258

Muda M., Boschert U., Dickinson R., Martinou J. C., Martinou I., Camps M., Schlegel W. & Arkinstall S. (1996a) MKP-3, a novel cytosolic protein-tyrosine phosphatase that exemplifies a new class of mitogen-activated protein kinase phosphatase. *Journal of Biological Chemistry*, Vol. 271, No. 8, pp. 4319-4326, ISSN 0021-9258

Muda M., Theodosiou A., Rodrigues N., Boschert U., Camps M., Gillieron C., Davies K., Ashworth A. & Arkinstall S. (1996b) The dual specificity phosphatases M3/6 and MKP-3 are highly selective for inactivation of distinct mitogen-activated protein kinases. *Journal of Biological Chemistry*, Vol. 271, No. 44, pp. 27205-27208, ISSN 0021-9258

Mukherjee A. & Soto C. (2011) Role of calcineurin in neurodegeneration produced by misfolded proteins and endoplasmic reticulum stress. *Current Opinion in Cell Biology*, Vol. 23, No. 2, pp. 223-230, ISSN 0955-0674

Mukherjee S., Poddar R., Deb I. & Paul S. (2011) Dephosphorylation of specific sites in the KIS domain leads to ubiquitin-mediated degradation of the tyrosine phosphatase STEP. *Biochemical Journal*, doi:10.1042/BJ20110240, ISSN 0264-6021

Munoz J. J., Tarrega C., Blanco-Aparicio C. & Pulido R. (2003) Differential interaction of the tyrosine phosphatases PTP-SL, STEP and HePTP with the mitogen-activated protein kinases ERK1/2 and p38alpha is determined by a kinase specificity sequence and influenced by reducing agents. *Biochemical Journal*, Vol. 372, Pt 1, pp. 193-201, ISSN 0264-6021

Newton A. C. (2003) Regulation of the ABC kinases by phosphorylation: protein kinase C as a paradigm. *Biochemical Journal*, Vol. 370, Pt 2, pp. 361-371, ISSN 0264-6021

Nguyen T. H., Liu J. & Lombroso P. J. (2002) Striatal enriched phosphatase 61 dephosphorylates Fyn at phosphotyrosine 420. *Journal of Biological Chemistry*, Vol. 277, No. 27, pp. 24274-24279, ISSN 0021-9258

Nishi A., Bibb J. A., Matsuyama S., Hamada M., Higashi H., Nairn A. C. & Greengard P. (2002) Regulation of DARPP-32 dephosphorylation at PKA- and Cdk5-sites by NMDA and AMPA receptors: distinct roles of calcineurin and protein phosphatase-2A. *Journal of Neurochemistry*, Vol. 81, No. 4, pp. 832-841, ISSN 0022-3042

Ohtsuki T., Matsumoto M., Kitagawa K., Mabuchi T., Mandai K., Matsushita K., Kuwabara K., Tagaya M., Ogawa S., Ueda H., Kamada T. & Yanagihara T. (1996) Delayed neuronal death in ischemic hippocampus involves stimulation of protein tyrosine phosphorylation. *American Journal of Physiology*, Vol. 271, No. 4 Pt 1, pp. C1085-C1097, ISSN 0363-6143

Okamoto S., Pouladi M. A., Talantova M., Yao D., Xia P., Ehrnhoefer D. E., Zaidi R., Clemente A., Kaul M., Graham R. K., Zhang D., Vincent Chen H. S., Tong G., Hayden M. R. & Lipton S. A. (2009) Balance between synaptic versus extrasynaptic NMDA receptor activity influences inclusions and neurotoxicity of mutant huntingtin. *Nature Medicine*, Vol. 15, No. 12, pp. 1407-1413, ISSN 1078-8956

Olney J. W. & Sharpe L. G. (1969) Brain lesions in an infant rhesus monkey treated with monsodium glutamate. *Science*, Vol. 166, No. 903, pp. 386-388, ISSN 0036-8075

Oyama T., Goto S., Nishi T., Sato K., Yamada K., Yoshikawa M. & Ushio Y. (1995) Immunocytochemical localization of the striatal enriched protein tyrosine phosphatase in the rat striatum: a light and electron microscopic study with a complementary DNA-generated polyclonal antibody. *Neuroscience*, Vol. 69, No. 3, pp. 869-880, ISSN 0306-4522

Ozawa S., Kamiya H. & Tsuzuki K. (1998) Glutamate receptors in the mammalian central nervous system. *Progress in Neurobiology*, Vol. 54, No. 5, pp. 581-618, ISSN 0301-0082

Pardo R., Colin E., Regulier E., Aebischer P., Deglon N., Humbert S. & Saudou F. (2006) Inhibition of calcineurin by FK506 protects against polyglutamine-huntingtin toxicity through an increase of huntingtin phosphorylation at S421. *Journal of Neuroscience*, Vol. 26, No. 5, pp. 1635-1645, ISSN 0270-6474

Paul R., Zhang Z. G., Eliceiri B. P., Jiang Q., Boccia A. D., Zhang R. L., Chopp M. & Cheresh D. A. (2001) Src deficiency or blockade of Src activity in mice provides cerebral protection following stroke. *Nature Medicine*, Vol. 7, No. 2, pp. 222-227, ISSN 1078-8956

Paul S. & Lombroso P. J. (2003) Receptor and nonreceptor protein tyrosine phosphatases in the nervous system. *Cellular and Molecular Life Sciences*, Vol. 60, No. 11, pp. 2465-2482, ISSN 1421-682X

Paul S., Nairn A. C., Wang P. & Lombroso P. J. (2003) NMDA-mediated activation of the tyrosine phosphatase STEP regulates the duration of ERK signaling. *Nature Neuroscience*, Vol. 6, No. 1, pp. 34-42, ISSN 1097-6256

Paul S., Snyder G. L., Yokakura H., Picciotto M. R., Nairn A. C. & Lombroso P. J. (2000) The Dopamine/D1 receptor mediates the phosphorylation and inactivation of the protein tyrosine phosphatase STEP via a PKA-dependent pathway. *Journal of Neuroscience*, Vol. 20, No. 15, pp. 5630-5638, ISSN 0270-6474

Pearce L. R., Komander D. & Alessi D. R. (2010) The nuts and bolts of AGC protein kinases. *Nature Reviews Molecular Cell Biology*, Vol. 11, No. 1, pp. 9-22, ISSN 1471-0080

Pelkey K. A., Askalan R., Paul S., Kalia L. V., Nguyen T. H., Pitcher G. M., Salter M. W. & Lombroso P. J. (2002) Tyrosine phosphatase STEP is a tonic brake on induction of long-term potentiation. *Neuron*, Vol. 34, No. 1, pp. 127-138, ISSN 0896-6273

Perez-Navarro E., Canals J. M., Gines S. & Alberch J. (2006) Cellular and molecular mechanisms involved in the selective vulnerability of striatal projection neurons in Huntington's disease. *Histology and Histopathology*, Vol. 21, No. 11, pp. 1217-1232, ISSN 0213- 3911

Pineda J. R., Pardo R., Zala D., Yu H., Humbert S. & Saudou F. (2009) Genetic and pharmacological inhibition of calcineurin corrects the BDNF transport defect in Huntington's disease. *Molecular Brain*, Vol. 2, pp. 33, ISSN 1756-6606

Poddar R., Deb I., Mukherjee S. & Paul S. (2010) NR2B-NMDA receptor mediated modulation of the tyrosine phosphatase STEP regulates glutamate induced neuronal cell death. *Journal of Neurochemistry*, Vol. 115, No. 6, pp. 1350-1362, ISSN 0022-3042

Pulido R., Zuniga A. & Ullrich A. (1998) PTP-SL and STEP protein tyrosine phosphatases regulate the activation of the extracellular signal-regulated kinases ERK1 and ERK2 by association through a kinase interaction motif. *EMBO Journal*, Vol. 17, No. 24, pp. 7337-7350, ISSN 0261-4189

Rangone H., Poizat G., Troncoso J., Ross C. A., MacDonald M. E., Saudou F. & Humbert S. (2004) The serum- and glucocorticoid-induced kinase SGK inhibits mutant huntingtin-induced toxicity by phosphorylating serine 421 of huntingtin. *European Journal of Neuroscience*, Vol. 19, No. 2, pp. 273-279, ISSN 0953-816X

Ravikumar B., Vacher C., Berger Z., Davies J. E., Luo S., Oroz L. G., Scaravilli F., Easton D. F., Duden R., O'Kane C. J. & Rubinsztein D. C. (2004) Inhibition of mTOR induces autophagy and reduces toxicity of polyglutamine expansions in fly and mouse models of Huntington disease. *Nature Genetics*, Vol. 36, No. 6, pp. 585-595, ISSN 1061-4036

Renna M., Jimenez-Sanchez M., Sarkar S. & Rubinsztein D. C. (2010) Chemical inducers of autophagy that enhance the clearance of mutant proteins in neurodegenerative diseases. *Journal of Biological Chemistry*, Vol. 285, No. 15, pp. 11061-11067, ISSN 0021-9258

Rosenstock T. R., Duarte A. I. & Rego A. C. (2010) Mitochondrial-associated metabolic changes and neurodegeneration in Huntington's disease - from clinical features to the bench. *Current Drug Targets*, Vol. 11, No. 10, pp. 1218-1236, ISSN 1389-4501

Runne H., Regulier E., Kuhn A., Zala D., Gokce O., Perrin V., Sick B., Aebischer P., Deglon N. & Luthi-Carter R. (2008) Dysregulation of gene expression in primary neuron models of Huntington's disease shows that polyglutamine-related effects on the striatal transcriptome may not be dependent on brain circuitry. *Journal of Neuroscience*, Vol. 28, No. 39, pp. 9723-9731, ISSN 0270-6474

Saavedra A., Garcia-Martinez J. M., Xifro X., Giralt A., Torres-Peraza J. F., Canals J. M., Diaz-Hernandez M., Lucas J. J., Alberch J. & Perez-Navarro E. (2010) PH domain leucine-rich repeat protein phosphatase 1 contributes to maintain the activation of the PI3K/Akt pro-survival pathway in Huntington's disease striatum. *Cell Death & Differentiation*, Vol. 17, No. 2, pp. 324-335, ISSN 1350-9047

Saavedra A., Giralt A., Rue L., Xifro X., Xu J., Ortega Z., Lucas J. J., Lombroso P. J., Alberch J. & Perez-Navarro E. (2011) Striatal-enriched protein tyrosine phosphatase expression and activity in Huntington's disease: a STEP in the resistance to excitotoxicity. *Journal of Neuroscience*, Vol. 31, No. 22, pp. 8150-8162, ISSN 0270-6474

Sancak Y., Peterson T. R., Shaul Y. D., Lindquist R. A., Thoreen C. C., Bar-Peled L. & Sabatini D. M. (2008) The Rag GTPases bind raptor and mediate amino acid signaling to mTORC1. *Science*, Vol. 320, No. 5882, pp. 1496-1501, ISSN 0036-8075

Sanna P. P., Berton F., Cammalleri M., Tallent M. K., Siggins G. R., Bloom F. E., Francesconi W. (2000) A role for Src kinase in spontaneous epileptiform activity in the CA3 region of the hippocampus. *Proceedings of the National Academy of Sciences USA*, Vol. 97, No. 15, pp. 8653-8657, ISSN 0027-8424

Sarbassov D. D., Guertin D. A., Ali S. M. & Sabatini D. M. (2005) Phosphorylation and regulation of Akt/PKB by the rictor-mTOR complex. *Science*, Vol. 307, No. 5712, pp. 1098-1101, ISSN 0036-8075

Sarbassov D. D., Ali S. M., Kim D. H., Guertin D. A., Latek R. R., Erdjument-Bromage H., Tempst P. & Sabatini D. M. (2004) Rictor, a novel binding partner of mTOR, defines a rapamycin-insensitive and raptor-independent pathway that regulates the cytoskeleton. *Current Biology*, Vol. 14, No. 14, pp. 1296-1302, ISSN 0960-9822

Schilling B., Gafni J., Torcassi C., Cong X., Row R. H., LaFevre-Bernt M. A., Cusack M. P., Ratovitski T., Hirschhorn R., Ross C. A., Gibson B. W. & Ellerby L. M. (2006) Huntingtin phosphorylation sites mapped by mass spectrometry. Modulation of cleavage and toxicity. *Journal of Biological Chemistry*, Vol. 281, No. 33, pp. 23686-23697, ISSN 0021-9258

Seta K. A., Kim R., Kim H. W., Millhorn D. E. & Beitner-Johnson D. (2001) Hypoxia-induced regulation of MAPK phosphatase-1 as identified by subtractive suppression hybridization and cDNA microarray analysis. *Journal of Biological Chemistry*, Vol. 276, No. 48, pp. 44405-44412, ISSN 0021-9258

Shamloo M., Soriano L., Wieloch T., Nikolich K., Urfer R. & Oksenberg D. (2005) Death-associated protein kinase is activated by dephosphorylation in response to cerebral ischemia. *Journal of Biological Chemistry*, Vol. 280, No. 51, pp. 42290-42299, ISSN 0021-9258

Sharma E., Zhao F., Bult A. & Lombroso P. J. (1995) Identification of two alternatively spliced transcripts of STEP: a subfamily of brain-enriched protein tyrosine phosphatases. *Brain Research Molecular Brain Research*, Vol. 32, No. 1, pp. 87-93, ISSN 0169-328X

Shehadeh J., Fernandes H. B., Zeron Mullins M. M., Graham R. K., Leavitt B. R., Hayden M. R. & Raymond L. A. (2006) Striatal neuronal apoptosis is preferentially enhanced by NMDA receptor activation in YAC transgenic mouse model of Huntington disease. *Neurobiology of Disease*, Vol. 21, No. 2, pp. 392-403, ISSN 0969-9961

Sheppeck J. E., Gauss C. M. & Chamberlin A. R. (1997) Inhibition of the Ser-Thr phosphatases PP1 and PP2A by naturally occurring toxins. *Bioorganic & Medicinal Chemistry*, Vol. 5, No. 9, pp. 1739-1750, ISSN 0968-0896

Shi Y. (2009) Serine/threonine phosphatases: mechanism through structure. *Cell*, Vol. 139, No. 3, pp. 468-484, ISSN 0092-8674

Shibasaki F. & McKeon F. (1995) Calcineurin functions in Ca(2+)-activated cell death in mammalian cells. *Journal of Cell Biology*, Vol. 131, No. 3, pp. 735-743, ISSN 0021-9525

Shibasaki F., Hallin U. & Uchino H. (2002) Calcineurin as a multifunctional regulator. *Journal of Biochemistry*, Vol. 131, No. 1, pp. 1-15, ISSN 0021-924X

Shields S. M., Ingebritsen T. S. & Kelly P. T. (1985) Identification of protein phosphatase 1 in synaptic junctions: dephosphorylation of endogenous calmodulin-dependent kinase II and synapse-enriched phosphoproteins. *Journal of Neuroscience*, Vol. 5, No. 12, pp. 3414-3422, ISSN 0270-6474

Shimizu K., Okada M., Takano A. & Nagai K. (1999) SCOP, a novel gene product expressed in a circadian manner in rat suprachiasmatic nucleus. *FEBS Letters*, Vol. 458, No. 3, pp. 363-369, ISSN 0014-5793

Shimizu K., Okada M., Nagai K. & Fukada Y. (2003) Suprachiasmatic nucleus circadian oscillatory protein, a novel binding partner of K-Ras in the membrane rafts, negatively regulates MAPK pathway. *Journal of Biological Chemistry*, Vol. 278, No. 17, pp. 14920-14925, ISSN 0021-9258

Shimizu K., Phan T., Mansuy I. M. & Storm D. R. (2007) Proteolytic degradation of SCOP in the hippocampus contributes to activation of MAP kinase and memory. *Cell*, Vol. 128, No. 6, pp. 1219-1229, ISSN 0092-8674

Shin D. Y., Ishibashi T., Choi T. S., Chung E., Chung I. Y., Aaronson S. A. & Bottaro D. P. (1997) A novel human ERK phosphatase regulates H-ras and v-raf signal transduction. *Oncogene*, Vol. 14, No. 22, pp. 2633-2639, ISSN 0950-9232

Snyder E. M., Nong Y., Almeida C. G., Paul S., Moran T., Choi E. Y., Nairn A. C., Salter M. W., Lombroso P. J., Gouras G. K. & Greengard P. (2005) Regulation of NMDA receptor trafficking by amyloid-beta. *Nature Neuroscience*, Vol. 8, No. 8, pp. 1051-1058, ISSN 1097-6256

Song C., Zhang Y., Parsons C. G. & Liu Y. F. (2003) Expression of polyglutamine-expanded huntingtin induces tyrosine phosphorylation of N-methyl-D-aspartate receptors. *Journal of Biological Chemistry*, Vol. 278, No. 35, pp. 33364-33369, ISSN 0021-9258

Springer J. E., Azbill R. D., Nottingham S. A. & Kennedy S. E. (2000) Calcineurin-mediated BAD dephosphorylation activates the caspase-3 apoptotic cascade in traumatic spinal cord injury. *Journal of Neuroscience*, Vol. 20, No. 19, pp. 7246-7251, ISSN 0270-6474

Stoof J. C. & Kebabian J. W. (1981) Opposing roles for D-1 and D-2 dopamine receptors in efflux of cyclic AMP from rat neostriatum. *Nature*, Vol. 294, No. 5839, pp. 366-368, ISSN 0028-0836

Strack S., Choi S., Lovinger D. M. & Colbran R. J. (1997) Translocation of autophosphorylated calcium/calmodulin-dependent protein kinase II to the postsynaptic density. *Journal of Biological Chemistry*, Vol. 272, No. 21, pp. 13467-13470, ISSN 0021-9258

Sun Y., Savanenin A., Reddy P. H. & Liu Y. F. (2001) Polyglutamine-expanded huntingtin promotes sensitization of N-methyl-D-aspartate receptors via post-synaptic density 95. *Journal of Biological Chemistry*, Vol. 276, No. 27, pp. 24713-24718, ISSN 0021-9258

Takaki M., Ujike H., Kodama M., Takehisa Y., Nakata K. & Kuroda S. (2001) Two kinds of mitogen-activated protein kinase phosphatases, MKP-1 and MKP-3, are differentially activated by acute and chronic methamphetamine treatment in the rat brain. *Journal of Neurochemistry*, Vol. 79, No. 3, pp. 679-688, ISSN 0022-3042

Tang T. S., Slow E., Lupu V., Stavrovskaya I. G., Sugimori M., Llinas R., Kristal B. S., Hayden M. R. & Bezprozvanny I. (2005) Disturbed Ca2+ signaling and apoptosis of medium spiny neurons in Huntington's disease. *Proceedings of the National Academy of Sciences USA* , Vol. 102, No. 7, pp. 2602-2607, ISSN 0027-8424

Thompson L. M., Aiken C. T., Kaltenbach L. S., Agrawal N., Illes K., Khoshnan A., Martinez-Vincente M., Arrasate M., O'Rourke J. G., Khashwji H., Lukacsovich T., Zhu Y. Z., Lau A. L., Massey A., Hayden M. R., Zeitlin S. O., Finkbeiner S., Green K. N., LaFerla F. M., Bates G., Huang L., Patterson P. H., Lo D. C., Cuervo A. M., Marsh J. L. & Steffan J. S. (2009) IKK phosphorylates Huntingtin and targets it for degradation by the proteasome and lysosome. *Journal of Cell Biology*, Vol. 187, No. 7, pp. 1083-1099, ISSN 0021-9525

Tian Q. & Wang J. (2002) Role of serine/threonine protein phosphatase in Alzheimer's disease. *Neurosignals*, Vol. 11, No. 5, pp. 262-269, ISSN 1424-862X

Torres-Peraza J. F., Giralt A., Garcia-Martinez J. M., Pedrosa E., Canals J. M. & Alberch J. (2008) Disruption of striatal glutamatergic transmission induced by mutant huntingtin involves remodeling of both postsynaptic density and NMDA receptor signaling. *Neurobiology of Disease*, Vol. 29, No. 3, pp. 409-421, ISSN 0969-9961

Valjent E., Pascoli V., Svenningsson P., Paul S., Enslen H., Corvol J. C., Stipanovich A., Caboche J., Lombroso P. J., Nairn A. C., Greengard P., Herve D. & Girault J. A. (2005) Regulation of a protein phosphatase cascade allows convergent dopamine and glutamate signals to activate ERK in the striatum. *Proceedings of the National Academy of Sciences USA* , Vol. 102, No. 2, pp. 491-496, ISSN 0027-8424

Walaas S. I. & Greengard P. (1991) Protein phosphorylation and neuronal function. *Pharmacological Reviews*, Vol. 43, No. 3, pp. 299-349, ISSN: 0031-6997

Wang H. G., Pathan N., Ethell I. M., Krajewski S., Yamaguchi Y., Shibasaki F., McKeon F., Bobo T., Franke T. F. & Reed J. C. (1999) Ca2+-induced apoptosis through calcineurin dephosphorylation of BAD. *Science*, Vol. 284, No. 5412, pp. 339-343, ISSN 0036-8075

Wang L. Y., Orser B. A., Brautigan D. L. & MacDonald J. F. (1994) Regulation of NMDA receptors in cultured hippocampal neurons by protein phosphatases 1 and 2A. *Nature*, Vol. 369, No. 6477, pp. 230-232, ISSN 0028-0836

Wang X., Zhong P. & Yan Z. (2002) Dopamine D4 receptors modulate GABAergic signaling in pyramidal neurons of prefrontal cortex. *Journal of Neuroscience*, Vol. 22, No. 21, pp. 9185-9193, ISSN 0270-6474

Warby S. C., Chan E. Y., Metzler M., Gan L., Singaraja R. R., Crocker S. F., Robertson H. A. & Hayden M. R. (2005) Huntingtin phosphorylation on serine 421 is significantly reduced in the striatum and by polyglutamine expansion in vivo. *Human Molecular Genetics*, Vol. 14, No. 11, pp. 1569-1577, ISSN 0964-6906

Warby S. C., Doty C. N., Graham R. K., Shively J., Singaraja R. R. & Hayden M. R. (2009) Phosphorylation of huntingtin reduces the accumulation of its nuclear fragments. *Molecular and Cellular Neuroscience*. Vol. 40, No. 2, pp. 121-127, ISSN 1044-7431

Weir D. W., Sturrock A. & Leavitt B. R. (2011) Development of biomarkers for Huntington's disease. *Lancet Neurology*, Vol. 10, No. 6, pp. 573-590, ISSN 1474-4422

Wera S. & Neyts J. (1994) Calcineurin as a possible new target for treatment of Parkinson's disease. *Medical Hypotheses*, Vol. 43, No. 3, pp. 132-134, ISSN 0306-9877

Westphal R. S., Tavalin S. J., Lin J. W., Alto N. M., Fraser I. D., Langeberg L. K., Sheng M. & Scott J. D. (1999) Regulation of NMDA receptors by an associated phosphatase-kinase signaling complex. *Science*, Vol. 285, No. 5424, pp. 93-96, ISSN 0036-8075

Wood A. M. & Bristow D. R. (1998) N-methyl-D-aspartate receptor desensitisation is neuroprotective by inhibiting glutamate-induced apoptotic-like death. *Journal of Neurochemistry*, Vol. 70, No. 2, pp. 677-687, ISSN 0022-3042

Wu H. Y., Tomizawa K., Oda Y., Wei F. Y., Lu Y. F., Matsushita M., Li S. T., Moriwaki A. & Matsui H. (2004) Critical role of calpain-mediated cleavage of calcineurin in excitotoxic neurodegeneration. *Journal of Biological Chemistry*, Vol. 279, No. 6, pp. 4929-4940, ISSN 0021-9258

Wu Z. L., O'Kane T. M., Scott R. W., Savage M. J. & Bozyczko-Coyne D. (2002) Protein tyrosine phosphatases are up-regulated and participate in cell death induced by polyglutamine expansion. *Journal of Biological Chemistry*, Vol. 277, No. 46, pp. 44208-44213, ISSN 0021-9258

Xifro X., Garcia-Martinez J. M., Del Toro D., Alberch J. & Perez-Navarro E. (2008) Calcineurin is involved in the early activation of NMDA-mediated cell death in mutant huntingtin knock-in striatal cells. *Journal of Neurochemistry*, Vol. 105, No. 5, pp. 1596-1612, ISSN 0022-3042

Xifro X., Giralt A., Saavedra A., Garcia-Martinez J. M., Diaz-Hernandez M., Lucas J. J., Alberch J. & Perez-Navarro E. (2009) Reduced calcineurin protein levels and activity in exon-1 mouse models of Huntington's disease: role in excitotoxicity. *Neurobiology of Disease*, Vol. 36, No. 3, pp. 461-469, ISSN 0969-9961

Xu J., Kurup P., Zhang Y., Goebel-Goody S. M., Wu P. H., Hawasli A. H., Baum M. L., Bibb J. A. & Lombroso P. J. (2009) Extrasynaptic NMDA receptors couple preferentially to excitotoxicity via calpain-mediated cleavage of STEP. *Journal of Neuroscience*, Vol. 29, No. 29, pp. 9330-9343, ISSN 0270-6474

Yamanaka T., Miyazaki H., Oyama F., Kurosawa M., Washizu C., Doi H., Nukina N. (2008) Mutant Huntingtin reduces HSP70 expression through the sequestration of NF-Y transcription factor. *EMBO Journal*, Vol. 27, No. 6, pp. 827-839., ISSN 0261-4189

Zala D., Colin E., Rangone H., Liot G., Humbert S. & Saudou F. (2008) Phosphorylation of mutant huntingtin at S421 restores anterograde and retrograde transport in neurons. *Human Molecular Genetics*, Vol. 17, No. 24, pp. 3837-3846, ISSN 0964-6906

Zeron M. M., Chen N., Moshaver A., Lee A. T., Wellington C. L., Hayden M. R. & Raymond L. A. (2001) Mutant huntingtin enhances excitotoxic cell death. *Molecular and Cellular Neuroscience*, Vol. 17, No. 1, pp. 41-53, ISSN 1044-7431

Zeron M. M., Hansson O., Chen N., Wellington C. L., Leavitt B. R., Brundin P., Hayden M. R. & Raymond L. A. (2002) Increased sensitivity to N-methyl-D-aspartate receptor-mediated excitotoxicity in a mouse model of Huntington's disease. *Neuron*, Vol. 33, No. 6, pp. 849-860, ISSN 0896-6273

Zhang Y., Venkitaramani D. V., Gladding C. M., Kurup P., Molnar E., Collingridge G. L. & Lombroso P. J. (2008) The tyrosine phosphatase STEP mediates AMPA receptor endocytosis after metabotropic glutamate receptor stimulation. *Journal of Neuroscience*, Vol. 28, No. 42, pp. 10561-10566, ISSN 0270-6474

Zhang Y., Kurup P., Xu J., Carty N., Fernandez S. M., Nygaard H. B., Pittenger C., Greengard P., Strittmatter S. M., Nairn A. C. & Lombroso P. J. (2010) Genetic reduction of striatal-enriched tyrosine phosphatase (STEP) reverses cognitive and cellular deficits in an Alzheimer's disease mouse model. *Proceedings of the National Academy of Sciences USA* , Vol. 107, No. 44, pp. 19014-19019, ISSN 0027-8424

Zhang Z. Y. (2002) Protein tyrosine phosphatases: structure and function, substrate specificity, and inhibitor development. *Annual Review of Pharmacology and Toxicolology*, Vol. 42, No. pp. 209-234, ISSN 0362-1642

21

BDNF in Huntington's Disease: Role in Pathogenesis and Treatment

Maryna Baydyuk and Baoji Xu
Georgetown University
USA

1. Introduction

Huntington's Disease (HD) is a neurodegenerative disorder characterized by motor, cognitive, and psychiatric abnormalities, and is inherited in an autosomal dominant fashion (Borrell-Pages et al., 2006). HD is caused by the CAG trinucleotide repeat expansion in the first exon of the gene encoding huntingtin (htt) (The Huntington's Disease Collaborative Research Group, 1993). This mutation is translated into a polyglutamine (poly Q) stretch near the amino terminus of htt, which results in a toxic gain of function (Gusella & MacDonald, 2000). Although htt is widely expressed in the human body and its mutation is not tissue-specific, the striatum is preferentially affected. The pathological changes in the striatum develop first in the caudate nucleus and then in the putamen, causing a 50-60% neuronal loss in these areas (Mann et al., 1993; Vonsattel & DiFiglia, 1998). Striatal atrophy is due to selective degeneration of medium-sized spiny neurons (MSNs), which comprise 90% of striatal neurons. Interestingly, the MSNs of the indirect pathway, responsible for inhibition of involuntary movement, are preferentially affected, causing motor symptoms of HD such as uncontrollable sequence of movements called "chorea". In the course of the disease, atrophy spreads to other brain regions, including the cerebral cortex, the globus pallidus (GP), and the thalamus (Mann et al., 1993).

The mechanism behind selective degeneration of striatal neurons remains to be elucidated, but it has been suggested that reduced trophic support renders striatal neurons more vulnerable to the toxic actions of mutant htt. Numerous *in vitro* and *in vivo* studies have shown that striatal neurons require brain-derived neurotrophic factor (BDNF) for their survival and function. A deficiency in BDNF-mediated signaling alone is sufficient to cause dendritic abnormalities and neuronal loss in the cerebral cortex and striatum (Baquet et al., 2004; Gorski et al., 2003). Moreover, reduced levels of striatal BDNF were detected in both HD animal models (Apostol et al., 2008; Gharami et al., 2008; Spires et al., 2004) and HD patients (Ferrer et al., 2000). These observations raise the possibility that reduced levels of striatal BDNF may significantly contribute to HD pathogenesis. In support of this view, the progression of HD is accelerated in *Bdnf* heterozygous mice (Canals et al., 2004). Furthermore, alterations of gene expression profile in the striatum have been shown to be similar in HD patients and mice in which the *Bdnf* gene is deleted in the cerebral cortex (Strand et al., 2007). Importantly, the receptor for BDNF, tropomyosin related kinase B (TrkB), is preferentially expressed in striatal MSNs of the indirect pathway, which may explain why this population of neurons is degenerated first in HD patients (Baydyuk et al., 2011).

BDNF found in the striatum is synthesized and anterogradely transported from the cell bodies located in the cerebral cortex, substantia nigra pars compacta, amygdala, and thalamus (Altar et al., 1997; Baquet et al., 2004). Since striatum does not produce BDNF but depends on it for proper function, abnormalities in anterograde transport and reduced gene expression in brain regions supplying BDNF to the striatum might contribute to neuronal dysfunction and death seen in HD (Gauthier et al., 2004; Zuccato et al., 2001). In light of these findings, efforts have been made to test whether increasing BDNF expression represents a valuable strategy for treatment of Huntington's Disease. Indeed, increasing striatal BDNF levels by a transgene, viral delivery, or stimulations that induce *Bdnf* gene expression have been shown to improve disease phenotypes in several HD mouse models (Cho et al., 2007; Gharami et al., 2008; Xie et al., 2010).

This book chapter will review these recent discoveries regarding the role of BDNF deficiency in the pathogenesis of HD and BDNF as a potential therapeutic agent for HD.

2. Wild-type but not mutant htt promotes *Bdnf* gene expression

The pathogenic mechanisms induced by mutant htt are not fully understood but are thought to involve the gain of toxic function and/or the loss of normal activities. Htt is a ubiquitously expressed protein, highly enriched in the brain (DiFiglia et al., 1995). While its exact functions are unknown, htt has been shown to be essential during embryogenesis and possess anti-apoptotic properties during adulthood (Dragatsis et al., 2000; O'Kusky et al., 1999; Rigamonti et al., 2000). Several mechanisms have been proposed for the neurodegenerative effect of the expanded polyQ tract in htt (Rubinsztein, 2002). Discovery of neuronal intranuclear inclusions in HD patients and HD mouse models led to the hypothesis that these protein aggregates might cause neuronal death. However, studies in mice and cultured neurons indicate that the formation of nuclear inclusions does not correlate with neuronal death (Hodgson et al., 1999; Kim et al., 1999; Laforet et al., 2001; Rubinsztein, 2002; Saudou et al., 1998). At present, the molecular basis for the toxic gain of function associated with mutant htt remains unclear.

The loss of a beneficial activity of normal htt has been proposed to contribute to the pathogenesis of HD. Wild type htt is known to regulate transcription of multiple genes, among which the gene encoding BDNF has received special attention (Zuccato et al., 2001). BDNF is a member of the neurotrophin family, which also includes nerve growth factor (NGF), neurotrophin-3 (NT-3) and neurotrophin-4/5 (NT-4/5) (Reichardt, 2006). BDNF has been shown to promote neuronal growth, survival, and differentiation by activating its TrkB receptor tyrosine kinase (Patapoutian & Reichardt, 2001). Upon binding to BDNF, activated full-length TrkB triggers multiple intracellular signaling cascades through protein-protein interactions (Chao, 2003). TrkB-initiated signaling pathways have been shown to promote cell survival by up-regulating the activity of survival genes and inhibiting the function of the proteins that lead to programmed cell death (Bhave et al., 1999; Encinas et al., 1999; Yamada et al., 2001). TrkB signaling pathways can also mediate various synaptic reorganization processes, including formation and maintenance of dendrites and dendritic spines (McAllister et al., 1999). In support of these in *vitro* observations, deletion of either the *TrkB* or *Bdnf* gene leads to cell atrophy, dendritic degeneration, and neuronal loss, as shown in the excitatory neurons of the dorsal forebrain (Gorski et al., 2003; Xu et al., 2000).

In rodents and humans, the *Bdnf* gene is transcribed from at least 8 discrete promoters, producing many different *Bdnf* mRNA species that encode the same protein (Aid et al., 2007). The different transcripts are generated in different tissues in a stimulus- and development-specific manner and may have differential subcellular localizations and targets (Metsis et al., 1993; Pattabiraman et al., 2005; Timmusk et al., 1993). Zuccato et al. have shown that wild-type htt enhances *Bdnf* transcription from promoter II, whereas mutant htt suppresses *Bdnf* transcription from promoter II as well as two other *Bdnf* promoters in cultured cells and the cerebral cortex of YAC72 transgenic mice expressing mutant htt with an expanded tract of 72 glutamines (Zuccato et al., 2001). The same group further investigated the mechanism underlying *Bdnf* gene regulation by wild-type and mutant htt, and found that wild-type htt promotes transcription of promoter II by sequestering the repressor element-1 transcription factor/neuron restrictive silencer factor (REST/NRSF) in the cytoplasm, thereby freeing the nucleus of the inhibitory complex and allowing transcription to occur (Zuccato et al., 2003). In contrast, mutant htt is unable to retain REST/NRSF complex in the cytoplasm, leading to aberrant accumulation of REST/NRSF in the nucleus and inhibition of *Bdnf* gene transcription. Interestingly, the effect of htt on *Bdnf* gene expression in cortical neurons is specific since the protein does not affect expression of two other neurotrophins, NGF and NT-3, in cortical neurons (Zuccato et al., 2003). In agreement with these findings, levels of *Bdnf* mRNA are reduced in the cerebral cortices of HD patients (Zuccato et al., 2008). It also has been shown that lower levels of BDNF are associated with higher numbers of CAG repeats in mutant *htt* alleles and correlate with the severity of the disease (Ciammola et al., 2007). It is important to note, however, that this autopsy data should be interpreted with caution. As mutant htt alters electrophysiological properties of cortical neurons (Cummings et al., 2009) and neuronal activity regulates *Bdnf* gene expression (Aid et al., 2007), we should not exclude the possibility that the observed reduction in cortical *Bdnf* mRNA levels may be secondary to neurodegeneration.

Although the contribution of suppressed *Bdnf* transcription to reduced BDNF levels in HD striatum is a widely accepted hypothesis, there are studies that contradict this idea. A reduction in *Bdnf* transcription would predict reduced levels of BDNF protein in cerebral cortices of both HD patients and mouse models. This prediction has been confirmed in one study (Zuccato et al., 2008) but not in another study (Gauthier et al., 2004) using post-mortem tissues from multiple control subjects and HD patients. Furthermore, *in situ hybridization* revealed normal levels of cortical *Bdnf* mRNA in aging YAC128 mice that express the whole human *htt* gene with 128 CAG repeats (Xie et al., 2010). Consistent with this observation, levels of cortical BDNF in YAC128 mice and R6/1 mice, another HD model, were found to be similar to those in WT mice (Gharami et al., 2008; Xie et al., 2010). Further analysis of *Bdnf* gene expression in other HD mouse models at various ages is necessary to clear the discrepancy. Despite the discrepancy in determining cortical *Bdnf* mRNA levels, a significant reduction in striatal BDNF has been consistently shown in both HD patients and animal models.

3. Htt promotes axonal BDNF transport

Studies by Gauthier et al. indicate that in addition to controlling *Bdnf* mRNA production in the cortex, wild-type htt may also regulate BDNF transport along the corticostriatal axes (Gauthier et al., 2004). The idea that htt is involved in intracellular trafficking arose from the subcellular localization of htt and its association with various proteins of molecular motors. Although present in the nucleus, htt is predominantly found in the cytoplasm, where it interacts with the huntingtin-associated protein-1 (HAP1), a protein involved in axonal transport via association with p150^{glued} subunit of dynactin, which is an essential part of the microtubule-based motor complex (Block-Galarza et al., 1997; Engelender et al., 1997; Li et al., 1998). While htt and other pathogenic polyQ-containing proteins have been shown to affect fast axonal transport (Gunawardena et al., 2003; Szebenyi et al., 2003), the link between deficient trafficking and selective neuronal degeneration has not been established. Gauthier and colleagues show that in normal condition wild-type htt promotes neuronal survival by facilitating the transport of BDNF-containing vesicles along microtubules. Consistent with a loss of function hypothesis, reduction in wild-type htt levels leads to attenuated BDNF trafficking. On the other hand, mutation in htt increases association of polyQ-htt and p150^{glued} via HAP1 and prevents efficient movement of BDNF-containing vesicles along microtubules. They also demonstrate that disruption of BDNF transport leads to decreased neurotrophic support and neurotoxicity, which can be rescued by wild-type htt (Gauthier et al., 2004).

BDNF synthesized in the cortex and transported to the striatum via corticostriatal projections provides the main support for survival of striatal neurons in the adult brain (Altar et al., 1997; Baquet et al., 2004). Importantly, it has been shown that BDNF levels are reduced in the striatum but not in the cortex of HD patients (Gauthier et al., 2004). These observations are in agreement with the notion that both mechanisms, suppressed *Bdnf* gene expression and deficient BDNF transport, might concomitantly contribute to reduced levels of BDNF in the striata of HD patients and mouse models, thus providing strong evidence for BDNF as a crucial factor in the pathogenesis of HD.

4. Possible effects of mutant htt on BDNF maturation

One additional cause for reduced neurotrophic support of striatal neurons in HD may be due to deficits in processing of proBDNF. The proBDNF is a 32-kDa precursor protein that is cleaved to generate the mature BDNF protein of 14 kDa. Whereas the mature form binds to its TrkB receptor and promotes neuronal survival, the uncleaved proBDNF preferentially activates the low-affinity neurotrophin receptor p75NTR (Hempstead 2006), which is a member of the tumor necrosis factor receptor subfamily and is known to induce neuronal death via apoptosis (Frade & Barde, 1998; Teng et al., 2005).

As discussed earlier, immunoblotting analysis has consistently found a reduction in striatal levels of mature BDNF in HD mouse models (Apostol et al., 2008; Gharami et al., 2008; Spires et al., 2004; Xie et al., 2010) and HD patients (Ferrer et al., 2000; Gauthier et al., 2004). However, studies using ELISA assays reported normal striatal levels of BDNF in R6/1 HD mice (Canals et al., 2004; Pang et al., 2006) and increased striatal levels of BDNF in YAC72 mice (Seo et al., 2008). As ELISA assays detect both mature BDNF and proBDNF, this discrepancy suggests that maturation of proBDNF may be impaired in the striatum, leading to accumulation of proBDNF in the striatum. Impaired proBDNF maturation could be detrimental to striatal neurons, because they lose the protective effect of mature BDNF via TrkB receptor signaling and are subject to the apoptotic effect of proBDNF via the p75NTR receptor (Teng et al., 2005). As current BDNF antibodies are still problematic in detecting proBDNF on immunoblots, utilization of tagged *Bdnf* knockin mice (Matsumoto et al., 2008; Yang et al., 2009) will help uncover the effect of mutant htt on BDNF maturation.

A recent study has suggested that proteins involved in proBDNF axonal transport might also play a role in BDNF maturation (Yang et al., 2011). As discussed earlier, htt facilitates axonal transport via its interaction with HAP1 and the mutation in htt alters the formation of proper motor complex and inhibits BDNF transport (Gauthier et al., 2004). Yang et al. found that proBDNF interacts with both HAP1 and sortilin, a binding partner of p75NTR, to form a complex that prevents proBDNF degradation and modulates proBDNF targeting to dendrites and axonal organelles. Furthermore, their data suggest that the complex of proBDNF-HAP1-sortilin might facilitate cleavage and release of mature BDNF (Yang et al., 2011). Thus, it is possible that mutant htt can affect both BDNF maturation and trafficking via its interaction with HAP1.

5. BDNF and selectivity of striatal degeneration

Striatal neurons are not uniformly affected in HD. Immunohistochemical studies using tissues from HD patients have shown a greater decrease in the number of neurons co-expressing the dopamine receptor D2 (Drd2) and enkephalin (Enk) (Reiner et al., 1988). These neurons comprise the indirect pathway, projecting to the external segment of globus pallidus. The indirect pathway acts to terminate basal ganglia associated movements or suppress unwanted sequences of movements (Bolam et al., 2000). Hence, the loss of the indirect pathway neurons leads to disinhibition of the thalamus and increased facilitation of the motor cortex, producing hyperkinesias in HD patients (Calabresi et al., 1996). On the other hand, the direct pathway neurons co-expressing the dopamine receptor D1 (Drd1) and substance P (SP) are less affected in HD. In contrast, striatal interneurons containing acetylcholine, somatostatin/neuropeptide Y, or parvalbumin are spared in patients with HD; a striking phenomenon considering the fact that these cell populations comprise only 5% of striatal neurons (Ferrante et al., 1987a; Ferrante et al., 1987b). These findings suggest that the Drd2/Enk neurons of the striatum may be more vulnerable to the deleterious effects of mutated htt. However, the precise mechanism of this selective neuronal loss is unknown.

Genetic studies using HD mouse models with altered levels of BDNF indicate that BDNF plays an important role in this specificity of degeneration. Depletion of BDNF using heterozygous or forebrain-specific knockout mice results in alterations of striatal gene expression profiling that more closely recapitulates human HD than any other HD models (Strand et al., 2007). Deletion of one copy of the *Bdnf* gene in R6/1 HD mice resulted in early onset of the disease, more severe motor dysfunction, and led to a significant and selective loss of Drd2/Enk striatal neurons (Canals et al., 2004). Data originated in our laboratory show that the loss of striatal neurons was associated with reduced levels of mRNAs for both Enk and Drd2 in YAC128 HD mice (Xie et al., 2010), indicating selective degeneration of striatal neurons in the indirect pathway. Our recent data suggest that selective vulnerability of striatal neurons in the indirect pathway is due to differential expression of the TrkB receptor among striatal neurons. We found that the majority of the TrkB receptor was localized in striatal neurons of the indirect pathway in the adult mouse brain and deletion of TrkB receptor in the developing striatum caused selective loss of this neuronal population (Baydyuk et al., 2011). Together, all these findings indicate that a decrease in striatal BDNF can lead to dysfunction and death of MSNs in the indirect pathway, producing severe motor phenotype as seen in HD. Hence, restoring BDNF levels in the striatum may delay or even stop disease progression.

6. Increasing BDNF expression rescues disease phenotypes in HD mouse models

The evidence discussed above clearly indicates that the reduction in striatal BDNF levels plays a pivotal role in the pathogenesis of HD. Therefore, it is not surprising that efforts have been made to examine whether increasing BDNF expression represents a valuable strategy for treatment of Huntington's Disease. Indeed, increasing striatal BDNF levels via stimulation that induces *Bdnf* gene expression (Duan et al., 2003; Peng et al., 2008; Simmons et al., 2009; Spires et al., 2004) or by viral delivery (Cho et al., 2007) has been shown to improve disease phenotypes in several HD mouse models.

Early symptoms of HD are manifested by cognitive and memory deficits that start before characteristic motor dysfunction (Ho et al., 2003; Lawrence et al., 1998). In HD mouse models, impaired learning and memory, measured as hippocampal long-term potentiation (LTP), occur prior to motor deficits and neuronal loss (Mazarakis et al., 2005; Murphy et al., 2000; Van Raamsdonk et al., 2005). LTP, a form of synaptic plasticity, is potentiated by release of BDNF. Thus, reduced levels of BDNF in HD patients and mice can disrupt synaptic changes important for learning and memory formation. Applying low concentrations of BDNF to hippocampal slices prepared from HD mice fully restores LTP (Lynch et al., 2007). Furthermore, up-regulation of endogenous BDNF levels with an ampakine, a positive modulator of AMPA-type glutamate receptors, rescues synaptic plasticity and reduces learning deficits in HD mice (Simmons et al., 2009).

Altered neurogenesis has been reported in HD mouse models and in human postmortem brains (Curtis et al., 2003; Gil et al., 2005; Phillips et al., 2005). It has been shown that in addition to promoting survival and inducing synaptic plasticity, BDNF also regulates adult neurogenesis (Bath et al., 2011; Henry et al., 2007; Scharfman et al., 2005). The adenoviral delivery of BDNF and Noggin (a known suppressor of gliogenesis) to the striatum of R6/2 HD mice resulted in induction of striatal neurogenesis (Cho et al., 2007). The majority of the newly born neurons differentiated to MSNs and became functional, leading to delayed motor impairment and prolonged survival in R6/2 mice. Similar improvements have been seen in the same HD mouse model after administration of the antidepressant sertraline (Peng et al., 2008). By increasing BDNF levels and stimulating neurogenesis, sertraline treatment resulted in improved motor performance, reduced striatal atrophy, and prolonged survival.

To more directly evaluate the effect of increasing cortical BDNF supply to the striatum on the progression of HD, we overexpressed BDNF in the mouse forebrain by employing a *Bdnf* transgene under the control of the promoter for Ca^{2+}/calmodulin-dependent protein kinase II alpha (Gharami et al., 2008; Xie et al., 2010). This transgene starts to express BDNF in the cerebral cortex in the first postnatal week and reaches plateau in the third postnatal week, as does the endogenous *Bdnf* gene (Huang et al., 1999). It also expresses at low levels in the striatum where the endogenous *Bdnf* gene is mostly inactive (Gharami et al., 2008; Xie et al., 2010). We found that the *Bdnf* transgene was able to greatly increase BDNF levels in the striata of R6/1 and YAC128 mice, indicating that overexpressed BDNF in the cortex is efficiently transported to the striatum despite expression of mutant htt. Importantly, BDNF overexpression reversed brain atrophy, normalized the expression of several important genes in the striatum, and ameliorated deficits in motor coordination in these two HD mouse models (Gharami et al., 2008; Xie et al., 2010). In addition, overexpression of BDNF in YAC128 mice prevented loss of striatal neurons, normalized spine abnormalities of medium-sized spiny neurons, and significantly improved procedural learning (Xie et al., 2010). In summary, these studies suggest that increasing striatal BDNF supply may have therapeutic potential for HD.

7. Conclusion

Many pathways have been proposed to contribute to the pathogenesis of HD. Recent studies have identified complex molecular mechanisms that mediate neuronal dysfunction and death; these include transcriptional dysregulation, excitotoxicity, impaired axonal transport, and altered synaptic transmission. The findings presented in this chapter support the hypothesis that reduced striatal BDNF plays a crucial role in HD pathogenesis. Currently, drugs used to treat HD act on the symptoms and do not slow or stop the disease progression. Attempting to restore striatal BDNF levels or activate downstream signaling pathways may have therapeutic potential in treating HD patients.

8. Acknowledgment

This work was supported by a grant from the National Institutes of Health (R01 NS050596) to BX.

9. References

Aid T, Kazantseva A, Piirsoo M, Palm K, Timmusk T. (2007). Mouse and rat BDNF gene structure and expression revisited. *Journal of Neuroscience Research,* Vol.85, No.3, (Feb 15), pp.525-535, ISSN 0360-4012

Altar CA, Cai N, Bliven T, Juhasz M, Conner JM, et al. (1997). Anterograde transport of brain-derived neurotrophic factor and its role in the brain. *Nature,* Vol.389, No.6653, (Oct 23), pp.856-860, ISSN 0028-0836

Apostol BL, Simmons DA, Zuccato C, Illes K, Pallos J, et al. (2008). CEP-1347 reduces mutant huntingtin-associated neurotoxicity and restores BDNF levels in R6/2 mice. *Molecular and Cellular Neuroscience,* Vol.39, No.1, (Sep), pp.8-20, ISSN 1095-9327

Baquet ZC, Gorski JA, Jones KR. (2004). Early striatal dendrite deficits followed by neuron loss with advanced age in the absence of anterograde cortical brain-derived neurotrophic factor. *Journal of Neuroscience,* Vol.24, No.17, (Apr 28), pp.4250-4258, ISSN 1529-2401

Bath KG, Akins MR, Lee FS. (2011). BDNF control of adult SVZ neurogenesis. *Developmental Psychobiology,* (Mar 22), ISSN 1098-2302

Baydyuk M, Russell T, Liao GY, Zang K, An JJ, et al. (2011). TrkB receptor controls striatal formation by regulating the number of newborn striatal neurons. *Proceedings of National Academy of Sciences U S A,* Vol.108, No.4, (Jan 25), pp.1669-1674, ISSN 1091-6490

Bhave SV, Ghoda L, Hoffman PL. (1999). Brain-derived neurotrophic factor mediates the anti-apoptotic effect of NMDA in cerebellar granule neurons: signal transduction cascades and site of ethanol action. *Journal of Neuroscience,* Vol.19, No.9, (May 1), pp.3277-3286, ISSN 0270-6474

Block-Galarza J, Chase KO, Sapp E, Vaughn KT, Vallee RB, et al. (1997). Fast transport and retrograde movement of huntingtin and HAP 1 in axons. *Neuroreport,* Vol.8, No.9-10, (Jul 7), pp.2247-2251, ISSN 0959-4965

Bolam JP, Hanley JJ, Booth PA, Bevan MD. (2000). Synaptic organisation of the basal ganglia. *Journal of Anatomy,* Vol.196 (Pt 4), (May), pp.527-542, ISSN 0021-8782

Borrell-Pages M, Zala D, Humbert S, Saudou F. (2006). Huntington's disease: from huntingtin function and dysfunction to therapeutic strategies. *Cellular and Molecular Life Science,* Vol.63, No.22, (Nov), pp.2642-2660, ISSN 1420-682X

Calabresi P, Pisani A, Mercuri NB, Bernardi G. (1996). The corticostriatal projection: from synaptic plasticity to dysfunctions of the basal ganglia. *Trends in Neuroscience,* Vol.19, No.1, (Jan), pp.19-24, ISSN 0166-2236

Canals JM, Pineda JR, Torres-Peraza JF, Bosch M, Martin-Ibanez R, et al. (2004). Brain-derived neurotrophic factor regulates the onset and severity of motor dysfunction associated with enkephalinergic neuronal degeneration in Huntington's disease. *Journal of Neuroscience,* Vol.24, No.35, (Sep 1), pp.7727-7739, ISSN 1529-2401

Chao MV. (2003). Neurotrophins and their receptors: a convergence point for many signalling pathways. *Nature Reviews Neuroscience,* Vol.4, No.4, (Apr), pp.299-309, ISSN 1471-003X

Cho SR, Benraiss A, Chmielnicki E, Samdani A, Economides A, Goldman SA. (2007). Induction of neostriatal neurogenesis slows disease progression in a transgenic murine model of Huntington disease. *Journal of Clinical Investigations,* Vol.117, No.10, (Oct), pp.2889-2902, ISSN 0021-9738

Ciammola A, Sassone J, Cannella M, Calza S, Poletti B, et al. (2007). Low brain-derived neurotrophic factor (BDNF) levels in serum of Huntington's disease patients. *Am J Med Genet B Neuropsychiatr Genet,* Vol.144B, No.4, (Jun 5), pp.574-577, ISSN 1552-4841

Cummings DM, Andre VM, Uzgil BO, Gee SM, Fisher YE, et al. (2009). Alterations in cortical excitation and inhibition in genetic mouse models of Huntington's disease. *Journal of Neuroscience,* Vol.29, No.33, (Aug 19), pp.10371-10386, ISSN 1529-2401

Curtis MA, Penney EB, Pearson AG, van Roon-Mom WM, Butterworth NJ, et al. (2003). Increased cell proliferation and neurogenesis in the adult human Huntington's disease brain. *Proceedings of National Academy of Science U S A,* Vol.100, No.15, (Jul 22), pp.9023-9027, ISSN 0027-8424

DiFiglia M, Sapp E, Chase K, Schwarz C, Meloni A, et al. (1995). Huntingtin is a cytoplasmic protein associated with vesicles in human and rat brain neurons. *Neuron,* Vol.14, No.5, (May), pp.1075-1081, ISSN 0896-6273

Dragatsis I, Levine MS, Zeitlin S. (2000). Inactivation of Hdh in the brain and testis results in progressive neurodegeneration and sterility in mice. *Nature Genetics,* Vol.26, No.3, (Nov), pp.300-306, ISSN 1061-4036

Duan W, Guo Z, Jiang H, Ware M, Li XJ, Mattson MP. (2003). Dietary restriction normalizes glucose metabolism and BDNF levels, slows disease progression, and increases survival in huntingtin mutant mice. *Proceedings of National Academy of Science U S A,* Vol.100, No.5, (Mar 4), pp.2911-2916, ISSN 0027-8424

Encinas M, Iglesias M, Llecha N, Comella JX. (1999). Extracellular-regulated kinases and phosphatidylinositol 3-kinase are involved in brain-derived neurotrophic factor-mediated survival and neuritogenesis of the neuroblastoma cell line SH-SY5Y. *Journal of Neurochemistry,* Vol.73, No.4, (Oct), pp.1409-1421, ISSN 0022-3042

Engelender S, Sharp AH, Colomer V, Tokito MK, Lanahan A, et al. (1997). Huntingtin-associated protein 1 (HAP1) interacts with the p150Glued subunit of dynactin. *Human Molecular Genetics,* Vol.6, No.13, (Dec), pp.2205-2212, ISSN 0964-6906

Ferrante RJ, Beal MF, Kowall NW, Richardson EP, Jr., Martin JB. (1987a). Sparing of acetylcholinesterase-containing striatal neurons in Huntington's disease. *Brain Research,* Vol.411, No.1, (May 12), pp.162-166, ISSN 0006-8993

Ferrante RJ, Kowall NW, Beal MF, Martin JB, Bird ED, Richardson EP, Jr. (1987b). Morphologic and histochemical characteristics of a spared subset of striatal neurons in Huntington's disease. *Journal of Neuropathology and Experimental Neurology,* Vol.46, No.1, (Jan), pp.12-27, ISSN 0022-3069

Ferrer I, Goutan E, Marin C, Rey MJ, Ribalta T. (2000). Brain-derived neurotrophic factor in Huntington disease. *Brain Research,* Vol.866, No.1-2, (Jun 2), pp.257-261, ISSN 0006-8993

Frade JM, Barde YA. (1998). Nerve growth factor: two receptors, multiple functions. *Bioessays,* Vol.20, No.2, (Feb), pp.137-145, ISSN 0265-9247

Gauthier LR, Charrin BC, Borrell-Pages M, Dompierre JP, Rangone H, et al. (2004). Huntingtin controls neurotrophic support and survival of neurons by enhancing BDNF vesicular transport along microtubules. *Cell,* Vol.118, No.1, (Jul 9), pp.127-138, ISSN 0092-8674

Gharami K, Xie Y, An JJ, Tonegawa S, Xu B. (2008). Brain-derived neurotrophic factor over-expression in the forebrain ameliorates Huntington's disease phenotypes in mice. *Journal of Neurochemistry,* Vol.105, No.2, (Apr), pp.369-379, ISSN 1471-4159

Gil JM, Mohapel P, Araujo IM, Popovic N, Li JY, et al. (2005). Reduced hippocampal neurogenesis in R6/2 transgenic Huntington's disease mice. *Neurobiology of Disease,* Vol.20, No.3, (Dec), pp.744-751, ISSN 0969-9961

Gorski JA, Zeiler SR, Tamowski S, Jones KR. (2003). Brain-derived neurotrophic factor is required for the maintenance of cortical dendrites. *Journal of Neuroscience,* Vol.23, No.17, (Jul 30), pp.6856-6865, ISSN 1529-2401

Gunawardena S, Her LS, Brusch RG, Laymon RA, Niesman IR, et al. (2003). Disruption of axonal transport by loss of huntingtin or expression of pathogenic polyQ proteins in Drosophila. *Neuron,* Vol.40, No.1, (Sep 25), pp.25-40, ISSN 0896-6273

Gusella JF, MacDonald ME. (2000). Molecular genetics: unmasking polyglutamine triggers in neurodegenerative disease. *Nature Reviews Neuroscience,* Vol.1, No.2, (Nov), pp.109-115, ISSN 1471-003X

Hempstead BL. (2006). Dissecting the diverse actions of pro- and mature neurotrophins. *Current Alzheimer Research,* Vol.3, No.1, (Feb), pp.19-24, ISSN 1567-2050

Henry RA, Hughes SM, Connor B. (2007). AAV-mediated delivery of BDNF augments neurogenesis in the normal and quinolinic acid-lesioned adult rat brain. *European Journal of Neuroscience,* Vol.25, No.12, (Jun), pp.3513-3525, ISSN 0953-816X

Ho AK, Sahakian BJ, Brown RG, Barker RA, Hodges JR, et al. (2003). Profile of cognitive progression in early Huntington's disease. *Neurology,* Vol.61, No.12, (Dec 23), pp.1702-1706, ISSN 1526-632X

Hodgson JG, Agopyan N, Gutekunst CA, Leavitt BR, LePiane F, et al. (1999). A YAC mouse model for Huntington's disease with full-length mutant huntingtin, cytoplasmic toxicity, and selective striatal neurodegeneration. *Neuron,* Vol.23, No.1, (May), pp.181-192, ISSN 0896-6273

Huang ZJ, Kirkwood A, Pizzorusso T, Porciatti V, Morales B, et al. (1999). BDNF regulates the maturation of inhibition and the critical period of plasticity in mouse visual cortex. *Cell,* Vol.98, No.6, (Sep 17), pp.739-755, ISSN 0092-8674
Kim M, Lee HS, LaForet G, McIntyre C, Martin EJ, et al. (1999). Mutant huntingtin expression in clonal striatal cells: dissociation of inclusion formation and neuronal survival by caspase inhibition. *Journal of Neuroscience,* Vol.19, No.3, (Feb 1), pp.964-973, ISSN 0270-6474
Laforet GA, Sapp E, Chase K, McIntyre C, Boyce FM, et al. (2001). Changes in cortical and striatal neurons predict behavioral and electrophysiological abnormalities in a transgenic murine model of Huntington's disease. *Journal of Neuroscience,* Vol.21, No.23, (Dec 1), pp.9112-9123, ISSN 1529-2401
Lawrence AD, Hodges JR, Rosser AE, Kershaw A, ffrench-Constant C, et al. (1998). Evidence for specific cognitive deficits in preclinical Huntington's disease. *Brain,* Vol.121 (Pt 7), (Jul), pp.1329-1341, ISSN 0006-8950
Li SH, Gutekunst CA, Hersch SM, Li XJ. (1998). Interaction of huntingtin-associated protein with dynactin P150Glued. *Journal of Neuroscience,* Vol.18, No.4, (Feb 15), pp.1261-1269, ISSN 0270-6474
Lynch G, Kramar EA, Rex CS, Jia Y, Chappas D, et al. (2007). Brain-derived neurotrophic factor restores synaptic plasticity in a knock-in mouse model of Huntington's disease. *Journal of Neuroscience,* Vol.27, No.16, (Apr 18), pp.4424-4434, ISSN 1529-2401
Mann DM, Oliver R, Snowden JS. (1993). The topographic distribution of brain atrophy in Huntington's disease and progressive supranuclear palsy. *Acta Neuropathologica,* Vol.85, No.5, pp.553-559, ISSN 0001-6322
Matsumoto T, Rauskolb S, Polack M, Klose J, Kolbeck R, et al. (2008). Biosynthesis and processing of endogenous BDNF: CNS neurons store and secrete BDNF, not pro-BDNF. *Nature Neuroscience,* Vol.11, No.2, (Feb), pp.131-133, ISSN 1097-6256
Mazarakis NK, Cybulska-Klosowicz A, Grote H, Pang T, Van Dellen A, et al. (2005). Deficits in experience-dependent cortical plasticity and sensory-discrimination learning in presymptomatic Huntington's disease mice. *Journal of Neuroscience,* Vol.25, No.12, (Mar 23), pp.3059-3066, ISSN 1529-2401
McAllister AK, Katz LC, Lo DC. (1999). Neurotrophins and synaptic plasticity. *Annual Reviews Neuroscience,* Vol.22, pp.295-318, ISSN 0147-006X
Metsis M, Timmusk T, Arenas E, Persson H. (1993). Differential usage of multiple brain-derived neurotrophic factor promoters in the rat brain following neuronal activation. *Proceedings of National Academy of Science U S A,* Vol.90, No.19, (Oct 1), pp.8802-8806, ISSN 0027-8424
Murphy KP, Carter RJ, Lione LA, Mangiarini L, Mahal A, et al. (2000). Abnormal synaptic plasticity and impaired spatial cognition in mice transgenic for exon 1 of the human Huntington's disease mutation. *Journal of Neuroscience,* Vol.20, No.13, (Jul 1), pp.5115-5123, ISSN 1529-2401
O'Kusky JR, Nasir J, Cicchetti F, Parent A, Hayden MR. (1999). Neuronal degeneration in the basal ganglia and loss of pallido-subthalamic synapses in mice with targeted disruption of the Huntington's disease gene. *Brain Research,* Vol.818, No.2, (Feb 13), pp.468-479, ISSN 0006-8993

Pang TY, Stam NC, Nithianantharajah J, Howard ML, Hannan AJ. (2006). Differential effects of voluntary physical exercise on behavioral and brain-derived neurotrophic factor expression deficits in Huntington's disease transgenic mice. *Neuroscience,* Vol.141, No.2, (Aug 25), pp.569-584, ISSN 0306-4522

Patapoutian A, Reichardt LF. (2001). Trk receptors: mediators of neurotrophin action. *Current Opinions in Neurobiology,* Vol.11, No.3, (Jun), pp.272-280, ISSN 0959-4388

Pattabiraman PP, Tropea D, Chiaruttini C, Tongiorgi E, Cattaneo A, Domenici L. (2005). Neuronal activity regulates the developmental expression and subcellular localization of cortical BDNF mRNA isoforms in vivo. *Molecular and Cellular Neuroscience,* Vol.28, No.3, (Mar), pp.556-570, ISSN 1044-7431

Peng Q, Masuda N, Jiang M, Li Q, Zhao M, et al. (2008). The antidepressant sertraline improves the phenotype, promotes neurogenesis and increases BDNF levels in the R6/2 Huntington's disease mouse model. *Experimental Neurology,* Vol.210, No.1, (Mar), pp.154-163, ISSN 0014-4886

Phillips W, Morton AJ, Barker RA. (2005). Abnormalities of neurogenesis in the R6/2 mouse model of Huntington's disease are attributable to the in vivo microenvironment. *Journal of Neuroscience,* Vol.25, No.50, (Dec 14), pp.11564-11576, ISSN 1529-2401

Reichardt LF. (2006). Neurotrophin-regulated signalling pathways. *Philos Trans R Soc Lond B Biol Sci,* Vol.361, No.1473, (Sep 29), pp.1545-1564, ISSN 0962-8436

Reiner A, Albin RL, Anderson KD, D'Amato CJ, Penney JB, Young AB. (1988). Differential loss of striatal projection neurons in Huntington disease. *Proceedings of National Academy of Science U S A,* Vol.85, No.15, (Aug), pp.5733-5737, ISSN 0027-8424

Rigamonti D, Bauer JH, De-Fraja C, Conti L, Sipione S, et al. (2000). Wild-type huntingtin protects from apoptosis upstream of caspase-3. *Journal of Neuroscience,* Vol.20, No.10, (May 15), pp.3705-3713, ISSN 1529-2401

Rubinsztein DC. (2002). Lessons from animal models of Huntington's disease. *Trends in Genetics,* Vol.18, No.4, (Apr), pp.202-209, ISSN 0168-9525

Saudou F, Finkbeiner S, Devys D, Greenberg ME. (1998). Huntingtin acts in the nucleus to induce apoptosis but death does not correlate with the formation of intranuclear inclusions. *Cell,* Vol.95, No.1, pp.55-66, ISSN 0092-8674

Scharfman H, Goodman J, Macleod A, Phani S, Antonelli C, Croll S. (2005). Increased neurogenesis and the ectopic granule cells after intrahippocampal BDNF infusion in adult rats. *Experimental Neurology,* Vol.192, No.2, (Apr), pp.348-356, ISSN 0014-4886

Seo H, Kim W, Isacson O. (2008). Compensatory changes in the ubiquitin-proteasome system, brain-derived neurotrophic factor and mitochondrial complex II/III in YAC72 and R6/2 transgenic mice partially model Huntington's disease patients. *Human Molecular Genetics,* Vol.17, No.20, (Oct 15), pp.3144-3153, ISSN 1460-2083

Simmons DA, Rex CS, Palmer L, Pandyarajan V, Fedulov V, et al. (2009). Up-regulating BDNF with an ampakine rescues synaptic plasticity and memory in Huntington's disease knockin mice. *Proceedings of National Academy of Science U S A,* Vol.106, No.12, (Mar 24), pp.4906-4911, ISSN 1091-6490

Spires TL, Grote HE, Varshney NK, Cordery PM, van Dellen A, et al. (2004). Environmental enrichment rescues protein deficits in a mouse model of Huntington's disease, indicating a possible disease mechanism. *Journal of Neuroscience,* Vol.24, No.9, (Mar 3), pp.2270-2276, ISSN 1529-2401

Strand AD, Baquet ZC, Aragaki AK, Holmans P, Yang L, et al. (2007). Expression profiling of Huntington's disease models suggests that brain-derived neurotrophic factor depletion plays a major role in striatal degeneration. *Journal of Neuroscience,* Vol.27, No.43, (Oct 24), pp.11758-11768, ISSN 1529-2401

Szebenyi G, Morfini GA, Babcock A, Gould M, Selkoe K, et al. (2003). Neuropathogenic forms of huntingtin and androgen receptor inhibit fast axonal transport. *Neuron,* Vol.40, No.1, (Sep 25), pp.41-52, ISSN 0896-6273

Teng HK, Teng KK, Lee R, Wright S, Tevar S, et al. (2005). ProBDNF induces neuronal apoptosis via activation of a receptor complex of p75NTR and sortilin. *Journal of Neuroscience,* Vol.25, No.22, (Jun 1), pp.5455-5463, ISSN 1529-2401

The Huntington's Disease Collaborative Research Group. (1993). A novel gene containing a trinucleotide repeat that is expanded and unstable on Huntington's disease chromosomes. *Cell,* Vol.72, No.6, (Mar 26), pp.971-983, ISSN 8458085

Timmusk T, Palm K, Metsis M, Reintam T, Paalme V, et al. (1993). Multiple promoters direct tissue-specific expression of the rat BDNF gene. *Neuron,* Vol.10, No.3, (Mar), pp.475-489, ISSN 0896-6273

Van Raamsdonk JM, Pearson J, Slow EJ, Hossain SM, Leavitt BR, Hayden MR. (2005). Cognitive dysfunction precedes neuropathology and motor abnormalities in the YAC128 mouse model of Huntington's disease. *Journal of Neuroscience,* Vol.25, No.16, (Apr 20), pp.4169-4180, ISSN 1529-2401

Vonsattel JP, DiFiglia M. (1998). Huntington disease. *Journal of Neuropathology and Experimental Neurology,* Vol.57, No.5, (May), pp.369-384, ISSN 0022-3069

Xie Y, Hayden MR, Xu B. (2010). BDNF overexpression in the forebrain rescues Huntington's disease phenotypes in YAC128 mice. *Journal of Neuroscience,* Vol.30, No.44, (Nov 3), pp.14708-14718, ISSN 1529-2401

Xu B, Zang K, Ruff NL, Zhang YA, McConnell SK, et al. (2000). Cortical degeneration in the absence of neurotrophin signaling: dendritic retraction and neuronal loss after removal of the receptor TrkB. *Neuron,* Vol.26, No.1, (Apr), pp.233-245, ISSN 0896-6273

Yamada M, Tanabe K, Wada K, Shimoke K, Ishikawa Y, et al. (2001). Differences in survival-promoting effects and intracellular signaling properties of BDNF and IGF-1 in cultured cerebral cortical neurons. *Journal of Neurochemistry,* Vol.78, No.5, (Sep), pp.940-951, ISSN 0022-3042

Yang J, Siao CJ, Nagappan G, Marinic T, Jing D, et al. (2009). Neuronal release of proBDNF. *Nature Neuroscience,* Vol.12, No.2, (Feb), pp.113-115, ISSN 1546-1726

Yang M, Lim Y, Li X, Zhong JH, Zhou XF. (2011). Precursor of brain-derived neurotrophic factor (proBDNF) forms a complex with Huntingtin-associated protein-1 (HAP1) and sortilin that modulates proBDNF trafficking, degradation, and processing. *Journal of Biological Chemistry,* Vol.286, No.18, (May 6), pp.16272-16284, ISSN 1083-351X

Zuccato C, Ciammola A, Rigamonti D, Leavitt BR, Goffredo D, et al. (2001). Loss of huntingtin-mediated BDNF gene transcription in Huntington's disease. *Science,* Vol.293, No.5529, (Jul 20), pp.493-498, ISSN 0036-8075

Zuccato C, Marullo M, Conforti P, MacDonald ME, Tartari M, Cattaneo E. (2008). Systematic assessment of BDNF and its receptor levels in human cortices affected by Huntington's disease. *Brain Pathology,* Vol.18, No.2, (Apr), pp.225-238, ISSN 1015-6305

Zuccato C, Tartari M, Crotti A, Goffredo D, Valenza M, et al. (2003). Huntingtin interacts with REST/NRSF to modulate the transcription of NRSE-controlled neuronal genes. *Nature Genetics,* Vol.35, No.1, (Sep), pp.76-83, ISSN 1061-4036

Part 7

Learning to Live with Huntington's Disease

22

Risk and Resilience: Living with a Neurological Condition with a Focus on Health Care Communications

Kerstin Roger and Leslie Penner
University of Manitoba
Canada

1. Introduction

There is little research examining the daily lived experiences of families in which one person is living with a neurological condition. Further, there is no research which examines how patients evolve and adapt in ways that allow them to have their needs met within the complex system of health care provision. This paper explores nine factors that emerged in original, empirical qualitative data, and how these factors are perceived by the participants to contribute towards increasing resilience when communicating with the health care system. The data were collected over a three year period, utilizing focus groups and interviews. Our primary focus in this paper will be on the participants living with Huntington's and Parkinson's, although other participants will be referred to as well.

Communication about care can be shaped by many factors such as a concern about being a burden to others, the proximity of family, family dynamics, personalities, stress related to the illness itself, and, socio-cultural belief systems about health and healthcare (in the example of persons with a collectivist versus more individualistic cultural background). It has become clear (Roger et al., 2010; Roger & Medved, 2010; Zloty et al., 2010) that contradictory and often unspoken expectations of care emerge between professionals, family and individuals. Because of this, errors, poor treatment, and misunderstandings can result, in addition to diminished well-being of caregivers and individuals. Further, the ways in which individuals and family members experience and perceive interactions with health care providers have been shown to affect the level of trust that the patients have in their care providers (Tarrant et al., 2003), and have been demonstrated to impact health care outcomes (Moreau et al., 2006; Safran et al., 1998). These outcomes can result in significant costs to individuals and the health care system. This paper will explore how the nine factors that emerged can contribute towards the development of tools that can be used to facilitate improved communication between patients living with neurological conditions and health care providers.

2. Ecological model

Bronfenbrenner's ecological approach (1979) is a suitable framework. His approach identifies how individuals, small groups and larger groups, as well as institutions, interact systemically and bi-directionally to shape relational patterns, norms and values.

Bronfenbrenner (1979) initially described four nested systems: micro-, meso-, exo-, and macrosystems, with family being the primary microsystem within which the individual develops (Bubloz & Sontag, 1993, p. 424). The mesosystem refers to interactions and relationships between the family and other systems where individual learning takes place. This is an important concept in relation to the current study because there is much health care teaching that individuals and their family members are exposed to when interacting with the health care system. Bronfenbrenner suggested that if the connections between families and the environments where learning takes place are healthy and positive, it would affect the individual in a positive way. For example, according to this theory, if individuals and their family members have open and positive connections with their neurologists, the health of these patients will be improved. The exosystem represents environments that adults participate in which, in turn, affect the individual and the family as a whole (e.g. work settings, recreational organizations, volunteer settings). Finally, the macrosystem, which includes cultural beliefs, customs and laws, is the outermost layer of the linked systems - it envelopes and influences the interactions of all system layers. Later in the development of his theory, Bronfenbrenner (1986) added the chronosystem, which refers to the influence of the person's development of changes over time in the context of the environments in which the person interacts. He suggested that life transitions were the simplest form of chronosystem, with the development of an illness being an example. This transition has the potential to affect the development of the individual (and family members) directly, or indirectly by affecting the interactions and functions of the family. The chronosystem includes the cumulative effects of evolving developmental transitions over the life of an individual. This concept is useful when exploring the impact that a chronic, deteriorating neurological illness has on the development of patients and their family members; despite being a difficult transition, it has the potential to instigate developmental change. This systemic approach is particularly useful here since the stories and narratives between couples and care professionals reflect important aspects of Bronfenbrenner's systems approach. The themes that emerged in this study moved from descriptions of very personal responses to situations, to daily communications and interactions among family members, to interactions with the larger network of resources and organizations.

Human beings are inherently social; they do not generally operate independently and in isolation. Therefore, it follows that the behaviour and responses of individuals cannot be fully understood without considering the environment within which they are embedded. Human ecological theory "….is concerned with interaction and interdependence of humans (as individuals, groups, and societies) with the environment" (Bubolz & Sontag, 1993, p. 421). A key concept in this theory is the process of adaptation — the way in which individuals and families attempt to cope with their ever-changing environments. Particular attention is given to communication and the underlying values which guide these decisions (Bubolz & Sontag, 1993). Resilient couples who interact with the health care system on a regular basis can and do find techniques and strategies which ensure effective and creative adaptation to the challenges they face. In this way, Bronfenbrenner's model is a helpful framework for the data to be discussed in this paper.

3. Background

The National Health Charities of Canada (NHCC) is an umbrella organization that works with government, researchers and the community to promote and support services and research to a number of related neurological disease conditions such as neurotrauma (e.g., acquired brain injury and spinal cord injury), neuromuscular disorders (e.g., cerebral palsy, epilepsy and spina bifida), degenerative demyelinating conditions (e.g., multiple sclerosis, Guillian-Barre syndrome), and movement and other neurodegenerative disorders (e.g., Parkinson's Disease, Huntington's Disease, Alzheimer's Disease and Amyotrophic Lateral Sclerosis [ALS]). This study recruited participants from conditions under this umbrella: to be discussed in this paper are participants with Parkinson's Disease (the majority of our sample) and Huntington's.

Huntington's Disease is an inherited neuropsychiatric disorder that causes brain cells to die, resulting in clinical features which present in a triad of movement disorder, cognitive dysfunction and psychiatric or behavioral disturbance (Sturrock & Leavitt, 2010). Because the disease is inherited as an autosomal dominant trait, each child of an affected individual has a 50% chance of inheriting the gene (Aubeeluck & Wilson, 2008). The average age of onset of Huntington's Disease is between 35 and 44 years (Paulsen, Ferneyhough Hoth, Nehl & Stierman, 2005), but it can present anytime between ages 2 to 85 years (Roos, 2010). Huntington's is a rare disease with a prevalence rate that varies between 5-10 per 100,000 in the American population (Nance & Myers, 2001) and 4-8 per 100,000 in the European population (Harper, 1992).There is no cure for Huntington's Disease at this time and patients die, on average, 20 years after onset (Paulsen et al., 2005).

Mild cognitive and personality changes can occur in the early stages of the disease (Sturrock & Leavitt, 2010; Walker, 2007). The early symptoms are often first noticed by family, friends and co-workers and may include disinhibited behavior, fidgetiness, irritability, anhedonia, obsessive behaviors, altered executive function, and slowed processing speed which manifests in decreased productivity (Sturrock & Leavitt, 2010). As the disease progresses, motor disturbances such as chorea, speech and swallowing difficulties, rigidity, bradykinesia and akinesia develop (Roos, 2010). Eventually, mobility is lost (Sturrock & Leavitt, 2010) and oral motor dysfunction leads to incoherence of speech and inability to eat (Sturrock & Leavitt, 2010).

Cognitive deterioration is progressive in Huntington's Disease (Sturrock & Leavitt, 2010). There is a decline in executive functioning that affects judgment, insight, and ability to organize (Roos, 2010; Sturrock & Leavitt, 2010) and there is a progressive deterioration in recall and complex intellectual functioning (Sturrock & Leavitt, 2010). Eventually, cognitive decline becomes global and all aspects of cognition become impaired (Sturrock & Leavitt, 2010). Psychiatric symptoms such as depression and anxiety are often present during the disease trajectory and are relatively independent of the motor and cognitive aspects of the disease (Paulsen, Ready, Hamilton, Mega & Cummings, 2001). As well, although less common, delusions and hallucinations can also as emerge during the course of the illness (Paulsen et al., 2001). There is an increase in suicidal ideation in individuals at risk for, and diagnosed with Huntington's Disease, particularly in the period immediately before receiving a diagnosis ,and in then again in the period where independence begins to diminish (Paulsen et al., 2005).

Parkinson's Disease is the second most common neurodegenerative disorder worldwide, second only to Alzheimer's Disease (de Lau & Breteler, 2006). It is commonly diagnosed in individuals over the age of fifty, with its prevalence affecting approximately three percent of the Canadian population over the age of 65 (Public Health Agency of Canada, 2000). Recent

Canadian research has shown that the prevalence of Parkinson's Disease is increasing, perhaps due to an aging population (Guttman, Slaughter, Theriault, DeBoer, & Naylor, 2003; Lix et al., 2010).

The most common symptoms of Parkinson's Disease are tremor, muscle rigidity and stiffness and psychomotor retardation (bradykinesia) (Clark, 2007; Heisters, 2011). As the disease progresses, postural abnormalities and instability can emerge (Clark, 2007; Nutt, & Wooten, 2005). These symptoms can interfere with ambulation and increase the risk of falls. There is a wide range of non-motor symptoms which pain, dementia, mood disorder, psychosis, apathy, sleep disorder and excessive daytime sleepiness, bowel and bladder dysfunction, excessive sweating, and sexual dysfunction (Clark, 2007; Heisters, 2011). Dementia is common, affecting approximately 30% of patients with Parkinson's Disease (Aarsland & Kurz, 2010). The severity and presentation of Parkinson's related dementia varies between individuals, but common characteristics are impairment in attention, memory, ability to plan, organize and problem solve, and impaired recall, personality changes, behavioral symptoms and hallucinations (Emre, 2003). Because non-motor symptoms affect several domains of functioning, and because they are difficult to treat, the impact on the patient's quality of life can be profound.

Both HD and PD are chronic neurodegenerative and progressive in nature (National Institute of Neurological Disorders and Stroke, 2008; Public Health Agency of Canada, 2000). As well, symptoms can affect all domains of functioning: physical, cognitive and emotional, resulting in complex health care needs.

Although each of these neurological conditions has their own etiology, they share symptoms which can impact everyday activities and the health and well-being of individuals and family members. Shared experiences include disrupted relationships and a reduction in participation in personally meaningful activities (e.g., employment, shared family activities) (Statistics Canada, 2007). While there has been some attention paid to the illness experience (Brody, 1987; Charmaz, 1991), little research was found that examines communication and the daily experiences of persons with neurological conditions in relation to their families and health care providers. The impacts in this context are multi-faceted including multiple changes to couples' roles and responsibilities over years, every day routines, marital relationships and financial status.

4. Methods

This was a qualitative pilot study conducted over three years using interviews and focus groups. Initially, community consultations were held with affiliated staff and organizations (see Table 1) to better understand organizational needs regarding care services to families and couples with long term neurodegenerative conditions (e.g. Huntington's, Alzheimer's, ALS, Multiple Sclerosis, and Parkinson's). This led to a clarification of the need for research on people's daily lived experiences and especially a need to better understand their interactions with the health care system. The research team was then successful in receiving funding to conduct Phase II in 2009. This explored decision-making between couples and care professionals. We received funding for Phase III where we were able to explore in more depth, given the same sample of participants, what had changed for them given their daily lived experiences in the last year. The data to be described in this paper emerged from this last sample in Phase III collected in 2010.

	2007-08	2009	2010
Event	PHASE I: Community-based consultations (3)	PHASE II: Interviews (16)	PHASE III: Follow-up interviews (8) and focus groups (2)

Table 1. Research Phases.

4.1 Ethics approval

The authors prepared an ethics protocol for the most suitable ethics review committee at the local university and this was approved. Components of this protocol included: i. a script for the research assistant to be used when discussing possible recruitment with relevant organizations and a script to be used when discussing the study with potential future participants; ii. a list of the interview questions; iii. a pledge of confidentiality for the research assistant and the transcriber; and, iv. a consent form that described the study, it identified the process that ensured the participant's confidentiality and how this would be maintained over time. The consent form addressed the treatment of the data once the study was to be completed, how the data will be stored, and when and how the data will be discarded.

4.2 Interview and focus group sample

We aimed for equal representation of individuals in the three proposed categories: the individual with a selected neurological condition, a familial support person, as well as a professional care provider (see Table 2). Fifteen people were interviewed in total over three time periods in the three years. Over the three years, some changes did occur to our sample due to divorce, moves out of province, and a willingness to participate in the study. We held two focus groups once the interviews were conducted and themes were analyzed. They were recruited in the same manner as described below for the individual interviews. The focus groups were comprised of six participants each (not the same persons as the interview sample) who were all formal caregivers working in any one of the affiliated organizations. Diversity for the focus group participation was sought in regards to the type of professional (e.g. nurses, social workers, administrators).

Participants with an interest to participate in the study were eligible if they met the following inclusion criteria:

i. They understood the primary goal of the study and were able to articulate their thoughts verbally on the topic;
ii. They fit on of the disease categories;
iii. They were able to provide consent at the beginning of the study by reading the consent form, asking questions about the study, and signing the consent forms;
iv. They were able to hold a full conversation in English;
v. They or a family support was affiliated with one of the selected and recognized institutions. Or, for the focus groups/consultations, they were a staff person in one of the affiliated organizations;
vi. A primary diagnosis of one of the conditions under the NHCC umbrella had occurred as confirmed by a key familial support person, a physician, social worker, nurse or patient care manager familiar with the participant's history;

vii. Participants for the interviews had engaged in communications regarding health care involving a person with a selected condition in the last 6 months; viii. Participants had to be over 18 years of age.

	PSEUDONYM	RELATIONSHIP	AGE	PROFESSIONAL BACKGROUND	PROFESSIONAL/ PATIENT WITH CONDITION
1	Neil	Married to 2			PD
2	Flora	*	65	SW	Support
3	Len	Brother to 4	79	Business	PD
4	Frieda	*	83	Secretary	Support
5	Estrella		2003 began	OT with MS	Prof
6	Sophie		1986 began	SW with HD/PD	Prof
7	Daniel	*	55	Welder	Support
8	Doreen		50	Computer analyst	MS
9	Jane	Married to Ken	59		MS
10	Ivan		1997 began	Warden/vol pall care	Prof
11	Leila		1999 began	nurse	Prof
12	Nettie		2005 began	Health care/MS	prof
13	Janelle	Married to 7	54	Bakery manager	PD
14	Margareta	*	81	banker	PD
15	Friederich	Married to 14	83	Power engineer	support

Table 2. Interview Sample.

Individuals for the interviews were asked whether they would provide one key family support person. A person with one of the selected neurological conditions could take part in this study even if they had no key family support person and wanted to select a second health care provider, or their key family support person declined but someone else was willing to be interviewed. The definition of the family support person was quite broad including family members, common law partners, neighbors who provide significant frequent care, or another relative doing the same. This individual would sign a consent form prior to being interviewed. Both the individual and their family support person received an honorarium for their participation. Confidentiality agreements as documented through the consent form applied to each participant. Professionals working in related areas and with populations who fit the criteria were recruited as well.

Once the consent forms were signed by participants, the interviews were held individually at a site selected by the participant, and lasted approximately two hours. A demographic section began the interview process including basic questions about a person's age, gender, professional affiliation if appropriate and so on. Semi-structured interview questions were then used to investigate the primary objectives of the study. Questions included asking about the wished for and perceived role of health care providers communicating on their behalf, the wished for and perceived impact of family members making decisions on their behalf, and the perceived changes and role of their own independence as their condition was diagnosed and as it progressed. New probes were developed as data was collected based on the findings in the ongoing interviews. All interviews were audio taped and then transcribed verbatim.

Once the consent forms were signed by the participants, two focus groups were held at locations convenient for the participants and lasted approximately two hours. Themes from the individual interviews had been compiled and were presented to the focus group members. Similarities between the themes and the participant's experiences, identified gaps and differences were discussed. Probes had been developed in the ethics protocol for this purpose. Detailed notes were taken in these two focus groups by a research assistant.

4.3 Analytic process

NVIVO8, a qualitative data management program, was used to code all transcribed interviews. Content analysis was used as a framework (Graneheim & Lundman, 2004) where constant comparison is possible between themes and sub-themes as they emerged. Once the researchers coded the transcripts and assessed the main and sub-themes, selected experts were provided with a sample of transcripts in order to code them. By applying the principles of qualitative content analysis, further exploration of the data was performed to encourage trustworthiness and credibility of the research findings. This approach allowed the researcher to "condense" the data into additional and/or comparative codes, followed by "aggregation" or the progressive interpretive process of thematic abstraction. The themes were then compared and similarities noted while differences will be documented in subsequent papers.

Rigor was determined according to principles set out for conducting qualitative research (Morse et al., 2002). For example, *consistency* was ensured by choosing participants who have experiences with the research topic and a genuine interest in taking part in the interviews. *Transferability* will be fulfilled by making certain that detailed information will be provided in future papers so that readers would be able to identify a similar situation in a similar context. *Credibility* was attained through editing of the interview transcripts as well as integration of field notes in line with what does exist in the literature. Inter-rater reliability throughout all phases of analysis solidified credibility. Results and interpretations were checked by members of the research team who then reviewed the analysis, obtaining consensual validation.

5. Findings

Three levels of the Bronfenbrenner ecological model were most apparent in our emerging nine factors: the microsystem, the mesosystem, and the chronosystem. The findings will be presented in two parts - the interview data first and the focus group discussions second.

5.1 Microsystem

Individual and family characteristics were included to represent the microsystem.

1. Manner of Communicating: Ability to Push the System
2. Self-Reflection about Characteristics Prior to Condition
3. Existing Social Supports
4. Education Levels
5. Gender of Primary Caregiver/ Professional

5.2 Mesosystem

The mesosystem includes the connections and interactions between the family and other systems where individual learning takes place. Here, the system where learning takes place is the health care system.

1. Health Care Literacy

5.3 Chronosystem

Bronfenbrenner's chronosystem is useful for examining the impact that the passage of time has on the individual. For example, the passage of time can impact the individual's physical state (eg. children maturing, individual's with chronic illness deteriorating), and this in turn, can affect the way in which they interact with the environment. As well, the passage of time can allow individuals to gain mastery and experience in coping with a difficult situation, for example. The chronosystem encompasses anything that has to do with the passage of time.

1. Length of Relationship to Primary Caregiver
2. Stage of the Illness
3. Age

5.4 Findings from interview data

5.4.1 Microsystem

5.4.1.1 Manner of communicating

Resilience research has often sought to understand specific personality traits or characteristics inherent in an individual - qualities which were described as being protective in nature (Earvolino, 2007; Johnson & Wiechelt, 2004; Richardson, 2002).The participants in this sample demonstrate personality traits that reflect their individual abilities regarding communication. For example, communicating about their care needs in a clear and articulate manner was for some a new skill and for others an existing long standing trait. Not being able to communicate well with care professionals in many cases reduced their ability to interact in effective and beneficial ways with the health care system.

It was evident that the manner in which a patient, family or professional caregiver communicated with others shaped how they perceived their care plans developed. Janelle talks about her ability to communicate to 'push' the system:

Janelle: *And the Pharmacare system out here is a lot stricter as well, and at first they weren't gonna cover me for a couple of my main medications. I had to, this is quite funny. Well it's funny but it wasn't. They made me get a letter from my doctor saying that it was absolutely necessary that I have these medications. And I said, well, of course I do, I have Parkinson's. So they said, well, isn't there something else you could use? And I said no, and I had to get a note from my doctor in order to have those two medications covered.* (Int.: "So you really had to push the system?") *Yes – I really had to work hard to get those medications!*

Janelle described that her ability to be persistent and forthright in her communications with care professionals was a skill she needed to learn, that she had previously been less assertive and unable to express her needs when interacting with professionals in larger institutions or organizations. Her illness 'taught' her this skill. Upon being diagnosed, she learned that her new found ability to 'push' the system by being more assertive was useful when advocating for herself. People are also aware that their ability to communicate on their own behalf will change as their condition progresses. Ironically, participants talked about how important it was to be polite and nice when interacting with health care professionals, to be courteous with care professionals in order to get ones care needs met. However, participants also spoke proudly about their ability to push the system and get what they needed when they were less polite. At times, they enlisted the support of health care professionals to achieve their goals. In cases where a care professional was not able to assist them in successfully 'pushing the system', it was clear that disappointment in the care professional occurred. There was an expectation that when patients had allies in health care, the patients felt better able to communicate within the system. This theme persists throughout the study and becomes important to our broader theme of social support promoting individual resilience. Those who felt supported by a social service professional also expressed higher levels of satisfaction with communications with health care professionals.

Troubling communications with people with neurological conditions are often attributed to the disease itself, while Janelle is articulate in stating that some of her characteristics, which may in another person be seen as a symptom of her condition, are part of her old normal self. Medicalizing personality traits that existed prior to the onset of an illness can lead to mis-understandings and inadequate care plans. In this sense, people's ability to communicate with health care professionals may be interpreted as part of their condition, but in fact, may be describing personal traits and qualities they have always had.

5.4.1.2 Self-reflection about personal characteristics prior to condition

The participants in this study demonstrated resilience in the example of adaptation and coping with extraordinary circumstances. While these may appear as minor or simple reflections of resilience, they must be understood in the context of challenging and usually complex lives. One approach to resilience research has been to seek out adaptive processes or a means of coping with various adversities and defining these as an opportunity to learn or improve upon an individual's protective qualities (Richardson, 2002). Research in this realm has helped to broaden the predominant mindset of problem-oriented approaches, where one would focus on the importance of prevention, encourage strengths and value human fortitude; ultimately, focusing on the basics of a strength-based lens (Krawlik et al., 2006).

For example, Neil says he has always had an optimistic attitude and that this now serves him well as he begins to live with some of his early limitations. He says it could have been anyone who received this diagnosis, so 'why not me?', and he states that he is fine even though he now has this additional condition in his life. His wife Flora also underlines how optimistic he remains after receiving his diagnosis. Janelle describes herself here:

Well, for instance, I'm terribly, terribly disorganized and always have been. Now, I don't write things down and I forget my appointments and like it causes everybody a lot of stress because I'm scrambling at the last minute to make arrangements and what have you, and she'll say to me, get a book. I'll buy you a book and write these things down and I'll help you with your appointments. I'll help you arrange them and I'll help you get there, you know, but you've got to make the effort to write them down and take note of them so, you know, we're not doing things at the last minute. And she said, I'm quite willing to take you and do anything you want, but you've got to at least be a little bit more organized. And it's true, like I need a kick in the butt. Its always been that way.

We know that self-reflection about personal characteristics can fade as disease trajectories of neurological conditions progress. A fading ability to recognize one's own style or interactions with others can lead to challenges for health care professionals with the goal of determining where a person with a neurological condition is at. Certainly, when health care providers have known an individual over years as their condition progresses, they are in a much stronger position to assess how an individual is doing – especially as they become less able to reflect on their own processes.

5.4.1.3 Existing social supports

While this factor (#3) does fit with other levels in the Bronfenbrenner system, it is being placed here as the primary location. Traditional research on resilience was rooted in psychiatric literature and focused primarily on children and adolescents capable of dealing with great difficulty (Garmezy, 1985, 1993; Masten, 1999; Werner, 1990; Werner & Smith, 1977). More current examples reflect essential components of social support when developing qualities associated with that traditional literature on resilience. It appears that social support is highly correlated to resilience and that networks are critical for a positive description of social support. A better understanding of how couples might experience or express resilience can directly and indirectly impact care provision, and a professional's inclination to provide better care for a particular dyad. If social supports result in a more resilient patient, it would be helpful to be able to assess this at intake.

It was clear that participants with strong and reliable family or friendship supports were more able to interact effectively with the health care system. Flora states,

We have a lot of friends who have Parkinson's, by coincidence, not that we've acquired them later on. A lot of people have Parkinson's. And the Parkinson's community now, we've seen it developed. When Nick was first seeing his doctor for Parkinson's, the specialist, there's one neurologist that was looking after it, it was him. Now you've got the Movement Disorders Clinic.....

Flora and Neil were among the most resilient couples, expressing satisfaction with each other and their communications with the health care system, stating in no uncertain terms that they were happy and getting their needs met. Their social network within the PD community contributed in particular to their perception of resilience, and this is mirrored in other research as well (Roger, 2007a,b; 2006a,b; Roger et al., 2010; Roger & Medved, 2010).

Jack lives in an intergenerational home that led to more supports than he would otherwise have: his wife's mother was living with him, and as an older senior who was cognitively well was supporting Jack in the care of her daughter (his wife with early onset dementia). Jack's grown daughter, who is now a mother as well, comes regularly to bathe and care for her own grandmother. This pattern of intergenerational supports clearly reduces the caregiving stress Jack has experienced.

However, alternately, another participant had a very different experience. Margareta, whom her husband described as quite a formidable force, did not find support in her own family. She said:

(Int.: "Do you talk to your kids about it at all?") No...No, they don't want to hear about it...They've got families of their own and they're all busy.

Margareta expressed disappointment that 'one must always do everything for oneself' in regards to her family. She also spoke disparagingly of the health care system as an institution.

Although Len did have a sister who lived near him, he was reluctant to over-rely on her for support. His children did not live in the same province, so when he went for what is typically a day surgery, he demanded extra support from the system to compensate for absent family support:

If I needed it, I'd damn well go after it...I would insist, as I did with the nurses 4 times a day. I told the doctor and his secretary, look, I don't have anybody. If I had somebody, I might not be here. I don't have nobody to do this for me. When I had my surgery, they want someone to stay with you overnight. You shouldn't be alone. I don't have anybody to stay overnight. No family. No relatives. Nobody that can stay the night. I want to stay in the hospital the first night cause I have nobody at home and that's what I want. (Int.:"And what did they do?") *They gave it to me...I told them I want it, that's it.*

It was apparent that when family supports were not available that participants depended much more strongly on the health care system and in this way also, our data suggests that they did not always feel the health care system was providing for them. Perhaps their expectation of imagined family assistance was superimposed onto the health care system in a way that could only disappoint. Furthermore, the health care system utilizes an individualistic approach which often leaves family support people "out of the loop". Over time, the rigid boundary which exists between the health care system and the family many have reinforced the tendency of individuals to over-rely on health care professionals, with expectations that the system should provide supports which are not possible in all circumstances.

5.4.1.4 Education levels

A study by Berkman and Syme (1979) identified that a greater social network and frequency of contact lead to decreased mortality of men and women across all ages, even when controlling for socioeconomic status, health status and health practices. However, interestingly, we found participants stated a perception that education and financial resources in fact would significantly improve their interactions with health care professionals. Neil suggests:

Yeah, probably. Cause I know how to write a letter. You know, that's the thing, when you're submitting your resume, you put a covering letter. The first 5 sentences determine where you're gonna go. I got good writing letters when I was a department head of science. There are ways of writing a letter of recommendation which are positive, and there are negative ways.

Later, he describes:

Well you know when you start speaking with an educated person or an uneducated person, you'll notice things in their conversation, grammatical structures and the like, that are more middle class or upper middle class, if you listen carefully, the subjunctive case if you were. If I were instead of if I was. That's what makes a more educated or less educated conversationalist. Now I don't want to be pedantic, but one, so I do make some really simple conversions, but there's different classes of language in English. If you listen carefully, you can tell a person's grade level.

Janelle's partner (who became her ex in the course of the study), was a skilled labourer and was intimidated not only of the health care system (as he said) but also by engaging with us as researchers in this study. He initially felt he had nothing important to say, which changed as he began to engage with us in the interviews.

It was apparent that other participants who felt confident about their ability to articulate their needs clearly and those with higher levels of education felt more confident in getting their needs met from the health care system.

5.4.1.5 Gender of primary caregiver/ professional

Upon examining ageing women with MS, Harrison and colleagues (2008) identified that women with higher levels of social support experienced higher levels of positive attitude towards their condition even when measured over a period of seven years. Janelle reflects here on her very close care relationships with her daughter:

My daughter and I...As close as you can get to a caregiver. (Int. "Okay. So I'm gonna just sort of, like the boys, are they more sort of a peripheral role?") *Yes.* (Int. "Okay. Okay. But your daughter is the one who is more primary?") *Yes, absolutely.* (Int. "Okay. Alright. So I'm just curious. Do you think that's a function of her being a woman or a nurse?") *Actually I think, it's hard to say, but I'd say as a daughter. More as a daughter.*

Another participant, Margareta, exclaimed in no uncertain terms that her girlfriends were her best supports, and regrettably she added, that may include ruling out her husband. Even Jack reflects on gender when he describes that he did not feel comfortable with the quick advice he received from his younger and male doctors:

Oh, the males were stereotypic doctors. They knew, they had all the answers and attitude and whatever. And the women were there to find out what was wrong with you.

Later, Jack says that his son was much less involved and even interested in the kinds of care needs he has with his wife and mother-in-law. Evidently, he thought gender played a role in how caregiving occurred and who was hands on in his household. It appears that the gender of the primary caregiver or the care professional shaped how communication occurred, in the minds of our participants.

5.4.2 Mesosystem

5.4.2.1 Health care literacy

Couples who had good 'health care literacy' were most able to effectively get their needs met. The relationship between health care provision and the couples we interviewed was described in particular in relation to the participant's ability to confidently interact with the system. Couples who were the most familiar with the health care system also appeared to have the most effective tools of communication.

For example, Flora and Neil had several experiences with illnesses (each) before Neil was diagnosed with Parkinson's Disease. This meant that they had had a previous entry into the health care system and how it worked and they were more able to translate this into assistance. In Flora's case, her ability to maneuver within the health care system was compounded by the fact that she had been a Social Worker and had firsthand knowledge of how 'care systems' and institutions operate. She speaks proudly of her high level of health care literacy, acknowledging that this has supported Neil's care plan over the years.

On the other hand, Janelle had not had previous health care concerns or interactions with health care, and so when she was diagnosed, she states in no uncertain terms that she found it very difficult to begin to understand the health care system in a way that provided her with the resources. She stressed that she had been a very independent person prior to being diagnosed, and that this new condition left her feeling more vulnerable and needy than previously. Her husband at the time similarly felt overwhelmed by potential interactions with health care professionals. It appears that being new to 'illness' and 'health care' was an additional challenge (and form of literacy) that some participants were less able to deal with.

In another example, Friedrich had been deaf for years, and that now that his elderly wife Margareta had Parkinson's, he said they just 'carried on as usual'. He said they were already familiar with health concerns and the health care system, they were older and had been married over 50 years, and he felt that they were simply able to step into the system and communicate their needs well within it.

In another similar example, Jack described that one of his parents had had a declining neurological condition and that he had grown up with awareness of this condition and the kinds of care and interactions with care professionals that it required. Now, he was caring for his wife and his mother-in-law. He stated that this gave him some insight and even potentially positive perspectives on how now to deal with his wife's young diagnosis in the context of the health care system.

A study by Wallace and colleagues (2001) suggested that resilience improved among individuals who had a sense of purpose and opportunities for communication particularly during times of personal duress. It was found that providing opportunities for communication could impact a person's compliance to medication, reduce errors in communication, and generally improve the experience of the disease trajectory. Further, those who had previous experience of stressful situations and had a sense of purpose might be better able to maximize previous successful social supports as new situations arose. It was evident that those participants in our study who had experienced previous health crises in their lives, and had already become familiar with the health care system, were much better equipped to handle changes now related to the new diagnosis. Clearly, health care literacy improved participant's interactions with health care providers.

5.4.3 Chronosystem

5.4.3.1 Length of relationship to primary caregiver

Weak social ties have been shown to be correlated with poor health and premature death (Berkman et al., 2000).This may be due to the fact that healthy support systems encourage good self-care. For example, men in happy marriages are less likely to have health problems, and both men and women in happy marriages are more likely to access health care services when required (Sandburg et al., 2009). Kiecolt-Glaser and Newton (2001) reviewed evidence from 64 articles published from 1991 to 2001, and concluded that poor marital quality negatively impacts health of individuals both directly and indirectly, through poor health habits and depression.

Papadatou (2009) states that social support shapes well-being between people over time, especially when they are dealing with exceptionally difficult life processes, and these may include: a long time committed and shared trajectory; shared responsibility; reinforcement of communication and involvement between committed partners; continuity in times of uncertainty and distress; and allowing for shared learning throughout the defining life process. Certainly, those participants who could be described as strong in their committed relationships appeared to be more resilient and excelled at interacting with the health care system. Thus, evidence suggests that the bond that couples have with each other can contribute to health-related benefits and improve upon an individual's capacity for resilience.

An important factor was how long couples had been together and whether they were married and had raised children together. It was clear that couples who had raised children together and/or had been married for a longer time, were more resilient in handling the issues that now arose. In fact, one of the younger (in age) couples who had only been together a few years and that we interviewed in the first Phase of the study, were no longer together in the second Phase of the study. It became clear in the second interview in Phase II, when we only interviewed one person in the couple, that the progression of the condition had contributed towards the ending of their relationship. In another case, however, a long time marriage led to a deepened sense of commitment to the husband who had one of the selected conditions.

I don't have trouble with that really. Well if anything's bothering me, I tell her right out. I don't care how they feel about it. In my mind, my concern is my husband. I don't care what you're doing with anybody else, but my husband, I want him looked after too. And they've been very, very good. (Marlene)

Marlene stresses that she knows her husband well and what his needs are and that this aids her in communicating more effectively with the health care professionals. She acknowledges that the long term care unit her husband is on is chronically lacking enough staff to adequately meet the residents' needs. However, her loyalty and commitment to her husband ensure she advocates for him so that he gets his needs met, despite a lack of resources:

I don't have trouble with that really. Well if anything's bothering me, I tell her right out. I don't care how they feel about it. In my mind, my concern is my husband. I don't care what you're doing with anybody else, but my husband, I want him looked after too. And they've been very, very good. (Marlene)

Length of time together is an indicator of relationship commitment and loyalty, and even when the patients cannot advocate for themselves, a committed and assertive partner can ensure they have their needs met, even in an overburdened healthcare system.

Margareta and Friedrich also stressed that their 50 + years of marriage had allowed them to grow more resilient as a couple over time in a way that was demonstrated by their ability to now cope. This length of time together, according to Friedrich, has positively impacted their ability to communicate with health care professionals.

5.4.3.2 Stage of the illness

It became evident over the three years of collecting data that the patient's stage of illness in tandem with existing social supports largely contributed to their sense of resilience (or lack thereof) and (in)ability to communicate well with health care professionals. Research has shown a consistent positive relationship between social connectedness, integration and stages of physical health (Cohen, 2004). Janelle became much less able to engage with the health care system after her relationship ended and her condition deteriorated. However, she then made the decision to move to another province where her children were living and this dramatically improved her health - as evidenced in her second interview with us. She felt more able to communicate about her needs, and she certainly had social supports around her.

Flora on the other hand, suggested that you need to prepare for times ahead when you know a condition will be deteriorating:

I think it's partly communication. Sometimes people just don't know what to ask for even in the beginning. Like you haven't been trained to know what to ask for. You know, people vary. Sometimes they just think, oh I'll just keep managing, or I'll just keep going. They wait until there's a crisis instead of saying, okay, we're on this path anyways now. But this is a chronic illness. What is it's path? What do we expect? You can't predict. You know, you could have Parkinson's and die of a heart attack. You have to start early and prepare for things down the road.

Other participants reflected on a 'future' when they may not be able to communicate as well as they do now. They were developing an awareness of the kinds of supports they might need to put into place for that point in time. When people are living with a neurological condition, a fact which clearly demarcates them from examples such as cancer, they know that their condition may progress over sixteen or more years. Their ability to consider a distant future means that they have a lot of time to consider changes that might occur to them. This makes their situation unique from other conditions. Their engagement with health care professionals will also remain in a more 'chronic' phase for many years, with room for many discussions about possible later stages. This can both be a strength and a vulnerability where people can prepare and plan, but also may become overwhelmed with realities. Health care professionals must have a clear understanding of this trajectory which includes emotional, cognitive and physical realms.

5.4.3.3 Age

While there were few quotes that specifically explored participant's reflections on age, it was apparent that 'age' was a constant presence in the discussions regarding health care communications. Jack stated how much he did not appreciate 'the younger physicians', who he felt had less experience and fewer insights into situations he might have questions about. Age shone through in another way as well - it was clear that the older couples in our sample were more resilient than the younger couples and that they were more prepared for eventualities related to neurological decline. They appeared to have more capacity when interacting with and communicating with the health care system as well. It was apparent that younger folk had a much harder time with their diagnosis, such as Janelle. She would not have anticipated this diagnosis at such a young age, and her stories indicate that she was having a harder time in general. Age also intersected with the long term couples to create a bond that promoted not just their social supports developed over time, but also their sense of resilience as individuals and a couple.

5.5 Findings from focus group data

The focus group data consolidated the importance of the nine factors found in the interviews that have been discussed above - bringing together key elements to this study and the emerging data. The above nine factors were presented in summary form to the focus groups to ascertain the extent to which these findings reflected the participant professional's experiences. Each of our focus group participants confirmed that these factors were highly relevant in their experience of intake with new individuals in order to determine how couples were handling difficult situations related to their diagnosis. They stated that the nine factors were central to how well the couples (and families) were doing and thus also how they would benefit from health care communications.

However, focus group participants also expressed concern that there was not a standard form or approach to this kind of intake assessment. Few tools if any existed to refine or better understand these nine factors. They underlined how relevant this information would be if it were being collected and analyzed more effectively by individual organizations / researchers.

One of the key discussions in the focus groups thus revolved around the preparation, implementation and compilation of their organization's intake forms. The participants in our focus groups stated that while intake forms were used to collect information related to the nine factors, the process of collecting information and the forms themselves were limited. For example, there were several different forms to be used at intake. Little training existed in how best to fill out the intake forms and what to do with the information professionals had collected. Further, our focus group participants described that any number of forms developed by their organizations might be used at intake and then filed away without being compiled or analyzed. Therefore, the information being collected was fragmented and lacked a cohesive structure or purpose after collection.

The focus group participants stressed that developing a unified form, and a comprehensive analysis of the data on these forms, could result in an important source of information for developing not only better assessment tools for intake, but also towards improved services for patients. Professionals stated they could better assess couples at risk and also provide more catered and effective care services if this information was made analyzed and compiled and then available to them. However, care professionals in our focus groups stated they had neither the time nor the skills to compile and analyze the data available to them through their own organizational intake forms.

6. Discussion

Since we know so little about interactions between family care providers and their loved ones in terms of health communications, and we know equally little about their interactions with health care professionals, it is even more significant that we also know so little about a large group of persons living with neurological decline. Living with a neurological condition is a reality which can affect individuals and their caregivers for well over fifteen years. The unique medical realities associated with the etiology of the neurological conditions encompassed within the mandate of the NHCC must be better understood.

Firstly, the value of social support has emerged in this data as central to the couples interactions with health care professionals, and this in turn, matches what the literature suggests about resilience. Assessing for social support and resilience early on in the intake procedure will benefit a well implemented care plan and improved well-being for all concerned. Social support has been shown to emerge not only through personal family and friendship relationships but also through professionals who have had a continuous relationship with the patient over time. When there is a high turnover and professionals have little education and training in dealing with neurological conditions, patients are impacted with less than optimal care conditions. Compliance to treatment plans and general well-being are affected.

In the context of Bronfenbrenner's ecological model, we can presume that a care professional's ability to communicate clearly with couples in a health care context depends on an understanding of factors that go beyond the medical condition itself. A couple's level of social support appears to be important to building their perception of their own resilience. Likely, social support is critical when we consider communication issues with health care providers as well - potential errors in developing care plans, misjudgment of what is required, and the patient's ability to successfully carry out what has been suggested. If professionals have a good understanding of social factors that surround a patient's daily life, these could be articulated and understood in order to benefit the medical aspects of the treatment. To separate out medical aspects of a condition from daily lived experiences (e.g. social support) may be setting the stage for poor care plans that may not be implemented properly – leading to potential for further crisis and poor health down the road.

The participant narratives demonstrate that the couples' perception of resilience was intertwined with factors associated with social support. Participants who were less able to 'push the system' also appeared to have fewer social supports in their life, they had less familiarity with the health care system, they appeared to have lower levels of education, or, they were simply farther along on the disease trajectory and depended on others to advocate on their behalf. A better understanding of the intersection of these nine factors is imperative.

The goal of this study has been to better understand how couples living with a neurological condition communicate and how they do so in the context of health care communications. In particular, the data discussed in this paper are intended to assist the development of tools that can also assist professionals in improving their understanding of individuals (e.g. couples and families who provide care) living with neurological conditions. Recommendations towards this goal follow.

7. Recommendations

Several recommendations are highlighted as a result of this study (see Table 3). It was evident that social support can be seen to improve the participant's sense of their own resilience when communicating about health and with the health care system. Intake could also be improved to assess for levels of social support and resilience in couples. The nine factors found in our sample reflect important aspects of social support and resilience. This information can lead the way to the development of a new tool that would provide more insight into how couples experience resilience and how this can benefit their communication with health care professionals and the health care system. This data leads to the recommendation that the nine factors be used to develop a new form which could be piloted in a selected set of organizations. These organizations could then participate in a second phase where the data collected over a specified period of time be compiled and analyzed with 'social supports as promoting resilience' in mind. This would have in mind a third phase with the purpose of establishing improved communication training (e.g. developing new training and education) between professionals and clients.

PILOT NEW FORM	NEW DATA COLLECTION UPON INTAKE	NEW TRAINING FOR PROFESSIONALS AND FAMILY
1. To develop a pilot of a new form that encompasses the nine factors identified here.	2. To collect data on each of the nine factors as these relate to patient intake. To support inhouse compilation and analysis of these data.	3a. This would include training for professionals in order to identify levels of risk and resilience. 3b. Selected mentors, advocates, andor a buddy system within health care could be created to train family members upon intake.

Table 3. Recommendations for New Research and Organizational Interventions.

8. Conclusion

This paper has discussed original, empirical data highly critical for health service provision given the anticipated increased diagnoses of persons with neurological conditions. Since we know that little research has been found examining the daily lived experience of families, where one person is living with a neurological condition, this study contributes to knowledge in this area. The paper has explored nine factors that emerged from the data. The authors suggest that communication about care, and the many factors that shape it, must be better integrated into our daily health care provision - expanding on what we know about medical aspects of conditions such as Huntington's and Parkinson's. Since contradictory and often unspoken expectations of care emerge between professionals, family and individuals, a better understanding of social factors that influence communication might reduce errors in care, poor treatment, and misunderstandings. These outcomes can result in significant cost savings for the health care system but also improved well-being of families and those affected by neurological conditions.

9. Acknowledgements

The authors would like to acknowledge the University Start-up Grant (University of Manitoba) and the University Research Grants Program each for funding this study over the three years.

10. References

Aarsland, D. & Kurz, M. (2010). The epidemiology of dementia associated with Parkinson disease. *Journal of the Neurological Sciences,* 289, pp. 18-22

Aubeeluck, A. & Wilson, E. (2008). Huntington's disease. Part 1: Essential background and management. *British Journal of Nursing,* 17, pp. 146-151

Berkman L. & Syme L. (1979). Social networks, host resistance, and mortality: A nine-year follow up study of Alameda County residents. *American Journal of Epidemiology,* 109, 2, pp. 186-204

Berkman, L., Glass, T., Brissette, I., & Seeman, T. (2000). From social integration to health: Durkheim in the new millennium. *Social Science & Medicine,* 51, pp. 843- 857

Brody, H. (1987). *Stories of Sickness,* Yale University Press, New York, USA

Bronfenbrenner, U. (1979). *The Ecology of Human Development: Experiments by Nature and Design,* Harvard University Press, ISBN 978-067-4224-57-5, Cambridge, Massachusetts, USA

Bronfenbrenner U. (1986). Ecology of the family as a context for human development research perspectives. *Developmental Psychology,* 22, pp. 723–742.

Bubolz, M. & Sontag, M. (1993). Human ecology theory, In: *Sourcebook of family theories and methods: A contextual approach,* P. G. Boss, W. J. Doherty, R. LaRossa, W. R. Schumm, & S. K. Steinmetz (Eds.), pp. 419-450, Plenum Press, New York

Charmaz, K. (1991). *Good Days, Bad Days: The Self in Chronic Illness and Time,* Rutgers University Press, ISBN 978-081-3519-67-8, New Jersey, USA

Clark, C. (2007). Parkinson's disease. *British Medical Journal,* 335, pp. 441-445

Cohen, S. (2004). Social relationships and health. *American Psychologist,* 59, pp. 676-684

Earvolino-Ramirez, M. (2007). Resilience: A concept analysis. *Nurse Forum,* 42, 2, pp. 73-82

Garmezy, N. (1985). Stress-resistant children: The search for protective factors, In: *Recent Research in Developmental Psychopathology,* J. Stevenson (Ed.), pp. Pergamon Press, ISBN 978-008-0308-28-9, Oxford, UK

Garmezy, N. (1993). Children in poverty: Resilience despite risk. *Psychiatry,* 56, 1, pp. 127-136

Graneheim U. & Lundman B. (2004). Qualitative content analysis in nursing research: Concepts, procedures and measures to achieve trustworthiness. *Nurse Educ Today,* 24, pp. 105-112

Guttman, M. Slaughter, P., Theriault, M., DeBoer, D., & Naylor, C. (2003). Burden of Parkinsonism: A population-based study. *Movement Disorders,* 18, pp. 313-336

Harper, P. (1992). The epidemiology of Huntington's disease. *Human Genetics,* 89, pp. 365-376

Harrison, T., Blozis, S., & Stuifbergen, A. (2008). Longitudinal predictors of attitude towards aging among women with multiple sclerosis. *Psychology of Aging,* 23, 4, pp. 823-832

Heisters, D. (2011). Parkinson's: Symptoms, treatments and research. *British Journal of Nursing,* 20, pp. 548-554

Johnson, J. & Wiechelt, S. (2004). Introduction to the special issue on resilience. *Substance Use and Misuse,* 39, 5, pp. 657-670

Kiecolt-Glaser, J. & Newton, T. (2001). Marriage and health: His and hers. *Psychological Bulletin,* 127, pp. 472-503

Krawlik, D., van Loon, A., & Visentin, K. (2006). Resilience in the chronic illness experience. *Educational Action Research,* 14, 2, pp. 187-201

Lix, L., Hobson, D., Azimaee, M., Leslie, W., Burchill, C., & Hobson, S. (2010). Socioeconomic variations in the prevalence and incidence of Parkinson's disease: A population-based analysis. *Journal of Epidemiology and Community Health,* 64, pp. 335-340

Masten, A. (1999). Resilience comes of age: Reflections on the past and outlook for the next generation of research, In: *Resilience and Development: Positive Life Adaptations,* M. Glantz & J. Johnson (Eds.), pp. 281-296, Kluwer Academic/Plenum Publishers, ISBN 978-030-6461-23-1, New York, USA

Moreau, A., Boussageon, R., Girier, P., & Figon, S. (2006). The "doctor" effect in primary care. *La Presse Medicale,* 35, pp. 967-973

Morse, J., Barrett, M., Mayan, M., Olson, K., & Spiers J. (2002). Verification strategies for establishing reliability and validity in qualitative research. *International Journal of Qualitative Methods,* 1, pp. 1-19

Nance, M. & Myers, R. (2001). Juvenile onset Huntington's disease—Clinical and research perspectives. *Mental Retardation and Developmental Disabilities Research Reviews,* 7, pp. 153-157

National Institute of Neurological Disorders and Stroke. (2008). NINDS Huntington's disease information page, 19.07.2008, Available from: http://www.ninds.nih.gov/disorders/huntington/huntington.htm

Nutt, J. & Wooten, F. (2005). Diagnosis and initial management of Parkinson's disease. *The New England Journal of Medicine,* 353, pp. 1021-1213

Papadatou, D. (2009). *In the face of death: Professionals who care for the dying and the bereaved,* Springer Publishing Company, ISBN *978-082-6102-56-0,* New York, USA

Paulsen, J., Ready, R., Hamilton, J., Mega, M., & Cummins, J. (2001). Neuropsychiatric aspects of Huntington's disease. *Journal of Neurology, Neurosurgery and Psychiatry,* 71, pp. 310-314.

Public Health Agency of Canada. (2000). Research on Alzheimer's care giving in Canada: Current status and future directions, *Chronic Diseases in Canada (CDIC),* 25, 3/4,, Available from:
www.phac-aspc.gc.ca/publicat/cdic-mcc/25-3/c_e.html

Richardson, G. (2002). The metatheory of resilience and resiliency. *Journal of Clinical Psychology,* 58, 3, pp. 307-321

Roger, K., Mary-Quigley, L., & Medved, M. (2010). Perceptions of Health Care and Familial in the Context of Parkinson's Disease and Multiple Sclerosis. *Journal of Communication in Healthcare,* 3, 2, pp. 124-137

Roger, K. & Medved, M. (2010). Couples manage change in the case of terminal medical conditions including neurological decline. *International Journal of Qualitative Studies on Health and Well-being,* 5, 2, pp. 5129

Roger, K. (2007a). It's a problem for other people, because I am seen as a Nuisance: Hearing the voices of people with dementia. *Alzheimer Care Quarterly,* 8, 1, pp. 17-25

Roger, K. (2007b). End-of-life care and dementia. *Geriatrics and Aging,* 10, 6, pp. 380-384

Roger, K. (2006a). Understanding social changes and the experience of dementia. *Alzheimer Care Quarterly,* 7, pp. 185-193

Roger, K. (2006b). Literature review on palliative care, end of life, and dementia. *Palliative and Supportive Care,* 4, pp. 1-10

Roos, R. (2010). Huntington's disease: A clinical review. *Orphanet Journal of Rare Diseases,* 5, 40

Safran, D., Taira, D., Rogers, W., Kosinski, M., Ware, L., & Tarlov, A. (1998). Linking primary care performance to outcomes of care. *The Journal of family practice,* 47, 3, pp. 213-220

Sandburg, G., Miller, R., Harper, J., Robila, M., & Davey, A. (2009). The impact of marital conflict on health and health care utilization in older couples. *Journal of Health Psychology,* 14, pp. 9-17

Statistics Canada. (2007). *Participation and Activity Limitation Survey 2006: Analytical Report.* Ministry of Industry, Catalogue no. 89-628-XIE, Ottawa, Ontario, Canada

Sturrock, A., & Leavitt, B. (2010). The clinical and genetic features of Huntington disease. *Journal of Geriatric Psychiatry and Neurology,* 23, pp. 243-259

Tarrant, C., Stokes, T., & Baker, R. (2003). Factors associated with patients' trust in their General practitioner: A cross-sectional survey. *British Journal of General Practice,* 53, pp. 798-800

Walker, F. (2007). Huntington's Disease. *The Lancet,* 369, 9557, pp. 218–228

Wallace K., Bisconti T., & Bergman C. (2001). The mediational effect of hardiness on social support and optimal outcomes in later life. *Basic Applied Social Psychology,* 23, 4, pp. 269- 279

Werner, E., & Smith, R. (1977). *Kauai's children come of age,* University of Hawaii Press, ISBN 978-0824804756, Honolulu, Hawaii, USA

Werner, E. (1990). Protective factors and individual resilience, In: *Handbook of Early Intervention,* S. Meisels & J. Shonkoff (Eds.), pp. 97-116, Cambridge University Press, 978-052-1387-77-4, Cambridge, UK

Zloty, A., Lobchuk, M., & Roger, K. (2010). A model for the development of caregiver networks. *WORK: A Journal of Prevention, Assessment & Rehabilitation,* in press.

23

Communication Between Huntington's Disease Patients, Their Support Persons and the Dental Hygienist Using Talking Mats

Ulrika Ferm[1], Pernilla Eckerholm Wallfur[2], Elina Gelfgren[2] and Lena Hartelius[2]
[1]DART Centre for Augmentative and Alternative Communication and Assistive Technology, Regional Rehabilitation Centre, Queen Silvia Children's Hospital, Sahlgrenska University Hospital,
[2]Institute of Neuroscience and Physiology, Division of Speech and Language Pathology, University of Gothenburg, Sweden

1. Introduction

Communication is at the heart of any health care situation and individuals, who have difficulties describing their problems and expressing their needs, are in danger of being misunderstood or mistreated (Bartlett et al., 2008). Persons with Huntington's Disease (HD) have cognitive, emotional and motor problems which affect their communication and they frequently need support to be able to communicate in their daily life in general and in health care situations in particular. This chapter describes an effort to enhance communicative effectiveness in a dental and oral health care situation using Talking Mats. Eleven individuals, their support persons and a dental hygienist volunteered to help in exploring the use of this method in a clinical situation.

2. Background

Several of the changes associated with the progression of HD affect communication. Cognitive and emotional changes lead to fewer communicative initiatives, word finding difficulties, grammatical errors and difficulties in keeping track of what is being said in a conversation (Jensen et al., 2006; Yorkston et al., 2004). Furthermore, difficulties in managing complex discourse, tasks that involve interpretation of ambiguous, figurative and inferential meaning, are common and can appear early in disease progression (Chenery et al., 2002; Saldert et al., 2010; Saldert & Hartelius, 2011). Changes in motor function affect speech and articulation and symptoms of dysarthria are common (Hartelius et al., 2003; Yorkston, et al., 2004). The most frequently occurring perceptual deviations found in continuous speech in the study by Hartelius, et al., were mainly related to speech timing and phonation and reflected the underlying excessive and involuntary movement pattern. Deviation related to speech timing were variations in speech rate, shortened phrase length, and prolongation of interword and intersyllable intervals. Phonation-related aspects included increased pitch, harsh and strained-strangled phonation, and decreased pitch variation. Imprecise consonant articulation was also prominent, but to a less severe degree (Hartelius et al., 2003).

Individuals with HD report that communication demands more concentration and is more tiring than before they had the disease (Hartelius et al., 2010). In the same interview study, family members and carers reported that the persons with HD had increased difficulties understanding complex information and that their personality changes also had led to decreased quality of communication with lack of in-depth talk, difficulties shifting focus in conversation, etc. One action to take to meet the communicative difficulties is to introduce different kinds of augmentative and alternative communication (AAC) strategies and tools. It is important that AAC, and communication aids in particular, are introduced early in the disease process, that they are simple to use and that a conversation partner is actively present to create structure and support in the use of the aid in real-life communication (Yorkston et al., 2004). Attitudes, skills and knowledge in conversation partners influence communication (Allwood, 2000; Kagan et al., 2004) and AAC-interventions relating to persons with HD should take the experiences of conversation partners into consideration (cf. Saldert et al., 2010).

Huntington's Disease is, in every sense of the word, a family disease, and significant others are often closely involved in care and communication surrounding the afflicted family member. The disease eventually leads to increasing need of health care and a multitude of health care contacts (McGarva, 2001; Roos, 2010). One of the health care professionals frequently in contact with persons with HD is the dental hygienist. Dental and oral health is of vital importance because of its effects on chewing, swallowing and speech and essential to avoid the increased risk of caries, gingivitis and periodontitis that comes with dental and oral care neglect (Kidd, 2005; Klinge & Gustavsson, 2011). Individuals with HD often need to increase the number of daily meals and the energy content in their food, they have decreased flow of saliva and frequently also anti depressant medication which create dry mouth. These factors all contribute to a danger of oral and dental health problems.

In dealing with dental and oral health care, significant others play an important role. Problems with fine motor control makes it more difficult for the person with HD to manage toothbrush, dental floss, fluoride tablets etc. and the cognitive problems are challenging when trying to follow instructions and remembering the appropriate use of different items (Gabre, 2009). The visit to the dental hygienist is not unique in this sense. Murphy (2006) investigated communication between health care personnel and persons with aphasia or cognitive disabilities. Both patients and personnel experienced misunderstandings because of communication difficulties. The patients had problems remembering what they wanted to say and following instructions from the doctor. Doctors also used words that the patients didn't understand. The patients expressed the need to have information given in writing and with supporting pictures. All social activities are related to certain procedures, goals and roles that influence communication in different ways. As far as the support of persons with cognitive and communicative disabilities is concerned, it is important to take these activity factors into consideration (cf. Ahlsén, 1995; Allwood, 2000).

Talking Mats™, TM (Murphy & Cameron, 2006) is a method used to enable persons with cognitive and communicative difficulties to express their opinions (see Figure 1). Talking Mats does not replace a person's communication aid but can be used without or together with a communication aid. The method consists of a textured mat on which relevant pictures are stuck in a structured way. There are three sets of pictures: a visual evaluation scale, a picture describing a topic and pictures associated with the different questions relating to the topic. The conversation partner (the person responsible for the TM conversation, e.g. a nurse, a speech-language pathologist, or as in this case, a dental hygienist) formulates open questions such as: "How does it work to use…?", "What to you think of…?" etc. In addition to the prepared picture-based questions, new issues written down on pieces of note paper or empty cards can also be added. The person answering the questions (the person with difficulties expressing themselves, in this case individuals with HD) places the picture representing a specific question below the picture in the visual evaluation scale that best matches his or her opinion, but can also point to the part of the evaluation scale where the picture should be put. At the end of the conversation, the conversation partner recapitulates the discussion and seeks confirmation regarding the opinions expressed by the person being interviewed (Murphy & Cameron, 2006).

Fig. 1. A mat (Talking Mats) including a visual evaluation scale at the top, a picture for the conversational topic at the bottom, and pictures for different questions relating to the topic in the middle. The figure includes Picture Communication Symbols © 1981-2011 by Mayer-Johnson LLC.

One aim of conversations using TM is increased communicative involvement. Murphy et al. (2010b) conducted a study where individuals with dementia and their significant partners were engaged in two types of conversation, with and without TM. The conversations were on everyday topics such as personal care and household work. Results showed that persons with dementia as well as their partners felt significantly more involved in the conversations using TM and the increase was significantly higher for the partners compared to the participants with dementia. Communication effectiveness was assessed to be significantly higher in the conversation using TM compared to without TM.

Ferm et al. (2010) compared unstructured and structured conversations with conversations using TM with five individuals in different stages of HD. Conversations were compared with respect to communication effectiveness as measured by EFFC (Effectiveness Framework for Functional Communication, Murphy & Cameron, 2008). Communication effectiveness was significantly higher in the structured conversation as compared to the unstructured and highest in the conversation using TM. The conversation partner expressed the view that the persons with HD showed a greater involvement in the conversation and also that it felt more natural to wait quietly for the participant's answer when using TM. Talking Mats as a communication support has also been tried successfully in group discussions for persons with HD (Hallberg et al., 2011) but not yet in real health care situations. The participants in the discussion group studied by Hallberg et al. were more effective communicating about diet and health when TM was used than when the questions around these topics were discussed without TM. The difference in communicative effectiveness between the conditions was significant on both individual and group levels. Another interesting finding of this study was that the group leader and some of the individuals with HD asked significantly more follow-up questions when using TM than when the group discussion was unaided. Over all the group members with HD were positive about using TM.

As mentioned earlier, one of the health care situations that persons with HD encounter is dental and oral care. In Gothenburg, Sweden, most individuals with HD visit the dental hygienist between once a month and once every third month, to create or to keep a good dental and oral health. During a typical visit, a good part of the communication is done when the patient receives his or her treatment, lying down in the dental chair. The dental hygienist starts by talking about general things to make the patient feel at ease and continues on to give instructions, frequently with the support of pictures. When the ability to care for their own dental and oral hygiene is decreased, the instructions are given to the person supporting the patient during the visit (e.g. family member, assistant or carer). Communicative support in the dental and oral health care situation is of great value (Lewis et al., 2008) and the development and evaluation of appropriate support methods is important.

The aim of the present study was to explore the use of Talking Mats in conversations with individuals with HD in the dental and oral health care situation. The specific research questions asked were: 1) is there a significant difference in communicative effectiveness between conversations where Talking Mats is used compared to conversations where TM is not used?, 2) Is there a significant difference in perceived communicative involvement between the two types of conversation on the part of the individuals with HD?, 3) Is there a difference in perceived communicative involvement between the two types of conversation on the part of the support persons? And 4) Does the dental hygienist perceive the use of TM as a beneficial support in the dental and oral health care consultation?

3. Method

The study was designed to compare two different types of conversations between persons with HD, their support persons and a dental hygienist using both quantitative and qualitative methodology. Data was collected during naturally occurring dental and oral health care consultations.

3.1 Participants

Twenty four persons participated in the study; eleven individuals with HD (seven men and four women, mean age = 52 years, range 24 – 75 years), twelve support persons and a dental hygienist. The same dental hygienist carried out all conversations. The individuals with HD and their partners formed eleven dyads. As can be seen in Table 1, dyad 1 included two assistants which meant it really was a triad (i.e., included three individuals). For the sake of simplicity, we will call it a dyad. The support persons were relatives and professionals that accompanied the person with HD to the dental hygienist. Ten of the participants with HD had continuous contact with the dental hygienist; one participant had met the dental hygienist a couple of times. The dental hygienist was trained in TM but had limited experience in its use.

Dyad	Participants	Age	Onset HD	Phase	Education	Used TM before	Length of relationship (years)
F1	F1	53	40	4	compulsory school	yes	
	F1(A)ass	55			high school	no	10 months
	F1(B)ass	28			high school	no	1 year
F2	F2	75	57	5	university	yes	>50 yrs
	F2husband	77			university	no	
F3	F3	64	58	3-4	university	yes	3.5 yrs
	F3ass	59			university	no	
F4	F4	58	50	4-5	university	no	5 months
	F4ass	19			high school	no	
M1	M1	24	20	4	high school	yes	6 months
	M1ass	45			high school	no	
M2	M2	28	22	3	high school	yes	3.5 years
	M2ass	57			compulsory school	no	
M3	M3	57	47	4	university	no	3 years
	M3ass	22			high school	no	
M4	M4	46	30	3	university	yes	2.5 years
	M4ass	50			high school	no	
M5	M5	52	50	3	compulsory school	no	1.5 years
	M5support person	46			university	no	
M6	M6	57	52	no info	compulsory school	no	a couple of years
	M6counselor	43			university	no	
M7	M7	57	54-55	2	compulsory school	no	29 years
	M7daughter	29			high school	no	

Table 1. Participant characteristics. F = female; M = male; Onset HD = Age of first symptoms of HD, Phase = TFC-phase (Shoulson et al., 1989) for the participant with HD according to the dental hygienist, Education = highest completed education, ass = personal assistant; Length of relationship = number of months or years the participants had known each other.

Invitation of participants was done by the dental hygienist. All participants were registered as clients at Mun-H-Center[1]. Inclusion criteria were HD, contact with the dental hygienist and interest and willingness to participate in a study about communication support. No formal cognitive or linguistic assessments were made. However, persons in the late stage of the disease were not invited to participate. All participants communicated through speech and no one used personal communication aids during the visit at the dental hygienist's. The participants' speech varied in intelligibility.

3.2 Ethical considerations

The study built on relevant research and was led by professionals with expert knowledge within the fields of HD and augmentative and alternative communication. Participation was voluntary and built on informed consent. The individuals with HD and their support persons all signed consent forms. Due to the cognitive and linguistic difficulties accompanying HD particular attention was given to the process of informing the participants with HD. The study was described in detail by the dental hygienist and the researchers on three different occasions. Simplified written information with pictures was also supplied. The participants were informed that they could withdraw from the study at any time without specific reasons and without personal consequences. They were also informed that their data would be treated with integrity and that no names would be used in the dissemination of the results. All participants were informed about the results of the study and participants with HD received photographs of their mats.

3.3 Material

Black textured mats (37 x 58 cm), five pictures (6 x 6 cm) representing a visual evaluation scale, a picture (5 x 5 cm) for the conversational topic oral hygiene and prophylaxis, and pictures (5 x 5 cm) of the questions relating to the topic were used. Velcro on the back of the pictures allowed these to be placed and moved around on the mat. Digital photographs and Picture Communication Symbols PCS (Mayer-Johnson™, 1981-2011) were used.

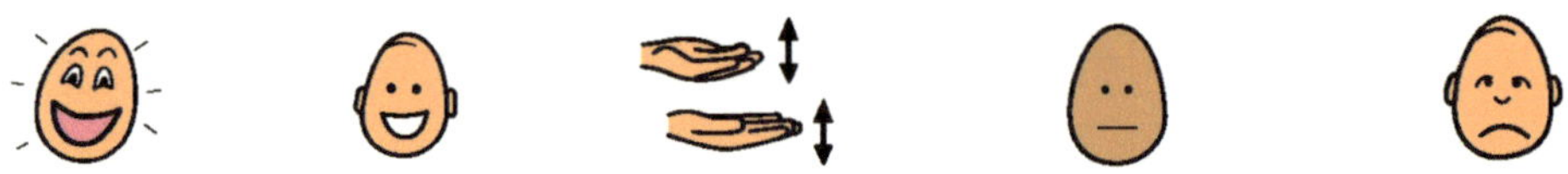

Fig. 2. The visual evaluation scale used with Talking Mats. The figure includes Picture Communication Symbols © 1981-2011 by Mayer Johnson LLC.

Twenty questions about oral hygiene and prophylaxis were developed by the dental hygienist and the researchers (Table 2). The questions formed two sets which included ten questions each. The questions were designed to be equivalent with regard to content and level of difficulty. Each dyad received both sets of questions; one set in the condition where TM was used and the other set in the condition where TM was not used. The order of the

[1]Swedish national oro-facial centre of expertise for rare disorders and national resource centre for oro-facial assistive devices, Gothenburg, Sweden.

question sets and conditions (TM and nonTM) were counter balanced (Table 3). The purpose of using two different but equal sets of questions in the two conditions was to create different but yet content wise similar conversations. In this way, the effects of TM, rather than of different questions, could be evaluated.

Question	Set 1	Set 2
1	How does teeth brushing work?	What would you say about getting help with teeth brushing?
2	How does it work brushing the inside of your teeth?	How does it work brushing exactly where you intend to?
3	How does it work using a regular toothbrush?	How does it work using a double toothbrush?
4	What would you say about getting help cleaning between your teeth?	How does it work cleaning in between your teeth?
5	How does it work using an interspace toothbrush?	How does it work using dental floss?
6	How does self cleaning work?	How does it work rinsing the mouth after the meal?
7	How does it work using toothpaste with extra fluoride?	How does it work rinsing the mouth with fluoride?
8	How does it work using fluoride chewing gum?	How does it work using fluoride tablets?
9	How does it work using gel against mouth dryness?	How does it work using spray against mouth dryness?
10	How does it work sitting in the chair?	How does it work lying in the chair?

Table 2. The two sets of questions used in the two conditions, TM and nonTM.

The participant with HD and the support person each filled out two questionnaires about the two different conditions. The questionnaire regarding the nonTM condition included seven questions (1 to 7 below). The questionnaire regarding the TM condition included the same seven questions (1 to 7 below) and one additional question (8). Questions 1 to 5 and 7 were similar to the questions used by Murphy et al. (2010b). Questions 6 and 8 were constructed for this study. The questions were: (1) Do you think that the questions asked were relevant for you? (2) Did the others listen to you in the conversation? (3) Were you able to express your opinions? (4) Did you have enough time to express your opinions? (5) Did you feel involved in the conversation? (6) Did it work well doing this together with NN? (7) How well do you think the conversation went? Circle the picture that best suits your opinion! (8) What do you think about using Talking Mats? Describe with your own words! A visual scale including four pictures of the concepts all/always, most/usually, a few/occasionally and none/never was used for questions 1 to 6. A seven point scale representing the continuum bad to excellent was used for question 7. The scales included pictures (Mayer-Johnsson™, 1981-2011) and were similar to the ones used by Murphy et al. (2010b).

The dental hygienist filled out two questionnaires for each dyad; one for the nonTM condition and one for the TM condition. The questionnaires included seven identical questions: (1) To what degree did the person with HD understand the questions? (2) To what degree did you get carefully considered answers to the questions? (3) To what degree did you feel listened to in the conversation? (4) How natural was the conversation? (5) How easy was it to stay on topic in this conversation? (6) How involved did you feel in the conversation? (7) How well do you think the conversation went? Circle the picture that best suits your opinion! Questions 1 to 6 and 7 were answered according to the same four and seven point scales that were used by the participants with HD and by the support persons.

After each consultation, a semi-structured interview was carried out with the dental hygienist. The interview included six open questions about the two conditions and about TM.

The consultations, with and without TM, were recorded using a Canon HD Legria HF S11 camera and the mats were photographed using a Panasonic Lumix DMC-TZ8.

The Effectiveness Framework of Functional Communication EFFC (Murphy & Cameron, 2008) was used to measure communicative effectiveness in the two conditions.

3.4 Procedure

Data collection was done during regular consultations with the dental hygienist at Mun-H-Center (9 dyads), at one participant's home (1 dyad), and at an activity centre (1 dyad) from November 2010 to February 2011. There were totally eleven consultations, one for each dyad. During each consultation two different conversations were carried out; one with Talking Mats (TM) and one without Talking Mats (nonTM). Consequently, there were 22 conversations in total. Each consultation started with repeated information about the study by the researchers and the signing of consent forms. A short conversation with TM was demonstrated. Thereafter the main researchers (second and third authors) left the room and the dental hygienist carried out the TM and nonTM conditions with the dyad. The dental hygienist was informed about the order of conditions and question sets for each dyad (Table 3) and about the fact that additional questions were allowed. The ten questions within each set were asked in the same order. Both conversations were recorded with a digital video camera. Towards the end of each session, some of the participants received dental treatment by the hygienist.

Dyad	Condition	Question set
M1, F2, M6	TM nonTM	1 2
M2, M4, M7	nonTM TM	2 1
M3, F3, F4	TM nonTM	2 1
F1, M5	nonTM TM	1 2

Table 3. Order of conditions and question sets for the eleven dyads.

After the completion of the two conversations the participants filled out the questionnaires. One researcher assisted the persons with HD who needed it by reading the questions aloud and by noting which picture in the visual scale the person pointed to. For question 8, the persons with HD were encouraged to describe their opinions about TM. These were written down by the researcher. The support persons filled out the questionnaires independently but a researcher was close by in case any of them had questions. The two questionnaires were answered in the same order as the two conditions had been carried out. The dental hygienist filled out the questionnaire on her own after the completion of the two conversations. The interview was carried out at the end of the consultation, that is, after the dental treatment. One researcher asked the questions and the other researcher took notes.

Data was compiled and communicative effectiveness as well as the participants' feelings of communicative involvement and satisfaction in the two conditions were examined and compared on group and individual levels.

3.5 Analysis

The communicative effectiveness of the persons with HD in the two conditions was evaluated by the two researchers who also assisted with data collection during the consultations. The evaluation was done using EFFC (Murphy & Cameron, 2008). Each conversation was evaluated according to four factors namely (a) the participant's understanding of the questions, (b) the participant's engagement in the conversation, (c) the participant's ability to keep to the questions discussed, and (d) the interviewer's (dental hygienist) understanding of the participant's views. The evaluations were based on the criteria set out by Murphy et al. (2010b), Murphy et al. (2010a) and Ferm et al. (2010) but also depended on thorough discussions taking place between all the researchers in this particular study. Each conversation was evaluated according to the four factors and using a 5-point scale representing low (0) to high (4) effectiveness. The evaluation of the *participant's understanding of the questions* was based on both verbal answers and body communication. To get a high score it should be obvious that the person with HD understood the questions. Lack of answers, irrelevant or inadequate answers resulted in low scores as did misunderstandings. A lower score was also given if it was difficult to understand the person's answers and hence to make the evaluation. The *participant's engagement in the conversation* concerned the social closeness that is a result of social interaction and which is maintained through different kinds of feedback and shared attention. Facial gestures and other body communication as well as verbal feedback were observed. High scores depended on active engagement and interest shown through eye contact, explicit feedback, and humour or by the participant's development of a topic. It was decided that to get one or more points, more than a short answer was needed. The *participant's ability to keep to the questions discussed* was based on the relevance of the participant's answers and on his or her ability to stay on track when answering and discussing the questions. A lower score was given if the participant changed or drifted away from the topic and if it, considering the person's communicative contributions, was difficult to make an evaluation of the factor. The *interviewer's understanding of the participant's views* was evaluated on the basis of the dental hygienist's reactions, verbal and through body communication, to the participant's answers.

The two researchers were trained in EFFC by evaluating video-recordings of conversations involving persons with HD that were not used in the study. Thereafter, the films of the 22 conversations were evaluated in a randomized order. First, the two researchers rated each recording independently. This meant that they looked at the recording together but did their own rating. Subsequently, the researchers discussed their ratings and reached a consensus score for each factor in each recording. The maximum score for each conversation was 16. Twelve points is the cut-off for an acceptable level of effectiveness (Murphy et al., 2010a). To check for interrater reliability, two independent external raters evaluated 30 % of the data using the same procedure. To check for intra-rater reliability, the two main researchers did a second evaluation of 30 % of the data a week after the first evaluation. The two conditions were also compared with respect to time and with respect to number of questions and follow-up questions that were asked.

The answers to the questionnaire items were transferred to a descriptive scale as follows: all/always (4), most/usually (3), a few/occasionally (2), and none/never (1). The individual scores for the six questions were added to form a total involvement score for each condition and participant. Means were calculated as well. Written comments in questionnaires were analysed and categorised with regard to content.

Statistical calculations were done using SPSS (version 19). Internal interrater reliability, measured using intra-class correlation (ICC) was 0.85. External interrater reliability, between the main two researchers and the external raters, was 0.64. The reliability between the researchers was higher (0.91) for the nonTM condition than for the TM condition (0.78). Intrarater reliability calculated on the basis of the researchers' consensus scores was 0.96. Differences in scores of communicative effectiveness and involvement as well as differences in the duration of the two conditions were analysed using Wilcoxon Signed Ranks Test ($p<0.05$).

4. Results

The inherent differences between the two types of conversations, with and without TM, had a few consequences that need to be kept in mind when interpreting the results. In the conversations using TM, the questions were introduced visibly using pictures presented to the participant. Also, in these conversations the participants had an opportunity to delete or add questions, while this was not an option in the conversation without TM. Several of the participants, particularly support persons, deleted as well as added questions both before the conversation started and during the conversations. Support persons were also more active in giving support during conversations using TM. After the TM conversations, the dental hygienist went through the answers together with the participants, who had an opportunity to change the answers. Some of the participants with HD chose to do so. In the conversations without TM, the questions were put to the participants with HD in the predetermined order, and the answers were not documented/written down, albeit recorded.

In the following, results will be presented according to the research questions being asked.

4.1 Is there a significant difference in communicative effectiveness between conversations where Talking Mats is used compared to conversations where TM is not used?

No statistically significant difference between the two types of conversation was found in this group (see Figure 3). Mean effectiveness score without TM (nonTM) was 12.27 (SD 3,26) and mean effectiveness score with TM was 11.45 (SD 2.98).

Individual ratings on the four different parameters of EFFC, together with the total score, are shown in Table 4. Six of the 11 participants with HD, were rated as more effective in their communication in conversations without TM and 3 were rated as equally effective in both conditions. Two individuals communicated more effectively using TM.

Additional qualitatively important aspects of communicative effectiveness are time (duration of the conversations) and number of questions and follow-up questions being asked. Table 5 shows that conversations where TM were used were significantly longer than conversations without TM. This is in part due to the fact that the TM conversations included significantly more follow-up questions, see Table 6.

Research questions number 2 and 3 concerned the perceived communicative involvement on the part of the individuals with HD and the support persons. These questions were answered using the previously described questionnaires.

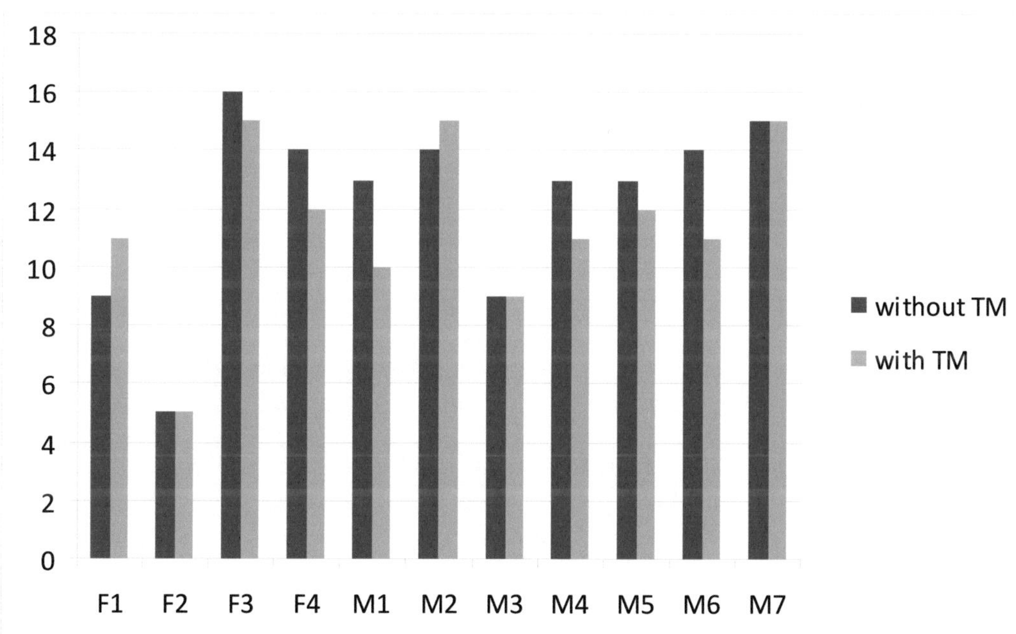

Fig. 3. Communicative effectiveness, as measured by EFFC, in conversations with (grey bars) and without Talking Mats (black bars). Acceptable communicative effectiveness cut-off value is 12 points.

Participant	Condition	Understanding	Engagement	Stick to the topic/questions	Interviewer's understanding	Effectiveness score
F1	nonTM	2	2	2	3	9
	TM	3	2	2	4	11
F2	TM	1	1	1	2	5
	nonTM	1	1	1	2	5
F3	TM	4	3	4	4	15
	nonTM	4	4	4	4	16
F4	TM	3	2	3	4	12
	nonTM	4	2	4	4	14
M1	TM	3	1	3	3	10
	nonTM	3	2	4	4	13
M2	nonTM	4	2	4	4	14
	TM	4	3	4	4	15
M3	TM	2	2	2	3	9
	nonTM	2	2	2	3	9
M4	nonTM	3	3	3	4	13
	TM	2	3	3	3	11
M5	nonTM	3	2	4	4	13
	TM	3	2	3	4	12
M6	TM	2	2	3	4	11
	nonTM	3	3	4	4	14
M7	nonTM	4	3	4	4	15
	TM	4	3	4	4	15

Table 4. Individual ratings and effectiveness scores of the eleven participants with HD. N.B.: nonTM = without Talking Mats, TM = with Talking Mats.

	nonTM	TM
Mean (SD)	3.7 (0.89)	12.8 (4.01)
Max	5.78	19.38
Min	2.47	7.68

Table 5. Duration of conversations in minutes, without (nonTM) and with (TM) Talking Mats. The difference is statistically significant (Wilcoxon Signed Rank test, p= .003).

	nonTM		TM	
	Questions	Follow-up questions	Questions	Follow-up questions
Mean (SD)	9.82 (0.41)	3.55 (2.12)	10.09(0.83)	9.64 (4.23)
Max	10	7	11	15
Min	9	1	9	3

Table 6. Number of questions and follow-up questions being asked by the dental hygienist during the two different types of conversations, without (nonTM) and with (TM) Talking Mats. The difference is statistically significant (Wilcoxon Signed Rank test, p= .005).

4.2 Is there a significant difference in perceived communicative involvement between the two types of conversation on the part of the individuals with HD? And is there a difference in perceived communicative involvement between the two types of conversation on the part of the support persons?

The participants with HD rated their communicative involvement significantly higher in conversations using TM compared to conversations without TM (nonTM). Differences between the two conditions were not statistically significant for the support persons or the dental hygienist, see Table 7.

	Participants with HD		Support persons		Dental hygienist	
Conversation type	nonTM	TM	nonTM	TM	nonTM	TM
Mean(SD)	21 (2.37)	22.45 (1.51)	21.25 (3.05)	21.50 (2.58)	20.27 (2.61)	20.82 (2.14)
Max	24	24	24	24	23	24
Min	18	20	15	15	16	16

Table 7. Ratings of perceived communicative involvement during the two different types of conversations, without (nonTM) and with (TM) Talking Mats. The difference is statistically significant for the participants with HD (Wilcoxon Signed Rank test, p= .048) but not for the support persons or the dental hygienist.

The participants' reactions to the use of TM were also collected and a general finding was that participants with HD as well as support persons were positive. A qualitative analysis of the notes generated four themes: *Understanding, Thinking and memory, Expressive function* and *The use of Talking Mats*. Themes and associated quotes are included in Table 8.

Theme	Illustrative quotes from participants with HD	Illustrative quotes from support persons
Understanding	*"people who do not understand sometimes people do not understand, then it would be good"* (M3) *"easier if you find yourself in a conflict situation, which I avoid, then it would be good"* (M3)	*"you don't always understand what he says, in that case he can show and point out"* (assistant to M4) *"I think the pictures are good, they help with understanding and being able to express oneself"* (assistant to F1) *"She understood more with the pictures"* (assistant to F1)
Thinking and memory	*"it's easier to think and understand when there are pictures"* (M7) *"made me think some more about the different stuff with oral care, that can be good"* (M5)	*"he has a fairly poor memory, and then you can take out the mat and show what we agreed on"* (assistant to M4) *"it can be useful to remember what you talked about even if the verbal communication does work"* (assistant to F3) *"she has difficulties making decisions and form opinions"* (husband to F2)

Expressive function	*"you can express feelings just by pointing to a face"* (M4) *"it is a bit easier to find the words with the mat"* (M2) *"it is a bit easier to talk about it with the mat"* (M2)	*"it clarifies when the words get muddy"* (assistant to M1) *"he talked more with the help from the Talking Mats"* (counselor M6)
The use of Talking Mats	*"damn good with pictures"* (M3) *"I didn't think it was going to be so easy"* (M3) *"it became a bit slow with the mat and better flow in the conversation without the mat" (F4)*	*"it was easier than I thought, and I don't think she had any difficulties with it either"* (assistant to F4) *"to think about using it during the right phase of the illness. To not be offended"* (support person to M5) *"a good "tool" to use in the future should it be necessary"* (daughter of M7) *"the mat was very good used like that regularly it will be great"* (assistant to M3) *"maybe clearer instruction before the interview about the use of Talking Mats about the pictures that was graded and the questions/conversation topic words"* (counselor to M6) *"felt a bit conflicting at times – several things in one picture that were opposites"* (assistant to F3) *"it is a bit difficult to say, I didn't know if I could expand, I was afraid to take over too much, I wanted to add more questions but didn't really know how much I was allowed to ask"* (assistant to M2) *"I didn't really know how much I was supposed to interfere in the conversation"* (assistant to K3)

Table 8. Themes and quotes from the interviews with participants with HD and support persons, concerning the use of Talking Mats.

The last research question concerned the perceived benefit of the use of TM as reported by the dental hygienist in the interview conducted after the consultation had been completed and both types of conversations implemented.

4.3 Does the dental hygienist perceive the use of TM as a beneficial support in the dental and oral health care consultation?

In the interview, the dental hygienist expressed the general view that the use of TM supported the counseling and treatment of persons with HD. Her opinion was, that all conversations using TM were superior to the conversations without TM. The qualitative analysis of her answers to the interview questions yielded six different themes: *Talking Mats as a method, Information and intervention, Individual adjustments, Memory, Understanding,* and *Naturalness.*

Talking Mats as a method. TM made the conversation obvious, transparent and concrete. One particular advantage of the method was that you were able to review the answers afterwards: *"the possibility to go back, add and comment on things in the conversation"*. The use of TM also gave a visible overview of the conversation. The pictures served as support for the memory both for the dental hygienist and the patient and they made the conversation concrete: *"easier to talk and discuss with the pictures as a support"*. The method Talking Mats and the mat in itself created a joint focus for the participants *"you have the whole conversation in front of you, it's there on the table"*. At times, the dental hygienist found it difficult to stick to the preset wordings of the questions: *"the questions are a bit tricky sometimes"*. It was evident from her answers to the questionnaire, that she found it difficult to engage in the conversations without TM – *"it feels as though you only want to get it over with"*, particularly if it was the second conversation: *"I don't feel as engaged when I'm going to ask almost the same questions again"*.

Information and intervention. One other positive aspect of the use of TM in conversation was that new and more in-depth information about several of the participants' oral and dental care appeared: *"I found out that she has difficulties remembering to brush her teeth but that she wants to be reminded"*, *"you can go deeper into the questions"*. Three participants with HD wanted to try new products or methods when they communicated using TM: *"it became apparent that the patient was interested in cleaning between his teeth, I thought this was impossible before, he has a strong sense of integrity"*. The dental hygienist experienced that two support persons contributed with new thoughts in the TM conversation: *"new thoughts appeared from the assistants"*.

Individual adjustments. The dental hygienist described that she would have wanted the opportunity to adjust the questions to particular individuals. In the present study, for the sake of consistency, she had to ask questions that were not relevant to all participants with HD: *"you have to individualize the picture material so that you can adjust the questions/pictures to the patients"*, *"I know that patient so it feels a bit silly to ask about things that are not relevant for that person"*.

Memory, Understanding and *Naturalness*. The dental hygienist saw TM as a support for her own memory: *"it supports the memory, you remember what the conversation has been about, what you have talked about"*, *"good when you go through the answers afterwards you remember what you talked about"*. Talking Mats seemed to make things more visible and clear both for her and for the participant with HD: *"easier for me to understand and probably also for the patient"*, *"it's good because it's difficult to understand what he says, you have to ask again, and in that case the mat helps"*. In two conversations using TM, with F4 and M7, the dental hygienist experienced that TM affected naturalness: *"it felt a bit repetitive, it did not have the same flow as earlier conversations using TM"*, *"it* (the TM conversation) *is not an ordinary conversation"*.

5. Discussion

We examined communicative effectiveness in conversations between persons with HD, their support persons and a dental hygienist, with and without Talking Mats. We also examined the participants' experiences from the two conditions as well as the dental hygienist's experiences from using TM in dental and oral health care consultations. To summarise, there was no increase in communicative effectiveness for the group when TM was used. Two individuals communicated more effectively with TM but more than half of the participants were evaluated as more effective when TM was not used. Three individuals were evaluated as equally effective in the two conditions. Importantly, the participants with HD experienced a significantly higher degree of communicative involvement when TM was used than when conversation was unaided. The support persons also experienced a higher degree of communicative involvement with TM than without TM but this difference between conditions was not significant. The dental hygienist was very positive about TM. In her view, conversations with TM worked better than those without TM. She received new and more comprehensive information when TM was used.

Previous studies have shown that TM leads to more effective communication for persons with dementia and for persons with HD in dyadic as well as group conversations (Ferm et al., 2010; Hallberg et al., 2011; Murphy et al., 2010a, 2010b). A number of differences between this and previous studies regarding the way data was collected, goals, roles and procedures of activities may have contributed to the different results obtained.

Individuals with HD have many medical and health related contacts in which communication is of central importance. A purpose of the present study was to examine if TM could be helpful in such consultations. Thus, it was important to collect data in situations that were as natural as possible. It was believed that it was more natural if all data from each participant was collected during one and the same consultation than if the participant had to come back to the dental hygienist a second time, just to finish data collection. Hence, all data from each participant was collected on one occasion. This procedure is different from that of previous TM studies of effectiveness and could have influenced the participants' behaviours and the rating of effectiveness. For example, it may have been unnatural as well as tiring for the participants to answer similar questions twice during one and the same visit. However, four of the six participants that were scored as more effective without the mat carried out this condition after they had used the mat. Perhaps it was even positive for these individuals, with varying levels of cognitive functioning, to hear the questions twice. The fact that the participants had answered questions before may have contributed to less hesitation and more concise answers. It is possible that these answers were seen as efficient and informative and thus rated as more effective by the outside observers.

Another important difference between this and previous studies is that previous studies examined activities where communication was a main goal. It has also been the case in previous studies, at least in the study by Ferm et al. (2010), that the activity was constructed for the study. Although communication is important in dental and oral health care consultations it typically is not the main goal of the activity and although the present activity was natural for the participants and not constructed, it was slightly changed as far as procedure and goals are concerned. Communication usually takes place when the patient lies in the treatment chair and the goal is to promote good oral health; the patient gives information about his or her dental and oral status upon questions and requests from the dental hygienist who, based on the information given by the patient, gives advice about care routines, aids and products. In this study, the consultation started by the table. Communication about dental and oral health was "lifted out" from the more "practical" dental treatment and in this sense got a different and somewhat more prominent function than the participants were used to. The change of the activity, and the fact that communication was more in focus than is often the case at the dental hygienist's, may have led to increased demands on the person with HD to, for example, give more nuanced answers than yes and no when TM was not used. Such changes in communication may have influenced the observers' ratings.

Dental hygienist consultations are typically associated with different roles which were slightly changed in this study. These changes, pertaining to the rights and obligations of support persons in particular (cf. Allwood, 2000), may have influenced the participants' communicative behaviours in ways that had effects on the ratings. The dental hygienist usually leads the activity and the patient, in this case the participant with HD, is supposed to do what the dental hygienist suggests. The support person usually does not accompany the person with HD into the treatment room but a prerequisite for participation in this study was that both the person with HD and a partner participated in the consultation. It is possible that the role of the partner was unclear. It was an unfamiliar situation for the partner both to discuss dental and oral health care and to participate actively in the conversation. Eight of the support persons were personal assistants whose role, apart from assisting the person they work for in relation to practical issues, involves promoting independence in that person. To argue against and even question the opinions of the person with HD during his or her "private" consultation was perhaps difficult. The analysis showed that overall, the support persons participated little in the conversations but the personal assistants interfered more often than others. The support person was supposed to have knowledge about the dental and oral health of the person with HD. This was not the case in all dyads. Some of the support persons managed everything that had to do with oral care in the person with HD; others didn't know anything about this daily issue. Not having the knowledge needed for participation in the study probably affected the support persons' behaviour and communication negatively. The dental hygienist's role was also changed. She would usually ask questions as she performed the actual treatment. In this study she talked with the patient before the treatment. She also used a communication method that was new to her.

All of the above mentioned factors could have influenced the participants' behaviours and, hence, the outside observers' rating of communicative effectiveness in the persons with HD. In future analyses, we will look more closely into these factors.

For two participants, F1 and M2, communicative effectiveness increased when TM was used. It is important to note that both of them had used TM before. For some people at least, communicative effectiveness with TM may be related to amount of experience as well. Both F1's own and the dental hygienist's understanding was higher in the TM condition. The TM conversation took a longer time which meant that F1's conversational space increased and, as a result, her and the dental hygienist's understanding. The dental hygienist asked F1 fewer clarifying questions (e.g., *And that works well?*) when TM was used which was interpreted as a sign for her better understanding. The situation was similar for M2. The TM conversation took a longer time and M2 had more room for developing his answers and showing humour, factors which, according to the criteria used, can have been a reason for why M2 was assessed as engaged in the TM conversation.

TM was assessed in three different ways; through outside observers' rating of communicative effectiveness, through the participants' responses to questionnaires about communicative involvement in the two conditions and through interviews with the dental hygienist. A most significant finding of the study is that the majority of the participants with HD and their support persons appreciated TM. Even if only two participants were more effective in conversations with TM, many participants appreciated the mat and thought it supported memory and word finding. Talking Mats also supported understanding in interaction. It was easier for the participants to make themselves understood and to understand others when TM was used. Interaction with other people is problematic for persons with HD who become less talkative and more isolated with the progression of the disease (Hartelius et al., 2010; Power et al., 2011). Finding ways to support participation in different social activities is important and the individual's ability to communicate in activities relating to own health should be prioritized. This study indicates that TM is one possible way of supporting communication between persons with HD and their conversational partners. The participants with HD felt significantly more involved, that is, experienced greater communicative involvement, when TM was used than when conversation was unaided. The support persons also experienced increased communicative involvement with TM but for them the difference between the two conditions was not significant. The support persons interfered more in the mat conversations. Some of them even assisted the person with HD, physically and psychologically, placing the pictures on the mat. Again, unclear instructions from the researchers and contradictions between the typical rights and obligations of assistants to promote independence in the person they work for and the expectations on assistants in this study, to converse with the person with HD on equal terms, may have contributed to the lack of significance between conditions as far as the support persons' feelings of communicative involvement is concerned. Still, the present findings are similar to those of Murphy et al. (2010b) where individuals with dementia and in particular their partners felt more involved in conversation when TM was used.

An all-embracing purpose of the study was to explore the use and function of TM for individuals with HD at the dental hygienist's and perhaps the most interesting finding is the fact that the dental hygienist experienced that TM conversations were better than conversations in which TM was not used. The dental hygienist also asked more follow-up questions when TM was used, indicating that TM stimulated conversation. A similar pattern was found in the study by Hallberg et al. (2011): In this study both the leader of the group discussions and the participants with HD asked more follow-up questions when they used TM than when they didn't have this support. By asking follow-up questions the dental hygienist could get more information about the dental and oral care situation of the person who has HD and as a result, she appreciated the situation. In her view, she got new and more comprehensive information from the dyads when TM was used, information that could lead to improved counselling and individual treatments of individuals who have HD. The dental hygienist's comment about individualizing questions and pictures is in line with the methodology of TM and would not be a problem in her future clinical work. For more comprehensive discussions in relation to follow-up questions and other queries that arise during conversations, she could use sub-mats (Murphy & Cameron, 2006).

Measuring communicative effectiveness is not easy and as has been shown in this study, ratings by outside observers must be complemented with measurements of the interlocutors own experiences. What then, is the difference between communicative involvement and communicative effectiveness? The questions about involvement used in this study were developed from Murphy et al. (2010b). It is possible that they reflect not only involvement but also effectiveness. Questions number 2, 3 and 4 in the questionnaire used in this study relate to communicative effectiveness; to be able to convey a message in an effective way and to be able to influence other people (Hustad, 1999). Question 4 focuses on time for expression of opinions. Perhaps the participants, in answering this question, considered both how much time they got from others and to what degree they were able to take their time in the conversations. In fact, it is reasonable to believe that the participants' rating of communicative involvement mirrored their perceptions of how effective they had been in the conversations.

5.1 Using the EFFC

Despite lots of training and discussion of criteria the researchers experienced difficulties using the EFFC in relation to these data. A re-analysis of consensus discussions, recordings and final scores shows that there were more disagreement between the raters in relation to the TM conversations than in relation to the unaided conversations, suggesting the former were more difficult to rate. Lack of experience in TM and EFFC as well as too vague criteria were obvious threats to agreement. The researchers' experiences and ideas about an ideal "effective" dental hygienist consultation also may have influenced their rating of the participants' communicative effectiveness. It also may have been the case, that the raters favoured oral expressions and treated these differently from body communication in their ratings. For example, the criteria used for rating of the participant's *understanding of the questions* and ability to *keep to the questions discussed* meant that higher points were given if the person was very explicit in her or his oral expression. It is possible that, unconsciously of course, a very short, adequate and concise utterance by a participant was valued higher and accordingly rated higher than a quiet placement of a picture on the mat. Some individuals with HD had less eye contact with their support persons and the dental hygienist when using TM. This seems to have influenced the observers rating of the participants' *engagement*.

5.2 Limitations

The conversations examined in this study were conducted by a dental hygienist who had limited experience in using TM. She was instructed about the order of question sets, conditions and individual questions but was free to formulate follow-up questions. Treatment integrity (Schlosser, 2003), that is, the degree to which the dental hygienist followed the procedure as planned, was considerable. Each participant received most of the questions and she was consequent in using the open question format. It is important to remember that this study was carried out in an authentic clinical situation and that time pressure and the dental hygienist's previous knowledge about the participants' dental and oral health may have influenced the conversations. Several of the participants had to catch transportation service at scheduled times which may have been stressful for both them and the dental hygienist. The dental hygienist's enthusiasm over TM and participation in the study certainly constituted threats to validity but were difficult to control and did not lead to higher ratings of TM conversations. Rather, it is possible that her enthusiasm affected her communicative behaviours in ways which had negative effects on the rating of effectiveness in participants with HD. Her satisfaction and hope in TM as a resource in her future clinical work may have influenced her ability to behave equally in the two conditions and to overestimate the benefits of TM in the interview.

A limitation of the study which, considering the purpose of exploring communication support in real life also is its strength, relates to the fact that each participant's data was collected on one occasion.

5.3 Strengths and clinical implications

The strengths of the study outweigh its limitations by far. A considerable set of interaction data involving as many as eleven individuals in different phases of HD and their support persons has been examined. The investigation of interaction in a natural health care situation is in itself unique. The fact that the intervention focused on the situation of individuals with HD, for whom communication is often complicated and related to the many other difficulties that come with the disease, makes the study even more interesting.

Both quantitative and qualitative methods were used and the fact the participants' own experiences were taken into consideration strengthens the ecological validity of the study.

Talking Mats is used by speech language pathologists (SLP), teachers and others who know of its benefits. The present findings suggest that TM could function as a communication support not only in dental and oral health care but also in other clinical care situations that are important for individuals with HD and those who care for them, for example in conversations with the physician, the dietician, the physiotherapist, the occupational therapist and the psychologist. With training and careful instruction to all people involved, TM could lead to increased communicative effectiveness and a feeling of communicative involvement for the person with disability as well as for conversation partners. Considering the strategies and experiences of conversation partners to individuals with communication difficulties is important (cf. Saldert et al., 2010).

More studies focusing the communication of individuals with cognitive and communicative disability in naturally occurring activities are needed. The present researchers' future contributions to the field include more comprehensive interaction analyses of the present data (Ferm & Saldert, 2011) as well as evaluations of TM in interactions between persons that have Parkinson's Disease and their partners at home.

6. Conclusion

Interactions between individuals with HD, their support persons and a dental hygienist have been examined regarding communicative effectiveness and perceived communicative involvement with and without Talking Mats (Murphy & Cameron, 2006). According to outside observers, TM may not lead to more effective communication for persons with HD during dental and oral health care consultations. However, a most significant finding is that the participants found it valuable using the mat. Both the participants with HD and their partners felt more involved in the TM condition than when conversation was unaided. For example, the participants commented that it was easier expressing feelings with the mat, that it was a good method for reflecting on oral health and that it was easier thinking and understanding with the mat than without it. Participants also reported that the pictures supported memory. The dental hygienist was positive as well. It was easier for her to understand the views of some of the participants when she used TM. For example, patients, who typically were inflexible as far as oral hygiene and prophylaxis is concerned, were more open minded and positive towards trying new methods and aids when discussing these issues with TM. According to the dental hygienist, TM has the potential to support communication in consultations involving persons with HD and their partners. Clinical activities in which TM could be useful include instruction and treatment planning, individual goal setting and follow up.

To date, few studies have investigated the use and value of augmentative and alternative communication for persons with HD and their partners in different activities. Research focusing communication support in care situations hardly exists. In this sense, and because it was conducted within an ordinary clinical practice, the findings of this study are important for the rehabilitation and treatment of individuals with HD and those who care for them.

7. Acknowledgment

Our greatest appreciation is to the individuals with HD and their support persons who accepted to participate in this research during one of many health consultations. Furthermore, the research had not been possible without the full participation by the dental hygienist. Her support was invaluable for the carrying out of the study. Her enthusiasm as far as communication support is concerned, and Talking Mats in particular, certainly will continue to be very important for families who are affected by HD. The writing of this chapter was partly funded by the Promobilia Foundation.

8. References

Ahlsén, E. (1995). Activity demands and communication ability in aphasia: A protocol and a case study. In: *Papers from the XVth Scandinavian Conference of Linguistics*, I. Moen, H. Gram Simonsen, & H. Lødrup, (Eds.), 1-12, ISBN 8291298025 9788291298023, Oslo: Oslo University, Department of Linguistics

Allwood, J. (2000). An activity based approach to pragmatics. In: *Abduction, Belief and Context in Dialogue: Studies in Computational Pragmatics*, H. Bunt & B. Black (Eds.), 47-80, ISBN 9027249830, Amsterdam: John Benjamins

Bartlett, G.; Blais, R.; Tamblyn, R.; Clermont, R.J. & MacGibbon, B. (2008). Impact of patient communication problems on the risk of preventable adverse events in acute care settings. *Canadian Medical Association Journal CMAJ*, 178(12), pp. 1555-1562, ISSN 0820-3946

Chenery, H. J.; Copland, D. A. & Murdoch B. E. (2002). Complex language functions and subcortical mechanisms: evidence from Huntington's disease and patients with non-thalamic subcortical lesions. *International Journal of Language and Communication Disorders*, 37(4), pp. 459-474, ISSN 1460-6984

Ferm, U.; Sahlin, A.; Sundin, L. & Hartelius, L. (2010). Using Talking Mats to support communication in persons with Huntington's disease. *International Journal of Language & Communication Disorders*, 205, pp. 523-536, ISSN 1460-6984

Ferm, U. & Saldert, C. (2011). Evaluation and revision in interactions with Talking Mats: Conversations between persons with Huntington's disease, their close others and a dental hygienist. *Manuscript in preparation*

Gabre, P. (2009). Strategies for the prevention of dental caries in people with disabilities: a review of risk factors, adapted preventive measures and cognitive support. *Journal of Disability and Oral Health*, 10, pp. 184-192, ISSN 1754-2758

Hallberg, L.; Mellgren, E.; Hartelius, L. & Ferm, U. (in press). Talking Mats in a discussion group for people with Huntington's disease. *Disability and Rehabilitation: Assistive Technology*, ISSN

Hartelius, L.; Carlstedt, A.; Ytterberg, M.; Lillvik, M. & Laakso, K. (2003). Speech disorders in mild and moderate Huntington disease: results of dysarthria assessments of 19 individuals. *Journal of Medical Speech-Language Pathology*, 11, pp. 1-14, ISSN 1065-1438

Hartelius, L.; Jonsson, M.; Rickeberg, A. & Laakso, K. (2010). Communication and Huntington's disease: quality interviews and focus groups with persons with Huntington's disease, family members, and carers. *International Journal of Language & Communication Disorders, 45*, pp. 381-393, ISSN 1460-6984

Hustad, K. C. (1999). Optimizing Communicative Effectiveness: Bringing it together. In: *Management of motor speech disorders in children and adults*, K. M. Yorkston, D. R. Beukelman, E. A. Strand, & K. R. Bell, (Eds.), 483-541, ISSN 0-89079-784-6, Austin, Texas: Proedition

Jensen, A. M.; Chenery, H. J. & Copland, D.A. (2006). A comparison of picture description abilities in individuals with vascular subcortical lesions and Huntington's disease. *Journal of Communication Disorders*, 9(1), pp. 62-77, ISSN 1873-7994

Kagan, A.; Winckel, J.; Black, S.; Felson Duchan, J.; Simmons-Mackie, N. & Square, P. (2004). A set of observational measures for rating support and participation in conversation between adults with aphasia and their conversation partners. *Topics in Stroke Rehabilitation*, 11(1), pp. 67-83, ISSN 1074-9357

Kidd, E. A. M. (2005). *Essentials of Dental Caries* (3d edition) ISBN 978-0198529781, Oxford: Oxford University Press

Klinge, B. & Gustafsson, A. (2011). *Parodontit: en introduktion* (4th edition), ISBN 9789172057609, Stockholm: Gothia Förlag AB

Lewis, D.; Fiske, J. & Dougall, A. (2008). Access to special care dentistry, part 7. special care dentistry services: seamless care for people in their middle years – part 1. *British Dental Journal, 205*, pp. 305-317, ISSN 1476-5373

Mayer-Johnson, 1981–2011, *The Picture Communication Symbols©*, Solana Beach, CA: Mayer-Johnson LLC,
http://www.mayer-johnson.com/

McGarva, K. (2001). Huntington's disease: seldom seen - seldom heard? *Health Bulletin, 59*, pp. 306-308, ISSN 0374-8014

Murphy, J. (2006). Perceptions of communication between people with communication disability and general practice staff. *Health Expectations, 9*, pp. 49–59, ISSN 1369-7625

Murphy, J. & Cameron, L. (2006). *Talking Mats a Resource to Enhance Communication*. Stirling: University of Stirling, Available from info@talkingmats.com

Murhpy, J.; Gray, C.M.; Cox, S.; Van Achterberg, T. & Wyke, S. (2010a). The effectiveness of the Talking Mats framework with people with dementia. *Dementia; International Journal of Social research and Practice* 9(4), pp. 454-472, ISSN 1741-2684

Murphy, J.; Oliver, T. M. & Cox, S. (2010b). *Talking Mats® and involvement in decision making for people with dementia and family carers*. Stirling: University of Stirling, Joseph Rowntree Foundation, Available from
http://www.jrf.org.uk/publications/talking-mats-decision-making

Murphy, J. & Cameron, L. (2008). The effectiveness of Talking Mats with people with intellectual disability. *British Journal of Learning Disabilities*, 36, pp. 232-241, ISSN 1468-3156

Power, E.; Anderson, A. & Togher, L. (2011). Applying the WHO ICF framework to communication assessment and goal setting in Huntington's Disease: A case discussion. *Journal of Communication Disorders*, 44(3), pp. 261-275, ISSN 1873-7994

Roos, R. A. C. (2010). Huntington's disease: a clinical review. *Orphanet Journal of Rare Diseases, 5*, pp. 1-8, ISSN 1750-1172

Saldert, C. & Hartelius, L. (2011). Echolalia or functional repetition in conversation – a case study of an individual with Huntington's disease. *Disability and Rehabilitation*, 33 (3), pp. 253-260, ISSN 1464-5165

Saldert, C.; Eriksson, E.; Petersson, K. & Hartelius L. (2010). Interaction in conversation in Huntington´s disease: An activity-based analysis and the conversation partner's view of change. *Journal of Interactional Research in Communication Disorders* 1(2), pp. 169-197, ISSN 2040-512X

Saldert, C.; Fors, A.; Ströberg S. & Hartelius, L. (2010). Comprehension of complex discourse in different stages of Huntington's disease. *International Journal of Language & Communication Disorders, 45*, 656-669, ISSN 1460-6984

Schlosser, R. W. (2003). *The Efficacy of Augmentative and Alternative Communication*, ISBN 978-0126256673, San Diego, CA: Academic Press

Shoulson, I.; Kurlan, R. A.; Rubin, A. J.; Goldblatt, D.; Behr, J.; Miller, C.; Kennedy, J.; Bamford, K. A.; Caine, E. D.; Kido, D. K.; Plumb, S.; Odoroff, C. (1989). Assessment of functional capacity in neurodegenerative movement disorders: Huntington's disease as a prototype. In: *Quantification of Neurologic Deficit*, T. L. Munsat, (Ed.), 271–283, ISBN 9780409901528, Boston: Butterworth

Yorkston, K. M.; Miller, R. M. & Strand, E. A. (2004). *Management of speech and swallowing disorders in degenerative diseases* (2nd edition), ISBN 0-89079-966-0, Austin, Texas: Proed